IMPENDING INQUISITIONS IN HUMANITIES AND SCIENCES (ICIIHS-2022)

About the Conference

In an era of increasing specialization, the need for cross-disciplinary dialogue demands an integrated approach that transcends the artificial boundaries between disciplines. "Impending Inquisitions in Humanities and Sciences" presents a groundbreaking tapestry of cutting-edge research across the spectrum of humanities and sciences. This volume presents a meticulously curated selection of research papers presented at the conference, a forum where scholars from diverse fields – English, Mathematics, Physics, and Chemistry – converged to engage in rigorous dialogue and push the boundaries of knowledge. From the nuanced interpretations of literary texts to the elegant formulations of mathematical models, from the awe-inspiring revelations of physics to the meticulous experiments of chemistry, each contribution challenges assumptions and provokes fresh perspectives. This collection serves as a valuable resource for scholars, students, and academic fraternity with an insatiable curiosity about the world around us.

Key Features:

- Interdisciplinary dialogue across English, Mathematics, Physics, and Chemistry
- Groundbreaking research challenging established paradigms
- Insights into the transformative power of interdisciplinarity
- Valuable resource for scholars, students, and inquisitive mind
- Invitation to embark on a journey of intellectual exploration

IMPENDING INQUISITIONS IN HUMANITIES AND SCIENCES (ICIIHS-2022)

Editors

Dr Mohan Varkolu
Assistant Professor, Department of Chemistry,
Koneru Lakshmaiah Education Foundation (KL Deemed to be University)
Off Campus Hyderabad

Dr. M P Mallesh
Associate Professor, Department of Mathematics,
Koneru Lakshmaiah Education Foundation (KL Deemed to be University)
Off Campus Hyderabad

Dr. Kranthi Priya Oruganti
Assistant Professor, Department of English,
Koneru Lakshmaiah Education Foundation (KL Deemed to be University)
Off Campus Hyderabad

CRC Press
Taylor & Francis Group
Boca Raton London New York

CRC Press is an imprint of the
Taylor & Francis Group, an **informa** business

First edition published 2024
by CRC Press
4 Park Square, Milton Park, Abingdon, Oxon, OX14 4RN

and by CRC Press
2385 NW Executive Center Drive, Suite 320, Boca Raton FL 33431

British Library Cataloguing-in-Publication Data
A catalogue record for this book is available from the British Library

ISBN: 978-1-032-78829-6 (pbk)
ISBN: 978-1-003-48943-6 (ebk)

DOI: 10.1201/9781003489436

Typeset in Times LT Std
by Aditiinfosystems

Dedications

We dedicate this book to the enduring power of research. Through rigorous inquiry and critical analysis, research illuminate the complexities of our world and empowers us to address the challenges of our time. We believe that this volume serves as a testament to the transformative power of research and its potential to benefit society.

This book is dedicated to the vibrant academic community, whose tireless pursuit of knowledge and commitment to rigorous research continue to shape our understanding of the world and drive progress across diverse disciplines. We are grateful for the collaborative spirit and dedication that fuel innovation and contribute to a brighter future.

We stand in awe of the collective effort of the academic community. Through knowledge sharing, you have contributed to the intellectual tapestry that enriches our lives and shapes our future. We are forever grateful for your dedication and unwavering pursuit of knowledge.

Impending Inquisitions in Humanities and Sciences (ICIIHS-2022) – Dr. Mohan Varkolu et al. (eds)
© 2024 Taylor & Francis Group, London, ISBN 978-1-032-78829-6

Contents

Impending Inquisitions in Humanities and Sciences (ICIIHS-2022) – Dr. Mohan Varkolu et al. (eds)
© 2024 Taylor & Francis Group, London, ISBN 978-1-032-78829-6

List of Figures

Impending Inquisitions in Humanities and Sciences (ICIIHS-2022) – Dr. Mohan Varkolu et al. (eds)
© 2024 Taylor & Francis Group, London, ISBN 978-1-032-78829-6

List of Tables

Impending Inquisitions in Humanities and Sciences (ICIIHS-2022) – Dr. Mohan Varkolu et al. (eds)
© 2024 Taylor & Francis Group, London, ISBN 978-1-032-78829-6

Acknowledgements

The success of any conference is the result of the combined efforts of many individuals and organizations. We express our sincere gratitude and are grateful for the contributions of the following individuals and organizations who contributed to the success of the conference and the subsequent publication of the proceedings and made this conference on "Impending Inquisitions in Humanities and Sciences" possible:

Organizing Committee

- Er. K. Satyanarayana, President and Chancellor, KLEF, Er. K. L. Havish, Vice President, KLEF, Er. K. Raja Hareen, Vice President, KLEF, Dr. K. S. Jaganatha Rao, Pro-Chancellor, KLEF, Dr. Partha Saradhi Verma, Vice-Chancellor, KLEF, Dr. A. V. S. Prasad, Pro-Vice Chancellor-I, KLEF, Dr. N. Venkatram, Pro-Vice Chancellor-II, KLEF, Dr. K. Subba Rao, Registrar, KLEF for their unwavering encouragement to conduct academic events; Dr. L. Koteshwar Rao, Former Principal, KLH Aziznagar campus and Principal, KLH Bowrampet campus for providing us his support and resources for conducting the conference and Dr. A. Ramakrishna, Principal, KLH Aziznagar campus and Director, Campus Placements for his constant motivation and encouraging us to publish the research papers.

Editorial Support

- Pillai, Vindhya H
- Nair, Margaserry Gopakumar

We would also like to acknowledge the invaluable contributions of all the participants who submitted papers, attended the sessions, and engaged in the critical discussions and reviewers who have been instrumental for their valuable comments in improving the quality of the manuscripts. Your insightful perspectives and lively debates formed the foundation for these proceedings.

We are particularly grateful to the authors who shared their research in this volume. Their courage and commitment to intellectual freedom in the face of adversity are truly inspiring.

Finally, we extend our deepest thanks to the publisher, Taylor & Francis, for their dedication in bringing this important collection of research papers to limelight. We believe that these proceedings will serve as a valuable resource for scholars, policymakers, and all those who are concerned about the future of the humanities and sciences.

Thank you once again to all who have made this publication possible!

Impending Inquisitions in Humanities and Sciences (ICIIHS-2022) – Dr. Mohan Varkolu et al. (eds)
© 2024 Taylor & Francis Group, London, ISBN 978-1-032-78829-6

Preface

The landscape of the humanities and sciences is shifting beneath our feet. Old certainties are fading, new challenges are emerging, and the very foundations of our disciplines are being questioned. As a community of scholars, we stand at a pivotal moment, poised on the precipice of both peril and possibility.

The International Conference on "Impending Inquisitions in Humanities and Sciences", gathered in these proceedings, serves as a testament to our shared commitment to navigating this complex terrain. We come together from diverse backgrounds and perspectives, united by a common desire to grapple with the pressing issues of our time and to chart a course for the future of our fields.

The papers presented within these pages offer a rich tapestry of insights, interrogations, and proposals. They span a broad range of topics, from the impact of technological advancements on our understanding of the human condition to the enduring questions of history, culture, and identity.

Among the key themes that emerge are the changing nature of knowledge production and dissemination in the digital age, the rise of new forms of scholarship and inquiry that challenge traditional methodologies, the growing interconnectedness of global challenges that demand interdisciplinary collaboration, the need to critically assess the impact of and political forces on our disciplines, and the vital role of the humanities and sciences in promoting critical thinking, responsibility, and global citizenship.

These proceedings are not simply a collection of individual voices; they represent a vibrant ecosystem of intellectual thought and exchange. They challenge us to re-examine our assumptions, to re-imagine the boundaries of our disciplines, and to embrace the dynamism that lies at the heart of academic inquiry.

Yet, our journey is not without its challenges. The humanities and sciences are facing unprecedented threats, from funding cuts and declining enrollments to attacks on academic freedom and open discourse. We must be vigilant in defending the values and principles that have long sustained our disciplines.

In the face of adversity, however, lies opportunity. The very uncertainties that we face present a fertile ground for innovation and change. We must seize this moment to re-invigorate our disciplines, to forge new connections, and to make our voices heard in the public sphere.

These proceedings are more than just a record of a conference; they are a call to action. They invite us to step outside the comfort of our disciplinary silos and to engage in meaningful dialogue with each other and with the world around us.

Let us use this platform to ignite a spark of intellectual curiosity, to challenge the status quo, and to build a stronger, more vibrant future for the humanities and sciences. For in the end, it is through the critical lens of our disciplines that we can illuminate the complexities of our times and contribute to a more just and equitable world.

The future of the humanities and sciences rests in our hands. Let us rise to the occasion and ensure that these vital disciplines continue to thrive as bastions of critical thought, creativity, and human understanding.

Dr Mohan Varkolu
Dr. M P Mallesh
Dr. Kranthi Priya Oruganti

Impending Inquisitions in Humanities and Sciences (ICIIHS-2022) – Dr. Mohan Varkolu et al. (eds)

About the Editors

Dr. Mohan Varkolu is currently associated with Koneru Lakshmaiah Education Foundation (KLH- KLEF off campus Hyderabad) as an Assistant Professor, Department of Chemistry. He has completed his Ph.D. degree in Chemistry from CSIR-IICT/Osmania University, Hyderabad, India in 2014. He has visited countries like USA, France, and South Africa. Dr. Mohan Varkolu has published 34 Articles (H-index 16) including 5 book chapters and 1 edited book so far. His area of research includes Valorization of Biomass, Heterogeneous catalysis, Photo/electrocatalysis, Single atom catalysis, Novel coupling reactions, organic synthesis, and Carbon Capture, Utilization and Storage (CCUS). Dr. Mohan Varkolu is editorial board member and reviewer of various national and international journals.

Dr. M P Mallesh received his Ph.D. degree from GITAM Institute of Science, Visakhapatnam in Applied Mathematics in 2017. He is currently working as an Associate Professor, Department of Mathematics, Koneru Lakshmaiah Education Foundation (KLH-KLEF Off-Campus Hyderabad). Dr. Mallesh's area of interest includes Finite Element Method, Finite Difference Method, Hybrid Nanofluids, Nanofluids, and Mathematical Modelling. He has published 16 Articles including 1 text book, 1 book chapter and 3 patents. He is an active member and reviewer for various technical societies and journals.

Dr. Kranthi Priya Oruganti is an Assistant Professor of English and the Head of the Department of English at Koneru Lakshmaiah Education Foundation (KLH- KLEF Off-Campus Hyderabad). She holds a Ph.D. in English Language Teaching from The English and Foreign Languages University, Hyderabad. Driven by a deep commitment to education, she is a passionate teacher actively involved in research. Her academic interests span Materials Development, Testing, and Evaluation. Dr. Oruganti has authored 7 publications, 3 patents and presented research papers at both national and international conferences. She is a seasoned workshop facilitator and has conducted numerous Faculty Development Programs (FDPs) as well as two international conferences. Her contributions extend to serving as a resource person for various workshops and FDPs. Among her significant achievements is the coordination of a team of 14 resource persons for a Teacher Training project under the auspices of The State Government of Karnataka and Sarva Sikhsha Abhiyan. Dr. Oruganti continues to make impactful contributions to the field of English language education through her teaching, research, and leadership.

Impending Inquisitions in Humanities and Sciences (ICIIHS-2022) – Dr. Mohan Varkolu et al. (eds)
© 2024 Taylor & Francis Group, London, ISBN 978-1-032-78829-6

Michael Foucault's Concepts of Parrhesia and Governmentality: A Study of Arvind Adiga's Selection Day

1

Varun Jain
Assistant Professor, Lovely Professional University,
Phagwara, India
Richa Arora*
Associate Professor, Lovely Professional University,
Phagwara, India

ABSTRACT Michael Foucault's thought process changes from the concept of sovereign power in the 1970's which dealt with territory to that of disciplinary power whose prime examples are prisons, schools, hospitals and finally to Governmentality in the 80's which is a positive form of power and does not require coercion but involves the active participation of those governed. To exhibit Governmentality, requires a certain form of Parrhesia which is the expression of free speech and it is dangerous in the sense that it talks the truth to power and shows it the mirror of its true self. By applying these concepts in Aravind Adiga's Selection Day, it is intended to understand a microcosmic structure that could then be applicable to the macrocosm.

KEYWORDS Governmentality, Parrhesia, Power, Foucault, Truth

1. Introduction

Parrhesia is a greek word which means saying everything without fear or inhibtion. According to Michael Foucault, "Parrhesia is the activity that consists in saying everything: pan rema. Parrhesiazesthai is 'telling all'" (Foucault, 2011) . It is an idea of true discourse. It does not bother about the consequences this free discourse would have on those who are in positions of power. It helps to bring about revolutionary changes in society. In other words, it leads to governmentality which is the art of governing in the rightful manner. It is different from exerting power in the sense that it does not force an individual rather regulates his/her conduct.

The word parrhesia appears for the first time in Greek literature in Euripides [c.484-407 B.C.] whereas "According to Foucault, western governmental rationality can be traced back to the Judeo-Christian tradition in general and to the Christian ideology and practice of the pastorate in particular" (Ojakangas 2012). As Foucault also mentions, "How to govern oneself, how to be governed, how to govern others, by whom the people will accept being governed, how to become the best possible governor - all these problems, in their multiplicity and intensity, seem to me to be characteristic of the sixteenth century" (Burchell 87)

Aravind Adiga is an Indian writer currently residing in Australia. He is the Man booker prize winner for his novel *The White Tiger*. His novels centre around oppression and power. In The White Tiger, Balram is the protagonist who murders his master to rise to the top. He has chosen the wrong path to succeed. Similarly, Amnesty is the story of an illegal immigrant residing in Australia, Danny who is finnaly deported back to his native country Sri Lanka. But he exhibits parrhesia and follows the right

*Corresponding author: Richa.21590@lpu.co.in

DOI: 10.1201/9781003489436-1

path of truthfullness to unravel a crime in his illegally occupied country. Heis actions lead to regulation of conduct towards illegal immigrants in Australia, thus bringing about Governmentality.

Selection Day is a novel by Aravind Adiga published in the year 2016 that went on to become a series on Netflix in the year 2018. In this novel, Tommy Sir is the head coach who spots talent in the world of cricket. He has been in service for 39 years and still he is made to clean up the grounds. That is the influence of the people sitting at the top who exert power. Tommy Sir leaves no stone unturned in exhibiting Parrhesia. He is bluntly honest to his players. As is said about him, "Tommy Sir was given to truth as some men are to drink" (Adiga, 2016). He would ruthlessly tell a player if he was not fit for cricket. He would not praise a batsman even after he has scored a century just to make him aspire for more.

The reality of India is also depicted very subtly through the novel. "Blue smoke rose from the garbage burning in a corner" (Adiga, 2016). It causes smog in the air. Pollution is a menace and with such actions and a callous attitude, it becomes even more deadly. Cricket is a frenzy in the country and in Mumbai especially, where people come from different parts of the country to make their lives and Mohan is one of them. He wants his two sons to become the number one and number two cricketers in the country.

These two brothers, Radha and Manju aspire to be the future cricketers of India but not by choice. They are forced to make cricket as their careers by their father who exercises power over his two sons. This kind of power is sovereign power wherein a man is coerced into following a particular regime. The sons are immature and so they unknowingly accept their father's dreams as theirs. Next, they are coerced into continuing to play cricket by the capitalist society which provides them a decent living, a house etc. so that their destinies are controlled by a businessman who sponsors them. Sir Tommy, their head coach, exhibits parrhesia and thus moulds the conduct of these two players positively. This type of influence is governmentality, a term coined by Michael Foucault. It is a positive form of power and can bring about positive results. In the novel, even though the elder one likes playing it, he is unable to be successful because of the adamant behavior of his father. Manju, the younger brother, is the only chance, but he is not interested in cricket but in science. Their father does not understand this and weilds his power over them. He shuts his eyes towards the reality and the aspirations of his two sons. He is unable to regulate their conduct. This spells doom for his two sons. Manju, his younger son, is also a homosexual, but he cannot come out of the barriers that society has put on him. Power dynamics are present at every point in the novel as the elder brother tries to prevent the younger one from getting selected for the national team if he, himself, is not selected. Their careers falter as they do not have the courage to exhibit parrhesia. It is not only presented through one's speech but also through one's actions.

2. Literature Review

Dr. Dnyanoba B. Mundhe analyses the representation of Caste-System, Crime and Religion in *Selection Day.* Anamika Suman studies about the aspect of social realism in the novel. J. Hilda Malar, R.Venkatraman and J. Amutha Monica also talk about India's social structure in the novel. Joshi Manohar Purushottam has done a comparative study on culture between the novels of Aravind Adiga and Kiran Desai.

3. Power Struggles

Manju is not able to express himself freely as he is subdued by his father. He uses his muscle power to tame them. He cannot allow his sons to have an affair with any girl and if he gets the slightest hint that they have been just looking at them, he will beat them, thus showing his power over their bodies. In fact, this is one of the power elements in Foucauldian discourse wherein individuals used to be and are even today in some cases disciplined by subjecting bodies through pain.

The Politicians' role in society is depicted in the novel as some of them, notorious criminals still find their way on billboards flying high in the city. The brazen use of power is exposed here when the narrator says, "There were so many medieval criminals whose grinning heads had been hoisted above a city gate" (Adiga, 2016). On the contrary, the abject state of poverty is depicted when a boy "offering or threatening to polish shoes for 5 rupees each" is spotted in the local train on which Manoj Kumar and his two sons are travelling. Thus, two contrasting Indias can be seen: On the one hand, there are the rich and the powerful like Bal Thackeray who is referred to as "the permanent boss of the city" and on the other hand, there are the slum dwellers who are poor and destitute. Apart from all this, there are some budding talents also in the country like Sachin Tendulkar who is referred to as the "living viceroy of God" (Adiga, 2016).

Mohan, to teach cricket to his two sons, Radha and Manju, has weaved a web around them which they cannot transgress. Radha feels they are "prison bars" (Adiga, 2016) as various restrictions are imposed on them like not shaving until 21 and not driving a car. It is a sort of disciplinary institution that is seen in Foucauldian power struggles, and it is one way of making people trained and efficient so that they can contribute effectively to their own lives and to the society around them. Mohan has had his own idiosyncrasies since childhood as once he remarked that India would be manacled at the hands of some Muslim powers just as they had overthrown the governments in Egypt and Syria. The mentality of Mohan also depicts the general tendency of bigotry among many people in the nation. There are also other problems depicted subtly through the novel as Aravind Adiga always does; the irony of huge billboards showcasing images of politicians along with broken houses portray a contrast between the ruling class and the ruled ones.

"Push and survive" (Adiga, 2016) is the mantra of Mumbai especially when one talks of the Mumbai local trains. Mohan Kumar's childhood was also tough. From earning three and a half rupees a day to preparing his sons for cricket, it has been a challenge for him. His life changes when he starts to read and write. So, education can be seen as a harbinger of change. When in Mumbai with his two sons, he faces an uphill task. Many times, he feels dejected and even resorts to cheating and fraud for making a living in Mumbai. He, however, gets caught and is put in jail and that in a country where the "big thief walks free "(Adiga, 2016) but "small thief gets caught" (Adiga, 2016).

Ultimately, he finds Mr. Anand Mehta who decides to sponsor his two children. Mr. Mehta is foreign returned, has high aspirations and loves to follow his own path even if it means suffering losses and failures. It is a perfect example of being parrhesiastic as Parrhesia is not only about speaking one's mind but also following it up with action that is of your doing. This is a character having its strengths and weaknesses, weaknesses being his visiting to 2-star hotels "that was melodious with moonlighting college girls" (Adiga, 2016), an indirect reference by the author towards prostitution. He has had his ups and downs. At one point in time, he was pessimistic about cricket as he came to know about the deep hands of match fixing in cricket. He wants his fellow countrymen to know "the truth about their most cherished national memories" (Adiga, 2016). This is a case of hiding the real identity of something to deceive people. In fact, the absence of Parrhesia does that. It gives room for manipulation and deceit. Even his wife does not support him in his endeavors. She is always skeptical of him as she would say, "what if the two sensations ran away from Mumbai with their money and went back to their village" (Adiga, 2016). Her trust in him or one would say mistrust is what leads him to lose many of his investments.

Mumbai is a city of dreams for many. The writer feels it is a dying city yet it is tenacious because of the integrity of its people. That is what keeps it going. Being honest and straightforward is what holds the city together. Mohan Kumar has promised that he would pay back whatever he owes to Anand who is sure about it. Parrhesia is the hallmark of the conduct over here. Mohan Kumar's honesty is what has led him to get a contract for his children, to etch out a decent living for them so that they can take care of themselves. This is what Foucault mentions in Parrhesia, being something that leads one to the care of oneself and ultimately to society. In this case, two young and budding cricketers will be born and who will glorify the nation on every front. But this road to success is not an easy one for many Indians. As the writer mentions, there are troubled waters before the 1991 reforms kick in. The socio-economic condition of people is very dim during this time. Most people cannot save anything. Mohan Kumar has been persistent in getting the due for his sons but not everyone is like that. When Mohan announces that they would be buying a new house with the money that they have saved every month, Manju, his younger son, is delighted and goes "bullocking down the bridge" (Adiga, 2016)) to a place where there are "a group of unemployed young men"(Adiga, 2016) smoking and listening to cell phone radio. This is the reality of India. Contrast is drawn here between the common life of citizens and someone like Mohan Kumar who rises from the ashes. Still, the journey for Mohan Kumar was not an easy one. In fact, it will not be easy for anyone who tries to change his standard of living. In a country where the rich become richer and the poor poorer, a person tries to break away from the shackles and achieve prosperity. He is somewhat successful in doing that; yet his plight shows on his face in the form of "a pre liberalization gaze" (Adiga, 2016) as Anand Mehta points out when he gives the cheque to Mohan on the first of every month. He has a lot of challenges in front of him as according to Aravind Adiga, this world runs with "the combination of money and influence" (Adiga, 2016) and this is necessary how things can be made to work here.

Governmentality refers not only to the regulation of the conduct of others but also the conduct of oneself. Governmentality is a term coined by Michael Foucault. Its purpose is to secure the "welfare of the population, the improvement of its condition, the increase of its wealth, longevity, health, et cetera" (Burchell et al., 1991). It involves willing participation of the governed. By being parrhesiastic , one can achieve governmentality as one can regulate the conduct of others. It is not easy getting into cricket. The writer gives the example of Eknath Solkar who goes to bat in cricket and ultimately wins the Ranji match for his Mumbai team on the very day he loses his father. He comes and plays and says that he must do his duty. "On a day of supreme personal pain, on a day rich with excuses not to do his job, he does his job" (Adiga, 2016). How Cricket has transformed the lives of millions of young men and women is a spectacle to be seen in rising India. This has all been possible because of the ultimate toil that players like Sachin Tendulkar and MS Dhoni have shown that has catapulted them onto the world stage. They have been successful in controlling the minds, the aspirations of 1.3 billion people and that too without any force. This is the power of governmentality. Care of the self leads to the care of others. In fact, instances like the 1983 cricket world cup bring in a deep sense of unity among the people at a time when there are riots in the country. The then prime minister declares a national holiday in view of the victory of the Indian cricket team.

Making a career in cricket is a risky proposition. Most of the people, after making an assessment, find it too risky and decide to go abroad (Adiga, 2016). In a land where the population is huge and the opportunities do not match it, it is conducive for someone to go abroad. As in the words of the writer, most of the "socialist factories" (Adiga, 2016) in India are already shut down so there is a huge unemployment crunch. Moreover, it is not unemployment but the lack of strong will that also determines the failure in this country. It is the will to enter the terrain of the unknown, the unchartered territories where no one has ever dared to dream. It is on following these naive paths that one achieves unparalleled success; an entrepreneur is created in the process, thus unleashing a score of opportunities for the youth. Anand Mehta is one such entrepreneur who has carved his niche in the tinsel town of Mumbai. He has invested in the two sons of Mohan and when finally, the elder one becomes famous after scoring a triple century in a school match, he finally feels his gamble is going to pay off and thus exhibits his emotions by "devouring all the bread in the Kumar home" (Adiga, 2016).

Radha thus becomes the cynosure of all eyes. His younger brother Manju is also not far behind. But there is one flaw in Manju's character. He does not have the courage to do things his way. His father wouldn't allow him and his brother to shave even and they would follow his instructions in a docile manner. This anomaly, as the readers of this novel find out later, becomes the graveyard for Manju. Mohan has control over his two sons until the time they come to Mumbai and join Ali Weinberg's academy. Here the power rests with Tommy Sir and it is he who decides the future of the two players. Their father is left in the lurches which he repents later. So, the power dynamics shift from the father to the coach and the sponsor who invests in the two players as the father realizes in the "step by step encroachment of his paternal rights" (Adiga, 2016) by Tommy Sir.

With power also comes criticism. One who is at the helm of affairs is bound to be ridiculed also. There is always the negative as well as positive side of the same coin. Karim Ali, who is the founder of the Ali Weinberg International school and many other prominent institutions in Mumbai, thus being a very prominent and powerful personality of Mumbai, finds himself in the same spot. He is loved as well as hated at the same time depending on person to person. But it is generally considered that he breaks some rules to expand his empire as the writer mentions that "this city's laws were meant to be broken" (Adiga, 2016). A similar situation arises in Aravind Adiga's *The White Tiger where* the protagonist, who is also the servant, kills his master to rise to fame and build an empire. So, some sort of fraudulent means is used to rise to power. Michael Foucault says that "Where there is power, there is resistance, and yet, or rather consequently, this resistance is never in a position of exteriority in relation to power" (Foucault, 2009). So, since Karim Ali is in no doubt at a position of power, he is despised by many people. At the same time, he is also liked by many people. This shows the subjectivity of truth. Truth is relative in nature depending on the person who sees the truth. Or one can say that Ignorance is the reason for the truth being hidden. Being truthful about one's failure is also very important as it gives a lesson to others to take the right step forward thus leading to governmentality. Tommy Sir loves studying about failure as it has a "tragic grandeur" (Adiga, 2016). Moreover, it inspires generation after generation when they study about the near victorious heroics of their ancestors. Sir Tommy is a true parrhesiast. Despite the game of cricket getting murkier with allegations of betting and match fixing, Sir Tommy sticks to his stand of ethics and morality and does nothing that would be injustice to cricket. In the process, he produces some of the finest players that cricket has seen. It is because of his and many other honest coaches' fair play techniques that India has seen raw talent budding to play the finest game in which India has excelled in the recent years.

Manju is depicted as a homosexual in the novel. His elder brother Radha does not like that and so expresses his disapproval straightaway. The reason for doing so is not the betterment of his brother but because Manju has overtaken him in terms of play on the field. So, Radha, in a way, mocks his brother. So, this situation cannot be termed as a parrhesiastic situation but nevertheless it helps Manju to focus on his game and momentarily forget about his friend Javed.

Power comes with knowledge and experience and one who has more experience exhibits greater power. In the case of Javed and Manju, who are getting close to each other in a homosexual relationship, Javed is already the superior one as he has got into one such relationship on his 16th birthday. So, he smiles at Manju condescendingly (Adiga, 2016). And with power comes resistance and so when Manju rebels against his very own father when he gets his beard shaved, his father can only look at him with astonishment. It is like the tables have turned and the weak have become powerful. This is what happens when one employs parrhesia. It leads to a reversal of positions.

Adiga mentions *Animal Farm* by George Orwell in which the animals' rebel against humans, cease power from them and ultimately spell their own misfortune. This is what is going to happen to Manju as well. After rebelling against his father, Manju is going to spell his own doom. Javed, who has instigated Manju into rebelling against his father, says his own father of being afraid of him. "All of them are. I told you, read The Animal Farm, Manju" (Adiga, 2016). Javed also provokes Manju into admitting that he is a gay and tries to make him talk about sex but Manju, out of fear, not necessarily knowing the origin of that fear, is not courageous enough to accept the truth that he is a gay and tries to run away from the realities of his life. But the truth comes to haunt him continuously and he is embroiled in a tussle with his inner self. It is a stage of parrhesia where he is trying to reveal the truth about himself but is unable to do so.

Foucault mentions that Parrhesia is that activity by which one nourishes oneself with true discourse" (Foucault, 2011). In fact, throughout the novel, there is a power struggle as Mohan wants both his sons to be cricketers despite the unwillingness of the younger one, Manju. Manju is forced to join cricket although he loves Science. Although he plays better than his elder brother, he is dissatisfied with what he is doing. He wants to break the shackles which his father has tied him with. If instead of exercising his power, if Mohan would have regulated their conduct as in the case of governmentality, which is defined as "the conduct of conducts" (Foucault, 2014), his sons would have done better.

Cricket is the game where one must be honest because even if one gets into the team by any means whatsoever, "when he/she stands at the crease, all three stumps fly" (Adiga, 2016). At the top echelons of the game, it is the talent and determination that counts, not nepotism of any sort. So, the right conduct is very significant and that is achieved by being parrhesiastic or truthful. Parrhesia is not only free and unrestricted speech but also fearless action because ultimately what one thinks is what one does. One should not conceal one's follies and performing little actions in the right direction ultimately led to larger-than-life stories. This is what is shown by Manju in the end when he is made to choose between Javed and cricket. Finally, he overcomes the devil in him to leave out Javed and returns to cricket. That is self-care, the ultimate form of Parrhesia. As Foucault mentions, "What is at stake in this new form of parrhesia is the foundation of ethos as the principle on which conduct can be defined as rational conduct in accordance with the very being of the soul" (Foucault, 2011). Ultimately, it is also the victory for their coach Tommy Sir who has also given up on any chance of success. With one success in cricket, many others rise and so is the case with right conduct. When one does the right work with the right intentions, millions of people come out of poverty and destitution. As Tommy Sir points out, "they do have characters in the slums" (Adiga, 2016). From a slum in Dahisar to the lanes of cricket, Manjunath has done it and with it appends the fortunes of Tommy Sir and Anand Mehta.

4. Conclusion

This paper concludes that if one does not exhibit Parrhesia or the courage to speak the truth as illustrated by Michael Foucault, one cannot make the best use of his/her life and must ultimately settle down for a mediocre life. Also, Governmentality can bring about the best in an individual without having to coerce or force someone into submission. If the father had exhibited Governmentality, his sons' careers would have taken a different but positive trajectory. Mohan tries to use his position to dominate his two sons. Rather than understanding the desires of his two sons, he is hell bent on fulfilling his own aspirations. He cannot be blamed because of the circumstances in which he has lived his entire life. For a poor person in India, options are very less and in the process of achieving something worthwhile in life, one often tends to overdo a particular thing. This is what happens in Mohan's behavior towards his two sons as he behaves autocratically with them. He wields power over them. The sons do not like it and they rebel ultimately to the chagrin. But now, one of them has developed a weight transfer problem due to which he cannot play cricket well and the other one, despite having all the talent, has left his home and cricket after a gay guy named Javed Ansari. So, their cricketing careers are virtually ruined. It is a sort of anti-parrhesia where partial or incomplete realisation of parrhesia ruined governmentality.

REFERENCES

1. Adiga, A. (2016). *Selection Day: Netflix Tie-in Edition*. Pan Macmillan.
2. Burchell, G., Gordon, C., & Miller, P. (1991). *The Foucault Effect: Studies in Governmentality : with Two Lectures by and an Interview with Michel Foucault*.
3. Foucault, M. (2009). *Security, Territory, Population: Lectures at the Collège de France 1977–1978*. Macmillan.
4. Foucault, M. (2011). *The Courage of Truth*. Springer.
5. Foucault, M. (2014). *On The Government of the Living: Lectures at the Collège de France, 1979-1980*. Springer.
6. Ojakangas, M. (2012). Michel Foucault and the enigmatic origins of bio-politics and governmentality. History of the Human Sciences, 25(1), 1–14. https://doi.org/10.1177/0952695111426654

Impending Inquisitions in Humanities and Sciences (ICIIHS-2022) – Dr. Mohan Varkolu et al. (eds)
© 2024 Taylor & Francis Group, London, ISBN 978-1-032-78829-6

Showcasing the Hidden Multitudes in Literary Paradigms

2

L. Santhosh Kumar*
Assistant Professor of English, Kristu Jayanti College (Autonomous),
Affiliated to Bengaluru North University, Bengaluru, Karnataka

ABSTRACT Literature in contemporary times traverses the unheard truths and foregrounds the veracity of the falsified and the original truths. Muted voices are brought to the limelight and create counter histories due to manipulating truths that happened deliberately due to political monopolisation. After the advent of cultural studies, literature underwent a paradigmatic shift which paved the way for a multidimensional approach to literary studies. Literature does not confine itself only to books; what history books cannot do, literature can do with its language. Socio-political discourses play a vital role, and to a greater extent, they create rampant discourse in the literary and social paradigms. History and Historiographies play a vital role in the literary sensibilities of Marginalised literature. The landscape and Mindscape of an individual have got a unique significance in the present scenario. Books are said to be the repository of cultures which will keep the wound alive. On the whole, this research article foregrounds the plight of the muted voices and also culls out their multifacetedness which is the need of the hour.

KEYWORDS Veracity, Political monopolization, Cultural studies, Socio-political, Multifacetedness

1. Introduction

Literature is said to be culture-specific production. Authenticity and documentation of facts play a vital role in contemporary discourses. Voices of the Marginalized are being muted by the power structures owing to political monopolisation. The supreme irony is that the element of subversion and containment are being state-sponsored. Academia emphasises the facts which are supposed to be authentic, and the truth is that what history books cannot do, literature can do miracles with its language. Marginalised voices challenge history and historiographies, and they also emphasise that history books from the good old past had spoken only about the importance of an individual. In contrast, they have not spoken about the plight of the marginalised voices.

Testimonios are said to be the social documentaries which act as a threat to power structures. This particular branch of literature originated in Latin America, where women folks in Latin America protested against the patriarchal structure of the government. In the Initial stage, academia considered that Testimonio belongs to one of the schools/waves of Feminism, and later it was understood that testimonios traverse the unheard and hidden truths so that the unknown becomes known. The Best way to learn Subaltern is through Testimonial Literature. (John Beverly 1989). This paper investigates the plight of the Marginalised voices and, most notably, the importance of oral tradition regarding Tribal discourses; muted voices are being explored. Books are said to be the repository of culture, which will keep the wounds alive and most importantly, they act as a social documentary. Poems from Tribal discourses talk about the element of displacement, dislocation, and disjunction, and the significant trajectory is on

*santhoshcool247@yahoo.com, lsanthosh@kristujayanti.com

DOI: 10.1201/9781003489436-2

the internal conflict, which reflects the disturbed mindscape and the landscape of the natives. The word development is said to be a negative term in tribal discourses, and they also condemn the capitalistic attitude of humankind. Their main emphasis is that man should not think he is the crown of all creations created by God. Paradigmatic Shift is one of the prominent features in literary studies, and it creates a rampant discourse in the text's literariness. Most importantly, it makes the literary world redefine literary sensibilities, which is the need of the hour.

2. Literature Review

In the tribal discourses, women-centred issues were brought to the limelight. Womenfolks is said to be the custodians of culture, and they believe that Nature is an entity interwoven in their daily lives (Dr Ajit Kumar Kullu 2020). Emancipation of Women's struggle and the exploitation of natural resources, westernised education in tribal schools paved the way for the degradation of tribal culture (Sanjay Nath, 2016). Every natural resource and landscape plays a vital role in the tribal discourses, and they believe every habitat is interwoven with their culture (Amrita Sengupta, 2016).

Tribal Discourses Showcases the Hidden Histories and also, at the same, Act as a Threat to Power Structures

Tribal Poetry

Poetry is literature in its most intense, imaginative, rhythmic forms. Literature in contemporary times showcases radical thoughts and also, at the same they, create counter histories in order to foreground the falsification of truths. The muted voices of Women and also the pathetic state of the tribal landscape are foregrounded in their poems. Literature is said to be a socio-political tool which vehemently protests against the power structures. Also, they blur the line between personal and political so that the hidden brutalities of political monopolisation are made known to the world. Tribal poetry, far from being squashed politically, has become perhaps *the* voice of the Yemeni revolution. Not coincidentally, perhaps, rap poetry is now at the forefront of protests against the government in Dakar, Senegal. Although social media are being vaunted as the communications backbone of street revolutions today, oral poetry is their flesh and blood. (Steve.S. Caton 2012).

Political Ramifications in Tribal Poetry

Literature has produced renowned masterpieces, and also to be noted that what history books cannot do, literature can do miracles with its language. History has become an inevitable phenomenon in the literary discourses which deal with the element of protcst. Histories have created many poets and influenced histories, traditionally archiving their histories through songs, folklore, and myths. Adivasis have aced their way through their ancestral accounts accumulation. Adivasis are folks who function their lives through flourishing flora and fauna encapsulated with enriching cultures and practices and lives among the areas of the Indian subcontinent. Adivasi is a common term coined in the 1930s to address the indigenous groups of India, while the legal term "scheduled tribes" is used in the constitution. They are the most prehistoric inhabitants of the sub-continent and are a heterogeneous group with diverse ethnicity and linguistics. Post the Aryan intervention; the Adivasis began to trade with people of the plains; it was during the mid-eighteenth century that the indigenous people of the east revolted against the political British and its intrusive regulation upon the mainlands of the indigenous folks. This renounced the defenceless position of the land and its people. Even today, the Adivasis encounter various forms of social discrimination, political power lash, and remain economically stagnant. (Anushri Muthusamy, 2022)

History and Political Ramifications Play a Vital Role in Tribal Poetry

Tribal Poetry: A Paradigmatic Shift

Tribal Writers

Literature and History are considered irreconcilable binaries, whereas it is not proper to a greater extent. Both are reconcilable binaries that academia fails to consider and acknowledge it. Tribal poetry emphasises on Landscape and Mindscape of the people. Historiography plays a vital role in Marginalised works of literature, and most importantly, they make the unknown known. The word development is said to be a negative term in tribal discourses, and it will be apt to say that tribals are the only people on earth who are in tune with Nature. Tribal poetry is said to be the epitome of cultural studies, which bridges the gap between Anthropology, politics, and social sciences. Owing to Tribal Poetry, Anthropologists and humanistic scholars have an intellectual thrive on give and take. Thus begins the Interdisciplinary approach to knowledge. The writings of Jacinta Kerketta showcases the hidden multitudes and also at the same time they foreground how power structures operates at various levels in order to get rid of political monopolisation. The word development is said to be a negative term in tribal discourses and the writer like Jacinta Kerketta vehemently protest against the institutions which act as an hindrance to one's own culture.

3. Methodology and Conclusion

The poems taken for study in this paper are the selected poems of Jacinta Kerketta and the selected poems of Mamang Dai. Both poets have a unifying element in their poems which talks about the element of despair and isolation. It is evident to say that the poems of Jacinta Kerketta and Mamang Dai are said to be the repository of cultures which will keep the wounds alive and also, at the same time, draw the line between personal and political. The element of protest is showcased in their poems which question the power structures, and also at the same time, they bring the veracity of the original and falsified truths.

Fleeing the roof over their heads, they often ask:

O, city!

Are you ever wrenched by the very roots

In the name of so-called progress? (OC 3-7)

The poetic lines of Jacinta Kerketta showcase the conflict between 'Roots Vs Routes'. It is a well-known truth that the tribals have an intense communion with Nature, and there is a strong belief that they associate every natural resource with their livelihood. Jacinta Kerketta expresses her protest through the mode of writing a poem. She condemns the capitalistic attitude of man, and she also emphasises that man should not think he is the crown of all creations created by God. The above poetic lines cull out the plight of tribals in their homeland. The Capitalistic society grinds the landscape and the mindscape of the tribals. The sons of the soil are forced to forget their ancestral roots and distance themselves, which makes them traverse the other routes. The developments happening in the current society by exploiting natural resources make Jacinta Kerketta undergo the pain of displacement, dislocation, and disjunction.

Now a dam to hold back the welling tears,

Now a dam to contain the seething rage.

These dams shall burst one day for sure,

When the boughs of sakura

From the hilltops in rebellion roar,

Sweeping out powers that destroy and displace.

And once again in the breeze will sway

The ears of paddy in their majesty,

Enclosed by mud mounds, no more by dams. (EOPTBD 23-31)

Jacinta Kerketta is a social rebel and activist with a sense of envisionment in her writings. She is a typical Adivasi, and she also registers the plight of her people in agriculture. For her people, agriculture is said to be the essential occupation, and they are completely away from the capitalistic society; they also believe that it is only through organic farming that their mode of life will survive. River beds and dams are said to be the cradles and the birthplace of a civilisation, and they have a strong unifying force in one's life. She is afraid and also in pain that the capitalistic society, known for cunningness, might exploit their dams for the welfare of others, and the people of the tribal region will be in despair for their livelihood.

Why did we think gods would survive

deathless in memory,

in trees and stones and the sleep of babies;

now, when we close our eyes

and cease to believe, God dies. (R 18-22)

Mamang Dai is an ardent lover of Nature, and most importantly, in her poems, she talks about the unification of natural resources. She is a mystic poet and firmly believes that Nature is a good repository for the people of her landscape and also they have a vital communion with Nature. Tribal discourses are known for their matriarchal legacy. They also believe that womenfolks are said to be the custodians of culture and are said to be the typical representatives of their own culture. Mamang Dai expresses her protest through the overhead lines and emphasises how the people of her own landscape were being dejected by the power structures owing to political monopolisation. Trees and other natural resources are said to be the essential elements in the tribal discourses, and the sound of a tree and the river flow can elucidate many meaningful and unheard voices which foreground the element of nostalgia which can be painful and also an ever-lasting one.

This paper on the whole, foregrounds the multifacetedness of tribal voices and simultaneously shows the element of literariness in the writers mentioned above. Voicing out the muted voices is the essential trait of contemporary literature, and literature is said to be socio-political discourse which encompasses all disciplines under one roof. Writers like Mamang Dai and Jacinta Kerketta they emphasis on the fact that man should give up his ego in order to achive the eco consciousnesses. Tribals are also human beings with different multitudes and also they draw the line between personal and political so that unheard truths are brought to the limelight. The poems of Jacinta Kerketta and Mamang Dai insists that the literariness of a text is decided by the Historicty of a text and the textuality of history. The adavantage is that what history books cannot do literature can do miracles with its language. The writers of tribal discourses make every individual to re-read and re-learn and revisit the text so that the cultural beauty is made known to the world and the power structures in the good old past made it to be unknown. The advent of culture studies has made literary texts to be Autotelic which is the need of the hour.

REFERENCES

1. Barla, A. G. (2015). *Indigenous heroines: A saga of tribal women of India.*
2. Dai, M. (2022). *The inheritance of words: Writings from Arunachal Pradesh.* Zubaan Books.
3. Devy, G. N. (2002). *Painted words: An anthology of tribal literature.* Penguin Group USA.
4. Giri, D. (2020). *Tribal perspectives in India: Critical responses.* Booksclinic Publishing.
5. Kerakettā, J. (2020). *Angor.*
6. Ramesh, M. (2018). *Tribal education in India: A case of lambada Tribe in Telangana.* Education Publishing.
7. *Recent trends in the tribal literature of the north-east India.* (2021).
8. Sen, A. K. (2017). *Indigeneity, landscape and history: Adivasi self-fashioning in India.* Taylor & Francis.
9. *Tribal literature and oral expressions in India.* (2021).
10. Yadav, A. (2017, 22). *The anger of Adivasis turns to poetry of anguish and hope in a young woman's hands.* Scroll.in. https://amp.scroll.in/article/808591/the-anger-of-adivasis-turns-to-poetry-of-anguish-and-hope-in-a-young-woman-s-hands

Impending Inquisitions in Humanities and Sciences (ICIIHS-2022) – Dr. Mohan Varkolu et al. (eds)
© 2024 Taylor & Francis Group, London, ISBN 978-1-032-78829-6

Language and Society 3

Vanishree Adoni*
Assistant Professor, KITS Singapur, Huzurabad, Karimnagar

ABSTRACT This article is intended to give an idea of what the language is and how it is related to society. As the society means, living beings and all living beings communicate in one way or another. And the way humans speak needs a language: must be understood and serve the purpose of the communication to progress in all fields of life.

It is through language that we communicate with the world, define our identity, express our history and culture, learn, defend our human rights and participate in all aspects of society.

If the above said facts are lacking or hindered, the growth of a language is gradually stopped. The same happened with Sanskrit and English in Bharath during the freedom movement.

During foreign invasion to dominate and rule, a situation was created so that Indians were curtailed using our own languages and prefer other foreign languages. Because there is a hidden treasure of knowledge regarding medicine, yoga, spirituality, science ect. in books which are in sanskrit. So English was forced into our society and in today's world it is mandatory to keep abreast to have knowledge and development across the globe.

KEYWORDS Language, Society, Communication, knowledge

1. Introduction

Language is essentially a system of communication where ideas and feelings express objects, events and thoughts. The history of language dates back to several thousands of years. As language is the key to human lives, it can be used as an instrument to transfer communication among people. Shelly (2022) quotes Malinowski in her article, as language is the necessary means of communication; it is the one indispensable instrument for creating the ties of the moment without unified social action is impossible.

Language can thus be assumed to be at the interior part of humanity. Language grows based on its usage. When one communicates regularly for almost all purposes in society like business, education, legal and medical, language grows. And if one is restricted from using it, its growth slowly deteriorates, and a stage comes, where society stops using it.

In this article Sanskrit and English language are taken as examples to show how Indians restricted use of Sanskrit only to rituals and English has become part and parcel of the society.

*vanimbnr@gmail.com, vani.adoni@yahoo.com

DOI: 10.1201/9781003489436-3

2. The Purpose of Communication

The main goal of communication is to express the feelings, emotions and thoughts and serve the purpose. And this need for communication was the main cause of language making. The old proverb "Necessity is the mother of invention" proves itself; it was to meet the needs of daily life that helped and speech was the result. Charles Winick (2021) defines language as, a system of arbitrary vocal symbols, used to express communicable thoughts and feelings and enabling the members of a social group or speech community to interact and to cooperate. It is the medium of oral expression.

So, it is obvious that one needs to acquire knowledge about that language in which the information is hidden for the prosperity of the society. In Vedas which are written in Sanskrit, there is a lot of information regarding all fields, like medicine, spirituality, science etc. This information through Sanskrit is used to find a solution in the society: the solution can be of any type like health, societal, legal, agricultural etc. The purpose of communication in other words is passing information from one to another in society to enrich one's life- which was served in those days.

When the purpose of using Sanskrit was not served during the foreign invasion, Hindustanis slowly started ceasing the use of this language. On the other hand Persian, Arabic and English were introduced in the educational system. Wikipedia (2022) clearly mentions that, "Islamic education became ingrained with the establishment of Islamic empires in the Indian subcontinent in the Middle Ages while the coming of the Europeans later brought western education to colonial India."

3. Effect of Language on the Society

Effect of language on society is a result of not one cause but of diverse aspects. It is, in fact, a social creation, a discovery, an innovation of a whole in itself. The language that we speak influences our cultural identities and our social realities.

The effect of other languages for example English started showing on all walks of life of Indians and still continues. It depends on the horizons the language plays on different fields like medicine, education etc. Sanskrit has some limitations when compared to English. For example in medical field, Dr. Anshu (2016), in her article, "Evolution of medical education in India: The impact of colonialism", states that, "There was a clash of cultures where the East was seen as weak against the powerful knowledge of the West." And "Allopathic practitioners saw themselves as modernizers and often treated their indigenous counterparts with contempt for their "inferior knowledge." Local knowledge was labeled unscientific or irrational."

Language is therefore a social phenomenon which is the indispensable instrument of society and on which it acts continuously.

Thus language creation is not the job of one person or of one period but it is an organization, on which hundreds of generations and umpteen individuals have contributed. Markandey Katju (2010) shares, "The various immigrants/invaders who came into India brought with them their different cultures, languages, religions, etc. which accounts for the tremendous diversity in India".

And further J. Mark (2020) remarks that, "This belief system developed uninterrupted until the rise of Islam in the north of India beginning in the 7th century CE which became pronounced by the 12th century CE. Islamic rule only gradually came to tolerate Hindu practices. A far more significant threat to the Vedic vision came later in the form of British colonialism and imperialism in the 18th-20th centuries CE. The British tried to convert the Indian people to Protestant Christianity and expended considerable effort in re-educating the populace and dismissing Hinduism as an evil superstition." Thus, Indians whether willing or not, were compelled to follow the Islamic rulers followed by the Britishers and neglecting our own languages and culture to a larger extent. The result is today's society where Indians cannot imagine a world without English and Sanskrit has become merely an optional Language.

4. The Role of Language in Society

Sanskrit is considered spiritual and Godly, whereas English is considered to earn livelihood. Therefore Sanskrit is elective in schools and English compulsory in our educational system. We hardly find a teacher to teach in schools and colleges but on the contrary is English. Everyone is chasing English . The people who converse are considered highly sophisticated and elite in modern society. Many movies, stories in novels also reflect the same.

According to Y. Lopamudra Ganguly (2014), "Sanskrit is vital to Indian culture because of its extensive use in religious literature, primarily in Hinduism, and because most modern Indian languages have been directly derived from, or strongly influenced by, Sanskrit." But according to Subhashree Das (2017), "There are several factors that make us learn English. First of all, the language has international standards. Everyone needs to learn English to get in touch at the international level. In the educational field, it is seen that much of the syllabus is written in English. Children are encouraged to learn English at the starting levels. And, accordingly, as they advance to the next levels they study almost all the subjects in English." Both are correct in their own perspective . But one should accept the fact that English assumes growing importance in India.

5. The Impact of Sanskrit Language on the Society

Sanskrit is the ancient language of the world. Knowledge of Sanskrit is a treasure of the world. It is considered as "Deva Bhasha" or the language of the Gods by Indians. This language is a sign of the great Indian heritage which has manifested complete freedom in the search of wisdom and has shown acceptances towards universal truth. This unique language contains not only a good account of wisdom for the people of this country, but it is also an unparalleled and right way to acquire knowledge and is thus significant for the people of the entire world.

Wikipedia (2022) cites, Sanskrit has been the predominant language of Hindu texts encompassing a rich tradition of philosophical and religious texts, as well as poetry, music, drama, scientific, technical and others. It is the predominant language of one of the largest collections of historic manuscripts. One can see the evidence of Sanskrit chronicles which manifests a comprehensive plan of the humans like spiritual, emotional, intellectual and material. Its philosophical composition confers a clearly organized way of appreciating our relationship to the rest of the world and lays out suggestions on how to lead life well. Writers and poets have artistically crafted the language to offer timeless perceptions of humans. Not only Indians but also people across the world value Sanskrit.

And according to the Dalai Lama, the Sanskrit language is a parent language that is at the foundation of many modern languages of India and the one that promoted Indian thought to other distant countries. In Tibetan Buddhism, states the Dalai Lama, Sanskrit language has been a revered one and called *legjar lhai-ka* or "elegant language of the gods. wikipedia.org (2022)

Even though many eminent personalities across the globe have referred to and defined Sanskrit in different ways, today its use is restricted to rituals. There are myriad reasons. According to Wikipedia (2022), Sanskrit declined starting about and after the 13th century. This coincides with the beginning of Islamic invasions of South Asia to create, and thereafter expand the Muslim rule in the form of Sultanates, and later the Mughal Empire. Further followed by other invaders and continued for more two hundred years.

But still Indians are inclined to this rich language and want their next generations to learn and benefit because of the hidden treasure. The following section discusses the problems of Sanskrit for being moderately used among Indians:

Four tools why Indians minimized use of Sanskrit.

Observation

Indians slowly started observing that during Muslim rule other foreign languages started dominating and gaining popularity for varied reasons like, compelling by the rulers and employment etc. According to Wikipedia (2022), "Muslim rulers patronized the Middle Eastern language and scripts found in Persia and Arabia, and the Indians linguistically adapted to this Persianization to gain employment with the Muslim rulers." So when society is compelled and discouraged not to use a language it loses its importance. Same happened with the Indianization of English also.

Consideration

Indians after careful thought considered accepting other foreign languages as superior and better, to have safe and secured status in the society. This continued throughout foreign rule on Bharath, right from Mughals to Britishers. Because, the main purpose of any society is smooth going and livelihood . And it is through language that we communicate with the world, define our identity, express our purpose, protect the rights and property and participate in all aspects of society. When this is curtailed by the rulers, one is forced to adapt to the ruler's language and adopt their culture too.

Need

There was a need to create an identity in the society by the individuals. The only means was to understand the invaders and learn their traditions and customs. Because language is also a distinctive and effective means to influence the individual's as well as the society's social, economic and educational conditions. Language can decide the potentials of an individual and a group as a whole, as well as play a crucial part in building recognition in the society. So, Indians have not only neglected their own native language and culture but converted and accustomed to foreign traditions.

Acceptance

India admitted the fact that Sanskrit which is the basis for many languages may not serve the purpose and has accepted Arabic, Persian and English for their survival and English continues till date. As Priyadarshi Dutta (2017), mentions in her article, Why Sanskrit remains confined, "There is dichotomy about that classical language. It remains a popular subject at higher secondary, graduation and post-graduation level. The only other place it exists outside the curricula is Hindu religious ceremonies (*karma kanda*) like pujas, marriages and shraddhs etc. But otherwise there is practically no journalism; no modern literature; no film industry; no television channel or music industry in the language."

6. The Impact of English Language on the Society

When the British introduced English to India in the 19th century, knowledge of Sanskrit and ancient literature continued to flourish as the study of Sanskrit but changed from a more traditional style into a form of analytical and comparative scholarship mirroring that of Europe.

Shelly Shah (2022) in her essay quotes Whitney; Language making is a mere incident of social life and of cultural growth. It is as great an error to hold that at some period men are engaged in making and laying up expressions for their own future use and that of their descendants, as that, at another period, succession shall find expression. Each period provides just what it has occasion for, nothing more. The production of language is a continuous process; it varies in rate and kind with the circumstances and habits of the speaking community, but it never ceases; there was never a time when it was more truly going than at present.

It is absolutely true regarding English in India. After independence the great leaders decided that English will be the national language for over a period of time but it never ceased, still we use and are privileged to use it. Because of the four tools the English language carries.

Observation

Even though many great leaders like Mahatma Gandhi, Tagore, Dayanand Saraswati and Gokhale supported vernacularisation medium of education, which did not work out because our own leaders like Raja Ram Mohan Roy who discouraged Indian languages and encouraged English. It shows that few leaders were concerned and selfless about society and their needs and could feel it by learning the English language to serve the purpose.

Macaulay's Minute 1835 clearly states that this group would be "Indian by blood and color, but English by tastes, opinions, morals, and intellect." As the Britishers wished to create a pool of Indians capable of serving them and remaining loyal. Here the need of the society particularly Indians who are deprived of using their own language, are compelled to learn the rulers language. And the needs are satisfied with the use of English, as it flourished in all the sectors of the society. Krishnaswamy aptly mentions that the English language and the Western system of education were only the means for a cultural and religious conquest of the Hindus, the ultimate aim being trade and political power. The master and pastor were to be used as tools to bind the empire, So, Grant pleaded that the British government forcefully introduced English in India. Further it changed from dominative to collaborative language in the society.

Consideration

After careful investigation about English over a period of time, Indians considered that by learning English the horizons are opened in all the sectors of the society. Indians started enjoying the privileges and felt it is the only way is to learn English. Sangeeta Arora (2013) clearly states in her article about the use of English and role in the society, "English is used by scientist, business organizations, internet, education tourism sector so it plays a dominant role in most of the fields of the present world. In short most of the world's communication is done in English and it has become principal language of communication." Unique Position of English in the Present Scenario"

Need

English was used by Indians as it was needed. If one assesses, today the demand of society is met by the use of English to a larger extent. It is a widely conserved language for all purposes. And one finds it difficult to find an immediate word in Indian languages when compared to English, for example Auto, Cigar, phone, wire, etc.

English is everywhere and in all domains, may be in education, medicine, economy, tourism and technology etc. People from different backgrounds use English. It has become a privilege to use English. Today it is necessary for society to keep abreast of the world when compared to any other Indian languages.

Acceptance

Those foreign languages which were forced on the Indians have become an indispensable part of the society. Indians accepted not only the English language but their culture and tradition. Conversions into Christianity were common during those days and continuing till today. English has become so popular in all fields because of its usage. One may not find a substitute word in our own mother tongue and to find a word in Sanskrit is far behind.

The belief is like when men think for a period of time about anything he starts feeling it and finally he implements it and that leads to habituation.

7. Conclusion

Language is an arrangement of symbols utilized by humans to converse or express ideas and thoughts in the society. The language refers to influences and effects of the society and vice versa. Therefore it is apt to say that the language and the society have a close association. This can be referred back to Bharath under foreign rule. Many foreign languages like Persian, Arabic, English etc. were forcefully introduced. And their tradition was also introduced in the society and till date it is followed. There is a parable that says that language indicates the nation.

Language eventually leads to the shape of civilization. It expands and blossoms when appreciated and contracts and shrinks, whether induced or forced in the society.

Finally, it is clear why Indians preferred English over Sanskrit, not only Sanskrit but many other languages across the globe too. The reason is that of foreign invasion or having less scope for development. As the languages and society has had a very large impact across all the earth, there was a need for any language to be a beacon in order to project itself towards the future society. As English language satisfies all criteria to be used, it is favored and promoted.

REFERENCES

1. Shah, Shelly. Article. (2022). *Language and its Importance to Society*, Essay.sociologydiscussion.com
2. Winick, Charles. (2021). *Sociology of Language: History and Society*, https://appliedworldwide.com
3. https://en.wikipedia.org/wiki/Sanskrit#Origin_and_development. (2022)
4. Anshu, A Supe. (2016), *Evolution of medical education in India: The impact of colonialism*. https://www.ncbi.nlm.nih.gov.
5. Katju, Markandey. (2010). *Sanskrit Literature and the Scientific Development in India*. Speech. Banaras Hindu University, Varanasi.
6. Mark, Joshua. (2020). Article. *The Vedas* - World History Encyclopedia. https://www.worldhistory.org
7. Ganguly, Lopamudra. (2021). *Importance of Sanskrit*. eduindex.org
8. Das. Subhashree. (2014). Article. *English assumes growing importance in India*. https://www.dailypioneer.com.
9. Dutta, Priyadarshini. Shah, Shelly. Article. (2022). *Language and its Importance to the Society*, Essaysocialogydicussion.com.
10. Swamy, Krishna, N. and Lalitha K.M. *The Story of English In India*, Google Books. Foundation Books. 11–13.
11. Arora, Sangeeta. (2013). *Unique Position of English in the Present Scenario*. www.ijhssi.org

Impending Inquisitions in Humanities and Sciences (ICIIHS-2022) – Dr. Mohan Varkolu et al. (eds)
© 2024 Taylor & Francis Group, London, ISBN 978-1-032-78829-6

Self Translation of Poetry by Agyeya: An Analysis

4

Manisha Gupta*
Research Scholar, School of Translation Studies and Training, IGNOU, New Delhi

ABSTRACT This research paper examines the English translation done by Agyeya of his four Hindi poems *Yaad* (याद), *Kitni Navon Mein Kitni Baar* (कितनी नावों में कितनी बार), *Saal-Dar-Saal* (साल-दर-साल), and *Chup-Chap Nadi* (चुप-चाप नदी). The importance of poetry translation is that it cannot be translated; it is always recreated. This study will analyse whether, being a poet himself, Agyeya has translated his poems without deviating from the essence of the poems. This paper will also highlight how as a translator Agyeya overcame the challenges of poetry translation, such as aesthetic, linguistic, and socio-cultural problems. This paper will also study that only a poet can translate poetry. This study will conclude its findings by analysing the characteristics of Agyeya's poetry translation.

KEYWORDS Agyeya, co-creation, cultural translation, literary translation, new writing, poetry translation, trans-creation, untranslatability

1. Introduction

Agyeya is the nom de plume of the legendary Hindi writer Sachchidanand Hiranand Vatsyayan. He was born on 7 March 1911 and died on 4 April 1987. He was a prolific author, poet, revolutionary, and freedom fighter. As an author and poet, he presented emotions, feelings, and truth to readers. His creations were heartfelt, and readers got easily attached to them. His writing skills made his novels, stories, poems, diaries, essays, memoirs, plays, and travelogues remarkable. Readers can feel the sufferings and desires of Agyeya in his works. Emotional aspects in his works are very crucial. He was a profound thinker. He laid the foundation of prayogvad or experimentalism in Hindi poetry with the publication of Tar Saptak. He has received the Sahitya Akademi Award, Jnanpith Award, and Golden Wreath Award. He has written both in Hindi and English languages. He has translated some Hindi and Bangla novels into English and English novels into Hindi. He has done self-translation also. He has translated his Hindi poems and novels into English. In this paper, we will study and analyse his four Hindi poems and their English translation. Four Hindi poems are *Yaad* (याद), *Kitni Navon Mein Kitni Baar* (कितनी नावों में कितनी बार), *Saal-Dar-Saal* (साल-दर-साल), and *Chup-Chap Nadi* (चुप-चाप नदी).

Yaad (याद) poem was written on 21 March 1969 at the Kabuki theatre in San Francisco. In this poem, Agyeya as a poet, has narrated his emotions and feelings of missing and remembering someone close to his heart. A pair of cranes is a symbol of love. Cranes make their pairs once in their life. If a crane in the pair dies, the other one lives alone, remembering its partner. Cranes form lifelong monogamous pair bonds. *Kitni Navon Mein Kitni Baar* (कितनी नावों में कितनी बार) poem is written on 30 May 1966 at Ljubljana (Yugoslavia). In this poem, Agyeya narrates the journey of a man in search of the truth and God. A

*sinha.mansi@gmail.com

DOI: 10.1201/9781003489436-4

man seeks God using religious traditions, cultures, and priests. A boat is a symbol of religious traditions, cultures, and priests. He narrates the bitter truths of materialism and the vices of a man. He showed us how a man is trapped in his earthly desires. *Saal-Dar-Saal* (साल-दर-साल) is written in September 1976 at Winsor. In this poem, Agyeya narrates the beauty of the cycle of seasons. He used the cycle of seasons to represent hope and a new beginning in life. *Chup-Chap Nadi* (चुप-चाप नदी) was written on 3 February 1976 at Heidelberg. Agyeya wrote this poem based on the primate human skulls and bones found near Heidelberg. His imagination gives momentum to the poem and wakes up humanity.

2. Poetry Translation

The word 'translation' is derived from the Latin 'translatio', which means carrying across. In general terms, translation is the act of transferring a text from the source language to the target language. According to J. C. Catford (1965), translation is "the replacement of textual material in one language (SL) by equivalent textual material in another language (TL)" (p. 20). Andre Lefevere considered translation as rewriting in his book 'Translation, Rewriting, and the Manipulation of the Literary Fame'. Translation studies focused on the 'cultural turn' later in the 1980s. Andre Lefevere and Susan Bassnett co-edited and published the book 'Translation, History, and Culture' in 1990. Lefevere and Bassnett emphasised the role of culture in translation work. They shifted the linguistic approach to the cultural approach in determining translation subjectivity and discussed the cultural influence in translation. Andre Lefevere, in 1975 argued on seven strategies of poetry translation.

Lefevere has stated the following seven strategies of poetry translation:

(i) Interpretation can be in two ways, i.e., version and imitation. The translator enjoys the freedom to alter the meaning and structure of the ST while using this strategy. With this strategy, the translator creates his own poem.
(ii) The phonemic translation is just like transliterating a text. When the translator attempts to reformulate the source text sounds, then the context and message of the original poem are lost and destroyed.
(iii) Rhyming translation is where the translator reproduces the source language poem's meter, rhyme, and rhythm in the target language. This is the most challenging rendition of the ST.
(iv) The metrical translation is the strategy where the translator aims to retain the meter of the original poem.
(v) Poetry into prose strategy is where the translator tries to convey the context and message of the original poem into the target language while changing the poetic form into prose.
(vi) Poetry into blank verse strategy is where the translator tries to reproduce metrical patterns without rhyme.
(vii) The literal translation is the strategy where the translator alters the structure, context, and meaning while doing word-for-word and group-for-group translations.

English Translation of the Hindi Poem *Yaad* (याद)

This poem in Hindi is beautifully crafted and written by Agyeya. In this poem, a pair of cranes symbolises love, which is forever, and the poet remembers his lost love. Agyeya translated the poem's title literally as 'memory' which perfectly conveys the poet's message and feelings. He translated the first line of the poem याद : सिहरन : उड़ती सारसों की जोड़ी as Memory: a shiver. / A pair of cranes in flight. The words 'shivering' and 'in flight' are not the appropriate selection of equivalent words that can depict the poet's true message and emotions. Tingling is the feeling, not shivering, we feel when we remember and miss our loved ones. The word 'flying' is a more appropriate equivalent than the word 'flight'.

He translated the second and third lines of the poem याद : उमस : एकाएक घिरे बादल में / कौंध जगमगा गयी as Memory: a closeness / Lighting across the heave of clouds. He translated उमस as closeness. Agyeya tried to represent the climate metaphorically, but this rendition is inaccurate. He translated the fifth line of the poem बादल में सारसों की जोड़ी ओझल as Cranes dissolving in the cloud. Here, Agyeya chose 'dissolving' for the word ओझल. This does not correctly convey and represents disappearing in the clouds.

English Translation of the Hindi poem *Kitni Navon Mein Kitni Baar (कितनी नावों में कितनी बार)*

Agyeya translated the title of the Hindi poem literally as How many times in how many boats. He has omitted the information on the place and date of creation as it is present in the original text of the source language. He has not translated अंतहीन सच्चाइयाँ / कितनी बार मुझे / खिन्न, विकल, संत्रस्त (the three lines before the last line of the poem). Due to this, the English translation of the Hindi poem will not be considered a complete translation.

This poem is about the human journey to know God and attain salvation. Agyeya has translated the third line of the poem मैं तुम्हारी ओर आया हूँ as I have voyaged towards you. The word 'voyaged' adequately conveys the original poem's context and message. He has translated the sixth line of the second stanza of the poem एक नंगा, तीखा, निर्मम पकाश as By a ruthless garish light. This translation puts stress on the central message of the poet.

Agyeya has translated the third line of the second stanza किन पराये देशों की बेदर्द हवाओं में as To what heartless winds of what alien lands. Agyeya chose 'alien' instead of 'foreign'. This dilutes the main context and message of the poem.

English translation of the Hindi poem *Saal-Dar-Saal* (साल-दर-साल)

The Hindi poem *Saal-Dar-Saal* (साल-दर-साल) is a poem on hope in life. Agyeya has translated this poem into English as Tree Calendar. The title of the Hindi poem *Saal-Dar-Saal* represents the life cycle, like the seasons. Seasons keep on changing and repeating themselves year by year. This change shows that life is not always the same. If there is sadness, then happiness will come. But the translation of the title into English as 'Tree Calendar' does not convey the same message as the original one.

Paar Saal (पार साल) is the first line of the first and second stanzas of the poem, but Agyeya has translated it as 'year before last' in the first stanza and as 'last year' in the second stanza. The beauty of repetition in the original poem is lost in the translation. In the fifth stanza, Agyeya translated लकड़हारों as 'woodsmen' and not as woodcutters. The word 'woodsmen' is not the correct equivalent for लकड़हारों.

In the last stanza, Agyeya translated the following lines as:

जिस पर	And over it hovering
मंडराएगी	Softly land
उतरेगी	A tiny humming-bird
पिद्दी-सी फूलचुही:	And with the boldness of love
प्यार से	Break into song.
ज़िद्द करती	
गाएगी!	

Agyeya translated उतरेगी as 'softly land' to convey the context and message of the poet. The word 'softly' serves the purpose, but the word 'land' does not seem appropriate to represent the action of sitting by a small bird. He chose the word 'boldness' for ज़िद्द. He translated गाएगी as 'break into song' and omitted the exclamation mark (!) in the rendition into English. He has also omitted the information about the poem's place and date of creation in the English translation. So, the translation does not truly represent the poet's feelings, emotions, and message.

English Translation of the Hindi Poem *Chup-Chap Nadi* (चुप-चाप नदी)

Agyeya has translated the title of the poem *Chup-Chap Nadi* (चुप-चाप नदी) as 'Soundlessly the stream. He has done literal translation of the first stanza of the poem. He used 'stream' and 'river' for the word नदी in the poem. This poem has only two stanzas. The last stanza is as follows:

और एक मेरे भीतर नदी है स्मृतियों की
जो निरंतर टकराती है मेरी दुरंत वासनाओं की चट्टानों से
जिस नदी की जिन चट्टानों से जिस टकराहट से
कोई शब्द नहीं होता
और जिस के प्रवाह में
बरफ़ की लड़ी की
छाया-सा
लटका हूँ मैं।

Agyeya has translated the last stanza into English as:

And there is this river of memories

Dashing themselves against the unending rocks –

Soundlessly

Above whose rush

Hangs an icicle's shadow

Me.

Agyeya has omitted मेरे भीतर in the English translation. He translated मेरी दुरंत वासनाओं की चट्टानों as 'the unending rocks'. He translated जिस नदी की जिन चट्टानों से जिस टकराहट से कोई शब्द नहीं होता as 'soundlessly'. He translated प्रवाह as 'rush'. These selections of words do not convey the original Hindi poem's context, emotions, and message. He omitted the information about the original poem's place and date of creation while translating it into English. The beauty, style, rhythm, and aesthetic of the Hindi poem are lost.

3. Conclusion

We can conclude that poetry translation interprets poetry in the source language and brings it to life in the target language. Translation of poetry is not only confined to form and meaning, but it also aims to preserve the beauty of the poems. Each language has its speciality, aesthetic, and rhythm. Translating poetry is a difficult path for translators. Robert Frost, an American poet, has considered poetry translation as an activity where poetry gets lost in translation. This is because poems have a rhythm, style, and shape because of the elements of poems. A poet expresses his thoughts, ideas, and feelings in a vivid and imaginative style, forming aesthetic tension. It is challenging for translators to preserve poems' aesthetics, style, rhythm, and shape. If a translator tries to get all the elements of ST poems in the TL, then the translation of poetry is impossible. If a translator tries to get as far as possible the aesthetic, style, rhythm, and shape in their translation, then only translation is possible. Despite difficulties attached to the poetry translation, it is the answer to the isolation and essential need for communication of SL in TL. Poetry translation enriches the target language literature, and a successful translation transcends boundaries.

Agyeya has carefully shaped his poems in Hindi to guide the reading process. He has given information about the place and situation of the creation of poems. So, his Hindi poems echoed his thoughts, emotions, feelings, and messages. But his English translations do not serve the purpose and many places. He has not provided any information about the place and situation of the creation of the poems as he did in the SL. He has done literal translations of poems. Because of this literal translation, the message is altered in many places. The word selection is not appropriate in many places. Due to this, we don't get the aesthetic and beauty of English poems. Even being a poet himself, it was not easy for Agyeya to translate the poems without losing some features. Still, these English translations are very important as poetry translation gives a new life to Hindi poems. Half a loaf is indeed better than none!

REFERENCES

1. Agyeya. (1981). *Nilambari*. Delhi: Clarion Books.
2. Barnstone, W. (1993). *The Poetics of Translation: History, Theory, Practice*. New Haven and London: Yale University Press.
3. Bassnett, S. and Lefevere, A. (eds.). (1990). *Translation, History and Culture*. London and New York: Pinter Publishers.
4. Boase-Beier, J. (2011). *A Critical Introduction to Translation Studies*. New York and London: Continuum International Publication.
5. Catford, J. C. (1965). *A Linguistic Theory of Translation: An Essay in Applied Linguistics*. London: Oxford University Press.
6. Corbett, J. and Huang, T. (eds.). (2020). *The Translation and Transmission of Concrete Poetry*. London and New York: Routledge.
7. Dalmia, V. (ed.). (2012). *Hindi Modernism: Rethinking Agyeya and His Times*. Berkeley: The University of California.
8. Diaz-Diocaretz, M. (1985). *Translating Poetic Discourse: Questions on Feminist Strategies in Adrienne Rich*. Amsterdam and Philadelphia: John Benjamins Publication.
9. Gibbs, R. (ed.). (2008). *The Cambridge Handbook of Metaphor and Thought*. New York: Cambridge University Press.
10. Khaitan, J. P. (2015). *Ajneya: Chetana Ke Seemanta*. Varanasi: Vishwavidyalaya Prakashan.
11. Johnsgard, P. (1983). *Cranes of the World*. Bloomington: Indiana University Press.
12. Lefevere, A. (1975). *Translating Poetry: Seven Strategies and a Blueprint*. Amsterdam: Van Gorcum.

12. Lefevere, A. (1992). *Translating, Rewriting and the Manipulation of Literary Fame*. London and New York: Routledge.
13. Ormsby, E. (2001). *Facsimiles of Time: Essays on Poetry and Translation*. Ontario: The Porcupine's Quill.
14. Paliwal, K. (ed.). (2011). *Ajneya Rachanawali* (Vol. 2). New Delhi: Bharatiya Jnanpith.
15. Raffel, B. (1998). *The Art of Translating Poetry*. University Park and London: The Pennsylvania State University Press.
16. Shah, R.C. (2012). *Ajneya Ka Kavi-Karm*. New Delhi: Vani Prakashan.
17. Sharma, S. (2021). *Ajneya: Vichar Aur Chintan*. Kanpur: Vikas Prakashan.
18. Singh, A. K. (2019). *Hindi Anuvad Vimarsh (Vol. 1-2)*. New Delhi: Sahitya Akademi.

Impending Inquisitions in Humanities and Sciences (ICIIHS-2022) – Dr. Mohan Varkolu et al. (eds)

Generational impact of Diaspora and Patriarchy in Jhumpa Lahiri's, "The Namesake"

5

Penamareddy Kavya Sai Sri

Department of English, St. Francis College for Women, Hyderabad, Telangana, India

ABSTRACT Jhumpa Lahiri's novel, "The Namesake" offers nuanced insights into the lives of those displaced from their homelands. Lahiri's characters in 'The Namesake' offer the readers glimpses of the diverse ways in which diasporic circumstances impact various aspects of one's life.

This research aims to analyse the two female characters of Jhumpa Lahiri's, "The Namesake", Ashima Ganguly and Moushumi Mazoomdar. The analysis points out the contrasting and diverse viewpoints, that Ashima and Moushumi provide the readers, in regard to their respective response to diaspora and patriarchal oppression in their lives. Lahiri has penned down a beautiful and subtle narration that offers the readers a peek into Ashima and Moushumi's personal dynamics in relation to diaspora and patriarchy. The research follows both the characters journey through the course of the novel and traces the collective impact patriarchy and diaspora has on these characters lives, their journey, their choices and their personalities.

This paper explores the varying impacts of diaspora on first generation immigrants as well as the second-generation immigrants. Ashima and Moushomi's characters portray these diverse impacts of diaspora. The constant longing, Ashima has for her homeland paired with the sense of loss of identity, the inability of Moushomi to claim a land, a country, a culture her own, symbolise the conflict ridden state brought about by diaspora. The paper also aims to shed light on the extent to which patriarchy plays the role of one of the primary driving factors in Ashima and Moushomi's lives and the choices they make throughout. The constant presence of patriarchal expectations has a tremendous impact in shaping the personalities and lives of both these female characters.

KEYWORDS Patriarchy, Diaspora, Impact, Personality, Choices

1. Introduction

"The Namesake", is an intimate tribute to all those whose lives have been split between lands. Lives that find themselves struggling to truly, wholly claim a land as theirs, as their home. The novel is a lyrical narration of the complexities and conflicts faced by a Bengali immigrant family in the USA. This paper focuses specifically on the influence, diaspora and patriarchy collectively has on the female characters of the novel

"The Namesake", Ashima and Moushumi. This collective influence of both patriarchy and diaspora have far reaching impacts on both the characters and can be traced in the choices, decisions and outcomes of their lives. The impact of diaspora and patriarchy, unquestioningly seeps into various aspects of these female characters lives, their personal lives, careers, personality and relationships.

*pkavya.2803@gmail.com

DOI: 10.1201/9781003489436-5

Diaspora, a word derived from the Greek word 'Diaspeirein' which means 'Scattered' or 'Dispersed'. Diaspora as a concept has come a long way, from its initial associations with the centuries long search for a homeland by Jews to its contemporary understanding. Diaspora in literature revolves around the idea of homeland. Longing for homeland coupled with the search for a 'sense of belonging' and identity in new lands, is what shapes diasporic literature. Today, diaspora has emerged as one of the prominent themes in Indian English Literature.

Jhumpa Lahiri, is an English-born American author, born to Bengali Immigrant parents, she was born in the UK and was raised in Rhode Islands, USA. Jhumpa Lahiri much like her characters was split between her Indian roots and American circumstances. Jhumpa Lahiri confessed that she neither felt Indian nor American:

> *"Like many immigrant offspring I felt intense pressure to be two things, loyal to the world and fluent in the new, approved of on either side of the hyphen" Looking back, I see that this was generally the case. But my perception as a young girl was that I fell short at both ends, shuttling between two dimensions that had nothing to do with one another" ("My Two Lives" 2006).*

Her experiences undoubtedly contribute to the subtle brilliance of her works. They enable her to illuminate the intricate intimacies of an immigrant family's experience. Jhumpa Lahiri has weaved a path for the readers through the means of the novel, that offers us an intimate and a sensual glimpse into the intricacies of the characters lives. A perspective often left unventured by many.

2. The Concept of 'ABCD'

The concept of 'ABCD' or 'American Born Confused Deshis' has recurrently emerged in popular media, literature and art. ABCD or 'American Born Confused Deshis' is a concept of identity often associated with second generation immigrants from India and other South east Asian countries like Bangladesh, Pakistan, Sri Lanka, etc. There has been growing research and interest towards the concept of 'ABCD', primarily in the context of diaspora and diasporic literature. ABCD captures the fragmentation and dilemma faced by most diasporas. Diasporas, especially second-generation immigrants are often left grappling with the question of their own 'Identity', they are split between their ancestral roots and their current circumstances in life. More often than not, Diapsoras are left with a yearning for a 'Sense of belonging', this yearning's root cause could be attributed to the fact that diasporic community cannot associate themselves to a single land or a single culture entirely.

The concept of 'ABCD' has made its appearance in Lahiri's novel "The Namesake", as a concept that the Protagonist stumbles upon, in a literary seminar during his college days at Yale. This encounter with 'ABCD' leaves the protagonist thinking about its's relevance and existence in his own life. Not just Gogol, but various second-generation characters in "The Namesake" can be seen reflecting this concept, that accurately captures the essence of conflict diasporas face in the context of their identity. These characters can be witnessed struggling to find their feet among various cultures in their lives. All the second-generation characters in the novel Gogol, Moushumi and Sonia, can be considered as portraying the'ABCD' identity in one way or the other.

While on one hand 'ABCD' captures the conflict of identity characteristic to many diasporas. On the other hand, in recent times there has been a in boom in content embracing diasporic diversity. Contemporary approach on the concept of diasporic struggle and 'ABCD' has evolved into a positive, inclusive form. Contemporary artists on various social media platforms are portraying a new stream of Indian, South-East Asian diasporic identity. A new stream of inclusive identity embracing the diversities of intertwining cultures of homeland and hostland. This new stream of diasporic identity breaks the stereotype of diasporic identity which was more often than not associated with the conflict and struggle to 'fit in' and brings forward a completely new identity that the current diasporic community has embraced with pride. An identity encompassing their roots as well their contemporary cultural circumstances. Many contemporary authors, artists and social media influencers are engaged in creating art that embraces their cultural intricacies and proclaim it to the world with pride.

3. Cultural Disparity

In the wake of Globalisation and the rise of immigration, "Cultural Disparity" is relevant in countless immigrant lives. Cultural disparity is the struggle to find a cultural identity, that one can claim as their own. A cultural ideology that brings about a sense of belonging amidst a sea of conflicting cultural identities, those learnt back home and those unknown in new lands. Cultural disparity focuses on the assimilation and clashes of culture. It is the struggle between one's loyalty towards their homeland, their roots and the need for survival, adaption to their contemporary circumstances in a foreign land. More often than not this struggle to balance between two contrasting ideologies leaves individuals conflicted and overwhelmed with a sense of loss, loss of identity, loss of belonging.

Ashima Ganguly's character is a literary representation of cultural disparity faced by diasporic women. Ashima Ganguly a young woman who spent her entire life in Calcutta, with her family and her siblings, growing up in the familiar surroundings of her childhood, is left grappling to her place a foreign land and to live with a man she barely knew. As a young bride of nineteen Ashima is seen accompanying her husband across seas, leaving behind all that was familiar. Leaving behind the familiarity of the culture she was taught and brought up in.

When Ashima finds herself in a foreign country with ways of life completely unknown to her, she struggles to find a balance between the deep rooted norms of her homeland and the current unknown culture she finds herself in. Ashima holds on the culture she grew up, the sarees she's seen her mother wear, the food she ate growing up, the music accustomed to her ears, the language of her loved ones, her mother tongue. Through the course of three decades she spends in the United States of America, she plunges herself into relentless efforts to make a home far away from her homeland, a home that resembles the culture she grew up, a culture akin.

Her efforts to imbibe the culture of her homeland in her American children is a symbolic representation of the cultural conflict faced by first generation immigrants. Ashima ensures to teach her children Bengali poems, the ones she grew up reciting and narrate to them stories of deities and characters from her childhood.

Lahiri has penned down the contrasting impacts of cultural disparity between the first and second generation immigrants. The loss of a sense belonging, the conflicts and struggle to find their identity is shared by both first and second generation immigrants due to their collective diasporic circumstances, but the same does not hold true in terms of cultural disparities. The difference in the struggle to find a balance between both the generations can be traced in the lives and choices made by the second-generation immigrants like Moushomi, Sonia and Gogol who claim their American identity. While the second-generation immigrants also ail with the conflict of finding their identity amid a sea of contrasting ideologies they claim their American identity and the culture around them with relative ease compared to the first generation immigrants.

In contemporary times cultural disparity becomes a concept prevalent in thousands of lives, as a result it finds its representation in literary narratives of diasporic writers like Jhumpa Lahiri, Chitra Banerjee, Anita Desai and many more.

The Impact of Diaspora and Patriarchy on the Female Characters in "The Namesake".

The impact of diaspora is far reaching on all the characters of the novel. While, diaspora's influence is inescapably present on all the characters, the female characters are also prone to patriarchy. Patriarchy is a significant driving force, in both the female protagonists lives. Ashima and Moushomi's lives are significantly influenced by the constant scrutiny of patriarchy and patriarchal expectations. Despite the differences in their backgrounds, origins, circumstances, personalities, careers or era that they belong to, they have been entrusted with a mould. A mould, sculpted by the society specifically for their gender. Fitting into which perfectly is considered their pious duty, failing to do which could lead catastrophic outcomes.

The patriarchal expectations entrusted on women's shoulders has had a strenuous impact on the entire womankind for centuries. These patriarchal expectations associated with women may have undergone certain reforms but still remain obsolete and oppressive, despite countless revolutions that have taken place over the years, to enable women their rightful share of equality. Despite the advances made in the context of opportunities in the professional and educational spheres, women are still susceptible to the age-old gender roles. Patriarchal expectations such as the autonomy of household duties, caretaker and nursing responsibilities are still intact. Ashima and Moushomi's, journerys in life, reflect the contrasting paths they have chosen to grapple with the constant pressure of patriarchy. Despite the differences in their journeys, both these women have done a commendable job in overcoming and establishing an identity for themselves. The contrasting choices and journeys of these two women reflect the diverse impact of patriarchy and diaspora.

The Influence of Diaspora and Patriarchy on Ashima's life

Through the course of three and half decades, Ashima from the "The Namesake", has come a long way. Ashima Bhaduri, a young, shy bride from Calcutta, who crosses seven seas, to spend the better part of her life with her husband in an unknown land, to Ashima Ganguli, a woman who finds her own feet in a foreign land and takes up a job in her late 50s.

Jhumpa Lahiri's character Ashima, is a beautiful character that leaves the readers with a sense of serenity. Ashima is a strong, resilient and an independent woman who had the courage to build a new life, a new home in an entirely foreign land not just for herself but her family too. The brilliance of Ashima's character is in the way her character has evolved from a homesick young bride filled uncertainty to a woman who ventures out to find her independence despite her age. The novel opens with a pregnant Ashima, filled with longing for her homeland.

> *"On a sticky August evening two weeks before her due date, Ashima Ganguli stands in the kitchen of a Central Square apartment, combining Rice Krispies and Planters peanuts and chopped red onion in bowl."*
>
> *–Jhumpa Lahiri, "The Namesake"*

This extract for the novel reflects the impact diaspora has on Ashima. Her craving for an Indian snack symbolises her longing for home. Due to her circumstances in life, Ashima Ganguli is led to lead her life in a foreign country far away from everything familiar, everything associated with home. Ashima Ganguli's longing of home is a collective representation of the yearning and homesickness that diasporas feel. Countless young brides and students are often left with a similar feeling, when they find themselves in a foreign land. Ashima's foreign surroundings which fill her with a longing for her homeland, also lead her to perform certain symbolic acts, as a means of holding onto to her roots.

Ashima's unfaltering devotion to Indian cooking throughout her life in America, the way she dresses in sarees and puts on bindis, her relentless attempts to teach her kids about the culture, religion and customs of her homeland, Ashima and Ashoke's search for solace in weekly gatherings of Bengali friends, friends honoured with the titles of Mashi and Masha and most importantly how closely she holds on to the Bengali magazine 'Desh', that accompanies her on her first journey to the USA, the magazine from her homeland, printed in alphabets from her mother tongue. All these choices of lifestyle, these acts that encompass the majority of her life are reflections of the longing and yearning that Ashima ails from for her homeland. This longing is a constant companion of Ashima despite her life in America for more than three decades, this yearning reflects the impact diaspora has on immigrants.

> *"For being a foreigner, Ashima is beginning to realize, is a sort of lifelong pregnancy – a perpetual wait, a constant burden, a continuous feeling out of sorts. It is an ongoing responsibility, a parenthesis in what had once been ordinary life, only to discover that previous life has vanished, replaced by something more complicated and demanding."*
>
> *–Jhumpa Lahiri, "The Namesake"*

Ashima was filled with emotional distress and disillusionment upon her arrival in the USA. She was filled with dread and uncertainty at the prospect of having to lead a life in an unknown land. Ashima's emotional distress could be attributed to drastic change in her circumstances of life, a radical shift from everything familiar.

> *"On more than one occasion he has come home from the university to find her morose in bed, rereading her parent's letters."*
>
> *–Jhumpa Lahiri, "The Namesake"*

Motherhood is a challenging responsibility provided the availability of support. But, motherhood in an entirely foreign land with no familial support, is all the more challenging. Ashima dreads having to raise a baby all by herself in an unknown country with no family to support or help her out. Ashima dreaded having to raise a baby in a foreign land, when she herself remains a stranger, lost amidst the unfamiliar.

> *"I'm saying I don't want to raise Gogol alone in this country, It's not right. I want to go back."*
>
> *–Jhumpa Lahiri, "The Namesake"*

These above lines reflect Ashima's unwillingness to raise a child, her dread and uncertainty at having to spend her life as a woman, a wife, a mother in an unknown country.

Despite Ashima's initial longing for home and dread at the life ahead in a foreign land, she begins making peace with the circumstances of her life and plunges herself into efforts to establish a new life for herself in the USA. Ashima moves ahead in life with resilience to find her own feet. Ashima's resilience to move forward despite her dissent towards her situation in life is a reflection of the strength of her character. After the birth of her son Gogol, Ashima lifts herself up and pours her entire strength into being a mother to her son. Her resolution to focus her entire energy in being a good mother could be viewed as an initial step towards, years long journey of her own independence, her own identity as a woman.

> *"She begins to pride herself on doing it alone, in devising a routine. Like Ashoke, busy with his teaching and research and dissertation seven days a week, she, too, now has something to occupy her fully to demand her utmost devotion, her last ounce of strength."*
>
> *– Jhumpa Lahiri, "The Namesake"*

The societal conventions may quite conveniently tag Ashima as an ideal Indian homemaker, wife and mother. Ashima may appear as an 'ideal woman', according to the stereotypes on the surface. But, despite the deceiving, deep rooted stereotypes, Ashima goes beyond the mould, she's a character filled with nuances. She undoubtedly is a homemaker, a wife, a mother, a daughter, a sister, a granddaughter who fulfils all the responsibilities and duties bestowed upon her by patriarchy and society. But, she is also the same woman who finds her own feet, her own independence, she is also the woman she finds herself a job in a foreign land at the age of fifty, she is also the woman who learns to drive and live all on her own, despite spending her entire life in the company of her family, husband and children, she is also the woman who makes peace with her children's independence, their lack of need of Ashima in their lives. She most certainly has come a long way not just as a wife, mother or a daughter but as a woman.

Ashima has travelled a long way, from a young bride who travelled across seas to accompany her husband to a widow who was travelling back to a homeland, not entirely home. Ashima loved and lost, yet she goes on. Ashima's, strength is in her silent, subtle rebellion, in her mundane acts of claiming her own identity. Ashima is a woman true to her name. Ashima, someone without boundaries or borders, someone limitless.

The Impact of Diaspora and Patriarchy on Moushumi's life

Moushumi Mazoomdar is a bold and a beautiful character crafted by Jhumpa Lahiri. Moushumi's character has evolved from being a shy, timid girl who always accompanied her Bengali parents to Gogol Ganguli's house for weekly gatherings of Bengali families, a girl always seen with glasses and a book in hand at these gatherings into a bold, independent woman pursuing her PhD in French Literature at NYU.

Moushomi's character is a representation of the norms and expectations enforced upon the second-generation immigrants, by their own parents, relatives and society as a whole. Indian immigrant parents can often be seen trying to both convince as well as force their American children into the cultures and customs of their homeland. The efforts of immigrant parents directed towards convincing their children to adopt their homeland's culture, is relentless. Immigrant parents try everything in their power to convince their children to adopt the culture and customs accustomed to them as a desperate effort to hold on to the culture familiar to them from a far-away home. While all second-generation immigrants are prone to constant pressure to claim the culture of their ancestral homelands, second-generation Indian-American women are also prone to patriarchal expectations of a woman's virtue, chastity, duty, responsibilities, etc. Moushumi Mazoomdar, like countless Indian-American women is faced with constant pressure of diasporic as well as patriarchal expectations. While she tries to find and claim an identity in a setting of diverse cultures, she is also under the constant pressure of patriarchal expectations. Throughout her life Moushumi is persistently reminded, not just by her parents but everyone around her of her duties as an 'Indian Woman'.

> *"She had always been admonished not to marry an American, as had he, but he gathers that in her case these warnings had been relentless, and had therefore plagued her far more than they had him. When she was only five years old, she was asked by her relatives if she planned to get married in a red sari or a white gown."*
>
> *–Jhumpa Lahiri, "The Namesake"*

Moushumi's entire life was a parade of proclamations of expectations out of her life, proclamations by her parents, relatives sometimes even strangers. Proclamations of her duties and responsibilities as an Indian daughter and most importantly as an 'Indian woman'. The impact of these countless, relentless patriarchal and diasporic expectations throughout her life undoubtedly seeps into various aspects of this character's life choices. The persistent eye of scrutiny plays an influential role in shaping Moushomi's personality, her career, relationships and life choices.

> *"By the time she was twelve she had made a pact, with two other Bengali girls she knew, never to marry a Bengali man. They had written a statement vowing never to do so, and spit on it at the same time, and buried it somewhere in her parent's backyard."*
>
> *–Jhumpa Lahiri. "The Namesake"*

The prominent influence of diaspora and patriarchal expectations from Moushumi throughout her life, can be traced in her Academic, Professional and Personal lives.

Moushomi went against her family's insistence to pursue Chemistry at Brown University, and further follow the footsteps of her father and work as a chemist. Instead Moushumi double majored in French and picked up everything she owned and moved away to France at the first opportunity. This academic rebellion of Moushumi is symbolic representation of her breaking away from years long culture and norms enforced upon her against her will. Moushumi's decision to choose French, an entirely new, foreign, unexplored language portrays her efforts to dismiss the decades long forced norms and identity, barely resonating her and embracing an entirely new territory, a new identity, something her own, untarnished by the struggle to fit in but rather a new beginning giving her space to be herself unabashedly.

> *"Immersing herself in a third language, a third culture, had been her refuge she approached French, unlike American or Indian, without guilt, or misgiving, or expectation of any kind. It was easier to turn her back on the two countries that could claim her in favor of one that had no claim whatsoever."*
>
> *–Jhumpa Lahiri, "The Namesake"*

The impact of diaspora and patriarchy can be traced in Moushomi's romantic relationships as well. Every relationship of Moushomi from Adolescence through Adulthood reflect her state of conflict, struggle of finding her identity due to diaspora as well as her dissent at patriarchal expectations. Her first romantic interest Dimitri is symbolic to Moushumi's first ever encounter with rebellion to dismiss years long conditions and traditions imposed upon her. Her involvement with Dimitri in her adolescence is symbolic rebellion against everything foreign imposed on her, and her act to claim her own identity. Her first romantic relation enabled Moushumi to look at herself in an entirely new light. Following this, her various casual indulgences and romantic, sexual relations with strangers are further symbolic to the moulding of Moushumi's personality into an uninhibited, bold, confident woman.

> *"In retrospect she saw that her sudden lack of inhibition had intoxicated her more than any of the men had"*
>
> *–Jhumpa Lahiri, "The Namesake"*

Moushumi's subsequent passionate and intense romantic involvement with an American, Graham also symbolises Moushomi's struggle with her diasporic identity and her efforts to dismiss the relentless expectations of her parents and relatives, which in turn could be seen as an act of claiming her own life. Failure of Moushumi and Graham's relationship post Graham's unsolicited remarks on India, which also was a trigger to their ultimate breakdown is also a symbolic representation of the Diasporic struggle in Moushumi's life.

Moushumi eventually ends up marrying Gogol, the protagonist who also happens to be a successful architect, more importantly an Indian- American with Bengali roots, the son of Moushumi's parents friends. Moushumi and Gogol, contrary to the belief that they held for the better part of their life, never wanting to marry an Indian, end up doing the same. Moushumi and Gogol's relationship and marriage turns out to be a calm, fulfilling part of their lives pushing them to make themselves better people and most importantly make peace with their decades long struggle to rebel against the norms enforced upon both of them. Despite the warm comfort of the love between them, Moushumi associated Gogol with resignation. Resignation towards the life that she fought so hard to leave behind.

> *"They had both sought comfort in each other, and in their shared world, perhaps for the sake of novelty, or out of fear that world was slowly dying"*
>
> *–Jhumpa Lahiri, "The Namesake"*

This peace turned to be short-lived for Moushumi, who finds herself finding her way back to Dimitri. Moushumi's marriage to Gogol, her subsequent extramarital relation with Dimitri symbolises the ongoing struggle, that has Moushumi with her own self.

> *"Sometimes she wondered if it was her horror of being married to someone she didn't love that had caused her, subconsciously, to shut herself off."*
>
> *–Jhumpa Lahiri, "The Namesake"*

Despite the dire conditions in her personal life, Moushumi still has come a commendable way as an individual, as a woman who claimed her own identity and established a successful career for own self. Moushumi is a strong, bold, independent woman who still manages to go forward despite the torments in her life. True to her name, she is indeed 'Moushumi' - A damp southwesterly breeze, a force of nature.

4. Conclusion

In the contemporary circumstances Diaspora has emerged as a pressing issue of debate and contemplation. Given the extensive rise in globalisation and migration for education and economic opportunities, the debate on the impact of diaspora on people becomes something of utmost importance. Diaspora undoubtedly seeps into every aspect of an individual's life and their choices. It so happens that diasporic women are not just prone to diaspora but also to the continual gaze of patriarchal expectations. The sad reality happens to be that, whether the female character is Jhumpa Lahiri's, Arundati Roy's or Virginia Woolf's, regardless of their nationality, roots or the age they belonged to women have been and will be, for at least the foreseeable future under the same scrutiny of patriarchal expectations. The forms, shapes and methods may change but it lurks in a woman's life without doubt.

REFERENCES

1. Lahiri, Jhumpa, The Namesake. London: HarperCollins, 2006. print.
2. Raina, J. (2017) 'Identity and cultural conflicts in Jhumpa Lahiri's novel, The Namesake'. *Universal Research Reports, 4 (6).*

Impending Inquisitions in Humanities and Sciences (ICIIHS-2022) – Dr. Mohan Varkolu et al. (eds)
© 2024 Taylor & Francis Group, London, ISBN 978-1-032-78829-6

Dislocation, Conflicts and Issues of Identity in Michael Ondaatje's *The English Patient*

6

Pooja Shankar*
PhD. Research Scholar,
Department of English, Central University of Himachal Pradesh, Dharamshala

ABSTRACT Humans have moved from their native countries and other geographic and cultural regions since the beginning of time, sometimes voluntarily, partially forcibly, or forcibly. Everyone wishes to emigrate from their own nation in hopes of finding wealth, possibilities for education and work, a lucrative career, or other privileges. It gives us a chance to see a variety of issues that immigrants face in different nations in regard to other people's cultures, races, identities, and religions, among other things. The bulk of newcomers struggle to acclimatize to their new environment. Some experiences in their chosen land bring them into contact with two opposed cultures and two competing universes. Migration causes psychological problems to the person. One of the major traumas of the modern diaspora is the sense of loss of identity. The greatest and most accurate approach to represent and talk about the problems faced by immigrants in exile and even in their own countries is through diasporic writing. Many writers are concerned with migrant conditions and the sense of alienation inherent with relocation but, Philip Michael Ondaatje's writing appeals to the wider international and globalized audience. He is one of the foremost writers, whose work has influenced an entire generation of writers and readers. This paper aims to analyze *The English Patient* by Michael Ondaatje in order to discuss the problem of identity in connection to war. It focuses on elements like violence and death as it explores the links between conflict and the formation of identity. Together with the World War II events that serve as its backdrop, Ondaatje explores individual stories in his novel. The way Kip views and experiences war and the way he lives through it lead one to believe that there are many internal tensions at play that either drive people apart or bring them together. The hunt for a new meaning of the self is an issue in *The English Patient*, when Hanna, Kirpal Singh, and Almásy make an effort to free the self by looking into the past. By focusing on the concepts of mobility and transformation as they are thematized in the war storyline and act as the basic mechanisms that determine identity, the novel thereby problematizes the complexities of human relationship. This paper will highlight the issues of identity, sense of loss, interpersonal and intrapersonal conflicts, alienation, homelessness etc.

KEYWORDS Philip Michael Ondaatje, Dislocation, Conflicts and issues of Identity, Diasporic literature, Self, War

1. Introduction

Born in Sri Lanka, Michael Ondaatje comes from a family of Sri Lankan Burghers. He has earned a special position among other renowned poets and novelists around the world. He is recognized for his work on topics including identity crises, cultural tensions, homelessness, and expatriation. The current study will be beneficial to all immigrants moving from one location to another and dealing with issues like identity crises and cultural confrontations. Most immigrants should be aware of the problems posed by intercultural migration. The writings of Michael Ondaatje appeal to a wide, more globalized public. Living

*poojashankarjnr@gmail.com

DOI: 10.1201/9781003489436-6

in Canada, he has a tremendous drive to build his own identity. Michael Ondaatje struggles with his allegiance to the host society and his yearning for the native culture. It is inevitable that the literature created by migrant or diasporic writers will portray the sorrow and difficult situations they confront in a foreign nation. It develops as a result of difficulties that arise and a specific type of self-analysis to survive in the global environment. Migrant or diasporic writers are compelled to express the agony and painful experiences they encounter in a foreign nation through their works of literature.

The writers from the Sri Lankan diaspora, such as Gertrude De Silva, Nira Wikramasinghe, Romesh Gunesekera, V.V. Ganeshnathan, Carl Muller, Shyam Selvadurai, Yasmine Goneratne, Michelle De Kretser, and Ru Freeman, to name a few, have addressed issues like intercultural conflict, identity crisis, sense of loss, alienation, homelessness, etc. In the Oxford Advanced Learner's Dictionary, the term "diaspora" is defined as "the movement of the Jewish people out from their own country to live and work in other countries". The term "diaspora" historically refers to a particularly particular circumstance, namely the expulsion of Jews from the Holy Land and their subsequent dispersion around the globe. And in the Britannica Encyclopedia the term "diaspora" is defined as "the dispersion of Jews among the Gentiles after the Babylonian exile (586 BC); or the aggregation of the Jews outside Palestine or modern-day Israel" (Vol. 3 of the Britannica Encyclopedia).

2. Literature Review

Avinash Jodha has published a book- *Michael Ondaatje's Fiction: Poetics of Exile*. This book concentrates on Ondaatje's multiple negotiations of identity.

Maduri Bhosale in *In Search of Centre: Identity Crisis in Michael Ondaatje's 'The English Patient'*. (A research article on Journal of Higher Education and Research Society: A Refereed International , ISSN 2349-0209, Vol.-4/Issue-2, Oct.2016.) - She shows different aspects which creates identity crises. We find that there is no stable identity in characters.

Lucian Blaga in a research article Identity and War in Michael Onaatje's The English Patient reflects the fundamental change that occurred in the colonial world following World War II. The novel examines this issue by investigating identity as a concept based on the link to the "other," whether distinct people or a prior embodiment of one's self. It is multilayered and has a polyphonic structure. Each character's identity is shaped by the changes brought on by the conflict, and it is based on the character's inquiry into the past as a means of understanding a discontinuous present. Thus, experiencing battle helps to demystify and revalue the past.

In Ondaatje's novels, the main themes of the study—dislocation, conflict, and identity crisis—are explored. Ondaatje, who is Dutch and Indian by descent and currently resides in Canada, was born in Sri Lanka, previously known as Ceylon, during World War II. 1992 Booker Prize winner Ondaatje's *The English Patient*, addresses the issue of displacement. The main subjects of the novel are how the Second World War affected the lives of the protagonists, who are all refugees living in the barren Tuscan Villa San Girolamo in Central Italy. All of these displaced people have distinct cultural backgrounds and ethnicities. In conflict, they suffer severe psychological or bodily wounds. In the privacy of the villa, they have established a community. The villa serves as an army hospital for the Allies throughout the conflict.

In Ondaatje's narrative the individuals have likewise journeyed much and have ended up in a ruined villa in a foreign nation where they are attempting to reassemble new identities. The Second World War's Italian campaign serves as the book's setting. Each person living in the villa is a symbol of their own ethnicity and culture. The patient is Hungarian, but due to his English accent and attitude, he is referred to as the English patient. Caravaggio and Hana are Canadian, while Kip or Kirpal Singh is Indian.

English is not spoken at the villa. The protagonists in this book, like the author, are compelled to flee their own nations due to conflict. They remain at a villa in a small Italian village, forging new bonds with one another. A few of the individuals who find themselves in Italy are Hana, a nurse, Almasy, a wartime mapmaker, Caravaggio, a robber, and Kip, a bomb defuser.The four protagonists are connected by their experiences with trauma during World War II, although they each arrived at Villa San Girolamo for reasons that are distinct. The four main characters all come to Villa San Girolamo for different reasons, yet they are all linked by World War II. The protagonists try to reconstruct their identities while residing in Italy while being influenced by both the world that surrounds them and their own ideas, feelings, and experiences. Tyson argues that culture plays a part in the construction of identity in addition to individual will and desire. According to Lois Tyson, a person's identity and their cultural environment "inhabit, contemplate, and define each other" (Tyson).

Even though the battle between the nations in the narrative is nearing to an end, the havoc it caused is still devastating the characters. They all had significant pain as a result of the experiences Caravaggio, Kip, Hana, and the English patient underwent during World War II. It is easier for the characters to ignore their trauma than to deal with it. The protagonists are first unable to confront their lost identities, loved ones, innocence, youth, significance, or honour on a mental or bodily level. Characters' attempts to escape their given names, bodies, and situations help to create and develop their identities. No matter how hard the protagonists attempt to hide their trauma, real-world events constantly bring it back.

Hana likes to take care of her English patient in this terrible situation since he is a "despairing saint" in her eyes and lives with a man whose entire identity was destroyed in a fire. Because of the war, Hana is psychologically damaged. Her mental fragility is caused by the deaths of her father Patrick in the war, her lover and the father of her unborn child, her abortion, and her loved ones as well as war victims. Hana is driven insane by all the deaths happening all around her. Without the mystery of long hair, her face depicts her personal loss as a result of World War II. "I worked very hard in the hospitals and treated from everybody around me. Except the child, who I shared everything with. In my head. I was talking to him while I bathed and nursed patients. I was a little crazy."(Ondaatje, 82). She was suffering with intrapersonal conflicts. She was continuously thinking about her past, she was also suffering with guilt consciousness. All these circumstances create identity crises in her life. However, Hana finds meaning of her life in caring of this burned Englishman, because Hana's father Patrick has died of burning in France when Hana was in Italy, nursing her patients who are the victims of the war. She is the victim of the sadness. She couldn't nurse her father who was burned in the dove cote. So she has decided not to leave the English patient and the villa San Girolamo.

A tent is set up close to the villa by the Sikh man, (Kripal or Kip) the sapper. "She (Hana) sees his shirtless brown body as he tosses water over himself like a bird using its wing." (Ondaatje p. 72). Hana also takes note of his wrist's "darker brown flesh." Racial identity is based on skin tone. A small nod of the head served as an indication of identification. The Sikh man seems excessively finicky in Caravaggio's opinion. He chuckles at the Sikh sapper's habit of always washing his hands. The sapper explains to Caravaggio about the Indian habits as "I grew up in India, Uncle. You wash your hands all the times. Before all meals. A habit. I was born in the Punjab"- (Ondaatje, 76). Teeth brushing is an outdoor practice in the East, particularly in India. Since he was a young child, he has always enjoyed being outside while brushing his teeth. (Ondaatje, 86). Hana and Caravaggio eventually adopt the eating and brushing routine of the Sikh sapper after a week. During meals, he spreads onions and other herbs that are wrapped in handkerchiefs. The way a sapper uses his right hand to carry food to his lips with his fingers irritates Caravaggio.

Eastern and Western cultural values are different and are reflected through interactions. Intermingling of any two cultures always lead to cultural conflict. The other three characters—Hana, the English patient, and Caravaggio—are Westerners who reside in the house, while the sapper, Kirpal Singh, also known as Kip, is the sole one from the East who lives outside the villa in a tent. In opposition of family custom, Kirpal went to England to fight in the war. As Kirpal has voluntarily adopted his new nickname, Kip, Lord Suffolk greets him there in England. Kip describes his memories of Westbury, England in the 1940s in his flashback. Upon arriving in England, Kip in his flashback recalls the memories of Westbury, England in 1940s. He remembers his arrival in England. As a twenty-year-old boy "knowing no one, distanced from his family," (Ondaatje, 187) in the Punjab, the only Indian among the applicants, Kirpal the second son of the family breaks the tradition of his family and joins a Sikh regiment to fight the war for England.

Ondaatje writes about Kip. All inhabitants in the villa continue to live through hardship or adversity and with their physical and psychological wound. There in England Kirpal is welcomed by Lord Suffolk, as Kirpal has willingly accepted his changed nickname Kip, "but the young sikh had been there by translated into a salty English Fish." (Ondaatje, 87). Kip faces racial discrimination in England. He is aware that he will never be able to fully integrate into the English culture. Despite this, he has moulded himself to fit into the culture of the host country and only trusts people who have befriended him. He felt identity crises because the Britishers refuse to communicate with Indians, yet they desire Indians to fight in their wars. The English patient is Count Ladislaus de Almasy, a European desert adventurer. The Cultural conflict has affected Almasy. When kip hears he news of dropping of the atomic bombs. He presses the English patient's chest with the rifle stock pointed in that direction. He doesn't have faith in him. He doubts him, that's why Almasy hates national identities.

3. Conclusion

In conclusion we can say that however, the story depicts the newly created imaginary community of all the people who were forced to flee and were injured during the conflict in the villa San Girolamo, which is a symbol of universal peace. Colonial domination is one of the reasons behind cultural conflict. They have a difficult time getting by because of their negative personal encounters in this strange country and society. The desert serves as a metaphor for their erratic national identities, which are divided and different. Because of that characters are suffering with identity issues, and cultural conflicts especially in the case of Almasy. And character like Hana is suffering with identity issues because of her personal reasons. But one thing is similar that all are creating artificial identities to make their self comfortable in the place where they are displaced.

REFERENCES

1. Adhikari, Madhumalati, "History and story: Unconventional History in Michael Ondaatje The English Patient and James A. Michener's Tales of the South Pacific," *History and Theory*. Vol. 41, No. 4, December 2002, pp. 43–55.
2. Bhosale, Madhuri. "Search of Centre: Identity Crisis in Michael Ondaatje's 'The English Patient". *Journal of Higher Education and Research Society: A Refereed International*, Vol.-4, Issue-2, Oct. 2016, pp. 722–26.
3. http:// literature.britishcouncil.org/Michael Ondaatje.
4. Jodha, Avinash. *Michael Ondaatje's Fiction: Poetics of Exile*. Rawat Publication, 2011.
5. Tyson, Lois. Critical Theory Today. 3rd ed., Routledge, 2014.
6. Ondaatje, Michael. *The English Patient*. Picador, 1993.
7. www.Oxford learner Dictonary.com

Impending Inquisitions in Humanities and Sciences (ICIIHS-2022) – Dr. Mohan Varkolu et al. (eds)

Feasibility of Using Newspaper Content in Classrooms to Reinforce the Vocabulary Learning among Students of Engineering

7

Challa Srinivasa Rao[1]
Assistant Professor, Dept. of Science & Humanities, Sreenidhi Institute of Science and Technology, Yamnampet, Ghatkesar, Hyderabad, Telangana, India

Karayil Suresh Babu[2]
Associate Professor, Dept. of Sciences & Humanities, Vasireddy Venkatadri Institute of Technology, Nambur, Guntur, Andhra Pradesh, India

ABSTRACT The present study, entitled "Feasibility of Using Newspaper Content in the Classrooms to Reinforce Vocabulary Learning among the Students of Engineering," aims to explore the feasibility of using authentic material like newspaper content to reinforce the learning of vocabulary among students from the Computer Science and Engineering stream. This study adopted both a qualitative and quantitative approach. The specific questions this study attempts to investigate are the feasibility of using newspaper content in the classroom and the effectiveness of the activity in achieving the target of reinforcing the vocabulary of the students. The study is conducted on a sample of 60 students from two different engineering colleges located in the two Telugu-speaking states of Andhra Pradesh and Telangana. Thirty difficult words identified by the students from the first two chapters of *Operating System Concepts* authored by Abraham Silberschatz, Peter Baer Galvin, and Greg Gagne are considered for the study. Newspaper items that contain the same words are identified and presented for reading through an instructional programme of 10 hours' duration with minimal teacher intervention. The performance of the students on the tests after intervention indicated significant improvement in receptive and productive vocabulary.
KEYWORDS Authentic materials, Vocabulary building, Communicative language teaching, Newspaper in education

1. Introduction

The current study is part of the effort to study the feasibility of embedding authentic material like newspapers into the classroom. The objective of embedding the newspaper reading activity is to study if it can enhance the rate at which the students learn new words and understand whether it significantly improves the effectiveness of how students learn new words. Two complete parametric research studies conducted by the current researchers have given the impetus to study all the possibilities of making English language learning at the tertiary level, especially in engineering education, more fruitful. The syllabus prescribed by the universities, duration available for teaching, motivation of the students, classroom size, willingness of the teachers to try new methods, availability of authentic material, and examinations are some of the aspects to be considered while devising any method. The prescribed English textbooks do not completely fulfill the language needs of the students in general and their vocabulary needs in particular. There is a need to guide students in learning vocabulary in an engaging and effective manner. Using interesting news articles is one of the most effective ways to learn vocabulary.

[1]chsrao17@gmail.com, [2]karayilsuresh@gmail.com

DOI: 10.1201/9781003489436-7

2. Literature Review

Ellis (1994) terms incidental learning "learning without any intention to learn." In other words, understanding and remembering the new word happens as a natural process rather than a conscious effort to learn it. Incidental vocabulary learning assumes that certain key conditions must be fulfilled for the learning to take place. According to Huckin, T., & Coady, J. (1999), accurate word recognition, good textual clues, a clear context, clear-cut interpretation, considerable prior vocabulary knowledge, and good reading strategies are some of the key conditions that enable learning.

A considerable amount of research was done in the field of authentic materials being used in English language teaching. The fundamental quality of any authentic material, especially printed sources that can be used in the classroom environment, should be the same as how it would be used in real-life situations. One of the chief characteristic features of communicative language teaching (CLT) is the use of authentic materials in classroom teaching. Larsen-Freeman (2000) and Jacobson et al. (2003) view authentic materials as printed materials that are used in classrooms in the same way that they would be used in real life. Authentic material can be prepared or designed for English language teaching from various sources. The use of authentic content gives teachers the chance to create innovative lesson plans and teaching strategies. "Authentic texts can be motivating because they are proof that the language is used for real-life purposes by real people." (Nuttall 1996).

Sourcing authentic material has been simplified by the Internet, which has a huge repertoire of audio, visual, and printable sources like newspaper content. Printable newspaper content from the internet, the most economical of all the sources, also provides more reliable and authentic content for structure and vocabulary instruction. Since newspapers contain the most current information connected to everyday life, students may not show resistance during the learning process. Moreover, the content may kindle interest among the students and support learning.

Use of Newspapers as Authentic Material

Using newspapers in the classrooms is an idea that is being put into practise because of the initiatives of newspapers like "The Times of India" and "The Hindu." Deep research in this field is needed to develop methods and activities that aid vocabulary learning. According to a research study by Manoj Kumar Nagasampige and Kavita Nagasampige (2016), Newspapers in Education (NIE) programme conducted at the middle school level was quite successful. Previous studies on the effectiveness of newspapers in the classroom have shown that they can be used to develop writing, speaking, and reading lessons (Lindsay Clandfield and Duncan Ford, 2011). Hwang, K., and Nation, I.S.P. (1989) conducted a study to look at different ways of selecting stories in newspapers and the effect that these ways have on the occurrence of low frequency words. It was found that reading running stories provided better conditions for reducing the learners' vocabulary load and also for the acquisition of vocabulary. Research done by Schmitt and Carter (2000) indicates that language learners can acquire vocabulary knowledge through reading newspaper articles. They compared the lexical profiles of two sets of topic-related and topic-unrelated stories in newspapers. Their survey suggests that learners can be quick to realise the benefits of narrow reading and are willing to have it become part of their coursework. Another advantage of using the newspaper in the language classroom is that it is a source of authentic language in a communicative classroom, thereby providing stimulation for learners to think, talk, and write about the things that they are familiar with and that matter to them (Grundy, 1993). Additionally, he recommends various material resource books and textbooks and offers comprehensive advice on how to choose newspaper items. Grundy also claims that news articles provide engaging conversational topics in the classroom. The objective of the current work is to evaluate a method where newspaper content is used in instruction to reinforce vocabulary items encountered by students in their domain textbooks. The proposed method will need to provide more reading opportunities that aid students in reinforcing their vocabulary. This can be ascertained by examining the effectiveness of learning.

3. Research Questions

(i) Can English newspapers be used in classrooms for vocabulary instruction?

(ii) Would students be able to perform better in recognizing more new words thereby improving their ability to understand original textbooks?

Research Process

The research process is divided into three stages. In the first stage, the words to be used for learning were identified. The second stage is the instructional stage. In the third and crucial stage, the acquisition traits of the students were tested using a test designed for this purpose. The data obtained is used to validate the research claim.

Population of the Study

The sampling frame for this study comprises engineering students who are in their second year of study (III Semester). Sixty students were selected from two different engineering colleges, Vasireddy Venkatadri Institute of Technology (VVIT) and Sreenidhi Institute of Science and Technology (SNIST), located in Andhra Pradesh and Telangana states, respectively. These students have a minimum of 13 years of English learning. All these students belong to the Computer Science and Engineering disciplines of SNIST and VVIT. All the students are drawn from different educational and social backgrounds.

Selection of the Target Words

Unknown vocabulary occurring in their *Operating Systems Concepts* textbook was identified. The words are selected from the first two chapters of *Operating System Concepts*, authored by Abraham Silberschatz, Peter Baer Galvin, and Greg Gagne. The students identified thirty words as tough words that they encountered while reading the first two chapters of the textbook.

Instruction

In the second stage of the research, the students were given news articles containing the difficult words that they identified from the textbook. The news items are sourced from the official websites of the three national English dailies 'The Hindu', 'The New Indian Express' and 'The Times of India.' Each news item is printed on an A4 sized sheet for the convenience of reading in the classroom. While selecting the news items importance was given in choosing items that may entice some interest among the students to read. The news items were not chronologically arranged. Multiple news items for each new word are created for the activity. The teacher's involvement was minimal. The students were encouraged to read the news items and guess the meanings of the unknown words. The teacher's role was restricted to a facilitator of the activity than that of a preceptor.

The Test

The test designed for this study has two parts, i.e., Part-A (choose the right answer) and Part-B (fill in the blanks). The two parts have 20 questions each. The test consisted of two varieties of items: multiple-choice questions and fill-in-the-blanks. The 20 questions of Part-A were based on 20 words randomly selected from the 30 words identified by the learners. The same pattern is followed for the 20 questions in Part B of the test. This 30-minute test was administered after ten hours of instruction. This test was conducted on an open-source online learning management system platform named Moodle. Online testing facilitated the smooth conduct of the examination and helped in the analysis of the test result.

4. The Result

The results are compiled for the statistical analysis. The mean, mode, median, standard deviation, and variance were the standard descriptive statistical tools used in the analytical part of this research work. Since this research design is not based on an entry-level and exit-level test, advanced statistical analysis was not used.

5. Discussion

Students who read the newspaper content showed a proclivity towards remembering and using newly acquired words. An examination of the data generated by the tests revealed that 78% of students who used newspaper items to learn words recollected the words. The result is produced after the students spent only ten hours of instruction. The instruction strategy followed is minimal interventional instruction. In other words, the students were encouraged to guess the meanings from the context on their own. The data collected after the test show a positive trend and emphasises that ten hours of focused reading activity with little or no help from the teachers could generate considerable learning. The results of the study further show that both the reception and production of vocabulary items are enhanced when newspapers are used as a tool in the classroom. It can be observed from Table-1 that 85% of the students were able to score between 21 and 60 percent marks in Part-A, i.e., the receptive part of the test, whereas 78% of the students were able to score between 21 and 60 percent marks in Part-B, i.e., the productive part of the test. 93% of the students scored between 21 and 60 percent in the aggregate of the test. This data very clearly states that there is a perceivable impact of the reading material used on the ability of the students to learn new words. 38 students, i.e., 63% of students scored between 31 and 60 percent in the aggregate.

The descriptive statistics obtained after the study also corroborate the researchers' claim that there is a significant impact of the reading activity on the students. The students scored an average of 16 marks over 40 which is equal to 40%. This is not a mean performance. It should be noted that the students took a test in which they answered questions based on words they claimed to be unknown to them.

Table 7.1 Frequencies of the performance of the students in the test

Frequency	Part-A	Part-B	Total
0-10	1	6	1
11-20	1	5	2
21-30	9	11	10
31-40	13	19	19
41-50	18	8	19
51-60	11	9	8
61-70	7	2	1

Table 7.2 Descriptive statistics of the test results taking marks as the basis

Stat	Part-A	Part-B	Total
Mean	9	7	16
Median	9	7	16
Mode	10	7	19
SD	3	3	4

Table 7.3 Descriptive statistics of the test results taking percentage as the basis

Stat	Part-A	Part-B	Total
Mean	45	36	41
Median	45	35	40
Mode	50	35	48
SD	14	15	11

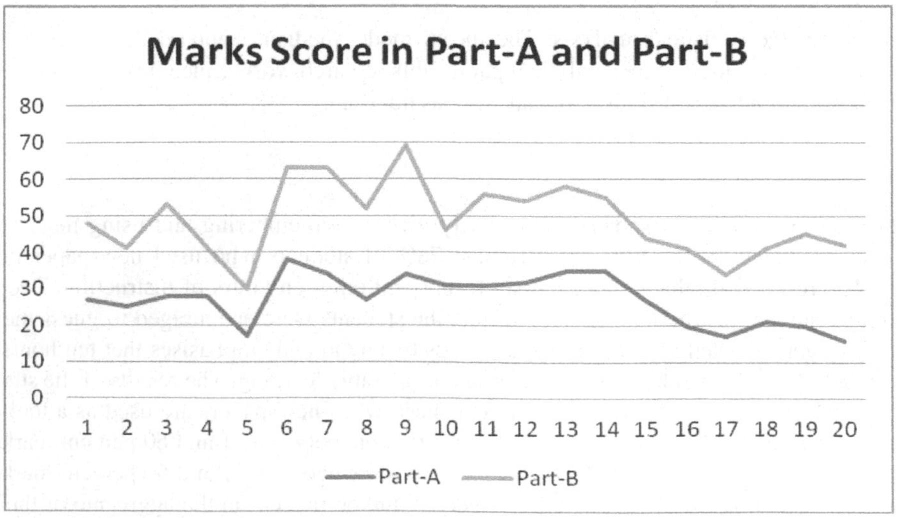

Fig. 7.1 Performance of the students on 20 questions in Part-A and Part-B

4. Conclusion

The statistical analysis of the results obtained after the tests proved that there is a significant improvement in the vocabulary learning ability of the students who were instructed using newspaper content. The results achieved after 10-hour minimal intervention instruction corroborated the researchers' claim that the newspaper content helped the students learn new vocabulary in a quick and effective manner. The newspaper of the current day can be used in the classroom for better results. News items sourced either in print format or from the websites printed or pasted on separate sheets can also be used in the classroom for good results. The researchers opine that further research is needed to be undertaken to examine the feasibility of embedding this activity into the English language classroom.

REFERENCES

1. Abraham Silberschatz, Peter Baer Galvin, et al. (2009). *Operating System Concepts*, John Wiley & Sons, Inc., USA.
2. Ellis, R. (1994). *The study of second language acquisition*. Oxford: Oxford University Press.
3. Grundy, P. (1993). *Newspapers*. Oxford: Oxford University Press
4. Huckin, T., & Coady, J. (1999). Incidental Vocabulary Acquisition in a Second Language: A Review. *Studies in second language acquisition*, 21,181.
5. Hwang, K. and Nation, I.S.P. (1989) Reducing the vocabulary load and encouraging vocabulary learning through reading newspapers. *Reading in a Foreign Language* 6, 1: 323–335.
6. Jacobson, E., Degener, S., & Purcell-Gates, V. (2003). *Creating authentic materials and activities for the adult literacy classroom*: A handbook for practitioners. USA: NCSALL.
7. Larsen-Freeman, D. (1986). *Techniques and Principles in Language Teaching*. New York; Oxford University Press.
8. Lindsay Clandfield and Duncan Foord. (2011). Classroom 1. In *Teaching materials: using newspapers in the classroom* 1/. http://www.onestopenglish.com/support/methodology/ teachingmaterials/teaching-materials-using-newspapers-in-the-classroom-1/ 146510.article.
9. Nagasampige, Kavita; Nagasampige, Manojkumar. (2016): Educational Quest; New Delhi Vol. 7, Iss.1 https://search.proquest.com/openview/fe4647ed321be7bf0eaf656e45290b6e/1?pq-origsite=gscholar&cbl=2032163.
10. Nuttall, C. (1996) Teaching Reading Skills in a foreign language (New Edition) Oxford, Heinemann.
11. Schmitt, N. and Carter, R. (2000). The lexical advantages of narrow reading for second language learners. TESOL Journal, 9(1), 4–9.

Appendix – 1 List of words selected for the study

1. delineated
2. explore
3. monopolize
4. concurrently
5. synchronize
6. trigger
7. invoke
8. stack
9. explicitly
10. immutability
11. cartridges
12. fetched
13. hierarchy
14. trade-off
15. relegated
16. volatile
17. deterioration
18. cluster
19. reliability
20. transition
21. enhancement
22. privileged
23. contemporary
24. primitives
25. manipulating
26. replica
27. constraints
28. prevalent
29. peers
30. retrieve

Appendix – 2 Question Paper

Max Marks – 40

Time 30 Minutes

Part – A

Choose an appropriate word that is closer in meaning to the word given in Capital letters.

S.No	Word	A	B	C	D	E	Ans
1	EXPLORE	search	devise	neglect	ignore	swift	
2	DELINEATED	unclear	confused	illustrated	developed	indistinct	
3	CONCURRENTLY	agreeably	unlikely	divergently	interestingly	simultaneously	
4	FETCH	arrange	get	buy	send	sell	
5	INVOKE	reply to	regret for	remember about	appeal to	retained by	
6	STACK	mine	rubble	debris	pack	margin	
7	EXPLICITLY	doubtfully	regularly	quietly	peacefully	clearly	
8	HIERARCHY	powerful	follow	ranking	inheritance	disorder	
9	TRADE-OFF	contest	dispute	disagreement	measure	deal	
10	DETERIORATION	downfall	comeback	ascent	decoration	improvement	
11	CLUSTER	split	chunk	disperse	divert	derive	
12	TRANSITION	beginning	transportation	temporary	transformation	conclusion	
13	IMMUTABILITY	breaking	ending	stability	irregularity	variability	
14	CONTEMPORARY	current	preceding	asynchronous	futuristic	old fashioned	
15	PRIMITIVE	advanced	modern	exact	actual	underlined	
16	REPLICA	converse report	different version	original side	carbon copy	irrelevant account	
17	PREVALENT	frequent	rare	abnormal	unusual	actual	
18	RETRIEVE	lose	recover	misplace	repeat	rebuild	
19	VOLATILE	durable	everlasting	uniform	fluctuating	dependable	
20	TRIGGER	stunt	activate	halt	destroy	kill	

Part – B

Fill in the blank in each sentence by choosing an appropriate word from the options given.

1) It may be said that Microsoft had nearly ____________ the business of the Operating systems and has pushed its rival companies to distant second position.

A) explored B) monopolized C) invoked D) stacked E) triggered

2) Most cell phones on the market have features that help users to ________________ with the contact list in their mail ids and with a personal computer.

A) monopolized B) clubbed C) synchronize D) clustered E) retrieved

3) As a senior system administrator he has all ________________ to access all folders in the system and to install or uninstall any application in the system.

A) privileges B) peers C) hierarchy D) opportunities E) access

4) Grabbing a fresh ________________ the policeman began to reload his automatic rifle though fumbling a couple of times before he charged at the attacking terrorist.

A) stacks B) peers C) triggers D) cartridge E) cluster

5) If Delhi Dare Devils loses again they may be ________________ from the top position to the list of insignificant teams of this edition of the premiere league.

A) invoked B) triggered C) disqualified D) banned E) relegated

6) While every effort has been made to ensure the ________________ of the information presented in this publication, Kranthi Publishers, a noted education publisher, does not guarantee the accuracy of the data used.

A) volatility B) prevalence C) immutability D) source E) reliability

7) All manufacturing units of the company have been consolidated and brough under single management, allowed the stakeholders to experience productivity ______________ and thereby enjoy improved profits.

A) enhancement. B) deterioration C) manipulation D) exploration E) relegation

8) Efficiency of the computer science and engineering solely depends on its efficiency in storing and ________________ data and information faster and more accurately than the previous system.

A) retrieving B) fetching C) constraining D) manipulating E) delineating

9) 9. My friend's family has decided to abandon a trip they planned to Singapore because of financial ______________.

A) cartridges B) trade-offs C) enhancement D) concurrence E) constraints

10) Among his ________________ , only Anant was more adept at deciphering the problem and suggesting a suitable solution, than Karan, though both have been working on the project for a long period.

A) privileges B) clusters C) peers D) relations E) friends

11) Virtual Reality aims to give us artificial worlds to ________________ outside normal space and time.

A) explore B) delineate C) relegate D) trigger E) explore

12) It is unlikely that the network will schedule all of those shows ________________ .

A) immutably B) concurrently C) reliably D) explicitly E) primitively

13) The UN threatened to ________________ economic sanctions if the talks were broken off.

A) invoke B) retrieve C) manipulate D) replicate E) fetch

14) Unless this is noticed quickly, rapid ________________ of water conditions can occur.

A) enhancement B) synchronization C) deterioration D) transition E) erosion

15) This training course aims to smooth the ________________ between education and employment.

A) migration B) reliability C) mutation D) transition E) exchange

16) The officers were ________________ instructed to involve themselves in local economic affairs, education, and medical care so that people can see the government's efforts in extending good governance.

A) thoroughly B) explicitly C) hierarchically D) covertly E) indirectly

17) These diseases are more ________________ among young children.

A) primitive B) contemporary C) vibrant D) prevalent E) explicit

18) Food and fuel prices are very ________________ in a war situation.

A) manipulated B) triggered C) volatile D) exploded E) reliable

19) In the ________________ western world, rapidly changing styles cater to a desire for novelty and individualism.

A) primitive B) monopolized C) volatile D) reliable E) contemporary

20) There was a little ________________ of admirers round the guest speaker.

A) replica B) cluster C) constraint D) stack E) members

Impending Inquisitions in Humanities and Sciences (ICIIHS-2022) – Dr. Mohan Varkolu et al. (eds)
© 2024 Taylor & Francis Group, London, ISBN 978-1-032-78829-6

Comparative Analysis of Tom Stoppard and Girish Karnad's Political and Historical Plays

8

Osheen Sharma*
Research Scholar, Department of English,
Central University of Himachal Pradesh, Dharamshala

ABSTRACT The late 20th Century post-modernist writing is particularly rich in content and diversity. Societal factors have a transitional role in literary developments, which in turn interact with the political, philosophical, and economical climates. As liberation moves to liberalism, there is a corresponding trend away from social realism towards more self-consciously hybrid genres in which the relationship between the historical, magical, and fictional is problematized. The Czech playwright Tom Stoppard portrays the disillusionment of his characters in the prevailing social scenarios and registers his protest against the abuses of Sovietism and Communism. Harboring similar courage as this great writer of the dissidents, many writers of this era, although in different landscapes but agnates in literary tendencies, took upon themselves very similar tasks. Girish Karnad, an Indian playwright and proponent of multi-culturalism and freedom of expression, deals with history and folktales in his plays to express the modern Indian contemporary political and social issues in the post-colonial scenario. Drawing on the political disillusionment of Soviet Communism in the historical plays of Tom Stoppard and the allegorical depiction of the Nehruvian era in the works of Karnad, this paper will attempt to map the theoretical shifts between stability and instability, political dependency and rationality, resistance and revolution, among other motifs.

KEYWORDS Stoppard, Karnad, Political, Historical, Modernism, Postmodernism

1. Introduction

The late 20th Century post-modernist writing is particularly rich in content and diversity. Societal factors have a transitional role in literary developments, which in turn interact with the political, philosophical, and economical climates. The focus on individual rights and freedom has also been present in drama, a genre where the individual and societal facets of life are taken into account. As liberation moves to liberalism, there is a corresponding trend away from social realism towards more self-consciously hybrid genres in which the relationship between the historical, magical, and fictional is problematized. However, it is evident that this generic instability is either emancipatory or treacherous, or both; relying on foreign traditions and genre norms while also developing site-specific ways of resistance. With the advent of modernism in literature in the early twentieth century, there was a rejection of traditional realism in favour of experimental forms of various kinds (Barry 79). Chronological plots were abandoned, narratives came out complex and proper beginnings and endings became a thing of the past. Post-modernism, then, took this even further with the adoption of fragmented forms of "parody and pastiche" (80).

Political movements contribute to wider structural issues and bureaucratic procedures, highlighting the cultural and historical components of social movement processes. This reflects the cultural turn in the social sciences and humanities more extensively.

*osheensharma1423@gmail.com

DOI: 10.1201/9781003489436-8

Literature itself is a part of society, and plays are representations of it. For the playwright, there are several tasks at hand like observing the never-ending power struggle in human history, taking into account social reality, considering the absolute necessity of an individual's freedom, and not ignoring the demands that the social system places on the individual. After the odyssey of postmodernism, the impact and influence of literary theories on literature grew multi-fold. To achieve such multi-accentualities and literariness of the text, the authors had to bring out something different each time. However, some authors rejected or rather objected such deliberate intricacies. The result is a return to traditional values, particularly after the 1950s, wherein moral authority, old values, and social responsibilities are favoured. This is evident in works of authors who do not fail to portray strong beliefs in terms of individual thought, and a critique of orthodox institutions.

Postmodernism, both cultural and artistic, is inextricably tied to shifting socioeconomic situations (Heuvel 217). Keeping in view the transitional role of societal factors in this period, this paper intends to explore how literary developments interact with the political, philosophical, and economic climates. It will draw upon the active strands of historical and political influences on two eminent personalities of this era, namely Tom Stoppard and Girish Karnad, and carry out a comparative analysis of their works. Their thematic concerns and motifs, the formation of the collective ideal of social, psychological, and political protest, will be examined. Drawing on the political disillusionment of Soviet Communism in the historical plays of Tom Stoppard and the allegorical depiction of the Nehruvian era in the works of Girish Karnad, this paper will attempt to map the theoretical shifts between stability and instability, political dependency and rationality, resistance and revolution, among other motifs.

Tom Stoppard's Political and Historical Plays

Czech playwright Tom Stoppard has a strong affinity for the intellectual, artistic, and ideological stances often associated with postmodern drama, while at the same time he rejects the radical notions and claims of this theory (Heuvel 213). Postmodernist notions provoke odd blending of traditional and non-traditional styles without any proportionality of meaning under the complex usage of exclusive idioms (216). However, Stoppard does not seem to favour the fractured nature of postmodern culture and theory in blending of political and historical genres. One particularly successful take on this dimension is manifested in the framing perspective of political protests in his works. His writings suggest that resistance movements are made up of members of sizable communities who are disillusioned or feel unimportant. These movements provide participants with a sense of empowerment and community that they would not otherwise have. In his dramatic comedies, Stoppard portrays the disillusionment of the characters in the prevailing scenario and registers his protest against the abuses of Sovietism and Communism. He does so with the aid of the historical and monumental, using it as a parable for the "arrogance and anxiety of a new generation moving as fast as it can" (Brantley).

Stoppard is a moral writer and opposes any political idea ejecting morality. He has proven that he is a man of human rights, both in his thoughts and his actions. His criticism of Marxism and Leninism is clear evidence of it. Of all the values held dear, morality has a significant place in his plays. He puts forward that ethical attitudes may show differences in different conditions. He states, "I have never been comfortable with the idea that words would have preferential status over and above the discomfort they may give others" (qtd. in Nadel 422-423). Even when confronted about his postmodernist tendencies, he avows the claims and clarifies that he favours structure and stability above everything else (Heuvel 227).

Conventionally, his trilogy, *The Coast of Utopia* (2002) is a kind of historical account of the origins of Russian Socialism. A dramatic biography of a group of Russian radicals in the nineteenth century, it is made up of three consecutive plays (*Voyage, Shipwreck*, and *Salvage*). It focuses on their attempts to bring about political change in a nation plagued by poverty, injustice, and centuries of authoritarian Tsarist rule. The main characters are Mikhail Bakunin (founder of anarchism and Russian populism), the socialist philosopher Alexander Herzen, the literary critic Vissarion Belins and the novelist Ivan Turgenev. All of them want to improve Russia and its condition, but their ideologies are different and constantly evolving. *Voyage* is set in serf-owning Russia during Tsar Nicholas I's brutally oppressive rule, *Shipwreck* is centered on the events of the 1848 European uprisings, and *Salvage* is based on the revolutionary idea of creating a utopia, which ironically never happens. Instead, the hell of the Bolshevik Revolution begins therefrom. Herzen is the spokesman for Stoppard and fully assumes the central role in at least two parts of the trilogy. He is a critic of Marxism, and through him, the author shows the impossibility of any utopia. The central idea that the play gives is that "no philosophy or political theory will direct the course of nations in an expected manner–that history, in turn, will reject a neat dialectic" (Choate 137).

Written in the late 1970s his political work, Professional Foul (1977) takes the Soviets and Eastern European Communist countries in the bowels of their repression, as the subject of criticism. Stoppard finds it difficult to accept anything as absolute or universal and stresses freedom of thought. The play is a contrast between personal and political approaches to the idea of liberty, portrayed in the experiences of a scholar of philosophy, who goes to Czechoslovakia to present a paper at a conference. Stoppard had written the play after his own experience in Moscow, where he and others had been searched and investigated by the K.G.B. on grounds of human rights in Russia. (Takkac 2) The arrest of playwright Vaclav Havel in 1977 for submitting a document on human rights to the Czech government was another reason for writing *Professional Foul* in which Stoppard directly attacks undemocratic systems, not in a specific country but in general. (3) The protagonist of the play, Anderson, and his friend McKendrick, travel to Czechoslovakia to present papers at a conference. Stoppard's concern about ethics and morality is carefully brought forth in the experience of Anderson in the totalitarian regime of Czechoslovakia, where he comes across his old student Hollar. Hollar wants Anderson to smuggle his thesis to England, as it is in no way publishable in the Czech where political systems control each and everything. Anderson's introspection over the true meaning of freedom, and his acceptance of ethics as a real-world and not an absolute or ideological aspect, is brought to light in his conversations first with Hollar and then his colleague McKendrick.

Travesties (1979) is set in World War I, at the peak of the brawl between communism and capitalism. All over eastern Europe, communist classes gained power and the formation of Soviet Communism in Russia took place. In the play, Stoppard portrays revolutionary individuals like Tzara and Lenin and bashes the politicians through dark parody and crude mockery. This play is a ridicule on political theories as being absurdities of reason. (Görmez 52) He targets the Russian political scene, by portraying an interaction between the Bolshevik leader Vladimir Lenin, the Dadaist poet Tristan Tzara, and the novelist James Joyce. Henry Carr, a British diplomat dreams about these influential personalities, discussing the state of affairs. Stoppard portrays Lenin with serious strokes, as a political dictator, as he is even opposed to free art forms like music. Carr's words show Stoppard's criticism clearly, "you're either a revolutionary or you're not, and if you're not you might as well be an artist as anything else" (qtd. in Görmez 54).

Girish Karnad's Political and Historical Plays

Harbouring similar courage as the great writer of the dissidents Tom Stoppard, many writers of this era, although in different landscapes but agnates in literary tendencies, took upon themselves very similar tasks. Girish Karnad, an Indian playwright and advocate for multiculturalism and linguistic as well as cultural diversity, deals with history and folktales in his plays to express the modern Indian contemporary issues in the post-colonial scenario. Karnad does not need any special recognition because he has a very prestigious place in Indian English drama. He is also a critic of religious fundamentalism. Although his plays were originally written in Kannada, and then translated into various languages, Karnad's critique of socio-cultural issues in his milieu, abridges the gap between different time frames and his commentary is often stark and pungent.

Bakhtin says that we need a cross-cultural dialogue to stop discourses from homogenising (Rao 1). Karnad's plays "re-invent the Indian tradition", by the use of versatile characters and stories across various Indian cultures (1). The most important contribution he has made to Indian English drama is his effort to get back the cultural and mythological rich traditions of the Indian past. He is a political reformer and puts attention to the social problems, caste system, gender discrimination, patriarchal family system, and situation of women; not only Karnad talks about basic problems of Indian society but also the belief of superstitions underlying these. Karnad is definitely modern in his sensibilities, nevertheless his works also incorporate the postmodernist elements of pastiche and parody. His plays deal with societal and personal issues, relying heavily on the past but discussing modern scenarios. He makes use of magic realism, a postmodernist element which refers to an amalgamation of fantasy and reality (Veeramani 214).

In his plays, he also incorporates historiography as a venture of postmodern counter discourse, using elements of fiction and literary imagination (Choudhary 80). His characters show the instability and fragmentation attributed to the postmodernist perspective. However, for his themes he relies heavily on the classics. Karnad's use of the ancient and medieval past is not only a compulsory choice but also a tribute to it. His familiarity and ease with the stories, stems from his strong feeling of cultural ties to India. It is rightly said that, "Karnad belongs perhaps to the last generation of the urban Indian writers who encountered the 'great' and 'little' traditions of myth, poetry, history, legend, and folklore at first-hand in his earliest childhood, and internalized them deeply, enough to have their about authorial selves shaped by them" (qtd. in Kumar 314).

His play, *Tale-Danda* (1990) is based on the Bhakti movement in the southern part of Karnataka in the 12th Century. The old historical tale takes a new meaning and significance when the issues of caste and class it raises are considered in the context of contemporary times. The distinction between caste-based and communal violence has been depicted by Karnad. The play's central character, Basavanna, a pious man, dreams of a casteless society and struggles to speak out against prejudice in his community. His followers, known as *sharanas* work to upend the advantages enjoyed by higher caste people and to promote social justice, freedom, and equality. (Wadikar 3) However, amidst this cultural movement, the marriage of a brahmin girl to a cobbler boy leads to a riot-like situation, massacre, and bloodshed. *Tale-Danda* exposes the ugly deformity of the society, the hollowness of idealistic preaching depicted through Basavanna's failed attempt to create a casteless utopian society. (6)

Tughlaq (1977) revolves around Muhammad Bin Tughlaq, the 14th Century Indo-Turkish ruler. His failed vision is a crafty analogy of the disenchantment of Nehruvian ideals. In these plays, Karnad takes the historical events and fuses them into a whole to portray the fatal evils of the bygone days prevalent even in the present days. Karnad also examines conscious historiography, which has produced a distorted account of Indian history. The eponymous ambitious ruler Tughlaq gets alienated from his society because of his overreaching idealistic thoughts which neither match the times nor the people's mindset. The drama highlights the dichotomy of the idealistic Tughlaq, who never experiences true peace and is instead constantly distrustful of everyone around him. Karnad uses this paradox to show his critical views of political idealism in the past as well as present times. The play raises some concerns about the socio-political reality of post-Independence India and then goes back to tradition and history to find an answer.

The Dreams of Tipu Sultan (1997), another historical play by Karnad, is set in Karnataka in the eighteenth century and seeks to highlight the imbalance between the greatest king's aspirations and actions, which started with the noble intention of establishing a glorious kingdom but ended in gory violence. In the play, Tipu Sultan has dreams of an independent India free from British colonial rule, and his dreams serve as manifestations of this goal. He wanted to revolutionize Indian society, improve trade and economy, and wipe out the colonial regime from the country. Karnad depicts Tipu as a politically sound leader, with fair knowledge of commerce and administration, however, his whims are way above his head. His alliances with other countries ultimately lead to his demise since they undermine his leadership. In his final, allegoric dream, Tipu forms forces with the Marathas and the Nizams to completely overthrow the British rule in India, although the reality is completely opposite. Karnad also gives his significant critique on historiography and shows how history is always in the hands of the powerful. In the play, the use of dreams and recollections creates a transition between the fictitious and the actual world. (Sharmila 7) The dramatist is successful in capturing the core of the Sultan's values as well as his hopes and aspirations as a cross-section of his life by blending the boundaries of the two states of awareness. (8)

Comparative Analysis of Stoppard and Karnad

The idealism of the resistance and revolutionary periods, the following disenchantment, and the search for new forms of protest and community all the way from Stoppard through Karnad has been illustrated above and shall be scrutinized further. There is no doubt about the use of historical and political themes by both these playwrights. Neither does Tom Stoppard sway from staunch criticism of Marxism and Leninism but rather is bold in his representation of the political characters he used in his plays. Girish Karnad is unapologetic in his treatment of religious and social fanaticism and through his allegorical use of myth, he leaves no stone unturned in satirizing the Indian contemporary scene.

Although written more than two decades apart, *The Coast of Utopia* by Tom Stoppard bears striking affiliations with Girish Karnad's *Tughlaq*. In the former Stoppard gives a pungent critique of the rise of Russian socialism in the 19th century, in the latter, Karnad directly attacks the failing political situation during the Nehruvian era in India. The main narrative of *Tughlaq* depicts the downfall of an ambitious autocrat, while the secondary plot centers on a common dhobi who takes advantage of the king's plans for his own gain. Similar contrasts are drawn by Stoppard between Herzen's existence and its smallness in *The Coast of Utopia* and the massive socialist political projects that were being built all throughout Russia at the time. The central idea is that idealistic politics fall short when confronted with reality, which both playwrights are trying to put forth. Karnad's portrayal of Tughlaq is of an uncompromising ruler, whose ambition gets the best of him. Stoppard too, depicts the Tsar in a similar fashion, as cruel, devout, and callous. The recurrent theme is again similar in both these dramas, of the superiority of human nature and life over politics, social activism, and utopian visions.

Similar to Stoppard's *Utopia* which is never achieved in Russia, the Nehruvian vision of idealized communalism given by Karnad in post-Independence India, is never attained. In Tughlaq uncompromising idealism is strongly critiqued, and so is chance and deliberation in *The Coast of Utopia*. The characters of Barani and Najib represent two opposite selves of Tughlaq, whereas Aziz is a treacherous manipulator who takes advantage of the Sultan. Likewise, Stankevich and Belinsky are a contrast to the character of Herzen, whereas Bakunin, is a passionless anarchist. The word "prayer" is repeated several times in *Tughlaq* and reverberates throughout the play as a necessity for the masses. In *The Coast of Utopia* "family" is used as a metaphor; Herzen cannot think above family, not even about important matters of the state. Last but not least, the theme of commonality that is presented in Herzen's character when he is only shown to be grieving the loss of his wife and son is also present in

Tughlaq's character when the emperor is reduced to a common man as he touches the feet of the cunning Aziz disguised as Ghiyas-ud-din Abbasid.

Published in the same year as Karnad's *Tughlaq*, Tom Stoppard's *Professional Foul* is a commentary on personal ethics and morality as opposed to that of political ethics and morality. All political actions, according to him, must be evaluated morally. In the face of political unrest, Professor Anderson is shown as being ignorant of human pain and freedom. Girish Karnad's *Tale-Danda* then, is a direct attack on the age-old caste-system prevalent in India of which Karnad has always been a staunch opposer. Are there any resemblances in the two? Let us examine. Karnad handles the societal issues directly by using metaphorical depictions of actual figures from the past, unlike Stoppard, who utilises fictitious characters to criticise false moralities in his play (despite the fact that it was based on real-life occurrences). At the time which Karnad wrote the play, there were already Mandal and Mandir conflicts going on in India which situates it as a contemporary piece of writing. He himself states, "the horror of subsequent events and religious fanaticism that has gripped our national life today have only proved how dangerous it is to ignore the solution they offered" (qtd. in Abidi 75). Stoppard also included contemporary events into his play by ilucidating the communist dictatorship in Prague and the rigid political system of Russia.

Both the playwrights use their intuitive feelings about ethics, justice, human rights, and morality in order to criticise the system of absolute principles. Karnad satirises the supremacy allotted to higher castes in India and Stoppard challenges the lack of freedom of thought and expression in a totalitarian regime like Russia. They are against the dogmatised systems which sacrifice human emotions and rights. In *Tale-Danda*, Karnad has depicted all strata of life through his colourful characters. There are Brahmins, cobblers, barbers and skinners who represent hierarchy of Indian Varna system. Stoppard has also used a wide range of dramatis-personae in his television play *Professional Foul*, namely professors, historians, scholars, writers, policemen, and clerks. Karnad also weaves another thread of social evil in his play, i.e., superstition, which can be compared to Stoppard's idea of ethics. In the actual world, where academics like Hollar are terrified of punishment by the law, Anderson's ethics—which he defines as being a good guest for the Czech Republic—are irrelevant. Similar to the superstition shown by Karnad in the villagers' dwelling, they are imaginative constructs of a blinded people. There is a similar tone in the following lines by Stoppard and Karnad, wherein both are scrutinizing the idealized societal constructs:

> Hollar: The ethics of the state must be judged against the fundamental ethic of the individual. The human being, not the citizen. I conclude there is an obligation, a human responsibility, to fight against the state correctness. (Fleeming 144-45)
>
> Bijjal: One's caste is like the skin on one's body. You can peel it off top to toe, but when the new skin forms, there you are again: a barber-a shepherd-a scavenger! (Karnad 14-15)

The Dreams of Tipu Sultan is a historical play by Karnad based on the life of monarch Tipu Sultan of Mysore who reigned from 1782-1799. In the play Karnad uses dreams as an important technique of narration. Tipu's dreams are sole determinants of his thoughts. It is also a symbolic representation of his downfall, as the message is loud and clear–dreams aren't always fulfilled. Stoppard's *Travesties* is "unreal, chaotic, devoid of coherence and impervious to sense" (Delaney 58). The play echoes history as it portrays the Russian ideologue Lenin, the Rumanian artist Tristan Tzara, and the Irish novelist James Joyce, as residents of Switzerland in 1916. Both the plays depict a crucial time for the respective nations, the colonization of India in the early twentieth century and the socio-political turmoil in Russia during World War I. However, this is not the only similarity between the plays. Where Stoppard portrays influential characters with flawed outlooks, and idealistic sensibilities, as a lost cause for the Soviet nations, Karnad pretty much does the same thing in the depiction of a pre-Independence ambitious ruler of India, who causes his own doom like Icarus. Stoppard is angry with politicians like Lenin, who abuse art or philosophy for their own ideologies. He is almost brutal in his character sketch of Lenin and offers serious criticism of his theories. Likewise, Karnad also offers bold judgment on the failure of Tipu as a monarch.

2. Conclusion

The aforementioned comparisons demonstrate how the two writers are not at all hesitant to express their opinions on social issues and to provide personal criticism on the shortcomings of their respective nations. Stoppard is critical of Marx and his theories, false moralities, political dictatorship, and idealism, which he has asserted clearly in his political plays. Karnad too shows his stance clearly against the flaws of Nehruvian ideals, ills of the Indian political system, caste-based prejudice, and racism. For voicing these bold opinions both playwrights adopt distinct genres, which suit their purpose, and are unique in their own way. Where Stoppard is a famous and successful dramatist of black comedies and absurdist plays, Karnad's plays are hybrids constructed with blended strokes of mythology, fiction, and history. There is an abrupt similarity in their genrefication, which appears as a conscious effort to create something new. This creation is not defined or regulated by any one literary movement, rather appears to be a blend of multiple literary discourses.

At the same time, there is also a striking difference between the dark and complex writing style of Stoppard and the symbolic but plain language of Karnad. Nevertheless, like true modernists, both playwrights depend upon political and social themes for their plays. The postmodern elements of historiography, pastiche, parody, categorical disillusionment and fragmentation are found in both Karnad's and Stoppard's plays. They have made use of historical hard facts and mingled them carefully with fictional touches to suit their purpose. The conclusion is that both Tom Stoppard and Girish Karnad can be attributed largely for adding versatility and substance to post-modern English drama and bringing about a change in society through their quality writings. Their literary prowess transcends the boundaries of time, space, and location and instead applies to a far wider context.

REFERENCES

1. Abidi, S.A. Raza. "Social and Political Conflicts in Girish Karnad's Tale-Danda." *International Journal of English Language*, Literature and Humanities, vol. V, no. II, 2017, pp. 74–83.
2. Brantley, Ben. "The Coast of Utopia: Part One- Voyage." Young, Restless and Russian, Devouring Big Ideas. *The New York Times*, 28 Nov, 2006. www.nytimes.com/2006/11/28/theater/reviews/28coas.html.
3. Choate, E. Teresa. "The Coast of Utopia (Voyage, Shipwreck, Salvage) review." *Theatre Journal*, vol. 60, no. 1, 2008, pp. 133–137. doi.org/10.1353/tj.2008.0072.
4. Choudhary, Kafeel Ahmed. "Things Fell Apart: A Re-reading of Girish Karnad's Tughlaq in Postmodern Perspective." *International Journal of Arts, Humanities and Management Studies*, vol. 1, no. 6, 2015, pp. 80–84.
5. Delaney, Paul. "Mortal Flesh in a Moral Matrix of Words: The Temporal and the Timeless in Travesties." *Tom Stoppard: The Moral Vision of the Major Plays*, Macmillan P Ltd., 1990, pp. 58–81.
6. Fleeming, John. *Stoppard's Theatre: Finding Order amid Chaos*, U of Texas P, 2001.
7. Görmez A. "Morality and Politics in Tom Stoppard." *Journal of Social Sciences Institute*, no. 22, pp. 45–58, 2007.
8. Heuvel, Michael V. "'Is postmodernism?': Stoppard among/against the postmoderns." *The Cambridge Companion to Tom Stoppard*, edited by K. Kelly, Cambridge UP, 2001, pp. 213–228.
9. Karnad, Girish. *Tale-Danda: a play*, Ravi Dayal Publisher, 1993.
10. Kumar, Laxman. "Perspectives on Thematic Concerns in Girish Karnad's Yayati." *The Creative Launcher*, vol. II, no. VI, 2018, pp. 313–320.
11. Nadel, Ira. *Double Act: A Life of Tom Stoppard*, Methuen, 2002.
12. Rao, T Eswar. "Re-Inventing Tradition: A Study of Girish Karnads Plays." *International Journal of Creative Research Thoughts*, vol. 6, no. 2, 2018, pp. 1-5.
13. Sharmila, K. "Girish Karnad's Retold History in The Dreams of Tipu sultan."
14. Takkac, Mehmet. "Alternative Theories of Ethics: Tom Stoppard's Professional Foul." *Journal of the Institute of Social Sciences*, Spring, no. 16, 2006, pp. 1–14.
15. Veeramani, S. "Multiaccentualities in Girish Karnad's Nagamandala- A Postmodern Reading." *International Journal of English Language, Literature and Humanities*, vol. 2, no. 3, 2014, pp. 213–217.
16. Wadikar, Shailaja B. "Tale-Danda: India: A Million Mutinies Now." *Pune Research*, vol. 5, no. 3, 2019, pp. 1–7.

Impending Inquisitions in Humanities and Sciences (ICIIHS-2022) – Dr. Mohan Varkolu et al. (eds)
© 2024 Taylor & Francis Group, London, ISBN 978-1-032-78829-6

Being on the Margins: Gender, Disability and Violence in Mridula Garg's *Kathgulab*

9

Shivanee*
Research Scholar, Department of Humanities and Social Sciences, NIT Hamirpur, Himachal Pradesh

Manoj Kumar Yadav[1]
Assistant Professor, Department of Humanities and Social Sciences, NIT Hamirpur, Himachal Pradesh

ABSTRACT The experiences of disability are deeply gendered. Disabled women tend to be undereducated and poorer than their male counterparts, and more likely to be at risk of sexual abuse and violence. This paper suggests that gender needs to become an integral part of disability studies for bringing the experiences and realities of women with disabilities. It analyzes Mridula Garg's novel, *Kathgulab*, which portrays how Smita becomes a rape victim by her sister's husband. It explores such prominent themes as sexuality and relationship, gender-based violence, family, survival challenges, poverty, resistance, and challenges to the dominant negative constructions of disabled women. Further, the paper shows how the intersectionality of gender and disability addresses women's experiences of disability; and how literary representations feed directly into cultural perceptions of disability.

KEYWORDS Gender, Disability, Discrimination, Violence, Women

Women with disabilities experience gender-based violence and discrimination and are more prone to vulnerabilities. The intersectionality of disability and women reveal different types of barriers in terms of their education, job, living, healthcare and justice experienced by them in their daily life. They also face various types of disproportionate violence, be it physical, psychological, sexual, economic or societal. The paper explores the marginalized voices of a disabled woman providing intricate dynamics of gender, disability, violence and marginalization. It shows the representation of a disabled woman, her agency of self and identity, experiences of discrimination and various forms of violence that she encounters. Writing about disability in India needs to recognize the social, cultural, religious, historical, and economic circumstances that shape the experiences of disabled people. Being a disabled woman, one has to face not just "double marginalization," which Gaytri Charkravarti Spivak refers to in the context of colonial and patriarchal oppression, but also another jeopardy of the hegemony of normalcy. Such complex discourses further contribute to marginalization and discrimination against them. UN Human Rights Council 47th Session (2021) observed that women with disability often "experience discrimination, exclusion, isolation and denial of their dignity and autonomy, both in public and within their own families" (np). However, the intersection of gender and disability creates new possibilities and opportunities for women and girls with disabilities. It critiques the male-centric representation of disability studies and subverts the prevalent notions of the "normal" and "natural" in the "abled" male world. In my analysis, I would use the methodology of feminist disability studies promulgated by Renu Addlakha, Anita, Ghai, and Rosemarie Garland-Thomson. It is explicit and unquestionable that the gradual development of Western feminist discourses and gender studies since the 1960s, from Simon de Beavoir's *The Second Sex* to Judith Butler's *Gender Trouble* modified and upgraded the representation of woman and womanhood but women with disabilities have not received much attention.

*Corresponding author: shivanee@nith.ac.in
[1]manojk@nit.ac.in

DOI: 10.1201/9781003489436-9

Being on the margins suggests an exclusion or being on the periphery in a group, society or mainstream norms. It also refers to the condition of being socially, economically, and politically marginalized. The people on the margins are excluded from the opportunities, benefits and power enjoyed by the mainstream people. They have limited access to resources and face unequal treatment. In this regard, the focus is on how people are excluded and disadvantaged in terms of gender and disability, and how such people are put outside the mainstream or dominant societal structures and norms. The paper deals with the intersection of gender, disability and violence, and how they contribute to the structural process of marginalization in Mridula Garg's novel, *Kathgulab*, published in 1996. Garg was awarded the eminent Vyaas Sanman Award in 2004 for *Kathgulab*. She has been imprisoned for writing about matters that are generally preferred to be glossed over such as reproductive rights, sexuality, and other such issues considered peripheral to the progress of a nation. This novel can be considered a radical feminist novel which articulates a didactical rhetoric of the male oppression of women. Garg's two major novels '*Chitkobra*' and '*Kathgulab*' have also been considered controversial due to her candid views on sexual violence, oppression and openness of sex in a man-woman relationship. *Kathgulab* represents a complex story of Smita being marginalized on the ground of her disability and gender and sheds light on her experiences of being on the margins of society. The novel embodies her simultaneously marginalization due to her gender and disability, along with the violence she encounters. The novel is a seminal text on marginalization, inequality, violence, and resilience in Hindi literature.

Though *Kathgulab* narrates the story of a young girl, Smita, it is not just her personal story. She embodies the story of all the disabled women who face suffering in society. Nothing is indeed purely personal and neither is purely social. Society has to be tested through the individual and the individual through the society. Needless to say, when this person is a woman, the interactions become more complex and interesting. The story in the novel, *Kathgulab*, is written about the life of women and their struggles. The plot exposes the exploitation, injustice and plethora of struggles of women prevailing in society. It further showcases how women are exploited, sexually harassed by men in their homes and society and thrown away after using them. Along with such issues, disability, its consequences and various other problems faced by women in their everyday lives are the centre of this novel. The novel taking stock of the lives of many such women finds out all the relevant and incompatible elements of life, which give birth to a continuously dynamic society by connecting with the individual and sometimes breaking it. This novel gives an opportunity for deep introspection and warns that the social system needs to transform itself. The novel is in a way living document of the pain and struggle of Indian women. In this novel, there is direct evidence of the complex texture of male-female relations in the context of the injustice, oppression and pain of women. The novel questions dominant cultural propositions about patriarchy, hierarchy and women's (both disabled and nondisabled) subordination in everyday experiences. The novel is ideologically and stylistically considered feminist because it promotes women's expression in the structure of the narrative itself, and denounces the oppression of females by men. The symbolic meaning of *Kathgulab* is '*jijivisha*' (willingness to live). In this work, Garg has underlined that women are not roses, which also blossom when they grow. They are woodrose, which needs little care to blossom.

In post-Independence Hindi literature, some novels have raised the problems of people with disabilities that they suffer with. They expose violence against women, taking advantage of their disability or physical weakness constitutes. The issues of rape or such other misdeeds have been dealt with by many writers in their fiction. Similarly, Mridula Garg, a very well-known writer of Hindi literature has prominently raised the problem of rape in her novel *Kathgulab*. The problem of rape is a serious issue in the Indian society. Many women become victims of rape. Sometimes the family members, sometimes the relatives, and sometimes the other members of society destroy their lives by raping them. Even women with hearing, visual, and mental disabilities are not spared. Women from all strata of society are subjected to such rape. If a woman opposes rape, she has to face social stigma and humiliation.

Smita, the heroine of the presented novel goes through a lot of problems because of her poor eyesight. She has weak eyesight at the age of twenty. She cannot see anything without her glasses, and cannot go anywhere. The narrator describes her situation that "to spend a single day without spectacles is like falling into a blind well" (Garg, 2019, p. 18; this and subsequent translations are my own). She lives with her sister and brother-in-law after the death of her parents, and they bear the expenses of her education. Her brother-in-law has always eyes on her; seeing no one around, he kisses her without her consent. She doesn't tell her sister about him because she has no other place to live. He keeps doing dirty tricks with her many times and wants to get rid of her and treats her like a burden in his house. But at the same time, he also wants to take advantage of Smita as long as she is in his house. He says that she is his sister-in-law, what a thing she is. If she is around, I keep drooling over her, she is my sister-in-law and half-wife; hence he has every right on her. When Smita resists him, she is being defamed. He complains to his wife (Smita's sister) about Smita: "If you are not going to pay for my admission, I will defame you." Namita slaps Smita's cheeks. In this incident, Smita's spectacles fall and shatter.

The oppression of Smita in a traditionally patriarchal society, represented by her sister and brother-in-law, is two-fold: she is exploited on the ground of being disabled and a woman. What is more fascinating is that the violence against Smita is committed by both her brother-in-law and her sister. In "Disabled Women: An Excluded Agenda of Indian Feminism," Anita Ghai observes that "disabled women occupy a multifarious and marginalized position in Indian society based on their disability and also on sociocultural identities" (2002, p. 49) of class, caste, community, etc. The life and daily experience of disabled women are extremely painful, perplexing and difficult. They have to undergo several types of ordeals, physical and emotional violence and discrimination daily. Smita's sister slaps her which leads to the breaking of her glasses, and consequently making her blind as she cannot see without the glasses. Her brother-in-law's fabricated complaint and consequent punishment by her sister in the novel suggest that disabled women are exploited, defamed, physically abused, blamed and thrown out of the house. Renu Addlakha's statement justifies apathy towards people with a disability:

> Historically in India as elsewhere in the world, there has been a deep-rooted cultural antipathy to persons with disabilities. Throughout the ages, the disabled have been looked down upon with disdain, almost as if they were sub-human. They have been portrayed as medical anomalies, helpless victims and a lifelong burden on family and society. (2007, p. 9)

The accusations by Smita's brother-in-law suggest such typical stereotypes about people with disabilities as deceit, mischief, unscrupulous actions, devilry, and other negative cultural identities. Smita represents the general condition of routine violence against disabled women in India. Such women are exploited by their family members and relatives, silenced, and left to suffer. When they are dependent on their violators, they bear everything silently. The poor economic condition of women with disabilities becomes a reason for violence against them in the name of some help. They resist but their resistance finds no support. Public consciousness remains averse to them. Ghai rightly states that "the meaning of disability in India is embedded in the basic struggle for survival" (2015, p. 51). Smita's initial dependence for the educational fees on her brother-in-law suggests what poor economic conditions can do to the people. *In Rethinking Disability in India* (2015), Ghai remark that "poverty de-individualizes, and alienates those affected from the mainstream of society" (pp. 101–02). Poverty puts a limitation on disabled women and curtails their quality of life making them helpless and hopeless. It affects them disproportionately. Ghai further remarks that poverty disables people and disabled people are often poor.

Thinking of leaving home, Smita arranges some money by going to her friend Asima so she had to stay two more days till her spectacles are repaired. Without spectacles, her life was like a person with blindness. Meanwhile, her demonic brother-in-law takes advantage of her weak eyes and rapes her brutally in the darkness of the night. Tying her hands to the cot, stuffs her mouth with a cloth, tears apart her clothes from the body, and spits on every part of her body. She tries very hard to lift her eyelids to look up but could only see the hazy shadow. The narrator says "Thankfully, without glasses, she could see so little that the havoc being inflicted on her body blurred everything. After consuming her by raping her, he laughed a suppressed, slow, cruel laugh and had gone out leaving her dumb, bound and naked" (Garg, 2019, p. 21). When her sister finds her lying naked and her mouth stuffed with cloth, she informs that they were looted last night. She tells a lie about the robbery but Smita was sure who her rapist was. He was her brother-in-law who covered his face and didn't utter a single word throughout the heinous act. She wanted to report to the police but her sister forbids in the name of insults, but implicitly same the perpetrator.

Smita has become the object of the male gaze to be exploited by her brother-in-law. She is a victim of domestic violence and sexual violation. Sexual violence against Smita splinters her subjectivity and identity. She was sexually assaulted and physically abused. She faced the worst type of human crime yet her sister tells a lie to protect her husband since she and Smita are dependent on him for survival. Further, the issues of honour, social status, stigma and shame forbid her to reveal the truth. Opportunities for survival and improving the life of rape victims and disabled women are non-existent in India. Above all, Smita is considered as a social and economic burden. Smita was already a subject of pity to her sister because of her physical and cognitive impairments, but now her rape has made her undesirable in the house. She is not provided with basic human treatment and, instead, bears the brunt of violence, violation, exclusion, and discrimination from both male and female members of the house. Smita had a feeling that there was no other but her brother-in-law because he knew that she cannot see anything without her glasses. The metaphor of broken glasses and consequent vision loss suggests that disabled women undergo the trials of violence but cannot narrate the same in this unjust society as if nothing has happened. Hence, she has to close her eyes to the violence inflicted upon her. Though, the narrative suggests that she couldn't see anything, the fact is that she did know everything. It is her sister who pretends not to see anything and it's her culprit brother-in-law who pretends nothing has happened. She has faced unprecedented pain, suffering and violation. The other meaning of the metaphor of the broken glass suggests that Smita, being a disabled girl, will not be admissible by society as a witness to the sexual and physical assault on herself. Similarly, the camouflage by her sister and disowning by her brother-in-law suggest a preoccupation with the hegemony of normalcy. Brian Watermeyer rightly observes the denial of disabled people's experience:

> the very real danger for disabled people is that this sphere of our human experience, of our sense of self, may be rendered less admissible or possibly banished entirely from view under the host of moral imperatives exercised by a medicalizing society – to overcome, normalize, disguise, defeat, disown, defy or otherwise void the perceived emotional trappings of disablement. (cited in Ghai, 2015, p. 17)

People with disabilities are not just negated but also denied respect. The cultural expectations constructed concerning the "abled" and "normal" human body do not value the narratives of disabled people. Non-disabled people construct, marginalize, and disempower disabled people, and make them internalize negative stereotypes about themselves. Renu Addlakha argues, "Pity, segregation, discrimination and stigmatization became normalized in the management of persons with disabilities" (2007, p. 4). Smita has to accept the practice of silencing disabled people or she has to face further structural exclusion, violence and social stigma. She has to survive through acceptance and denial: acceptance of her suffering and denial of her pain by others.

Smita keeps sparring to avenge her rape and, despite being raped, she does not have any guilt, but she is angry at this injustice done to her which burns the fire of vengeance in her. She engages herself in the task of gathering strength for resistance. She tries to improve herself in every way to carry out her vengeance. Hence, she leaves her sister's house to study in Kanpur on her own then moves to America after getting a job. Where she describes herself as sad and lonely and sees trees as her only friends. But during this time her sense of vengeance does not diminish. Her journey for education and a job is not easy. Disabled women face challenges in education, job, marriage and motherhood; and they get exploited everywhere. The problems in carrying conventional roles of women, disabled women are bracketed with "rolelessness," a concept used by Michele Fine and Adrienne Rich, "for their social invisibility and cancellation of femininity" (Ghai, 2002, p.55). Smita's decision to leave her sister's house, seeking help from her friend, continuation of studies and moving to the USA show, to use Arthur Frank's words, a "quest narrative" which embodies her efforts to derive meaning and purpose to her life by rising to the circumstances. In a "quest narrative," the character shares the lessons learned or the deeper meaning represented in the experience: "The quest narrative tells self-consciously of being transformed; undergoing transformation is a significant dimension of the storyteller's responsibility" (Frank, 1995, p.118). She defies the general fate of blind people, who generally become beggars, and pursues her education. Despite her mental breakdown, she copes with her experiences and grows.

During her stay in America, she gets married to her psychotherapist, named Jim Jarvis in Boston, a horrible egotistic male chauvinist and a psychopath. Smita again becomes a victim of marital rape. Jarvis wanted love, affection, trust, need, affinity, friendship, and satisfaction, all to be expressed only in words. If she could not, she would have only one option left, sex. Her every 'silence' would make him invent a new way of enjoying her body, more despicable than ever, humiliating and indecent. Jarvis abuses and beats her with belts and rapes her causing a miscarriage and due to which she can never conceive a child in future. He keeps torturing her mentally and physically. To escape from this sting, Smita reaches an Institution named RAW (Relief for Abused Women). She also demands a divorce from Jarvis, but Jim Jarvis's devious mind leads to Smita being declared a lunatic and a victim of psychosis in court and Jarvis being relieved of her maintenance. Jarvis's deceitful scheme does not allow Smita to win. Her brother-in-law also had a similar heinous plan against Smita, where Namita, her real sister goes against her. Jarvis' behaviour was a mindset of "normate" who made himself appear like a part of "polite society" but his real nature gets exposed soon. Jarvis' engagement with Smita during her psychotherapy was his sense of pity and fascination but soon her disability implicitly becomes a matter of repulsion and an unwelcome reminder. He further tortures her with sexual and physical assaults. For similar kinds of violence, she has left her sister's house but again she faces the same suffering.

The story of Smita, her failures and success, defy the norms of "normalcy" with "ability" and "abnormal" with disability. She challenges the social process of disabling and changes the perception of the disability. Her all efforts to make a space for herself in the unjust society are an attempt at the integration of the people of disabilities into the larger community without discrimination. Her disability, to borrow the words of Helen Deutsch, in the novel becomes "a vehicle for self-reflection, self-representation, and self-legitimation" (cited in Wang, 2023, p.46). It invites people to introspect about disabled people, to question the negative portrayal of disability and to assert their selfhood and identity. The experience and encounters of Smita at different stages of her life suggest what Rosemarie Garland-Thomson (1997) calls "the rules of social engagement" between the disabled and non-disabled: "the nondisabled person may feel pity, fascination, repulsion, or merely surprise, none of which are expressible according to social protocol" (p.12).

The problem of education for people with disabilities is a persistent concern. Garg makes a subtle marking of the myriads of problems and obstacles faced by people with disability like Smita in "*Kathgulab*" while pursuing education. She completes her B.Sc. somehow, but when she asks her brother-in-law for money for the admission fees, she has to be humiliated many times and suffers physical pain also. In the name of help, he often demands sexual and other bodily pleasure. In a way, he compensates for the financial help with his sexual advancements. Being an orphan and disabled child, Smita asks for help from her sister and her husband but her brother-in-law rapes and molests her in favour of letting her stay at his home. One day a man is called to see Smita for marriage. Then Smita says "I don't want to marry, neither to a middle-aged nor to a young. I want to study (Garg, 2019, p.15)". She wants to pursue M.Sc. while living in a hostel. But due to a lack of financial income, she remains worried about the expenses. During her education, her brother-in-law always tries to fix a match for her marriage saying who will marry this girl. Hence, most of the would-be grooms that he brings are unsuitable and in the guise of fixing her marriage, he tries to molest her. His description of Smita is based on gender prejudices against disabled women which constructs disabled women's bodies as *asexual*. Her marriage was a counter-discourse to such a type of representation. Garland-Thomson considers the disabled body as "extraordinary rather than abnormal" (1997, p.137). While there are efforts to mainstream the concerns of women by empowering them but women with disability remain on the margins. There are many grave issues which people with disabilities are confronted with in their lives, such rape, mismatched marriages, social neglect, etc.

Mridula Garg's novel, *Kathgulab*, depicts the intricate intersectionality of gender, disability, and violence. Through the character of Smita, the novel brings the lived experiences of disabled women who often find themselves oppressed within society. It highlights the multifaceted nature of discrimination and the resulting violence. Disabled women experience double and triple marginalization. Ghai states that "they fare worse than either disabled men or nondisabled women socioeconomically, psychologically and politically. Disability compounds their already marginal status as women" (2002, pp. 57–58). Smita's life suggests that she fights against both disability and gender. She fights against the stigma of disability and copes with gendered violence in order to assert her identity, re-emphasize humanity, and create her own space in society. She exposes the oppressive nature of society. In *Extraordinary Bodies*, Garland-Thomson asserts that disability requires accommodation rather than compensation, and to shift our conception of disability from pathology to identity" (1997, p.137). Garland-Thomson questions the "lack of agency" associated with a disability because of the female body's inherent contingency and vulnerabilities. They are not helpless objects but visible subjects in society. Smita often confronts physical and attitudinal barriers, limited access to education, healthcare, and employment opportunities. The novel exposes ableism ingrained in society and the various forms of discrimination Smita encounters. The life experiences of Smita affirm a need for inclusivity, accessibility, and the recognition of disabled individuals' ts to dignity and full participation in society. The novel endeavours to promote the empowerment of women (disabled and nondisabled) within the space of the narrative. Thus, it delves into the intricacies of disability and gender and also interrogates the cultural specificities of "disabled gender trouble' in the Indian context. By representing a disabled girl, Smita, as the protagonist, Garg challenges societal norms and the systemic barriers faced by those on the margins, promoting empathy and social change.

REFERENCES

1. Addlakha, R. (2007). "Gender, Subjectivity and Sexual Identity: How Young People with Disabilities Conceptualise the Body, Sex and Marriage in Urban India". occasional Paper Series no. 46, Centre for Women's Development Studies, New Delhi.
2. Frank, A. (1997). *The Wounded Storyteller: Body, Illness, and Ethics*. Chicago: University of Chicago Press.
3. Garg, M. (2019). *Kathgulab*. New Delhi: Bhartiya Gyanpith.
4. Garland-Thomson, R. (1997). *Extraordinary Bodies: Figuring Physical Disability in American Culture and Literature*. New York, Columbia University Press.
5. Ghai, A. (2002). "Disabled Women: An Excluded Agenda of Indian Feminism". *Hypatia*, Vol. 17, No. 3, Feminism and Disability, Part 2, pp. 49–66.
6. ---. (2015). *Rethinking Disability in India*. London: Routledge.
7. "Violence against women and girls with disabilities". United Nations Human Rights. (2021). https://www.ohchr.org/en/statements/2021/07/panel-1-violence-against-women-and-girls-disabilities. Accessed on 22 June 2023.
8. Wang, Fuson (2023). *A Brief Literary History of Disability*. London: Routledge.

Impending Inquisitions in Humanities and Sciences (ICIIHS-2022) – Dr. Mohan Varkolu et al. (eds)

Culture in the Contours of Language and Literature

10

Bh V. N. Lakshmi[1]
Professor, Department of English, S R K R Engineering College (A),
Bhimavaram, Andhra Pradesh, India
Satish Kumar Nadimpalli[2]
Associate Professor, Department of English, S R K R Engineering College (A),
Bhimavaram, Andhra Pradesh, India

ABSTRACT In the artistic gallery of language and literature a writer is a painter of words and his iconic works are poetry in colors. Shedding flood lights on the language and literature enables one to experience the power and beauty of words. Reiterating Ngugi's reflections on the relationship of language to human experience, human culture and the human perception of reality, in this paper the researchers made an attempt to critiquing the language of a few great literary artists -- Toni Morrison, R.K.Narayan, Salman Rushdie and Paulo Coelho—highlighting their classic expressions that etched a mark in the society. Thinking is energy and it is words that dress every thought. Words are articulative, interactive and even suggestive and reminiscent. Indispensable to human life, language not only intertwines with class, gender, race and culture but also keeps changing from time to time. It is interesting and even essential to observe how age defines the definition of tragedy. Moving from classical drama to modern drama, it is identifiable that the aspects such as morality, male dominance, royalty etc., shifted towards realism, commonality, psychoanalysis, women empowerment, etc. Evidently, Shakespeare's writings are more of tragedy of royalty, Arthur Miller's are tragedy of middle classes and Thomas Hardy's are tragedy of peasantry. Common man's tragedy might appear trivial and ridiculous, but still it is a tragedy. Every age has its own definition, all cannot be measured with the same cup. A glimpse at the spectrum of writers across the globe with variegated historical vicissitudes such as colonization, wars, renaissance, industrialization, science and technology, etc., as a backdrop fits the bill for enunciating the fact that human creativity is a blend of culture, literature and language.

KEYWORDS Culture, Language, Literature, Reality, Percept

1. Introduction

Art and literature are standards of culture. Literature is art in words whereas painting is poetry in colors. From primitive cave paintings to modern abstract creations, the artistic trajectory reflects mankind's progressive knowledge and creative outcomes in all walks of life. Since language is integral to human life, it reflects the significant aspects of the seminal journey of humanity. The chronicles of human history mirror not only the changes in human life but also changes in literature. Undoubtedly, every age has its own literature and definitions. Tracking the language gradation following the literary footprints enables one to experience the aesthetics of creative thinking and power of words. Reiterating Ngugi's reflections on the relationship of language to human experience, human culture and the human perception of reality, in this paper the researchers made an attempt to critiquing the language of a few great literary artists highlighting their classic expressions that etched a mark in the society.

[1]drbhvnlakshmi1@gmail.com, [2]saketarama77@gmail.com

DOI: 10.1201/9781003489436-10

Thinking is energy and it is words that dress every thought. Evidently right word at the right time is the key for wit and wisdom. Sketched from the linguistic palette, words can be articulative, interactive and even suggestive and reminiscent. Sometimes due to multipronged nature of words what one says might be different from what the saying suggests. Also, the distant echo of words might occasionally refresh a faraway expression or statement. The potpourri of expressions such as literal, figurative, phrasal, idiomatic etc., are sometimes even quizzical. The variance of sense and essence of human expression to some extent typifies the cultural diversity. Indispensable to human life, literature and language intertwine with class, gender, race, and culture and keep changing from time to time. The trajectory of these relational changes becomes visible if one probes into the annuls of literature and language. Keeping this as a backdrop, it is interesting and even essential to observe how age defines the definition of tragedy. Moving from classical drama to modern drama, it is identifiable that the aspects such as morality, male dominance, royalty etc., shifted towards realism, commonality, psychoanalysis, women empowerment. Evidently, Shakespeare's writings are more of tragedy of royalty, Arthur Miller's are tragedy of middle classes and Thomas Hardy's are tragedy of peasantry. Common man's tragedy might appear trivial and ridiculous, but still, it is a tragedy. Undoubtedly the problem of holding to the old definitions will not suit the present time.

2. Tragedy from Classical to Modern Drama

Most of the traditional dramas have their philosophical roots in Senecan tragedy, and typically revolve around nobles who possess a tragic flaw (hamartia) which ultimately leads to a dramatic reversal of their fortune (peripeteia). Later, carrying the legacy of the ancient dramatists Shakespeare flourished as an actor cum writer during The English Renaissance. However, the eternal effect of his works lies in 'his capacity to speak to his contemporaries as also to the succeeding generations'(Srinivas Iyengar). In his poem 'To the Memory of My Beloved, the Author Mr. William Shakespeare,' Ben Johnson passionately expressed that Shakespeare's relevance transcended his own era and was timeless, enduring for all generations. And today his plays are performed all over the world in almost every medium and, even many TV shows and films adapted from his works. Critiquing Shakespeare, the genius, Srinivas Iyengar says that "he was able to see the universal behind the particular, the timeless beyond the contemporaneous and no wonder his portraits of religious bigotry, political chicanery, recruiting vagaries and heroic attitudinisations strike us still as astonishingly 'modern.' Coleridge describes it as Shakespeare's 'prophetic' quality.

Stratford and London groomed Shakespeare's religious and political knowledge and his imaginative vision added a new dimension to it. Despite many theatrical limitations of his times, he superlatively understood his play-loving audience and gave them what they wanted simply reflecting their own thoughts and feelings in his plays. When plague-ridden London suffered from the immediacy of death, of treason and public execution, of invasion and possible conquest, Shakespeare could still make his dramas a noble gesture of defiance and he could make death itself glorious and beautiful. He could write memorable lines about the pathos of the human situation. For instance, it is Ross speaking in Macbeth (IV, iii, 164):

Alas, poor country,
Almost afraid to know itself! It cannot
Be call'd our mother, but our grave; where nothing,
But who knows nothing, is once seen to smile;
Where sighs, and groans, and shrieks, that rent the air,
Are made, not mark'd; where violent sorrow seems
A modern ecstasy; the dead man's knell

His tragic heroes are from real history, the royal figures - kings, princes, or military generals. He reshaped them with more effective plots and a wider philosophical outlook. Thus, either prince Hamlet, King Lear, Macbeth, or Othello has tragic flaw and contributes to his own destruction before the equilibrium is restored. And the cathartic effect is akin to the Aristotelian purification. However, with the advent of time the focus on Shakespeare's tragedies marked new understanding, away from the Aristotelian pattern. The new ages reinterpreted old tragedies viewing with the lens of psychoanalysis. For instance, Shakespeare's Othello can be seen as a modern psychological drama about a psychopath who manipulates everyone around him just for fun. Therefore, it is more of disgust for Iago rather than pity for Othello. Salman Rushdie in his novel Fury speculates that Desdemona's death is an 'honor killing.' Othello loves only himself and Desdemona is just his doll.

Commending the universality of Shakespeare's writings, one of the online sources aptly addresses Shakespeare as the Michelangelo of literature. In the hands of Shakespeare, Iago is a fully realized psychological character just as the statue of David in the hands of Michelangelo is a fully realized man physically. Interpreting Shakespeare's plays Wilson Knight in his book *Wheels of Fire*, offers a fresh perspective arguing that character is merely a role that we play, not our true nature. Shakespeare delves beneath the surface of character and describes the fundamental motivations that are common to all people. That is why his plays have eternal appeal. Wilson even identifies 'Shakespearean Progress' and says that 'if Macbeth is Hell and Antony and Cleopatra Paradse, King Lear is Purgatory.'

It is true that times changed, contexts turned different, yet the writings of Shakespeare always fit the bill. He provided us with exceptionally vivid means to articulate feelings of hope and despair, sorrow and anger, love and desire. It's nearly unavoidable to refrain from quoting him, even for those who have never acquainted with his writings. For instance, one might be in a dilemma - "to be or not to be" (*Hamlet*), sometimes allow oneself to "gossip" (*A Midsummer Night's Dream*), never forgets that "nothing will come of nothing" (King Lear), aims for "the be-all and end-all" (*Macbeth*), expects "fair play" (*The Tempest*) and always tries to fight "the green-eyed monster" (*Othello*). Undoubtedly Shakespeare's presence is indelible in all walks of life. Shakespeare's genius, as reiterated my many, lies not only in creating right word at the right time, but also in choosing the right character to voice it. It's not surprising that critic Harold Bloom titled his book from 1998 about Shakespeare as *The Invention of the Human*. If being a great writer is defined by continued reading of their works, then perhaps being a genius is marked by the fact that their words are still in circulation through speech as well.

Life is a blend of culture, literature and language. Invariably changes in life mirror in literature and language. In the past imagination is the major source for writing whereas, reality is the major source for modern fiction. Subsequently, even the definition of a tragic hero changed. Tragic heroes no longer have to be only nobles, or only men. The works that best exemplify this transition is that of Thomas Hardy's. He is a social critic who portrays the low standards of life that the poor lived during the period of industrial revolution. However, he incorporates his new ideas synchronizing with the traditional tragic views. With this new vision of tragic literature, he creates modern tragic heroes, like Henchard, Tess, ClymYeobright etc. In *The Mayor of Casterbridge*, Hardy creates tragedy with Henchard's downfall and intensifies it with his ironic twists. For instance, Henchard regains his virtuous ways only after his decline. This appears to follow the pattern of a Shakespearean tragedy in which the tragic hero reclaims his honor, albeit too late to rescue himself.

Hardy is a staunch believer in Church and learned a lot about the Bible in his formative years. Evidently biblical allusions are aplenty in his works especially *Jude, The Obscure*. In it Hardy traces the odyssey of Jude, showing that at important turning points in his life Biblical references serve as guideposts marking his direction. Later, Charles Darwin's *Origin of Species* triggered the advancement of science and the Victorians were caught in the catch 22 situation. They could not believe either God or science and Hardy slipped into doubt and pessimism. He resorted to Paganism and Hellenism. This is obvious in his character Angel in *Tess of the D'Urbervilles*. Hardy concludes Tess' story by stating that "'justice' had been served, and the President of the Immortals, as Aeschylus might put it, had concluded his game with Tess." Tess, much like Prometheus, appears to have been a plaything of the moral and religious deities of Victorian England, and she had to be offered up as a sacrifice for the betterment of humanity. Every aspect of Tess' existence can be attributed to either chance, destiny, or the interference of divine forces. Hardy frequently employs elements of chance, misfortune, and happenstance as key narrative tools. In works such as *The Mayor of Casterbridge, The Return of the Native, Jude the Obscure*, and *Tess of the d'Urbervilles*, chance takes on a broader significance, symbolizing Thomas Hardy's personal outlook. According to Karl (196), the prevalence of happenstance in his literary world mirrors his pessimistic philosophy, which developed in response to the prevailing optimism of the Victorian era. Hardy strongly believed in the inherent disorderliness of the observable world. Consequently, various "isms" have been associated with Hardy, including pessimism, fatalism, naturalism, humanism, and meliorism.

3. Victorian Age

Moving away from the strict principle of 'art for art's sake' to 'art for moral or life sake,' the great literary works of the Victorian Age are realistic mainly concerned with practical problems and solutions. Despite great economic, scientific, and social revolutions, this was more like the age of pessimism and confusion. Writers of that time such as George Eliot, Thomas Hardy and Charles Dickens, appropriated many of the scientific arguments of their day into their works. Born in a middleclass family with financial difficulties during the time of industrial revolution, Dickens is always sensitive to the suffering of poor children. His works such as David Copperfield, Bleak House, Little Dorrit etc. are autobiographical presenting his social concerns. His own financial crisis and plight of poor children as industrial workers initiated his writing of A Christmas Carol. During the Great Depression some saw the story as what Davis calls "the Christmas of the common man."

A Christmas Carol made the expression "Merry Christmas" widely known among the people of the Victorian era, and the name "Scrooge" was adopted to describe a miser. In fact, it was officially included in the Oxford English Dictionary with this meaning in 1982. The Santa Claus, who was an emblem of 'good cheer' in Victorian age, even turned into a popular image of Coco Cola advertisement in modern times. It even became an established tradition after the publication of the poem 'A Visit from St. Nicholas', known today as '*The Night Before Christmas*'. Though Christmas is affiliated with happiness, it is not happy or merry to the poor. Dickens and Robert Frost both have suffered the pangs of festival as they cannot afford to buy Christmas gifts to their children. However, in the later stage their writings have become fountain head of their popularity and prosperity. Frost penned his poem "*Stopping by the Woods on a Snowy Evening*" during a harsh winter in New Hampshire. He wrote it after a challenging day when he was on his way home, grappling with the realization that he couldn't afford Christmas gifts for his children due to an unsuccessful trip to the market. This situation left him deeply disheartened, causing him to halt his horse at a curve in the road where he allowed himself a moment of tears. However, after a brief pause, the jingling of the horse's harness bells served as a reminder of the responsibilities awaiting him back home, prompting him to continue his journey. Thus, he writes in his poem:

> The woods are lovely, dark and deep
> But I have promises to keep
> And miles to go before I sleep
> And miles to go before I sleep.

During the Great Depression times economics dominated politics and the American Dream shattered. Perspectives changed and Arthur Miller's '*Tragedy and the Common man*' became the tag line of the social drama. Profoundly affected by the human tragedies of his era, Miller reinterprets what Aristotle referred to as 'hamartia' in modern terms. He views it as the hero's inherent reluctance to accept a passive role when faced with what he perceives as a challenge to his dignity and rightful place in society. In this reinterpretation, Miller shifts the tragic flaw onto society and makes the individual a victim of this flaw. His renowned play *Death of a Salesman* symbolizes the commodification of individuals in the modern capitalist society. The protagonist, Willy Loman, along with his sons Biff and Happy, must summon the courage to resist the temptation to act unethically in pursuit of 'the American Dream. Willy wants to refabricate the broken pieces of American Dream through selling talcums and lipsticks when people are barely managing for their essentials. He is bound to fail, and he cannot accept it. He lives in a fool's paradise claiming to his wife that all women are swarming around him wherever he goes. His two sons, Biff and Happy, cannot even secure employment. Lowman is too old to hold his job anymore. Miller created Willy as a product of society, a contemporary tragic figure who vocalizes the inequities within a broken dream. At the end of the drama, the audience see him planting seeds of hope for the future but his wife cries at his grave that all their life they went on paying the installments for their apartment and now there is no one to live in it.

4. Impact of Colonialism

Another significant aspect that changed human thinking across the globe is colonization. Colonialism isn't a recent occurrence; throughout world history, there are numerous instances of societies gradually expanding by annexing neighboring lands and establishing their populations on newly acquired territory. This practice was observed among various cultures, including the ancient Greeks, Romans, Moors, Ottomans, and many others. Thus, colonialism isn't confined to any one era or location. Eventually, British Empire emerged powerful spreading its political wings across Americas, Australia, and parts of Africa and Asia. This initiated crisscrossing of cultures and communications. British Raj established its governance in India in 1858 and colonial rule impacted Indian literature and education to that extent that Indian writers began to write poetry, short stories, and novels in English. Thus, Indo-Anglian Literature prospered and the writings of the three literary giants namely R. K. Narayan, Mulk Raj Anand and Raja Rao enunciate the hues, tones, and shades of the intersection of cultures and perspectives. R. K. Narayan's rendition of soul in all its nudity and utter frankness, Mulkraj Anand's Leftist ideologue and Raja Rao's indianized English language stand significant in this aspect. Narayan's imaginary town of Malgudi undergoes a transformation both before and after independence, mirroring the shifts in colonial and post-colonial India. This transformation signifies the shift from innocence to experience, the overshadowing of the subtleties of human nature by the noise of urbanization, the overpowering of traditional values by the forces of change, and the loss of stability and certainty. Highlighting the social context, he creates comic-tragedies using illiteracy, superstitions, astrology, marriage, nationalism, karma etc., and his Brahmanical perspective echoes throughout. Characters like Chandran, Raju, Rosie, and Margayya, after enduring a series of life's ups and downs, ultimately find themselves back in the encompassing world of Malgudi, having gained wisdom and experienced setbacks. Narayan's humor-filled perspective, infused with the distinctive essence of India, resonates universally. Even Western critics emphasize that Narayan's achievement lies in his ability to use one cultural context to capture the ambiance of an entirely different culture.

Unlike R K Narayan, Mulk Raj Anand carries the echo of social reform in his writings depicting the lives of the poorer castes in traditional Indian society. Even though he was born into a privileged caste, his objective is to demonstrate to the Western world that the Orient possesses a richness beyond what can be gleaned from the works of writers like Omar Khayyam and Li Poor Kipling. Consequently, he places characters such as the vulnerable Munoo in *Coolie*, the untouchable Bakha, and the indentured laborer Gangu at the heart of the narrative, highlighting the pervasive cruelty and exploitation that that held India in its vicious grip. According to Srinivas Iyengar, Anand is a veritable Dickens in his ability to depict the injustices and peculiarities of the contemporary human condition candidly and precisely. He was not a historian, but the history of India seeped into his narrative and transferred it. He embraces the language of immigrants to portray untouchables and laborers, and to convey the spirit of one's own and created a lasting impression in the literary firmament. He openly acknowledges the unique challenges he faces, carrying the weight of both European literary traditions and the deep history of India. He positions himself as a significant figure bridging the gap between imagination and the actual social and cultural landscape of that era in India.

Raja Rao makes a remarkable triad with Mulk Raj Anand and R K Narayan in the choice of themes but not in writing style. His style is exceptionally unorthodox because of his attempt to adapt in English the idiom, the rhythm, the tone, the total distinctness of vernacular speech. Consequently, his novel, *Kanthapura,* thematically may appear like 'Gandhi and our village,' but the narrative style transforms it into more of a 'Gandhi Purana.'

It offers a wonderful paradigm of the synthesis of the cross-cultural experience. Since language is culture specific, the act of writing necessitates conveying meaning from one culture to another through a complex process of assimilation, interpretation, and transformation. Raja Rao achieved this by infusing his language with a unique Indian expression. In doing so, he deftly manipulated both the symbolic and literal aspects of language, causing rigid distinctions to blur and allowing different sound patterns to interact. This approach aimed to articulate a distinct essence, representing the essence of Indian poetics—a holistic response encompassing all aspects of aesthetics—in a unified, all-aesthetic experience.

Later in post-colonial times the diasporic writer, Salman Rushdie, referring to himself, his characters and his fellow Indian writers says that 'we are translated men.' In his sense writing means being translated. He carved a niche with his new language. He is a sorcerer of words, and the words are not of just one language. It's as though multiple languages, summoned as needed, respond readily to the magician's command. Describing the impact of Rushdie's 'chutnification,' Rustom Baharucha suggests that it's akin to transforming the Queen's English by infusing it with the flavors of Indian cuisine, essentially deep-frying it in ghee and immersing it in curry. Baharucha argues that Rushdie's voracious appetite for words isn't just about imbuing his writing with an Indian essence but ultimately aims to free Indian English from its artificial puritanism and insincere refinement. Additionally, Rushdie asserts that utilizing language in his writing serves the purpose of liberating the entire culture from the shackles of colonial domination.

Passing through various historical and cultural vicissitudes the semantics of literature keeps modifying. Obviously, every age has its own definition and its own Shakespeare. The past reflects in the present in a new perspective. In this aspect, the autobiographical novel *My Feudal Lord* by the Pakistani writer, Tehmina Durrani, is a Shakespearean tragedy in modern times and Tehmina herself is a Desdemona. Othello's inferiority complex reflects in her husband, Mustafa Khar, who despite being a charismatic political hero and a Cupid to many women spares no effort to disfigure physically and psychologically his wife, Tehmina after their love marriage. He justifies his position saying that "A woman was like a man's land – 'The Koran says so.'" (107)

Colonization, cultural confrontation, and system of slavery are major ingredients in the world of letters in America and Africa. Black verbal art as stated by Ellison is more than a sum of its brutalization. Slavery was more than a labor system; it also influenced every aspect of colonial thought and culture. Right from 1619 the American colonies took many black people to America as indentured slaves and blacks suffered cultural shock and a complex form of economic and psychological oppression. Chinua Achebe's *Arrow of God* describes how the Africans suffered many terrible and lasting misfortunes in their long encounter with the Europeans. The cultural difference enforced the whites to behave brutally whilst the blacks' pagan culture took them very near to nature which has semblance to animal life. Hence, the whites looked down on them as animals. As the black community suffered increasing brutality, they became even more animalistic. Thus, Cholly Breedlove in Toni Morrison's *The Bluest Eye*, rapes his daughter Pecola and Bigger Thomas in Richard Wright's *Native Son* ends the life of his white employer's daughter and rapes his love Bessie to prevent her from telling the truth. Cholly and Bigger did not possess inherent criminal tendencies from birth. Instead, they were individuals shaped by American culture, influenced by the pervasive violence and racism deeply ingrained within it. Ralph Elison's *Invisible Man* and Morrison's *Tar Baby* carry the folktales of Brer Rabbit, Tar Baby etc., projecting the meaning of blackness and the role of Black oral tradition.

The end of the Civil war and Emancipation Proclamation did not mean the end of the racist oppression of African Americans. Economic and political claims of newly freed blacks fanned pervasive racial fears and hatred, spawning a segregated Southern, Jim Crow culture with lynching, vigilante justice and limited educational opportunities. During the post-slavery era, black individuals faced a challenging task of reconciling their newfound freedom with the loss of their cultural heritage. Consequently, writings from this period often illustrate their efforts to both document their past and embark on a fresh start as though nothing had transpired. However, the profound scars left by slavery inevitably surface, leading to manifestations of trauma, much like what is depicted in Toni Morrison's novel *Beloved.* Morrison employs elements such as blues and jazz, references to Tar women, the Seven Days group, the myth of flying Africans, and African storytelling techniques to connect the Black community's folk heritage with the broader Black American literary tradition. For Morrison and other contemporary Black women writers, the attention to the role of women in passing on the tradition comes directly from their African and American foremothers. It begins with the ancestral lineage of a woman and then expands to encompass all the revered figures in female culture, including folk healers and spiritual practitioners referred to as root workers and Obeah women. It also includes influential political activists like Sojourner Truth and Ida B. Wells, who have left their imprint on different eras of Black history. Morrison's achievement in creating characters very distinctly as if she were the creator who is above the trivialities of gender significance justly made her the first black woman Nobel laureate. Thus, she can be aptly addressed as a black Tiresias.

The black culture essentially depends on mutual respect of human emotion and relation. But the encounter with white world turned them into materialistic, killing the very essence of their existence. In Morrison's Jazz Joe kills his eighteen-year-old lover, Dorcas, "just to keep the feeling going on." And his wife Violet let loose the birds in the cage including the parrot that says, 'I love you.' Morrison symbolically reflects on many occasions the way human relations degenerated because of the dominant culture depriving the nourishment of the minority culture. The dominance of this cultural authority over uneducated and vulnerable individuals naturally gives rise to a sense of confusion regarding their own identity. This crisis deteriorates to such a profound extent that they begin to doubt whether they are alive or dead. Pecola, Cholly, Sethe, Son are only a few characters to name in Morrison's novels that represent the extremity of cultural bankruptcy.

African American women occupy a distinctive position in both American society and the world of literature. Morrison has drawn a larger landscape; she has cast a wider net. She calls the historical ability of black women to keep their families and their households together the 'tar quality.' Thus, her characters – Eva Peace, Pilate, Baby Suggs, Ondine, Therese, Consolata etc. – are audacious in times of misery and it is in the development of this 'tar women' that Morrison herself engages in the kind of ethnic cultural feminism that she advocates. She aims to authentically convey the cultural significance that underpins the historical context of her work. This includes the emotions, behaviors, values, and memories that influence and envelop the actions of her characters. She believes that hidden within historical experiences are mythical meanings that can rejuvenate the spirit and hold the wisdom for survival. Unlike Wole Soyinka and Chinua Achebe who projected the intrusion of the white evil into the black tradition, ethos and life, Morrison comes out successfully in presenting the cultural violation, interaction and adaptation without sacrificing the human values. It is sure that the American blacks cannot reclaim their African heredity, nor can they recreate the African milieu to see their culture flourish. Morrison offers a blend where the blacks can live like blacks without losing their identity and the essentials of their cultural heredity in consonance with the white wide world.

The black experience in the white world has sown the seeds of Harlem Renaissance which marked a turning point for African American literature as it started percolating into the American culture. It has been the breeding ground for creative endeavors by authors, poets, and artists. Anew literary art form called jazz poetry emerged and Langston Hughes' jazz poem *The Negro Speaks of Rivers* provides unity for the African American history; and Countee Cullen's poem *Black Christ* promotes equality, condemning racism and injustice, and celebrates African American culture and spirituality. Fashion in the 1920s was another avenue through which jazz music impacted popular culture. Jazz contributed to the advancement of the Women's Liberation Movement, offering a means of rebelling against established societal norms. This period marked the first instance where the culture of a minority group became widely sought after by the majority. Furthermore, the Harlem Renaissance played a pivotal role in showcasing African American dance heritage, which drew inspiration from various African cultural traditions. In 1921, the African American musical '*Shuffle Along*' made its breakthrough on Broadway, capturing the American imagination and energizing the Harlem Renaissance. This production not only provided new opportunities for talented black artists but also demonstrated that substantial profits could be earned from locally rooted dance forms. Black visual artists, whether they be sculptors, painters, photographers, directors, or illustrators, have experienced an explosion in ideas and energy during Harlem Renaissance and made a name for themselves throughout history.

Colonialism was an ugly thing and by mid 20th century people started fighting back. Postcolonial writers from Africa, South Asia, the Caribbean, South America, and other places started engaging in a significant act of responding to the empire through their literature and triggered the process of decolonization. Four prominent thinkers of this time – Frantz Fanon, Edward Said, Home Bhabha and Gayatri Chakravorty Spivak – hailing from diverse geographical origins, nationalities, and social contexts, have significantly contributed to the field of postcolonial literature with their remarkable works. Their literature emerged from the efforts of colonized communities to secure justice, equality, and liberty. Chinua Achebe's novel, *Things Fall Apart,* stands as one of the most impactful novels within the realm of postcolonial literature, delving into the dynamics between traditional African society and British colonial forces. J. M. Coetzee's novel, *Disgrace,* portrays the efforts of both colonizers and the colonized for reconciliation in post-apartheid South Africa. Throughout many of his works, he skillfully depictshis own alienation from his fellow Africans. Jamaica Kincaid articulates her strong disapproval of British colonial practices imposed upon the colonized in her famous novel, *A Small Place*. Also, Filmmakers such as Satyajit Ray, Deepa Mehta, Mira Nair and Shyam Benegal have contributed to post colonialism depicting colonial and postcolonial predicaments in their films. Music in postcolonial nations also serves as a reflection of cultural identity and values, with a prime illustration being Ravi Shankar's fusion of classical Indian music with Western elements in the realm of popular music.

Though English language is used to connect people globally, each country takes pride in its own language. Due to colonization a colonized country is forced to use colonizer's language and ironically even after independence it fails to decolonize the language and thus suffers 'colonial alienation.' According to Ngugi "colonial alienation," is a state where the mind and body are divided, existing within two unrelated linguistic realms within the same individual. On a broader scale it can be a society with bodiless heads and headless bodies. Ngugi firmly believes that the selection and utilization of language play a pivotal role in how a people define themselves in relation to the entire universe. Thus, he renounces writing in English and writes in his native language Gikuyu.

Through his book Decolonizing Mind, Ngugi appeals all African writers to come back to their native languages in their writings by saying that it is an appeal to rediscover the universal language of mankind – the language of struggle. Struggle is the force that shapes history and defines our existence. Our history, our language, and our very essence encompasses struggle. This struggle exists wherever we are and in everything we undertake. That's when we join the ranks of the countless individuals whom Martin Carter once saw sleeping not to dream but dreaming to transform the world.

5. Significance of Translation

Translation bridges the linguistic and cultural barriers. The popularity of the Brazilian writer, Paulo Coelho, exemplifies this. Today he is a man of many quotable words and an internet celebrity with millions of followers. His writings are in Portugese, and the translations of his books sold worldwide in millions. He represents a global literary phenomenon that transcends the origin of culture and occupies the realm of "world literature."Speaking of culture, Coelho says that culture makes people understand each other better. His writing philosophy is straightforward and effective. Instead of engaging in complex linguistic acrobatics or deep psychological examinations, he provides readers with skillfully crafted narratives coupled with self-help advice. He is an inspiration and spiritual guru for many, and his stories have become the themes of symphonies. Berthold Zilly, a translator and literature professor, asserts that Paulo Coelho is an author with a global appeal. He contends that the themes found in Coelho's books are universal and could easily resonate with readers from Europe, North America, or the Arab world.

6. Music and Art

The art forms can complement one another well. In his article, "When Books and Music Meet, Flawlessly and Otherwise" Tobias Carroll describes how the intersection of pop music and literature can be a wonderful and mutually beneficial. Evidently, Bob Dylan's song lyrics and composition style transformed the landscape of rock and pop music, a contribution that ultimately led to his recognition and the award of the Nobel Prize in Literature in 2016.

> Come writers and critics
> Who prophesize with your pen
> And keep your eyes wide
> The chance won't come again

Bob Dylan's lyrics and songwriting changed rock and pop music and earned him the Nobel Prize for Literature in 2016. His songs draw inspiration from the rich tradition of American folk music and bear the imprint of modernist poets and the beatnik movement. Through his song "*The Times They Are A-Changin'*," he urges the influential to grab the moment of social change that has come. His impact on contemporary culture is deep and enduring. Also, Art and architecture have been complementary to literature. All the works of Dan Brown are associated and integrated with art and architecture. Historically Cubism, Surrealism and Abstract arts inspired writers in the West to visualize their creative thinking and the epics such as the Ramayana and the Mahabharata are sources for many artists in the East.

7. Conclusion

Summing up the understanding of how culture impacted literature and language, it comes clear that in the early times it is human emotions and actions and primitive beliefs that triggered literary creativity, then it is wars, religious beliefs and colonialism followed by revolutions, isms, scientific discoveries, and inventions that crept into literary world. Absorbing all these variations, literature like a chameleon changes its colors and contours adapting to its time and human atmosphere. The literary journey of the present times acquired modernization and intertwined with digitalization, communication, and globalization.

CITATIONS

1. A Christmas Carol. https://en.wikipedia.org/wiki/A_Christmas_Carol#CITEREFDavis1990a
2. Aristotle and Shakespeare. http://www.stjohns-chs.org/english/Shakespeare/classical/poetics.html
3. Lenny Wiza M. A., Music and Modern Literature. https://www.enotes.com/homework-help/compare-what-bob-dylan-suggests-writing-music-1176721
4. Post Colonialism. https://www.wattpad.com/511005217-psyphilit-magic-12-post-colonialism
5. Shakespeare Tragedy Plays, https://www.nosweatshakespeare.com/shakespeares-plays/play-types/tragedy-plays/
6. Shakespearean Tragedy, https://en.wikipedia.org/wiki/Shakespearean_tragedy

REFERENCES

1. Achebe, C. (2013). Arrow of God. Penguin, UK.
2. Anand, M. R. (1994). Coolie. Penguin Books India.
3. Anderson, Hephzibah. (21 Oct 2014). How Shakespeare Influences the way we speak. http://www.bbc.com/culture/story/20140527-say-what-shakespeares-words
4. Bădulescu, D. (2012). Rushdie's Sorcery with Language. PhilologicaJassyensia, 8(2 (16)), 129–142.
5. Carroll, Tobias. "Pop Lit: When Books and Music Meet, Flawlessly and Otherwise."
6. Davis, Paul (1990a). The Lives and Times of Ebenezer Scrooge. New Haven, CT: Yale University Press. ISBN 978-0-300-04664-9.
7. Dickens, C. (2003). A Christmas Carol and Other Christmas Writings. Penguin UK.
7. Domeneck, Ricardo. (24-8-2017). "Why Paulo Coelho, now 70, isn't a typical Brazilian writer."
8. Durrani, T., Hoffer, W., & Hoffer, M. (1995). My feudal lord. Random House.
9. Ellison, R. (2016). Invisible man. Penguin UK.
10. Frost, R. (1998). Stopping by Woods on a Snowy Evening. Hastings W.-Nw. J. Envtl. L. &Pol'y, 5, 135.
11. Hardy, T. (2007). Tess of the D'Urbervilles. Broadview Press.
12. https://hazlitt.net/feature/pop-lit-when-books-and-music-meet-flawlessly-and-otherwise
13. https://www.dw.com/en/why-paulo-coelho-now-70-isnt-a-typical-brazilian-writer/a-40217950
14. Hardy, T. (2004). The mayor of Casterbridge. OUP Oxford.
15. Iyengar, K. S. (1985). Indian writing in English. Sterling Publishers.
16. Karl, Frederick R. "The Mayor of Casterbridge, A New Fiction Defined," Modern Fiction Studies, VI(3), 1960, 195–213.
17. Knight, G. W. (2005). The Wheel of Fire. Routledge.
18. Miller, A. (1996). Death of a Salesman: Revised Edition. Penguin.
19. Morrison, T. (2015). Tar baby. Christian Bourgois.
20. Morrison, T. (1994). The Bluest Eye. 1970. New York.
21. Morrison, T. (1993). Jazz. 1992. London: Picador.
22. Morrison, T. (1988). Beloved. 1987. New York: Plume.
23. Murta, A. P. (2018). Paulo Coelho: Transnational Literature, Popular Culture, and Postmodernism.

24. Rao, R. (1967). Kanthapura (Vol. 224). New Directions Publishing.
25. Reese, M. M. (1958). Shakespeare: his world and his work. Arnold.
26. Rushdie, S. (2010). Fury: a novel. Vintage Canada.
27. Ryan, Kierman. (15 Mar 2016). An Introduction to Shakespearean Tragedy. https://www.bl.uk/shakespeare/articles/an-introduction-to-shakespearean-tragedy
28. Shakespeare, W. (2001). The tragedy of Macbeth (Vol. 2). Classic Books Company.
29. Sherman, G. W., & Sherman, G. W. (1976). The Pessimism of Thomas Hardy. Fairleigh Dickinson Univ Press.
30. Showalter, E., Baechler, L., &Litz, A. W. (Eds.). (1993). Modern American women writers. Simon and Schuster. p 209.
31. Thamarana, S. (2015). Significance of studying Postcolonial literature and its relevance. Research Journal of English Language and Literature, 537–541.
32. Venkatesan, Priya. (7 Dec 2007). Of Dickens and Darwin. https://www.the-scientist.com/daily-news/of-dickens-and-darwin-45733
33. WaThiong'o, N. (1998). Decolonising the mind. Diogenes, 46(184), 101–104.
34. Willis, S. (1987). Specifying: Black Women Writing the American Experience. Univ of Wisconsin Press. p 6.
35. Wright, R. (2016). Native son. Random House.

Impending Inquisitions in Humanities and Sciences (ICIIHS-2022) – Dr. Mohan Varkolu et al. (eds)

Discourse Analysis of Implied Heroism Through Politeness Strategies in Telugu Films

11

V. Ashok Kumar*
Department of English, Anurag University, Venkatapur,
Ghatkesar, Medchal–Malkajgiri district, Hyderabad, Telangana, India
Department of English, Koneru Lakshmaiah Education Foundation, Hyderabad, Telangana, India

B. Chandrashekar
Department of English, Anurag University, Venkatapur, Ghatkesar,
Medchal–Malkajgiri district, Hyderabad, Telangana, India

ABSTRACT People use certain linguistic strategies when communicating with others to achieve their goals, but there is a chance that conflict will arise when people don't use appropriate politeness strategies in their conversations. Therefore, the politeness phenomenon is of great importance in the fields of discourse analysis and pragmatics. This paper examines how speakers tend to employ various politeness strategies when communicating with their interlocutor(s) and how such politeness strategies contribute to the construction of implied heroism with reference to three South Indian (Telugu) films directed by SS Rajamouli namely Baahubali: The Beginning, Magadheera, and Baahubali: The Conclusion was taken to find out both polite and impoliteness strategies used by the important characters in the films.

KEYWORDS Politeness strategies, Implied heroism, Frequency of politeness strategies

1. Introduction

Generally, people communicate with each other to fulfil their desires or to reach their goals. Therefore, an interaction never takes place without any intention. Hence, our intentions can be traced by using spoken/written (verbal communication) or by body language, gestures, signs (non-verbal communication). Basically, a person's age, gender, class and context affect any interaction. However, how, when, why and where to communicate with people must be learned from the society itself. In addition, our emotions, feelings, attitudes play a vital role in any interaction and the language that we use in society always shows our personality, status, knowledge, nature and social behaviour. Many scholars, who are associated with the study of linguistic phenomenon, and have tried to define politeness but couldn't succeed or satisfy the critics of linguistics, for instance, *"we might resort to expressions like the language a person uses to avoid being too direct' or 'language which displays respect towards or consideration for others'. Once again we might give examples such as language which contains respectful forms of address like sir or madam'. 'Language that displays certain 'polite' formulaic utterances like please, thank you, excuse me or sorry or even 'elegantly expressed language' (Watts, 2003:1)* Some scholars feel that using polished language in society is politeness but for other scholars it would be the use of socially accepted human behaviour or using certain tools to meet their wishes or desires. Many books and articles were published on the concept of politeness after the recognition of the formal discipline of politeness as linguistic behaviour during the 1970s. But there are a few books which came out on the concept of

*Corresponding author: vemulashokkumar@gmail.com

DOI: 10.1201/9781003489436-11

impoliteness. Some of the scholars like Lakoff, Grice, Brown and Levinson, and Leech have contributed their work on both the areas i.e politeness and impoliteness phenomenon in linguistics. When Brown and Levinson published an article on politeness, many scholars from sociolinguistic and anthropology tried to find out how this politeness phenomenon is realized in different linguistic fields and languages. Hence I am interested in exploring politeness strategies to analyse the conversations among film characters for my present study.

2. Review of the Literature

Many scholars conducted their research on politeness in the field of pragmatics based on different languages and cultural settings. For instance, Sabine Walper and Renate Valtin (1992) conducted a research which is titled "*Children's Understanding of White Lies*" to see how the acquisition of white lies varies for different age groups.

This research focused on how children understand politeness or polite behaviour and honesty in their communication and found that 'the understanding of white lies among the children who are ten years old has become more prominent in realization of polite behaviour which respects and supports the hearer's feelings or face (ibid. p. 248–249). Saeko Fukushima and Yuko Iwata (1985) conducted a research on politeness which is entitled "*Politeness in English*" in Japanese society. They worked on two objectives; one is to see how politeness is being used in English by advanced level EFL students whose native language is Japanese and another one is to see what kind of hurdles the advanced EFL students are facing to use polite expressions in English. Interpretation and analysis of the data showed that the participants used same expressions with their friends (classmates) and with their teachers as well (ibid. p. 11–12), but students had good awareness in using the language to show status difference especially in using polite expressions. However, they did not succeed in expressing their feelings or thoughts in English. Sachiko Ide also contributed something to the concept of politeness in her book "*The Concept of Politeness: An Empirical Study of American English and Japanese*". In this research, she tried to find out how the linguistic forms of politeness are different from American English and other languages. In other words, her research focused on how people understand politeness and its forms/strategies from one language to other language. And the result ofher study showed that there was a difference in the use of 'Politeness' from American English speakers to Japanese speakers. Perception of being friendly or polite was similar or common for the American speakers whereas those two notions were different for Japanese in their culture. Therefore, the findings of this research have demonstrated that the meaning of politeness varies from culture to culture across the world (ibid. p. 292).

3. Theoretical Framework

We come across many approaches to study politeness and impoliteness strategies in the field of pragmatics. Therefore, this chapter discusses some of the approaches that are useful for the study of (im)politeness and can be considered as the framework for the analysis of my present study. I have used one of the most successful and widely accepted approaches of Brown and Levinson (1987) as it is a classical model for the interpretation of politeness. In addition, I have also drawn on Culpeper's model of impoliteness to get an understanding of the politeness phenomenon.

Face and Conversation: The term 'face' has taken from Goffman (1967) to support Brown and Levinson's politeness theory. In this regard, Cameron (2001:79) has treated the term 'face' as a person's "social standing or esteem". And it has also being commented by Culpeper (2001:238) that face refers to one's "reputation, prestige and self-esteem" or it is treated as 'public self-image'. Hence, Brown and Levinson (1987) have shared that face is "something that is emotionally invested, and that can be lost, maintained or enhanced, and must be constantly attended to in interaction" (ibid. p.61). Therefore, they assume that interactants in general understand the individual's face in any communication and the observation of face in any interaction is based on the mutual vulnerability of face. And maintenance of one's face normally depends on other's face. Hence, face is concerned as basic want of a person. Therefore, it is a general assumption that people maintain their faces when they participate in an interaction and maintaining one's face is a kind of person's self-respect or privilege. Similarly, Ukosakul (2005) stated that "one must avoid behaviour that may cause shame to another" (ibid. p.120). Therefore, people from different languages or cultures perceive politeness strategies to express their feelings and to get whatever their faces want. Hence, "face is a universal characteristic across cultures that speakers should respect each other's expectations regarding self-image, take account of their feelings and avoid face threatening acts" (Cutting, 2002:45). However, people often want to defend their face when somebody threatens their face, and in defending, they might threaten the other's face as well. And people often ignore the other's face wants inemergency or social boycott situations. However, Brown and Levinson (1987) said that when a person requests someone, he or she threatens the other person or addressee's face. In addition, they (1987) also have mentioned that the act of face threat is performed at two levels; based on people's beliefs, feelings or attitudes in relation to the particular situation or context. They are namely 'positive face' and 'negative face'. And definitions of these two types are framed by Brown and Levinson (1987). **Positive Face:** appreciation or approval of a person's desire or self-image by others refers positive face. In other words, desires of a person must be understood, ratified, approved, admired or liked by others. For instance: An old lady who is passionate about gardening spends most of the time in her rose garden to look after them. And she loves the roses very much. She expects the visitors of the rose garden to admire/appreciate her. So she delights whenever she looks at the visitors' expressions like 'what a beautiful rose!' I too want my garden like this and how do you make it?' all these expressions imply that they too want what she wanted and achieved it. So here visitors' appreciation/admire towards her work enhances her positive face wants. **Negative Face:** Everyone wants his or her actions should not be impeded by others and this can be claimed as negative face. In other words, freedom from impositions or freedom of action implies negative face. For instance; take the above example; if the gardener says

> don't cut off the rose, when a visitor tries to cut it off. Here, the utterance of gardener attacks the negative face of the Hearer (visitor(s)) as the visitor doesn't want anyone to stop him or her from cutting off the rose." (ibid. 1987:61)

Politeness Strategies by Brown and Levinson;Brown and Levinson (1987) argued that no one can escape from the FTAs as communication becomes an integral part of life. Any conversation which takes place between two persons can face at least some degree of threat to both the participants or any one of the speaker's or hearer's face. But most of the people try to escape from the FTAs or try to reduce the FTAs by using conversational strategies to reach their objectives. In this regard, Brown and Levinson (1987) have proposed four strategies which people often use them to reduce or minimize the face threat to H. Therefore, strategies like Bald-on-record, Positive politeness, Negative politeness, and Off-record show how people try to minimize the FTAs.

4. Analysis and Interpretation

Extract-I (Film: BAAHUBALI, The beginning), Telugu Version:

1. **Sanga:** Shivudu..! Aapara..! oddura..! kallupotayira..! aapu…! Aapara…! Aapu…! Shiva…! (With lot of disappointment) (Then Shivudu Succeeds in digging Shivalinga and takes it to keep under water falls)
2. **Shivudu:** Amma…! Nee Shivudikiveyyikadu, gadiyagadiyakitalatanale. Oppena?
3. **Sanga:** She smiles (She feels very happy as her son did a good job)

English Version:

1. **Sanga:** Shivudu…! Please stop it, stop it. You will become blind if you won't stop it. Stop it I say, stop it. (While Shivudu is digging the Shivalinga). (Then Shivudu Succeeds in digging Shivalinga and takes it to keep under waterfalls)
2. **Shivudu:** Mother….! Your lord Shiva will get full water on him every time. Is it okay?
3. **Sanga:** She smiles (She feels very happy as her son has done a good job)

Analysis and interpretation: the moment Sanga listened to the news from one of the villagers; she thought that her son is doing something wrong and finds that her son is breaking the Shivalinga. So she thinks that it is not good for him as well as for the people who are living there. So she rushes towards her son and beats him with a stick by saying '*Shivudu..! Aapara.!oddura! kallupotayira! aapu! Aapara! Aapu!'* in turn-1, shows the negative impoliteness as she attacks the negative face of the addressee by belittling him. And in this scene, the non-verbal communication or action which is done by Shivudu employs a threat to the negative face of the hearer. Hence, the utterance and uncontrollable negative emotion falls under the negative impoliteness strategy for frightening the hearer. Since, she thought that it's not possible to stop him from breaking the Shivalinga, she sits down and prays to Lord Shiva to help her and stop him from wrong doing by her helpless expression 'Shiva…!' with low voice, which shows her anxiety and desperation. However, non-verbal action/communication frightens the hearer/audience which fall under Negative Impoliteness, as he threatens the negative face of the people. Meanwhile, Shivudu digs the Shivalinga and carries it by putting on his shoulders and puts it under the water fall. This non-verbal action shows his respect and politeness towards lord Shiva as well as towards her mother in saving her from pouring water on Shivalinga by climbing up and down. When he places Shivalinga under water, he feels very happy that he has saved his mother and says '*Amma…! Nee Shivudikiveyyikadu, gadiyagadiyakitalatanale.Oppena?*' in turn-2, which falls under the positive politeness (strategy 15: Give gifts to H (goods, sympathy, understanding, cooperation). In other words, the speaker made an effort to satisfy hearer's positive face by fulfilling her wants. Hence, Sanga feels very happy that her positive face has been enhanced by Shivudu in turn-2. Hence she too enhances the positive face of Shivudu by showing her smile and hands showering her blessings on him in turn-3. So, both the turns indicate that the S and H are cooperative towards each other and Shivudu wants to satisfy his mother's wants. Therefore, the non-verbal action employs the positive politeness strategy - 10 (Be optimistic). In addition, the action which Shivudu has done in this scene can be the construction of heroism as he maintained positive politeness among the people, his mother and showed respect to Lord Shiva. And he didn't hurt his mother as well as Lord Shiva and his spiritual values. So, positive politeness shows his heroism as he enhanced everyone's positive face in this scene. His determination to please his mother and go against the social rule also shows his heroism and courage. And mother's negative politeness shows her worry and anger towards her stubborn son but in the end she also praises him for his action.

Extract – II, Telugu Version:

1. **General Secretary:** Bhalladeva…! Aabhiaabhi.
2. **Bhallaladeva:** Dunnapothunukhalimathakubhaliistadu. Adharpanaaaa…!
3. **General Secretary:** Baahubali…! Aabhi..aabhi.
4. **Soldiers:** Baahubali…! Baahubali…!
5. **Baahubali:** Dunnapothubhaliniaputhadu.
6. **Kattappa:** bhahu…! Idi tharatharalugaostunnaaacharam. Bhaliivvakapothesainikulukeedunisenkistaru.
7. **Soldiers:** avunuyuvaraja
8. **General Secretary:** yuddanikivellemundubhalivakkapote? Amma... aagrahistundi.
9. **Baahubali:** bhaliivvadanikimoogajanthuvupranamendukumantri..? urakaluvestunnanaaraktamsiddangaundi. Jai Mahishmathi…!

English Version:

1. **General Secretary:** Bhallaladeva..! Come..!
2. **Bhallaladeva:** cuts buffalo to pay tribute to Godess Khali. Invincible..! Invincible..!
3. **General Secretary:** Baahubali..! Come..!
4. **Soldiers:** Baahubali…! Baahubali…!
5. **Baahubali:** will not cut buffalo (non-verbal action)
6. **Kattappa:** Baahu..! It is an ancestral culture. If you don't sacrifice the animal, soldiers feel it as a bad omen.
7. **Soldiers:** Yes…! My Prince..!
8. **General Secretary:** If we don't offer the blood before war, Goddess will be angry.
9. **Baahubali:** Why should we kill the innocent animal to sacrifice the blood, Secretary? Here, I have my own active blood to offer. Hail Mahishmati..! (he cuts his hand and offers blood)

Analysis and Interpretation: When the General Secretary invites Bhallaladeva to sacrifice buffalo to Goddess Kali, Bhallaladeva takes the sword and cuts the buffalo's neck which is not meant for the prince; instead he would have granted the permission to cut it. The non-verbal behaviour of Bhallaladeva shows that he is very rude and cruel towards animals and doesn't care for them. Since, the non-verbal behaviour frightens the H's negative face; it can be treated as an example of negative impoliteness. But when the General Secretary invites Baahubali to sacrifice buffalo, he denies obeying the ancestral culture and offers his own blood to lord Kali instead of the innocent buffalo's blood. He shows concern towards the dumb animals and this shows that the S is caring about the H which employs positive politeness (Strategy 15: Give gifts to H (goods, sympathy, understanding, cooperation). In addition, if we compare both the princes' characters in verbal and non-verbal behaviour; one (Bhallaladeva) doesn't care for others and the other one (Baahubali) cares, understands and shows sympathy for others. Therefore, we can say that the strategy 15 of positive politeness shows the construction of heroism in Baahubali and villainy in Bhallaladeva. It shows that courage, decision making, patriotism, care for soldiers, are all aspects of the protagonist's heroism.

Extract-III (Film: MAGADHEERA), Telugu Version:

1. **Sharekhan:** Naakupranabikshapedtava..? kudhauhe…! Nee ontloroshamunteneekadhalonijamunte… namanishulniondhamandhinipampistadhaniontipaicheyipadakundaapu. Ee rajyaniaaraninineekeappajebutha.
2. **Bhairava:** vennuchupinchaniveerulniennukonimaripampinchu.
3. **Sharekhan:** vallanuchustene nuvvu sagamchastav ra.

4. **Bhairava:** ekkuvainaparvaledulekkathakkuvakakundachusuko.
5. **Sharekhan:** ondhalookkadumigilina nuvvu odinattera.
6. **Bhairava:** okkokkannikadhusharekhanondhamandhini oke sari rammanu.
7. **Sharekhan:** rahaath..! neepogarumattallokadurachetallochupinchu.

English Version:

1. **Sharekhan:** You have granted me life, right? kudhauhe..! If you have power and truth in your story, I will send 100 soldiers, stop them from manhandling her. I will surrender this kingdom and the princess to you.
2. **Bhairava:** You can even select the bravest soldiers and send them.
3. **Sharekhan:** You will be half dead if you see them.
4. **Bhairava:** No matter if you send more army but see that the counting is not decreased
5. **Sharekhan:** You may lose race if you leave a single soldier without killing.
6. **Bhairava:** Not one by one Sharekhan, send 100 soldiers at a time.
7. **Sharekhan:** Rahaath..! Don't show your proud words. Show it practically.

Analysis and Interpretation; seeing the mercy for his life, Sharekhan gets more angry and conveys his offer that if Bhairava has courage and fact in his ancestor's story of killing 100 soldiers in a war is true, he tells Bhairava that he will send 100 soldiers and asks him to stop them from touching Mithrabindha and he also extends his offer that if Bhairava wins the challenge, then he would surrender the kingdom as well as the princess to Bhairava. Since the utterance implies the expression of challenge, it impinges upon H's positive face wants. Apart from that the utterance also shows that the S is trying to minimize FTA on H's positive face wants by making an offer or making a promise. And the S tries to fulfil H's wants. Therefore, the utterance falls under the positive politeness strategy 10: (Offer, promise). However, Bhairava accepts the challenge and conveys to Sharekhan to send 100 selected soldiers who are brave enough to fight and who don't escape from the battle field and send them at a time to attack in turn - 2 '*you can even select the bravest soldiers and send them.*' indicates that Bhairava is not scared of the enemies in the battle field and also wants to show his courage in the battle field. That's why Bhairava offers him to send the selected 100 brave soldiers to fight with him. Thus, the utterance observes an FTA on H's positive face wants. And H feels some pressure to accept the challenge. However, since it is a direct and unambiguous dialogue, it falls under the impoliteness Bald-on-record strategy. Looking at the confidence and courage of Bhairava, Sharekhan warns him that he would be scared and die just by seeing the dangerous army in turn - 3 '*you will be half dead if you see them*'. In other words S frightens the H's by making an implicature. Here, S attacks the positive face of the H. Therefore, the utterance falls under the positive impoliteness strategy for expressing an uncontrollable emotion. But subsequently, the following utterance by Bhairava in turn - 4 '*No matter if you send more army but see that the counting should not be decreased*' shows that Bhairava is not at all scared of Sharekhan's dangerous army. So he tells him that it doesn't matter if he sends more army than 100 but asks him to see that the counting while Bhairava is killing Sharekhan's army should not be decreased. In other words, S shows that he is powerful than the H, therefore, the S expects the H to accept and obey his words. And moreover, ignoring the H's positive face falls under the positive impoliteness strategy to insult or snub the enemy.

By looking at the pride of Bhairava, Sharekhan warns him in turn - 5 '*you may lose race if you leave a single soldier without killing.*' that if Bhairava fails in killing the 100 soldiers or if he leaves a single soldier without killing, Bhairava will be treated as a loser of the challenge. In other words, S ignores the H's positive face wants and frightens H, so H's feels some pressure on his positive face. Therefore, the utterance falls under the positive impoliteness strategy to insult or snub. But by accepting the challenge Bhairava tells Sharekhan in turn - 6 '*not one by one Sharekhan, send 100 soldiers at a time.*' Bhairava doesn't want to fight one by one or doesn't want to kill one by one. That's why he asks Sharekhan to send hundred soldiers together for the fight. The utterance also shows Bhairavas energy and courage to fight with hundred enemy soldiers. In other words, S provokes the H's positive face by making a series of challenges. So the H feels some pressure on his positive face wants. Since it is a violent or strong insult, it falls under the positive impoliteness strategy to frighten the other. Subsequently, feeling tired by Bhairava's provocation, Sharekhan asks him to show his pride practically; means by killing hundred soldiers, not by words, in turn - 7 '*Rahaath..! Don't show your pride in words but show it practically*' His feeling is, words will not work in the battle field, the way a soldier shows his courage tells his bravery. So he asks Bhairava to show his courage in the battle field. Since it is an aggressive verbal exchange, here we see the pride, courage and determination of the hero Bhairava who sacrifices all to protect the kingdom and his love implies the construction of heroism through these actions and dialogues.

5. Conclusion

The analysis of the extracts highlights the construction of implied heroism (loyalty, patriotism, love for the nation and respect and care for women) by both male and female protagonists. The results are not generalizable but indicate how filmmakers construct implied heroism through politeness strategies and recreate the social world through the dialogues and setting. And more frequently uttered dialogues by the protagonist falls under the positive politeness. However, understanding and interpretation of any conversation may change from one participant to another participant or from one situation/culture to another one. Further work can be done in this area by using other tools of discourse analysis to study films.

REFERENCES

1. Brown, Penelope. and Stephen C, Levinson. *Politeness: some universals in language Usage*. Cambridge university press, 1987.
2. Camaron, Deborah. *Working with Spoken Discourse*. Sage Publications, 2001.
3. Cutting, Joans. *Pragmatics and Discours: A resource book for students*. Routledge, 2002.
4. Culpeper, Jonathan. "Towards an anatomy of impoliteness"journal of pragmatics 25(3):North-Holland, 1996/3/1, pp. 349–367.
5. --- *Language and characterisation: people in plays and other texts*. Longman, 2001.
6. --- *Impoliteness: Using Language to Cause Offence*. Cambridge University Press, 2011.
7. Elia, Norbert. *The Society of Individual*. The Continuum International Publishing Group, 1991.
8. France, P. *Politeness and its Discontents: Problems in French Classical Culture*. Cambridge University Press, 1992.
9. Fraser, Bruce. *Whither politeness*. Benjamin publishing house, 2005
10. Grice, Herbert Paul. *Logic and conversation*. Academic press, 1975.
11. Halliday, M. *Language as a Social Semiotic: the Social Interpretation of Language and Meaning*. London: Edward Arnold, 1978.
12. Leech, Geoffrey Neil. *Principles of Pragmatics*. Longman, 1983.
13. --- *Explorations in Semantics and Pragmatics*. John Benjamins Publications, 1980.
14. Locher, Miriam A. and Bousfield, Derek. *Introduction: impoliteness and power in Language*. Mouton de Gruyter, 2008.
15. --- *Power and Politeness in Action: Disagreements in Oral Communication*. Mouton de Gruyter, 2004.
16. Lakoff, Robin Tolmach. *Language and woman's place: Text and commentaries*. Harper and Row Publishing Company, 1975a.
17. ---"The limits of politeness: therapeutic and courtroom discourse" A journal of *Multilingua*,Vol 8, issue 2-3, 1989, pp. 101–29.
18. --- and Sachiko, Ide. *Broadening the Horizons of Linguistic Politeness*. John Benjamains Publishing Company, 2005.
19. Mills, Sara. *Gender and Politenesss*. Cambridge University Presss, 2003.
20. --- *Gender and impoliteness*. Journal of Politeness Research, Vol 1, issue 2, 2005/07/27, pp. 263–280.
21. Thomas, J. *Meaning in Interaction: An introduction to Pragmatics*. Longman, 1995.
22. Watts, Richard J. *Politeness*. Cambridge University Presss, 2003.
23. Yule, George. *Pragmatics*. Oxford University Press, 1996.

Web-References

1. SS, Rajamouli. "Baahubali, the beginning" en.wikipedia.org/wiki/Baahubali:_The_Beginning
2. --- "Magadheera" en.wikipedia.org/wiki/Magadheera
3. --- "Baahubali, the conclusion" en.wikipedia.org/wiki/Baahubali_2:_The_Conclusion

Impending Inquisitions in Humanities and Sciences (ICIIHS-2022) – Dr. Mohan Varkolu et al. (eds)
© 2024 Taylor & Francis Group, London, ISBN 978-1-032-78829-6

Theme of Cultural Discrimination and Immigration Issues in Nadine Gordimer's *The Pickup*

12

S. P. Sekhara Rao*
Department of Basic Sciences & Humanities,
GMR Institute of Technology, Rajam, Andhra Pradesh, (India)
S. P. Jnaneswari
PG Scholar, Padmalaya Mahila University, Tirupati, Andhra Pradesh, (India)

ABSTRACT Nadine Gordimer (1923-2014) was the first woman novelist of South Africa who received the Nobel Prize for literature for her magnificent epic writing in 1991. She has published fifteen novels, more than two hundred short stories and many essays over fifty-year period of her literary career. Her principal novels were *The Conservationist (1974). Burger's Daughter (1979), July's People (1981), A Sport of Nature (1987), My Son's Story (1990), The House Gun (1988) and The Pickup (2001).* She received honorary doctorates from fifteen academic institutions. Her work has been translated into 31 languages. Indeed, most of her works illustrate the impact of apartheid on the lives of native South Africans. Her novels such as *The Lying Days (1953), A World of Strangers (1958), A Guest of Honour (1972), The Conservationist (1974). Burger's Daughter (1979), July's People (1981)* were banned because the apartheid government had felt those novels offensive on moral, religious and political grounds. In fact, she is a great supporter of the National Congress (ANC). Gordimer's earlier novels focus on the struggles of her characters to attain freedom from political and racial problems and her later works tend to concentrate on personal freedom. Gordimer in her later novels also dealt with different themes like the issue of violence, crime identity, dislocation, globalization, multicultural problems, exile, hybridity, theme of forgiveness and reconciliation. This paper mainly focuses on the cultural differences between the two main characters Julie Summers and Abdu in Gordimer's thirteenth novel, The *Pickup*. The novel depicts an illegal immigrant from an unnamed Islamic country, is out of place in the world of a "global city" and then by making the woman the outsider, as she follows him, when he is deported to his homeland.

KEYWORDS Nadine Gordimer, Apartheid, Cultural discrimination, Identity, Immigration, Globalization and the Pickup

Gordimer, was born in a country where people were not allowed to express their views quite freely during the period of apartheid. She was a kind of writer never afraid of criticizing the policies of apartheid regime in her writing and always stressed apartheid free government and harmony among the people. Gordimer, through her imagination, has changed the perception of the reader about reality of the life and its relationships under apartheid in South Africa. She was able to portray successfully how the cruel system of apartheid turned innocent blacks into criminals. Readers, admirers of Nadine Gordimer can notice that the evolution of change in her literary career of fifty long years. Since 1980 there have been extraordinary changes in the South African historical setting. In her later novels like *The House Gun, Telling Times, No Time Like the Present and The Pickup*, setting of most of her novels was Johannesburg, South Africa's 'Golden City." It is the city in which the rich and the poor live and work together almost side by side, where black and white came into contact daily.

*Corresponding author: sekhar.sp@gmrit.edu.in

DOI: 10.1201/9781003489436-12

This paper also narrates how Gordimer evolves as a writer with different themes in her novels. Therefore, this article throws some light on some of Gordimer's works which were penned during the reign of apartheid government in her country, South Africa. In the beginning of Gordimer's writing career, she mainly focused on the external reality of South Africa in which she depicts the comfortable existence of middle-class white people. Gordimer's debut novel *The Lying Days (1953)* was more autobiographical. Her own personal exploration is brought forth through the central character Helen Shaw, who gains political consciousness through an affair with a social worker, Paul. Though the novel appears to be autobiographical, it shows Gordimer's gift for creating individual truths that reflect more generally, the public truth.

A World of Strangers (1958) was Gordimer's second remarkable novel. What it narrates indicates the dilemma of the whites and shows the role whites may play in South African liberation movement. Gordimer expresses her point of view on this through one of her characters, Tobby Hood, an Englishman and outsider who came to replace a publisher's agent and he tries to unite both white and black. In *Occasion for Loving (1963)* she explores the moral dilemma confronting South African white liberals. *The late Bourgeois World* (1966) explores the hypocrisy of South African bourgeois mentality. *A Guest of Honor (1970)* is about the return of an exiled African after Independence. *The Conservationist (1974)* could be seen as Gordimer's version of Waste Land as it fictionalizes collapse of familiar sources of power in South Africa. In *Burger's Daughter* (1979) Gordimer's outstanding work Burger's Daughter written in (1979) portrays the impingement of political developments on the personal lives in a transitional society. In *July's People (1981)* she traces the end of the white regime and the impact of the sudden reversal of power relationships on both white and blacks. *My Son's Story*, (1990) colored family ravaged by white human rights advocate. It describes progression as a political activist and the toll it takes on him and his family. Thus, Gordimer's works focus on race, identity, place, and suppressive political system imprinted themselves onto the lives of her characters.

Nelson Mandela's release from prison was a watershed year in the history of South Africa. He was imprisoned for 27 years. In 1994, the country's new constitution paved the way for general elections. Mandela was South Africa's first black president, wiping out the last remnants of the apartheid system.

People began to ask what Gordimer would write about when apartheid was over. Gordimer replied, "Life doesn't end with apartheid; new life begins". To everyone's surprise she started writing like her contemporary authors Andrew Brink and J.M.Coetzee on the issues of violence and crime as well as theme of forgiveness and reconciliation in her novels like *The House Gun and No Times like the Present*. Salman Rushdie quote- "Despite the caveat in her Nobel Prize speech, I was interested to find out Nadine Gordimer was responding to the changes in South Africa since democratic elections of April 1994.

In one of her interviews, Gordimer expressed her views about her thirteenth novel *The Pickup* and its inner logic in the following words: "I've just finished a new novel, The Pickup, which shall come out in September this year; I surprised myself because, when I look at my fellow writers, they're all writing from their middle age having its own disillusion. My novel's characters are two young people, 28 and 30 years old. It's a story of two young people coming from very different back grounds and thematically, it is about what we've been talking on and off: the question of people coming to a country as migrants, the old question of economic refugee whether you come from Ireland during the Potato feminine to America, or from Russia, as a poor Jew, to America or to South Africa. It seems it has been repeated generation after generation. Particularly in our country, these people who come with hope. They come in, some illegally and they've got to live an incredibly difficult life here, taking another identity. My character is a Muslim from some unnamed Arabian country living here illegally and the novel is about what happens to him and to a young white woman who has every advantage of being in South Africa. These two young people are my protagonists but I see that the theme is really about exile, displacement". Thus, Gordimer presents many global issues in her new novels and proved that there is literary life after apartheid.

Gordimer's short fiction, *The Pickup*, concentrates on the problems of immigration and discrimination. Like her 2005 novel *Get a Life*, it sets in South African city Johannesburg but moves beyond the borders of South Africa. Gordimer tries to throw light on the problems faced by individuals in South Africa which are now global concern. Gordimer's heroine, Julie Summers, is a white rich girl. She meets Abdu, an illegal Muslim immigrant and unqualified mechanic in a garage when her car breaks down. She approaches him to repair her damaged car. She falls in love with him. She finds that his visa had expired and was forced to leave the country. She decides to go with him to his country which eventually turns out to be a home away from home. Gordimer, crosses South Africa's boundary to imbibe the cultural ethos of Muslim's community in the Arab nation. As Illeana Dimitriu in one of her studies "*Postcolonial sing Gordimer: The Ethics of 'Beyond" and significant Peripheries in the Recent Fiction*" had mentioned greatly. She gave up her focus completely on South Africa. She was considered to have been in cultural isolation'...Now writers feel less moral pressure, she is keen to offer literary solutions to questions 'How does each country go about to create a culture that will benefit self and others?' (159-60) Thus, Gordimer had expressed her views about the universal acceptability of her writing in one of her earlier interviews. She says, "quite a lot my writing could have come about absolutely anywhere" (Bazin and Seymour 35)

Gordimer's earlier works speak against policies in a strong way for which some of her works were banned. Policies of racial discrimination and apartheid to separate the people in a society are rarely successful. Indeed, Gordimer has not been only worried about the problems faced by her own country's people but the problems of the world. Gordimer crosses the boundary of South Africa to absorb and assimilate the ethos of an unknown Arab Country. In her novel *The pickup*, she focuses on internal issues happening in newly formed democratic South Africa by ANC. The need to adapt to globalization is also stressed in the post-apartheid era to improve the living conditions of her country's people. In the words of Ileana Dimitriu, Gordimer even presents the burning issues of local people under the government of Africa and the global problems raised in postcolonial era: identity and dislocation, migration, exile and hybrid all steeped in the tension between 'centre and periphery' as a global phenomenon after apartheid" (Shifts 90). Gordimer, wants to wipe out the tag to her novels, mostly related to themes of resistance to apartheid. She proved herself that she is not only capable of writing inner problems of her country but also the problems of the globe. Karina Magdalena Szczurek's observes, The Pickup does not have a complex plot like the other plots of Gordimer's novels but "in its character development, scope and narrative complexity it can certainly be considered a novel." (235) if anyone glances over quickly, understands that the novel is about a boy meets a girl tale by Gordimer but anyone who delves deep into it to understand what is happening in the rainbow nation of South Africa.

Thabo Mbeki became the second president of democratic South Africa in 1999. He listed the main problems faced by the country in his inaugural speech to the nation on 16 June 1999. These problems continue: many people live in utter poverty. We cannot sleep easily when children get permanently disabled, both physically and mentally. No night can be restful when people have no jobs, there is no food. People have to beg or steal to survive. Majority of our population are afflicted by HIV/AIDS. Many women and children fall victim to rape and other crimes of violence. Terrible deeds of criminals and their gangs cause safety and security problems. Corruption robs the poor of what is theirs and corrodes the value system. Liberation cannot be realized until our people are freed from oppression and dehumanizing legacy of depravation inherited from our past.

South Africa became officially an independent nation on April 26, 1994. It had been racially divided nation for a long period. The new government was elected in a democratic way declared that there would not be racial discrimination in the rainbow nation. But the dreams cherished by Nelson Mandela and many more black leaders are still far from being achieved. The legacy of apartheid government is still prevalent in the society of South Africa. Gordimer has said repeatedly in her novels that any nation cannot be developed only through multicultural society. Apartheid related problems are still haunting like a ghost. Exact opposite has happened to the promises made by Nelson Mandela and Mbeki at the time of taking oath of office. Corruption is mounting. Their leadership was unable to uplift the population above poverty, reduce corruption and mitigate or minimize social disorder. South African people are divided by economic disparity. The gap between the poor black and the rich white people has not been reduced but has widened and cannot be bridged in the foreseeable future. It remains as it was even after the demise of apartheid. The first president, Mandela and his successor have tried their best to fulfil the desire of Rainbow nation, but it is nowhere near their planning.

The present novel, The Pickup by Gordimer depicts, how the rural black is entering into areas exclusively reserved for whites who lived in rich places in Johannesburg. Racial segregation is slowly breaking down. People are coming from rural areas and working side by side with white people in cities. Fiction written by Gordimer during the apartheid period narrated that there was a stark difference between the black township and the white residential areas. This novel tells that one can move anywhere freely. She explores that those who enter into these places illegally engage in less paid jobs in the suburban areas of big cities like Johannesburg. Thus, we can see the immigrants from Nigeria, Congo, Senegal and Arab countries.

Gordimer's pickup narrates the growth of two protagonists from differing cultural backgrounds. Julie Summers is a white privileged girl of rich parents in South Africa. She hates the materialism of her affluent, privileged background and lives in a rented cottage in northern suburb area of Johannesburg. She spends her leisure with friends who are intellectuals and free thinkers and meets them at a café called 'The Table'. Abdu, later known as Ibrahim Ibn Musa hailing from an Arab nation enters into Johannesburg illegally and works as a mortar mechanic despite having a degree in economics from his native country. Meeting between Julie Summers and Abdu happens in a strange manner. Her car breaks down; she seeks Abdu's help to repair the car. She befriends him as he has strong muscular body. He likes her as she is the daughter of a rich banker. Gordimer is very clever in introducing the characters who have different cultural backgrounds. The motive of both persons is different. Abudu comes to Johannesburg to settle well and lead a westernized life. He lives in utter poverty in his native country and wants to escape from poverty. In his country, employment opportunities are very few. Western world is offering opportunities to grow in life. He comes to Johannesburg, South Africa as an unqualified mechanic. Though she is a rich lady, she is fed up with materialistic life. Her father divorced her mother and married another rich lady of his class. Her mother also married a rich man and lives in America and runs a casino. She is free to go anywhere and she has many relationships but she wants a permanent partner in her life. Gordimer's plots of her novels are very interesting not because of her characters are genuine sufferers but because some good things are revealed in their suffering. Despite the differences in their motives, they become lovers.

Nodine Gordimer is chiefly known for her writing full of complexities. In the beginning, relationship between Abdu and Julie Summers is highly bonding and a happy one. One colonizer meets another colonizer in a dramatic way. In the post- colonial South Africa, there is no apartheid class distinction. Both the black and the white move freely in bigger cities but there was restriction during the apartheid period. People are coming to the bigger cities form the third world countries for better living, In the case of Abdu, it is the same. He has come to Johannesburg with the hope of getting a suitable job. The irony of Abdu is that he is working as a mechanic as if he was in his home country. The fact is that people coming from poor nations to prosperous countries are provided only with menial jobs such as washing dishes, cleaning the tables and working in restaurants. Whites rarely do such kind of things. People who have not found employment in their country move to the foreign nations to get employment.

The wages for immigrants are always lower than that is paid to the whites and local population. They are not given high perks as are provided to European nationals. They are not even insured and cannot claim other benefits that are generally entitled due to employment. Now the relationship between the two Characters becomes complex. Author narrates in the beginning that Julie is not enamored by the materialistic life provided by her parents. She was entitled to all the privileges that white community possess in South Africa. Abdu, wants all the privileges that are enjoyed by the white man. Both of them never express their feelings clearly. They want to escape from the country to settle well. People generally desire what they do not possess. One of them wants to lead a rich and glorious life that is probably seen in the modern westernized world and the other aspires for happy, peaceful and full of care and love. In the observation of J.M Coetzee, South African novelist, and winner of Nobel Prize finds that the cause behind accompanying Abdu even if his deportation lies in sex. He says," Since sex with Abdu continues to be greatly satisfying; there must be deeply hidden potential to the relationship". (qtd. in Szczurek 248). Though the saying by the author in the above quote seems true, those things are never clearly elucidated or expressed and it is only to be assumed by the readers. In fact, Abdu never shares such feelings anywhere in the novel.

Julie is attracted to Abdu's muscular body and Abdu is mainly impressed by her rich father's background. She rejects the bohemian life and is free to move on in a multicultural society. She is fed up with the life she has been living aimlessly and now wants a permanent and fulfilling life. She is satisfied with her new relationship with Abdu.

Trouble starts when Abdu receives an official order from the Department of Home Affairs to leave the country within fourteen days or face deportation. He wants extension to stay in the country. Julie's father has already heard about their affair and cuts off her allowance. Abdu insists on meeting her father, who is an influential person and a famous banker. However, she doesn't want a meeting with her father as she thinks that her father's reputation may be damaged. Instead, she likes to meet her friends to seek their help in this regard. Abdu asks her to meet her father's friend, Motsamai, an eminent lawyer. Julie, on his request seeks the lawyer's help. Motsamai expresses his inability to help them in this regard. Though the lawyer is black like her lover Abdu, he never thinks of color of her friend, Abdu. He belongs to disadvantaged group in South Africa. He was also one of the victims during apartheid regime. The lawyer is looking for an opportunity to become rich. He doesn't even show sympathy towards them. He disapproves of their relationship. Gordimer says that the past still lingers on in post-apartheid South Africa. When all attempts fail, Julie precipitates a commitment which was not anticipated by buying two airlines' tickets to Abdu's country and insists on accompanying him. When Abdu knows about this, his reaction is:

> *"Who asked you to buy two tickets? You said nothing to me. Don't you think you must discuss? No, you are used to making all decisions, you do what you like, no father, no mother, nobody ever tell you. And me- what am I, don't speak to me, don't ask me – you cannot live in my country, it's not for you, you can't understand what it is to live there. You can wish you were dead, if you have to live there. Can't you understand? I can't be for you-responsible." (95)*

For a while, Abdu is dumbstruck about Julie's decision to accompany him to his native country. She is an independent woman in her own country. She has absolute freedom in this newly formed rainbow nation but the same will not be available in Abdu's country. It is a male dominated society where people are habituated to live their lives as per the commandments of Koran. Abdu expresses his inability to take her to his country as she is no better than a whore. He knows that his parents are orthodox Muslims. They will never allow a woman to live with their son without marriage. Now they marry hurriedly in a register office (against the wishes of their parents). Now his real name is Ibrahim ibn Musa. Gordimer depicts the problems of multicultural society in big cities of the world. In this novel, both the protagonists hide their identity. She feels ashamed of being a daughter of a rich business man and she abandons the rich life style. She lives in a small cottage and moves with black and colored people and enjoys their company. Many of her friends have progressive ideas whereas Abdu is ashamed of his origins. After reaching Abdu's country, Julie adjusts quite nicely and quickly with family members. She mingles with the females of the family. She becomes very close to Abdu's sister, Maryam in a short span of time. She even starts taking English classes for children and other female members of Abdu's family. Indeed, she is not a qualified teacher but she teaches satisfactorily. She even expresses that she couldn't get this kind of satisfaction by serving others in her earlier profession as PRO in the entertaining field. She studies the religious book, Koran. She is a helping hand in Ibrahim's house hold. She tries to become a good daughter-in-law. At dawn she goes out into the desert which is a few blocks away to sit at the edge of the desert, allowing the desert to enter her. Once she visits a rice field in the oasis, which inspires her to drill wells for an agricultural project with the money from the trust fund set up for her. She is so contended and happy in Abdu's house. On the contrary, Abdu is dissatisfied at not getting the permanent visa to go to a foreign country.

The irony of the novel is that, Abdu is trying to leave his country to find settlement in one of the rich nations of the world, whereas Julie leaves all her comforts and home trying to settle in Abdu's desert place. Abdu is trying very hard to get visas for emigration to those endowed countries of the world where he can make a living and live in comfort. He wants to get away from his country where a corrupt government is ruling and people are suffering from abject poverty. Despite his university degree in economics, he has not got a job in his nation. Every time he tries to sneak out from his country, he fails and his application to get a visa is rejected. After persistent efforts he succeeds to get a visa for United States. Now Abdu is very happy. Julie's mother is living in USA. Abdu dreams to go into information technology or else to into casino business with the help of Julie's stepfather. On the other hand, Julie is unwilling to go to America. The young woman is deprived of family bonding in her country. She has fond love, happiness and a purpose in the Arab village. Ibrahim is shocked to find her decision not to go with him to the USA. Thus, Abdu leaves his country for the USA in his utter disappointment and he hopes she will join him later.

Gordimer is exceptionally talented writer in the world of literature. Her chief concern of writing this novel, The Pickup' is to remove the tag that has been carried with her since her inception of her writing of her first novel. No doubt, most of her works move around apartheid related themes. Gordimer's post-apartheid works have mainly concerned about the external issues rather than internal. Thus, Gordimer exposes the real issues of post-apartheid period. She tries to project the basic problems in multicultural society. She emphasizes how apartheid problems such as discrimination and racial issues still persist in the post- apartheid period. She says that the past is still haunting the people of South Africa. Though the leadership is in the hands of black people, problems from the past still need to be addressed. This, she projects though one the characters of her novel by name Motsamai, a famous defense lawyer who becomes corrupt to mingle with rich people. She even projects how people who are in economic distress are forced to do indecent things. This is presented through Abdu. She even tries to project how people have shown adaptability in different situations. It is narrated through the character of Julie Summers. She stresses globalism. She advocates for socio-political harmony, pluralism and multi-culture in society so that racial discrimination can be eliminated completely in the society.

REFERENCES

1. Anthony Sampson, The observer Review of Nadine Gordimer, Living in Hope and History, 1997.
2. Bazin, Nancy Topping, and Marilyn Dallman Seymour, eds. Conversations with Nadine Gordimer. Jackson: U.P Weekly of Missipi, 1990. Print
3. Boston Globe interview by David Mehergan http://www.boston.com /globe/search/stories el/1994d. html
4. Clingman, Stephen. The Novels of Nadine Gordimer: History from the Inside. 2nd ed. Amherst: U of Massachssetts P, 1992. Print.
5. Diala, Isidore. "Interrogating Mythology: The Mandela Myth and Black Empowerment in Nadine Gordimer's post-Apartheid Writing." Novel: A Forum in Fiction 38.1 (2004): 41–56. Print
6. Dimitriu, Ileans. " Postcolonialising Gordimer: The Ethics of 'Beyond' and Significant Periheries in the Recent Fiction." English in Africa 33.2 (2006): 159–80. Print
7. Gordimer, Nadine. Burger's Daughter. London: Bloomsbury, 2000. Print
8. Gordimer, Ndine. The Conservationist. London: (Jonathan Cape, 1974)
9. Gordimer, Ndine. The House Gun. New York Penguin, 1978. Print.
10. Gordimer, Ndine. July's People. (Penguin, 1970)
11. Gordimer, Ndine. The Late Bourgeois' World. London: Jonathan Cape, 1996
12. Gordimer, Ndine The Lying Days. (Jonathan Cape, 1953).
13. Gordimer, Ndine. My Son's Story. (Delhi: Doaba, 2001)

14. Gordimer, Ndine. Occasion for Loving. London: Bloomsbury, 1994. Print.
15. Gordimer, Nadine. The Pickup (London-New Deli-New York-Sydney bloomsburypbks, 2001)
16. Gordimer, Nadine. A World of Strangers. London: Bloomsbury, 2002. Print
17. Jaggi Maya. "Culture-crossed Lovers in The South Africa." Rev. of The Pickup. The Gurdian Weekly 165.13 (September 20-26 2001): 21
18. Jonson, Walton R. "education: Keystone of Apartheid." African Education and Social Stratification. Spec. Issue of Anthropology and Education Quarterly 13.3 (1982): 214–37. JSTOR. Web. 21 July 2011
19. Szezurek, Karina Magdalena. Truer than Fiction: Nadine Gordimer Writing Post-ApartheidvSouth Africa. Saarbrucken, chriften Germany: Sudwestdeutscher verlag fur Hochschuisschrifen, 2008. Print.
20. Sue Kosse. Living in Hope: an Interview with Nadine Gordimer, Commonwealth 23.2.
21. Boston Globe interview by David Mehergan http://www.boston.com /globe/search/stories el/1994d. htm)
22. http://www.boston .com/ globe/search/stories/nobel/1991jhtml

Impending Inquisitions in Humanities and Sciences (ICIIHS-2022) – Dr. Mohan Varkolu et al. (eds)
© 2024 Taylor & Francis Group, London, ISBN 978-1-032-78829-6

Fighting Patriarchy through Literature: An Analysis of the 21st Century Indian Mythological Fiction

13

Vurity Mounika*
Assistant Professor, Koneru Lakshmaiah Education Foundation, Vijayawada

ABSTRACT India is a land of languages, cultures and stories. Indian mythologies are known for their values and morals that promote integrity among commoners, loyalty in women and valor in men. Quite often referred to as the foundation stone for patriarchy in India, these stories are now witnessing a change in 21st century fiction. Irrespective of the gender of the author, women in mythologies are taking up arms, involved in decision making and standing up to their rights. This paper analyzes popular titles of the 21st century under mythological fiction written by Indian authors to study the role of women with respect to the accepted norms in the society.

KEYWORDS Feminism, Gender stereotypes, Women authors, Literature

1. Patriarchy and Indian Mythologies

Patriarchy is a form of societal system where the dominion of men over women and children is a norm. Patriarchy, the Greek word, directly translates to "Father's Rule". Research on patriarchy and literature can be traced back to Kate Millet's Sexual Politics, 1970. Millett developed the notion that men have institutionalized power over women, and that this power is socially constructed as opposed to biological or innate. She analyzed the patriarchal myths in literature through the works of four renowned authors DH Lawrence, Henry Miller, Norman Mailer, and Jean Genet. Millet laid the foundation to recognizing cultured inequalities in the guise of natural occurrences through literature (Millett, 1969). According to Kate Millet, Patriarchy in a society exists in two forms. 1) Male subordination of females, 2) Older male subordination of younger males. Millet's radical feminism is followed by Marxist feminist literature emphasizing the notion of class superiority in relation to female subordination. Psychoanalysis and Feminism of Juliet Mitchell analyzes the system of exchanging women within families and its impact on the psychological status of the 'inferiorized' women (Mitchell, 1974). Integrating the central ideas of radical feminist thought with those pivotal for Marxist or socialist class analysis, Zillah R. Eisenstein analyzed patriarchy with respect to motherhood, reproduction and male supremacy through a socialist feminist historical analysis (Mitchell, 1974).

Patriarchy in its current form relates to any form of domination or superiority meted out by men upon women. Social science research considered patriarchy as a form of inequality patronized over centuries like racism and casteism (McKinney, 1985). However, patriarchy is still considered natural and unavoidable (sometimes unquestionable) despite unrelenting research in the field. Patriarchy is unrelenting and rigid as ever in the society. Though research points to the decline of patriarchal tendencies in current times, parallel research is going on regarding the return of patriarchy as well across the globe (Haulman, 2007). Patriarchy is promoted in society through various fields and institutions. While the family system is one key institution that promotes patriarchy, religion comes next in line.

*Corresponding author: mounikassc@gmail.com, v.mounika@kluniversity.in

DOI: 10.1201/9781003489436-13

Mythologies over the world are strict upholders of patriarchal values with their warrior princes and devoted princesses. Indian mythology, for example, is full of tales of kings and princes and emperors with their adventures and wars fought for, over and because of women. Retellings over centuries have further intensified the patriarchal stand of the society (Kaur, n.d.). In recent times, mythologies have become a weapon for moral policing for godmen and sexist misogynists. In the year 2012, the Bombay High Court observed that a wife should be like Goddess Sita, while hearing divorce petition by a man against his wife's disapproval to relocate with him to his new workplace. Mythology as a tool for moral policing is not uncommon anywhere in the world, especially in India. Instances of God men accusing women of not following their mythological role models as a reason for rape and sexual assaults against women are repeatedly witnessed in India (The Associated Press, 2015). This repeated references to mythological figures as watchdogs of patriarchy is being countered by 21st century Indian writers through literature. Indian writers feel patriarchy underplays Indian mythology by ignoring the status of women in the mythological stories. Retellings of the age-old tales are accused of promoting patriarchy and suppressing gender rights promoted and upheld in mythology (Chadda, 2016). Devdutt Pattanaik, the author of Sita, retold the story of Ramayana from the perspective of Sita. Amish Tripathi, titled the literary popstar of India, penned the Shiva Trilogy, portraying Sati as a warrior queen fighting till her last breath as a warrior should. Kavitha Kane, a feminist mythological writer, is relentlessly penning down the stories of the women in mythology to explore the world of Treta Yuga and Dwaparayuga through the eyes of a woman. Chitra Banerjee Kulkarini, known for her Palace of Illusions is another prominent Indian writer working on promoting gender equality through mythological retellings from the perspective of women. Many other young writers are taking up to literature to give voice to women through mythological retellings. The fight against patriarchy through mythological retellings is however, not a novel concept. Ahalya Draupadi Kunti Tara Mandodari tatha / Panchakanya smaranityam mahapataka nashaka is a Sanskrit sloka that is promoted among women as a shield against sin and all evil. The sloka glorifies the five strong women in Indian mythology, Ahalya, Draupadi, Kunti, Tara and Mandodhari; all women who faced strong opposition from male chauvinists throughout history and mythology. There is an argument that any woman who remembers the fate of these five women shall abstain from taking a wrong path in her life. However, feminists laud the sloka for glorifying the five wronged women from Indian mythologies as epitomes for freedom and gender rights.

2. Method and Analysis

There are many works of feminism on patriarchy and violence against women but patriarchy there is no popularly accepted theoretical framework for measuring patriarchal elements in literature. Gwen Hunnicutt in his article Varieties of Patriarchy and Violence Against Women: Resurrecting "Patriarchy" as a Theoretical Tool tried to build a theoretical framework for patriarchy. However, the research ended in identifying methods to analyse factors for violence against women (Hunnicutt, 2009). Patriarchy, for the purpose of the research, is classified into five categories, viz., financial patriarchy, social patriarchy, psychological patriarchy, physiological patriarchy and political patriarchy. Though the classification is not earlier used, references to these factors as a cause for patriarchy are found in social sciences research (Cho, 2003).

For the purpose of the study, Indian books published in English on mythological fiction have been chosen as the population. 30 books written by four authors are purposefully picked as the sample for the study owing to their ratings in the concerned genre. This paper is a work of narrative analysis of the sample to study the presence of financial, social, psychological, physiological, and political superiority of men over women. Narrative analysis refers to a cluster of analytic methods for interpreting texts or visual data that have a storied form (Figgou & Pavlopoulos, 2015). Narrative analysis is said to have emerged from the narrative turn in psychology and the seminal work of Jerome Bruner more than 20 years ago. Narrative analysis is concerned with the structure, content and intent of the story (Carolin Demuth, 2015). In this research, the overall story is taken into purview to identify the presence of patriarchal supremacy based on the lead characters in the stories.

Financial patriarchy can be defined as superiority of men over women with respect to monetary matters. Being the key earners of the family as well as the rightful inheritors of property, men tend to have an upper hand over women owing to their financial supremacy (Basile & Harriss-White, 1999). In India, it is a norm for men to deal with all financial matters of the family, irrespective of the woman's working status. This leads to women taking a second seat when it comes to financial decision making. The sample is analysed to identify the working status of women and their role in making financial decisions.

Social Patriarchy refers to men showing dominance over women in social circles. Men are said to have social dominance orientation over women due to various factors. Research is rampant in understanding the reasons behind the social dominance orientation of men over women (Schmitt & Wirth, 2009). This paper, however, tries to identify the presence of social dominance of men over women in the given sample by studying the hierarchal relationships between the key characters in social situations with reference to their gender.

Psychological patriarchy refers to a form of society where the male mind is considered to be superior to female mind. Women are in many instances not given a position that requires spontaneity as they are considered to be prone to mood swings and blackouts due to various factors like menstruation and ovulation (Koren, 2017). Women are psychologically thus considered to be inferior to men. In this paper, the sample is studied to analyse the presence of such scenarios where women are considered to be weak at mind, when compared to their male counterparts.

Physiological patriarchy talks about the notion that men are naturally strong when compared to women. Their physiological strength gives them the natural authority to be superior to women (Goldberg, 1989). Research, however, points out that there is always a difference between nature and nurture. Men may not be naturally stronger than women. Multiple factors lead to their strength, mostly owing to their nurture. In this paper, the presence of physiological patriarchy is analysed by studying the key characters and their physiological traits.

Political Patriarchy is a scenario where the administration lies in the hands of men, irrespective of their capabilities and merit. Many patrilineal societies consider men as the rightful heads of families, communities as well as states and countries. Women rarely take after their fathers or husbands as rulers and administrators. In this paper, the role of women as administrators is analysed by studying the characters and their political status in the stories.

3. Results and Discussion

Sita, the most loyal and devoted wife according to Hindu Mythology, is portrayed as the Prime Minister of Mithila in Amish's Sita - Warrior of Mithila. Sita's mother Sanaina is said to have handled the finances of Mithila while her husband, King Janak, is a philosopher. Sita, as prime minister, handles the economic fall down of Mithila after her mother's death. Similar references to women taking up financial decisions in times of emergency are seen in most of the works. Tara, in Anand Neelkantan's Vanara, realizes that a city cannot be built without money. She does not think like Baali, who demands his people to work for him through poverty and famines. She sends Sugreeva to get money for the Vana Nara people. Women in mythologies are usually not portrayed as handlers of the financial crisis. Millennial mythological fiction, however, is giving them that role, by clearly waylaying the rules of financial patriarchy.

In Ram, the Scion of Ikshvaku, Ram looks forward to meeting a woman who shall make him bow down to her in respect. Ram sees Sita for the first time when she tackles a criminal on the road. Her keen eye for detail, her warrior skills as well as her concern for the safety of the common public make him bow low in respect. Amish's introduction of Ram as someone bowing low to a woman is in contrary to the mythological tales that always speak of Ram's prowess at Sita's Swayamvara. Songs of praise are sung of Rama's unwavering focus towards the task on hand. He is said to have never once looked at Sita until he won the contest. Sita is portrayed as a timid bride who walks to him with her head bent low, a garland in her hands, accepting him as her husband for winning the Swayamvara. Sita, who could single handedly lift the bow, is rarely appreciated for her physical prowess, but is instead portrayed as a timid little girl grown up under her father's shadow. Kavitha Kane also breaks the rule in Sita's Sister, where she lets Sita take a stand when Ram decides to leave her in the palace while he goes on an exile. The way Bali asks Tara to talk to his people as he trusts she has a better way with dealing with the masses than him, clearly confirms that millennial mythologies are fighting social patriarchy through their characters.

Ravan, in Amish Tripathi's War of Lanka, narrates the story of Vedavati's death to Sita, with complete faith in her ability to take the news. Sita is called brave and wise by Kumbhakarna, Ravan's younger brother. The novel talks about Vishnu, as a position given to either men or women who show excellence in their qualities both mentally and physically. A similar relationship is seen between Shiva and Sati; where Shiva considers Sati to be able to take better decisions than him. Kavitha Kane's Queen of Lanka and Sita's Sister talk about women who could take decisions in times of crisis to protect their kingdoms and their people. Context within these stories prove that the millennial mythologists attempt to portray women as equal in psychological aspects when compared to that of men.

In an open challenge to Tarak, the most heinous warrior in Meluha, Sati calls for a test of fire; a one-on-one duel to death. Contrary to popular assumptions, Sati fights the villain, while Shiva watches the fight with complete faith in her. Shiva is nowhere shown as a better warrior than Sati throughout the trilogy. Till her last breath, Sati fights like a warrior and she dies a warrior's death. Sita, in Ram Chandra Series, as well, is portrayed with war scars unlike the usual beautiful unblemished Sita. Kavitha Kane, however, does not give them the warrior appearance. In Kavitha Kane's works, the women are bold and brave but not warriors with battle scars. Chitra Banerjee Kulkarni as well goes by this stand through the portrayal of her characters. Anand Neelkantan also does not give his queens and princesses a warrior-like trait. However, his works do not reflect the age-old portrayal of women as obedient and devout wives. Though it is a good start, Indian writers are still finding it challenging to portray their women as warriors on par with men. Sati, in the Shiva Trilogy, leads a war front while Shiva leads the other front. Sita, in Sita's Sister, takes after her father in helping him in his administrative works. In the Ram Chandra Series, Sita is the Prime Minister of Mithila, next in line in administration to her father. Urvi, Karna's wife, supports her father in administrative matters, and later supports Karna in his decision making. Mandodari, the Queen of Lanka, has a similar stand in dealing with administrative matters. The women of the 21st century mythological fiction are far more involved in administration than their previous versions in earlier narrations. However, there still is a long way to go to see women as true decision makers in society. Women are, in many instances, portrayed as the brain behind the crown, but they are not shown as able leaders and rulers in broad daylight.

4. Conclusion

Mythological fiction is witnessing a revival in India through literature and cinema alike. 21st century Indian movies are mostly concentrating on retelling mythological tales, quite often adaptations of literary works. Portraying strong women characters in these fictional works is restructuring the patriarchal tendencies in Indian society. Moral policing through references to mythology are facing a backlash through these retellings. The millennial mythologists are thus fighting patriarchy in India through stories. The study has its limitations with respect to its methods and results. The concept of fighting patriarchy through literature is not new. Lots of research was conducted in the field earlier. This study builds on the earlier research to throw some light on the Indian retellings of mythological tales in a completely altered perspective. Narrative analysis is often criticized for its inability to provide coded data that can be statistically tested. The current study is also a narrative analysis that concentrates on the overall story rather than on coding the characters, situations and the text within the stories. A quantitative analysis is also possible in this particular study to identify the presence of patriarchal tendencies through numbers; for example, the number of working women in a story to the number of non-working women. However, qualitative analysis is a better method for this study as it moves beyond the boundaries of numbers and figures and tries to qualitatively analyze the overall message sent out by the author. Further research can be done on contextual analysis and discourse analysis of the sample to study the presence of hidden stereotypes within the content. The work can also be a comparative study between earlier mythological fiction to understand the change in the portrayal of the character on patriarchal lines.

REFERENCES

1. Basile, E., & Harriss-White, B. (1999). The Politics of Accumulation in Small town India. *IDS Bulletin, 30*(4), 31–38.
2. Carolin Demuth, G. M. (2015). Qualitative Methodology in Developmental Psychology. In J. Wright (Ed.), *International Encyclopedia of the Social & Behavioral Sciences* (pp. 668–675). Elsevier.
3. Chadda, S. (2016, June 3). *Indian Patriarchy underplays Indian Mythology*. Retrieved from Woman Endangered.
4. Cho, U. (2003). *Global Capital and Local Patriarchy: the Financial Crisis and Women Workers in South Korea*.
5. Eisenstein, Z. R. (Ed.). (1979). Capitalist Patriarchy and the Case for Socialist Feminism. Monthly Review Press.
6. Figgou, L., & Pavlopoulos, V. (2015). Social Psychology: Research Methods. In *International Encyclopedia of the Social & Behavioral Sciences (Second Edition)* (pp. 544–552). Elseveir.
7. Goldberg, S. (1989). The Theory of Patriarchy. *International Journal of Sociology and Social Policy, 9*(1), 15–62.
8. Haulman, K. (2007). The Return of Patriarchy. *Reviews in American History, 35*(4), 483–489. Retrieved 11 5, 2022, from https://muse.jhu.edu/article/225328
9. Hunnicutt, G. (2009). Varieties of Patriarchy and Violence Against Women: Resurrecting "Patriarchy" as a Theoretical Tool. Violence Against Women, 15(5), 553–573. https://doi.org/10.1177/1077801208331246
10. Millett, K. (1969). Sexual Politics. Doubleday.
11. Mitchell, J. (1974). Psychoanalysis and Feminism. Pantheon Books.
12. Kaur, D. A. (n.d.). *Indian Patriarchy – An Intersection of Caste, Class and Gender*.
13. Koren, M. (2017, March 10). *Why Women Weren't Allowed to be Astronauts*.
14. McKinney, A. &. (1985). Review of Money, Sex and Power: Toward a Feminist Historical Materialism., by N. Hartsock. *The Journal of Politics*, 1298–1301.
15. Schmitt, M. T., & Wirth, J. H. (2009). Evidence that Gender Differences in Social Dominance Orientation Result from Gendered Self-Stereotyping and Group-Interested Responses to Patriarchy. *Psychology of Women Quarterly, 33*(4), 429–436.
16. The Associated Press. (2015, March 4). *Blaming women for rape a widespread issue in India*.

Impending Inquisitions in Humanities and Sciences (ICIIHS-2022) – Dr. Mohan Varkolu et al. (eds)

Research Study: Ponniyin Selvan as a Historical Fiction and its Adaption in Modern Cinema

14

K. Jayasree*
Assistant Professor, PG & Research Department of English,
Srimad Andavan Arts and Science College, Thiruvanaikovil, Tiruchirapalli

ABSTRACT Bringing Kalki Krishnamurthy's literary classic Ponniyin Selvan to the big screen has been a labour of love for Mani Ratnam for at least two decades. Published in 1955 as a five-volume series, Krishnamurthy's novel is a sweeping work of historical novels inspired by genuine figures and events from the eleventh century. The first volume of Ponniyin Selvan (the second will be published in 2023), titled Ponniyin Selvan: I, condenses the central conflict of Krishnamurthy's massive epic: the Chola empire is threatened by disaffected courtiers and past rulers of the country. In a story rife with malicious whispers and massive conflicts on land and water, the protagonists and antagonists are introduced in the first part. The story depicts a period of turmoil in the Chola Empire. Being one of the most iconic works in Indian Tamil literature the filming itself was a massive task with the cast shooting in different national and international locations for lengthy periods. The cast consists of some of the most popular faces in Indian cinema. The film released on 30th September 2022 in Indian theatres and worldwide will be discussed. It was a huge success and is considered the highest-grossing Tamil Film in 2022.
KEYWORDS Ponniyin Selvan, Ponniyin Selvan: I, Mani Ratnam, Kalki Krishnamurthy

1. Introduction

Multiple filmmakers have tried their hands at adapting Ponniyin Selvan to the big screen, beginning with M. G. Ramachandran in 1958. He paid 10,000 (equal to 810,000 or US$10,000 in 2020) for the film rights to the novel and declared that he would produce, direct, and star in the adaptation alongside Gemini Ganesan, Vyjayanthimala Bali, Padmini, and Savitri. Ramachandran had an accident before filming could begin; because it took six months for the wound to heal, he was unable to move forward with the project even when the rights were renewed four years later. Entertainer Kamal Haasan along with director Mani Ratnam adapted the story into a film in the late '80s. Ratnam admitted that he and Kamal Haasan, who already had purchased the rights to the novel through Ramachandran, had worked on a screenplay for the film but had ultimately decided to scrap the project due to its lack of commercial viability.

Soon after, in the early 1990s, Kamal Haasan collaborated with screenwriter Ra. Ki. Rangarajan adapted the novel into a forty-episode television series, but production on the show was eventually shelved. The book was supposed to be adapted into a web series by Jeyamohan as well as Soundarya Rajinikanth, who was hired as a creative producer by Eros International in 2016, however, the project never began production. Mani Ratnam officially picked up shooting again on his movie in the first few months of 2019. An all-star cast, including Vikram, Aishwarya Rai Bachchan, Jayam Ravi, Karthi, and Trisha, appeared in both of his film adaptations. And by September 2021, shooting had wrapped on both segments (SOONTHODU, and WAHAB,

*jayasree@andavancollege.ac.in

DOI: 10.1201/9781003489436-14

2021). On September 30, 2022, the film was made available in five different Indian languages. Furthermore, in this research study, we will research the historical background as well as the movie and theatre adaptation of Ponniyin Selvan.

2. Discussion

The Son of Ponni, or Ponniyin Selvan, is the Tamil title of a historical fictional book by Indian writer Kalki Krishnamurthy. From October 1950 through May 1954, it appeared in weekly instalments in the Tamil magazine Kalki, before being collected into five volumes the following year. The life of Chola emperor Arulmozhivarman is chronicled in this 2,210-page work (Rumsby, 2018). Kalki travelled to Sri Lanka on three separate occasions to learn more and draw inspiration from the culture. If you're looking for an excellent Tamil novel, look no further than Ponniyin Selvan. The series gained such a large following that it helped boost Kalki's circulation to 71,366 copies per issue, a substantial number in the newly independent India. Even in the present day, individuals of all ages continue to hold the book in high esteem, and it has even attracted a cult following. Ponniyin Selvan's portrayal of the Chola empire's machinations as well as the desire for power throughout the 10th century, as well as its tightly knit plot, vivid narration, and witty language, have won the novel widespread critical acclaim. Indian director Mani Ratnam is now working on a film adaptation of the novel. Published on September 30, 2022, was Ponniyin Selvan: I, the first installment.

Ponniyin Selvan: Historical Background

No other novel had as much appreciation while also being written as "Ponniyin Selvan", which was published as a serial in Kalki. Fans of the Kalki magazine began lining up at train stations on May 16, 1954 and continued to do so for the next three and a half years in order to get their hands on a copy of the publication (Muthuvel, and Kumaravel, 2021). Even the novels written by "Alexander Dumas" were first published in serial form in journals. Additionally, there were echoes of Dumas in a number of other areas. Milady de Winter, from "The Three Musketeers", served as the inspiration for "Beautiful Nandini". Even "Vanthiyathevan" had a striking resemblance to "D'artagnan". The suspense would build to a climax in the final words of each episode, which piqued the audience's curiosity and encouraged them to discuss the episode up until the next one. However, serialization was not without its challenges. Due to time constraints, Kalki was unable to correct the factual inaccuracies he made. For instance, a character who is first presented as mute will end up speaking at various points throughout the story.

Ponniyin Selvan included a large number of regular people, and Kalki illustrated what daily life was like for Chola citizens, which was something that had never been done before. The audience was able to visualize the story because of the breath taking graphics that Maniam created, which contributed to the story's overwhelming success. The sales of Kalki magazine reached unprecedented heights, which sent its competitors, such as "Ananda Vikadan", into a state of panic.

The narrative was bolstered by remarkable historical particulars that had only just been uncovered at the time of writing. But Kalki made a shrewd choice by selecting a hero who appeared only twice in the edicts that stretched over a mile and a half within the Tanjore large temple. Rajaraja was an important figure who should not have been trifled with. He was unable to frolic romantically among the woods. Characterized as the kid next door, "Vanthiyathevan" role in the story was limited to that of a courier for the heir presumptive (Rajendran and Kumar, 2019).

It just so happens that Vanthiyathevan spends the night at the sprawling "Kadambur" stronghold when he is traveling to Tanjore. He overhears the chieftains of the Chola realm discussing among themselves who the future Chola monarch should be in order for the crown to be placed on their head. An unexpected onslaught of treachery and murder had begun to pursue the Cholas, and an incensed "Vanthiyathevan" assisted his people in maintaining their position. In the background, among all of the intrigues, there was a multitude of peaceful relationships, and Kalki closes the novel with a flower-picking child as the emperor-to-be and a boat maiden as his princess, which serves as a reminder to the people that liberty was here to remain.

Ponniyin Selvan is not an accurate history in any way, shape, or form; but, the narrator of the tale certainly has the right to follow whichever makes the greatest sense to him. In the year 1954, Kalki performed something that was unimaginable. He finished the book but failed to provide a satisfying conclusion. The principal characters never got married, and the person who gave the book its name was never crowned. Ponniyin Selvan has been published in book form for the past 65 years, and it continues to be a best-seller. MGR was able to acquire the film's rights, but he, together with "Kamal Haasan" and "Mani Ratnam", were unable to produce the film. When that point came around, the reader's mental images of the personalities were solidified. In the past half-century, there have been at least three sequels released.

Movie Adaptation of Ponnyin Selvan

Vandiyathevan, the protagonist, is a charming, courageous, and clever young man who travels across the Chola nation to deliver a message from crown prince Aditya Karikalan to the Monarch and Princess. The narrative alternates between the adventures of Vandiyathevan in Chola territory and those of Prince Arulmozhivarman (formally renamed "Raja Chola") in Sri Lanka. The story follows Kundavai, his sister, as she tries to persuade Arulmozhivarman to return and restore political stability in a country where vassals and small chieftains appear to be plotting civil war. After Parantaka Chola's first son Rajaditya was killed in combat, his younger son Gandaraditya took his place. Since Gandaraditya's son Maduranthaka was only two years old when their father passed away, the throne was passed to Gandaraditya's brother Arinjaya. The son Parantaka II (Sundara Chola) succeeded him as king after Arinjaya's death. Adithya Karikalan and Arulmozhivarman were his sons, and Kundavai was the name of his daughter. At the start of the novel, Emperor Sundara Chola is sick and bedridden. Aditya Karikalan, his son, is the Northern Command's general and a resident of Kanchi. Arulmozhivarman, the youngest child (and future Raja Raja Chola I) is currently engaged in battle in Sri Lanka. Kundavai Piratti, their sister, resides in the Pazhayarai palace of the Chola king and queen.

Story Plot

The plot develops when suspicions spread about a scheme involving Sundara Chola as well as his children. Vanar Kula Veeran Vallavarayan Vandiyathevan, a warrior, happens to see the Pandya conspirators somewhere at the palace of his buddy Kandhamaaran. Most of the novel's characters—including protagonist Nandhini (the chief conspirator) and protagonist Arulmozhivarman (the prince who everyone loved) are introduced to us by Vandiyathevan. Aditya Karikalan had been in passion about Nandhini when he was young, but after he killed Veerapandiyan, she became vengeful and swore to bring down the Chola dynasty. After hearing about the scheme, we also encounter Kundavai Piratti, who dispatches Vandiyathevan to Sri Lanka with a message for Arulmozhivarman. There are also secondary characters like as Sundara Cholar's Prime Minister and all-seeing snoop Aniruddha Brahmarayar's son Gandaraditya Thevar, the conspirators' target for coronation as king. Along the way, he encounters Brahmarayar's spy, Azhwarkadiyan Nambi, a gentleman who travels the country challenging others to disputes. He is constantly near Vandiyathevan and has come to his aid in perilous situations; he gathers intelligence for the Prime Minister (Sankaravelayuthan, 2019).

Poonkuzhali, the boat woman who paddles the heir to the throne to Lanka, Mandakini, the deaf and mute parent of the legendary Maduranthaka Chola, as well as Vanathi, a Kodumbalur royalty (the woman who will become Arulmozhi's wife afterward) are all featured. Nandhini is the most famous of the women in the story, and her attractiveness is considered to be enough to sway even the stubbornest male. Manimegalai, sister of Kandhamaran (Kadamboor prince), aids Nandhini without realizing that she is the conspirator, and Kandhamaran betrays his best friend, Vandhiyathevan. Vandiyathevan, meantime, is aided in his journey to Sri Lanka by Poonkuzhali, where he eventually meets and befriends Arulmozhivarman. Arulmozhivarman, while in Sri Lanka, learns that his father was with a deaf and mute girl who was born on an island close to Sri Lanka. Through her drawing, he learns that she and his father have sired two children. Are those kids legitimate heirs to the throne, and who are they? Later, Vandiyathevan witnesses Nandhini and the Pandya conspirators putting a young child on a throne in the Thirupurambayam forest and making a pledge before him. However, the main question remains "What makes this young man so deserving of the monarchy, anyway?"

Alignment of Story and Movie

Arulmozhivarman vanishes in a cyclone while returning from Sri Lanka. He is said to have died, although he actually lives at the Buddhist monastery of Choodamani Viharam. Once again, the scattered relatives begin to reassemble. Meanwhile, the plotters settle on a single day for the triple murder of the monarch, his two sons, and their own father. Meanwhile, Nandhini summons Aditya Karikalan to the Kadambur Palace for a meeting about the kingdom's destiny. Karikalan goes to the Kadambur palace to see Nandhini, despite the fact that he knows doing so puts his life in grave danger. Next, Aditya Karikalan is killed at the palace of Kadambur. Arulmozhivarman, in meantime, is on the mend and has returned to Tanjore, in which he was initially coerced into accepting the throne. Eventually, he pulls a fast one and succeeds in having his relative Uthama Chola crowned king. Tyaga Sigaram, the zenith of sacrifice, is an appropriate title for the book's fifth section (Mariyappan, 2020).

3. Conclusion

The aim of this segment will be to conclude the overall report. The report was focused on the discussion and evaluation of a fictional character and the historical and movie background has been described. The demonstration and the illustration of the facts and the information have been accurate and thus the identification of the concepts has been clear and precise. Thus all the aims and objectives of the report have been fulfilled with precision and accuracy.

REFERENCES

1. Mariyappan, K.R., 2020. The Journey of the Hero in Nigel Tranter's The Bruce Trilogy and Kalki's Ponniyin Selvan. PalArch's Journal of Archaeology of Egypt/Egyptology, 17(9), pp. 5927–5932.
2. Muthuvel, K. and Kumaravel, K., 2021. Une Etude Caractéristique Des Romans Tamouls Réalisés Par Un Ecrivain Remarquable De Renaissance. Annals of the Romanian Society for Cell Biology, 25(6), pp. 10996–11004.
3. Rajendran, V. and Kumar, G.B., 2019. A robust syllable centric pronunciation model for Tamil text to speech synthesizer. IETE Journal of Research, 65(5), pp. 601–612.
4. Rumsby, J.H., 2018. Otherwordly others: racial representation in fantasy literature.
5. Sankaravelayuthan, R., 2019. Lessons from Translation of a Historical Novel from Tamil to English. Translation, Nation and Knowledge Society, p. 73.
6. Soonthodu, S. and Wahab, I.N., 2021. Exploring Niche Tourism The Indian Perspective.

Impending Inquisitions in Humanities and Sciences (ICIIHS-2022) – Dr. Mohan Varkolu et al. (eds)
© 2024 Taylor & Francis Group, London, ISBN 978-1-032-78829-6

Soft Skills for Engineering Undergraduates: A Brief Study

15

K. Geeta*
Assistant Professor, Department of English, Anurag University, Hyderabad

ABSTRACT Soft skills are personal traits that drive interactions, improve job performance, and improve a person's chances for a better job and a better job. The lack of soft skills and values is usually what causes people to be demotivated, insubordinate, unpredictable, dishonest, conflict, or not trust each other. Soft skills are learned behaviours that need to be practised and used in a specific way. Soft skills will help people who have a strong conceptual and practical framework to build, develop, and manage teams at work. In the long run, this will help them get a better job because they will have a better overall personality. Companies require their employees to have more than just knowledge in technical and analytical skills. They also need them to be able to deal with clients, and government officials. Soft skills are what make us who we are. They include our attitudes, habits, and how we interact with other people, to name a few. They aren't as tangible as hard or technical skills, but they are still important to have. These skills can't be learned by going to a training class. These skills can, however, be learned via educational and work experiences, as well as in life.

KEYWORDS Soft skills, Personal traits, Employment skills, Analytical skills, Attitudes, Habits, Life skills etc.

Soft skills are currently gaining increased attention from a range of age groups in conjunction with occupational education programmes to better prepare individuals for future jobs. Nonetheless, communicating such concepts in a manner fit for higher education presents a substantial challenge. Soft talents are defined in a variety of ways. In general, education programmes emphasising soft skills strive to raise or decrease the number of unemployed graduates and to efficiently match graduates with employers, not just technically, but also in terms of company values. Borrego and Bernhard (2011) assert that the bulk of the skills that future engineers must study in class and hone during their professional practise are soft skills. These are transferrable skills that can be used in a number of settings throughout life, most notably in extremely competitive job environments. They are necessary for employment market entry and have gained in prominence in engineering professional contexts, alongside hard and technical skills.

While today's engineering graduates are equipped with an abundance of technical knowledge, they frequently lack the critical social skills for today's professional situations, such as leadership, communication, and teamwork. One significant topic of engineering education research is the design of higher education engineering courses with the purpose of predisposing future engineers to be competent, autonomous, and decision-makers in order to meet labour market demand for highly educated professionals.

Engineering has traditionally been characterised by a strong emphasis on its technical components. This is because engineering is more removed from interpersonal interactions than other fields. In these disciplines, the outcome is paramount, and personal concerns are not required to achieve success. However, the traditional image of an engineer working alone on a customised product has shifted. Businesses undertake numerous projects involving a large number of individuals. This means that the

*geetakalakuntla@gmail.com

DOI: 10.1201/9781003489436-15

relationships between the various stakeholders in a project are critical to its success. As a result, certain personality traits, which we like to refer to as soft talents, such as cooperation or leadership, have begun to be recognised. Historically, these skills have been overlooked in engineering curriculums. However, these soft skills requirements are taken into account, particularly given that engineers work in a project-oriented setting.

The study of soft skills is a future trend in rising sectors of engineering, such as information and communication technologies (ICT). Today, people in all walks of life seem to be paying more attention to soft skills than they have ever before. Soft skills are the positive traits that people look for in someone who is good and honest. They have to be learned as a habit. All of a man's good qualities and the values he holds dear are thought to fall under the term "soft skills." It is thought to be a sign of intelligence when a person is able to express his feelings in the best way possible. Thus, one can analyse how important soft skills are as it enables him/her to understand and react according to the needs of the situations and people.

In the workplace, the term "soft skills" refers to an individual's emotional intelligence, which is a collection of personality traits, social graces, communication, language proficiency, personal habits, emotions, optimism, and collaboration that characterise interpersonal connections. While soft skills are difficult to define, they are critical for personal and professional success. Our future success depends on our ability to communicate fluently and accurately in the English language, according to education providers, academics, human resources departments and corporations. As a result, little is known about how soft skills affect business and the current level of soft skill attainment and English language proficiency among Indians.

According to Backlund, G. & Sjunnesson, J. (2012), the importance of teaching students at the undergraduate level soft skills has grown steadily over the previous decade. For example, Backlund and Sjunnesson argue that technical competence is a small part of the set of skills needed to succeed on the job, but many engineering students leave college without having been taught the skills needed to work in a business context, such as empathy, self-awareness, and the ability to understand others.

University graduates, no matter how competent they may be, often lack basic communication and interpersonal skills, making them ill-equipped for the workplace. Most courses and assessments do not emphasise soft skills despite their obvious usefulness and necessity. Furthermore, the majority of soft skills cannot be assessed objectively using summative approaches. Soft skills are highly prized by employers as a sign of potential employees or job candidates. Learning soft skills helps students land jobs more easily and increase their prospects of future success.

The author, Elena Spirovska (2015), claims that: "Students would be expected to 'apply their language skills and knowledge when conversing through e-mail, presenting or listening to presentations.' To accomplish this, a large number of them are required to solve problems or come up with new ideas. Working in an English-speaking environment necessitates the development of LSRW skills while still in the undergraduate stage of engineering education. Additional training is needed in self-management and interpersonal skills.

Academic achievement is not solely based on students' ability to acquire knowledge. Soft skills are just as important as hard ones in order for students to compete and be self-sufficient or independent. This includes an understanding of learning and the ability to apply knowledge and skills to real-world issues. Developing students' soft skills is an essential part of preparing them for the challenges they will confront in the real world. The findings of the study led to the selection of seven soft skills for implementation across the nation's higher education institutions. Critical thinking and problem techniques, management skills, and professional skills are only a few among these skills.

Student-centered teaching practises are underutilised in higher education because academic success is prioritised over personality. Ethics, values, and professionalism are just a few of the soft skills that need to be instilled in children as early as possible (Mangala, 2010). Teaching professionals had to interact with students, teachers and administrators on a daily basis; whereas during their training, they communicated mostly with their peers and lecturers. It is true that good interpersonal communication and the capacity to work well with others go hand in hand. All programmes and curriculum should include soft skills as a separate topic. In order to improve one's soft talents, one needs to demonstrate them. It is not enough if only soft skills are included in the curriculum but it is to be made sure that they are given the attention they need by incorporating hands-on practise sessions. Many students and teachers alike see soft skills as a non-core subject, and as a result, they don't give it the importance it deserves. Soft skills are critical complements to "hard skills," which are job and many other activity-specific criteria. They are easily observable, quantifiable, and quantifiable. They include credentials, degrees, employment titles, technical expertise, and general technical and administrative procedures. They are the soft capabilities that enable institutions of higher education and the workforce to advance professional development and personal improvement. Additionally, they generate new opportunities and are motivated by factors other than financial gain.

Soft skills, it has been proposed, may be more valuable in the long run than vocational skills in a number of occupations. In legal, educational, and even technological professions, an individual's ability to interact with others effectively and graciously can determine his or her professional success, which can ultimately result in the success of an organisation or institution. As a result, employers increasingly value soft skills in addition to traditional (technical) certifications, and they may be vital in a variety of occupations.

People who have these skills can use them in a lot of different jobs. They are personal characteristics and attitudes that can help us collaborate more effectively and have a good effect on both the organisations for which we work and on ourselves. That is not to say that technical skills and knowledge are unimportant in today's job market. While technical skills can be acquired rapidly, they cannot be acquired as quickly as soft skills. This is because technical skills may be easily acquired. While educational degrees are necessary, developing soft skills is more about how to interact with others than it is about how to do things. These are critical components of a dynamic workforce and are always in high demand, making them critical. According to Wolfe and Johnson (1995),

> From their past jobs, responsibilities, life experiences and other things they've done in the past, everyone learns new skills. It is possible that we even have skills that we don't know about. When these skills are found, they can be added to our resume and make us a better candidate for the job. (p.178)

There are hard and soft skills. It is possible to learn hard skills. Hard skills are the skills that are required to do a job. Everyone who has the same level of education and experience should be able to do the same things with their hands, it's been thought. If one wants to be good at hard skills, he needs to know how to use a machine, computer programmes, safety rules, financial systems, technical analysis, and sales administration. Hard skills are easier to see, measure, and quantify than soft skills.

In this scenario, soft talents trump hard skills. They are necessary for an individual to be a decent person and perform successfully at work. Our social status in a society is determined by our social skills, which make it simpler to engage with other people. An individual's credentials and status will be meaningless in the absence of soft skills. Finally, people will always admire someone who is cheerful and friendly. If a person has demonstrable hard skills, h/she may be invited to an interview. According to mythology, your soft skills are what will get you hired. They suggest that firms would likely perform better if they hired individuals with excellent soft skills and then taught them how to hone their hard talents in their field of employment.

Because soft talents are intangible, they are frequently difficult to teach. Soft skills are a set of advantageous personal characteristics and skills that contribute to the enhancement of relationships, job performance, and market value. Soft skills are necessary if one wishes to listen effectively, communicate effectively, be positive, manage conflict effectively, collaborate effectively with everyone, accept healthy criticism, manage pressure, while possessing decent attitude and behaviour. They are more likely to be the result of individuals' communication styles, motivations, social relationships, and work ethic. The majority of employers use soft skills to separate candidates in a competitive job search. To differentiate oneself from the mass of job searchers, an individual must first establish his/her own soft skills and then communicate them to prospective employers.

The relative importance of each talent is defined by its vocation. Communication skills (particularly listening skills), corporate communication skills, problem-solving skills, and teamwork skills are all recognised as vital and difficult to perfect in the information technology services field (Conard, 2005). Communication, selling, promotion and distribution, and customer service are the primary skills required in organised retail. This holds true for the financial services business as well. The following may be considered a few such skills that are very important for a student to be acquired.

1. Effective Communication: This is the most frequently listed characteristic on person specifications for job openings; effective communicators readily capture their colleagues' attention. Additionally, subordinates listen to and comprehend orders and convey their message without becoming enraged or upset. Additionally, they can adapt their communication style as required by their employment from time to time. This can be extremely beneficial in a variety of scenarios, from resolving conflict to convincing a buyer of the merits of purchasing your goods. If a person possesses effective communication skills, he or she should be able to establish productive working connections with colleagues and benefit from constructive criticism. Active listening, presenting, and outstanding writing talents are all examples of communication skills.

2. Decision making: Effective decision making involves gathering critical facts, seeking advice, considering the bigger picture, exploring alternatives, and being aware of the consequences. By avoiding emotions like wrath and jealousy, individuals can conserve energy and make sound decisions.

3. Work commitment: Employers value employees who are dependable, practical, eager, and committed to their work. Committed employees require less supervision and motivation to perform at their best, ultimately leading to higher productivity.

4. Adaptability and flexibility: In rapidly changing work environments, being adaptable and flexible is crucial. Employers appreciate individuals who are willing to step out of their comfort zones, try new things, and embrace change. Having an optimistic and can-do mentality is highly valued.

5. Time management: Effective time management involves prioritizing critical tasks, determining the most productive actions, and meeting deadlines. Prioritizing and allocating time efficiently helps ensure tasks are completed on time and with maximum output.

6. Leadership skills: Good leaders motivate and inspire others, continuously strive for personal growth, and maintain a positive mindset. They understand when to take initiative and when to follow instructions, balancing independence and collaboration.

7. Team player: Being a good team player means focusing on the team's objectives, collaborating with others, providing constructive feedback, and actively listening. Working well with others and maintaining open communication is essential for success.

8. Self-confidence: Employers value individuals who take pride in their work and have the confidence to present their results. The ability to maintain composure and confidence in the face of adversity is also highly regarded.

9. Performing under pressure: Some individuals thrive under pressure, exceeding expectations and performing exceptionally well. Being able to maintain performance and deliver results in high-pressure situations is a valuable skill.

10. Computer and technological literacy: Proficiency in computer software and technological skills is becoming increasingly important across various industries. Demonstrating competence in these areas can enhance your career prospects and advancement opportunities.

11. Interpersonal skills: Working effectively in teams, resolving conflicts, and having good communication skills are highly valuable in the workplace. The ability to collaborate and interact positively with others is crucial for career advancement.

12. Project management skills: Efficiently organizing, planning, implementing, and evaluating projects and tasks is a valuable skillset. Even if not in a formal project management role, possessing these skills can contribute to successful project outcomes.

13. Problem-solving skills: The ability to creatively and effectively address challenges using reasoning, experience, information, and available resources is highly sought after. Problem-solving skills save time and contribute to organizational success.

14. Emotional intelligence: Demonstrating emotional intelligence, which involves understanding and managing emotions in oneself and others, is highly valued. It can be conveyed through strategic resume writing and showcasing relevant experiences.

15. Highlighting soft skills: During interviews, provide concrete examples of how you have utilized your soft skills in professional or personal contexts. Sharing experiences of time management, working under pressure, innovation, and effective communication can make your soft skills stand out.

16. Networking skills: Building a strong network can facilitate various tasks such as finding job opportunities, seeking advice, connecting with business partners, and locating customers. Networking enhances career opportunities and professional growth.

Developing and improving these skills can greatly enhance your professional prospects and contribute to your overall success in the workplace.

Self-Management Skills are concerned with how an individual views himself and others, how h/she regulates his\her emotions, and how h/she responds to bad situations (Bernd, 2008). They enable us to achieve success in both our professional and personal lives. These skills include a growth mindset, self-awareness, emotion control, self-confidence, stress management, resilience, forgiving and forgetting skills, persistence and perseverance, patience and perceptiveness, and persistence and perseverance. Expertise in a particular field of expertise is no longer adequate. The competition is strong, and the soft skills will set the person apart. Businesses are increasingly recognising that in order to maintain a competitive edge, they must guarantee that their staff understand proper workplace behaviour and how to deal with customers and peers. Soft skills are necessary in all occupations, not just those that involve direct consumer interaction. They are, however, critical for all employees in a business. Consider the leaders in your profession and ask yourself, "Did they rise to the top due to their hard skills or their soft skills?" When the salary structure and categorization of employees are specified, it is instantly apparent that people at the top of the pay scale excel at soft skills.

1. Conclusion

Soft skills have a significant impact on the correctness and fluency of the UG learners' English language skills at the engineering as well as other levels. This clearly demonstrates the link between soft skills and LSRW English language skills. People, businesses, and society all depend on the right mix of skills. People with higher levels of education and experience are more likely to have higher incomes and better physical and mental well-being. Employees who have advanced degrees have a higher rate of production and innovation than those without. Human capital and economic growth are clearly linked in economics.

Current employees are expected to meet these qualifications. Personal attributes and "soft skills" are just as important as hard skills in obtaining success. According to the World Economic Forum, several of the 'crucial proficiencies in the 21st century' are non-technical. Employability and entrepreneurship are only two of the many words for soft skills that may be used in a multitude of sectors and professions. In addition to teaching communication and teamwork, these courses also cover topics such as problem solving and emotional judgement, as well as professional ethics and global citizenship.

It is thus crucial to understand the critical role soft skills play in your team and to aim to enhance them not just within your organisation, but throughout. Numerous characteristics, including personal accountability, degree of collaboration, interpersonal negotiation skills, conflict management, adaptability, flexibility, communication clarity, inventive thinking, and individual inclusion, must be discovered in order to do this. All of these factors have an effect on how an individual interacts with clients, customers, employees, supervisors, and other stakeholders. The more optimistic an individual is, the more successful the connections will be. This results in exceptional team performance and motivates individuals to make major contributions to the organization's vision and strategy.

REFERENCES

1. Ackerman, P. And Heggestad, E. (1997), "Intelligence, Personality and Interests: Evidence for Overlapping Traits", *Psychological Bulletin*, 121, 219–245.
2. Almlund, M. Et al. (2011), "Personality Psychology and Economics". *IZA Discussion Paper* (No. 5500). http://ftp.iza.org/dp5500.pdf.
3. Barrick, M. and Mount, M. (1991), "The Big Five personality dimensions and job Performance: A Meta-Analysis", *Personnel Psychology*, 44(1), 1–26.
4. Backlund, G. and Sjunnesson, J. (2012), Training Young Engineers to See, *AI & Society*, 27(4), pp. 509–515.
5. Borrego, M. and Bernhard, J., 2011, The emergence of engineering education research as a globally connected field of inquiry., Journal of Engineering Education, 100. 1, pp. 14–47. http://dx.doi.org/10.1002/j.2168-9830.2011.tb00003.x
6. Bernd S., (2008), "The Importance of Soft Skills: Education Beyond Academic Knowledge", *Journal of Language and Communication*, Vol. 2(1), pp. 146–154.
7. Conard, M. A. (2005), "Aptitude is Not Enough: How Personality and Behavior Predict Academic Performance", *Journal of Research in Personality*, Vol. 40, pp. 339–346.
8. Tevdovska, Elena Spirovska (2015). Integrating soft skills in higher education and the EFL classroom: Knowledge beyond language learning. Seeu Review 11 (2):95–106.
9. Mangala Rani. (2010), "Need and Importance of Soft Skills in Students", *Open Journal Academic System*, Vol. 2, Jan-June, pp. 1–6.
10. Wolfe, R. and Johnson, S. (1995), "Personality as a Predictor of College Performance. *Educational and Psychological Measurement*", Vol. 55(2), pp. 177–185.

Impending Inquisitions in Humanities and Sciences (ICIIHS-2022) – Dr. Mohan Varkolu et al. (eds)

Gender Difference in Emotional Choices: A Study on Willa Cather's "A Wagner Matinee" and O. Henry's "A Retrieved Reformation"

16

Jeeva M.*
Assistant Professor, Kristu Jayanti College, Bengaluru, India,
V. Vinesh Raj[1]
Verbal Trainer, 6th Sense Training Academy, Tiruchirappalli, India
S. Kalaivani[2]
Assistant Professor, School of Distance Eduaction, Coimbatore, India
Bhavani Sushma Garlapati[3]
Assistant Professor, Department of English,
Koneru Lakashimiah Education Foundation, Hyderabad

ABSTRACT This paper tries to explore the emotions and choices of different gender (male and female) in decision-making for their future. Choices decide destiny and it differs from person to person based on their culture, gender, and background. But the emotional part is very peculiar while making decision in life. Daniel Goleman's *Emotional Intelligence* theory helps to explore the literary characters and their decisions in life, based on their culture and gender. This paper particularizes the gender difference in emotional choice in the personal life of individuals. For the study, two different writers and their works from American Literature are included. They are Willa Cather's "A Wagner Matinee" and O. Henry's "A Retrieved Reformation". These two stories deal with the concept of marriage, but the end is different from each other. In the above stories, the theory of Emotional Intelligence is applied to understand the reason for their differing ends.

KEYWORDS Gender, Marriage, Choices, Awareness of self, Emotional intelligence

1. Review of Literature

A study of the literature on emotions and EI reveals that the gender differs significantly in areas of the emotional world. The feminine gender, in particular, has long been associated with the emotional dimension of human beings to a larger extent than the male gender, which experiences positive and negative emotions more intensely than the male gender (Grossman & Wood, 1993).

This "feminist vision of emotions" has been explained using both biological and social elements (Nolen-Hoeksema & Jackson, 2001). According to the biological explanation, women's biochemistry is better suited to regard one's own and other people's emotions as essential components in survival. Women may have greater emotional processing brain regions than men in some cases. (Baron-Cohen, 2002; Gunning-Dixon, Bilker, & Gur, 2002).

*Corresponding author: jeevamlit@gmail.com
[1]vineshraj89@gmail.com, [2]kalaivani.lit@gmail.com, [3]bhavani.sushma@klh.edu.in

DOI: 10.1201/9781003489436-16

In addition, to prevent interpersonal relationships from degrading and to create fulfilling social networks, women spend more time socially interacting with the emotional world and are more concerned with preserving the good tone of their own and others' emotions. (Nolen-Hoeksema & Jackson, 2001).

"Acts of Love (and Work): Gender Imbalance in Emotional Work and Women's Psychological Distress," by Strazdins, Lyndall, and Dorothy H. Broom, was published in the journal "Acts of Love (and Work): Gender Imbalance in Emotional Work and Women's Psychological Distress," and it discussed gender imbalance. The disparity between the sexes put women's health at risk and contributed to the understanding of gender differences in psychological distress brought on by marriage failure.

Lisa Eklund (2013) described the concept of ecology and partner in his work "Marriage squeeze and mate selection: The ecology of choice and implications for social policy in China." In order to demonstrate how many social processes and practises affect who can marry, with whom they can marry, and under what circumstances, the concept of the ecology of choice in mate selection is used in this work.

Emotional intelligence is defined as the ability to effectively identify, appraise, express, and monitor one's own and others' feelings and emotions, as well as the ability to use such information to guide one's thinking and behavior. The study examines Daniel Goleman's idea of emotional intelligence to show how emotional choices impact gender-based disparities in character and behaviour. Emotional intelligence has an impact on emotional decisions. It illustrates the obvious differences between how men and women make emotional decisions in their lives.

Emotional intelligence is defined as the capacity to recognize, control, and evaluates emotions. In his Emotional Intelligence theory, Danial Goleman defined five dimensions to understand people's emotional intelligence. The fundamental notion is self-awareness, which serves as the foundation for all other domains. Self-awareness is considered the mother of all life skills. It includes emotional awareness, accurate self-evaluation, and self-assurance. All three principles are important for a better understanding of oneself and can be used together.

The second step in achieving emotional intelligence is self-regulation. Self-control, trustworthiness, conscientiousness, adaptability, and originality are some of the levels used to assess emotion management. It is not necessary to understand all of the key concepts to understand self-regulation. Three or four notions are needed to recognize that the person is struggling with self-control.

Motivation is a term that is well-known and expected in the outside world. That can be defined as an external factor, as well as our goal and objective. The notion is linked to motivation, commitment, initiative, and optimism. This is where society can gain an understanding of oneself. Personal competency encompasses all three of these factors.

Empathy and social skills are both included in the category of social abilities. Empathy is the ability to relate the outside world to one's self. It is also known as empathy, development, and service orientation. Emotion is the idea of balancing one's feelings with those of others. Influence, communication, leadership, collaboration, and cooperation are some of the primary phrases used to describe social abilities.

Gender plays a significant impact in making emotional decisions. "A Wagner Matinee" and "A Retrieved Reformation" are the titles of the two stories that follow. Marriage is a central subject in both books, and it has the power to change one's life forever. However, the options differ depending on gender. Two women of different genders make life-changing decisions in different ways that lead to distinct outcomes. In the first story, "A Wagner Matinee", a woman named Aunt Georgiana makes an emotional decision to marry Howard Carpenter, who is unemployed and without hope. The writer correctly observed, "Inexplicable infatuation was that she eloped with him, that their love is like infatuation" (02).

"A Wagner Matinee" is a story written by Willa Sibert Cather. The story revolves around a character called Georgiana. The story opens with the letter received by Mr. Clark, the nephew of Aunt Georgiana. The letter tells him that his aunt is coming to Boston. When he reads the letter, he begins to recollect the experience that he had with his aunt. Clark remembers the past and tells the readers that Aunt Georgiana was a music teacher in Boston before she got married. Mr. Howard carpenter falls in love with Georgiana. They both elope and get settled in Nebraska Frontier. Life in Nebraska Frontier was all the more difficult. Georgiana starts working on the fields, taking care of the children, cattle, and doing the household things. For thirty years, she has accustomed herself to this kind of life in Nebraska Frontier. One day it has become necessary for Georgiana to visit Boston after thirty years. Clark, knowing the past of Georgiana, has arranged a surprise matinee show for her aunt. Aunt Georgiana is not able to accept the fact that she is in Boston. Clark taker her to the symphony orchestra. In the beginning, Georgiana feels ashamed of her attire in the crowd. As there is progress in the orchestra, Aunt Georgiana begins to recollect her past being a music teacher in Boston. At the end of the orchestra, she bursts into tears. The story ends with the sad note that she does not want to go back

"A Retrieved Reformation" is a story written by O. Henry. At the beginning of the story, a character called Jimmy Valentine is imprisoned because of robbing a bank in Springfield. Jimmy Valentine is a professional safecracker. Having spent ten months in the prison, he now walks out of prison. He meets Mike Dolan, who helped him to come out of prison. He then goes to his room and checks whether his burglary tools are safe. After that, he resumes his robbing business in different parts of the city. The series of burglaries triggers the police investigation. Ben Price, the detective investigates this case and finds out that Jimmy Valentine is the one who has robbed all these banks. Therefore, Ben Price starts to look for him. On the other hand, Jimmy Valentine goes to Elmore to rob a bank. Accidently, he meets a girl called Annabel Adams with whom he falls in love. He changes his name to Ralph Spencer in order to hide his previous identity.

Later, he starts up the shoe business and becomes rich. He also wins the love of Annabel Adams. They are about to get married in two weeks. Jimmy Valentine decides to quit his robbing business and so he writes a letter to his friends. He wants to give his burglary tools to his friends as he is starting a new life in weeks. Mr. Adams, the father of Annabel Adams wants to show everyone about the vault that has been newly installed in Elmore Bank. Ralph Spencer on his way to give the tools to his friends, is also invited for the inspection. Annable's married sister also comes to the spot with her two children May and Agatha. Mr. Adams takes pride in explaining the different features of the new safe. After some time, May playfully closes the door of the safe with Agatha inside but she is not able to open it. Mr. Adams understands that time lock has been enabled. Thus, saving Agatha is an impossible task. Annable Adams implores the help of Ralph Spencer. Ralph without any hesitance saves Agatha with the help of tools that he is supposed to deliver to his friends. Ben Price who has also come to Elmore watches Jimmy Valentine opens the safe. At the end of the story, Ben Price acts rather strange that he does not know Jimmy Valentine.

In the story "A Retrieved Reformation", Jimmy Valentine is very well aware of his internal states and intuitions. Self-awareness also includes emotional awareness, accurate self-assessment, and self-confidence. Jimmy Valentine, when arrested for the Springfield bank robbery case, he is very confident that he will be released soon from prison. He believes in his self-worth and capabilities.

> It was a complete set, made of specially tempered steel, the latest designs in drills, punches, braces and bits, jimmies, clamps, and augers, with two or three novelties, invented by Jimmy himself, in which he took pride (140).

This statement gives a clear picture that Jimmy Valentine believes in himself in his potential and strength. It is self–confidence that drives him to rob a series of banks in the city even after his release from prison. When he goes to Elmore to rob a bank, he falls in love with Annable Adams. He recognizes his own emotions and effects. He also decides to changes his name and to quit his old business. Apart from robbing a bank, he has learned to stitch shoes in prison. Therefore, he decides to open up a shoe store.

In the story, A Wagner Matinee, Georgiana recognizes her emotions when she meets Howard. But, she fails to recognize the effects of her emotions. Georgiana, being a music school teacher elopes with Howard. She does not think about the consequences in the future. Georgiana suffers in Nebraska Frontier leaving behind the comfortable life in Boston. They both suffer a lot.

> They built a dugout in the red hillside, one of those that cave-dwellings inmates usually reverted to the conditions of primitive savagery. The water they got from the lagoons where the buffalo drank, and their slender stock of provisions was always at the mercy of bands of roving Indians (5).

The second component, self-Regulation is about managing one's own internal states, impulses, and resources. Self-Regulation also includes self-control, trustworthiness, conscientiousness Adaptability, and innovation. Jimmy Valentine in "A Retrieved Reformation", regulates his internal states when he sees Annable Adams. Jimmy Valentine changes his name to Ralph Spencer. He can adapt himself to a new place, Elmore. His flexibility in handling things is very remarkable. Being a stranger to Elmore, he enquires about the business. He said he had come to Elmore to look for a location to go into business, "How was the shoe business, now, in the town? He had thought of the shoe business. Was there an opening?" (142). Jimmy can think about his strengths and what he knows apart from robbing. He is also comfortable with new ideas and approaches as he comes to Elmore.

Georgiana in "A Wagner Matinee" is also adapting to the new place, Nebraska Frontier. The transition from city life to village life is not easy. But, Georgiana being a stranger to all kinds of fieldwork, adapts herself to this new kind of life. On the contrary, she is not able to regulate her internal states and impulses. At one point she says, "Don't love it so well, Clark, or it may be taken from you. Oh! Dear boy, pray that whatever your sacrifice be it is not that" (6). This statement gives a thought that she is into an internal conflict that sparks her past life. She regrets the fact that she is not able to be play music at present. The music also connects the way she lived in Boston.

The third component, motivation is about the way of achieving the goal. Jimmy Valentine in "A Retrieved Reformation", finds the purpose of his life when he comes to Elmore. Until he meets Annable Adams, his purpose remains in robbing banks. It is Annable Adams who drives him to achieve something in his life. Jimmy's commitment to winning the love of Annable Adams is interesting. He takes an initiative to changes his name to Ralph Spencer and he knows the challenges that he has to face while setting up the shoe store in Elmore. After a few days, his business flourishes like a palm tree. He has earned the respect of the community and has won the love of Annable Adams. "I want to make you a present of my kit of tools… say, Billy, I've quit the old business a year ago…I am going to get married in two weeks… I tell you, Billy, she is an angel (144)". Surrounded by positive vibes, he decided to quit his old ways. He explains in a letter that he is about to get married and he wants to change his way of living. He also acknowledges that Annable Adams is the reason behind all this sudden change. On the other hand, in the story of "A Wagner Matinee", Georgiana is not able to reach her goal. It is a drastic change for her to move from Boston to Nebraska Frontier. But, in terms of achieving her goal, this is a complete deterioration. She is completely blinded about her future when they decide to elope together. There was no motivating factor in her life. She does all other household work because she has made her choice. Life in Nebraska is monotonous for her.

The fourth component is empathy. It is being aware of others' feelings, needs, and concerns. Jimmy Valentine in "A Retrieved Reformation", understand the need of others. Though he was a safecracker before he met Annable Adams, he is also filled with empathy. Toward the end of the story, May, a little girl playfully shuts the safe door with Agatha inside it. Mr. Adams knows that it can never be opened because the time lock has been enabled. The mother of Agatha cries for help. That is when Annable Adams asks Ralph Spencer to save her. Ralph Spencer knows for sure the effects of saving her. He is also carrying a suitcase which is full of burglary tools. He is about to give it to his friends. He is aware that if he saves the child, he might have to lose Annable Adams and the people around him will name him a professional safecracker. But, despite all these, he goes to the safe and carefully opens it, and saves Agatha. He truly understands the feelings of others. On the other hand, Georgiana, being immersed in her internal conflict has failed to show empathy.

The fifth component is social skills which are proficient at inducing desirable responses in others. Change catalyst is one of the components of social skills. Jimmy Valentine initiates the change after coming to Elmore. From the very beginning, he has a good taste for society. The following sentences speak about Jimmy.

> Ben Price knew Jimmy's habits. He had learned them while working on the Springfield case. Long jumps, quick get-always, no confederate and a taste for a good society—these ways had helped Mr. Jimmy Valentine to become noted as a successful dodger of retribution. (141)

Even after going to Elmore, without hesitance, he approaches a boy who was loafing on the steps of the Elmore Bank. He learns more about Annable Adams from that boy. In Planter's hotel, he enquires about the business that he can start. Jimmy once committed to Annable Adams, he even goes to the extreme of doing anything for her. Georgiana, in "A Wagner Matinee", fails to handle relationships. Her internal conflict is reflected when she comes to Boston. She feels timid to venture into the world that was shut for thirty years. While attending the orchestra along with Clark, she tends to remember her past. She is also becoming aware of her surroundings. "From the time we entered the concert hall, however, she was a trifle less passive and inert and seemed to begin to perceive her surroundings" (7).

Dr. Peter Sifneos, psychiatrist coined the term 'alexithymia', which means the result of emotional intelligence imbalance. He rightly defines the problem, "They give the impression of being different, alien beings, having come from an entirely different world, living in the midst of a society which is dominated by feelings" (75). Here, Aunt Georgiana, feels left out from the luxurious world in Boston. She feels inferior in the music hall and feels not belonging to the crowd. This is the place, where readers can find and understand the emotional intelligence imbalance of Aunt Georgiana. She burst out into tears by saying, "I don't want to go, Clark, I don't want to go!" (12).

Emotional intelligence refers to the ability to perceive, control, and evaluate emotions. Jimmy Valentine in "A Retrieved Reformation" can channel his emotions and make rational decisions. He perceives his emotion of love for Annable Adams and takes a rational decision of reforming his life by getting involved in the shoe business. On the other hand, in "A Wagner Matinee", Georgiana submits to her emotion of infatuation and take impulsive decisions.

Daniel, in his theory Emotional Intelligence, discusses two different mental state to become competent in both personal and professional life. They are rational state of mind and emotional state of mind. In professional life, all the people are able to balance both mental states to achieve in their career. But, in personal life, some characters failed to balance the both, as a result, could not take right decision at the right time. In the above discussed two characters, are best example to understand the emotional intelligence imbalance and the consequences? Jimmy Valentine is the character, who balances his emotional intelligence in his personal life, which leads him to become successful business man. In the other hand, Aunt Georgiana, who fails to balance her rational mental state and emotional mental state? Her emotions dominate the decision to elope with Howard.

REFERENCES

1. Grossman, M., & Wood, W. (1993) Sex Differences in Intensity of Emotional Experience: A Social Role Interpretation. Journal of Personality and Social Psychology, 65, 1010–1022.
2. Gur, R.C., Gunning-Dixon, F., Bilker, W., & Gur, R.E. (2002). Sex Differences in Temporo-limbic and Frontal Brain Volumes of Healthy Adults. Cerebral Cortex, 12, 998–103.
3. Nolen-Hoeksema, S., & Jackson, B. (2001). Mediators of the Gender Difference in Rumination. Psychology of Women Quarterly, 25, 37–47.
4. Nolen-Hoeksema, S., Larson, J., & Grayson, C. (1999). Explaining the Gender Difference in Depressive Symptoms. Journal of Personality and Social Psychology, 77, 1061–1072.
5. Nolen-Hoeksema, S., Wisco, B., & Lyubomirsky, S. (2008). Rethinking Rumination. Perspectives on Psychological Science, 3, 400–424.
6. Sánchez Núñez, M. T., Fernández-Berrocal, P., Montañés, J., & Latorre, J. M. (2008). Does Emotional Intelligence Depend on Gender? The Socialization of Emotional Competencies in Men and Women and its Implications. Electronic Journal of Research in Educational Psychology, 15, 455–474.
7. Rivers, S. E., Brackett, M. A., Salovey, P., & Mayer, J. D. (2007). Measuring emotional intelligence as a set of mental abilities. In G. Matthews, M. Zeidner, & R. D. Roberts (Eds.), *The science of emotional intelligence: Knowns and unknowns* (pp. 230–257). Oxford University Press.
8. Mayer, J.D., Salovey, P., & Caruso, D. (1998). Competing models of emotional intelligence. In R. J. Sternberg (Ed.), Handbook of human intelligence (2nd Ed.). New York: Cambridge University Press.
9. Bhalla, S. & Nauriyal, D.K. (2004). Emotional intelligence: The emerging paradigm in personal dynamics. Psychological Studies, 49(2–3), 97–106.
10. Mayer, J. D., Roberts, R. D., & Barsade, S. G. (2008). Human abilities: Emotional intelligence. Annual Review of Psychology, 59, 507–536.
11. Mayer, J. D., & Salovey, P. (1997). What is emotional intelligence? In P. Salovey, & D. J. Sluyter (Eds.), Emotional development and emotional intelligence: Educational implications (pp. 3–34). New York: Basic Books.
12. Salovey, P., Mayer, J. D., Caruso, D., & Yoo, S. H. (2009). The positive psychology of emotional intelligence. In S. J. Lopez & C. R. Snyder (Eds.), *Oxford handbook of positive psychology* (pp. 237–248). Oxford University Press.
13. Rivers, S. E., Brackett, M. A., Salovey, P., & Mayer, J. D. (2007). Measuring emotional intelligence as a set of mental abilities. In G. Matthews, M. Zeidner, & R. D. Roberts (Eds.), *The science of emotional intelligence: Knowns and unknowns* (pp. 230–257). Oxford University Press.
14. https://www.penguinrandomhouse.com/books/69105/emotional-intelligence-by-daniel-goleman/9780553804911/readers-guide/
15. https://static1.squarespace.com/static/4ff4905c84aee104c1f4f2c2/t/5084d976e4b0f4598aaa459f/1350883702078/Salovey+Pizarro+2003+%5Bchapter%5D.pdf

Impending Inquisitions in Humanities and Sciences (ICIIHS-2022) – Dr. Mohan Varkolu et al. (eds)

Mother Tongue Influence on English Language Learning

17

Mallesham D., Boddu Chandrashekar*
Department of English, Koneru Lakshmaiah Education Foundation,
Hyderabad, Telangana, India

ABSTRACT Indian languages differ from English in many aspects, be it morphology, syntax, grammar or pronunciation. These differences result in the impact of First Language (L1) on Second Language (L2), which in turn pushes the learners to make mistakes in using the second language fluently. Though these differences are not similar for all, there are multiple reasons associated with Mother Tongue Influence (MTI) on English language learning. This paper attempts to discuss the role of languages and their impact on each other by explicitly giving examples of language in use. It further presents a clear analysis of the reasons mentioned for MTI. This paper also highlights on the problems of MTI among Indian speakers, especially Telugu Speakers and talks about the effects of MTI. The paper also presents a few possible solutions and ways of overcoming MTI and attempts to categorize the differences in L1 and L2 by giving examples in aspects of phonological, syntactic, semantic, morphological, lexical, grammatical, contextual code switching, usage to literal translational differences.

KEYWORDS Mother tongue influence, First language, Second language, Pronunciation, Telugu speakers, Syntax, Morphology

1. Introduction

In India, many people are bilinguals. They use their mother tongue to communicate in their everyday life and also learn a second language for various purposes. However, the influence of L1 on L2 cannot be ignored. It affects how we process and use words in our second language, often resulting in errors or mistakes. In this article, we will explore how MTI affects the learning of a second language in India and discuss ways to overcome it.

According to Manjula. R (2022), "Influence of mother tongue means interference of child's first language in learning the pronunciation of otherlanguage" (Manjula, 2022, p.66). Mother tongue influence (MTI) is an important factor in language learning, especially in India where many people grow up speaking more than one language. The impact of MTI can be seen in the way people speak, write and understand their second language. It affects the pronunciation, grammar, syntax and even vocabulary of a person's second language. This has a direct effect on how people communicate with each other and how they express their thoughts and feelings. MTI has both positive and negative implications for those learning a second language, but it are important to recognize its importance in order to make the most out of it. Avery & Ehrlich (1992) also stated that the sound features of the learners' first language are transferred into the target language that results in mispronunciation. The mispronunciation of words by non-native speakers reflects the rules, stress, intonation and sound influence of their native language. Mother tongue influence (MTI) is the impact of one's mother tongue on the second language; for example, English.

*b.chandrashekar@klh.edu.in

DOI: 10.1201/9781003489436-17

The ethnicity/nationality of a person with the way he/she uses the dialects that interferes with English speaking. This is very common in India.

2. Role of Languages

Language is an important part of identity, and it is no surprise that the way people speak reflects their ethnicity or nationality. The way a person speaks English is often influenced by their ethnicity or nationality. This is especially true in India, where different dialects and languages are spoken by different ethnic groups. For example, an Indian person may use words or phrases from their native language when speaking English. This can make it difficult for people who are not familiar with the language to understand what is being said. Furthermore, this phenomenon can lead to misunderstandings and miscommunication between people of different ethnicities or nationalities in India. As a result, many people in India have difficulty speaking English fluently due to interference from their native language. This can create communication barriers between people from different backgrounds and make it difficult for them to understand each other. It also makes it difficult for non-native English speakers to communicate effectively with native English speakers.

L1 is the language we grew up speaking. This means the language we learn as we start speaking: the language we speak since we were a child. There is a well established notion that in India, language changes at every ten miles. This makes it difficult for us to speak in one standard accent as all of us have our own way of speaking a second language such as English. This is what is called Mother Tongue Influence- where the effect of mother tongue on English becomes evident.

English has become the language of choice for most Indians; however, it is not always spoken in a standard accent. This is known as MTI (Mother Tongue Influence) and it can make it difficult for many to communicate effectively. This can be seen in the way people pronounce words, use slang or local dialects and even how they structure sentences. As a result of this, there is a need to understand the nuances of English language as used by Indians.Most Indians find it difficult to speak in a standard accent as each of us has our own way of speaking English. MTI can be seen in the way we pronounce words, use sentence structure and even grammar. As a result, many Indians struggle to understand and be understood by non-native speakers of English.

The difference in pronunciation and usage can be accounted to the difference between their native language and English, which can make it difficult for them to accurately pronounce words in English. Fortunately, there are ways to improve pronunciation for such individuals and help them become more confident in their ability to communicate in English.

English is one of the most widely spoken languages in the world, and many South Indians struggle with its pronunciation. This is especially true for certain words that are commonly mispronounced, such as 'temple' as '*temble*' and 'Ramesh' as '*Ramess*'. The problem of mispronunciation can be attributed to a lack of exposure to English language, accent differences between English and native languages, and even cultural differences. Fortunately, there are several ways to help Indian speakers improve their pronunciation of English words.

How does MTI Develop?

It is quite common to hear people using words from their native language while speaking in English. This phenomenon has been studied extensively, and the evidence suggests that mother tongue influences the way Indian speakers use English, both linguistically and pragmatically.

English is a language with a rich history, and it has been influenced by many different cultures over the years. One of the most notable influences on English is the influence of mother tongue on English speakers in India. This influence has had an undeniable impact on how Indian English is spoken today. From subtle changes to major differences, the evidence of mother tongue influence on English is very obvious in Indian speakers.

The evidence of mother tongue influence on Indian speakers of English is very obvious. This manifests in the form of incorrect pronunciation word choice and grammar. Pronunciation error may be due to many issues. Many people from different natives cannot pronounce many words correctly. Some of the words that most Indians find difficulty in pronouncing are: 'measure', 'pleasure', 'treasure', 'support', 'college', 'bus', 'school', 'zero', 'pleasure', 'treasure', 'smart'. The words 'measure', 'pleasure', 'treasure' are pronounced as *'mezur', 'plezur', 'trezur', and* 'college' *becomes 'collige' and* 'zero' *becomes 'jiro' or 'jeero'.*

MTI among the Telugu Speakers

These sections present a brief discussion of a few reasons for MTI among theTelugu speakers:

There have been distinct errors that can be categorized based on the type of MTI among the speakers of English form the southern part of India. First language influences can range from phonological, syntactic, semantic, morphological, lexical, grammatical, contextual code switching, usage to literal translation.

Moreover, with regard to the mother tongue influence on second or foreign language, an example of a morphological influence could be suffixing sounds such as 'aa' or 'uu' at the end of certain words. For example, words such as seen, run, fire, car, girl etc when pronounced become seenu, runu, fireu, caru, girlu and words such as want, hint, flight, hint, sad become wanta, hinta, flighta, hinta where an indication of a question mark is inserted along with the word. Also, sounds such as /s/ /z/ and /j/ are mispronounced. For example, 'zoo' becomes 'joo' and words starting with sounds such as /e/ or /a/ are precede with a /y/ sound and so 'every' becomes 'yevery', 'easy' becomes 'yeasy' and 'elbow' becomes 'yelbow'. Apart from this, there are sounds like /r/ and /l/ which are often pronounced more than required. For example, 'worst' is pronounced as 'warrust', and the sound /l/ in 'calmer' is pronounced even when it is silent.

Similarly, MTI can be clearly seen in the usage of words like 'small', 'big', 'light', 'first' and so on which become 'small-small', 'big-big', 'light-light', 'first-first' which replicate the mother tongue usage of words such as 'china-china' or 'pedha-pedha' (as used in Telugu).

Other errors that come under lexical errors such as "My professor was very serious today (to imply strict or formal)" or code switching such as 'keep leave' (leave pedtha) for 'apply leave', 'over-action' to imply hyper-activeness or exaggeration, 'mind absent' to imply 'absent mindedness'

Moreover, MTI influences can be clearly seen in how grammatical errors are carried due to the literal translation of sentences. Examples like saying "why because"(endhuku ante), what you are looking for is near me' (nadeggiraundhi) instead of 'what you are looking for is with me' (natho)", 'come at 9, 10 like that' (9 10 ki atlara), to sentences such as 'frustrated on me' (namedha) instead of 'frustrated with me' (natho), and finally, I will say you later instead of I will tell you later (Transitive/ Intransitive verb errors). Further, errors such as "angry on me' (namedhakopama) for 'angry with me', 'He went without telling anyone' (evarikicheppakundavelladu) to imply 'he went without informing anyone', and 'she made the baby to sleep' for she put the baby to sleep'.

The above errors though present a sample of MTI on the usage of English and mother tongue influence on second or foreign language can be seen among the speakers of Telugu. MTI can be attributed to various reasons such as guesswork or vagueness of the correct form of a word or sentence, a general ineptness of the language could be the reason of mispronunciation, vernacular (for example, Telugu) medium schools, or children being introduced to L2 (English) at a later stage in childhood.

The following section presents a clear analysis of the reasons mentioned above.

1. Guesswork or vagueness of the correct form of a word or sentence:

MTI can be seen in both written and spoken forms, with incorrect spellings, wrong usage of words, incorrect grammar, etc. similar to the ones presented above. The underlying cause of this problem is the lack of understanding of English grammar rules and how they should be applied in different contexts. As a result, Indians often find themselves making mistakes when writing or speaking in English.Many Indians struggle with MTI when speaking and writing in English due to the guesswork or vagueness associated with the correct form of a word or sentence. This is because of the lack of understanding and knowledge about the nuances of English grammar. As a result, many Indians are unable to express themselves correctly, leading to confusion and misunderstandings.

2. A general ineptness of the language could be the reason of mispronunciation:

Language is an important part of communication and the ability to communicate effectively can be hindered by mispronunciation. Mispronunciation can be caused by a general ineptness of the language, which could lead to mistakes in pronunciation or even MTI. This is especially true for those who are not native speakers of the language they are trying to learn. In order to overcome this problem, it is important to understand the basics of the language and learn how to pronounce words correctly. This can be done through practice, studying grammar rules and also listening closely to native speakers. However, many people find it hard to pronounce words correctly due to their lack of knowledge or understanding of the language. This could be attributed to a general ineptness in the language, which can lead to mispronunciation and MTI.

This is where technological assistance can come into play. By using different tools, we can easily identify mispronunciations and mistakes in interactions. This way, they can help correct these errors and ensure that their content is accurate and free from any errors. Additionally, AI writing tools can also suggest alternative words or phrases that may better suit a particular context.

3. Vernacular medium schools:

Majority of Indians suffer from Mother Tongue Influence (MTI) due to the large number of Vernacular medium schools in India. Vernacular medium schools are those that teach in a language other than English, such as Hindi, Bengali, Marathi etc. This has resulted in a huge gap between the English-speaking and non-English speaking population of India. MTI can be seen in the way people speak and write English, which often results in miscommunication and misunderstanding.

Vernacular medium schools are a common sight in India, with a majority of students attending them. This has led to MTI and is a major problem faced by many Indians, especially those from rural areas, where vernacular medium schools are more common.

4. Introducing English (L2) to Children at a later stage:

It is essential for children to be introduced to the language at an early age. Unfortunately, in India, English is often introduced at a later stage in childhood, leading to Mother Tongue Influence and low confidence when speaking and writing in English. This is especially true for those who come from non-English speaking backgrounds which can lead to difficulties in school, work, and other areas of life. As such, it is important to ensure that children are given the right support and guidance when learning English so that they can develop their confidence and proficiency in the language.

Indian Speakers and Problems of MTI

Indian speakers often struggle with the correct pronunciation of English words, as they are more accustomed to speaking their native language. Additionally, they tend to use words and phrases that are more common in their native language than in English. As a result, they end up speaking neither a hybrid language which is neither English nor their native language. This can lead to confusion among other English speakers as well as cause difficulty in understanding each other's messages.

Indian English is a unique variety of English that has its own set of pronunciation rules and conventions. This can make it difficult for Indian speakers to pronounce words correctly in the English language. Pronunciation errors are common among Indian speakers, and this can be attributed to a number of different issues. For instance, some may have difficulty with the phonetic alphabet or have trouble distinguishing between similar sounding words. These pronunciation errors may be due to many issues, such as lack of exposure to native speakers, incorrect pronunciation of certain sounds, or difficulty with the phonetic spelling system. All these factors can lead to an Indian speaker making mistakes when pronouncing words in English.

Additionally, cultural differences in pronunciation can also lead to confusion and mispronunciation. Furthermore, lack of exposure to the language being spoken can also lead to errors in pronunciation. Ultimately, it is important for Indian speakers to recognize these issues and take steps towards improving their pronunciation skills.

3. Observations

Many studies show that children, who begin to learn a second language in their childhood, learn it faster. Learning at a later stage makes it difficult becausethe child by then has already internalized the sounds of his/her mother tongue and in turn applies the same to English. This is where the problem starts. Lastly, it could also be because of lack of exposure to English.

Recent studies have shown that children, who are exposed to a second language at an early age, tend to learn it faster and more efficiently than those who start at a later age. This is due to the fact that young children have an innate ability to absorb new information quickly and easily. Additionally, their brains are more malleable and can adapt to new concepts much faster than adults. This is due to the fact that young minds are more receptive and adaptable, making they better suited for learning a new language.

However, this in turn can lead to MTI, which is when the child's native language affects their ability to learn a new language. It is important to understand how MTI can affect the process of learning English and how it can be managed in order to ensure successful learning outcomes.

Effects of MTI

MTI can be hindering the confidence of a person in a number of ways. A persons inability to express in a second or foreign language due to the interference of mother tongue can have multiple effects. The following are a few that can be the effects of MTI.

- People ridicule others at the expense of those who speak with strong MTI.
- But, it is not right to do so because people who speak English with MTI know that they speak wrong. They know that people make fun of them when they speak English and so, they refrain from speaking English altogether.
- Lack of confidence in speaking English
- Embarrassment in communicating with others publicly
- Inability to speak English properly due to internalization of sounds from the mother tongue
- Slow career/academic growth
- Most of the learner expressway MTI is such a big issue.
- Are the various accents in the world, such as American and British accents, not an example of MTI?

Mother tongue influence has a profound effect on students' ability to learn and understand other languages. It can also affect their overall academic performance, as well as their social and emotional development. Research shows that students who are exposed to their mother tongue from an early age tend to have better language proficiency, better cognitive development, and higher self-esteem. Furthermore, mother tongue influence can help students develop a deeper understanding of the cultures associated with the language they are learning. In short, having a strong mother tongue influence is beneficial for students in many ways.

It is important to note that mother tongue influence can vary depending on the student's cultural background and language proficiency. Therefore, it is essential for teachers and parents to be aware of the potential effects of having mother tongue influence on students' learning outcomes. This will help them create an environment where students can gain maximum benefit from their exposure to their native language while also gaining proficiency in other languages.

Technically speaking various accents is also an example of MTI. But the difference is that these are universally accepted and uniform. In India, there is no consistency. We do have a standard Indian accent which we aim to make popular, but there is still a long way to go. Are the various accents in the world, such as American and British accents, not an example of MTI?

As of now, the northern part of India speaks English in a different manner; the southern part has its own pronunciation and way of speaking and so on. Not just that, there are variations within the state as well! This is why MTI is an issue in India.

How to Get Rid of MTI?

Mother tongue influence is a common problem faced by many writers, especially those who are not native English speakers. It can be difficult to write in a language that is not your own, and this can lead to mistakes that can be difficult to detect. Fortunately, there are a few simple steps that students can take to reduce the impact of their mother tongue on their language. By understanding and applying the right techniques, writers can ensure their work is free from any errors due to mother tongue influence. Though it is difficult to get rid of something you have internalised. However, it is not impossible.

Hence it can be said that one can get rid of MTI by doing the following:

- Investing time and effort
- Practicing with dedication
- Practicing tongue twisters to improve focus while speaking English.
- Reading a text aloud and note down the words that you pronounce incorrectly to practice later.
- Listening to podcasts to observe and understand the correct sound of each syllable.
- Listening to an English news channel every day to improve pronunciation
- Watching English movies and listening to English songs to find out how native English speaker speak English.
- Recording yourself speaking English to find out which words you pronounce incorrectly.

4. Conclusion

All in all, I'd say that we need not be ashamed of how we speak. But, we need to neutralize our MTI in order to achieve a standard Indian accent that will be universally acceptable for intelligibility. The need to neutralize our Mother Tongue Influence (MTI) in order to achieve a standard Indian accent that is universally acceptable for intelligibility has become increasingly important. This would involve reducing the influence of regional dialects and other language influences that are present in the Indian accent. Standard Indian accent will help individuals communicate more effectively and confidently with people from different backgrounds, cultures, and countries. To achieve this, it is essential to understand the various factors that contribute to the formation of an Indian accent and then use techniques such as voice training and practice speaking in English with a neutralized MTI in order to attain the desired accent.

To conclude, mother tongue influence is an issue that many people face when trying to communicate in a language other than their native one. It can also be a can be a challenge for many people, especially when they are trying to learn a new language. For instance, learners can focus on learning the structure of the language instead of its vocabulary and use materials such as books, audio recordings and online courses to practice the language in a more effective way. However, with the right strategies and techniques, this can be overcome. Additionally, by immersing oneself in the target language through conversations or engaging with media such as films and music, learners can become more familiar with its nuances and reduce mother tongue influence.

REFERENCES

1. Avery, P. Ehrlich, S. (1992). 'Problems of Selected Language Groups'. In Teaching American English Pronunciation. Oxford University Press. ISBN:978-0194328159.
2. Burt, M, & Kiparsky, C. 1974. Global and local mistakes. In Schumann, J. H., and Stenson, N., (Eds.), New Frontiers in Second Language Learning. Rowley, MA: Newbury House.
3. Carter, R., & Nunan, D. (2001) The Cambridge Guide to Teaching English to Speakers of Other Languages.
4. Darcy et, (1953) "A Review of the Literature on the Effects of Bilingualism upon the Measurement of Intelligence", A Journal of General Psychology.
5. Garrigues, S. (1999). Overcoming Pronunciation Problems of English Teachers in Asia. (on-line) Available: http://asianbridges.com/pac2/presentations/garrigues.html.
6. Gelvanovsky, G.V. (2002). Effective Pronunciation teaching: Principles, factors and teachability. In P.V. Sysoyev (Ed.), Identity Culture and Language Teaching. USA: REEES.
7. Harmer, J. (2007). The Practice of English Language Teaching (4th ed.). Longman Handbooks for Language Teachers: Pearson Longman.
8. Julia, G. (2002). Introducing English rhythm in Chinese EFL Classrooms: A Literature Review. Published by the Faculty of Education at the University of Melbourne, Australia. 3(1), 26–42.
9. Manjula, Reddivari. (2022). 2021-Mother tongue Influence in English Pronunciation-Problems of Learnig English As a second language in India. PourhoseinGilakjani, A. (2012). A Study of Factors Affecting EFL Learners' English Pronunciation Learning and the Strategies for Instruction. International Journal of Humanities and Social Sciences, 2 (3), 119–128.
10. Rao, S Jayasrinivasa. (2012). To Correct or Not to Correct--Usual and Unusual Errors among Telugu Speakers of English [Language in India, Volume 12 : 7, July 2012] [ISSN 1930-2940] (Pp. 313–322).
11. Tickoo, M. L.(2003). Teaching and Learning English : A Source book for Teachers and Teacher Trainers, Delhi: Orient Longman, P.349.
12. Yates, L., & Zielinski, B (2009). Give it a go: Teaching Pronunciation to Adults. AMEP Research Centre, Department of Immigration and Citizen ship, Macquarie University, Sydney, Australia.

Impending Inquisitions in Humanities and Sciences (ICIIHS-2022) – Dr. Mohan Varkolu et al. (eds)
ISBN 978-1-032-78829-6

Cryptocurrency Market—An Overview

18

Balanagalakshmi B.*
Assistant Professor, Business School,
Koneru Lakshmaiah Education Foundation, Hyderabad, Telengana, India

Swetha A. B.
PG Student, Faculty of Technology, CEPT University,
Ahmedabad, Gujarat, India

Siddharth A.
Student, Department of CSE, IIITDM Kurnool,
Andhra Pradesh, India

ABSTRACT Cryptocurrency is a new type of digital currency creating by using the cryptography technology. Cryptocurrencies are not controlled by a single person or a government, it follows the decentralized control system. The first cryptocurrency was Bitcoin, created by Satoshi Nakamoto who built its decentralized control system also. It was mined by millions of people from different parts of the world. These Bitcoins were open to all, and its transactions were supposed to be anonymous. At the time of inception, the promise was given that, Bitcoins were the Universal Electronic Currency that passed around the world in minutes. Bitcoins have the quality of to treat this as a coin and a store of value and a network of payments. In this paper an attempt is made to explore the working of cryptocurrencies, its different types, benefits, drawbacks, ways to protect from digital frauds and the perspectives of Cryptocurrency at the global level in detail.

KEYWORDS Crypto currency, Cryptography, Decentralized Control, Digital currency, Electronic mode

1. Introduction

Cryptocurrency is a new type of digital currency created by using the cryptography technology. It is used online to buy or sell goods or services using online ledger. To secure online transactions, online ledger with the strong Cryptography is used. Cryptocurrencies are not controlled by a single person or a government, it follows the decentralized control system. The companies specifically create their own currency in the form of Tokens, the clients must make use of the same to transact with the company. The customers or clients must get the real currency in exchange of cryptocurrencies. The interest behind the usage of Cryptocurrency is to trade for profit, but for speculators it is to drive the prices skyward. The appeal for the technology is the security what it provides.

*Corresponding author: balanagalaxmi@gmail.com

DOI: 10.1201/9781003489436-18

Table 18.1 Important Types of Crypto Currency

Types	Meaning	Coin Images
Bitcoin	Bitcoin is a decentralized currency, not administered by a single person or by a bank. The transactions are between the users of the currency on peer network basis. There is no involvement of intermediaries.	
Ethereum	It is a decentralized and native cryptocurrency, blockchain which is open source with smart contract functionality. It is the second largest and the most actively used blockchain after Bitcoin in the market.	
Cardano	Cardano is a part of blockchain/cryptocurrency. It is also a decentralized digital currency, and a proof of ownership is also given. In this also there is no intermediary's involvement in the transactions.	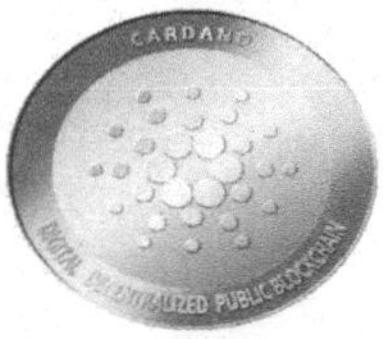
Dogecoin	Billy Markus and Jackson Palmer who were the software engineers have created the Dogecoin as a cryptocurrency. This coin was created out of fun with wild speculation in cryptocurrencies. though it is satirical in nature, it is also considered as a legitimate investment.	
Binance Coin	This type of currency is issued by Binance exchange with its own symbol (BNB). Initially it was followed Ethereum blockchain but now it is following its own Binance chain. In terms of market cap Binance coin is just behind the performance of Bitcoin and Ethereum.	
Tether, USDT	Tether is a controversial but most stable crypto currency backed by the traditional currencies like Dollar, Euro and Japanese Yen. It is using USDT symbol and traded with blockchain technology.	
Bitcoin Cash	One Bitcoin Cash value is around Rs. 35, 887, the transaction in this currency is very fast. The network used is also without any congestion. It is a simple, peer-to-peer electronic cash. It is more acting as a investment vehicle than as a currency.	
Zcash	It is a cryptocurrency created with cryptography technology. It provides privacy for the users comparing to other crypto currencies. It is based on Bitcoin's base code.	

Namecoin	It is a cryptocurrency forked from bitcoin software. It is based on the code of bitcoin but not intended to use as a currency but developed as a decentralized DNS. Like Bitcoin, it is also mined up to a limit of 21 million. The important aspect of this currency is that it can't be seized or censored.	
Bitcoin Gold	BTG is a cryptocurrency using bitcoin fundamentals but mined using GPU technology. It is also a open source, digital and decentralized currency using peer-to-peer transaction without involving intermediaries for exchange.	
Litecoin	One Litecoin value is around Rs. 10,102. By considering the market capitalization, it is the 6th largest cryptocurrency. It produces a greater number of coins and transaction speed is also fast when comparing to Bitcoin. So, it gives encouragement and enthusiasm to the investors and doesn't impact on the value of in its usability.	

Source: https://www.bitdegree.org

2. Journey of Cryptocurrency

1998 – B-Money and Bit Gold, Digi Cash, were the first formulated concepts of cryptocurrencies but not yet developed in full format. These currencies laid the base for Cryptocurrency but came into use for the public after 11 years of its inception.

2008 – Bitcoin which is a peer-to-peer electronic cash which was posted to the public mailing list based on cryptography technique.

2009 – Bitcoin which was added to the mailing list of the public was made available for the first time and the mining of Bitcoin was also started.

2010 – Bitcoin was initially bartered because it can't be traded but mined. So, there exists inherent difficulty in assigning the monetary value for the cryptocurrency.

2011 – Due to the increasing acceptance and popularity for the Bitcoin, the new currencies were also started introduced such as Namecoin and Litecoin. They are considered as the improved versions of Bitcoin.

2016 – Ethereum rose in its values and came closer to the values of Bitcoins during 2016. It has used the Ether platform to facilitate blockchain-based smart contracts and apps. The arrival of Ethereum was marked by Initial Coin Offerings (ICOs). These platforms provide opportunity for the investors to trade in the form of stocks or shares in start-ups, the same way they invest in and trade cryptocurrencies.

2019 – Bitcoin reached $10,000 and continued to grow: the revaluation and the recognition of the cryptocurrency market began gradually during this period and it is assumed that more money was flowing into the Bitcoin and crypto coin ecosystem. In this period the market capital of all crypto coins rose to an enormous rate from $11bn to $300bn. Various banks came forward to work with cryptocurrency markets.

2021 – Pandemic and Bitcoin: Today there are around 5,392 cryptocurrencies. Around 2.9 to 5.8 million people were the active users of cryptocurrency wallet at present.

2022 – There are more than 12000 cryptocurrencies at present. The Global crypto owners reached 1 billion by the end of 2022. It was 178% increase from the year 2021.

3. Cryptocurrency: Its Working and Mining

Cryptocurrency can be used just like PayPal or bank credit and the working is just equal to online banking services. The online transactions are recorded in a public digital ledger called Blockchain. The recorded currencies can be accessed through the software, wallets. With the help of online exchanges these digital currencies can be traded or exchanged. There is no central power authority instead is controlled in a decentralised manner by the concerned traders who have created these currencies to trade with them.

The first cryptocurrency was Bitcoin, created by Satoshi Nakamoto who built its decentralized control system also. It was mined by millions of people from different parts of the world. Those Bitcoins were open to all and its transactions were supposed to be anonymous. At the time of inception, the promise was given that, Bitcoins were the Universal Electronic Currency that passed around the world in minutes. Bitcoins have the quality of to treat this as a coin and a store of value and a network of payments.

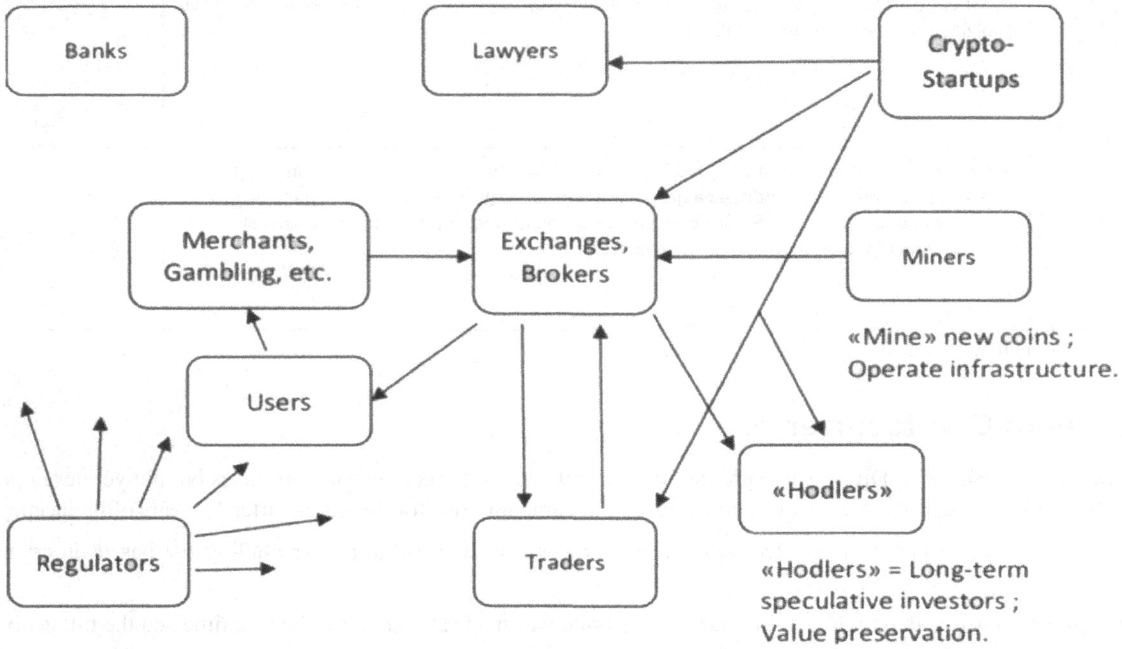

Fig. 18.1 Working of Crypto Currency

Source: CoinMarketCap.com

Bitcoin is a decentralized currency. In general, if a transaction is done using bitcoin two addresses would be there in the transaction details which are the sender's and receiver's addresses and the information like how much amount was transferred etc. are noted in a ledger which in this case is called a blockchain. In 2008, Satoshi Nakamoto invented blockchain_which is used as the public transaction ledger of the cryptocurrency bitcoin. These details are processed by many computers running special software. A blockchain keeps on changing for many times in a single day. Details or addresses of each transaction goes through all the computers which run the special software. For example, somehow a person hacks and changes the details or increases the number of bitcoins one has might think he succeeded by getting few bitcoins which is not true as details of each transaction or blockchain goes to all the computers. If one computer shows different value than all the others it reverts which means he didn't succeed in getting a few bitcoins and to succeed, he must change the blockchain in each computer individually which is impossible. So, you get an increased protection while using bitcoins.

3.1 Bitcoin Mining

Whenever a transaction is done using bitcoin the senders and receivers' addresses are noted in blockchain which keeps updating multiple times a day which is sent to each computer which processes bitcoin multiple times in a network to ensure that each computer uses the same/correct copies of the blockchain, cheating or hacking a computer to change wrong values is virtually impossible. This verification process is done by people who are connected using various devices such as a PC, a laptop etc. are known as bitcoin miners. The software run by these miners is done by grouping the recent transactions into blocks and it is accepted only if these transactions are hashed correctly which requires the computer to find correct numerical values. Whenever a computer correctly processes a block, it is added to a blockchain and a bitcoin is added to the miner's digital wallet which is literally a computer making money to the miner, but mining consumes a lot of power and electricity. This process of earning a bitcoin is known as bitcoin mining. Mining a bitcoin is not advisable for everyone because of the fluctuation of the value of bitcoin and the electricity costs in some parts of the world are really high that some people trying to make money using mining actually losing money instead of making a profit. As of the values of 29th June 2021, a person in Hyderabad using a NVIDIA GTX 1650 graphics card can make a profit of around 0.98-1.5$ a day which is approximately 0.000034 bitcoins and a person there using NVIDIA GTX 1060 graphics card can make a profit of around 1.34-1.8$ a day which is approximately around 0.000043 bitcoins.

4. Tips to Invest in Cryptocurrency Safely

According to Consumer Reports, Investments in Cryptocurrency is one of the riskiest avenues. CNBC has forecasted that, at the end of 2018, the value of these digital currencies would go up to $1 trillion dollars and it has offered fours important tips for the investors to invest in this Cryptocurrencies.

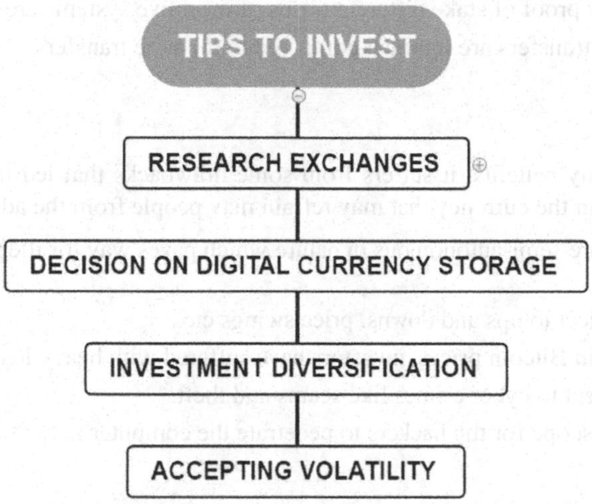

Fig. 18.2 Tips to invest in Cryptocurrency

Source: Statista Global Consumer Survey

Research Exchanges

The investor should collect details about the cryptocurrency exchanges before start investing. According to Bitcoin.com., there exists 500 exchanges available to provide the means to buy and sell these currencies. So, before moving ahead, the investor should do his/her own basic research about the investment type, collect reviews and discuss with the field experts if needed before the final decision.

Digital Currency Storage

The investor should take the decision on where to store the digital currency. It can be stored either in an exchange available or in a digital 'wallet' depends on the requirements and security demanded by the type of cryptocurrency what the investor is choosing.

Diversify Your Investments

The success of any investment is on the strategy diversification and the same is applicable to Cryptocurrency investments also. It is suggested that, instead of investing in the single options, the investors can choose among the number of alternative cryptocurrencies available to minimize the risk and maximize the returns.

Prepare for Volatility

As the cryptocurrency market is volatile, the investors should be prepared to accept the ups and downs in the market. There may be dramatic price swings in the market. The investment portfolio and the mental well-being of the investor should be strong enough to select this digital currency as an investment option.

5. Advantages

Cryptocurrencies have become popular in the recent years since 2018 and more than 10000 types are traded publicly. Cryptocurrencies are continued to proliferate and raises money through Initial Coin Offerings. The increasing acceptance of these digital currencies are due to the below mentioned reasons.

i. It is easy to transfer the funds between two parties through these Cryptocurrencies.
ii. No interference of involvement of intermediaries like banks or credit card company.
iii. Transfers are enabled with the help of either public key for the wallet access and the private keys known only for the owners and is used to sign the transactions are highly protected.

For example: Bitcoin gives poor choice for conducting illegal business and gives way for forensic analysis enable the authorities to identify, arrest and prosecute criminals.

iv. Based on the proof of work or proof of stake different forms of incentive systems are also.
v. The processing fees for Fund transfers are minimal compared to the wire transfers.

6. Drawbacks

Though Cryptocurrencies offer many benefits, it suffers from some drawbacks that led to huge losses for few people. The following are the obstacles present in the currency that may refrain may people from the adoption of this technology.

i. Cryptocurrency transactions are semi-autonomous in nature which paves way for illegal activities like money laundering and tax evasion.
ii. Highly volatile in market subject to ups and downs, price swings etc.
iii. In 2013, due to sudden crash in Bitcoin prices, investors have suffered with heavy losses.
iv. Being digital currencies, subject to cyber crimes like scams and theft.
v. Cryptocurrency market gives scope for the hackers to penetrate the computer networks and access others digital wallets.

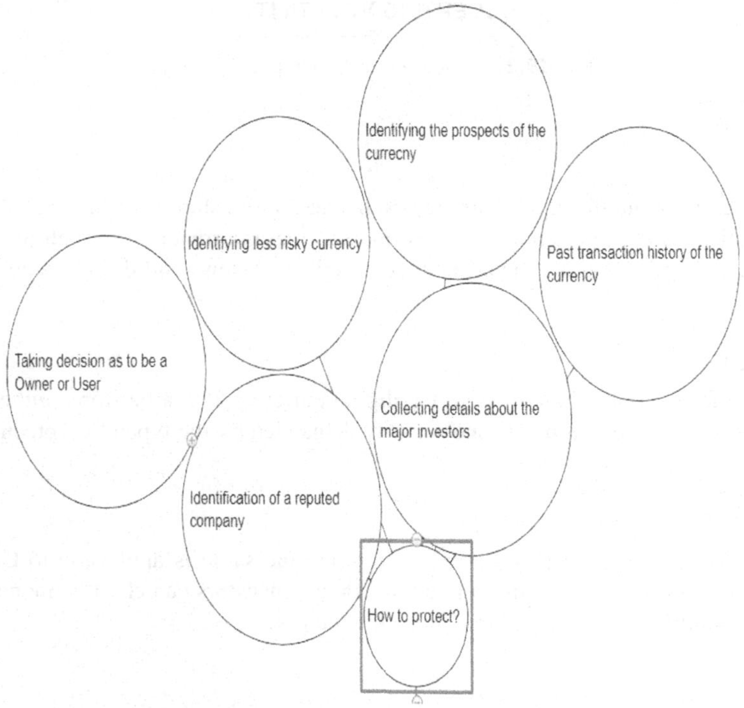

Fig. 18.3 How to safeguard from digital frauds?

Source: https://www.nerdwallet.com

7. Cryptocurrency – Global Perspective in General

According to the data derived from Statista Global Consumer Survey that, the people from Africa, Asia and South America are interested and having ownership of these Cryptocurrencies than those in North America, Australia and Europe. Especially only single percentage people from U.S. owned and used these digital currencies. Africa is the king of Cryptocurrency both in Ownership and in Usage. Its increasing population, poverty, high usage of mobiles, less transaction charges attracted to own and use Cryptocurrencies. It is evident from the report from Bitcoin.com that, 32% of the people in Nigeria own and use cryptocurrency as against 6% from America. The world's Top 10 Crypto countries as per **Analytics Insight** are given below:

Table 18.2 Top Ten Crypto Countries in 2022

Name of the Country	% of Participation	Position
Thailand	20.1%	1
Nigeria	19.4%	2
Philippines	19.4%	3
Turkey	18.6%	5
Argentina	18.5%	5
Indonesia	16.4%	6
Brazil	16.1%	7
Singapore	15.6%	8
South Korea	13.4%	9
Malaysia	13.2%	10

Source: Analytics Insight

As per the recent study on 'Which countries are using Cryptocurrency the Most?' done by Andrew Lisa (2021, June 29) is that currently, Bitcoins are not the hotcakes in market which is grown only upto 113% those it is the original and biggest digital currency as against Dash (198%), Ethereum (324%), Maker (760%), Dogecoin (7,555%).

According to the estimation done by the company Triple A (Cryptocurrency and Blockchain Technology Company) it is given that there are 300 million people are using digital currencies across the world. On an average 3.9% is the cryptocurrency ownership rate of countries and 18,000 business across the world are accepting the payments through digital currencies. It is also found from their study that out of every five users three are male members and they are young & educated in comparison.

According to CoinMarketCap survey, $1.7 billion is the total market value of the cryptocurrency and they are the front and center in the transactions of everyone including the non-investors. The non-investors are the people playing the vital role in making these digital currencies through celebrity tweets in social media, finance articles, memes and news in headlines. The popular misconceptions about the cryptocurrencies such as Cryptos are medium used for illegal and criminal dealings, Cryptos are outlawed and governments will ban in future, and the transactions in Cryptos are very much complicated were now become meaningless and it is understood with the help of market share and increased acceptance from the people all around the world.

The total market capitalization of cryptocurrency was peaked at over $2.9 trillion by the end of 2021 but at the same time it was $789 billion by the end of 2022. It showed that the market witnessed a fall during the year 2022 though there is an increase in the users and owners of cryptocurrency. In the year 2023, the investors of cryptocurrency market are expecting to rebound its position. For this the experts are having number of conversations about cryptocurrency regulations to establish laws and guidelines to safeguard the investors and have less appealing to the cybercriminals to avoid trafficking and terrorism sponsorship and etc.

8. Conclusion

Though Bitcoins are receiving acceptance from the world countries especially African countries, China's Central Bank (PBOC) has recently released an announcement that, there is ban of Cryptocurrency trading and mining in China. This decision was taken due to deepened crackdown on cryptocurrencies. The PBOC in Beijing urged institutions to launch a thorough check on client's account to identify the involvement and of digital currencies and take strict steps to cut down their payment channels.

The PBOC even stated that, this decision was taken in state cabinet or council meeting. The experts in different fields of finance expressed that this depression in China can be overcome by having a good digital asset management solution. Despite all discussions, it can be understood that Cryptocurrency is still in its infancy stage. Investment in new avenues always brings few challenges to be faced. So, to avoid issues the investors should prepare themselves to do research and try to start investing conservatively.

Co-founder and CEO of Unocoin crypt exchange Mr. Sathvik Vishwanath said that the involvement of corporates at present will not let the market to become a bear market and there are bright chances of having bull market in the upcoming periods. So, the crypto market is expecting a glory in the year 2023.

REFERENCES

1. Corbet, S.; Lucey, B.; Urquhart, A.; Yarovaya, L. Cryptocurrencies as a financial asset: A systematic analysis. *Int. Rev. Financ. Anal.* **2019**, *62*, 182–199.
2. Eyal I. and Sirer E.G. Majority is not enough: Bitcoin mining is vulnerable. 2013. arXiv: 1311.0243.
3. Gerlach, J.C.; Demos, G.; Sornette, D. Dissection of Bitcoin's Multiscale Bubble History. *SSRN Electron. J.* **2018**, 18–30.
4. Kristjanpoller, W.; Bouri, E. Asymmetric multifractal cross-correlations between the main world currencies and the main cryptocurrencies. *Phys. A* **2019**, *523*, 1057–1071.
5. Kristoufek, L.; Vosvrda, M. Cryptocurrencies market efficiency ranking: Not so straightforward. *Phys. A* **2019**, *531*.
6. Kroll J.A., Davey I.C. and Felten E.W. The economics of Bitcoin mining, or Bitcoin in the presence of adversaries, Mimeo. 2013.
7. Krugman P. Adam Smith hates Bitcoin. NYTimes blog. 2013. http://krugman.blogs.nytimes.com/2013/04/12/adam-smith-hates-bitcoin
8. Laurie B. Decentralised currencies are probably impossible (but let's at least make them efficient). 2011. (http://www.links.org/files/decentralised-currencies.pdf)
9. Satoshi, N. Bitcoin: A Peer-to-Peer Electronic Cash System. 2008. Available online: http://bitcoin.org/bitcoin.pdf
10. Stosic, D.; Stosic, D.; Ludermir, T.B.; Stosic, T. Multifractal behavior of price and volume changes in the cryptocurrency market. *Phys. A* **2019**, *520*, 54–61.
11. Urquhart, A.; Zhang, H. Is Bitcoin a hedge or safe haven for currencies? An intraday analysis. *Int. Rev. Financ. Anal.* **2019**, *63*, 49–57.
12. Wei, W.-C. Liquidity and market efficiency in cryptocurrencies. *Econ. Lett.* **2018**, *168*, 21–24.
13. Zięba, D.; Kokoszczyński, R.; Śledziewska, K. Shock transmission in the cryptocurrency market. Is Bitcoin the most influential? *Int. Rev. Financ. Anal.* **2019**, *64*, 102–125.
14. https://www.statista.com/global-consumer-survey
15. https://www.bitdegree.org
16. https://www.nerdwallet.com
17. https://www.statista.com
18. https://coinmarketcap.com

Impending Inquisitions in Humanities and Sciences (ICIIHS-2022) – Dr. Mohan Varkolu et al. (eds)

Narrative Unbound: Textuality of "Influence" in Ian McEwan's The Child in Time

19

Pravin K. Patel*
Assistant Professor, Department of English (MMV),
Banaras Hindu University, Varanasi

ABSTRACT This paper explores the concept of "influence" as textuality and relationship between the texts in Ian McEwan's novel, *The Child in Time*. By examining the intricate interplay between the characters' subjective experiences of the loss of child and the textual construction of the narrative, it suggests how the idea of influence permeates the novel on multiple levels. The analysis explores how influence shapes the idea of the text in terms of child loss and different shades of time. This study unravels the complex and dynamic nature of influence in the novel.

KEYWORDS Influence, Textuality, Child-disappearance, Time

1. Introduction

This paper focuses on the crucial concerns of "anxiety of influence" in Ian McEwan's *The Child in Time*. McEwan offers a critique of the author as the originator of the text which, in his words, is "rented or borrowed;" in other words, he counters the humanistic concept of literary texts presuming as a creation of the author's imagination, intellection and experience. The novel embodies a pool of writers and texts disseminating logocentric structures of the story and narrative towards intertext and influence. In other words, the concept of "influence" challenges the social and literary authority, and lays bare the structure of story, plot and narration. The novel incorporates tradition and transformation through experimental and metafictional techniques along with the strong moral sensibility of 19th-century canonical English novels. The novel narrates the life and time of the relationship of the couple Stephen and Julie, the mysterious abduction of their three-year-old daughter from the supermarket, and their painful relationship after the disappearance. The protagonist of the novel, Stephen Lewis, tries to heal from the trauma of his daughter's disappearance, the consequent breakdown of his marriage and subsequent identity crisis. Further, it depicts the complexities of time. Stephen and Julie, in their efforts to find out the lost child, encounter wild, distressing, warm and humorous moments. In so doing, the novel incorporates several writers' views on child, childhood, and time.

Objectives

This paper aims to explore the "relationship" between the texts through Harold Bloom's idea of "anxiety of influence". The critical reading of the text challenges the notions of mimetic criticism that a text is a representation of the world outside, expressive criticism that a text is a creation of the mind of the author, and liberal humanism that the meaning of the text is natural and universal. It affirms that *The Child in Time* does not embody self-sufficient, independent or autonomous meaning, and owes its existence to mighty precursors. Further, it puts an emphasis not on *what* is said but on *how* it is said, that is, its major focus goes to the texture of expression than the content. The novel represents Bloomian "relationships between the texts"

*pravin.patel@bhu.ac.in

DOI: 10.1201/9781003489436-19

in terms of its thematic and stylistic concerns. In this context, I shall focus on the issues of the disappearance of the child, the concept of time and the relations between the personal and the public.

3. Theoretical Approach

Bloom's concept of text bears resemblance to Jacques Derrida's notion of the text as "a fabric of traces referring endlessly to something other than itself, to other differentiated traces" (1979, p. 84). Bloom's "influence" like Derrida's "trace" turns towards itself or moves outside to dislocate its origin, structure, meaning and unity. However, unlike the "trace," "influence" directly incorporates the historical and psychological relation of the text. The Nietzschean "internal recurrence" and Freudian "oedipal relation" leads Bloom to defend the rhetorical will of the original creation against its paternal precursor, the literary mighty father. In the preface to *The Anxiety of Influence*, Bloom (1997) writes that "the anxiety of influence" is not a simple Freudian Oedipal rivalry (p.xxii) or a narrative of the struggle between one poet (author) and his precursor which is a naïve and weak application of belatedness. However, Bloom does emphasize a direct "borrowing" or assimilation of textual features and materials from the precursors but this evokes anxiety in the belated authors for becoming "original," and for this reason, they read the parent-text "defensively" by stimulating drastic distortion and ambivalence in their creative interpretation.

The act of reading makes an intersection between the author's text and the canon of literature. The literary influence is perceived as a conscious misreading and the process of misprision is the creation of new text, which means all reading is misreading and all interpretation is misinterpretation. The opening lines of *A Map of Misreading* make it obvious that reading is a "belated and all-but-impossible act" (1975, p. 3). The quest for meaning and deciphering of truth in the text is a misreading, a strong meaning is a strong misreading because literary language undermines the literary meaning. In "The Necessity of Misreading" he further writes that the act of reading and interpretation is an exercise in belatedness and a necessary misprision (2002, p. 70). Hence, literary influence is an intentional and essential misinterpretation of the preceding writers as "creative correction"[1]. Bloom's theory assumes that a work is perpetually a "misprision" of already existing great works of both male and female writers[2]. Bloom's idea of influence suggests that "there are *no texts*, but only relationships *between texts*. These relationships depend on a critical act, a misreading or misprision, that one poet performs upon another, and that does not differ in kind from the necessary critical acts performed by every strong reader upon every text he encounters" (1997, p. 3).

The Return of the Text of the Past

The Child in Time incorporates a vast array of the texts of McEwan's predecessors. The novel attempts both the continuation of preceding authors' views on child, childhood and the disappearance of the child and the disruption of the same. The "anxiety of influence" in the novel is represented in the form of "latent" and "manifest" relationships between texts. The former can be viewed for and against the Romantics' conception of a child and the latter is a direct reference to *Dream Babies* by Christina Hardyment, *Wholeness and Implicate Other* by David Bohm and *Magical Seeds* by Chilton Pearce. In other words, the latent anxiety of influence is a "strong misreading" in which the "revisionary rewriting" of the precursors is not to be easily identified and it gives the reader diverse ways to interpret the text; while the manifest anxiety of influence is a "weak misreading" in which the sources can smoothly be recognized and at the same time regulate the interpretation(s) of the text in a particular direction. In this way, the novel embodies the return of the past texts, that is, the texts and the writers of the past that have influenced the novel.

The Bloomian reconceptualisation of tradition engraved in the novel dislocates the text to different contexts to expose, what Blooms regards, the "four largest illusions" about the text: "the *religious* illusion" that the text "possesses or creates a real *presence*," "the *organic* illusion" that the text "possesses or creates a kind of *unity*," "the *rhetorical* illusion" that the text "possesses or creates a definite *form*" and "the *metaphysical* illusion" that the text "possesses or creates *meaning*" (2002, pp. 84–85, emphasis in the original). The text does not possess "presence," "unity," "form" or "meaning" on its own as they are a matter of faith, construed mistake or a lie, metaphor and arbitrary interpretation. *The Child in Time* constructs its presence through an organic form with the birth of a baby, the motifs, child and time. The reader interprets a predominant meaning that the child is the nucleus of a family and its loss devastates the same. The novel exists in "misprision" or "misreading," without which it is the tennis coach's description of Stephen:

> You're passive. You're mentally enfeebled. You wait for things to happen, you stand there hoping they're going to go your way. You take no responsibility for the ball, you're making no active calculations about the next move. You're inert, spineless, you're half asleep, you don't like yourself . . . *You're not all here. Even as I'm speaking to you now you're not all here*. (1998, p. 157, emphasis added)

Just like Stephen, the text by itself remains passive and does not take any responsibility for meaning. It is "inert, spineless … [and] half asleep" without a "strong misreading" and its meaning does not arise by itself, instead derive from other texts ("you're not all here").

Search for the Lost Child

The novel deals with the tragedy experienced by Stephen and Julie as they cope with the missing of their daughter and the consequential changes and developments in their lives for the next two years. The novel shows how the married life of Stephen Lewis, a famous writer of literature for children and a member of the Government Commission focused on Child-care, and his wife, Julie, gets spoiled when their daughter goes missing in a London supermarket. This leads to the disintegration of their marriage. In the beginning, they came closer and shared their grief, though it was only in "rhetorical questions" and theories without any definite and substantive conclusions, but later they drifted apart[3]: "Stephen and Julie had clung to one another, sharing dazed rhetorical questions, awake in bed all night, theorizing hopefully one moment, despairing the next. But that was before time, the heartless accumulation of days, had clarified the absolute, bitter truth. Silence drifted in and thickened" (1998, pp. 23–24).

The disappearance of Kate, which causes grief to the parents, emphasizes the purposelessness and hurting vacuum left in the family of adult people. Stephen, unable to cope with the loss, unable to repress his love for her, unable to cease his desire for her presence, imagines her in the children on the street. His family gets devastated after Kate's disappearance, which he tries to balance through the act of creating the Childcare Manual. In "A Loss Beyond Imagining," Margarida Morgado (2002) argues that the novel seems to integrate and decipher the idea of the family and child evolved throughout the last two hundred years. However, such notions of family are confronted in the present time and challenge the middle-class ideals of patriarchy where the father goes outside for work and the mother and children at home.

After the disappearance of Kate, Stephen and Julie begin to drift apart: "Their loss had set them on separate paths" (McEwan 1998, p.52) because "[b]eing together heightened their sense of loss" (p.53). However, 'It was not a divorce, of course, nor a separation, but "a time apart"' (p.53). The search for the child becomes the search for the parents' identity, Stephen transgresses himself into an unborn baby requesting his mother not to abort him (p.60) and Julie too goes through distressing moments and regresses into childhood. She resides in a house which a child can imagine:

> It was a house such as a child might draw. It was box-shaped with its front door dead centre, four small windows near each comer, and constructed of the same red bricks as The Bell. A path made out of left-over bricks made a shallow S-shape between the gate and the front door. (pp. 66–7)

Similarly, Stephen is too engrossed in the innocent moments of childhood by constructing a sand castle along with Kate and Julie. They wanted to entertain their daughter but they got absorbed in the game and forgot about the approaching tide:

> When everything was completed and they had walked round their achievement several times, they squeezed inside the walls and sat down to wait for the tide. Kate was convinced that their castle was so well built it could resist the sea. Stephen and Julie went along with this deriding the water when it simply lapped around the sides, booing it when it sucked away a piece of wall. (p.106)

The regression of the parents into childhood is a kind of search for a child in every human being, and by emphasizing the image of child, McEwan highlights "the innocence and purity of childhood" which is "quickly and almost completely forgotten in adulthood" (Slay 1996, p.118). The other characters in the novel like Charles and his wife, Thelma, are also searching for the child-self within themselves. He tells Stephen that "childhood is timeless" and Thelma also tells Stephen that Charles wants to embrace childhood for its innocence, frailty and compliance. Childhood to him is freedom from responsibilities, plans, economic securities, etc. The projection of purity and innocence through childhood is derived from a long tradition of writers, particularly in the Romantic period. The Romantics, like William Blake and William Wordsworth, have represented children as pure and innocent. Thus, the loss of Kate represents the loss of childhood into the adult world. In "The Necessity of Misreading" (2002), Bloom writes, "If there are no texts, then there are no authors – to be a poet is to be an interpoet" (p.80). However, the successive generations of writers approach their predecessors with repressive and antagonistic perspective to escape from the "accumulating anxiety" that result from the burden of great past forms and accomplishments. Hence, McEwan offers an alternative perspective on the Romantics' conceptions of child. There are instances in *The Child in Time* which draw an opposite meaning of child and childhood by portraying them amid intersecting narratives and interpretations that dismantle the notion of innocence and progress associated with children and childhood. As McEwan (1998) says, "It should be remembered that childhood is not a natural occurrence. There was a time when children were treated like small adults. Childhood is an invention, a social construct, made possible by society as it increased in sophistication and resource" (p.93).

The Subject of Child Disappearance

Bloom's emphasis on "relationships between texts" or "influence" in the pattern of defensive misreading proposes the theory of belatedness and consequent revisionism in the composition of the text. It does not make current poets (authors) less original; instead, it shows immense anxiety of belatedness from and indebtedness to their mighty precursors. Paul de Man in his review of *The Anxiety of Influence* argues that all (mis)reading is a power play in which the later texts strive to usurp and displace the preceding texts.

The tradition of writing a fictional text about child disappearance does not begin only with McEwan's *The Child in Time*. Genevieve Jurgensen's novel, *The Disappearance*, deals with the story of two children who go missing. The novel emphasizes the deep emotional turmoil, pain, grief and anguish of parents over haunting memories of their children's death. Similarly, Nicholas Roeg's film, *Don't Look Now* (I973), presents the devastating loss of Christine, a young girl who tragically drowned in a pond. Her father's "crazed persistence of desire," desperation and wild hope appears similar to that of Stephen. The unspeakable loss of a child, by abduction or death, leads to the worst nightmare, tragic, and overwhelming emotional impact on the parents. The theme of child disappearance is also portrayed in such contemporary picture books for children as Maurice Sendak's *Outside Over There* and Jon Sciezska and Lane Smith's *The Stinky Cheese Man and Other Fairly Stupid Tales*. Sendak narrates the abduction of a baby girl by goblins in the absence of her parents. The narrative of the kidnapping girl child, as in R. L. Stevenson's *Kidnapped* (1886), is there for centuries for the interest of an individual or community. The story of stolen or missing children, who come back later, is a frequent concern in thrillers and mystery stories of Mary Kelso's *Abducted*! (1988), James Grippando's *The Abduction* (1998), Jacqueline Mitchard's *The Deep End of the Ocean* (1999), etc.

Loss of Child: Juxtaposing the Personal and the Public

One of the important aspects of the novel is the juxtaposition of personal and public or personal and historical incidents. The relationship between personal and public appears to be linked with oedipal reactions. What Bloom states about the author's revisionism also pertains to one's own two or more spatiotemporal concerns in which the author either tries to close the gap between them or psychologically de-idealizes one over the other. It is an internal narrative struggle in the structure of the text which is simultaneously unconscious and deliberate act dictated by the individual poetic psyche. Peter V. Zima (2002) regards that "[w]hat is true for the poet's psyche is . . . also true for the psyche of the average reader whose reception also follows the principles of misreading" (p.154).

The national history and political changes are very clearly reflected and paralleled in the private life of Stephen. During his discussion with the Assistant Secretary of the Prime Minister, Stephen points out: "I resent what the Prime Minister's been doing in this country all these years. It's a mess, a disgrace". The secretary asks him, "Then why did you accept first time?" and he replies that "I was a mess too. Depressed. Now I am not" (McEwan 1998, p.154). The intersection of personal and public in the novel manifests during the private child care of parents and the role of government in child care. In this regard, McEwan is inspired by Christina Hardyment's *Dream Babies: Three Centuries of Good Advice on Child Care* which is a kind of "history of child care manuals". McEwan represents the impact of prevailing societal values throughout the ages. The impact of the Victorian notion of controlling children implies strict adherence to norms and suppression of a child's will in the present era. Similarly, the Edwardian influence suggests an emotionally-driven and sentimental approach to children, and the decade of the 1920s and '30s represents a pseudo-scientific behaviourist approach to childcare. The public and the private spheres in the novel are dealt with the help of three major components: first, Stephen's quest to find out his missing daughter and re-building the connection with his estranged wife; second, his involvement in the government child-care initiatives, and the last, his association with a former government minister who has worked as a publisher. Further, the novel shows a kind of parallel between the nation and the parents: and accordingly, there is an analogy between the "family" and the "nation" where the latter is depicted as a larger version of the former. Thus, the head of the nation is described as the "nation's parents" and the parents in the family are "embodiments of society" (p.93). Moreover, the importance of family brings faithfulness to the nation. As McEwan writes in the epigraph to Chapter 4, "[F]rom love and respect for the home we derive our deepest loyalties to nation" (p.69).

Loss of Child and Recovery in Time

The rise of the stream-of-consciousness novel at the beginning of the nineteen twenties reflects the inwardness of life consequent upon the break-down of accepted values. Further, even the influence of Henry Bergson was so enormous that even the writers who did not read him were influenced by his idea of *la Durée* suggesting time as a continuous flow. Bergson's emphasis on "inner time" or *Durée* and "clock-time" or mechanical time categorizes time into two aspects. "Inner time" represents psychological time which is considered a continuous flow and fluid stream while mechanical time represents artificial time and incorporates the division of time into past, present and future. As Charles R. Schmidtke insightfully notes:

> *Durée* . . . is not a metaphysical link (or point or now) between the past and the future. . . . The past is not stored; it continually flows through the present in a cumulative process. The future is not an object at the end of a string of presents. . . . Past-present-future are not three separate points or areas on a line; rather, for Bergson, the past is really memory flowing through consciousness, the present is continuous perception with its characteristic *durée*, and the future is the creation, newness and unforeseeability of experience. (Cited in Barnard 2012, p.30)

The concept of *durée* provides freedom from rigid notions of time and thus blurs demarcations into "now" and "non-now" of time. In *The Child in Time*, the concept of time is not solely associated with theft and recovery but also carries a sense of relativity. The motifs of relativity are replete in the novel. For instance, the huge traffic jam that paralyzes London conveys to pavement crowds "a sense of relative motion, of drifting slowly backwards[5]" (McEwan 1998, p.7). After some time, the narrator understands that the way time is perceived can be inaccurate and distorted, as he says: "But time—not necessarily as it is, for who knows that, but as thought has constituted it—monomaniacally forbids second chances. *There is no absolute time . . . no independent entity. Only our particular and weak understanding*" (p.14, emphasis added). In this case, McEwan seems to borrow Bergson's idea of *durée*.

Later, the narrator directly engages the reader in the novel: "Only when you are grown up, perhaps only when you have children yourself, do you fully understand that your parents had a full and intricate exercise before you were born" (p.48), and thus follows mechanical time. In this way, the novel foreground the idea of limitations and incompleteness of any representation or description. The novel dramatizes movement between different times during Stephen's recollection of various past events in his committee room. He recalls his missing daughter and his relationship with his parents. Bloom rightly says, there is no end to the "influence" of one writer or literary tradition upon the later authors who simultaneously adopt and alter the content and style of earlier writer/s. For instance, Bloom considers that the Shakespearean characters, like Brutus, believe in the "influence" of stars on their fate, and Shakespeare employs the term "influence" in his plays and sonnets to convey the sense of "inspiration".

The Psychological Time, Childhood and Healing

The novel represents childhood as a continuous flow of time which remains uninterrupted by humdrum realities. It is not a linear succession of time and progress of events. According to Thelma, Charles "wanted the security of childhood, the powerlessness, the obedience, and also the freedom that goes with it, freedom from money, decisions, plans, demands. He used to say he wanted to escape from time, from appointments, schedules, deadlines. Childhood to him was timelessness, he talked about it as though it were a mystical state[4]" (pp. 200–1). One of the major concerns in the novel is "references to the seeming instability of time". Even the divisions in time are shown as superficial, for example, Stephen constantly turns into the memory through structured dreams enabled transition from present to the past and vice versa, facilitating shifts in temporal perception. The form and pattern of time in the novel help to unite the work: "like the image of the child, images of time also serve to unify the novel," says Jack Slay. The theft of Kate created a rift in the parents but Stephen's visit to his wife rejuvenates their relationship and reminds him of his parents before his birth. This particular episode challenges the idea of mechanical time and represents the redemptive experience of time.

However, the novel has a great deal of focus on mechanical time. The subservience of the child and the clock in the novel are fairly balanced after the birth of another baby. The novel opens with a baby and ends with another baby whose gender is not revealed. The birth of the baby unites the alienated parents, as Julie signals and the narrator comments, "'A girl or a boy?' And it was in acknowledgement of the world they were about to rejoin" (McEwan 1998, p. 220).

Theft of Child as Vandalizing Force of the Clock-time

The novel, *The Child in Time*, commences and concludes with "inner time" through the metaphor of the child, but its major emphasis was laid on movement between different times: Stephen's remembering of past events, like his relationship with his parents and the abduction of his daughter, in his committee meetings. The narrative attempts to recuperate the loss and trauma through memory. Stephen tries to remember most of the past events. According to David Malcolm (2002), 'The motif of historical loss is matched by motifs of a general and widespread loss and transience. This is part of the meaning of "in time" in the novel's title: time is loss, time is transience' (p. 103). The novel reveals, to use John Keats' phrase, "the weariness, the fever, and the fret" of the contemporary culture and society, which is represented in the form of the disappearance of the child, evaporation of hopes and destruction of ambitions of ill-paid teachers, taxi-drivers, cleaners, licensed beggars, etc.

Stephen and Julie's loss of their daughter changes the course of time in their life for the next two and half years. As the narrator states, "It was not principally a search, though it had once been an obsessive hunt, and for a long time too. Two years on, only vestiges of that remained; now it was longing, a dry hunger. There was a *biological clock, dispassionate in its unstoppability* ..." (McEwan 1998, p.7, emphasis added). Commenting upon the intricate nature of the time, Jack Slay (1996) writes, "McEwan creates a sense of time that is malleable, wondrous, infinitely complex. Time is a vandal: it is the essence that can make one forget the child, the youthful joy of life. Simultaneously, time is also vandalized: characters experience periods that stall in slow motion, that pass in a blur of quickness, that is even altered with the past coming round to the present. Time then serves as an emblem for the complexities and difficulties that exist in everyday contemporary society" (pp. 115–16). Stephen fights against the vandalizing force of time, a power which will erase the memory of his daughter, by purchasing gifts which Kate used to like. On Kate's birthday when there was nobody, "he began to sing Happy Birthday" and bring a gift walkie-talkie for her (McEwan 1998, p.129). Thus, she was always present in time with machine proximity (p. 129).

3. Conclusion

The Child in Time shows the intricate web of textual influences intermingling a diverse range of texts in "the ways in which the new must grow out of the old". The novel transcends the conventional narrative boundaries through interconnectedness of texts. The analysis of the novels shows that the heterogeneity of the text can be defined, but its univocal meaning cannot be justified. It suggests that the text is not a unified whole, and has no existence in themselves. It shows that meaning arises from conceptual and figural implications rather than from authorial intentions. It shows that the novel shatters the commonsensical idea that the meaning of a text lies inside the work, and exemplifies the assertions of Roland Barthes (1977) that a text is composed of anonymous and untraceable "quotations without inverted commas," or of "a tissue of quotations drawn from the innumerable centre of culture" or of previous works of literature, or a text is a "multi-dimensional space in which a variety of writings, none of them original, blend and clash" (p.146). The study reveals that the meaning of fiction is not self-sufficient and independent, instead, it claims, no text is self-born and every text of the present is a repetition of and an intertext with something happening with it and a text of the past. The concept of "influence" functions as the invention and intervention of the parent texts that disseminate the text from its unified whole to multiple other texts. The notion of "repetition" and "influence" in McEwan's fiction deconstructs the concept of the "author" as the "originator" of the texts.

END-NOTES

1. Ezra Pound proposes a completely different and directly opposed version of the influence of predecessors on belated poets. The notions of "belatedness," "revisionism," "distortion," "misprision," "misreading," and reading the precursor "defensively" are crucial to Bloom in his theory of the anxiety of influence, whereas "invention" or "discovery" is central to Pound. He does not consider it as the "burden of the past" in the manner of belatedness or oedipal struggle. Poetic influence for him is not simply a revisiting of past works or mighty predecessors but is instead a more positive and conscious process for broader interaction between the predecessor and the successors in terms of epistemological assumptions and cultural attitudes vis-à-vis literary form and language.
2. In *The Madwoman in the Attic*, Sandra Gilbert and Susan Gubar argue that women writers do not experience the "anxiety of influence" like their male counterparts because their precursors are exclusively males.
3. Like Ian McEwan's *The Child in Time*, Dunmore's *Mourning Ruby* (2003) is filled with the emotional lives of the English middle classes and is centred on a moment of terrible loss – the death of a child in a road accident. This tragedy drives apart the child's parents, Rebecca and Adam.
4. The fiction of McEwan presents the children as perceptive and the adults as sluggishly unobservant. *The Child in Time* demonstrates a child's gaze as a magnifying glass whereas an adult's is a world-weary observer. The child possesses the quality of how to observe things closely, and the adults lose it. In the following passage, the narrator shows Kate's close observation of the natural world and contrasts it with his own: "The wood, this spider rotating on its thread, this beetle lumbering over blades of grass, would be all, the moment would be everything. He needed her good influence, her lessons in celebrating the specific; how to fill the present and be filled by it to the point where identity faded to nothing. He was always partly somewhere else, never quite paying attention, never wholly serious" (McEwan 1998, p.105).
5. The time has become destructive in the novel and offers a dystopic vision of Thatcherite Britain. The "throbbing cars," traffics and "[t]he steady forward press of the pavement crowds" in the city of London at barely nine-thirty in the morning remind a scene of crowds passing over London Bridge in T.S. Eliot's "The Waste Land". In the poem, Eliot describes how people are passing over the bridge in a winter morning: "Unreal City, / Under the brown fog of a winter dawn, / A crowd flowed over London Bridge, / so many, / I had not thought death had undone so many. / Sighs, short and infrequent, were exhaled, / And each man fixed his eyes before his feet".

REFERENCES

1. Barthes, R. (1997). "Death of the Author". *Image, Music Text*. London: Fontana. pp. 142–148.
2. Barnard, G.W. (2012). *Living Consciousness: The Metaphysical Vision of Henri Bergson*. Albany: State University of New York.
3. Bloom, H. (1997). *The Anxiety of Influence: A Theory of Poetry*. New York: Oxford University.
4. ---. (1975). *A Map of Misreading*. New York: Oxford University.
5. ---. (2002). "The Necessity of Misreading". *The Georgia Review*. Vol. 56, No. 1. pp. 69–87.
6. Carter, A. (1982). "Notes for a Theory of Sixties Style". *Nothing Sacred: Selected Writings*. London: Virago, 85–90. Print.
7. Derrida, J. (1979). "Living on, Border Lines" in Harold Bloom et al eds *Deconstruction and Criticism*. New York: Seabury Press, pp. 75–176.
8. Malcolm, D. (2002). *Understanding Ian McEwan*. Columbia: University of South Carolina.
9. McEwan, I. (1998). *The Child in Time*. London: Picador.
10. Morgado, M. (2002). *"A Loss beyond Imagining: Child Disappearance in Fiction"*. The Yearbook of English Studies, Vol. 32, Children in Literature, pp. 244–259.
11. Slay, J. (1996). *Ian McEwan*. London: Twayne Publishers.
12. Zima, P.V. (2002). *Deconstruction and Critical Theory*. Trans. Rainer Emig. London: Continuum.

Impending Inquisitions in Humanities and Sciences (ICIIHS-2022) – Dr. Mohan Varkolu et al. (eds)
© 2024 Taylor & Francis Group, London, ISBN 978-1-032-78829-6

Translating *Toba Tek Singh*: From the "Sacredness" of the "Original" to the "Profaneness" of the Adaptation

20

Manoj Kumar Yadav*, Maneri Sundeep
Assistant Professor, Department of Humanities and Social Sciences, NIT Hamirpur, Himachal Pradesh

ABSTRACT *Toba Tek Singh* (1955) is an iconic short story written by Saadat Hasan Manto (1912–1955) that intricately captures the aftershocks of partition; trauma, mass hysteria and absurdity prevalent in the border areas. The story is set in a mental asylum which represents how lost and confused people were in a world that was simmering with madness and violence all around. The short story comments ironically on the insane decision of the then-political leaders to divide the country resulting in widespread bloodshed, displacement and anarchy. Since the publication of the short story, it has transcended the boundaries of media and genres and has been adapted into various art forms. The current article looks at a film adaptation of the short story directed by Ketan Mehta that was released in the year 2018.

This article explores the complex process of mediation that a story undergoes from its written version to visual recreation. In the process, it closely analyses the visual text vis-a-vis the written text and focuses on techniques/tools that the film uses to recreate the short story on screen. As a mode of translation, the adaptation of written text into visual text, has often been seen as an aesthetic challenge involving the movement across two differing media. Still, the relocation of written text onto a cinematic space, cannot be appreciated solely in its formal dimensions; instead, it should be seen within a larger, millennial movement across cultural spaces. The article will argue that while analysing the two texts across different media, the autonomy of the works of art is of primary significance. Hence the typical discourse of "fidelity" and "betrayal" towards the source text must be negated to explore the "autotelic" nature of the two works.

KEYWORDS Fidelity, Loss in translation, Adaptations, Partition, Source-text vs target-text

1. Introduction

Saadat Hasan Manto (1912–1955) was probably the most significant, controversial and provocative writer writing during the progressive writers' movement of the early nineteenth century. Although he is equally popular today for film scripts, novels, and essays, he was best known for his short stories. His stories are known for the portrayal of strong human emotions, complex internal conflicts, precision of narrative, and stark realism.

Living under the shadow of an authoritarian and abusive father, Manto had a difficult childhood. He would be punished by his father for his minor mischief. One such incident turned exceedingly traumatic for him as he had to jump from the rooftop to save himself (Hashmi, 2012). Even in the later part of his life, several instances can be pointed out that suggest that Manto might have suffered from severe depressive disorders. These moments in his life were: the years following his father's demise when he stated to have been "constantly dissatisfied" and "restless"; the period following his sister's miscarriage when he contemplated committing suicide; and the years between 1951–1952 when he had to be admitted to Punjab mental hospital

*Corresponding author: manojk@nit.ac.in

DOI: 10.1201/9781003489436-20

for his alcohol-induced psychosis. Eventually, he died at the age of forty-four due to liver cirrhosis possibly caused by excessive alcoholism.

Toba Tek Singh is a short story written by Saadat Hasan Manto. The story is set in the time just before the partition of India and Pakistan. It tells the story of a Sikh man who is living in an asylum for the mentally ill. The man, whose name is Toba Tek Singh, is from the town of Toba Tek Singh in Punjab. He does not remember how he got there, or even his own name. The only thing he knows is that he is from Toba Tek Singh, the village he left behind.

The story revolves around the theme of religious intolerance. Toba Tek Singh is a Sikh, and he is living in an asylum with Muslims, Hindus, and Christians. All of the patients in the asylum are from different religious backgrounds. The staff at the asylum tries to keep the patients segregated by religion, but Toba Tek Singh does not understand why they are doing this. He just knows that he wants to go home to Toba Tek Singh. The story also explores the theme of partition. The partition of India and Pakistan was a very contentious issue at the time, and it led to much violence and bloodshed. The character of Toba Tek Singh represents all of those who were displaced by partition. He does not know which side he belongs to, and he just wants to go home.

Film Analysis

Ketan Mehta's film *Toba Tek Singh* (2018) is the most recent recreation of Manto's short story of the same name. The film begins with a prelude by the director:

> Bringing together creative talents from India and Pakistan on the same platform and sharing their creative visions through an exchange of ideas and inspirations can go a long way in bringing peace and understanding in this subcontinent (Mehta, 2018, 0:0:09)

The film begins with a scene capturing the dawn on the premises of an asylum in Lahore in the year 1947. In the backdrop, the sound of birds chirping coincides with the sounds of prayers of namaz and of bells ringing in the temples. The first shot captures a man standing behind bars and staring blankly without blinking. After a few moments, he speaks gibberish and goes quiet again. Later, the same morning, when the asylum attendant Hamid Miyan opens the gate of the confinement, we come to know that the man is Bishan Singh. Bishan Singh (also known as Toba Tek Singh) has been there in the asylum for the last 10 years. Bishan Singh is also addressed as *darakht insan* (tree-man) as he has neither lied nor sat down in the last 10 years. The film narrative recreates the year 1947 on-screen by taking recourse to a few apparent audio-visual markers. For instance, when the newly appointed asylum superintendent Sadat Hasan reaches the asylum on his bicycle to join his duties, the subtitle "Lahore 1947" appears on screen, and in the background, the audience can hear the song "afsana likh rahi hoon…" from the film *Dard* (1947) which was a major musical hit of that year. A couple of shots later, another popular song *zawan hai muhabbat, haseen hai zamana* [The Love is Young and the World is Beautiful] from the film *Anmol Ghadi* (1946) is employed in the background as Saadat Hasan cycles to his residential quarters.

Two important characters are introduced early in the film that play an important role in how the film version of the short story shapes up. First is the character of Saadat Hasan Manto, an aspiring writer, and second, is the character of the outgoing asylum superintendent Mr. Malik. Both of these characters are not present in the source text but they are significant in the narrative scheme of the visual text. The outgoing superintendent appears to be disillusioned and hopeless about the future of the country. Superficial conversations also appear to bear deep connotations. When Manto laughs at Malik's casual banter of calling him an "idiot", Malik quips: "Laugh….till the time laughter strangles you". Malik also wonders why would any sensible person take the charge of the Lahore asylum: "I was helpless when I took this job….but why did you come to this hell willingly" (Mehta, 2018, 0:06:19). Malik feels, as an outgoing superintendent, it was his duty towards the new officer in charge to introduce him to all the lunatics in the asylum. Malik has seen all sorts of lunatics in his tenure, as a result, he is barely affected by even the strangest occurrences. When Malik walks Manto through the premises of the asylum, he is taken to an open area where all the lunatics are accommodated. Malik introduces Manto to some of the lunatics including Toba Tek Singh. When they are about to leave the area, a fight breaks between two of the asylum inmates, the scene captures two contrary reactions. Manto is shocked and scared whereas Malik is seen laughing. He has been accustomed to such incidents and does not take them very seriously. Malik represents a system of the old-world order; with all its problems and complications, giving us an idea of an undivided India. Whereas the character of the new superintendent represents a new world order; trepidations and apprehensions of the chaos that is approaching day by day. The scene of the lunatics' infighting in the asylum and the scene capturing the plight of women in the women's ward envisages an impending catastrophe. The linear narrative of the source text is replicated in the film text as he starts recapitulating the recent events of violence and insanity prevalent in the country:

> Since the beginning of 1947, a mad wind is blowing over India ... a hot wind like the burning wind of the desert. People become violent on the smallest occurrences, their minds infuriate, and their souls burn (Mehta, 2018, 0:09:24).

The scene that follows, captures a lunatic sitting in his gloomy chambers of the asylum and singing the famous poem of the mystic poet Bulleh Shah (C. 1680–C. 1757):

> Chal Bulleya chal uthay chalyae jithay saray anay
>
> Na koi saadi zaat pachane
>
> Na koi sanu manne
>
> Chal bulleya chal uthay chalyae jithay saray anay (Mehta, 2018, 0:09:50).
>
> [O Bulleh Shah let's go to a place where everyone is blind
>
> Where no one asks us about our lineage, and no one revers us
>
> Neither does anyone holds me in high respect
>
> O Bulleh Shah let's go to a place where everyone is blind]

At this point, the reader questions the basic premise of the film, the madness of the asylum inmates. Because many asylum inmates appear to convey the pithiest, meaningful, and mystical conversations. There are two particular scenes in the film that capture two major events in the world beyond the asylum walls; the first is a shot capturing the aftermaths of the partition and the second, is a shot that arrests the freedom of Pakistan and India on the 14th and 15th of August respectively. As opposed to the rest of the film, both of these scenes are filmed in black and white and are employed as a subtext to the principal plot. The first scene intends to recreate the communal violence that follows the partition. It shows burning public properties and buildings, destroyed buses, factories on fire, people running around with sticks and guns in their hands, dead bodies lying on the roads, and injured people (mostly children and women) being carried on stretcher-turned cots. The last few shots in this scene also successfully capture mass migration and displacement.

Fig. 20.1 A scene of an overcrowded railway station and a train showing the mass displacement after the partition [7]

The 58-second-long scene is backed by the narrative voice of the asylum superintendent Manto in the background:

> It was as if everyone was baying for blood. Violence and bloodshed rampaged everywhere. Brutality crossed all limits. On both sides of the border, hatred reigned supreme. Lakhs of people were killed and millions became homeless. Generations were destroyed (Mehta, 2018, 0:38:26).

The quick coverage of the monochrome past in black & white provides the director the space to familiarize the audience with a series of events that have occurred at the national level and will have more to do with the fate of the asylum inmates in the coming days. However, this scene fails to convey whether or not the families of the lunatics were affected in the aftermath of the partition. The audience is also not informed whether or not the families of the asylum staff members were safe and secure during the communal violence. The director does not seem to contextualize the events occurring in the world beyond the asylum walls to the asylum inmates.

The second scene vividly captures the freedom and the birth of twin countries i.e. Pakistan & India. Once again, several shots within 20 seconds give us a glimpse of the freedom march of the two armies, the unfurling of the national flags, and the scene where Mohammad Ali Jinna and Pandit Jawaharlal Nehru are seen addressing two different gatherings. While the audience is left wondering about the content of Jinnah's speech, a part of Jawaharlal Nehru's famous Tryst of Destiny speech, delivered on August 15, 1947, is very clearly audible. But all these celebrations appear to be ironic because of stark reality as appended through the narrative voice: "And then on the 14th and 15th of August, the twin nations were born…soiled in, blood, wounded with fear and hatred, burdened with shock and guilt" (Mehta, 2018, 0:46:53).

Fig. 20.2 A Scene capturing the unfurling of the Indian national flag for the first time at the Red Fort

Notable Omissions and Additions

There are several interesting additions and omissions that contribute significantly to the shape and identity of the film. The omission of *Toba Tek Singh*, the village, in the target text is significant in the first place. The narrator informs the audience that Bishan Singh was a prosperous landlord in the village but went mad when the issue of the division of property with his brothers cropped up. However, in the film, the audience is left wondering about the village, Bishan Singh's family, and the trauma that made him go mad. When Bishan Singh's family members visit him in the asylum, they inform him about the impending partition [*batwaaraa*] of the country. Bishan Singh gets occupied by the word *batwaaraa* (partition) and is seen mumbling the word several times on the asylum premises. The audience is left wondering whether it partition of property in the village still haunts him. The allusion is ambiguous and the audience is left wanting to know more about the village.

Another significant change in the film is the absence of the father-daughter relationship in the target text. The source text describes that when Bishan Singh was confined first, he left behind an infant daughter (Roop Kaur), now a pretty young girl of fifteen years. It is also described in the source text that she would visit Bishan Singh in the asylum occasionally and would sit in front of him for hours with tears rolling down her cheeks. But the father-daughter relationship, which constitutes an important part of the source text narrative, is subdued is unexplored in the target text. In the film, only one scene captures the father and the daughter in a single frame. Appalled at her father's mental health, she is seen crying and turning her face away. In the source text, the last mention of Roop Kaur occurs when Fazal Din, a neighbour of Bishan Singh from his village, informs him about the migration of his family across the border. Fazal Din appears a bit hesitant to inform about the girl: "Your daughter Roop Kaur....—he hesitated—She is safe too....in India". The pauses create doubt in the mind of the readers and indicate the possibility of Roop Kaur becoming a victim of post-partition brutality and violence. But this dimension is not articulated well in the film as it does not create a similar impression.

The inclusion of a "women's ward" in the asylum is another major change made in the film version of the text. Given the fact that women and children were at the receiving end of the post-partition violence, the inclusion of two scenes on the "women's ward" appears to be a meaningful addition. The first scene provides a basic introduction to the women's ward. The second scene shows a traumatized victim of rape and violence brought to the asylum premises by a nun. The asylum officials appear to be shocked and dismayed at her condition. With her clothes torn and her dupatta soaked in blood, the girl is in a state of shock and is not able to utter a word. The asylum attendant Ashraf miya wants to know about her religious orientation, to which the nun says that she is a girl and that was enough. She is eventually given shelter in the asylum premises. This incident, as viewed through Manto's own eyes, intriguingly incorporates a scene from Manto's previous short story titled *Khol Do*, in which a traumatized rape victim instinctively opens her salwar whenever someone utters the phrase "Khol Do". While the allusion appears relevant, the audience still feels that the scene is not very impactful owing to the irksome lack of genuine emotions. With this one scene, the film effectively conveys in the minds of the audience the veracity of the violence and abuse meted out to women in the aftermath of the partition.

2. Conclusion

Ketan Mehta's film *Toba Tek Singh* in many ways is an important work of adaptation tailor-made for the millennial audience. The question of fidelity to the original text is something that has been discussed at length by critics and reviewers. The objective of this article is not to essentialize the idea of "true" portrayal instead it focuses on the process of transformation that the source text undergoes in its visual recreation. The film seems to fluctuate between the domains of fidelity and creativity and in the process strives to create its own identity. While the central conflict of the division of India remains in the background, the film does not offer enough insight to contextualize the parallel developments during the partition to the actions of the lunatics in the asylum. The film raises questions not only about the exclusionary politics of the partition narratives but also offers an intervention by drawing attention to the characters' feelings of confusion and uncertainty as a ploy to recreate the insanity prevalent at the time of partition.

The conventional understanding is that adaptation involves adapting one form of expression of representation into another. But as we see in the case of the film *Toba Tek Singh*, adaptation opens a text to endless possibilities. And in the process, language becomes a privileged site of exchange between a source text and its translation into a visual image, music, or theatre. No sooner one pauses on this idea, it becomes quite apparent that apart from the adaptation of words into other symbolic and representational forms, there are many more mechanisms by which adaptation serves as a ground for the production of meaning.

REFERENCES

1. Bharati, J. (2019). Existentialism of Manto: Absurdity of Human Condition in "Toba Tek Singh" and "Khaled Mian".
2. Das, S. (2005). Space of the Crazy in Saadat Hasan Manto's Toba Tek Singh. *South Asian Review*, 26(2), 202–216.
3. Fleming, (1977-8). "Riots and Refugees: The Post-Partition Stories of Saadat Hasan Manto" *Journal of South Asian Literature* 13, 1–4.
4. Hashmi, Ali M. (2012). "Manto: A Psychological Portrait." Social Scientist 40 (11/12): 5–15.
5. Ispahani, M. (1988). Saadat Hasan Manto. *Grand Street*, 183–193.
6. Manṭo, Saadat Ḥasan. (1987). *Kingdom's End and Other Stories*. India: Verso.
7. Mehta, Ketan (Director). (2018). *Toba Tek Singh* [Film]. Zee Entertainment Enterprises.
8. Saxena, A. (2022). Reading Nearby: Teaching Sa'adat Hasan Manto's "Toba Tek Singh". In Parui, A. (2015). Memory, nation and the crisis of location in Saadat Hasan Manto's 'Toba Tek Singh'. Short Fiction in Theory & Practice, 5(1–2), 57–67. Teaching Literature in Translation (pp. 35–43). Routledge.

Impending Inquisitions in Humanities and Sciences (ICIIHS-2022) – Dr. Mohan Varkolu et al. (eds)

Use of Mother Tongue in English Classroom: An Evidence Based Strategy to Implement National Education Policy-2020

21

Satish Kumar Nainala*
Assistant Professor, Dept. of English
Gandhi School of Humanities and Social Sciences
GITAM (deemed to be) University, Hyderabad, India

ABSTRACT The changing status of English form a colonial imposition to an aspirational language proposes a change in the understanding the spread of English across the globe and rise of its status to the global language. The steady demand for English and subsequent growth of the English users, however is seen to be a blessing in disguise. The policies and practices that are proposed and put into practice to meet the English demand seemingly is exerting a negative influence on the status and learning of other languages let alone their survival. Even though there have been many policy measures suggested to bridge this gap, very little evidence of practicable measures of teaching is presented. Encouraging results from the recent studies in the English Language teaching using L1 for various roles as a resource for the development of target language, in most cases English, and its various language skills and the knowledge through the means of transfer provide an impetus to use Indian languages in the classroom. This evidence also shows that all languages and the proficiency developed are conceptually related to each other dialectically. Going further ahead the paper argues that a simultaneous use of many enabled languages, the mother tongues, encourages the teachers to put mother tongue into use in classrooms at various levels. There by it tries to offer a practical evidence based suggestion to use Indian languages in the classroom as proposed by National Education Policy, 2020 sustaining the multilingualism.
KEYWORDS National education policy, Common underlying proficiency, Mother tongue, Interference, Scaffold, Resource, Monolingual and multilingual

1. Introduction

A recent survey published in a French based website (https://www.statista.com/statistics/266808/the-most-spoken-languages-worldwide/) shows that English has become the most used language in the world surpassing Mandarin which enjoyed the status of most used language earlier. English, with a cumulative user population of 1.35 billion people worldwide either as native users or as a second language speakers, exceeded the 1.12 billion Mandarin users. The growth rate of English as envisaged by the 'World English Language Project' (Graddol, 2006), indicates that in a span of a few years, the possibility of 2 billion people simultaneously learning English could see a rise. In other words, the percentage of the population trying to learn English could account for ⅓ of the entire world population. On the other hand, the British council estimate in 2000 clearly indicates that almost 750 million and 1 billion people were learning the English language. It further goes on to estimate that the emerging economic scenario of urbanisation and increase in the middle class hints at the 'world English project' which if succeeded is estimated to have 3 billion speakers by 2040 who use the English language. The cause of the increase of the English users,

*snainala@gitam.edu

DOI: 10.1201/9781003489436-21

unlike earlier colonial times, is now willingly adopted as the component of basic education by the parents particularly in a country like India. This however shows a glaring contrast in the trend of the spread of English; from earlier being a colonial imposition to a willing adoption by the people in the post colony.

This however has also come with a price. Renowned literary critic and activist Ganesh Narayan Devy who set out to map the linguistic diversity of India through People's Linguistic Survey had documented almost 780 languages of India in 2010. He estimated that of all the languages around 600 of them were dying. He established that in Independent India within a span of 60 years almost 250 language had died. He quotes that the census of 1991 and 2001 show not more than 122 languages which evidently proves that the rest of the languages can potentially called as endangered. This reveals that the rest of the languages face a potential threat of endangerment. To address the language endangerment and to preserve the languages, there have been constant efforts to promote language preservation through language use in educational institutions and educational practices. In a similar line, National Education Policy, 2020 (henceforth referred as NEP, 2020) also expresses a concern for the endangerment of languages. Describing India as a treasure trove of culture with multiplicities of culture and language, it proposes to preserve the 'incredible India' by proposing to 'positive cultural identity and self-esteem' (53). NEP acknowledged that within a span of half a century Indian lost over 220 languages. Quoting UNESCO, NEP reiterates that 197 Indian languages as 'endangered'. To address the grim situation of Indian languages in educational practices, it proposes a mandate for the promotion of Indian languages in the educational practices as one of the measures.

However, the established teaching practices and the myths perpetrated about the use of Mother tongues in the classroom, especially English classroom would prevent the use of the mother tongue. In order to debunk the myth of interference and other myths around it a strong evidence based strategy (using it as post method pedagogic strategy, Kumaravadivelu, 2006) rooted in the updated research needs to be supplemented. The following section tries to explore the myths surrounding the use of mother tongue in English language classroom in particular.

The Myth of Mother Tongue as an Interference

In the educational context, in the past, use of two or more languages has not been encouraged as the existence of another language in English classroom has been seen as an 'interference'. This had successfully demotivated the teachers from using the enormous resource, mother tongue, in the English classroom. In fact, it went to the extent of preventing the use of L1 not just in the classroom but on the premises of the school or any educational institutions. In Phillipson, R (1992) words the monolingual fallacy that 'English is best taught monolingually,' (p.185) using English itself has gained prominence pushing the other languages including the home and local languages out of the learning environment. There have been many reported instances where children were punished for using the mother tongue for conversations on the school premises. Thus, it can be concluded that the fallacy actively derecognised and criminalised the use of L1.

Another pretext for the monolingual classrooms was the assumption that languages exist in separate compartments as separate entities in a human mind. The languages exist and should be kept in separate insulated compartments. If these by any chance get mixed then will result in confusion in the mind. So efforts were to keep the languages apart reinforcing the myth of mother tongue as a hindrance to learning of another language, in this case English. The myth also assumes that proficiency of languages exists in the separate and insulated compartments in human memory. And so each has to be learnt separately without disturbing the other compartment. This model of understanding of language proficiency is termed as SUP (Separate Language Proficiency). This model gave rise to insulated 'single language' classrooms which prevent skills transfer across languages.

C.U.P (Common Underlying Proficiency)

This model strengthens the argument that the experience of any language learning can contribute to developing the underlying proficiency, provided adequate motivation and exposure to the languages in school or in society. This model as presented by Cummins is compared to "a dual iceberg in which common lingual proficiency underlies the obviously different surface manifestations of each language" (Cummins & Swain, 1986–87). In general, surface features of L1 and L2 are linguistic features that are less cognitively demanding and constitute the label of a concept whereas the underlying proficiency is that involved in cognitively challenging and demanding activities.

Facilitating a base for learning languages, CUP gets expanded, which can exert beneficiary effects on both languages. Cummins (2000), in this context, states: "Conceptual knowledge developed in one language helps to make input in the other language comprehensible." In other words, if a student is familiar with and understands the concepts in one language i.e., in her/his own language, it will not be hugely difficult for her/him to learn the labels in another language whereas it can create hurdles if (s)he has to learn both the concept as well as the label in the other language (L2). This theory explains successfully why it is easier for adults to learn additional languages. From the above discussion, it can be inferred that the CUP is not language-specific. This theory asserts that cognitive and literacy skills in the mother tongue or L1 can be transferred across languages which is the primary argument of this project. This representation of bilingual proficiency would also suggest that a continued conceptual and linguistic development in the first language helps in learning a second language or another language.

The myths thus perpetrated did not let the policy makers and practitioners recognise the gigantic resource that a child brings to the classroom for learning. These myths augmented by the English demand could successfully create an English only environment and impacted the language and language education policy so that the demand for English would be served by teaching of English as a language as well as medium of instruction. This thus actively institutionalised the languages policies and practices that does not encourage the use of vernacular languages. Breaking this myth researchers working in the domain of English language pedagogy could successfully argue and prove that use of mother tongue is resource that facilitates the learning of another language. The evidence thus gathered paved a way for looking at the possibility of creating a multilingual classroom.

Use of Mother Tongue in English Classroom: Evidence for a Possibility of a Multilingual Classroom

The Indian reality poses a challenge to the monolingual fallacy and myths thus perpetrated as multilingualism is a grassroots phenomenon here. India is considered as sociolinguistic giant with language changing as the popular saying goes after a mile. In other words, India is a linguistically diverse country with the coexistence of multiple languages as a daily reality. This nature of Indian raises challenges to language as well as education policy makers to address this in the teaching of languages. The colonial proclamation concerning the teaching of English that it should be taught monolingually stands at crossroads because the use of only one language as a communicative tool would be not just uneconomic but absurd (Pattanayak 2014).

Especially in ESL contexts, the myth has made people to believe that languages bear no relation to each other and are and should be taught and learnt separately. In Indian ESL classrooms the teaching practice governed by English only principle while negating the use of the mother tongue is negating the fact that a child brings along a bank of knowledge with him in mother tongue. The 'immediate atmosphere' of the mother tongue in which children grow up is denied access as a result the children in ESL classroom are abandoned from a wealth of resource. In other words, the skills and world knowledge acquired through L1 cannot be transferred to L2 and vice versa. This myth was reinforced by the prescriptive nature of this myth was promoted the monolingual based methodology. This thus reinforced the myth of mother tongue as a hindrance.

Challenging the myth and recognising the fact that the learners are socialised in their L1 opens to the fact that their early knowledge construction happens using L1 which can be accessed and retrieved using the same language.

Arguing for a Positive Role of Mother Tongue

An individual's mother tongue is associated with various factors of human dignity and self-assertion. An L1, along with the language skills, also brings a variety of elements of assertion of a distinct culture and also self-respect of human beings to fore front in the classroom. So, the mother tongue brings with it cultural, societal and psychological features associated to it in the classroom in general and in ESL classroom in particular needs to be considered, recognised and used for the benefit of the learners.

The mother tongue is considered as one of the elementary and primary means of expression of a human being. A student, in the early years of the life, functions in the immediate environment, culture and the immediate society using the mother tongue for the fulfilment of all basic human societal and cultural needs. This is instrumental in grounding of firmly rooting the child deeply into a societal cultural identity rooted in set of social relations and sociocultural practices. This bond between a child and the mother tongue is strengthened in the early years of the child. Even at a later stages of the life in adulthood and old age people identify or prefer to identify themselves with their mother tongue only. And thus the mother tongue anchors child the land and its culture and identity.

As a child grows up in an immediate atmosphere of the mother tongue, the early knowledge and concept formation in the child happen in the L1. In the later stages of life, by means of 'accommodation' and 'assimilation,' the conceptualised world knowledge including the knowledge of concepts related to family, society and culture around of the children extends as they are exposed to the world. During the processes, old concepts formed through L1 might be negated completely, or modified, reformulated, restructured and reshaped into new concepts and new knowledge through the primary means of communication and expression. Moreover, it is very important to remember that in the early learning years of a child, the entire system of language and the knowledge is operated in the L1. The earlier foundation laid by the mother tongue acts as a base for further learning and also to build the superstructure of language system.

If the L1 is recognised and prominently used in the language classroom in general and English or any other L2 classroom in particular, all the features associated with become valuable pedagogic resources. Such recognition of the learner with special reference to the identity, background world knowledge and ethnic, linguistic and cultural identity results in an emotional satisfaction of the students. Mohanty (2007) highlights that the use of mother tongue maximises the utilisation of cultural, linguistic, and conceptual resources in an ESL classroom. In this regard, UNESCO (1958) recommends that a child learns faster and better through the use of mother tongue.

The mother tongue also increases learner comfort by reducing the affective factors and facilitates easy and free communication between students and also with the teacher enabling better learning of any language in general, a language skill (Listening, Speaking, Reading, Writing) in particular. The use of mother tongue also supplements the gaps in learning another language. Thus, the use of L2 gives a 'greater access to education' and 'equal opportunity to participate in national reconstruction' (Pattanayak, 2007: V).

As opposed to the traditional approached to teaching languages that negated any kind of use of L1 as an interference, the emerging body of research suggest a changing attitude and a positive tone. It is time and again reminded that if L1 access is restricted in the learning environment it is considered as a violation of the linguistic human rights (Skutnabb-Kangas and Philipson, 1995.). It not allowed can impedes the processes of 'accommodation' and 'assimilation.

Research based Evidence for the use of Mother Tongue

Earlier research in the use of mother tongue has described the use of it as incidental and it was used mainly for translation, explanation, or classroom management (Atkinson 1987; Cook 2001; Ramanujam 2003). Many studies explored the role of L1 as a scaffold (Bhattacharya 2001; Chimirala 2013; Ghosh 2010; Kumar 2011; Mahanand 2010; Mukhopadhyay 2003; Nirmala 2008; Pathak 2005, 2013). Some studies have used L1 to develop writing skills (Bhengre, 2005; Kedala 2013), to enrich vocabulary (Pavani 2010; John 2017), to teach English in general (Gupta 2009; Meher 2011), to tap the procedural knowledge (Victor 2007) and to transfer that capability to English (Ghosh 2010) (tertiary level of education, second year undergraduates).

It needs to be highlighted here that the subjects of the above studies ranged from school level to tertiary level of education. This suggests that mother tongue can be used as a resource across all levels of education.

With regard to the strategy transfer across languages, a body of literature suggests that strategies of reading as well as writing can be transferred from L1 to L2 or the other language. With particular reference to productive skills, i.e., writing previous studies say that writing in English is 'a comfortable bilingual activity' since a child's concept formation happens in their primary means of expression-the first language and by means of 'accommodation'/'assimilation' extends to the second language. Moreover, in the early learning years of a child, the entire system of language (syntactic, semantic, phonological, etc) is operated in her/his L1. The learners enter the ESL classroom with a more enabled language that can serve as a valuable pedagogic resource.

Role of L1 in ESL Writing Instruction

Research in ESL composition has already proved that students use their L1 while composing in L2 (Akeyl, 1994; Cohen & Brooks-Carson, 2001; Friedlander, 1990; Lay, 1982; Qi, 1998; Uzawa & Cumming, 1989; Wang, 2003; Wang & Wen, 2002; Wolfersberger, 2003; Woodall, 2002). However, the nature of L1 investigated in these studies is largely incidental and self-initiated by students. Using think aloud protocols research studies in the last decade proved successful in investigating the purposes for which L1 is used. Some of these purposes are; generating ideas (Akeyl, 1994; Friedlander, 1990; Wang, 2003; Wang & Wen, 2002; Wolfersberger, 2003; Woodall, 2002), planning and organisation of texts (Wang, 2003; Wang & Wen, 2002; Woodall, 2002), evaluating the text produced, process controlling, back tracking either to generate more text or, alternatively to check back on the success of the match between expression and the intended meaning(Wang, 2003; Wang & Wen, 2002; Wolfersberger, 2003), or solving linguistic problems while formulating text (Wang, 2003; Whalen & Menard, 1995). This paper grounding on the findings of these studies further proposes the deliberate use of L1 for the said purposes above to teach writing in the second language-English. Both, teacher and the students will use the L1 in a planned manner in different phases of writing: generation and organisation of ideas, discussion while writing, and revision.

This proposition is tested in an MPhil study with the title 'Use of L1 as a resource in ESL writing instruction' (2011)' with the above-mentioned purposes/stages of L1 in ESL writing has yielded better scripts with many a sign of growth. Growth in writing capability can be seen broadly as general growth in writing proficiency, and particularly in syntactic and semantic gains, or increased complexity of content. These categories of growth included fluency (text length and number of T-units), syntactic complexity (MTUL: Mean T-unit length), semantic complexity (introductory/stance making statements, reasoning statements, and concluding statements), and linguistic complexity and variety (intra-sentential elements, noun phrases and verb phrases)

It was concluded that unlike the role of the mother tongue (hence forth referred as L1) in teaching English (hence forth referred as L2) writing that is normally restricted to the translation of sentence structures and grammatical rules, the pedagogic value of the cognitive aspect of L1 can be effectively used to generate ideas, organise, plan and (re)write to produce better quality essays with reference to the number of ideas, organisation, vocabulary and language use. In other words, the use of L1 during the stages of prewriting i.e. brainstorming, for generating ideas, organizing, planning, and composing and also for consciousness raising would allow the students to generate more elaborate ideas. Consequently, this would produce more cohesive and qualitatively better L2 texts with enhanced language use.

This paper concludes with a further suggestion that the purposes for which the L1 as pedagogic resource is used can be extended to the other productive skill - speaking as it as a process also involves the same procedure for generating ideas, organizing, planning and composing. However, this suggestion exists with a limitation that this can only have fruitful results in terms of planned speech but not unplanned speech.

Connecting the Evidence for Implementing National Education Policy 2020

As mentioned earlier, National Education Policy-2020 recognising the threat to the Indian languages propose to encourage their use in the educational practices. In addition to many other suggestions of making 'a steady stream of high-quality learning and print materials in these languages including textbooks, workbooks, videos, plays, poems, novels, magazines, etc.' the policy also suggests to use languages extensively for 'conversation and for teaching-learning.' (p. 55). It also proposes to use home/ local languages in teaching making the learning more experiential with an inclusion of local knowledge.

The above suggested research as an evidence provides practicable strategies to implement the policy articulations in the classroom encouraging the use of India languages. It needs to be remembered that these strategies can be used across the curriculum and across all the levels of education. This evidence provides inputs for the curriculum, syllabus and materials developers, and teacher educators for developing programs that actively includes the Indian languages in the English classroom teaching and learning practices in particular and in teaching and learning in general. This also provides input for contextualising, in other words, particularising (Kumaravadivelu, 2006) the teaching learning in a sociocultural context acting as a post method strategy.

REFERENCES

1. Akeyl, A. "First Language use in EFL Writing: Planning in Turkish vs, planning in English." International Journal of Applied Linguistics (1994): 169–176.
2. Atkinson, D. 1987. The mother tongue in the classroom: A neglected resource? ELT Journal 41(4), 241–247.
3. Bhattacharya, T. 2001. Peer Evaluation of L2 Learners' Writing as a Pedagogic Tool. Unpublished MPhil Thesis. Hyderabad: CIEFL.
4. Bhengre, J.J. 2005. The Use of Translation as a Strategy to Develop Writing Skills in English. Unpublished MPhil Thesis. Hyderabad: CIEFL.
5. Bruner, J.S. 1985. Vygotsky: A historical and conceptual perspective. In J.V.Wertsch, (ed.), Culture, Communication, and Cognition: Vygotskian Perspectives, 21–34. Cambridge: Cambridge University Press.
6. Chimirala, U.M. 2013. Exploring Learners' Use of L1 in Collaborative Writing. Unpublished PhD Thesis. Hyderabad: EFLU.
7. Cohen, A. Brooks. & Carson. "Research on direct vs, translated writing processes: Implications for assessment." Modern Language Journal 85(2) (2001): 169–188.
8. Cook, V. 2001. Using the first language in the classroom. Canadian Modern Language Journal. 57(3), 402–423.
9. Cummins, J. 1979. Cognitive/academic language proficiency, linguistic interdependence, the optimal age question and some other matters. Working Papers on Bilingualism 19, 197–205.
10. Cummins, J. 1980. The cross-lingual dimensions of language proficiency: Implications for bilingual education and the optimal age issue. TESOL Quarterly 14(2), 175–187.
11. Cummins, J. & Swain, M. Bilingualism in education: Aspects of theory, research and policy. London: Longman, (1986).
12. Cummins, J. 1996. Negotiating Identities: Education for Empowerment in a Diverse Society. Los Angeles: California Association for Bilingual Education.
13. Cummins, J. 2000. Language Power and Pedagogy: Bilingual Children in the Crossfire. Cleveland: Multilingual Matters.
14. Dodson, C. J. 1967. Language Teaching and the Bilingual Method. London: Sir Isaac Pitman and Sons Ltd.
15. Durairajan, G. 2013. Realisations and perceptions of preferred languages in Indian classrooms: Teacher education implications. In P. Powell-Davies and P. Gunashekar (eds.), English Language Teacher Education in a Diverse Environment: Selected Papers from the Third International Teacher Educators Conference, 39-44. Hyderabad: British Council and EFLU.

16. Durairajan, G. 2017. Using the first language as a resource in english classrooms: What research from India tells us. In H. Coleman (ed.). 2017. Multilingualisms and Development. London: British Council.
17. Felix, P. 1981. Towards the Design of Self-Instructional Bilingual Reading Materials for Non-Formal Learners of English who are Literate in Tamil. Unpublished MLitt Thesis. Hyderabad: CIEFL.
18. Freire, P. 1970. Pedagogy of the Oppressed. (Translated by M.B.Ramos.) New York: Seabury Press.
19. Friedlander, A. "Composing in English: Effects of first language on writing in English as second language." Second language writing: Research insights for the classroom. B. Kroll. Cambridge: Cambridge University Press, 1990. 109–125.
20. Gazette of India. 2009. The Right of Children to Free and Compulsory Education Act No 35. New Delhi: Legislative Department, Ministry of Law and Justice. http://eoc.du.ac.in/RTE%20-%20notified.pdf
21. Ghosh, D. 2010. Transfer of L1 Literary Capability to L2: A Study at the Tertiary level. Unpublished MPhil Thesis. Hyderabad: EFLU.
22. Graddol, D. 2006. English next: Why global English may mean the end of 'English as a Foreign Language.' British Council: The English Company (UK).
23. Gupta, T.S. 2009. The Role of the Local Dialect in Teaching English through the L1: An Exploratory Study. Unpublished MPhil Thesis. Hyderabad: EFLU.
24. John, S.V. 2017. Unleashing potential in multilingual classrooms: The case of Bastar in Chhattisgarh State, India. In H.Coleman (ed.), Multilingualisms and Development, 181–188. London: British Council.
25. Kedala, K.S. 2013. Enhancing Writing Skills Using Local Folktales. Unpublished MPhil Thesis. Hyderabad: EFLU.
26. Kumar, N.S. 2011. Using L1 as a Resource in ESL Writing Instruction. Unpublished MPhil Thesis. Hyderabad: EFLU.
27. Lay, N. "Composing processes of adult ESL learners." TESOL Quarterl, 16(3) 1982: 406.
28. Mahanand, A. 2010. Using indigenous oral narratives as scaffolding in L2 learning. Languaging 2, 142–158.
29. Meher, S. 2011. Using Learners' Home Language in Teaching English at the Secondary Level in West Odisha. Unpublished MPhil Thesis. Hyderabad: EFLU.
30. Messick, S. 1989. Validity. In R.L.Linn (ed.), Educational Measurement, 13-103. (3rd edition.) New York: Macmillan Publishing Company for American Council on Education.
31. Mohanty, A.K. "Psychological Consequences of Mother Tongue Maintenance and Multilingualism in India." Multilingualism in India. Ed. D.P.Pattanayak. New Delhi: Orient Longman Private Limited, 2007. 54–66.
32. Mohanty, A. 1994. Bilingualism in a Multilingual Society: Psycho-social and Pedagogical Implications. Mysore: Central Institute of Indian Languages.
33. Mohanty, A. 2009. Multilingualism of the unequals and predicaments of education in India: Mother tongue or other tongue? In O.Garcia, T.Skutnabb-Kangas and M.E.Torres-Guzman (eds.), Imagining Multilingual Schools: Languages in Education and Glocalization, 262–283. Hyderabad: Orient BlackSwan.
34. Mohanty, A. 2017. Multilingualism, education, English and development: Whose development? In H.Coleman (ed.), Multilingualisms and Development, 261–280. London: British Council.
35. Mukhopadhyay, L. 2003. L1 as a Scaffolding Device in the Learning of L2 Academic Skills: An Experimental Study. Unpublished MPhil Thesis. Hyderabad: CIEFL.
36. Nation Education Policy, (NEP), 2020. New Delhi: Ministry of Human Resource Development, Govt of India.
37. Nirmala, Y. 2008. Teaching Writing Using Picture Stories as Tools at the High School Level: The Movement from Other Regulation to Self-Regulation. Unpublished MPhil Thesis. Hyderabad: EFLU.
39. Pathak, M. 2005. Using Knowledge of L1 as a Resource in L2 Teaching: An Exploratory Study. Unpublished MPhil Thesis. Hyderabad: CIEFL.
40. Pathak, M. 2013. The Role of L1 in L2 Reading Comprehension: Evidence from a Study on Strategy Instruction. Unpublished PhD Thesis. Hyderabad: EFLU.
41. Pattanayak, D.P. Multilingualism in India. New Delhi: Orient Longman Private Limited, 2007.
42. Pattanayak, D.P. 2014. Language policies in multilingual states. In D.P.Pattanayak, Language and Cultural Diversity: The Writings of Debi Prasanna Pattanayak, Volume 1, 721–729. Hyderabad: Orient Black Swan.
43. Pavani, P. 2010. Enriching the Vocabulary of Regional Medium Students: A Bilingual Approach. Unpublished MPhil Thesis. Hyderabad: EFLU.
44. Qi, D. S. "An inqury into language-switching in second language composing proceses." The Canadian Modern Language Review, 54(3) (1998): 413–435.
45. Rajagopal, G. 1992. Developing a Bilingual Method to Promote Reading Skills in English by Using the Learners' L1 Resources: An Experiment. Unpublished MLitt. Thesis. Hyderabad: CIEFL.
46. Ramanathan, V. 2005. The English-Vernacular Divide: Postcolonial Politics and Practice. Hyderabad: Orient Longman.
47. Ramanujam, P. 2003. The use of the mother tongue in the ESL classroom. The Journal of English Language Teaching (India) 39(4), 30–35.
48. Rao, R. 1996. Developing Writing Skills in L2 Using the L1 Abilities in a Bilingual Context. Unpublished MPhil Thesis. Hyderabad: CIEFL.

49. Selinker, L. 1974. Interlanguage. In J.Richards (ed.), Error Analysis: Perspectives on Second Language Acquisition, 31–54. London: Longman.
50. Sen, A. 1993. Capability and well-being. In A.Sen and M.Nussbaum (eds), The Quality of Life, 30–53. Oxford: Clarendon Press.
51. Simpson, J. 2015. Moving from monolingual models to plurilingual practices in African classrooms. Paper presented at the 11th International Language & Development Conference, 'Multilingualism and Development'. New Delhi.
52. Skutnabb-Kangas and Robert Phillipson. Linguistic Human Rights: Overcoming Linguistic discrimination. Berlin: Mouton de Gruyter. 1995.
53. Taylor, L. and Khalifa, H. 2014. Assessing students with disabilities: Voices from the stakeholder community. Paper presented at the ALTE 5th International Conference, 'Language Assessment for Multilingualism'. Paris.
54. Tharu, J. 1981. Measuring small gains. Paper presented at The National Seminar on Evaluation in Language Education. Mysore: CIIL (Central Institute of Indian Languages).
55. UNESCO. The Use of Vernacular Languages in Education. Paris: UNESCO. 1958.
56. Uzawa, K. &. "Writing stragies in Japanese as a foreign language: Lowering or keeping up the standards." The Canadian Modern Language Review, 46(1) (1989): 178–194.
57. Victor, S. 2007. Envisioning Language for the Technical Workforce: An ITI-Based Study. Unpublished PhD Thesis. Hyderabad: EFLU.
58. Wang, L. "Switching to first language among writers with differing second-language proficiency." Journal of Second Language Writing 12 (2003): 347–375.
59. Wang, W. &Wen, Q. "L1 use in the L2 composing process: An exploratory study of 16 Chinese EFL writers." Journal of Second Language Writing 11 (2002): 225–246.
60. Whalen, K. &Menard. "L1 and L2 Writers' Strategic and Linguistic Knowledge: A Model of Multiple-Level Discourse Processing." Language Learning 45:3 (1995): 381–418.
61. Wolfersberger, M. "L1 to L2 Writing Process & Strategy Transfer: A Look at Lower Proficiency writers." TESL-EJ 7:2 (2003).
62. Woodall, B. "Language Switching: Using the first Language while writing in a second language." Journal of Second Language Writing. 11 (2002): 7–28.

Impending Inquisitions in Humanities and Sciences (ICIIHS-2022) – Dr. Mohan Varkolu et al. (eds)

22 The Journey of Becoming and Unbecoming: Configuration of 'Identity' and 'Self' in Abdulrazak Gurnah's Paradise

Sunil Kamal*
Assistant Professor, Department of English,
Manav Rachna International Institute of Research Studies, Faridabad, India

ABSTRACT The current paper examines the colonial problematics of South Africa, particularly Tanzanian life from a postcolonial viewpoint in Abdulrazak Gurnah's Paradise. It fathoms into the magnitude collectively formed by colonial yoke and pre-colonial ill-practices within the region. It seeks to examine identity transformation in association with Raymond William's discourse categorizing it into three realms- dominant, residual, and emergent; however, this will not be separately studied on this praxis, but rather collectively with a postcolonial perspective as it posits serious threat of repetition.

Slavery and colonization are closely studied in consonance with each other as the former is a pathway to the latter. The characters' location vis a vis place and space is evaluated in the second phase as to how it contributes to the shaping and reshaping of identity in relation to Williams' conception. Besides, most of the characters have been affected by their displacement and dislocation- geographical, psychological, emotional, and cultural; which keep their identity in a loop of constant change; thus, a due section has been devoted to space and place-based analysis of cultural identity. Stuart Hall's correlation between past and present as 'becoming' and 'being' is closely associated with William's dominant, residual and emergent; then used as a tool to closely evaluate the multi-pronged facets of transformation of cultural identity in Gurnah's novel. The dimensions of space and time, place, and people are tested upon these variables.

KEYWORDS Slavery, Identity, Postcolonial, Colonialism, Becoming, Displacement, Residual, Dominant, Emergent

The inception of colonialism may roughly be traced back to the fifteenth century, although, European industrialization and so to say, its modernization set it on a global journey to exercise its power on vulnerable countries and colonized them. Postcolonialism and postmodernism are critical of, evaluate, and deconstruct colonialism in different forms and lights by scholars from innumerable parts of the world, especially by those who bore its scarring brunt. Both power and knowledge were brought into practice to strengthen their hold over foreign land; to ridicule and downplay indigenous knowledge and culture. Michel Foucault, the famous postmodernist scholar in Discipline and Punish (1975), aligns parallels between power and knowledge in the course of subjugation vis a vis power relation- "Power and knowledge directly imply one another... There is no power relation without the correlative constitution of a field of knowledge, nor any knowledge that does not presuppose and constitute at the same time power relations" (27).

In the postcolonial phase, the political, economic, cultural, and linguistic consciousness thwarted the established colonial residual of ill-notions, and the scholars from across the colonies started deconstructing the dominant appropriation of their culture and identity. The colonial views started being questioned and toppled by reclaiming and celebrating their original culture, tradition, and ethnicity. Predominantly, scholars from Asia, Africa, Australia, and several other continents and colonies

*snlkml1990@gmail.com

DOI: 10.1201/9781003489436-22

took over the responsibility of redeeming their identity. Through fiction and non-fiction, philosophical and political discourses, the writers and philosophers depicted colonial predicaments and refuted colonial ideology. Abdulrazak Gurnah's works are one of those postcolonial thinkers who shoulder the responsibility to redeem their cultural and national identity. In addition, interestingly, Gurnah locates inter-cultural and inter-religious problematics within the South-African locales of his setting and blends them with colonial ones in order to weigh the accurate magnitude of the sufferings of the general masses.

Gurnah's conception of identity is that it is not static and fixed, but rather is something that is constantly in flux and evolving. He argues that identity is formed and shaped by the experiences and situations that individuals encounter throughout their life. He believes that identity is a product of the many different places and cultures that a person inhabits. He argues that identity is not something that is predetermined or predetermined by the external world, but rather something that is constantly being shaped and reshaped by the individual's experiences and physical, psychological, and emotional journeys undertaken like his characters in Paradise. Gurnah's narrative of colonization is marked by a dual principle of internal and external forces, necessitating its rigorous analysis in terms of the internal dynamics of colonization of the indigenous populace by their own people and the subsequent colonization of the region by foreign powers, which was facilitated by the foundation of pre-existing colonial set-up. Consequently, Gurnah's narrative is distinguished by its vivid depiction of pre-colonial Tanzania, delving into the structure of the local administration under the leadership of tribal headmen and merchants, which engendered a system of slavery that was further exploited by Europeans by incarcerating the natives. In Gurnah's Paradise, the plight of slavery in pre-colonial Tanzania is narrated through the story of Yusuf, who is sold by his father to a businessman in the south as a form of compensation for a failed hotel venture. Throughout the narrative, Yusuf interacts with Khalil, another slave who works for Aziz and takes care of his sister, who was brought with him and eventually became Aziz's mistress.

Aziz leads trade caravans to the mountains and beyond to trade with the local people in exchange for valuable products. He puts together a team of porters and armed guards to accompany him. If someone fails to repay a loan to him, he takes their children as payment and makes them work as slaves. Yusuf is taken along on his last and most daring journey, far beyond the sea. Everywhere they go, they find the Europeans have already arrived, so trade is difficult because the people are wary of foreigners. When the caravan returns, Aziz has lost a great deal of money and his prospects for further trading seem dim. The lives of the slaves are left in an uncertain state, as their families may have died or been displaced.

Gurnah's narrative reveals the detrimental effects of both slavery and colonization, which are used to strip characters of their natural identity and replace it with a manufactured one. These two oppressive structures are closely linked in their role in the formation of identity. The wealthy merchants and business class see slavery as an acceptable means of repayment for debts, typically involving children who have not yet come to understand their place in the world. In this way, their enslavement gives those in power the time and opportunity to psychologically colonize them and create an identity that is convenient for them. This has resulted in a large slave diaspora within the country. In the present postcolonial era, we observe a diaspora of people scattered around the globe, who were formerly enslaved or taken as indentured labourers during colonial times. In Fiji, over thirty-five percent of the population are descendants of Indian people brought over as indentured labourers by the colonial authorities and left there. They were promised wages and lands, but this turned out to be a ruse to trap them and they were never given a free return passage after the end of their ten-year contract. This happened in other places around the world, including South Africa, Australia, and other European colonies. Even now, it is remarkable that Britain still has many overseas territories, a remnant of its colonial past.

Gurnah's saga of slavery in Tanzania offers insight into the archetypal nature of enslavement, how the area is initially a pre-colonial or partially colonized nation, but then quickly becomes an epicenter for the export of people as slaves to the benefit of Europeans. Narratives like Twelve Years a Slave (1853) a story of Solomon Northup, authored by David Wilson, tells the harrowing tale of an artist held captive for a dozen years, demonstrating the cruelty of this practice; from human labor to natural resources, how slaves were exploited mercilessly, leaving the colonies in a state of destitution after having been robbed of their assets.

Slavery and colonization are two distinct stages of a process. While slavery is a tool of colonization, colonization is a broader ideological process manifested in slavery on a smaller scale. In the novel, both slavery and colonization shape and dismantle the identities of characters by creating dominant social structures. Aziz is representative of the capitalist class which has power and resources over the oppressed, but the Europeans come in and re-establish the dynamics of identity and power. This is exemplified in the novel when the Europeans are said to possess a ring with which they can command the spirits of the land: "The European had the power over the chiefs of the savage tribes, whom he nonetheless admired for their cruelty and implacability . . . It was said that the European possessed the ring with which he could summon the spirits of the land to his service" (62, 63).

The ones who had been in the topmost echelon of the power structure, are now facing a crisis of identity, while the residual of their former power only serves to make their pain more acute. Meanwhile, the colonizers attempt to gain control of every business and piece of land, converting them all to their own advantage. Yusuf sees first-hand how the Europeans have taken over the farms, with their trucks and oxcarts, and a look of disdain for the locals, "European farmers came into the town in their trucks and oxcarts, for supplies and to conduct their mysterious business. They had no eyes for anyone, and strode about with a look of loathing" (69)

Inferior identity here on the pretext of external colonization is the realization as well as acceptance of a higher power structure as B. Dirks, in his essay, "Colonialism and Culture", argues that colonialism was not only a consequence of superior arms and political power but also a product of cultural technology. He explains that colonialism was a project of control, which imposed its own cultural effects, leading to the acceptance of a higher power structure by the colonized natives due to their fear of technological, scientific, and ethnic superiority. This ultimately resulted in the formation of an inferior identity characterized by xenophobia:

> Although colonial conquest was predicated on the power of superior arms, military organization, political power, and economic wealth, it was also based on complexly related varieties of cultural technology. Colonialism not only has had cultural effects that have too often been either ignored or displaced into the inexorable logic of modernization and world capitalism, but it was also itself a cultural project of control. (Ashcroft et al 58)

Thus, in Paradise, Gunrah delineates colonization in two phases: internal and external. The internal phase is led by the dominant native class, such as merchants whose profits rely on slavery, external phase is when European forces colonize South African countries, governing them from abroad. By exploring these two forms of colonization, Gunrah exposes the dynamics of colonialism, as well as its performative role in establishing colonial empires, regardless of geography, color, or ethnicity. B. Dirks' thoughts on this provide further insight.

The use of "cultural technologies" to portray foreign high culture as superior to local culture and traditions, and to depict locals as "savage", is a culturally constructed concept that is used to further the agenda of modernization and civilization. European colonialism has employed ethnicity and religious conversion as a means of control, propagating the idea that these tactics are the only path to salvation. This has resulted in the institutionalization of these beliefs, and in their acceptance by the masses.

Interestingly, Gurnah's narrative offers us an angle to evaluate internal colonialism in the same way as European colonialism. The subjugation of the powerless by the powerful, of the poor by the wealthy, and of the uncivilized by the civilized is the same with the only difference being that the Europeans are replaced by locals. Aziz's trade expeditions into the wild mountainous regions give us insight into how this works- with wealth and power, Aziz buys weapons and an army to help him in his cause, then forms a class of micro-businessmen and borrowers, eventually robbing them of their children as slaves and their daughters as his mistresses. The internal corruption depicted by Gurnah is just as ruthless as a colonial yoke of persecution. His supremacy is largely established by the power of cultural legacy, which leads people in the south to view those from the mountains as 'savage others' - just as colonizers do. Modernization and civilization are seen as capitalist cults and the prerogative of the dominant class.

Coming back to Raymond Williams' conception of identity vis a vis colonization phenomenon on the discussed praxis, it works on dual principles with the same configuration. Khalil's and Yusuf's fathers aspired business success to be their emergent identity, a construct to get rid of their poverty-stricken residual; is eventually dominated by their present life crisis in the form of business failure. Their progeny on the other hand is meted out with pre-ordained 'dominant' in the garb of slavery. The residual of their family determines their 'dominant', wherein, their 'emergent' as slaves is a perpetual uncertainty, for they have no say of their own. The whole course of their identity is carved out by their masters; in fact, their life, by and large, is their earned property.

The dominant structures are often located to be in a perennial conflict for ensued power struggle- the merchants struggle to negotiate as well as win over the tribal heads for free and profitable trade; on the other hand, the tribal heads try to curb their trade and seek more and more presents to satiate their lust for wealth. More than often, they resort to violence to take control over one another and to benefit from the enterprise. However, collectively, their identity configuration takes the same course, whereby their residual-their legacy of dominance, strengthens their dominant, their stature as controlling powers in society. Nevertheless, their 'emergent' is their tragedy caused by the rupture of their dominant in the wake of the Europeans' arrival. Therefore, the identity configuration in Paradise, more or less, is structured on the basis of class consciousness, for class factor in the novel is a major determinant with regards to one's position in society.

The Europeans' colonialist attitude is often seen as a form of cultural destruction and misrepresentation, and despite being acknowledged by the natives, it is still unstoppable due to the Europeans' exercise of political, economic, and material power. This lust for power is fuelled by the Europeans' acknowledgment that the natives are vulnerable to being ruled. In the novel, the African colonization takes place despite the natives' prior knowledge of its ill intentions and repercussions. The conversation between Hussein, Hamid, and Kalasingha manifests these fears and apprehensions vividly as they translate the uncontested subjugation by the Europeans as the marked end of their cultural heritage and personal freedom. Hussein, expressing his fears and concerns over their shadowed emergent in the wake of colonialism, "Everything is in turmoil. These Europeans are very determined, and as they fight over the prosperity of the earth they will crush all of us. You'd be a fool to think they're here to do anything that is good. It isn't trade that they are after, but the land itself. And everything in it . . . us" (86). Hussein's fear of cultural destruction is well-founded when he says, "When they write about us, what will they say? That we made slaves" (88), as the western powers have a long history of misrepresenting and appropriating the cultures of colonized nations. Historiography has been used as a tool to depict these cultures as uncivilized and to justify their subjugation and exploitation. Gurnah's novel is a powerful representation of the effects of Western colonization on African culture, including the re-writing of literature and the invocation of existing evils such as slavery.

In Paradise, the effects of displacement and dislocation are evident in the plot, as well as in the narrative of Gurnah. As an expert on colonial and postcolonial discourses, Gurnah deliberately incorporates these elements to demonstrate the complexities of the subject matter and the various forms of oppression experienced by his characters. The narrative explores the effects of colonization in South Africa, both prior to and after colonization. During pre-colonial times, slavery was embraced as a cultural indulgence by the ruling class, and when colonization occurred, this tendency was perpetuated by the colonizers for their own benefit, resulting in slaves being exported to European countries. Yusuf's journey comes full circle when he returns to tending the garden, and his disobedience is further demonstrated when he falls in love with Aziz's younger mistress, Amina. In the end, the reader is left wondering what punishment Yusuf will receive for his disobedience. The different types of banishment he experiences shape and redefine his identity in terms of his dominant, residual, and emergent traits, with a particular focus on his emergent identity, which is constantly changing.

Similarly, Khalil and Amina like many other slaves suffer their banishments for the disobedience of their father. Yusuf's mother's trade for the sake of goats and beans is her father's disgraceful act that her husband often taunts her for. The greatest banishment of all comes with the colonization that thwarts the stalwart businessmen and tribal heads into a frenzy of everlasting exile within their homes and country. The agony of banishment is best conveyed through an episode of Yusuf's departure on his first second expulsion: "He felt as if he was being banished, and felt accused of a betrayal he did not comprehend. Khalil pulled him near and held him in a long embrace and then let him go. 'it's better for you,' he said" (53).

Everybody's dominant is always in a constant emergent; it keeps shaping and restructuring with a perpetual shift in place and space, although, their 'residual' is the end product of a historical phenomenon, which cannot be seen in isolation. The past intervention is significant in the making of the present dominant, thus, the present identity is the construction of its past. Stuart Hall elaborates on this historical construct of cultural identity precisely in "Cultural Identity and Diaspora", defining how the present condition is a matter of 'becoming'- which again is a historical process:

In the second sense, is a matter of 'becoming' as well as of 'being'. It belongs to the future as much as to the past. It is not something that already exists, transcending place, time, history, and culture. Cultural identities come from somewhere and have histories but like everything which is historical, they undergo constant transformation. (Ashcroft 435)

Thus, Hall draws a connection between three dimensions of time, namely past present, and future which collates with Raymond William's dominant, residual and emergent. The same may be linked to geographical as well as emotional dimensions as Hall suggests. While 'becoming' is essentially an evolution, a residual seeping into dominant and emergent; 'being' is essentially the end culmination, a 'dominant'; however, in the perpetual transformation as an 'emergent'. Certainly, cultural identities have a past and history involved, which is a dominant construct of 'being'.

Yusuf's life experiences have had a tremendous impact on the development of his character and identity. Through the move from a place viewed as uncivilized and wild to learn about the ways of a "civilized Muslim life" from Hamid, Yusuf's wisdom and understanding of his own character and identity grow. He changes from a timid, inexperienced boy to a more opinionated individual, capable of standing up to Aziz and plotting to run away with Amina. This transformation is the result of the journey he has taken throughout his life.

Aziz, on the other hand, is shaped by the past and his relocations, from an ordinary self into a revered as well as envied being. in spite of Zulekha's physical ailment, he marries her for the sake of wealth and prosperity. Taking advantage of the wealth he incurs, he grows his trade manifold and becomes a famous trader in the farthest region. Similarly, Khalid and Amina are both the victims of their past, whereby they have been bonded for their life; so are all other slaves. Their displacement is the transformant of their identity, reconfiguring it tremendously. Likewise, the past and displacements have a significant role to play in configuring and disfiguring cultural as well as individual identities in the novel. Abdulrazak Gurnah uses them as metaphors and as prominent constructs of cultural identity.

However, Yusuf is portrayed as a contemplative character who rarely speaks up about anyone's personality or identity, instead, he reflects on his own journey and examines people and their lifestyles. As he is relocated more and more, his nostalgia changes. In the beginning, it is his mother and her fragrance that causes him to weep in the night; later, at Hamid's, he remembers Khalid, the beach, and his customers; and ultimately, it merges with Amina's. Thus, the nostalgia for his many displacements continuously puts Yusuf in a perpetual battle to cope with his own existence in a state of limbo; he is always a secondary citizen wherever he goes.

Nervous and combative, Yusuf is hemmed in from all sides and dependent in a state of displaced identity. Like H. W. Longfellow's poem, "The Slaves Dream", Yusuf too is trapped in a dream-like state of being, longing for the memory and dream of home and freedom. This nostalgia works on him like an alcoholic hangover, intensifying his repulsion and revolt against the subjugation of Aziz. Like the protagonist of Longfellow's poem, Yusuf too becomes fearless of the consequences of disobedience and is ready to confront Aziz and blossom an affair with his mistress. Even when Aziz offers a truce, Yusuf is still determined to mutter the words of disobedience that he wants to speak- "I want to take her away. It was wrong of you to marry her. To abuse her as if she has nothing which belongs to her. To own people the way you own us" (241).

Yusuf's speech is a manifestation of his pent-up resistance to his master's authority, and a recognition of his own oppression and that of his beloved ones. His journey of self-discovery has been enabled by his displacements and dislocations, which have enabled him to explore the world and his own sense of self. His aspirants of paradise are those living a more compromised life, unlike wealthy people like Aziz and headmen who use wealth and power to strive for material advancement. Gurnah's intertextuality, both metaphorical and ethnic, provides a dual ethnic perspective that paradise is something to be sought and not just a world of wealth and power.

Conclusion

In his work, Abdulrazak Gurnah examines the pre-colonial African experience, particularly the practice of slavery, and how it contributed to the eventual colonization of the continent by Europeans. He scrutinizes the complex mix of diverse narratives and stories to explore how class structures within the African continent were used to the Europeans' advantage. This paper provides a postcolonial perspective on Gurnah's work, tracing the history of colonization through the pre-colonial period of slavery and European occupation, and helps understand how these hierarchal class structures shift with time, shifting with them the paradigms of identity too. The constant, never-ending journeys of characters on the other hand, and their permanent displacement remain a cosmetic force in perennially building and dismantling identity at the individual level. Everybody's journey is a saga of everybody's becoming and unbecoming.

REFERENCES

1. Achebe, Chinua. *Things Fall Apart*. Penguin, UK, 2001.
2. Conrad, Joseph. Heart of Darkness. Legend Press Ltd, 2020.
3. Jacobs, J. U. "Trading Places in Abdulrazak Gurnah's *Paradise" English Studies in Africa*, Tylor &Francis Online, vol. 52, no. 2, 2009, pp. 77–88. https://www.tandfonline.com/doi/abs/10.1080/00138390903444164. Accessed 22 May 2021.
4. Gurnah, Abdulrazak. Paradise. The New Press, 1994.
5. ______________. *Afterlife*. Bloomsbury Publishing, 2020.

6. ________________. *Gravel Heart*. Bloomsbury Publishing, 2017.
7. ________________. By the Sea. Bloomsbury Publishing India Private Ltd, 2002.
8. ________________. Memory of Departure. Random house, UK, 1987.
9. McQueen, Steve, director. *12 Years a Slave*. YouTube. https://www.youtube.com/watch?v=YdWTGFUnVKs. Accessed on May 15, 2021.
10. Achebe, Chinua. "Colonial Criticism". *The Post-Colonial Studies Reader*. Eds. Bill Ashcroft, Gareth Griffiths and Helen Tiffin. London: Routledge, 1995. 73–79.
11. Anderson, Benedict. "Imagined Communities". *The Post-Colonial Studies Reader*. Eds. Ashcroft, Grifiths and Tiffin. London: Routledge, 1995.
12. Ashcroft, Bill. Gareth Giffiths and Helen Tiffins, eds. *The Post-Colonial Studies Reader*. London: Routledge, 1995.
13. Dirks, Nicholas B. "Colonial Culture". *The Post-Colonial Studies Reader*. Eds. Ashcroft, Griffiths and Tiffin. London: Routledge, 1995.
14. Foucault, Michel. *Discipline and Punish*. New York: Vintage, 1995. p 27.
15. Longfellow, Henry Wordsworth. "The Slave's Dream". https://www.hwlongfellow.org/poems_poem.php?pid=77. Accessed on May 15, 2021.

Impending Inquisitions in Humanities and Sciences (ICIIHS-2022) – Dr. Mohan Varkolu et al. (eds)
© 2024 Taylor & Francis Group, London, ISBN 978-1-032-78829-6

Translation and Reflections: A Study of the Challenges Associated with the Translation of Vamsi's *Maa Diguva Godaari Kathalu*

23

M. Sundeep[1]
Research Scholar, Department of English,
Acharya Nagarjuna University, Guntur, Andhra Pradesh, India

Madupalli Suresh Kumar[2]
Professor, Department of English,
Acharya Nagarjuna University, Guntur, Andhra Pradesh, India

ABSTRACT Every literary text is a textual representation of culture: geography, landscape, lifestyle, social milieu and language, which is unique in its own way. Vamsi's *Maa Diguva Godavari Kathalu* (2011) is a collection of short stories that presents the 'authentic' culture of the Godavari delta region with its meticulous and lively description, thus successfully connecting the text to the readers of the source language. On the other hand, this profound authenticity makes it challenging to translate the text into English, and yet retaining the authenticity. It becomes further challenging when the target readers are not familiar with the culture of the source text as language remains the only medium to communicate all the aspects of culture to the target readers through the target-text.

This article looks into the practical challenges associated with the translation of these stories into English. These challenges are categorised into: (i) Linguistic (ii) Idiomatic (iii) Syntactic (iv) Cultural. The article further attempts to anticipate the possible implications of these challenges in the reception of the target-text by the target readers.

KEYWORDS Translation, Culture, Challenges, Source-text, Target-text

1. Culture, Literature and Translation

As Terry Eagleton (2016) says, "Culture is an exceptionally complex word in the English language… four major senses of it stand out. It can mean: (1) a body of artistic and intellectual work, (2) a process of spiritual and intellectual development, (3) the values, customs, beliefs, and symbolic practices by which men and women live, (4) a whole way of life". This complex nature of the culture influences the language, themes and settings of the literature produced in a society. The literature thus produced lays emphasis to represent the culture and thus influences the society. C.S. Lewis says, "Literature adds to reality, it does not simply describe it. It enriches the necessary competencies that daily life requires and provides; and in this respect, it irrigates the deserts that our lives have become." Similarly, Translation tries to re-present the cultural aspects presented in the literature to the target audience. As Anthony Burgess opines, "Translation is not a matter of words only: it is a matter of making intelligible a whole culture". Thus, Culture, Literature, and Translation are not only closely inter-related but also complementary with each other, and they are rather inseparable. Hence, the role of a translator while translating a literary text from the source language

[1]sundeepmaneri@gmail.com, [2]msk4anu@gmail.com

DOI: 10.1201/9781003489436-23

into a target language is challenging as he/she has to consider the cultural aspects involved in the source text to be presented to the readers of the target text. This challenge has raised curiosity to attempt the translation of Vamsi's much celebrated work, '*Maa Diguva Godaari Kathalu*'.

Vamsi's Writings and *Maa Diguva Godaari Kathalu*

Though Vamsi is more popular and familiar to the Telugu audience as a film director and writer, he is prolific writer as well. He is known for his popular works like *Manchu Pallaki, Mahallo Kokila, Pasalapudi Kathalu*, and *Maa Diguva Godaari Kathalu.* His works are the best representation of the Godavari delta region – geography, landscape, traditions, social milieu, idioms, dialect and lexicon. His fondness for the Godavari region is unparalleled as he made many of his films in the Godavari backdrop. The setting, themes, characters, events… everything presents the Godavari flavour in them. This can be seen at the highest level in *Maa Diguva Godaari Kathalu.* This is a collection of fifty-two stories that were published in the Swathi magazine. These stories present an impeccable and meticulous representation of the Godavari culture. Such representation brings the authenticity to this collection. B.S. Rama Rao in his foreword to this book aptly says, "The lower Godavari region leaves an indelible impression on the minds of the readers. It gives an experience of visiting the region."

Vamsi names the characters in these stories with their surnames. These names reflect the nature of that characters. Similarly, the stories give us the taste of the recipes of the region. The villages, islands, roads, streets… all are real, with no difference in their geography or topography, as B.S. Rama Rao writes. These qualities enhance the authenticity of this work, and thereby makes it challenging for translation. *Maa Diguva Godaari Kathalu* was received well by the Telugu readers, but the authentic description and its lexicon distance it from the Telugu readers of the Telangana and the Rayaleseema regions because of the unfamiliarity with the culture. Such unfamiliarity is more pronounced for the readers of when it is translated into the foreign language like English.

Challenges in Translation of *Maa Diguva Godaari Kathalu*

The profound authenticity generates curiosity to take up the translation of this work. Simultaneously, it poses various challenges during translation. It turns to be even more challenging if the target readers are not aware of the culture of the source text. These challenges can be categorized into (i) Linguistic (ii) Idiomatic (iii) Syntactic (iv) Cultural or Thematic. It is to be noted that certain challenges can be categorized under more than one category. This paper attempts to present the various challenges faced during translation with few examples from the source text.

2. Linguistic Challenges

The foremost linguistic challenge is the selection of word equivalents in the target language during translation. This arises due to the writer's choice of words within the source language, which is faced in the title of the source text itself. The writer prefers to call it as '*Godaari*' instead of 'Godavari' in the title '*Maa Diguva Godaari Kathalu*'. The word 'Godaari' is not familiar to the target readers. So, 'Godavari' is preferred to '*Godaari*' in the translation as 'Tales of Our Lower Godavari'. Another example of such choice by the writer is the word '*Bigisthaa*' to mean 'drink/consume alcoholol'. This expression is very colloquial to the source culture; the expression has a different meaning in formal usage as 'tighten'. One more example is '*Kaapuralu*' instead of '*nivaasam*' to mean 'to live / to reside'. Such word choices by the writer makes it challenging to the translator to find the right equivalent to retain the sense of the source text.

Another linguistic challenge is the pun and rhyming of the source text. In one of the stories, the writer uses the expression '*Mandeyinchana*' in a punny manner to convey two different connotations in a context, 'to fetch a drink' and 'to get some medicine'. These connotations are unfamiliar to the target language readers, and it is difficult to find expressions to generate the pun in the target language during translation. Rhyming in the source text is another linguistic aspect that makes it challenging to find pairs of words or expressions to in the target language. For example, Vamsi uses an expression '*ekkada penchaalo… ekkada thunchaalo*' to describe the negotiating abilities of a character. But such a figurative rhyming is not possible in the target language in translation.

The next linguistic challenge is the use of target language words by the characters in the source text but conveying a differing meaning or being redundant. For example, one of the characters says '*serious avvaku*' to mean 'not get angry', where serious does not mean actually angry in the target language. Another example is the use of the target language redundantly along with a source language word conveying the same meaning. One of the characters says '*whole mottham*', where the target language word 'whole' and the source language word '*mottham*' have the same meaning and thus being redundant. This is a deliberate attempt by the writer for the authentic representation of the characters.

Inflexions and compound verbs in the source language are the other challenges faced during translation to the target language as the target language does not allow such linguistic privileges. For example, '*evarintikoccharu*' which is a combination of pronoun, noun and verb. Such a compounding is not common in the target language. Few more examples include 'alludnayipothe' and '*mukkuladiripothunnayi*'. This can be seen as a conscious attempt by the writer to evoke the linguistic authenticity.

Another prevalent linguistic challenge is the onomatopoeia. The writer of the source text uses onomatopoeia to present an authentic description of events and emotions. '*chillobollomantu*', '*phelaphela chappullu*', '*malamala maadipothunnaru*', '*gayyigayyimantu*', and '*Ttinguna boni avadam*' are some examples of the use of onomatopoeia that pose challenges in finding an appropriate expression in the target language during translation.

Idiomatic Challenges

Every language has its own corpus of idiomatic expressions that are specific to its culture. They are related so closely with a group and rooted so deep in the culture that Saeed (2003) aptly defines idioms as "words collocated together happen to become fossilized, becoming fixed over time." Such rootedness of idioms makes it very challenging for translation into the target language, because literal translation of these idioms without the awareness of their usage and of the culture of the source text result in distortion of meaning. Similarly, Vamsi's use of authentic Godavari idioms makes it difficult for the translator. For example, Vamsi uses '*Kallaki kattianattu*' to mean 'very clear and lucid'; '*mukkusootiga matladatam*' to mean 'to be frank'; '*pottana pettukodam*' to mean 'to take the lives'; '*thalalo naalukaipoyindi*' to mean 'to be reliable and trustworthy.' The literal meaning of these idioms translates to: 'like tied to the eyes', 'talk directly to the nose', 'keep in the tummy', 'become tongue in the head' respectively, thus leading to the complete distortion of the text.

Syntactic Challenges

Syntax of the source text (Telugu) is inherently different from the target language (English). In addition, the colloquial language preferred by Vamsi in this text makes the syntax of source language even more distinct to that of the target language. This profound distinction is due to the usage of run-on sentences, fragments, informal dialogues and conversations, thus making it challenging for the translator. For example, Vamsi uses this run-on sentence: "*meerannattu pandolichi thonalu thiyyalante pedda thathangame... ye oornichochharu? Evarintikoccharu... ye pani meedocchaaru?*" Such long sentences with colloquial expressions in the source text makes it challenging to retain the informal tone of the source text in the translated version. Another example of such colloquial syntax is '*owner... owner laagunte maataadaleka poyevaadnemo...*' Such run-on sentences with ellipses are not common in the target language. Vamsi also uses sentences without verbs that is usually seen in the source language. For example, '*Gattuki itupakka godaavari. Atupakka diguvalo thathapudi*'.

Cultural Challenges

Culture is the determining aspect in the translation since it encompasses various aspects such as social hierarchy, religion, caste, class, food, customs, traditions, etc. The most common problem is the honorific suffixes like *ayya, andi* which is are commonly seen in the source culture, but the target language does not have equivalents for it. If words like 'sir' 'lord' are used as equivalents, the rustic authenticity gets lost in the process of translation. The next cultural challenge is Vamsi's choice of language for certain classes of people. He prefers to use a different kind of slang or dialect or informal language, which is rudimentary and un-refined, for the working class and underprivileged class. For example, one of his characters who is a labourer uses '*seminchandamma*' instead of '*kshaminchandamma*' to mean 'forgive me'. While such usage is common in the source language (Telugu) in the form of *prakruthi* (standard form) – *vikruthi* (slang/rudimentary form), such a distinct expression is not possible in target language (English). Some more examples are '*manedana*' instead of '*mano vedana*', '*poddothe*' instead of '*poddu pothe*', '*ujjogam*' instead of '*udhyogam*'. The next cultural challenge is the events and traditions that are specific to the source culture but are alien to the target culture. In such cases, it is difficult to find equivalent word in the target language, and the translator's responsibility is not only suggesting an appropriate word but also familiarising the culture to the target readers. For example, Vamsi gives a reference of '*bhogam melam*' which is a dance troupe that performs in villages on festive occasions and fairs, usually seen in the source culture. There is no such practice in the target culture and hence it is challenging to find an appropriate word during translation. Some more such examples include '*thaamboolam*' to mean 'beetle leaf pan given customarily to women on special occasions', '*aadapaduchu*' to mean 'husband's sister or a woman closely attached to one's family'. Such words make it challenging to the translator to preserve the authenticity in the process of translation.

Implications of Challenges

Apart from posing challenges to the translator, these problems have various implications which impact the reception of the target text and appreciation by the target readers. This section lists the implications of the various challenges discussed in the previous sections:

(a) loss of authenticity of the source text due to the lack of equivalents and cultural distance
(b) change of tone of the narration due to the syntactic and linguistic challenges
(c) genesis of ambiguity due to unfamiliarity of the culture
(d) emotional dissociation in the target readers due to cultural distance

REFERENCES

1. Eagleton, Terry. (2016). Culture. Yale University Press.
2. Lawrence, Venuti. (2012). The Translation Studies Reader. London: Routledge.
3. Saeed, J. I. (2003). Semantics. Oxford: Blackwell.
4. Vamsi, Nallamilli. (2011). *Maa Diguva Godaari Kathalu*. Sahithi Publications.

Impending Inquisitions in Humanities and Sciences (ICIIHS-2022) – Dr. Mohan Varkolu et al. (eds)
© 2024 Taylor & Francis Group, London, ISBN 978-1-032-78829-6

Shifting Paradigms in English Language Teaching: An Analysis of the 'Innovations

24

Kranthi Priya Oruganti*
Assistant Professor, Department of English,
Koneru Lakshmaiah Education Foundation, Hyderabad

ABSTRACT With the emergence of new theories of language teaching and learning, there have been many innovations in pedagogy. With the varied educational opportunities accessible to the present learner, new trends in ELT that emphasize more on quality language teaching and catering to the learner needs have turned out to be major factors in the ongoing phase of transition. Recent trends in ELT have laid enough emphasis on shifting the traditional classroom to a more flexible, effective and learner-centric one in order to overcome the limitations of the traditional methods of language teaching. Rather than learning in a traditional classroom setting, a classroom that uses different methods such as, Task-Based Language Teaching, Flipped Method and so on, provides space for a hands-on, problem-solving learning environment.

To get a comprehensive understanding of the shifts in language teaching and to identify their effectiveness in the classroom, this paper is divided into two sections. The first section presents a bird's eye view of the different theories of teaching and learning along with the different trends in ELT that have been prevalent in the recent times. The second section provides a description of the three trends in ELT which are Task Based Language Teaching (TBLT), Flipped Classrooms and Blended Teaching using Computer Assisted Language Learning (CALL)/Technology Enhanced Language Learning (TELL) with reference to the teacher's role and usefulness in the classroom. A brief description of the methods and the analysis of data from classroom findings will also be provided in this section.

KEYWORDS Recent trends, Language teaching theories, Methods and approaches, Task based language teaching, Flipping the classroom, Computer assisted language teaching

1. Introduction

English Language Teaching (ELT) in India has undergone a tremendous refurbishment over the past few decades. Innovative trends in ELT have been gaining significance in education systems around the world. Major changes have taken place in ELT over the last one decade. Language teaching in the twentieth century has undergone many stages which include a decline of popular methods and approaches that ruled the classrooms followed by emphasizing top-down and bottom-up approaches to teach the four skills and so on.

In order to present the shifts in language teaching and their effectiveness in the classroom in a more comprehensive way, this paper is divided into two sections. The first section highlights a few theories in education prevalent over the past few decades.

*priyavankalapati@gmail.com

DOI: 10.1201/9781003489436-24

2. Section I: Methods and Approaches in English Language Teaching

While some trends in ELT are quite noticeable, a few new ones are yet to make their impact visible. The Grammar-Translation Method (GTM) where the focus was majorly on teaching grammatical rules was replaced by Communicative Method of teaching where the focus shifted to meaningful and contextualized discourse. While the black board was the major tool for teachers while teaching using GTM, new trends have emphasized on using technology and other tools for effective teaching to take place.

However, it can be said that, the necessity for the evolution of theories, methods and approaches usually occurs when there arises a need to change the existing teaching methodologies; change the teaching content and test design; update and incorporate new approaches; and reflect on existing teaching-learning practices to acquire desirable learning outcomes.

Many methods and approaches in English Language Teaching have emerged over the years. Though theories like Gardner's theory of Multiple Intelligence (MI), Chomsky's Language Learning versus Language Acquisition theory, Krashen's theory of Second Language Acquisition (SLA) were predominantly successful, Asher's Total Physical Response (TPR), Whole Language (WL) approach, Neuro linguistic Programing (NLP), Cooperative Learning (CL) and Blooms Taxonomy have majorly influenced language teaching and learning immensely.

The following diagram (Fig. 24.1) presents a few of the methods and approaches that were used to improve teaching and learning opportunities.

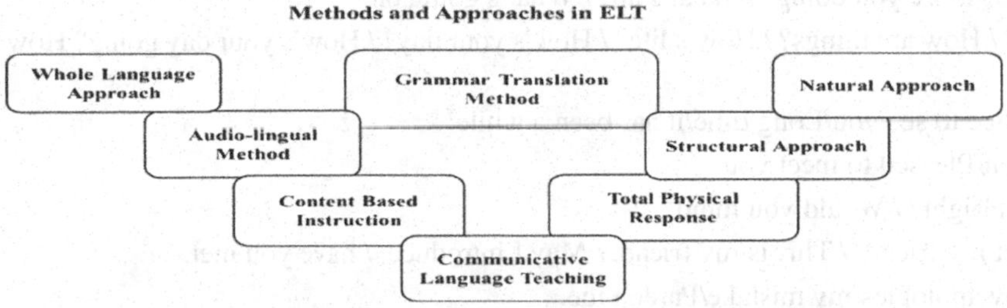

Fig. 24.1 Methods and approaches in ELT:

The below figures, (Fig. 24.2 and Fig. 24.3) are a representation of the two theories that dominated ELT over a period of time.

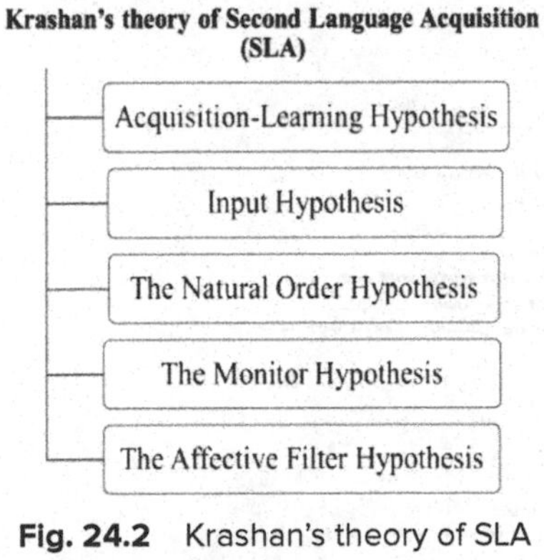

Fig. 24.2 Krashan's theory of SLA

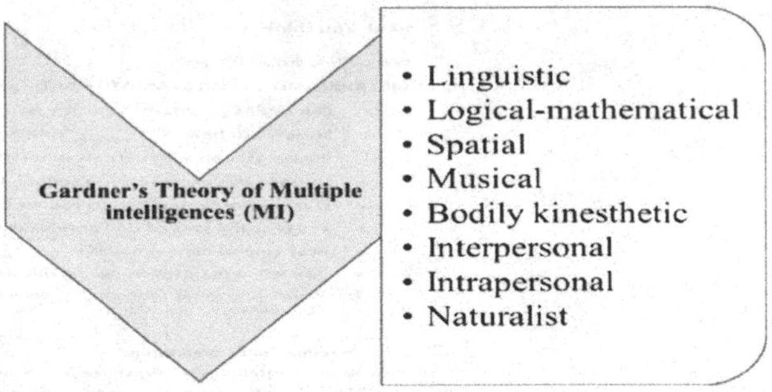

Fig. 24.3 Gardner's theory of Multiple Intelligences

On the other hand, recent innovations in English Language Teaching such as Content and Language Integrated Learning (CLIL), Computer Assisted Language Learning (CALL)/Technology Enhanced Language Learning (TELL), Information and Communication Technology (ICT), Web-Based Learning, Task Based Language Teaching (TBLT), E-Learning, Flipped Classrooms, Blended Teaching and Using Technology (Blogs, Email, Facebook, Skype, WhatsApp and so on) are the recent trends in ELT that are appealing to the new age learner. These trends not only encourage the ELT trainers to modernize their classroom arrangements but also ensure the learner's active participation in classroom activities.

The next section presents a detail three trends used in the classroom with reference to the teacher's role and their usefulness in the classroom for the 21st century language teacher.

3. Section II: Trends in English Language Teaching

This section provides a description of three trends in ELT that were used in the classroom at the tertiary level (B.Tech.) and proved to be effective in assisting language learning. Task Based Language Teaching (TBLT), Flipped Classroom and Blended Teaching using Computer Assisted Language Learning (CALL)/Technology Enhanced Language Learning (TELL) are the three approaches that will be discussed wherein the description of the approaches, teacher's role and usefulness of the approaches will be presented in this section.

Approach 1: Task Based Language Teaching (TBLT)

This approach focuses on the use of authentic language in the classroom where learners are involved in performing meaningful tasks using the target language. This approach attempts to develop the target language fluency by making the learners use language in real life contexts. This approach was used in an English Language Lab to teach communication skills. The learners were provided with language cues and sample role play and situational dialogues. They were later grouped into pairs where they were asked to prepare a role play using the language cues and sample tasks. The following are a few of the language cues and sample task used in the classroom:

Verbal Cues:

- Hey Man! / Good morning/Good afternoon/Good night.
- How's it going? / How are you doing? / What's up? / What's going on?
- How's everything? / How are things? / How's life? / How's your day? / How's your day going? How have you been? / How do you do?
- Good to see you/Nice to see you/Long time/it has been a while.
- It's nice to meet you/Pleased to meet you.
- Are you ok? / You alright? / Would you mind…?
- I'd like you to meet my friend. / This is my friend. / May I introduce. / have you met.
- Sorry about that/my apologies/my mistake/Pardon me.

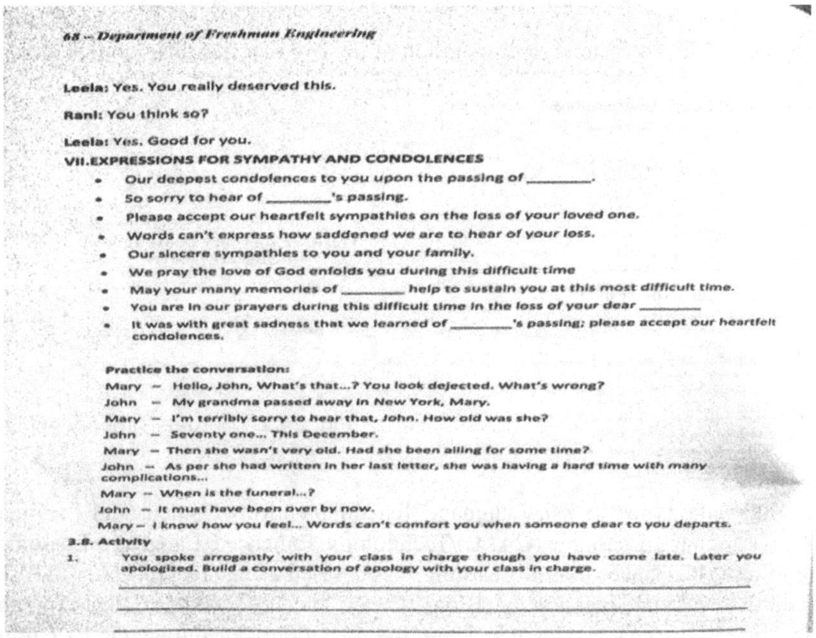

68 – Department of Freshman Engineering

Leela: Yes. You really deserved this.

Rani: You think so?

Leela: Yes. Good for you.

VII.EXPRESSIONS FOR SYMPATHY AND CONDOLENCES

- Our deepest condolences to you upon the passing of ________.
- So sorry to hear of ________'s passing.
- Please accept our heartfelt sympathies on the loss of your loved one.
- Words can't express how saddened we are to hear of your loss.
- Our sincere sympathies to you and your family.
- We pray the love of God enfolds you during this difficult time
- May your many memories of ________ help to sustain you at this most difficult time.
- You are in our prayers during this difficult time in the loss of your dear ________
- It was with great sadness that we learned of ________'s passing; please accept our heartfelt condolences.

Practice the conversation:

Mary – Hello, John, What's that...? You look dejected. What's wrong?
John – My grandma passed away in New York, Mary.
Mary – I'm terribly sorry to hear that, John. How old was she?
John – Seventy one... This December.
Mary – Then she wasn't very old. Had she been ailing for some time?
John – As per she had written in her last letter, she was having a hard time with many complications...
Mary – When is the funeral...?
John – It must have been over by now.
Mary – I know how you feel... Words can't comfort you when someone dear to you departs.

3.8. Activity

1. You spoke arrogantly with your class in charge though you have come late. Later you apologized. Build a conversation of apology with your class in charge.

Fig. 24.4 Sample language task from laboratory handbook on English for communication

Source: Vardhaman, Laboratory Handbook on English for Communication, pg: 68, Lab Manual, June, 2021.

The learners were given time to prepare their roleplay and perform. The focus here was to enable the learners to plan their activity and speak without hesitation. Their performances were also recorded to further check their performances and reflect on their language use. The following are two audio transcripts of their performance in the classroom.

Audio Transcript 1:

Student 1: You have broken my window. Can't you see?

Student 2: We are so sorry uncle. We were playing cricket. We never thought the ball would hit the window.

Student 1: Haven't I told you not to play in the street? Why don't you go to the park?

Student 2: We know we shouldn't play here. We are very sorry. We will not repeat this again.

Student 1: That's okay. But I don't want you to play in the street again.

Audio Transcript 2:

Student 1: Good afternoon. Can I help you?

Student 2: I hope so. I bought this television last month, but the sound and picture quality are awful. The picture is never clear and there is a dark line down the left-hand side of the screen. Also there is an annoying sound in the background.

Student 1: Do you have an outdoor antenna?

Student2: Yes, I do.

Student1: Have you tried adjusting the antenna?

Student 2: Many times.

Student 1: Hmmm. I'll get our engineers to have a look at it.

Student 2: My friend's cousin also bought the same model here and had exactly the same problems. I want a refund.

Student 1: If that is the case, let me take you to our manager sir. Can you please come this way?

Teacher's role:

The teacher had to plan the list of verbal cues and role play samples prior to the class and acted as a monitoring body during the activity. The role of the teacher was a passive one where occasional support was given to the learners along with constructive feedback on learner performance.

Usefulness:

Using Task Based Language Teaching in the classroom was helpful in motivating the learners to perform the task with more interest. It provided space for more meaningful communication among the learners. The tasks were more productive in terms of language use and it improved the learner's confidence in using the target language. Using this approach proved to be effective in terms of giving opportunities to the learner and making them speak fluently.

Approach 2: Flipping the Classroom

In this approach, classroom time is used to explore topics to a greater depth and create meaningful learning opportunities. Learners are actively involved and are encouraged to participate in discussions, pair-work, group-work and problem-solving activities. When intended for a class to teach content reading and critical thinking skills, this approach was planned in a way where the content delivery and introduction of topics was done prior to the class through reading texts and supporting videos. The text was divided into five parts and so were the learners into five groups. Each group had twelve learners and they were asked to read one part of the text assigned to them along with watching the supporting video shared to them. During class time each of the learner provided few points about what they had read and also gave their opinion about the text. Two learners from each group were assigned the task to write down the points put forth by their group members which was later summarized by one learner from the group.

The following is an example of the script presented by two groups on two different occasions where a flipped classroom was implemented.

Sample Script 1: Lesson - Presidential Address

Group 1:

Student 1: APJA Kalam 1931-2015, chief of Indian Research and Development Program. He is called as people's president and won Bharataratna.

Student 2: Talks about cooperative federalism and requirement to develop competitive strengths for the states so that they can excel at national level and global level.

Student 3: He says that India is still a developing country but not a developed country and with the power of knowledge we can empower India at various levels. So, when people are with unity the country will be a developed country.

Student 4: The President speaks about the ideal qualities that a person should have and he says to have future planning to become a developed nation.

Student 5: He talks about the responsibilities of the citizen to de developed and about the important elements of a nation.

Student 6: he talks about the ideal qualities that a nation should possess which are disease free, wealth, high productivity, harmonious living and strong defense.

Sample Script 2: Lesson – Good manners

Group 3:

Student 1: Do not underestimate the people who have disabilities.

Student 2: Unless we are confident about a situation we should not speak. Don't say badly about the person when you are back of him. Arguing unnecessary is not good you cannot understand that what he is saying.

Student 3: Different people has different perspectives and good manners comes from having sympathy towards others.

Student 4: to acquire good manners treat others as we want to be treated. Think about others and the truth we speak is not the actual truth there us something else (original truth).

Student 5: We should understand the real meaning of the speaker and we should have patience to hear and when we speak something it should be clear to the listener. Different people speak in different ways it is our responsibility to make them understand clearly. We should not be overconfident about our truth.

Teacher's role:

The teacher's role in a flipped classroom is very minimal. Though a proper plan to execute a flipped classroom is required for its successful administration in the classroom, the teacher is merely a passive observer in the classroom where the learners have to perform the tasks on which they are instructed. The teacher also can provide immediate feedback to the learners in the classroom.

Usefulness:

This approach proved useful in terms of promoting a collaborative learning environment where learners were willing to participate. It was found that the learners had less fear or inhibition in communicating their ideas and they were capable to identify their own errors and corrected each other in the process. The learners were also interested in knowing the alternative interpretations of the same content from their peers. There was better scope for the teacher to provide immediate and effective feedback to the learner.

Approach 3: Blended Approach using Computer Assisted Language Learning (CALL)/Technology Enhanced Language Learning (TELL)

A Blended Approach is one in which the teacher uses the aid of technology to provide the learner a better learning atmosphere. CALL/TELL is a study of applications of the computer in language teaching and learning. It acts like a tool that helps the teacher to facilitate the language learning process and promotes learner autonomy. This approach was used in the English Language Communication Skills (ELCS) Lab to teach phonetics. During one of the ELCS Lab classes, the learners were introduced to speech sounds and their symbols during this class. The learners with the help of a software (K-VAN Solutions) heard the sounds and related them with the place and manner of articulation. The learners were assisted with a lab manual in which they had to enter the symbols and its corresponding word.

Figure 24.5 is an sample of the contents page of the list of topics that were part of the English Language Lab that were taught using the blended approach:

Teacher's role:

The teacher in a Blended Approach classroom acts more as an support for the learner in using technology and monitors learner's understanding and performance in the given task. The teacher helps the learner in the process of acquiring knowledge of the content on their own, thus by promoting learner autonomy.

VARDHAMAN COLLEGE OF ENGINEERING
(AUTONOMOUS)
Shamshabad, Hyderabad – 501218

LIST OF MODULES		
Module. No. 1	CALL: Introduction to Phonetics - Speech Sounds – Vowels and Consonants ICS: Ice-Breaking activity and JAM session	
Module. No. 2	CALL: Past Tense Marker and Plural Marker – Syllable Structure – Consonant Clusters - Minimal Pairs	
Module. No. 3	ICS: Situational Dialogues – Role-Play – Expressions in Various Situations: Greetings: Self-introduction and Introducing others – Apologies – Requests – Complaints– Congratulating – Expressing sympathy/condolences	
Module. No. 4	CALL: Basic Rules of Word Accent – Stress Shift – Weak Forms and Strong Forms	
Module. No. 5	ICS: Asking for and Giving Directions – Giving Instructions – Seeking Clarifications – Thanking and Responding – Agreeing and Disagreeing – Seeking and Giving Advice – Making Suggestions	
Module. No. 6	CALL: Neutralization of Mother Tongue Influence-Common Indian Variants in Pronunciation – Differences between British and American pronunciation	
Module. No. 7	CALL: Intonation Patterns-Types of Tones - Sentence Stress	
Module. No. 8	ICS: Social and Professional Etiquette - Telephone Etiquette	
Module No. 9	ICS: Oral Presentation Skills (short presentations) - Making a Presentation-Prepared –Extempore	
Module No. 10	ICS: Listening-Types of Listening-Steps to effective Listening –Business Listening Comprehension exercises	

Fig. 24.5 Index from laboratory handbook on English for communication

Source: Vardhaman, Laboratory Handbook on English for Communication, pg: Index, Lab Manual, June, 2021.

Usefulness:

The usefulness of this approach lies in the fact that it appeals and challenges the 21st century learner. It was effective in ways that it engaged the learner and learning was an active process. It also enabled the learners to take responsibility for their own learning and helped the learner understand the mechanics of speech in a better way. Further, in a phonetics lab class, this approach was helpful in assisting the learner comprehend the symbols and the sounds of phonetics more effectively.

The above mentioned three trends which were used in the classroom were to a great extent useful to grab the attention of the learner and motivated the learner to participate in the classroom activities. These trends were instrumental in involving the learner in the process of teaching and learning in a more active way which provided ample scope for creating a learning environment that was more learner-centered.

4. Conclusion

Based in the above discussion, it can be summed that the 21st century teacher in the classroom has to play a multifaceted role of being an assessor, facilitator, organizer, observer, guide, tutor, resource person, participant, and a promoter of knowledge. The following chart shows the role of the 21st century teacher.

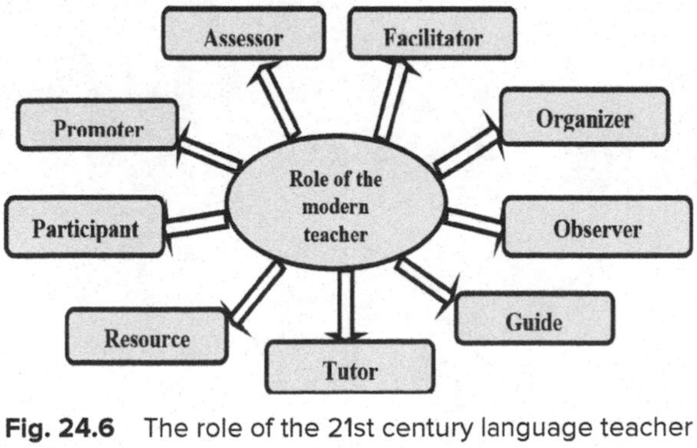

Fig. 24.6 The role of the 21st century language teacher

Source: Author

Moreover, language teachers need to update themselves with the current trends in language teaching and develop a positive attitude towards continuous professional development. In addition to updating themselves, they need to pace up with the fast-changing world by innovating new techniques for language teaching.

In conclusion, English Language Teaching has undergone a lot of changes and most of which were a result of the need for catering the English classroom to the learner's needs. In an attempt to provide ample learning opportunities to the learner, teachers of English need to update themselves with the recent trends in ELT and adapt their teaching methodologies in ways to help the learner learn better. While the new cultural contexts and diverse learner groups pose a challenge to the teachers, it is in these contexts that the new theoretical frameworks are devised based on successfully tested teaching practices. The current trends in ELT such as Flipped Classroom, TBLT and Blended Approach that were discussed above are a few of the approaches to teaching English that can prove to be useful in ensuring learner progress. Based on the above analysis of various classroom scenarios, it can be safely argued that the 'innovations' and the every-day practices that effect these 'innovations' are not the two entities in isolation, they are the two interconnected realities that help evolve each other.

REFERENCES

1. L. Cameron. Teaching Language to young learners. Cambridge: Cambridge University Press 2005.
2. J. Cummins, & C. Davison. "International handbook of English language teaching". Vol. 15, Springer 2007.
3. A. Maley. Global issues in English Language Teaching. Practical English Teaching, 13 (2), 73, 1992.
4. A. Adrian Underhill. "Trends in English Language Teaching Today" in Trends in English language teaching today, April 2004.
5. MED Magazine, issue 18. Retrieved September 15, 2007 www.macmillian.com. (Accessed: September 20, 2019)

Impending Inquisitions in Humanities and Sciences (ICIIHS-2022) – Dr. Mohan Varkolu et al. (eds)

Interrogating Causal Links Between Government Expenditures and Funding Sources, Catalyst for Private Sector in PPPs. A Case of Malawi

25

Raphael K. Munthali
PhD Commerce Scholar, School of Commerce and Management Studies,
DMI-ST. Eugene University, Lusaka, Zambia
K. Logasakthi[2]
Associate Professor, Universal Business School, Universal AI University, Karjat, Maharashtra, India

ABSTRACT This empirical study examines the causal links between monthly government expenditures and tax revenues, grants and loans at consolidated national level for Malawi Government using data sets from 2004 to 2007. The data and models were subjected to statistical tests. Findings show no expenditures to tax revenues link hence parameters are decided independently while loans and grants cause expenditures. One month lagged autoregressive variables for expenditures and tax revenues impact current amounts Unidirectional loans and grants causalities to expenditures infer that government budgets and expenditures are dependent on loans and grants. Here, it is suggested that budget deficits can get closed by intensifying collection of tax revenues and practicing prudent expenditure decisions. Studies of this nature can assist in establishing government's financial capacities to fulfil financial obligations in Public-Private Partnerships which can also guide players in private sector to decide on extents of their participation.

KEYWORDS Government-expenditures, Tax-revenues, Causality, Public-private partnerships

1. Introduction

Gounder *et al.* (2007) proclaims that governments aspire to achieve economic growth with low debts and friendly fiscal policies. Increase in taxes for accommodating government spending affects the capital investments by investors as they fear subsequent higher taxes. Malawi experienced growth in expenditure as a percentage of GDP since its independence attributed to growths in government revenue, government services driven by population growth, services containing unemployment and increased government employment. Goetz (1977) established that different states' ideologies, systems and theories underpin fiscal policies regarding government taxes and expenditures. Moisio (2001); Baffes and Shah (1994) in their studies found empirical inconsistencies to substantiate ideologies and theories backing government fiscal policies moderating taxes and expenditures hence the need for more research to attain consistency. Variations imply that states are prone to divergent findings hence there is no blanket application to them all.

There are various interlinks between states' ideologies and economic, political and financial systems. Fiscal Illusion theory links government growth to convoluted tax systems that mask the costs of public goods (Goetz, 1977). However, empirical support for these theories has variations leading to some loss in their quick progress hence warranting more research so that consistency is achieved (Moisio, 2001; Baffes and Shah, 1994). Understanding the government budgetary process and rational influence on fiscal choices are dependent on the ability to predict the impacts of policy on the behavior of government sectors.

[1]rolf_77bypass@hotmail.com, [2]dr.logasakthi@gmail.com

DOI: 10.1201/9781003489436-25

This can offset influences of temporal policy measures like temporal cuts in grants, changes in taxation rules and changes in business tax revenue sharing among government sectors. Nevertheless, understanding causality links and relationships between government expenditures and adjustments in various funding sources may help in making informed decisions when it comes to policy formulation and implementation. Most literature in the area of government spending has focused on factors attributing to growth in government expenditures. Empirical studies have shown that in most resource endowed countries; the tax revenue-spend hypothesis usually holds since they enjoy high proceeds from foreign exchange earnings that determine their spending pattern. For least developed countries, the links and relationships have not been fully established to substantiate the rapid growth in expenditures and policy orientation so that knowledge gaps are bridged through formulation and implementation of pertinent fiscal policies that can win private sector's trust to invest in public-private partnerships and re-invest their returns.

The major purpose of this study is to examine the causal links and relationships between Malawi government expenditures and revenue sources, that support financial year budgets; to inform expenditure plans, taxation structures, annual fiscal budgets and fiscal policy for the country.

This paper tries to examine and establish policies (spend-and-tax; tax-and-spend or fiscal synchronization) as they apply to Malawi and establish relationships between government expenditures and various government funding sources that support the government budget. For this, consolidated monthly national financial time series data is collected from 2004 to 2007 (the time span Malawi was categorized the second fastest growing economies) and analyzed to establish prevailing causalities. Bivariate and multivariate models are formulated with grants and loans as complementary variables that make the budget complete.

The rest of the paper is structured as follows: Section-2 reviews the existing literature and underlying hypotheses. Section-3 describes the sample and variables. Section-4 details the research methodology. Secttion-5 discusses the empirical findings. Section-6 concludes the paper and Section-7 provides references.

2. Literature Review

Tax and Spend Relationship

The theory was founded by Milton (1982) and is based on tax-expenditure causality implying adjustments in expenditure can be supported by revenues. The appetite for expenditure may undermine efforts to reduce debt, in turn creating budget deficits, as it implies incremental revenues support increased expenditures. It may also be regarded as a careful budgetary policy since the funds are gathered before spending. Saeed and Somaye (2012) used Granger causality test and Vector Autoregressive (VAR) approach in oil exporting countries, 2000–2009, and established a positive unidirectional causality from revenue to expenditure, revenue-spend hypothesis. In Nigeria, Aregbeyen and Ibrahim (2012a and 2012b) did a study using ARDL bound test and Cointegration approach, 1970–2008. Revenue-spend relationships existed. Ogujibuba and Abraham (2012) used the VECM, 1970–2011, for the Nigerian economy where the study supported the revenue-spend hypothesis. The study by Obeng (2015) for Ghana used the VAR causality test, 1980–2013, and found revenue-expenditure hypothesis for that economy. The hypothesis based on theory is as below:

H_1: *Tax Revenues unidirectional cause Government Expenditures.*

Spend and Tax Relationship

The theory was founded by Peacock and Wiseman (1961) and Barro (1979). It advocates expenditure then collecting tax revenues. It is a causality link where increase in tax and borrowing are as a result of increased government expenditure mostly triggered by crisis (exogenous disturbances or crises). its persistence after the crisis or without can be regarded as careless budgetary policy since expenditures are raised before the funding level is determined. The scenario and problems of budget deficits can be resolved by downsizing spending so that revenues collected may contain deficits as in the Tax and Spend. Nwosu and Okafor (2014) used the VECM, 1970–2011, for the Nigerian economy where the study results cemented the spend-revenue hypothesis.

H_2: *Government Expenditures unidirectional cause Tax Revenues.*

Simultaneous Spending and Taxation

The theory was founded by Black (1958) and advocated by Musgrave (1966), and Meltzer and Richards (1981). It embraces the idea of fiscal synchronization of revenues and expenditures with theoretical background in Lindahl's model of benefit taxation and the median voter rule. This budgetary behavior is considered efficient since both revenues and expenditures have a causal effect on each other, variables have feedback relationship enabling public to determine levels of government spending and taxation as they trade off government capital/services and costs. Many states consider it a benchmark. In Nigeria, Aregbeyen and Ibrahim (2012a and 2012b) did a study using Granger causality test, 1970-2006 whereby bi-directional relationships existed. Takumah (2015) for the Ghana economy, 1986–2012, used VECM and established a bi-directional relationship between expenditure and revenue endorsing the fiscal synchronization hypothesis. Baharumshah *et al.* (2016) did a study for the South African Economy using the Threshold Autoregressive (TAR) and Momentum Threshold Autoregressive (MTAR) models, 1960-2013, and found neutrality. Conversely, using the same models on 1960–2016 quarterly data, Phiri (2016) observed a bi-directional causality, fiscal synchronization hypothesis, for that economy. The hypothesis is as outlined below:

H_3: *Tax Revenues and Government Expenditures bidirectional cause each other*

Changes in Revenues and Expenditures are Independent of each other

The idea is typical of budget processes being seriously affected by interests and agendas that are divergent in nature as per findings by Hoover and Sheffrin (1992) in the United States of America for 1970–1991. Reactive actions to this include attempts to create causal interdependences between spending and tax mechanism decisions. Such attempts are regarded most unwanted since controlling expenditures is very difficult. Baharumshah *et al.* (2016) did a study for the South African Economy using the Threshold Autoregressive (TAR) and Momentum Threshold Autoregressive (MTAR) models, 1960–2013, and found neutrality. The hypothesis based on this theory is as below:

H_4: *Tax Revenues and Government Expenditures are independent of each other*

Grants, Loans and Government Expenditures

Much as grants and loans are taken on board as they support government expenditures and make the budget complete, there are high likelihoods that they, somewhat, have causalities on either other funding sources or government expenditures. For small economy governments, it is necessary to include all funding sources when analyzing tax revenue – expenditure causality links (Sobhee, 2004).

H_5: *Grants and Loans unidirectional cause Government Expenditures.*

In summary, the literature has portrayed that the hypotheses showing the relationship between government tax revenues and expenditures have no uniform array for all countries; the results from the studies are dependent on the data and estimation techniques employed by the author. No study has been conducted in the least developed and weak economy countries like Malawi to empirically establish improvements attained from reforms in fiscal policies.

3. Data and Variables

Study Period, Sample and Data

The study used Republic of Malawi online longitudinal aggregated monthly, time series financial data and sources of funding government expenditures as explanatory variables at National level. The study period span from January 2004 to August 2007 representing a sample size of 44 observations. Data for variables of Government Expenditure (G), Government Tax Revenue (TR), Grants (GR), and Loans (L) is collected.

Dependent Variables

The main dependent variable in multivariate model is Government Expenditure (G) though the rest take turns in bi-variate models when establishing whether causalities exist between the two variables under consideration.

Independent Variables

The main independent variable in multivariate model is Government Tax Revenue (TR) though the rest take turns in bi-variate models when establishing whether causalities exist between the two variables under consideration.

Control Variables

Bivariate and multivariate models include grants (GR) and loans (L) as additional explanatory variables, being funding sources for Malawi government expenditures. Their inclusion is backed by Sobhee (2004) who recommended inclusion of necessary funding sources for small economy governments to avoid the situation of fiscal deficit. Grants and loans are also used as control variables in the multivariate models.

4 Methodology and Model Specifications

This study used bivariate and multivariate Granger causality models adopted from the works by Darrat (1998) in Turkey and Moalusi (2004) in Botswana. This acts as a benchmark to avoid the problem of omission of variables, a persistent problem in the earlier studies. Granger test is used because it is a simple test and widely accepted as a reliable test for causality between two variables. Bivariate and multivariate models include grants and loans as additional explanatory variables, being funding sources for Malawi government expenditures. Their inclusion is backed by Sobhee (2004) who recommended inclusion of necessary funds for small economy governments to avoid the situation of fiscal deficit.

In both bivariate and multivariate models, selected model specification tests like stationarity, granger causality, autoregressive, autocorrelation-1st order (Durbin-Watson), F-statistics and Wald Chi-square ($\chi 2$) tests are applied to check the serial correlation and over-identification issues. The insignificant autoregressive terms (AR) of Durbin-Watson test indicates the absence of Serial Correlations. The insignificant p-values of Sargan test indicate no over-identifications issues while Durbin-Watson test with significant p-value implies the overall robustness of the model results.

Model Specifications

Bivariate Distributed Lag models

Adopted from L.M. Koyck approach (1954), are defined as in Equation 1 below: -

$$G_t = \beta_0 + \sum_{i=1}^{n1} \beta_{1i} G_{t-i} + \sum_{i=1}^{n2} \beta_{2i} TR_{t-i} + \varepsilon_1 \tag{1}$$

Where: G = government expenditure; TR = government tax revenue; and ε_1 = white noise error term. The equation (1) represents causality from tax revenue (TR) to government expenditure (G) in presence of one month lagged government expenditure. H_0: TR does not granger cause G. The hypothesis be accepted if $\sum \beta_{2i}$ are not significantly different from zero simultaneously. The model for the other H_0 : G does not granger cause TR (TR being dependent variable) in presence of one month lagged TR, a second hypothesis, is as in Equation (2) below: -

$$TR_t = \phi_0 + \sum_{k=1}^{m1} \phi_{1i} TR_{t-k} + \sum_{k=1}^{m2} \phi_{2i} G_{t-k} + \varepsilon_2 \tag{2}$$

Where ε_2 is a white noise error term. The H_0 be accepted if $\sum \phi_{2i}$ are not significantly different from zero simultaneously.

Multivariate Distributed Lag models

Adopted from L.M. Koyck approach (1954), are defined as in equations 3 and 4 below: -

$$G_t = \beta_0 + \sum_{i=1}^{n1} \beta_{1i} G_{t-i} + \sum_{i=1}^{n2} \beta_{2i} TR_{t-i} + \sum_{i=1}^{n3} \beta_{3i} GR_{t-i} + \sum_{i=1}^{n4} \beta_{4i} L_{t-i} + \mu_t \tag{3}$$

$$TR_t = \phi_0 + \sum_{i=1}^{m1} \phi_{1i} TR_{t-k} + \sum_{k=1}^{m2} \phi_{2i} G_{t-k} + \sum_{k=1}^{m3} \phi_{3i} GR_{t-k} + \sum_{k=1}^{m4} \phi_{4i} L_{t-k} + v_t \tag{4}$$

Where: GR = Grants; L = Loans; μ_t and ν_t = white noise error terms. For equation (3), H_0: TR does not Granger cause G in presence of GR and L. H_o be rejected if $\sum \beta_{2i}$ were significantly different from zero as a group. For equation (4), H_0: G does not Granger cause TR in presence of GR and L. The null hypothesis be rejected if $\sum \phi_{2i}$ were significantly different from zero as a group.

Maddala (2008) and Gujarati (2006) urge that stationarity tests be done before performing causality test to avoid lagged, dynamic and non-stationary relationships between dependent and independent variables in regression models. Non-stationary data set be differenced to the appropriate level or proven to co-integrate/at equilibrium with another data set/variable data. E-Views software is used to perform the analyses.

5. Empirical Results

Test for Data Stationarity

The respective stationarity test models for the variable time series in this study (G, TR, GR, and L) were formulated and run through Unit Root test (Augmented Dickey-Fuller/DF Test as per Dickey and Fuller, 1979 & 1981) as in Table 25.1 below.

Table 25.1 Unit root test results

Series Code	ADF Value	Prob.	Reject H_0	ADF Value	Prob.	Reject H_0
G	−2.964720	0.0474	** and ***	−8.821088	0.0000	* ** and ***
TR	1.497296	0.9880		−15.07087	0.0000	* ** and ***
GR	−4.302667	0.0015	* ** and ***	−6.789933	0.0000	* ** and ***
L	−5.089485	0.0002	* ** and ***	−7.378363	0.0000	* ** and ***
1% Critical Value: −3.592462; 5% Critical Value: −2.931404; 10% Critical Value: −2.603944						
*: Significant at 1%; **: Significant at 5%; ***: Significant at 10%						

Source: Author's compilation

The results of data series for the variables indicate to be stationary. ADF test at levels show that the null hypothesis is rejected for G at 5% and 10%; TR not rejected; GR at 1%, 5% and 10% and L at 1%, 5% and 10% levels of significance. ADF at first difference indicate that all variables are stationary at all level of significance. In general, all variables are stationary at both levels and first difference except for TR, only stationary at first difference. Therefore, co-integration test was not run and the error correction model (ECM) not used in the study.

Granger Causality Tests

As in Gujarati D.N. (2006) and Maddala G.S. (2008) the tests were run on both bivariate and multivariate models with government expenditure and tax revenue as dependent variables alternatively while grants and loans used as control variables in the multivariate models. Results are in Table 25.2:

Table 25.2 Granger causality test results for bivariate and multivariate models

	Lags: 1			Lags: 2		
Null Hypotheses, HO	**F-Statistics**	***p*-Value**	**Reject H_0**	**F-Statistics**	***p*-Value**	**Reject H_0**
Bivariate Models:						
Dependent Variable: G						
TR does not Granger Cause G	0.72236	0.40043	No	0.1138	0.89275	No
Dependent Variable: TR						
G does not Granger Cause TR	0.00429	0.94813	No	0.9274	0.40458	No
Multivariate Models:						
Dependent Variable: G						
TR does not Granger Cause G	0.72236	0.40043		0.1138	0.89275	No
GR does not Granger Cause G	9.91918	0.00309	Yes	6.07639	0.00523	Yes
L does not Granger Cause G	20.1463	0.000059	Yes	8.18667	0.00114	Yes
Dependent Variable: TR						
G does not Granger Cause TR	0.00429	0.94813	No	0.9274	0.40458	No
GR does not Granger Cause TR	1.23504	0.27307	No	1.43408	0.25127	No
L does not Granger Cause TR	2.40698	0.12867	No	3.24699	0.05021	Yes, at 10%

Source: Author's compilation

The results show that both bivariate and multivariate models do not have enough evidence to reject the two null hypotheses that tax revenue (TR) does not Granger Cause government expenditure (G) and government expenditure (G) does not Granger Cause tax revenue (TR) on specification of lags 1 and 2 hence they are independent of each other. Adjustments in grants and loans tend to have some granger causality on government expenditure (G) at lags 1 and 2.

Bivariate and Multivariate Distributed Lag Models' Results

Model Estimation and Interpretation

Bivariate distributed lag (1) or autoregressive model results for government expenditures (G) and Tax Revenues (TR) - equations (1) and (2) are provided as below: -

$$G_t = 2354.843 + 0.202405TR_{t-1} + 0.611242G_{t-1} \quad (1)$$

$$\text{se} = (1306.708) \quad (0.238146) \quad (0.127793)$$

$$\text{t} = (1.802119) \quad (0.849917) \quad (4.783055)$$

$$R^2 = 0.448183 \quad \text{d} = 1.897733$$

$$TR_t = 1172.987 - 0.003936G_{t-1} + 0.789776TR_{t-1} \quad (2)$$

$$\text{se} = (614.7964) \quad (0.060126) \quad (0.112046)$$

$$\text{t} = (1.907928) \quad (-0.65464) \quad (7.048674)$$

$$R^2 = 0.597698 \quad \text{d} - \text{stat.} = 2.726642$$

For bivariate models: From Equation (1), government expenditure (G) at time t is positively and significantly affected by own expenditure and total revenue (TR) at one period lag ($t - 1$) but the effect is not statistically significant. For total revenue (TR), Eq(2), there is negative relationship between one period lagged ($t - 1$) government expenditure, G (−0.003936) and total revenue (TR) at time t, inferring that a unit increase in government expenditure (over-expenditure) at one period lag ($t - 1$) would reduce total revenue by a multiplier of 0.003936 at time t, though the G effect is statistically insignificant. For multivariate distributed lag (1) or autoregressive model results for government expenditure (G); Tax Revenue (TR); Grants (GR); and Loans (L) – equations (3) and (4) are as shown below: -

$$G_t = 2460.037 + 1.097029G_{t-1} - 0.327065TR_{t-1} - 0.171163GR_{t-1} - 1.012416L_{t-1} \quad (3)$$

$$\text{se} = (1145.617) \quad (0.264192) \quad (0.257720) \quad (0.234263) \quad (0.337775)$$

$$\text{t} = (2.147347) \quad (4.152393) \quad (-1.269071) \quad (-0.730645) \quad (-2.997312)$$

$$R^2 = 0.641630 \quad \text{d} - \text{stat.} = 2.05317$$

$$TR_t = 1184.877 + 0.666138TR_{t-1} + 0.112910TR_{t-1} - 0.046026GR_{t-1} - 0.237264L_{t-1} \quad (4)$$

$$\text{se} = (641.0693) \quad (0.144216) \quad (0.147838) \quad (0.131090) \quad (0.189014)$$

$$\text{t} = (1.848282) \quad (4.61902) \quad (0.763745) \quad (-0.351101) \quad (-1.255276)$$

$$R^2 = 0.630414 \quad \text{d} - \text{stat.} = 2.568616$$

For Eq. 3, negative relationships exist with tax revenue, grants, and loans signifying that a unit increase in each of the cited explanatory variables at time lag of one month (t–1) lead to respective coefficient-based decreases in G in the subsequent month. Loan is the sole variable that is statistically significant. In Eq. 4, only one period lagged (t–1) tax revenue has a significant impact on tax revenue during the subsequent month(t). The results in both bivariate and multivariate models with one period lagged dependent variables, are alike and support the hypothesis that there is no causality of any direction between government expenditures (G) and tax revenues (TR). Causality is only within a variable with current changes depending on the same variable at one period lag. This might suggest the controlled nature of fiscal system or policies that are ceiling based.

Test for Autocorrelation - Model Robustness

Bivariate and multivariate models underwent Autocorrelation (Durbin – Watson) Test, the Difference Method, and results are as presented below:

Bivariate Models

$$G_t = 1757.577 + 0.78272G_{t-1} + 0.030304TR_{t-1} + [AR\ (1) = -0.162483]$$
$$se = (1241.353)\ (0.134558)\quad (0.216489)\ (0.176293)$$
$$t = (1.415856)\ (5.816964)\quad (0.139979)\ (-0.921663)$$
$$R^2 = 0.508629 \quad D - W\ stat. = 2.065459 \quad Prob\ (F\text{ - }stat) = 0.000000$$
$$h - statistics = -0.48 > -1.96$$

$$TR_{t-1} = 374.5787 + 0.9319TR_{t-1} + 0.008072G_{t-1} + [AR\ (1) = -0.537466]$$
$$se = (375.6089)\ (0.077378)\quad (0.040621)\ (0.142269)$$
$$t = (0.997257)\ (12.04352)(0.198721)\ (-3.777824)$$
$$R^2 = 0.687533 \quad D - W\ stat. = 2.69263 \quad Prob\ (F\text{ - }stat) = 0.000000$$
$$h - statistics = 2.677 > 1.96.$$

Both bivariate models proved not having enough evidence for problem of autocorrelation (neither positive nor negative) as computed h – statistics is greater than that from tables (–1.96 or 1.96), hence the models are robust and capable of forecasting.

Multivariate Models

$$G_t = 2003.937 + 1.166173G_{t-1} - 0.415828TR_{t-1} - 0.114258GR_{t-1} - 0.872929L_{t-1} + [AR\ (1) = -0.240012]$$
$$se = (995.6871)\ (0.249177)\quad (0.22571)\quad (0.232349)\quad (0.319097)\quad (0.165945)$$
$$t = (2.012617)\ (4.680102)\quad (-1.84231)\quad (-0.49175)\quad (-2.735624)\quad (-1.446329)$$
$$R^2 = 0.690362 \quad D - W\ stat. = 2.178861$$

$$TR_t = 551.1355 + 0.928689TR_{t-1} - 0.030593G_t + 0.064987GR_{t-1} + 0.003979L_{t-1} + [AR\ (1) = -0.532631]$$
$$se = (425.1608)\quad (0.110005)\quad (0.123831)\quad (0.119668)\quad (0.157133)\quad (0.147991)$$
$$t = (1.296299)\quad (8.442282)\quad (-0.247059)\quad (0.543061)\quad (0.025325)\quad (-3.599067)$$
$$R^2 = 0.697704 \quad D - W\ stat. = 2.587316$$

Both multivariate models proved to not having enough evidence of problem of autocorrelation (neither positive nor negative) – D – W stat around 2 hence they are robust and capable of forecasting.

6. Conclusion

This study interrogated the causal links between government expenditures and funding sources, a scenario for Malawi, in line with the methodology employed by Darrat (1998) and Moalusi (2004). Besides bivariate models, Koyck Approach (1954), distributed lag multivariate models were formulated. In the Distributed Lag multivariate models, grants (GR) and loans (L) are treated as control variables for ease of causality link analysis and their inclusion serves the purpose of avoiding the problem of omission of variables (funding sources supporting government expenditures). Time series aggregated monthly financial data at national level was used. E-Views statistical package was used to run the analyses.

Findings from bivariate and multivariate models provide alike results of no causal link between tax revenues and expenditures in the scenario of Malawi for the study period (2004/01 to 2007/08). This supports the hypothesis that there is no causal link/ interdependence between government expenditures and tax revenues. Such findings are in line with results from the study done by Hoover and Sheffrin (1992) for the United States for the data series spanning 1970 to 1990. There are unidirectional causalities from grants and loans to expenditures in multivariate distributed lag (1) or autoregressive models implying changes in lagged loans and grants affect current expenditures. Here, the results infer that government expenditures are dependent on exogenous support.

In practice, the negative link in multivariate model of government expenditures, as in Buchanan and Wagner (1978), suggests that deficits in Government budget can be corrected by raising more funds from taxes. Collecting more tax revenues and complying with expenditure ceilings may, in the long-run, reduce overdependence on grants and loans besides over-expenditure syndrome. Independent changes in revenues and expenditures (inconsistent with the fiscal synchronization hypothesis) infers to situations where budget processes are seriously affected by divergent interests and agendas or attempts to create causal interdependence between spending and taxing decisions with no tangible results.

Establishing government expenditures to revenues links and other funding sources enables displaying government's financial capabilities to fulfil financial obligations in Public-Private Partnerships which also guides private sector players in deciding extents of their commitment and participation. The study suggests that government can gradually shift from exogenous financial and budgetary support by tightening tax evasion loopholes, engaging private sector players in PPPs for efficiency gains in managing existing public assets while providing services at a fee or creating green assets and manage products or services based on an agreed viable business model.

Future research studies are recommended on long span time series and longitudinal data to establish prevailing government fiscal policies and theories practiced overtime for ease of forecasting the future scenarios.

REFERENCES

1. Al-Zeaud, H. (2014). The causal relationship between government revenue and Expenditure in Jordan. *Global Journal of Management and Business Research*, 14(6): 1–10.
2. Baharumshah, A., Jibrilla, A., Sirag, A., Ali, H. and I. Muhammad (2016). Public Revenue-expenditure nexus in South Africa: are there asymmetries. *South African Journal of Economics*, 84(4): 520–536.
3. Darrat A.F. (1998). Tax and Spend, or Spend and Tax? An Inquiry into the Turkish Budgetary Process; Southern Economic Journal; 64(4): 940–956.
4. Dickey, D.A. & W.A. Fuller (1981). Likelihood Ratio Statistics for Autoregressive Time Series with a Unit Root. *Econometrica*, 49: 427–432.
5. Elyasi, Yousef and Mohammed Rahimi (2012). The Causality between Government Revenue and Government Spending in Iran. *International Journal of Economic Sciences and Applied Research*, 5(1): 129–145.
6. Gujarati D.N. (2006). Essentials of Econometrics: Third Edition, McGraw-Hill, New York.
7. Maddala G.S. (2008). Introduction to Econometrics, Third Edition.
8. Meltzer A.H. and S.F., Richard (1981). A rational theory of the size of the Government; *Journal of Political Economy*, 89: 914–927.
9. Moalusi D.K. (2004). Causal Link between Government Spending and Revenue: A case of Botswana.
10. Moisio A. (2001). On Local Government Spending and Taxation Behaviour; Government Institute for Economic Research Helsinki, Finland.
11. Nwosu, D. and H. Okafor (2014). Government revenue and expenditure in Nigeria: A Disaggregate analysis. *Asian Economic and Financial Review*, 4(7): 878–890.
12. Obeng, S. (2015). A causality test of the revenue-expenditure nexus in Ghana. *ADRRI Journal of Arts and Social Sciences*, 11(1): 1–18.
13. Ogujiuba, K. and T. Abraham (2012). Testing the relationship between government Revenue and expenditure: Evidence from Nigeria. *International Journal of Economics and Finance*, 4(11): 172–180.
14. Phiri, A. (2016). Asymmetries in the revenue-expenditure nexus: new evidence from South Africa. *MPRA Working Paper*.
15. Saeed, P. and S. Somaye (2012). Relationship between government spending and revenue: Evidence from oil exporting Countries. *International Journal of Economics and Management Engineering (IJEME)*, 2(2): 94–97.
16. Sobhee, S.K. (2004). The causality between tax and spend of the public sector in maturities: A VECM approach. *International Journal of Applied Econometrics and Quantitative Studies*, 1(3): 115–130.
17. Takumah, W. (2015). The dynamic causal relationship between government revenue and government expenditure nexus in Ghana. *MPRA Working Paper*.

Impending Inquisitions in Humanities and Sciences (ICIIHS-2022) – Dr. Mohan Varkolu et al. (eds)

Exploring Institutional Use of Performance Measurement Instruments and Indicators for Success, Reinvestments, Partnerships and Survival of MSMEs: Perspectives from Agro-cooperatives in Malawi

26

Raphael K. Munthal[1]
PhD Commerce Scholar, School of Commerce and Management Studies,
DMI-ST. Eugene University, Lusaka, Zambia
K. Logasakthi[2]
Associate Professor, Universal Business School,
Universal AI University, Karjat, Maharashtra, India

ABSTRACT This study reviewed and established current practices in use of Performance Measurement (PM) frameworks in MSMEs based on their relevance, areas requiring improvement and practical recommendations for use. Four cooperatives are sampled for descriptive non-financial primary data obtained through semi-structured interviews and financial secondary data obtained from financial records, covering 2016 to 2019 financial years. Moderating variables of specialized trainings, access to market, stakeholders' management, member involvement in business activities and data management are measured by non-financial data. Independent variables of cost management, sales management, revenue management, profit management, asset utilization, and managing sources of financing are determined by financial data. Solvency and profitability are dependent variables that measure capabilities of MSMEs to reinvest, partner and survive in the business environment. SPSS and Microsoft Excel statistical packages are used to analyze the data collected. Study findings indicate that no cooperative redesigned conventional PMs though accepted the need to do so and acknowledged their associated benefits. MSMEs' financial performance (independent variables) has direct influence on their solvency (ability to meet financial obligations and earn income) and profitability (ability to earn profits) as dependent variables and precursors for success and capabilities to reinvest, partner and survive. It is observed that MSMEs that progressed well in moderating (non-financial) variables showed positive internal operations' results for efficiency and effectiveness as well as outshining in overall capabilities of solvency and profitability.

KEYWORDS MSMEs, PM instruments, Financial & Non-financial performance, Partnerships

1. Introduction

In Malawi, MSMEs contribute 24% of employment and US$6.8 billion to GDP. 32% of MSMEs are micro-businesses (K5 million turnover) and associated with annual losses. Their growth is constrained by factors constituting internal and external operation environments (FinScope Malawi, 2019). Malawi Country Private Sector Diagnostic (2021) states that MSMEs lack the means to signal their creditworthiness and the channels to provide transparent information about their financial activities. Malawi adopted cooperative prototype in 1940s yet remains underexploited for benchmarking interventions (ILO, 2017). A study by Zidana, R. (2015) found that most MSMEs' business owners in Malawi treated business as a temporally shelter while waiting for employment opportunities and not responding to opportunities in the economic environment. This mindset in business undermines government's industrious mechanisms of financing MSMEs besides commercial banks and equity

[1]rolf_77bypass@hotmail.com, [2]dr.logasakthi@gmail.com

DOI: 10.1201/9781003489436-26

investment schemes. Some MSMEs have strived to gain competitiveness or conducive internal environment by reorganizing into associations and cooperatives as to broaden their capital and acquire business skills. In spite of these external and internal efforts, little or no study has been done to explore whether MSMEs have so far developed own non-financial and financial capabilities to succeed, re-invest, partner and use any set of criteria to detect extents of their financial and non-financial dimensions.

Improving performance measurements (PMs) has been key by organizations globally alongside aligning its systems with strategic goals (Kaplan, 1983; Gregory, 1993). Numerous frameworks have benchmarked on Balanced Scorecard (Kaplan and Norton, 1992) but not balancing financial and non-financial measures and excluded MSMEs (Storey, 1994). Use of PMs in MSMEs has depended on employees' and consultants' viewpoints and not strategic goals and Key Performance Indicators (KPIs) agreed upon by owners, employees and stakeholders (Thakkar *et al.*, 2009). Bhagwat and Sharma (2007) recommended that performance measurement metrics for MSMEs should be easy to comprehend, while its data and information should be collected and analyzed with ease and in most economical way so that there is sustainability. This necessitated the study to explore possible and simplified frameworks that MSMEs can use to signal their performance (financial and non-financial or their drivers of performance) towards success, creditworthiness, strategy orientation and decision making for reinvestments, partnerships, sustained innovations and entrepreneurship.

This paper investigates effects of non-financial critical success factors (CSFs) on financial CSFs then in turn on the overall enterprise success or directly on the overall enterprise success, capabilities to reinvest and partner by analyzing enterprise's solvency and profitability. Study PM conceptual framework is developed benchmarking on Balanced Scorecard (BSC), DuPont Pyramid of Financial Ratios (Warren *et al.*, 2005) and substantiated by preceding studies. Four cooking oil producing cooperatives, representing MSMEs, are selected from which both primary and secondary data are collected then analyzed to establish extents of performance measurement dimensions.

Objectives: to explore the dimensions of MSMEs' financial and non-financial capabilities; to identify befitting and sustainable strategic PM tools and indicators for institutionalizing measurement of the identified dimensions in agro-cooperatives; to validate and recommend strategic PM framework for ascertaining agro-cooperatives' financial and non-financial capabilities.

The rest of the paper is structured as follows: Section-2 reviews the existing literature. Section-3 describes the sample and variables. Section-4 details the research methodology. Secttion-5 discusses the descriptive and empirical findings. Section-6 concludes the paper and Section-7 provides references.

2. Literature Review

Theoretical Literature Review

This study is underpinned by organization theory alongside others like goal theory, open system theory and stakeholder theory. Each theory expounds crucial phenomenon of the enterprise relative to performance measurement. Henri (2004) advocates for use of more than one theory in a study in cases where a single theory is perceived unable to capture the occurrences under study. This enhances holistic performance measurement approach to an enterprise.

Organization theory perceives an organization as a collection of individuals, aggregated into departments, structures and contexts, that do things together to achieve common organizational objectives. It is directly applicable to top-management concerns of setting goals, developing strategies, interpreting the external environment, and deciding organization structure and design; and middle- management concerns like overseeing major departments and how they relate to the rest of the organization while fitting with technology, power and politics, information and control systems (Daft, 2007). It strives to study the structure, functions, behavior of individuals and performance of organizations so as to understand how organizations should function and be managed, both in their current and foreseeable settings (Pugh, 1984). This study considers MSME an organization that constitutes individuals (owner/manager and employees) who work together to attain the enterprise's objective of financial success and wealth maximization.

Open system theory (Kleiner, 1986) is hatched on the idea that human entities right from individual to giant enterprises can be regarded as a system interacting with its environment, both internal and external. The internal environment constitutes people operating internally (employees, shareholders and managers) while the external environment is likened to people outside the enterprise like intermediaries, suppliers, competitors and customers (Henri, 2004). It assumes an existing boundary between the enterprise operating system and its environment as the system receives inputs from the environment, process the inputs, and release products as outputs back to the environment (Robbins, 1990). Studies by Katz and Kahn (1966) urge that there is a two-way causal relationship between environment and enterprise as they affect each other. This research, therefore, strives to establish the aforementioned relationship within the context of performance measurement in MSMEs, agro-cooperatives, by measuring inputs, internal system processes, and output elements of the open system.

Stakeholders' theory (Freeman, 1984) contests that an enterprise ought to take on board the interests of all the groups it interacts with besides shareholders. It recognizes customers, suppliers, employees, state and non-state institutions, communities, microfinance institutions and trade unions as some of the stakeholders besides shareholders (Freeman, 2010; Miles, 2012). Much as there is no consensus on the standard definition of stakeholders, a cross section of literature advocates for MSMEs to cultivate and sustain mutual relationships with their external stakeholders (Yu, 2011; Hutchinson et al., 2015). The theory compels this study to conduct performance measurement inclusive of stakeholders as to determine extents to which an enterprise can meet their needs, in the wider domain of stakeholder community, while enhancing performance of the whole enterprise. Therefore, the theory underpins the scope and framework of performance measurement.

Conceptual Review for Performance Measurement Frameworks and MSMEs

Several studies have unearthed such concepts and frameworks that have subscribed to performance measurement discipline since 1990s. From the pool of literature, few have ascribed some impact based on their dominance in the literature for the discipline. In spite of their dominance, each one has merits, demerits and areas requiring improvement.

Kaplan and Norton (1992) cofounded Balanced Scorecard (BSC) that considers both financial and non-financial measures to measure enterprise performance while aligning the measurement system to the company's vision and strategy. A survey by Bain & Co., (2003) indicated that 62% of large firms use the BSC because of its ability to reveal the underlying causes of financial performance, while assisting managers consider the short- and long-term implications of their decisions. Such abilities with large firms to use BSC and other tools do not exist with MSMEs but can get devolved to attempt bridging such gaps. However, Neely *et al.* (2005) contends that the framework leaves other stakeholders, is biased to internal measurement other than external for competitiveness and benchmarking sake.

A Stakeholder Approach to Strategic Performance Measurement (SASPM), as contained in Atkinson *et al.* (1997), considers the enterprise's relationship with its stakeholders besides other parameters. The Business Excellence Model (BEM) as contained in European Foundation for quality Management (2003), clarifies enablers of performance (policy and strategy, processes and resources, leadership and people management) with results being people satisfaction, customer satisfaction, and impacts on both business and society (Neely, 2007). The Performance, Development, Growth Benchmarking System (PDGBS), as contained in St-Pierre and Delisle (2006), focuses on SMEs while undertaking performance measurement exclusive of benchmarking perspective.

The concept of dissecting the financial statements of a firm to assess its financial condition through the DuPont ratio pyramid was founded by financial executive of E.I. DuPont De Nemours & Co. in 1919 (Courtis, 1978; Bayldon *et al.*, 1984 and Warren *et al.*, 2005). It merges the income statement and balance sheet into two summary measures of profitability, thus ROA and ROE. The financial ratio has a clear hierarchical structure where corrective measures can be linked to different levels of an organization. DuPont ratio has two major contributions towards financial analysis that it helps (1) to identify the sources of strength and weakness in current performance, and (2) to focus attention on 'Value Drivers.' The DuPont formula decomposes Return on Equity (ROE) into three components as it provides a summarized and focused look at how ***profitability, leverage and efficiency*** are combined by multiplying them together to determine the return on equity (ROE) of a business. Financial ratios have gained recognition by financial analysts as acceptable performance evaluation techniques for business firms. Much as it could be challenged that the credibility of ratios relies on the reporting behavior of firms as to whether relevant and accurate information is disclosed or not, it remains a systematic approach (Warren *et al.*, 2005).

Review of Performance Measurement in MSMEs

Business success is linked to good performance and performance is a generalized term used for some or all activities of a firm in a specific period. This is oftentimes confined to large scale businesses and remains not easily defined in small businesses (Wach *et al.*, 2016; Gerba and Viswanadham, 2016). Performance may entail determining selected measures that can measure the success of a business by its growth and in making profits (Kusumadewi, 2017). Performance of a firm is, often times, linked to effectiveness and efficiency in carrying out activities within specified time period. It is also contested that success may at times be associated with social aspirations (recognition, satisfaction, job creation) other than economic (wealth creation) in nature, especially in MSMEs (Wach *et al.*, 2016). This lack of consensus compels small businesses to devise own, relevant and contextual based definition of success other than relying on outsiders (Simpson, 2012). Financial performance can be analyzed with different financial analysis tools to establish the business's financial and non-financial status over specified duration (Bogomin *et al.*, 2016). Kotane *et al.* (2016) contends that it is infeasible to solely depend on financial indicators to accurately predict the firm's financial stability. Furthermore, Dejene & Getachew (2015) state that there are a number of factors whose combination portray success of MSMEs, cooperatives in this study, in various areas like commitment of members, members' participation, member education, communication, managerial factors, structural and external factors. Much as there is an attempt to understand patterns portrayed by non-financial factors, this study recognizes the ultimate success as measured by business growth, profitability, solvency and capability of MSME to keep on operating through indicators of financial ratios.

Review on Critical Success Factors (CSFs) and Key Performance Indicators (KPIs) in MSMEs

Key or critical success factor (CSF) is anything that enhances an enterprise to acquire business (Masocha and Charamba, 2014). This study endeavors to review and identify CSFs relevant to MSMEs in the Agro-Cooperative sector influencing their business performance and success of devised performance measurement framework.

Non-Financial Critical Success Factors for MSMEs

a) Member Education, Specialized Trainings and Success of MSMEs

Aribawa (2016) found that financial literacy was one of the factors influencing MSMEs performance and sustainability. There was enormous challenge for creative MSMEs to have more knowledge about financial literacy. Lusardi and Mitchell (2013) defined financial literacy as person's ability to manage information about the economy, make plans in finance, and make better decisions about the accumulation of wealth, pensioners and debt they have. One need to amass the capacity to gather pertinent and significant information, while differentiating between diverse financial choices, discussing monetary and financial issues, planning and determining decisions about the use of finances. Bogomin *et al.* (2017) contends that problems with unbalanced information on financial services hindering the success of MSMEs to compete is attributed to limited financial literacy. Anania & Rwekaza (2018) urges that cooperative education and trainings are essential in promoting better performance of savings and credit cooperative societies. Quartey *et al.* (2017) portrayed that only respondents from the banking sector have high financial literacy compared to other sectors hence MSMEs, backbone of the economy, must acquire high financial literacy and financial inclusion to enable the public in obtaining financial services in economic activities. The above findings attest the need for MSMEs in Malawi to acquire financial literacy and skills in specialized trainings.

b) Commitment of Managers, Employees, Members and Success of MSMEs

Commitment of the management is associated with recording high return on assets and investments, and decisions resulting in long-term success of the business since managers are agents of change by communicating strategies alongside respective performance measures, tailored trainings (Berko *et al.*, 2016). This also holds in making a performance measurement framework effective and successful (Amir, 2011).

Employee commitment determines success of any enterprise. This commitment is measured by proxies like their participation in decision making, autonomy, employee motivation, job satisfaction, employee loyalty, feedback, employee learning and professional growth (Valaei and Rezaei, 2017).

Levels of commitment by managers and employees interrelate with other CSFs and the vice versa. This underlines the need to consider both manager and employee commitment since the vice versa may lead to non-use of performance measurement frameworks perceived to lack decision-making purposes; high staff turnover, absenteeism, and unwillingness to go extra mile hence low productivity (Biggart *et al.*, 2010; Hutchinson *et al.*, 2015).

c) Member Participation in Business Planning and Success of MSMEs

Strategic planning in business is linked to growth, success and survival though argued that it is not viable in MSMEs since they cannot withstand unpredictable business environment hence MSMEs avoiding formal planning (Parnell *et al.*, 2015; Pekkola *et al.*, 2016). Business planning enhances the enterprise to develop, collaborate, implement, and amend its strategies as to attain the set performance objectives while meeting needs of key stakeholders (Talib *et al.*, 2014).

Member participation refers to member's involvement in the activities taking place within an organization. Active member participation (cooperative governance) helps management in carrying out their responsibilities since their involvement maintains the direction of the cooperatives towards enhancing cooperative's performance hence long run survival of a cooperative (USDA 2011). This CSF can be displayed in proxies like taking part in policy (strategy) making process through attendance at annual general meetings and patronage on cooperative's own products and services. Amini & Ramezani (2008) contend that members' active participation and loyalty determine the success of cooperatives since members are also the consumers, employees and leaders. The CSF leads to effective monitoring and control of activities (managerial actions) by enhancing performance through provision of platforms for feedback, interests and expectations from different groups, guidance and monitoring managers.

d) Managing Information, Innovations, Customers, Suppliers, Competitors; Other Stakeholders and Abiding by Regulations.

Managing key and strategic information is vital for success of any enterprise with instances of market intelligence that enables MSMEs to explore new opportunities in processes, products and services through seeking and sharing both internally and externally with stakeholders. This will enhance meeting customer preferences, needs and satisfaction. Performance measurement is considered a component of information management (Bayraktar, 2015; Zerfass & Winkler, 2016).

Innovation is a prerequisite for sustained long term enterprise performance on which its success and survival depends (Bulak *et al.*, 2016; Saunila, 2016). Some studies proven a positive relationship between MSMEs' business performance and scope of innovation (Kotey, 2014; Forsman & Temel, 2011). MSMEs are oftentimes resource constrained hence they need sustained innovations in marketing strategies, internal processes, and delivery of goods and services to compete with large enterprises (Masocha and Charamba, 2014).

Customer management develops customer loyalty and trust since loyalty leads to customer retention. A loyal customer buys from the enterprise in spite of better alternative goods and services offered by competitors. Retaining current customers and acquiring new ones lead to expanded market share and higher market performance. This requires positive interactions with customers through market research to know future needs (Hutchinson *et al.*, 2015; Laukkanen *et al.*, 2014).

Supplier management is one of the vital drivers of financial performance since it controls costs for quality management, ordering and storing goods and services so that uninterrupted procurement and supply of inventory is sustained economically and profitably. Studies have urged enterprises to develop mutual relationships with its suppliers to attain a competitive advantage and sustained organizational performance (Shi and Yu, 2013; Talib *et al.*, 2014; Bulak *et al.*, 2016).

Knowing competitors in the enterprise domain, their operations, performance and products enables MSMEs to differentiate their products and services hence surviving in the market, benchmarking. This can be done internally when performance is compared against own set standards or externally when performance is compared against competitors' standards set as best standards for the industry (Taschner, 2016). MSMEs may also share knowledge, information, and resources with competitors by forming partnerships (Gunawan *et al.*, 2016).

MSMEs are required to abide by regulations set by different government regulatory authorities like those on taxes, standards, environmental issues, and monitoring compliance for them not to face penalties, closures and operate profitably (Jitmaneeroj, 2016).

Financial Critical Success Factors for MSMEs

(a) Managing Pool of Resources

Managing pool of resources like tangibles and intangibles or soft (brand quality, knowledge and skills, relationships, reputation, patents, technology, capabilities) is anchored by Resource-Based Theory which suggests that performance and growth of a business is triggered by resources possessed by that business and how they are used (Kotey, 2014; Yazdanfar & Ohman, 2015). Intangible resources may not be imitated hence giving an enterprise a competitive advantage over competitors.

(b) Managing Costs, Revenues, and Profits

Cost management leads to efficient operations in a business, improved profitability, and recover a business in times of reorganization (Laitinen, 2011)

Revenue management consists of collecting (at point of sale) and analyzing data (using various techniques like demand forecasting and programming, cost-volume analysis) to get information about the trends, habits, and demand patterns of customers as to assess enterprise profitability (Ng *et al.*, 2013).

Most MSMEs have measured their performance based on levels of profits as one way of appraising unit or divisional managers based on outputs of their responsibilities (Drury, 2004; Halabi *et al.*, 2010). However, it has been argued that profitability is a measure of outcome of performance and cannot control or drive performance (Otley, 2007).

(c) Managing Sources of finance

Access to cheaper sources of finance is one of the crucial elements for performance and success of MSMEs. Survival of most MSMEs is hindered by high costs of borrowing that lead to high cost of doing business (Ramukumba, 2014). Therefore, it is imperative for MSMEs to compare costs associated with various sources of finances, internally and externally, to minimize costs of doing business so that they attain success and survival.

In summary, the chapter has touched on business performance, theories anchoring the study, its measurement frameworks, concepts, CSFs and KPIs. There is no consensus on the general and most befitting performance measurement, especially for MSMEs and specifically in the Agro-processing sector with perspectives of CSFs and KPIs. Furthermore, there seem to be no any literature ascertaining the most essential CSFs for MSMEs in Agro-processing sector, especially in developing countries like Malawi. The study considers performance as the capability of an enterprise to attain its envisioned outcome in an effective and efficient way. The discussed theories have shown the need for holistic and augmented approach. Conceptual PM Frameworks have displayed the need to utilize both financial and non-financial measures in the performance measurement matrices. This has been substantiated with advances in accounting discipline while the marketing discipline has customer at the core of performance measurement alongside non-financial measures and treating financial measures as output measures. Operations management inclines on measurement of the efficiency and effectiveness of internal processes while supply chain management looks at the flow of goods and services between supplier and user. These disciplines and concept-based measurement practices are related hence the need to conceive a harmonized performance measurement framework that considers an enterprise as a business firm with goals to be attained while meeting stakeholders' needs, thus considering operations in both internal and external environments. The study benchmarks its conceptual and performance measurement frameworks on the Balanced Scorecard (BSC) as it augers well with theories underpinning the study and accommodative of other frameworks by considering both financial and non-financial measures, measuring performance using success factors for internal and external environments, and linking measures to the enterprise's vision and strategy. This is augmented by DuPont ratio pyramid which has assisted to analyze and link the individual financial ratios that are selected for the purpose of this study. The pyramid that measures the firm's financial capabilities has been identified through literature review and case studies then domesticated to Agro-processing MSMEs.

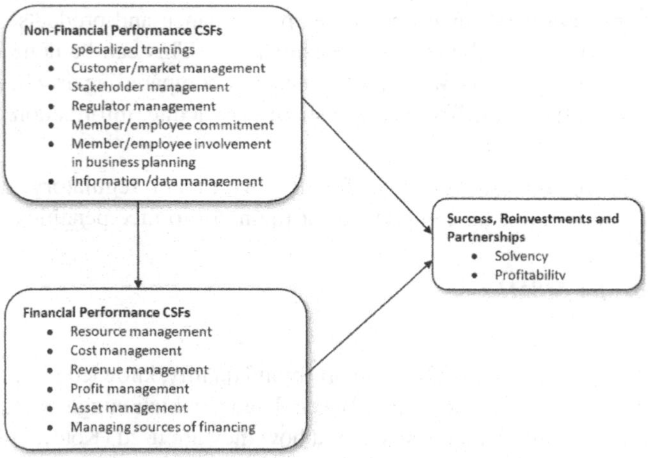

Fig. 26.1 Conceptual Framework for the Study

Source: Authors

3. Data and Variables

Study Period, Sample and Data

The study used quantitative secondary (feedback) data covering three financial years from 2016 to 2019 and descriptive primary (feed-forward) data on two non-financial variables. Financial performance measures are performed on the aforementioned period which is categorized as past performance and perceived not useful by some scholars, it is argued that they are equally important since feedback information is actual, reliable, objective and form a base for forecasting future performance, based on trends (Hegazy & Hegazy, 2012).

To derive the sample, the study purposively sampled active formal cooperatives that are into cooking oil processing in Mchinji district, thus the population for participating cooperatives was equal to the number of those deemed active and into those products.

The sampling for participating members on secondary data was as that of cooperatives, equaled total membership, while a simple random sampling with a sampling interval was used in soliciting primary data for two non-financial variables. These cooperatives are Machichi, Mikonga, Kamwendo and Mthiransembe. A longitudinal sample of 80 out of a population of 367 was derived.

Within the scope of performance measurement, key performance measure entails collection of data used to gauge the performance of an enterprise in line with a particular parameter deemed key to the success of an enterprise, focusing on the critical aspects of the business (Hegazy & Hegazy, 2012). Identifying KPIs for MSMEs is essential since they are resource and expertise constrained to define and measure KPIs. Hegazy & Hegazy (2012) urges MSMEs to focus on measuring only KPIs that reveal the state of affairs of the enterprise and that owners/managers ought to use the information and patterns derived from such KPIs for decision making. The study opted for performance measures with both feedback and feed-forward information attributes since forward looking performance measures are proven to be subjective and not easy to measure by MSMEs, expectations/probabilistic of future performance (Yu, 2013). Financial data was collected from net operating cash flows, financial records and balance sheets from sampled cooperatives and cross-checked with the database at district level. The data assisted in determining dimensions like sales growth, profit leverages, return on investment (ROI), return on assets (ROA), return on equity (ROE), earnings per share (EPS), return on sales, and return on capital employed and ultimately profitability and solvency (Al-Matari *et al.*, 2014). Non-Financial data was collected on Attendance for Cooperative Member Education and Specialized trainings, Member and Employee Participation in Business Planning, Access to Markets and Coordination with Stakeholders. Data for the last two parameters was primary and obtained through semi-structured questionnaire.

Dependent Variables

Solvency: It is a popularly used measure to establish the ability of a business to pay or meet its financial obligations/credits as they are due and to earn income.

Profitability: It is a widely used measure to quantify the ability of a business to earn profits depending on effectiveness and efficiency of its operations as well as the resources available to it. It focuses on the relationship between operating results and the resources available to an enterprise.

Independent Variables

The study considers internal processes, operating results that affect efficiency, effectiveness and financial position as key measures and independent variables for solvency and profitability. These are effectiveness in utilization of assets (net sales to assets), profitability/rate earned on total assets or asset turnover, profitability of the investment by stock/shareholders (rate of income earned on stakeholders' equity), quick ratio (instant debt-paying ability of a firm), and margin of safety for creditors (ability of the business to withstand adverse conditions). These, mostly, take the form of financial measures.

Moderating Variables

Certain performance CSFs neither have direct influence on enterprise's solvency and profitability nor do not but indirectly influence internal processes that in turn trigger such dependent variables. This study considers non-financial performance measures constituting innovation and learning and those in external environment as moderating variables. The study includes specialized trainings, access to reliable markets, coordination with stakeholders, member/employee commitment, participation in business affairs and information/data management to gauge their effect.

4. Methodology and Methods

The research method starts with the literature review to identify drivers that influence the business performance of MSMEs alongside means of measuring such drivers. This becomes a basis for comparison and identification of weaknesses warranting improvements and generating new knowledge. This is followed by establishing the current performance measurement practices by MSMEs. This enabled gathering what is done by MSMEs as to identify weaknesses of existing practices and propose possible amendments. Thirdly, attempt was made to develop a financial and non-financial performance measurement framework for MSMEs based on literature review and preceding empirical and descriptive studies. The general model was established, anchored by the framework, linking the identified financial and non-financial issues/indicators. Strategic financial and non-financial indicators and the framework are validated on sampled MSMEs (cooperatives) to ascertain their practical usefulness, thus to establish their likelihood in forecasting business success, survival, reinvestments and partnerships.

Research Questions

From a cross-section of literature, preceding studies have established that the aforementioned independent and moderating variables, respectively or collectively, have influence on enterprise's solvency and profitability that are proxies for its success and capabilities to survive, access capital credits, reinvest and form business partnerships. Based on what is established, the following research questions are developed.

(i) Do identified independent variables affect the success and capabilities of an MSME (agro-cooperatives)?
(ii) Do identified moderating variables affect the success and capabilities of an MSME (agro-cooperatives)?

The measurement linkages of all these variables are depicted in figure 1 for study conceptual framework.

5. Study Results

Non-financial Parameters Analysis

Trainings attended by cooperative members

The results are graphically presented in Fig. 26.2 below.

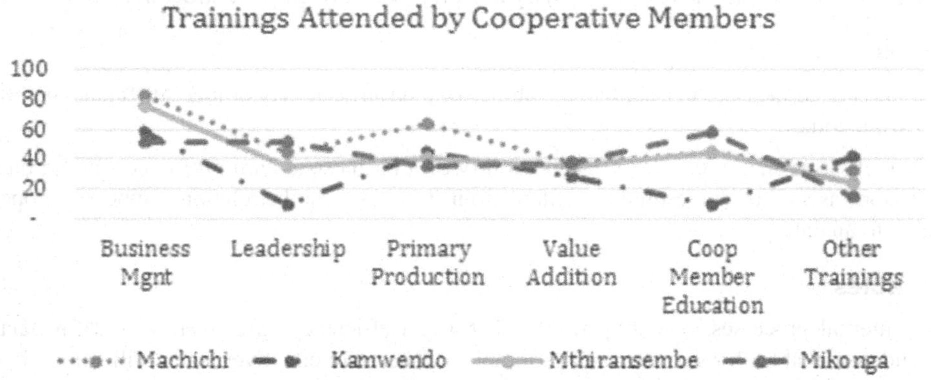

Fig. 26.2 Trainings attended by cooperative members

Source: Author's compilation

On this parameter, Mikonga cooperative seem to have done fairly well by having 60% (3 out of 5) of the specialized trainings attended by over 50% of its members and employees, followed by Machichi at 40% then Mthiransembe with around 40% of members in attendance and finally Kamwendo being the most fluctuating.

Member patronage for business meetings and group activities.

The results are graphically presented in Fig. 26.3 below.

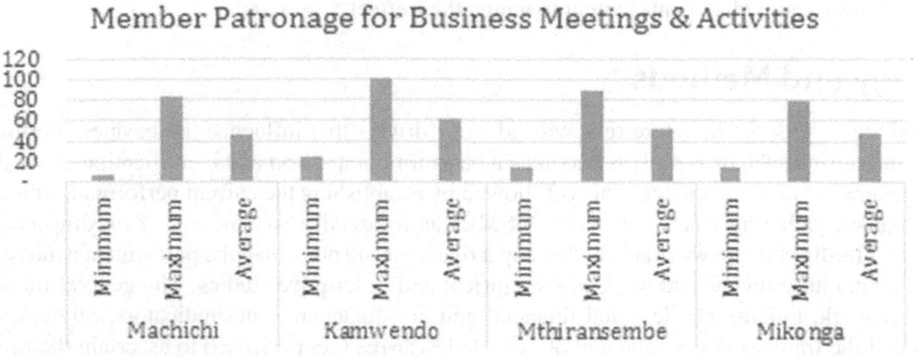

Fig. 26.3 Member patronage for business meetings and group activities

Source: Author's compilation

The bar graph shows Kamwendo having the highest average for member patronage (60%), followed by Mthiransembe (57%) then Mikonga (48%) and finally Machichi (44%).

Member and employees' patronage for business meetings and group activities.

The results are graphically presented in Fig. 26.4 below.

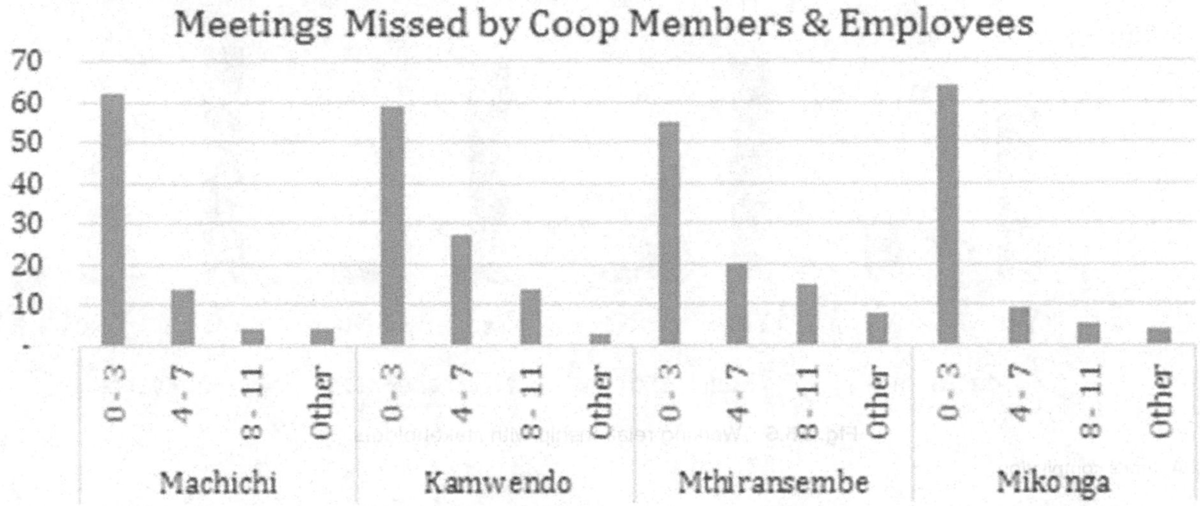

Fig. 26.4 Member and employees' commitment

Source: Author's compilation

The bar graph shows all cooperatives having more than 55% of cooperative members with least missed business meetings and group activities with Mikonga at 65%, Machichi at 62%, Kamwendo at 59% and Mthiransembe at 56%.

Access to reliable markets.

The results are graphically presented in Fig. 26.5 below.

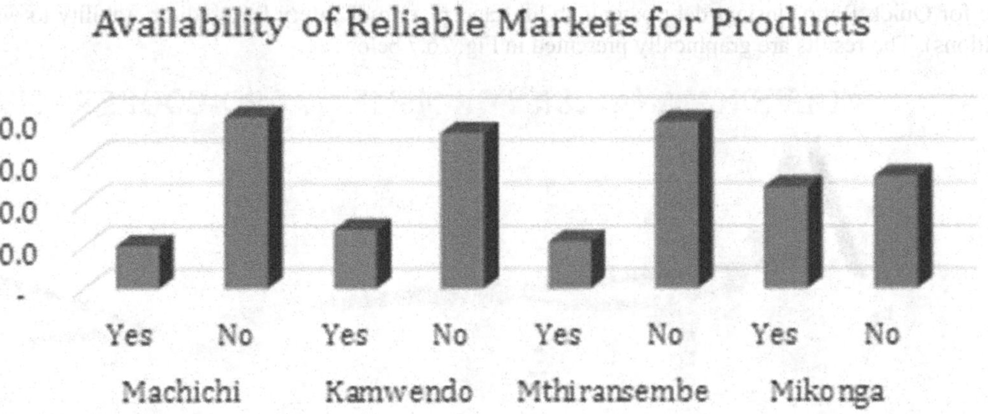

Fig. 26.5 Access to reliable markets

Source: Author's compilation

The above bar graph depicts Mikonga as having higher positive responses on access to reliable markets (42%) followed by Kamwendo (22%) then Mthiransembe (19%) and Machichi (18%). Conspicuously, Kamwendo cooperative is the only one with refinery, capacity to fortify with vitamin A, linkages to retail outlets outside the district and certified by Malawi Bureau of Standards since 2013. The other three cooperatives were linked to Kamwendo cooperative for its unique value addition services and competitive advantage at a fee.

Working Relationship with Stakeholders.

The results are graphically presented in Fig. 26.6 below.

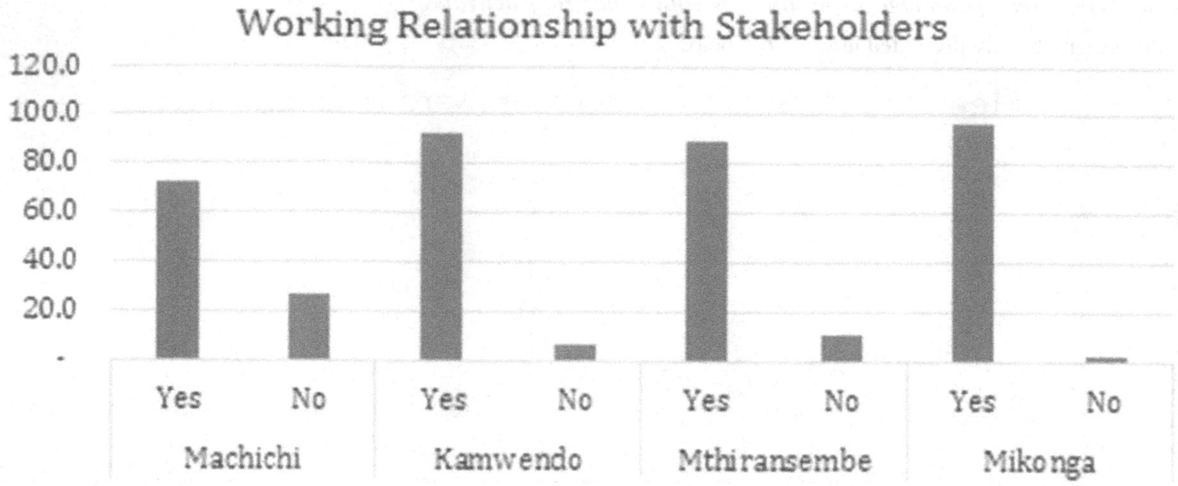

Fig. 26.6 Working relationship with stakeholders

Source: Author's compilation

The above bar graph depicts all cooperatives having goodness of working relations with stakeholders above 70% with Mikonga leading (98%) followed by Kamwendo (95%) then Mthiransembe (90%) and Machichi (75%). Conspicuously, Kamwendo cooperative is the only one with refinery, capacity to fortify with vitamin A, linkages to retail outlets outside the district and certified by Malawi Bureau of Standards since 2013. The other three cooperatives were linked to Kamwendo cooperative for its unique value addition services and competitive advantage at a fee.

Financial parameters analysis

Solvency analysis results.

The results are for Quick Ratio (instant debt paying ability) and margin of safety for creditors (ability to withstand adverse business conditions). The results are graphically presented in Fig. 26.7 below.

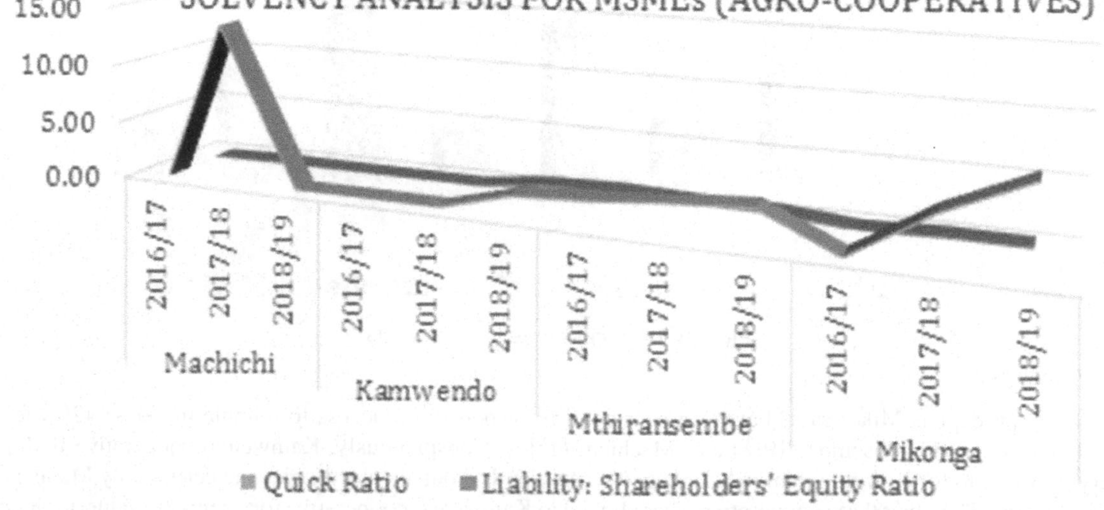

Fig. 26.7 Solvency Analysis Results

Source: Author's compilation

Mikonga displayed increases in instant debt paying ability (Quick Ratio) over the study period while for Mthiransembe the increase was steady, for Machichi it kept on fluctuating and for Kamwendo it was slightly above zero for the starting two years followed by an increase in the third year. For ability to withstand adverse business conditions, Mthiransembe and Mikonga are relatively better off. Therefore, it is implied that Mthiransembe and Mikonga passed the solvency test, having the abilities to meet financial obligations once due, withstand adverse business conditions and earn incomes.

Profitability analysis results

The results are for efficiencies, effectiveness, turnovers, operating ratios, and measures of financial position for an enterprise as presented in Fig. 26.8 below.

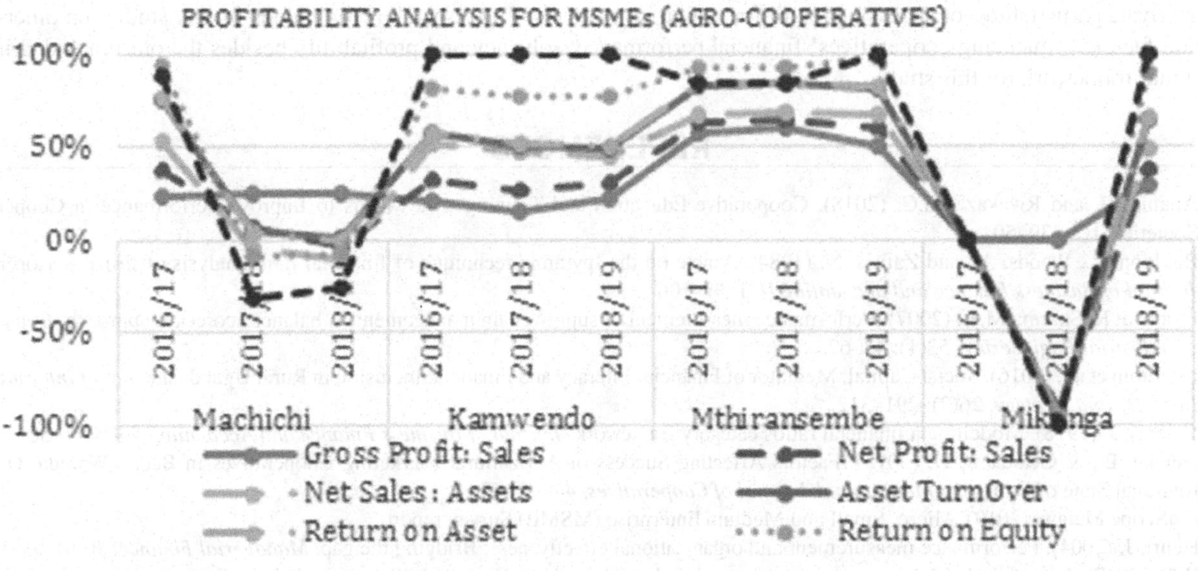

Fig. 26.8 Profitability analysis results

Source: Author's compilation

As displayed in Fig. 26.8, Kamwendo and Mthiransembe have all the profitability measure results above 0% during the study period. Such results imply that the two cooperatives had the ability to earn profits as evidenced by their effectiveness and efficiencies in entrepreneurial operations relative to available resources. Machichi and Mikonga displayed instabilities in most profitability parameters as evidenced by around zero and below zero or negative results in some of the years. This calls for use of multiple indicators to triangulate establishment of the enterprise's current financial position and capabilities to predict its future stand in the business domain.

6. Conclusion

The study has explored, framed and proposed performance measurements (PMs) for success, reinvestments and partnerships of MSMEs in Malawi, with cooking oil producing Agro-cooperatives representing MSMEs. Both financial and non-financial performance measurement dimensions are considered to determine their capabilities. Non-financial and financial parameters are identified by benchmarking on Balanced Scorecard (BSC) and DuPont Pyramid of Financial Ratios (Warren *et al.*, 2005) and substantiated by preceding studies. Four cooperatives are sampled for descriptive and non-financial primary data and financial secondary data covering 2016 to 2019 financial years. Primary data is obtained through semi-structured interviews while empirical data for financial dimensions is obtained from MSMEs' financial records and databases. SPSS and Microsoft Excel packages are used to analyze the data.

Study findings indicate that none of the cooperative redesigned conventional PMs in spite of accepting the need to do so and acknowledging their associated benefits. There are inconsistencies across MSMEs regarding their performance in non-financial and financial dimensions. There are no clear empirical direct causalities between results of non-financial (moderating) and financial (independent) performance variables much as they both, somewhat, influence solvency and profitability that affect enterprise's success, survival, reinvestments and partnerships. However, it is evidenced that MSMEs' financial performance (independent variables) has direct effect on their solvency (ability to meet financial obligations and earn income) and profitability (ability to earn profits) as dependent variables. MSMEs that produced positive results in moderating (non-financial) parameters retained similar internal operations' results for efficiency and effectiveness as well as outshining in overall capabilities of solvency and profitability. Here, it is inferred that there are linkages between non-financial, financial and overall enterprise performance. Overall, this study suggests that understanding financial and non-financial drivers of performance and critical success factors (CSFs) is paramount to MSMEs as evidence points that outstanding performance in all non-financial dimensions does not guarantee the same in financial dimensions and that of overall enterprise success and capabilities to survive, reinvest and form partnerships.

The study framework and methodology can be used by MSMEs to assess their worthiness for either inter-enterprise partnerships, public-private partnerships, or accessing credit for capital investments. The research recommends further studies on other non-financial factors influencing cooperatives' financial performance, solvency and profitability besides the ones included in the conceptual framework for this study.

REFERENCES

1. Anania, P., and Rwekaza, G.C. (2018). Cooperative Education and Training as a Means to Improve Performance in Cooperative Societies, 1(2): 39–50.
2. Bayldon, R., Woods, A., and Zafiris, N. (1984). A note on the 'pyramid technique of financial ratio analysis of firms' performance. *Journal of Business Finance and accounting II*, 1: 99–106.
3. Bhagwat R, Sharma M.K. (2007). Performance measurement of supply chain management: A balance scorecard approach. *Computers & Industrial Engineering*, 53(1): 43–62.
4. Bogomin et al., (2016). Social Capital: Mediator of Financial Literacy and Financial Inclusion in Rural Uganda. *Review of International Business and Strategy*. 26(2): 291–312.
5. Courtis, J. (1978). Modelling a financial ratios category framework. *Journal of Business Finance and Accounting* 5(4):371–386.
6. Dejene, E., & Getachew, D. (2015). Factors Affecting Success of Agricultural Marketing Cooperatives in Becho Woreda, Oromia Regional State of Ethiopia. *International Journal of Cooperatives, 4*(1): 9–17.
7. FinScope Malawi (2019). Micro, Small and Medium Enterprise (MSME) Survey report.
8. Henri, J. (2004). Performance measurement and organizational effectiveness; Bridging the gap. *Managerial Finance*, 30(6): 93–123.
9. ILO. (2017). *Use of statistics on cooperatives for national policy making*. Geneva: International Labour Organization.
10. Kotane, I., & Kuzmina-Merlino, I. (2017). Analysis of Small and Medium Sized Enterprises' Business Performance Evaluation Practice at Transportation and Storage Services.
11. Malawi Country Private Sector Diagnostic (2021). Creating Markets in Malawi. The Road to Recovery: Turning Crisis into Economic Opportunity. International Finance Corporation.
12. Munthali Thomas, C. (2019). ICAM 2019 Conference Paper: Positioning for Africa's Economic Renaissance - Malawi in The Equation. *New National Development Agenda for Malawi – Spreading Wealth Creation for All*. National Planning Commission, Malawi.
13. Quartey, P. and Abor, J. (2010). Issues in SME Development in Ghana and South Africa. International Research Journal of Finance and Economics. ISSN 1450-2887. Issues 39 @ EuroJournals publishing, inc.
14. Sushila, D., Nurizah, N., Mohd, S., Rafedah, J., and Farahaini, M. (2010). Success factors of Cooperatives in Malaysia. *Malaysian Journal of Cooperative Studies*, 6:1–24.
15. Turkkahramn, M. (2012). The Role of Education in the Societal Development. *Journal of Educational and Instructional Studies in the World*, 2(4): 38–41.
16. Warren, C. S, Reeve, J. M., and Fess, P. N. (2005). Managerial Accounting Book, 8th International Student Edition. Thomson South-Western, 5191 Natorp Boulevard, Mason, Ohio 45040: 267–522.
17. Zidana, R. (2015). Small and Medium Enterprises (SMEs) Financing and Economic Growth in Malawi: Measuring the Impact between 1981 and 2014. *Journal of Statistics Research and Reviews*, 1(1):1–3.

Impending Inquisitions in Humanities and Sciences (ICIIHS-2022) – Dr. Mohan Varkolu et al. (eds)
© 2024 Taylor & Francis Group, London, ISBN 978-1-032-78829-6

Managing Working Capital: Does it Impact the Profitability of HFCs?

27

D. Srinivasa Kumar*
Department of Basic Sciences and Humanities, GMR Institute of Technology, Rajam, Andhra Pradesh, India
T. Ramprasad[1]
Department of Mathematics, Vasavi College of Engineering (A), Hyderabad, Telangana, India

ABSTRACT The goal of this investigation is to precisely discover the association between Working Capital Management (WCM) and the profitability of Indian Housing Finance Companies (HFCs). Working Capital Management is foreseen to upgrade the estimation of the business and enrich its profitability. During this investigation, Return on Assets (ROA) is used as a dependent variable, and a few independent variables are used to measure the profit of organizations. The financial information collects from annual reports of HFCs for a period of ten years from 2011–2020 and consider for investigation as a secondary source. The regression analysis technique utilizes to analyze the whole financial information of selected housing finance companies. The end results show that there is no correlation between profitability and working capital management among the housing finance companies in India.

KEYWORDS Working capital, Housing finance companies, Return on assets, Current assets, Current liability, Total asset

1. Introduction

The primary goal of this investigation is to concentrate on exploring the various components of working capital management of the organisations and discovering their association with the profitability of organisations among Indian Housing Financing Companies (HFCs) during the study period from 2011 to 2020. The working capital management (WCM) of the firms covers not only the current assets administration, in tune with the administration's attitude towards risk and presenting itself at an acceptable level of current assets that balances liquidity and profitability standards, and the management to invest the selected level of current assets, once again taking into consideration the elevation of management towards risk. Deloof Marc has used an empirical test for sample companies. It is seen that many organisations have been investing a huge amount of money in the management of working capital (Deloof, 2003). The working capital management decision absolutely implies the decision on a firm's profitability (Adekola et al., 2017; Mannori and Mohammad, 2012; Deloof, 2003.

Previous studies and discoveries have affirmed the trade-off between the amount of working capital and its profit. Besides, there is an unfavorable relationship between the return on assets and components of working capital, while the return on assets is insistently related to the stock period of change and development of sales (Evci and Sak, 2018).

Proficiently, the management of working capital creates a remarkable activity in the overall business technique to produce an incentive for the financial expert. Working capital is considered as a result of the value deferral between the acquisition of raw

*Corresponding author: srinivasakumar.d@gmrit.edu.in
[1]rammaths.vasavi@gmail.com

DOI: 10.1201/9781003489436-27

materials and the collection of finished goods. Research investigates the extent to which the development effectiveness and cost effectiveness of 145 organisations can be improved by improving the exam boundaries. The exam tests the association between the liquidity score and the profitability estimated based on the compensation for liquid assets and the income from the average of total assets (Harsh and Sukhdev, 2013).

On the off chance that you are thinking about a review of the organization's financial strategies, the organisation can utilise short-term or long-term debt to fund business activities. The organisation has followed moderate financing strategies by utilising long-term debt to assist with its business activities, whereas the organisation uses progressively current liabilities to fund its tasks. The organisation follows a fantastic financial strategy, and the related relationship is employed as a substitute for estimating the finance strategy. The financing strategy is measured through Total Current Liabilities with respect to Total Assets. The most noteworthy extent suggests that it reasonably uncovers the aggressive policy to point out the impact of working capital management approaches on the organization's profit and is used to investigate return on assets (Soukhakian and Khodakarami, 2019). It will likewise be seen that manufacturing firms have a significant connection between various components of working capital and profitability. Working capital has a larger impact on the profitability of a large business (Anandasayanan *et al.*, 2014).

2. Objectives of the Study

- To check the association between firm management of working capital and its Profitability.
- To assess the impact on the ratio of Current Assets to Total Assets (CATA) and on its profitability.
- To evaluate the impact on the ratio of Current Liabilities to Total Assets (CLTA) on its profitability.
- To determine the impact on the ratio of Debt-to-Equity (DE) on its profitability.
- To analyze the Firm's Size (Ln) in any way with its profitability.
- To evaluate the impact of Net Operating Profit Margin Ratio (NOPMR) on its profitability.

3. Theoretical Framework

Return on Asset (ROA)

Return on Assets could be a profit index that assesses the entire total income made by all total assets for an amount of contrastive net revenue with average total assets. ROA estimates how well an organisation will handle its advantages in generating profit over some stretch of time. This indicator will permit both management and investors to perceive how well an organisation is transforming its investments into profits. The return on assets quantifies the firm's most recent profitability, which is a measure of how satisfied the assets are with being profitable; if this rate is higher, the performance is good (Gitman, 2004; Edmond et al., 2012). The higher ROA is better for investors as it indicates that the organisation can earn more net income if it properly manages its assets. Generally, a positive rate of return on assets also indicates an upward trend in earnings. It was discovered that there was a positive influence of working capital on the Return on Assets of the organization. If you must invest enough working capital, have enough profit for assets, which influences the increase in profitability of organisations (Kumar, 2017).

Ratio of Current Assets to Total Assets (CATA)

The higher magnitude relationship of current assets to total assets will indicate a reduction in profits quickly because the profit of current assets is lower than that of fixed assets. In addition to that, a few current assets (like an account receivable) may cause bad debts, which will eventually be discounted, causing reduced profits. The hostile investment policy variable was estimated from the TCA/TA proportion, validating that the association between traditionalist investment strategies and the profitability of the company shows a negative relationship. Therefore, the decision to adopt traditional investment strategies will negatively affect the company's profitability (Vahid *et al.*, 2012).

H_1: *Changes in CATA positively influence ROA.*

Ratio of Current Liabilities to Total Assets (CLTA)

The quantitative relationship of current liabilities to total assets displays the extent to which an organisation has used its short-term financial obligations to finance its total assets. This is the sum financed by creditors on the total assets of the organization's enterprise. This indicator is used by investors to assess whether the organisation has enough money to meet its entire short-term financial obligations and whether it can pay a reasonable return on its investment. When talking about a company's financial strategies, a company can use them to fund both short-term and long-term commitments to its activities. When using a long-term commitment to fund its activities, the company follows the traditional financing strategy. However, the company gradually uses its current obligations to finance its activities. At this point, the company follows a strong financing strategy, and the related index proxy is used to quantify the financing strategy. The financing strategy is evaluated by the ratio of current liabilities to total assets (CLTA).

H_2: *Changes in CLTA positively influence ROA.*

Ratio of Debt-Equity (DE)

Debt typically has a lower cost of capital compared with equity, fundamentally due to its rank on account of liquidation. Along these lines, numerous organisations may want to utilise debt over equity for capital financing. Sometimes, the debt-to-equity estimation might be constrained to incorporate only short-term and long-term debt. Frequently, it only incorporates some type of extra fixed payment. Together, the complete debt and total equity of an organisation combine to equal its total capital, which is additionally represented as all total assets. As with any ratio, the debt-to-equity ratio has additional meaning and understanding compared to similar calculations for different historical financial periods. This expanding impact increases additional risk to the organisation and increases costs due to higher interest and debt costs. The debt-to-equity ratio can be confusing unless it is used along with industry averages and financial information to decide how an organisation uses debt and equity compared to its industry. Organizations that take capital seriously may have a higher debt-to-equity ratio, while service firms will have lower ratios. A progressive sign indicates that if a DE ratio is established, it will increase the ratio of CATA. Since both are liquidity ratios, the consequences indicate that companies will be prepared to use their own assets to satisfy their current assets (Zafar *et al.*, 2016). There is a negative relationship between borrowed capital and profitability (Shin and Soenen, 1998) and it was also found that there is a negative correlation between ROA and DR (Akinlo, 2012).

H_3: *Changes in DE positively influence ROA.*

Size of Firm (Ln)

The size of a firm affects the profitability of its investment in working capital management. Small firms endure the effects of high debt interest rates; they may exhibit a more delicate positive and spread a negative trend in the U-shaped relationship. In addition, these organizations' break-even points may emerge at a good time and must be less than the initial investment goal of the full sample. In any case, large corporations recognise less interest and invest in cutting-edge ventures. Thus, their working capital management and U-shaped relationship can trigger a positive trend and decrease a negative trend (Mahmood *et al.*, 2019). A firm's size means how much an organisation holds in assets. The firms' volume and degree of their capacities in creating and developing countries is enormous to the organisations working in developing markets. Organizations in developing markets are frequently small and have little access to long-term capital markets. Those organisations are probably going to rely on extra genuine owner financing, trade on debt, and short-term debt subsidies to offset their attractive savings, cash, receivables, and inventories.

H_4: *Changes in Ln positively influence ROA.*

Net Operating Profit Margin Ratio (NOPMR)

When net operating income is an appropriate measure of organisational profit, it implies that the performance is acceptable and, therefore, profits will increase in the organization. In any case, if the profit is less, it implies that the indirect expenses of the organisation will increase. Numerous methods can be utilised to determine the profit of an organization, such as net operating margin, net income, and return on assets. The overall revenue indicates the distinctive financial plans adopted by organisations and is not an issue in the operating procedures. In this way, if the overall revenue in an organisation is low, this may prompt a high pace of profit for the owners' investments on account of the financial influence. Then again, if the pace of return is high, this demonstrates a decent presentation of the organization, high overall gains, and remarkable development. It is inferred that the financial performance of the chosen organisations as evaluated through the Return on Assets ratio is halfway acceptable (Najib *et al.*, 2021). If there should be an occurrence between the organisations, it is not adequate to get a huge profit for their assets in the event of a year-end investigation. However, it is not significant according to the net profit ratio and return on assets. The chosen organisations are neglecting to get their estimated net profit during the study period, and the associations moreover neglect to get a decent return on their assets.

H_5: *Changes in NOPMR influence ROA positively.*

Sample Design

In this study, five Indian Housing Finance Companies (HFCs) are selected, namely Life Insurance Company Housing Finance Co., Ltd. (LICHFL), Dewan Housing Finance Co., Ltd. (DHFL), Housing Development Finance Co., Ltd. (HDFCL), Can Fin Homes Limited (CFHL) and GIC Housing Finance Co., Ltd. (GICHFL). These five HFCs are playing a vital role in the housing business in India and have occupied various positions in the top ten HFCs in doing housing finance business in India. In this investigation, descriptive and correlation analysis have been used to find an empirical model and test the relationship between the WCM and the profitability of HFCs.

4. Methodology and Empirical Model

Methodology

This paper study adapts the various components of working capital management (WCM) and attempts to set up relationships between these components and the profitability of Indian housing finance companies (HFCs). The study period has been considered for 10 years, that is from 2011 to 2020. The whole information considered in this examination of secondary sources, mainly from annual reports of selected HFCs, which are published on respective organisation websites.

Empirical Model

In this regard, we developed one empirical model equation to consider the impact of WCM on HFC profitability. The practise of working capital management from the perspective of Sri Lanka was utilized. They utilised multiple regression analysis techniques to accurately determine the limits of industry best practises and measure the company's degree of separation from that limit. The target of the investigation was to seek after extra exploration instead of to uncover all the components related to WCM in the Sri Lanka perspective. The authors accept that the asset requirement might be a significant hindrance to the use of working capital by firms (Bei and Wijewardana, 2012). The regression coefficient among the self-employed group shows that the adjustments in the measure of housing finance are affected by the thought about independent variables. Only at the 5% level is the fitted regression model critical. Because of consolidated information, there are no significant factors that impact the measure of home finance. It shows that an expansion of one unit in the above factor won't increase the measure of housing finance. The change in housing finance is improved by the considered independent variables. The regression model is not fitted at the 5% level (Srinivasakumar, 2020).

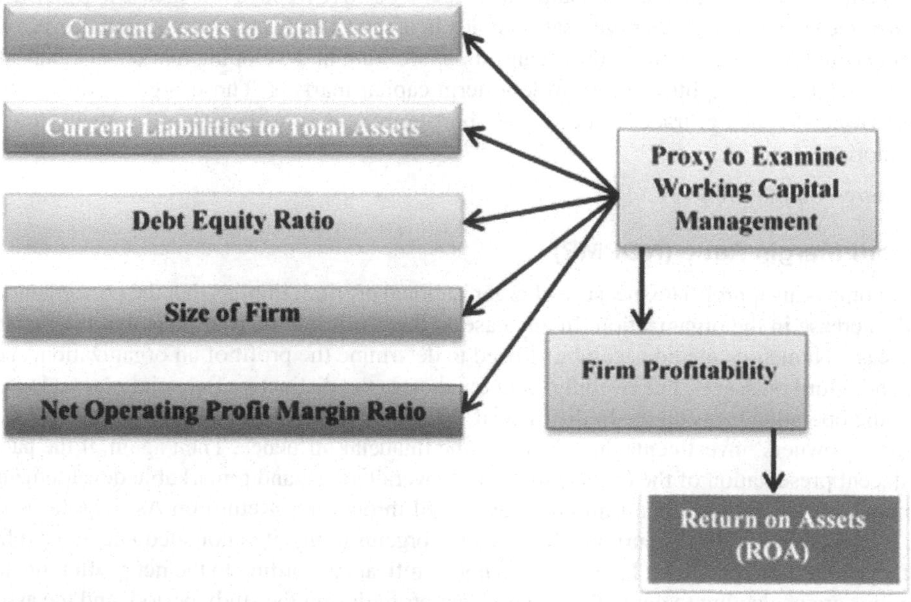

Fig. 27.1 Relation between dependent variable and independent variable

Fig. 27.1 shows the relationship between working capital management components and the firm's profitability. In this model, we test the variations in organizations' profitability (or ROA) by varying independent variables like Current Asset to Total Assets (CATA), Current Liabilities to Total Assets (CLTA), Debt to Equity ratio (DE), Size of Firm (Ln) and Net Operating Profit Margin Ratio (NOPMR). The regression analysis is a significant approach and typically attempts an investigation of how at least one variable has an effect or influences another variable. In this regression analysis, the level of significance is considered at a rate of 5% to test hypotheses, the following empirical research models are developed.

Model Equation

$$ROA_{it} = \beta_0 + \beta_1 CATA_{it} + \beta_2 CLTA_{it} + \beta_3 DE_{it} + \beta_4 Ln_{it} + \beta_5 NOPMR_{it} + \varepsilon_{it}$$

Here β_0 is intercept, β_1, β_2, β_3, β_4, and β_5 are the coefficients of variables

ε_{it} = Random error term for the ith firm at t time

ROA_{it} = Return on Asset for the ith firm at t time

$CATA_{it}$ = Current Assets to Total Assets Ratio for the ith firm at t time

$CLTA_{it}$ = Current Liabilities to Total Assets Ratio for the ith firm at t time

DE_{it} = Debt to Equity Ratio for the ith firm at t time

Ln_{it} = Firm Size for the ith firm at t time

$NOPMR_{it}$ = Net Operating Profit Margin Ratio for the ith firm at t time

Empirical Results and Discussion

Table 27.1 Descriptive statistics

	Mean	Median	SD	Min.	Max.
ROA	0.0157	0.0155	0.0056	-0.01	0.03
CATA	0.3522	0.1214	0.40776	0.01	1
CLTA	0.2123	0.1996	0.14259	0	0.87
DE	7.075	8.02	2.75694	0	11.86
Ln	10.467	10.645	1.67124	7.69	13.4
NOPMR	15.1539	14.8349	5.61792	-7.69	33.82

Source: Author's compilation

Table 27.1 shows the average mean of the ROA is 0.015 of total assets only and the SD is 0.0056. It is demonstrated that there is exceptionally less deviation in ROA among all factors. The ordinary assessment of the investment strategy was estimated through the ratio of current assets to total assets (CATA) with a mean of 0.35, recommending that the HFCs crush the preservationist investment strategy as they were investing more on current assets. The normal value of the financing strategy derived from the average estimate of the ratio of current liabilities to total assets (CLTA) is 0.21, indicating that HFCs have a sensible financing strategy by using the ratio of CLTA. The SD of the ratio of CLTA is 0.14, which also shows a small deviation. The normal mean debt-to-equity ratio is 7.08. It estimates that HFCs contribute more outsider funds to their assets than their own funds, and it also discovers that HFCs require less working capital for their operations. The mean value of firm size is 10.4. It implies the size of a firm shows an impact on its working capital to measure profitability. The normal mean of Net Operating Profit Margin Ratio (NOPMR) is 15.15 percent. It demonstrates that return on total sales is 15.15 percent. And since SD is showing 5.61, it shows that there is immense variation in profits in all HFCs during the examination period.

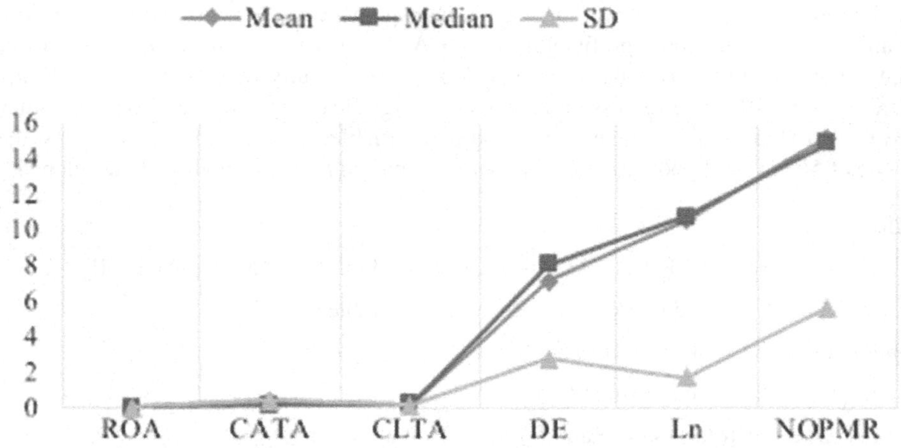

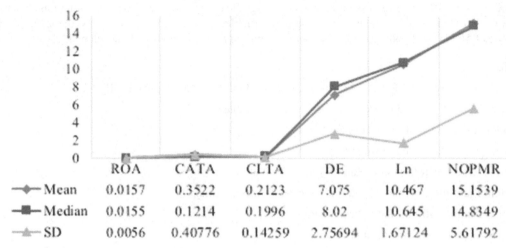

	ROA	CATA	CLTA	DE	Ln	NOPMR
Mean	0.0157	0.3522	0.2123	7.075	10.467	15.1539
Median	0.0155	0.1214	0.1996	8.02	10.645	14.8349
SD	0.0056	0.40776	0.14259	2.75694	1.67124	5.61792

Fig. 27.2 Descriptive statistics

Table 27.2 Correlation matrix of between the variables

	ROA	CATA	CLTA	DE	Ln	NOPMR
ROA	1					
CATA	−0.180	1				
CLTA	0.332*	−0.254	1			
DE	−0.525**	0.123	−0.767**	1		
Ln	−0.082	0.115	0.198	−0.167	1	
NOPMR	0.908**	−0.069	0.131	−0.242	−0.248	1

Source: Author's compilation

**Significant at the 0.05 level (2-tailed).*

***Significant at the 0.01 level (2-tailed) and N = 50*

Table 27.2 shows how the model data were tested for multicollinearity before evaluating the empirical model. As a result, it is critical to assess the degree of relationship between the variables used in this study. The independent variables should be correlated, low-correlated, or uncorrelated with each other to check for increased multicollinearity. In this examination, ROA and CATA, DE, and Ln are negatively related. In any case, ROA is positively correlated with CLTA and NOPMR, with correlation coefficients of 0.332 and 0.908, respectively. A muticollinearity issue happens if the correlation between two variables exceeds 0.90 (Hair *et al.*, 2006). The ROA and NOPMR are powerfully correlated with a correlation coefficient of 0.908, which will cause a positive multicollinearity issue if both are remembered for a similar regression model. Lastly, while in perception, CLTA positively correlated with all variables except CATA, because CLTA negatively correlated with CATA.

Table 27.3 Regression model summary[b]

R	R^2	Adjusted R^2	SE	Change Statistics					Durbin-Watson
				Change in R^2	Change in F	df1	df2	Change in Sig. F	
0.971a	0.944	0.937	0.0014	0.944	146.97	5	44	0	1.526

Source: Author's compilation

[a]Predictors: (Constant), CATA, CLTA, DE, Ln, NOPMR

[b]Dependent Variable: ROA

The variation rate of the dependent variable (ROA) is measured by R^2. This variation can be predicted from the independent variables like CATA, CLTA, DE, Ln, and NOPMR shown in table 27.3. It is indicated that the determined coefficient value of R^2 is 0.944. It is revealed that all independent variables, variation caused by changes in profitability by 0.944, the change in R^2 statistic is the same as the predicted model R^2 value. The Durbin Watson statistics value is 1.526, which is less than 2, which indicates positive autocorrelation between the variables.

Table 27.4 Results of regression coefficients for study model

Variable	Unstandardized Coefficients		Standardized Coefficients	T	Sig.	Collinearity Statistics	
	B	SE	Beta		0.05	Tolerance	VIF
(Constant)	0.006	0.002		2.989	0.005		
CATA	-0.002	0.001	-0.116	-3.057	0.004	0.896	1.116
CLTA	-0.005	0.002	-0.119	-2.045	0.047	0.376	2.657
DE	-0.001	0	-0.38	-6.561	0	0.384	2.607
Ln	0	0	0.102	2.628	0.012	0.854	1.171
NOPMR	0.001	0	0.849	21.896	0	0.853	1.172

Dependent Variable: ROA

Source: Author's compilation

The hypotheses of the study have been tested and analysed through t value. The significance level in this study is 0.05 (2-tailed test). Thus, the table value (critical value) of *t* is ±1.96. To test whether the hypothesis accepts or rejects the decision rule, compare the calculated t value and the critical value of t, and see if the calculated t value is greater or less than the table value, as shown in table 27.4. The above table reveals the relationship between the dependent variable and the independent variable. These estimations indicate that the measure of increments in ROA score would be predicated by a one-unit increment in the dependent variables (CATA, CLTA, DE, Ln, and NOPMR).

The hypothesis (H1) that there is a positive influence of changes in CATA on the profitability of HFCs (measured by ROA) is rejected on the grounds that the absolute determined value of *t* for CATA is 3.057 (since two-tailed) and the critical value of *t* in this investigation is 1.96. Thus, the determined value shows more than the critical value, so H1 is rejected. Hence, there was no significant impact of changes in CATA on HFCs' profitability (estimated by ROA). In another manner, additionally, with a level of significance (P-value) considered in this examination is 0.05, with a determined level of significance is 0.004 (P-value), so the determined value is less than the considered P-value, so H1 is rejected with this technique also. The regression coefficient of CATA is –0.002, which is statistically not significant, and it indicates that there is a negative relationship with the determined probability (P-value) during this examination.

H2 is that there is a positive influence of changes in CLTA on the profitability of HFCs (assessed through ROA) is rejected. The absolute determined value of *t* for CLTA is 2.405 (since two-tailed) and the critical value of *t* in this assessment is 1.96. Hence, the determined value is greater than the critical value. This hypothesis is therefore rejected. It was concluded that any changes in the CLTA ratio have no significant influence on the profitability of HFCs. In other words, the P-value being considered in this investigation is 0.05 and the determined value of the P-value for CLTA is 0.047. Consequently, the determined P-value is less than the considered P-value of 0.05, so H2 is also rejected in this situation. In this manner, the regression coefficient value of CLTA is –0.005, which is statistically not significant. In this manner, it was reasoned that there is a negative correlation with the determined P-value.

H3 is that there is a positive impact of changes in DE on the profitability of HFCs (as measured by ROA), which is also rejected in table 27.4. It is clear from the above outcomes that the absolute determined estimation of *t* for DE ratio is 6.561 and the critical value is 1.96, so the determined estimation of *t* more prominent than the critical value, and subsequently H3 is also rejected. It reveals that any changes in the DE ratio have no significant effect on the profitability of HFCs. Similarly, it was tried in another procedure additionally for a better conclusion with P-value. The determined P-value is 0.00 and the considered P-value is 0.05, so the determined P-value is less than the considered P-value. Hence, H3 is rejected in this case also. The result of the regression coefficient of DE is 0.001, statistically not significant, and it shows a negative correlation with the probability of the sampled HFCs for the investigation. It concludes that there is no significant influence of changes in DE on the profitability of HFCs.

H4 is that there is the positive influence of change in Ln (Size of Company) on the profitability of HFCs (measured through ROA), also rejected in the table 27.4. It is evident from the above results that the determined *t* value of Ln (Size) is 2.628 and the critical value of *t* is 1.96, so the determined value of t is greater than the critical value, hence, H4 is rejected. It shows that any change in Ln will not have a significant influence on the profitability of HFCs. Similarly, the calculated P value of Ln is 0.012, and the considered P-value is 0.05, so the calculated P-value is less than the considered P-value of this study, so H4 is also rejected in this technique. The result of the regression coefficient shows that Ln is insignificant with a probability value of 0.012 in this study. It concludes that there is no significant influence of changes in Ln on the HFCs' profitability.

H5 is that there is the positive influence of changes in NOPMR on the profitability of HFCs (assessed through ROA) is rejected. The calculated value of t for NOPMR is 21.90 and the critical value of t in this examination is 1.96, so the determined value is greater than the critical value. Hence, the H5 hypothesis is also rejected. It shows that any changes in NOPMR will not have a significant influence on HFC's profitability. In another way, this hypothesis is tested with a P-value. The predetermined P-value is considered in this examination is 0.05, and the calculated P-value for NOPMR is 0.00. Hence, the calculated P-value is less than the predetermined P-value. Therefore, this hypothesis is rejected. The regression coefficient of NOPMR is 0.001, statistically not significant even though there is a positive relationship with the calculated P-value.

In a regression analysis, VIF means variance inflation factor, which quantifies how much the variance of the estimated coefficients is inflated (Farhan *et al.*, 2020). A value of Variance Inflation Factor (VIF) greater than 5 may indicate that a particular variable has multicollinearity (Studenmund and Cassidy, 1992; Nastiti *et al.*, 2019), but if the VIF value shows less than 5, then it is a satisfactory value. It indicates that there is some multicollinearity in our data. In this examination, all independent variables' VIF value is less than 3. Therefore, all independent variables have strong multicollinearity.

5. Limitations

This paper ponders some components of working capital to explore the company's profitability. As a result, the estimates based on the variables only used to analyse working capital management and Housing Finance Company (HFC) profitability should have some degree of correlation. The present examination utilised published secondary information and, subsequently, the confinement of the published annual reports of respective companies may have been suitable in this investigation. This study concentrated on a few Housing Finance Companies (HFCs) only. Commercial banks are also playing a vital role in the Indian housing finance business, but commercial banks are not taken into consideration for this study. Because of the huge data taken into consideration to analyse the regression results, it might have increased the variability of end results.

6. Conclusion

This article examines how Housing Finance Companies (HFCs) manage their working capital to generate profitability. The result of this inspection is helpful for the short-term investment of HFCs and their money arrangements. As the results of this survey show, all the independent variables considered in this survey and the changes in their profitability have no significant impact. The predetermined hypotheses were rejected in this survey. There is no relationship between working capital and the profitability of HFCs, so profitability is estimated based on total revenue. The most influential factor in working capital management when it comes to Housing Financing Companies (HFCs), is that less money must be invested in fixed assets. However, a significant amount of money must be invested as working capital. Finally, the size of the company determines its working capital needs. Large-scale operations necessitate more working capital than smaller operations, so this is one of the most important factors influencing working capital. Another important influencing factor is operational efficiency. Businesses that improve their operational efficiency must invest less money in working capital. In contrast, organizations with lower operational efficiency require more funds to be invested in working capital. HFCs can significantly expand their profits by adequately managing their organization's working capital. Finally, it is prescribed that HFC attempt to keep up a decent balance between assets and liabilities, which is essential for HFC to force executives of any organisation to make money.

REFERENCES

1. Adekola, A., Samy, M., & Knight, D. (2017). Efficient working capital management as the tool for driving profitability and liquidity: A correlation analysis of Nigerian companies. *International Journal of Business and Globalisation, 18(2)*, 251. https://doi.org/10.1504/ijbg.2017.081957.
2. Akinlo, O. O. (2012). Effect of working capital on profitability of selected quoted firms in Nigeria. *Global Business Review*, 13(3), 367–381. https://doi.org/10.1177/097215091201300301.

3. Anandasayanan, S., Raveendran, T., & Raveeswaran, M. (2014). The relationship between working capital management and profitability of listed manufacturing companies in Sri Lanka. *SSRN Electronic Journal*. https://doi.org/10.2139/ssrn.2385949.
4. Bei, Z., & Wijewardana, W. (2012). Working capital policy practice: Evidence from Sri Lankan companies. *Procedia - Social and Behavioral Sciences, 40*, 695–700. https://doi.org/10.1016/j.sbspro.2012.03.251.
5. Deloof, M. (2003). Does working capital management affect profitability of Belgian firms? *Journal of Business Finance & Accounting, 30(3-4)*, 573–588. https://doi.org/10.1111/1468-5957.00008.
6. Edmond, T. P., Cindy, D. E., Bor-Yitsay, P. R & Nancy, W.S. (2012). *Fundamental managerial accounting concept*, Irwin: McGraw-Hill.
7. Evci, S., & Şak, N. (2018). The effect of working capital management on profitability in emerging countries: Evidence from Turkey. *Financial Management from an Emerging Market Perspective*. https://doi.org/10.5772/intechopen.70871.
8. Farhan, N. H. S., Tabash, M. I., & Yameen, M. (2020). The relationship between credit policy and firms' profitability: Empirical evidence from Indian pharmaceutical sector. Investment Management and Financial Innovations, 17(2), 146–156.
9. Gitman, L. J. (2004). *Principles of managerial finance (10th ed.)*. Prentice Hall.
10. Hair, J.F., Black, W.C., Babin, B.J., Anderson, R.E. & Tatham, R.L. (2006). *Multivariate Data Analysis (6th ed.)*. Upper Saddle River, NJ: Pearson University Press.
11. Harsh, V. K & Sukhdev, S. (2013). Managing efficiency and profitability through working capital: An empirical analysis of BSE 200 companies. *Asian Journal of Business Management*, 5(2), 197–207.
12. Kumar., D. S. (2017). Effect of working capital on firm's profitability: A pragmatic study with reference to pharmaceutical companies in India. *International Journal of Advanced Research, 5(7)*, 668–672. https://doi.org/10.21474/ijar01/4767.
13. Mahmood, F., Han, D., Ali, N., Mubeen, R., & Shahzad, U. (2019). Moderating effects of firm size and leverage on the working capital finance–profitability relationship: Evidence from China. *Sustainability*, 11(7). 1–14. https://doi.org/10.3390/su11072029.
14. Mannori E & Mohammad, J.(2012). The determinant of working capital management of manufacturing companies: case of Singapore firms. *Research Journal of Finance and Accounting 3 (11)*, 15–23.
15. Najib H.S. Farhan, Fozi Ali Belhaj, Waleed M. Al-ahdal & Faozi A. Almaqtari (2021). An analysis of working capital management in India: An urgent need to refocus, Cogent Business & Management, 8:1, 1924930, https://doi.org:/10.1080/23311975.2021.1924930.
16. Nastiti, P. K., Atahau, A. D., & Supramono, S. (2019). Working capital management and its influence on profitability and sustainable growth. *Business: Theory and Practice*, 20, 61–68. https://doi.org/10.3846/btp.2019.06.
17. Shin, H.H. and Soenen, L. (1998). Efficiency of Working Capital Management and Corporate Profitability. *Financial Practice and Education*, 8, 37–45.
18. Soukhakian, I., & Khodakarami, M. (2019). Working capital management, firm performance and macroeconomic factors: Evidence from Iran. *Cogent Business & Management, 6(1)*, 1684227. https://doi.org/10.1080/23311975.2019.1684227.
19. Srinivasakumar, D. (2020). Attributes influencing home loan borrowers in selecting housing finance companies in India. *International Journal of Scientific and Technology Research, 9(10)*, 4059–406.
20. Studenmund, A. H., & Cassidy, H. J. (1992). *Using econometrics: A practical guide*. Addison-Wesley Educational Publishers.
21. Vahid, T. K., Elham, G., Mohsen, A. K., & Mohammadreza, E. (2012). Working capital management and corporate performance: Evidence from Iranian companies. *Procedia - Social and Behavioral Sciences, 62*, 1313-1318. https://doi.org/10.1016/j.sbspro.2012.09.225.
22. Zafar, S., Nazam, M., Hanif, A., Almas, I & Sana, N. (2016). Impact of working capital management on firm's profitability: A case from food Sector of Pakistan. *European Journal of Accounting, Auditing and Finance Research, 4*(10), 8–58.

Impending Inquisitions in Humanities and Sciences (ICIIHS-2022) – Dr. Mohan Varkolu et al. (eds)

Chemical Reaction Effects on MHD Freeconvective Flow Past Parabolic Started Vertical Plate in the Presence of Viscous Dissipation and Variable Temperature

28

B. Shankar Goud
Department of Mathematics, JNTUH College of Engineering, Hyderabad, India
Pudhari Srilatha*, J. Suresh Goud
Department of Mathematics, Institute of Aeronautical Engineering, Hyderabad, India
Amraj. Srilatha
Department of Mathematics, Hyderabad Institute of Technology and Management, Hyderabad, India

ABSTRACT Research has been conducted in order to investigate the chemical reaction effect and viscous dissipation impacts on free convective MHD flow past a parabolic plate with varying temperature. The controlling boundary layer equations are changed into a set of nonlinear ODEs that may be elucidated numerically employing the finite element technique utilizing similarity conversions. The flow fields are discussed for several flow factors such as mass Grashof (Gc), mass Grashof (Gr), magnetic field (M), chemical (Kr), permeability parameters (k), Schmdt (Sc), Eckert (Ec), Prandtl numbers (Pr), and time (t). Its variations are shown graphically, whereas Skin friction values, Nusselt values, and Sherwood values are presented for various parameter values.

KEYWORDS MHD, Mass diffusion, Finite element method, Variable temperature, Viscous dissipation

1. Introduction

The combined issue of heat and mass transfer has become more significant in both the field of research and the field of technology. In many areas such as nuclear reactor cooling, aerodynamics, energy storage systems, cooling towers, accelerators in soil science and geophysics. Several studies of MHD fluid flows, such as Chamkha [1] investigated MHD flow via a vertical permeable surface influence of chemical reaction along with heat generation/absorption and chemical reaction. Ibrahim et al. [2] analyzed the passage of radiation absorption and chemical reaction to non-stationary MHD flow by free convection with a heat source/suction. Kandasamy et al. [3] examined combined impacts MHD flow heat and mass transport on via a vertical surface. Mansour et al. [4] considered the effects of chemical reaction and viscous dissipation on MHD flow on a vertically extended surface with steam and temperature stratification. Ramana and kumar [5] reported the impact of a chemical reaction on the MHD-free convection flow through an inclined channel. Ibrahim [6] explored the impact of a radiation and chemical reaction exposure on MHD natural convection movement along a stretching surface with heat generation and dissiaption. Pramod Kumar et al. [7] considered Soret number impacts on MHD Jeffrey fluid flow via a vertical permeable moving plate. Postelnik [8] deliberate the impact of magnetic field on heat and mass transmission during free convection on the vertical surface of permeable media. Anjalidevi and Kandasami [9] used a semi-infinite vertical plate to examine the behaviour of chemical reaction on heat and mass transmission on laminar flow. Radiation Impacts on MHD

*Corresponding author: pudhari.srilatha@gmail.com

DOI: 10.1201/9781003489436-28

stagnation point flow on a stretching sheet with slip boundaries was reviewed by Goud [10]. Goud et al. [11] regarded into the behavor of suction on the MHD flow and heat transmission via a permeable shrinking sheet that included a heat source/sink. S Goud et al. [12] investigated the impact that a sheet that was exponentially extending had on the boundary layer flow of an MHD fluid.

Therefore, the main perseverance of this exertion is to discuss the impacts of chemical reactions and MHD-free convective on a parabolic starting plate with mass diffusion and variable temperature. The numerical solution is carried out according to the Galerkin standard method and presented for various flow factors. The characteristics of flow fields, friction, Nusselt and Sherwood number were mentioned for the changes in the control factors.

2. Problem Formulation

Assume approximately conductive, viscous, incompressible fluids that undergo unstable free convection to transport heat and mass. These fluids flow over infinite vertical plates, maintain a constant temperature T'_w, undergo chemical reactions, and experience viscous dissipation. The axis is occupied perpendicular and x'-axis vertically up to the plate, and the uniform magnetic field of force B_0 is utilised to the plate in a constructive way. Initial fluid temperature and plate temperature are the same, and for At $t' > 0$ the plate begins to move at a velocity $u = u_0 t'^2$. As the temperature of the plate increases uniformly and linearly with time t, mass diffusion from the plate to the fluid similarly increases. The terms of the governing equation are independent of x' each other, & there will be no movement path in the y' direction. The equations that are used to govern the typical Bussiness assumption of the boundary layer are as follows:

$$\frac{\partial u'}{\partial y'} = 0 \tag{1}$$

$$\frac{\partial u'}{\partial y'} = \nu \frac{\partial^2 u'}{\partial y^2} + g\beta(T' - T'_\infty) + g\beta^*(C' - C'_\infty) - \left(\frac{\sigma B_0^2}{\rho} + \frac{\nu'}{k'}\right)u' \tag{2}$$

$$\frac{\partial T'}{\partial t'} = \frac{k}{\rho c_p}\frac{\partial^2 T'}{\partial y^2} + \frac{\mu}{\rho c_p}\left(\frac{\partial u'}{\partial y'}\right)^2 \tag{3}$$

$$\frac{\partial c'}{\partial t'} = D\frac{\partial^2 c'}{\partial y'^2} + K'_r(C' - C'_\infty) \tag{4}$$

and the subject to the boundary circumstances

$$\left.\begin{array}{c} t' \le 0 : T' = T'_\infty, u' = 0, C' = C'_\infty \ for\ all\ y' \\ t' > 0 : u' = u_0 t'^2, T' = T'_\infty + (T'_\infty - T'_w)At', C' = C'_\infty + (C'_\infty - C'_w)At' y' = 0 \\ u' \to 0, T' \to T'_\infty, C' \to C'_\infty\ y' \to \infty \end{array}\right] \tag{5}$$

And we add the non dimensional variables

$$U = u'\left\{\frac{u_0}{\nu^2}\right\}^{\frac{1}{3}}, t = \left\{\frac{u_0^2}{\nu}\right\}^{\frac{1}{3}} t', y = \left\{\frac{u_0}{\nu^2}\right\}^{\frac{1}{3}} y', Gc = \frac{g\beta^*(T'_w - T'_\infty)}{(\nu u_0)^{\frac{1}{3}}}, \theta = \frac{T' - T'_\infty}{T'_w - T'_\infty},$$

$$\Pr = \frac{\mu c_p}{k}, Sc = \frac{\nu}{D}, Gr = \frac{g\beta(T'_w - T'_\infty)}{(\nu u_0)^{\frac{1}{3}}}, C = \frac{c' - c'_\infty}{c'_w - c'_\infty}, Ec = \frac{\left\{\frac{u_0}{\nu^2}\right\}^{\frac{1}{3}}}{c_p(T'_w - T'_\infty)} \tag{6}$$

$$k = k'\left\{\frac{u_0}{\nu^2}\right\}^{\frac{1}{3}}, M = \frac{\sigma B_0^2}{\rho}\left\{\frac{\nu}{u_0^2}\right\}^{\frac{1}{3}}, A = \left\{\frac{u_0}{\nu^2}\right\}^{\frac{1}{3}}$$

The governing Eqs. are moderate to in the perspective of equations (2) to (4).

$$\frac{\partial u}{\partial t}=\frac{\partial^2 u}{\partial y^2}+Gr\theta+GcC-\left(M+\frac{1}{k}\right)u \tag{7}$$

$$\frac{\partial \theta}{\partial t}=\frac{1}{Pr}\frac{\partial^2 \theta}{\partial y^2}+Ec\left(\frac{\partial u}{\partial y}\right)^2 \tag{8}$$

$$\frac{\partial C}{\partial t}=\frac{1}{Sc}\frac{\partial^2 C}{\partial y^2}-KrC \tag{9}$$

The boundary circumstances are initiated

$$\left.\begin{array}{l} t\le 0:\theta=0, u=0, C=0 \ for\ all\ y \\ t>0: u=t^2, \theta=t, C=t \ \ y=0 \\ u\to 0, \theta\to 0, C\to 0 \ \ as\ y\to\infty \end{array}\right] \tag{10}$$

3. Problem Solution

Following the application of the Galerkin finite element technique across the element (e) in equation (7), is:

$$y_j\le y\le y_k \ is: \int_{vj}^{vk} N^{(e)}\left\{\left[\frac{\partial^2 u^{(e)}}{\partial y^2}-\frac{\partial u^{(e)}}{\partial t}-Bu+R_1\right]\right\}dy=0 \tag{11}$$

Where $R_1=Gr\theta+GcC, B=\left(M+\dfrac{1}{k}\right)$

The first part in equation (11) integrating by parts

$$N^{(e)}\left\{\frac{\partial u^{(e)}}{\partial y}\right\}_{yj}^{yk}-\int_{yj}^{yk}\left\{\frac{\partial N^{(e)}}{\partial y}\frac{\partial u^{(e)}}{\partial y}-N^{(e)}\left[\frac{\partial u^{(e)}}{\partial t}+Bu^{(e)}-R_1\right]\right\}dy=0 \tag{12}$$

Oversight the first term in equation (12):

$$\int_{yj}^{yk}\left\{\frac{\partial N^{(e)}}{\partial y}\frac{\partial u^{(e)}}{\partial y}-N^{(e)}\left[\frac{\partial u^{(e)}}{\partial t}+Bu^{(e)}-R_1\right]\right\}dy=0$$

Assume the linear piecewise estimation solution is $u^{(e)}=N^{(e)}\phi^{(e)}$ across the element (e), $y_j\le y\le y_k$, where $N^{(e)}=\left[N_jN_k\right], \phi^{(e)}=\left[u_ju_k\right]^T$ and $N_j=\dfrac{y_k-y}{y_k-y_j}, N_k=\dfrac{y-y_j}{y_k-y_j}$ are the The frame work defined as:

$$u^{(e)}=N^{(e)}\phi^{(e)}$$

$$\int_{y_j}^{y_k}\left\{\begin{bmatrix} N_j'N_j' & N_j'N_k' \\ N_j'N_k' & N_k'N_k' \end{bmatrix}\begin{bmatrix} u_j \\ u_k \end{bmatrix}\right\}dy+\int_{y_j}^{y_k}\left\{\begin{bmatrix} N_jN_j & N_jN_k \\ N_jN_k & N_kN_k \end{bmatrix}\begin{bmatrix} u_j \\ u_k \end{bmatrix}\right\}dy+B\int_{y_j}^{y_k}\left\{\begin{bmatrix} N_jN_j & N_jN_k \\ N_jN_k & N_kN_k \end{bmatrix}\begin{bmatrix} u_j \\ u_k \end{bmatrix}\right\}dy=R_1\int_{y_j}^{y_k}\begin{bmatrix} N_j \\ N_k \end{bmatrix}dy$$

After making it easier, we get

$$\frac{1}{l^{(e)}}\begin{bmatrix} 1 & -1 \\ -1 & 1 \end{bmatrix}\begin{bmatrix} u_j \\ u_k \end{bmatrix}+\frac{l^{(e)}}{6}\begin{bmatrix} 2 & 1 \\ 1 & 1 \end{bmatrix}\begin{bmatrix} \dot{u}_j \\ \dot{u}_k \end{bmatrix}+\frac{Bl^{(e)}}{6}\begin{bmatrix} 2 & 1 \\ 1 & 1 \end{bmatrix}\begin{bmatrix} u_j \\ u_k \end{bmatrix}+\frac{R_1l^{(e)}}{2}\begin{bmatrix} 1 \\ 1 \end{bmatrix}$$

Where dot and prime are respectively refers to diff. with res. to t and time y. The assembly of the element eqs for two successive elements

$$y_{i-1}\le y\le y_i \text{ and } y_i\le y\le y_{i+1}$$

obtained as follows:

$$\frac{1}{l^{(e)}}\begin{bmatrix} 1 & -1 & 0 \\ -1 & 2 & -1 \\ 0 & -1 & 1 \end{bmatrix}\begin{bmatrix} u_{i-1} \\ u_i \\ u_{i+1} \end{bmatrix}+\begin{bmatrix} 2 & 1 & 0 \\ 1 & 4 & 1 \\ 0 & 1 & 2 \end{bmatrix}\begin{bmatrix} u_{i-1} \\ u_i \\ u_{i+1} \end{bmatrix}+\frac{B}{6}\begin{bmatrix} 2 & 1 & 0 \\ 1 & 4 & 1 \\ 0 & 1 & 2 \end{bmatrix}\begin{bmatrix} u_{i-1} \\ u_i \\ u_{i+1} \end{bmatrix}=\frac{R_1}{2}\begin{bmatrix} 1 \\ 1 \end{bmatrix}\begin{bmatrix} 1 \\ 2 \\ 1 \end{bmatrix} \tag{13}$$

Now put row to zero referring to the node i, from equ. (13) the transformation forms with $l^e = h$ is:

$$\frac{1}{h^2}\left[-u_{i-1}+2u_i-u_{i+1}\right]+\frac{1}{6}\left[u_{i-1}+4u_i+u_{i+1}\right]+\frac{B}{6}\left[u_{i-1}+4u_i+u_{i+1}\right]=R_1 \tag{14}$$

Following Crank – Nicholson process equations are obtained after applying the trapezoidal rule:

$$A_1u_{i-1}^{n+1}+A_2u_i^{n+1}+A_3u_{i+1}^{n+1}=A_4u_{i-1}^{n}+A_5u_i^{n}+A_6u_{i+1}^{n}+R^* \tag{15}$$

Where $A_1 = A_3 = 2 - 6r + Bk, A_4 = A_6 = 2 + 6r - Bk$

$$A_2 = 8 + 12r + 4Bk, A_5 = 8 - 12r - 4Bk, R_1 = Gr\theta + GcC$$

Now from equations (8) and (9), below equations are acquired:

$$B_1\theta_{i-1}^{n+1}+B_2\theta_i^{n+1}+B_3\theta_{i+1}^{n+1}=B_4\theta_{i-1}^{n}+B_5\theta_i^{n}+B_6\theta_{i+1}^{n}+R^{**} \tag{16}$$

$$C_1C_{i-1}^{n+1}+C_2C_i^{n+1}+C_3C_{i+1}^{n+1}=C_4C_{i-1}^{n}+C_5C_i^{n}+C_6C_{i+1}^{n} \tag{17}$$

$$B_1 = B_3 = Pr - 3r, B_2 = 4Pr + 6r, B_4 = B_6 = Pr + 3r, B_5 = 4Pr - 6r$$

$$C_1 = C_3 = 2Sc + kScKr - 6r, C_2 = 8Sc + 4kScKr + 12r, C_4 = C_6 = 2Sc - kScKr + 6r, C_5 = 8Sc - 4kScKr - 12r$$

Here, $r = \frac{k}{h^2}$ where k, h are mesh size in the lines of time t and y respectively. Index i –indicates the space, and n –indicates the time (t). In equations (15)-(17), using the boundary condition (10), then the following equation system can be acquired:

$$A_iX_i = B_i \qquad i = 1, \ldots .m \tag{18}$$

Where A_i's are matrix of order m and X_i,B_i column matrices that have m components. The above equations can be solved by applying the Thomas algorithm to flow fields. Running the MATLAB application also yields the numerical results. The finite element Galerkin technique is stable and convergent since there was no discernible change in the flow areas with the decreasing values of h, k. For this kind of boundary movement, the physical parameters skini friction, Sherwood number, and Nusselt number are essential.

The dimensionaless form formed by the friction factor with the plate $C_f = \left(\frac{\partial u}{\partial t}\right)_{y=0}$. The heat transfer coefficient rate (expressed by Nusselt number as a dimensionless form) $N_u = -\left(\frac{\partial \theta}{\partial t}\right)_{y=0}$.

The mass transfer coefficient rate (expressed in a dimensionless form according to the Sherwood number) is defined as: $S_h = -\left(\frac{\partial c}{\partial t}\right)_{y=0}$.

Tables 28.1-28.4 display the impacts of the thermal (*Gr*) and Mass Grashof (*Gc*), Pramdtl (*Pr*), Schidmt number (*Sc*) and Chemical Reaction parameters (*Kr*) on the friction factor C_f, Sherwood (*Sh*), Nusselt numbers (*Nu*). $Gr = 2$; $Gc = 5$; $M = 1$; $Sc = 0.6$; $Kr = 2$; $Ec = 0.001$; $Pr = 0.71$; $K = 1.0$; $t = 0.4$.

Table 28.1 Effects of *Gr* and *Gc* on C_f

Gr	*Gc*	C_f
2	5	0.350162698
5	5	0.762891667
2	10	0.813812698

Table 28.2 Effects of *Pr* on C_f and *Nu*

Pr	C_f	*Nu*
0.71	0.350163	0.482018
1.0	0.349738	0.552118
7.0	0.232938	2.275468

The quantity of skin friction enhances when the Gr or Gc rise, as seen in Table 28.1. Table 28.2 demonstrates that as the Prandtl number and Nusselt number increase, skin friction decreases. Skin friction reduces as the Schmidt number enhances, whereas the number of Sherwood rises, as seen in Table 28.3. According to Table 28.4, the parameter for skin friction reduces when the *Kr* parameter rises, whereas the Sherwood parameter upsurges.

Table 28.3 Effects of *Sc* on *Sh*, C_f

Sc	C_f	*Sh*
0.3	0.411589	0.395041
0.6	0.350163	0.560908
1.0	0.315738	0.706493
2.01	0.258813	1.019718

Table 28.4 Effects of *Kr* on C_f and *Sh*

Kr	C_f	*Sh*
2	0.350163	0.560908
5	0.333738	0.711043
7	0.317438	0.810993

4. Result and Discussion

The velocity, temperature and concentration of different non-dimensional bounds are determined and illustrate with the help of figures.

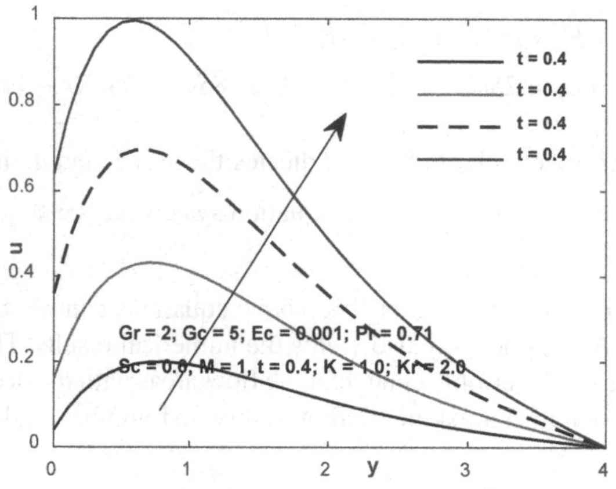

Fig. 28.1 Velocity for distinct *t*

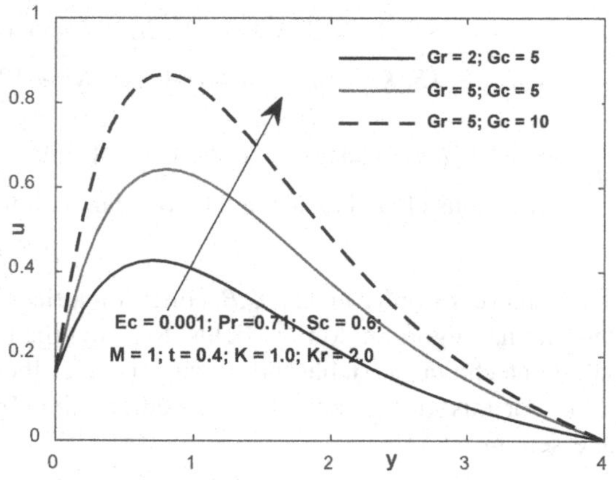

Fig. 28.2 Velocity for distinct *Gr* and *Gc*

The outcome of the time on the velocity, concentration and temperature is shown in Fig. 28.1, 28.9 and 28.14. According to the above figures with time changes, the velocity, concentration and temperature increase. The velocity for specific thermal (Gr) and mass Grashof number (Gc) values is exposed in Fig. 28.2. The effect of very leading thermal Grashof number, since plate temperature is assumed to be uniform. As the graph reflects the change in velocity as the amount of the thermal or the mass Grashof numberrises. Fig. 28.3 dealt with the impact on velocity of the magnetic parameter (M). When M escalations the velocity falls. The outcome of a transverse magnetic area on an electrically liquid result in the presence of a resistive form of force named the Lorentz force, which is opposed to the fluid flux.

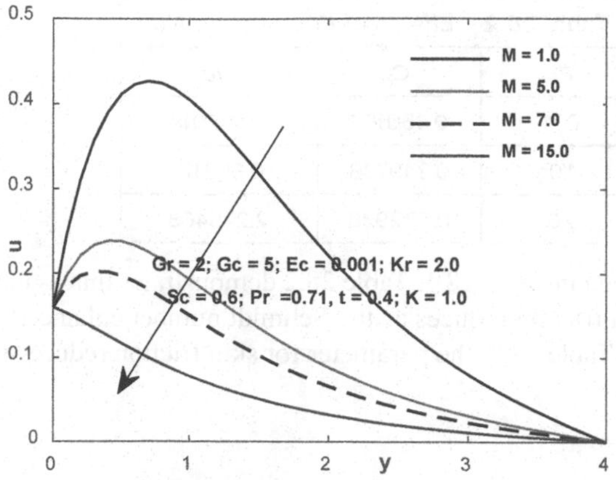

Fig. 28.3 Velocity vs M values

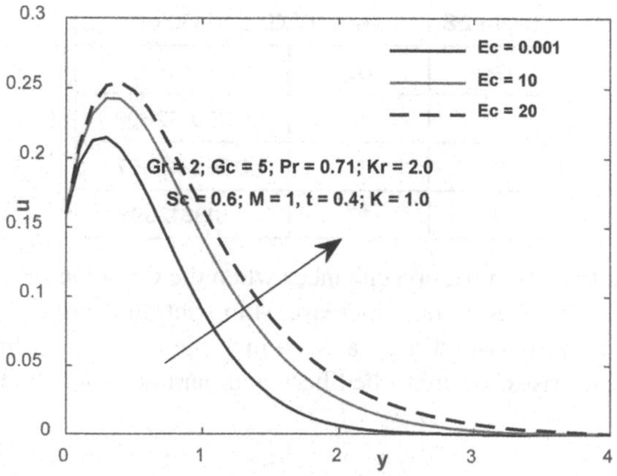

Fig. 28.4 Velocity for distinct *Ec* values

Fig. 28.4 represent the velocity for distinct values of the Eckert number is exposed. It is used to characterize the dissipative process. It is obvious *Ec* increases the fluid velocity. Hence, we acknowledged that the velocity enhances by raising in Eckert number. Fig. 28.5 shows the Schmidt number (*Sc*) impact on the concentration curves. It is clear that flow velocity declines with an upsurge in Sc. Fig. 28.6 depicts velocity for distinct values of Prandtl number (*Pr*). As the graph represents the velocity diminution by an enhance of *Pr*.

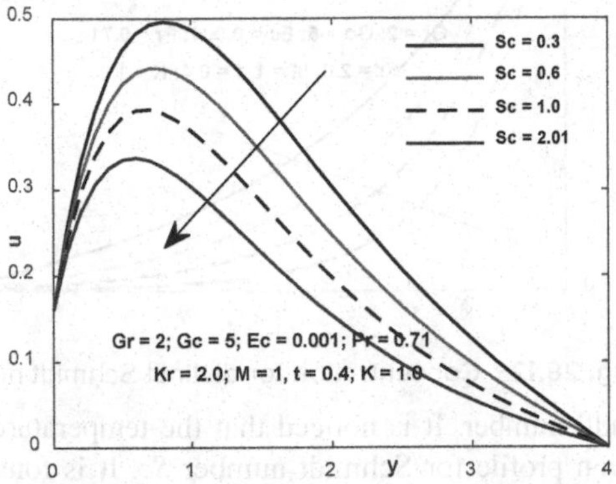

Fig. 28.5 Velocity for distinct Schmidt number

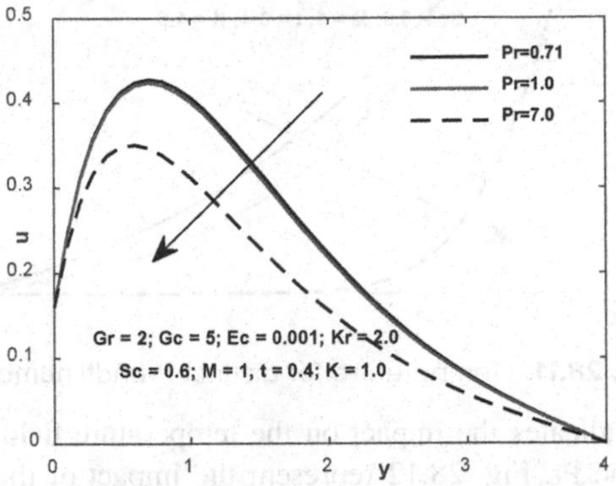

Fig. 28.6 Velocity for distinct Prandtl number

Fig. 28.7 indicates the impacts of the chemical reaction factor (*Kr*) on the velocity profile. There is evidence that indicates that the velocity decreases, as the *Kr* values increases. The velocity profile for distinct estimates of the permeability parameter (*K*) is shown in Fig. 28.8. It is found that the fluid velocity rises on permeability parameter upturns. Fig. 28.10 displays the distribution of temperature by means of the Eckert number effect. The temperature profile is found to increase by increasing the Eckert number values.

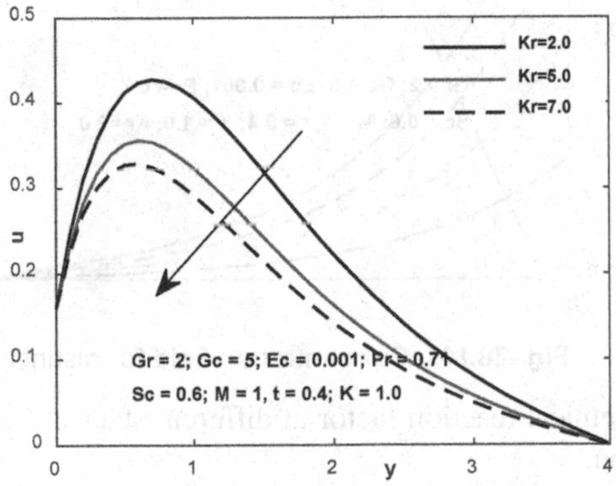

Fig. 28.7 Velocity for distinct chemical reaction parameter

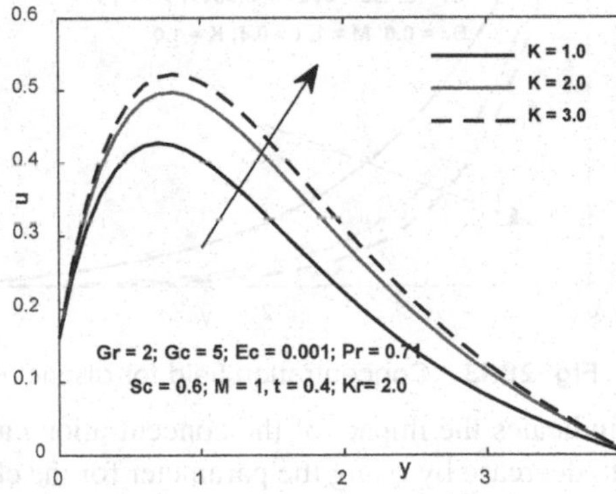

Fig. 28.8 Velocity for distinct permeability parameter

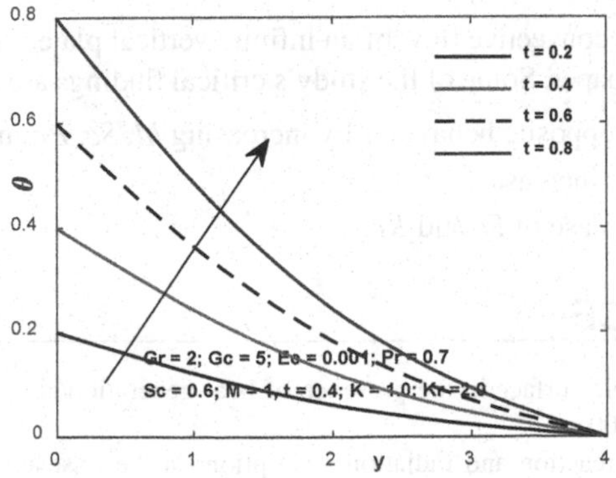

Fig. 28.9 Temperature for distinct of Eckert number

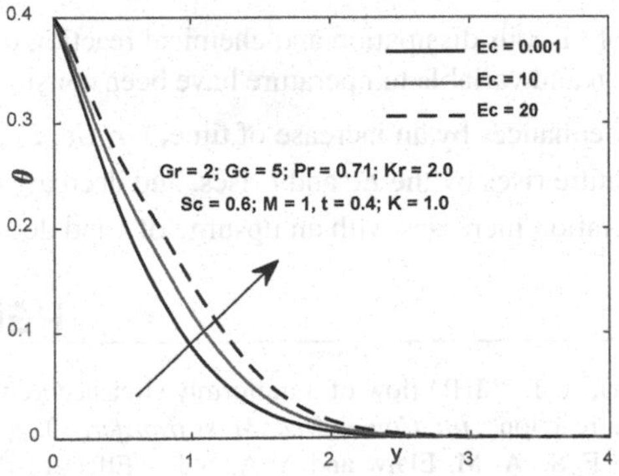

Fig. 28.10 Tempereture for distinct

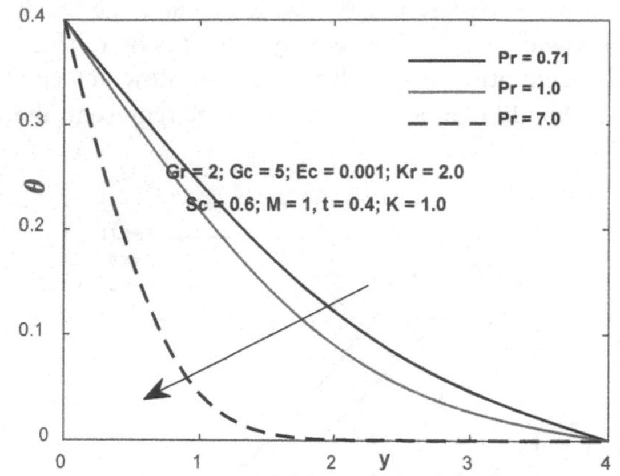

Fig. 28.11 Tempereture for distinct Prandtl number

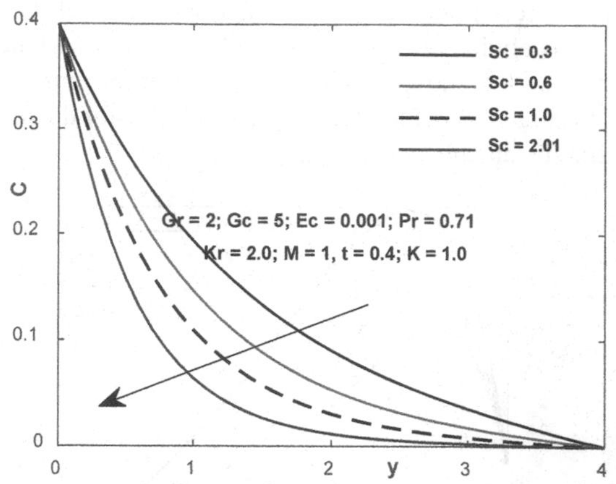

Fig. 28.12 Concentration for distinct Schmidt number

Fig. 28.11 indicates the impact on the temperature field of the Prandtl number. It is noticed that the temperature drops by increasing the Pr. Fig. 28.12 represent the impact of the concentration profile for Schmidt number *Sc*. It is found that the concentration declines as *Sc* increases.

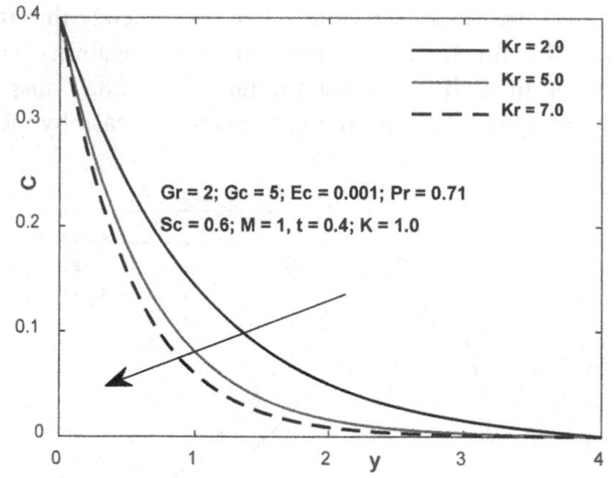

Fig. 28.13 Concentration field for distinct *Kr*

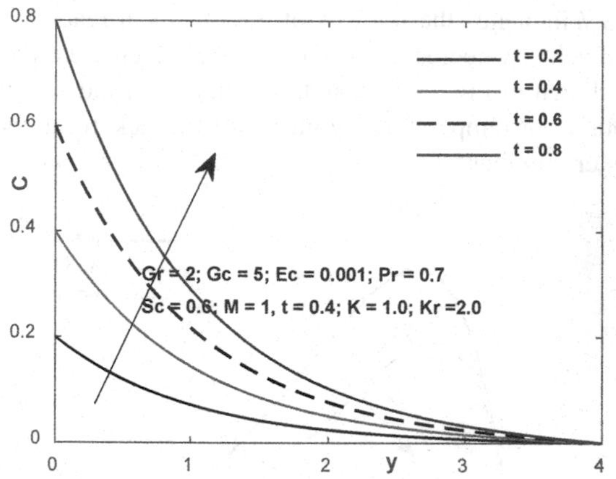

Fig. 28.14 Concentration field for distinct *t*

Fig. 28.13 indicates the impact of the concentration curves on the chemical reaction factor at different stages. Concentration was found to decrease by rising the parameter for the chemical reaction.

5. Conclusions

The impact of viscous dissipation and chemical reaction of MHD free convective flow of an infinite vertical plate with variable mass diffusion and variable temperature have been considered in this paper. Some of the study's critical findings are as follows:

- Velocity enhances by an increase of time, *Gr*, *Gc*, *Ec*, and *K* and opposite behaviour by increasing *M*, *Sc*, *Pr* and *Kr*.
- Temperature rises by the *Ec* and *t* rises, and decreases with the *Pr* increase
- Concentration increases with an upsurge of *t* and decrease by increase of *Sc* and *Kr*.

REFERENCE

1. Chamkha, A. J. "MHD flow of a uniformly stretched vertical permeable surface in the presence of heat generation/absorption and a chemical reaction", *Int. Comm. Heat Mass transfer*. 30, pp. 413–422, 2003.
2. Ibrahim, F. S., A. M. Elaiw and A. A. Bakr "Effect of the chemical reaction and radiation absorption on the unsteady MHD free convection flow past a semi-infinite vertical permeable moving plate with heat source and suction", *Comm. Nonlinear Sci. Numer. Simulation*, 13, pp. 1056–1066, 2008.

3. R. Kandasamy, K. Periasamy, K.K.S. Prabhu, “Chemical reaction, heat and mass transfer on MHD flow over a vertical stretching surface with heat source and thermal stratification effects”, *Int. Jour. Heat and Mass Trans.* 48, pp. 4557–4561, 2005.
4. M. A. Mansour, N.F. El-Anssary, and A.M. Aly, “Effect of chemical reaction and viscous dissipation on MHD natural convection flows saturated in porous media with suction or injection”, Int. Journal of Applied Mathematics and Mechanics, 4(2), pp. 60–70, 2008.
5. R. Mohana Ramana, J.Girish kumar “Chemical reaction effects on MHD free convection flow past an inclined plate”, Asian journal of current Eng. and maths, 3(6), pp. 66–71, 2014.
6. S. Mohammed Ibrahim, “Chemical reaction d radiation effects on MHD free convection flow along a stretching surface with viscous dissipation and heat generation”, Advances in applied science Research, Vol.4(1), pp. 371–382, 2013.
7. P. Pramod Kumar, B. Shankar Goud, Bala Siddulu Malga “Finite element study of Soret number effects on MHD flow of Jeffrey fluid through a vertical permeable moving plate”, Partial Differential Equations in Applied Mathematics, 1 (2020) 100005.
8. A. Postelnicu, “Influence of a magnetic field on heat and mass transfer by natural convection from vertical surfaces in porous media considering Soret and Dufour effects,” Int. Journal of Heat and Mass Transfer, 47(6-7), pp. 1467–1472, 2004.
9. S. P. Anjalidevi and R. Kandasamy, “Effects of chemical reaction, heat and mass transfer on laminar flow along a semi-infinite horizontal plate”, Heat and Mass Transfer, 35(6), pp. 465–467, 1999.
10. B. Shankar Goud “Thermal Radiation Influences on MHD Stagnation Point Stream over a Stretching Sheet with Slip Boundary Conditions”, International Journal of Thermofluid Science and Technology (2020), 7(2), Paper No. 070201.
11. B. Shankar Goud, Pudhari Srilatha, MN Raja Shekar “Effects of Mass Suction on MHD Boundary Layer Flow and Heat Transfer over a Porous Shrinking Sheet with Heat Source/Sink”, International Journal of Innovative Technology and Exploring Engineering, 8(10), pp. 263–266, 2019.
12. B. Shankar Goud, Pudhari Srilatha, P. Bindu, and Y. Hari Krishna “Radiation effect on MHD boundary layer flow due to an exponentially stretching sheet”, Advances in Mathematics: Scientific Journal, 9(12), 2020, pp. 10755–10761.

Note: All the figures and tables in this chapter were made by the authors.

Impending Inquisitions in Humanities and Sciences (ICIIHS-2022) – Dr. Mohan Varkolu et al. (eds)
© 2024 Taylor & Francis Group, London, ISBN 978-1-032-78829-6

Thermal Stratification Impacts On MHD Non-Darcian Flow due to Horizontal Stretching Sheet Embedded in a Porous Medium with Suction/Blowing

29

B. Shankar Goud*
Department of Mathematics, JNTUH University College of Engineering Hyderabad Kukatpally, Hyderabad, TS, India

Pudhari Srilatha
Department of Mathematics, Institute of Aeronautical Engineering, Hyderabad

D. Mahendar
B V Raju Institute of technology, Narsapur Medak-Dist

Kanti Sandeep Kumar
Department of Mathematics, Hyderabad Institute of Technology and Management, Hyderabad, Telangana, India

ABSTRACT The purpose of this work is to examine the non-Darcian impacts on the unsteady nonlinear MHD flow of viscous, electrically conducting, and an incompressible fluid across a horizontally stretching sheet immersed in a permeable medium with thermal stratification and suction and blowing. The dimensionless governing equations have been numerically solved by employing the MATLAB inbuilt solver bvp4c approach. The impacts of relevant factors on velocity and temperature are visually represented and thoroughly described.

KEYWORDS Non-darcian flow, bvp4c, Suction/blowing, Thermal stratification, MHD

1. Introduction

An investigation into how a fluid acts in its boundary layer near a surface that is continually stretched can find numerous significant applications in process development. Polymer extrusion, and paper, the formation of liquid films during the polymerization reaction, and a variety of other uses are a few of these uses. The ever-increasing uses in these manufacturing applications have resulted in a resurgence of concern in the fluid study flow and heat transmission via a stretching sheet. In noteworthy years, a significant amount of study has been done in order to uncover the stream, heat transfer, and mass transmission in fluid flow that occurs beyond a extending surface. Nakayama and Koyama (1989) investigated buoyancy induced pours across nonisothermal curved surfaces in thermally stratified permeable media. Heat and mass transmission were premeditated by Acharya et al. (1999) through an extending surface by a heat source and suction/blowing. The stimulus of suction/blowing on natural convective boundary layers on vertical surfaces subject to a constant heat flow was investigated by Chaudhary and Merkin (1993). Muthucumaraswamy(2002) describes how a chemical reaction affects a moving isothermal vertical surface exerting suction. Heat transport in a viscous fluid across a stretched sheet was investigated by Vajravelu and Hadjinicolaou (1993). Heat transfer mixed convection over a vertically going surface was investigated by Ali (2006), who

*Corresponding author: bsgoud.mtech@gmail.com

DOI: 10.1201/9781003489436-29

found that the influence of varying viscosity varied. Heat transport in a viscous fluid across a stretched sheet was investigated by Vajravelu and Hadjinicolaou (1993),who found that the influence of varying viscosity varied. The stagnation point flow with heat production was studied by Shankar Goud et al. (2019). This article by Goud (2020) discusses the affect of suction and injection on micropolar fluid, including the generation and absorption of heat. The effect of thermal radiation on a stretching sheet under slip boundary circumstances was investigated by B. S. Goud (2020). Effect of heat source & velocity slip on an electromagnetic surface with Soret outcome and variable suction/injection was studied by Asogwa and Goud (2022). The MHD heat and mass impact across a stretching sheet determined by thermal radiation was investigated by Reddy and Goud (2022).

In the past decade, the study of flow in porous media has become increasingly significant in a variety of subfields of engineering, including petroleum and groundwater hydrology. When the flow rate is large, the inertia impact as well as turbulence become significant, which results in non-Darcian flow. Darcy's law moreover explains the behavior of liquid flow in pores, which is accurate in a restricted assortment of low velocities. This particular kind of flow in aqueous media is useful for many different kinds of processes, cooling by geothermal process, and biomechanical activity. There have been a lot of different attempts through to research the non-Darcian flow. Studying NonDarcian assorted convection down a erect plate encased in a porous material, Chen and Chen (1990) drew some interesting conclusions. Mixed convection, characterized by suction and injection, was studied by Elbashbeshy (2003) along a vertical plate immersed in a non-Darcian porous medium. It was investigated by Taklifiand Aghanajafi (2011) how MHD non-Darcian flow behaves while passing via a non-isothermal vertical surface encased in a porous medium with radiation. The MHD non-darcian flow caused by a horizontally extending sheet immersed in a permeable medium using thermal stratification impacts was investigated by Banshiwal & Goyal (2018).

Incorporating the aforementioned studies, this work describes an examination of the effects of thermal stratification on non-Darcian MHD flow caused by a horizontally stretching sheet implanted in a permeable medium with suction and blowing. Employing the bvp4c yields numerical outcomes. Numerical values of friction factor, rate of heat transfer, and other relevant parameters are provided in tabular form, and their effects on fluid flow and thermal fields are illustrated with graphical demonstrations.

2. Mathematical Formulation

In the existence of a heat source and thermal stratification, assume the flow of a laminar, incompressible, two-dimensional boundary layer of an electrically conducting liquid across a horizontal porous elongating sheet implanted in a non-Darcian fluid flow. The induced magnetic field owing to the velocity of an electrically conducting fluid when intensity magnetic field B_0 is enforced across to the x-axis is insignificant. The velocity $U(x) = ax$ of the sheet stretching is assumed. In addition, T_w is the constant temperature of the sheet and T_∞ is the temperature of the surrounding fluid. The mass flux velocity is v_0 ($v_0 < 0$ for suction) & v_0 (($v_0 > 0$ for injection). Assuming the physical conditions described above, the Darcy–Forchheimer model's governing continuity and energy equations are

$$\frac{\partial u}{\partial x}+\frac{\partial u}{\partial y}=0 \quad \text{... (1)}$$

$$u\frac{\partial u}{\partial x}+v\frac{\partial u}{\partial y}=v\frac{\partial^2 u}{\partial y^2}-\frac{\sigma B_0^2 u}{\rho}-\frac{v}{k}u-\frac{C_f}{\sqrt{k}}u^2 \quad \text{... (2)}$$

$$u\frac{\partial T}{\partial x}+v\frac{\partial T}{\partial y}=\alpha\frac{\partial^2 u}{\partial y^2}-\frac{Q}{\rho C_p}(T-T_\infty)+\frac{\mu}{\rho C_p}\left(\frac{\partial u}{\partial y}\right)^2 \quad \text{... (3)}$$

These are the boundaries that are being applied to the flow.

$$\left.\begin{aligned} u &= U(x)=ax,\ v=-v_0,\ T=T_w(x)\ as\ y=0 \\ &=0, T=T_\infty(x)=(1-n)T_0+nT_w(x)\ at\ y\to\infty \end{aligned}\right\} \quad \text{... (4)}$$

Given a constant stretching rate, denoted by "a," and a constant number, "n," which is "$0 \le n \le 1$." If "n" is the thermal stratification factor which is specified as $\frac{m_1}{m_1+1}$ (Nakayama & Koyama (1989)) wherever m_1 is a constant. Similarity conversions are employed to figure out the Eqs. (for more Acharya et al. (1999)) $\psi=(vxU(x))^{1/2}f(\eta), \eta=\left(\frac{U(x)}{vx}\right)^{1/2}y$.

The components of velocity can be expressed as $u=\frac{\partial\psi}{\partial y},\ v=-\frac{\partial\psi}{\partial y}$ ... (5)

The following are the definitions of the dimensionless parameters:

$$\theta(\eta)=\frac{T-T_\infty}{T_w-T_\infty},\ Re_x=\frac{Ux}{\nu},\ Pr=\frac{\mu C_p}{k}=\frac{\nu}{\alpha},\ M=\frac{\sigma B_0^2}{\rho a},\ Q=\frac{Q}{\rho C_p a},\ Ec=\frac{U^2}{C_p(T_w-T_\infty)},\ k_1=\frac{\nu}{ka},$$

$Fs=\frac{C_f}{\sqrt{k}}x$, $S=-\frac{v_0}{\sqrt{a\nu}}$ ($S>0$ corresponds to suction, $S<0$ for injection). It is possible to express the wall shear stress, denoted by T_w relating to of the local skin friction factor, denoted by "C_f" as $C_f(Re_x)^{1/2}=f''(0)$. The numeric value of the local Nusselt scale, which is defined as $Nu=\frac{xq_w}{k(T_w-T_\infty)}$ where "q_w" is the amount of heat that is transferred from the sheet via: $q_w=-k\left(\frac{\partial T}{\partial x}\right)_{y=0}$. In equations, we obtain the dimensionless variables specified as $Nu/(\mathrm{Re}_x)^{\frac{1}{2}}=-\theta'(0)$.

We also take into account the fact that surfaces have different temperatures $T_w-T_\infty=Nx^n$ and As a result of this equation transformation, we get

$$f''+ff'-(f')^2-(M+k_1)f'-Fs(f')^2=0 \quad \text{... (6)}$$

$$\theta''-Prnf\left[\theta+\frac{n}{1-n}\right]+fPr\theta'+QPr\theta+PrEc(f')^2=0 \quad \text{... (7)}$$

As the boundary circumstances change,

$$\left.\begin{array}{l} f(0)=S,\ f'(0)=1,\theta(0)=1,\ as\ \eta=0\\ f'(\infty)=0,\theta(\infty)=0\ at\ \mu\to\infty \end{array}\right\} \quad \text{... (8)}$$

The friction factor, the local Nusselt quantity, and their dimensionless counterparts are the three numbers of importance from an engineering perspective as $Re_x^{1/2}Cf_x=f''(0)$ and $Re_x^{1/2}Nu=-\theta'(0)$.

Solution of the Problem

The altered governing equations (6) - (7) and boundary circumstances (8) are integrated. The BVP value is first turned into an IVP and solutions are prevailed by employing the bvp4c MATLAB in built solver. Follow the pattern of subsequent replacements

$$\{f,f',f'',\theta,\theta''\}=\{y_1,y_2,y_3,y_4,y_5'\}$$

$$y_3'=-(y_1*y_3-(y_2)^2-(M+k_1)y_2-Fs(y_2)^2)$$

$$y_5'=-\left(-Prny_1\left[y_4+\frac{n}{1-n}\right]+y_1Pry_5+QPry_4+PrEc(y_2)^2\right)$$

In response to shifting boundary circumstances, $\left.\begin{array}{l} y_1(0)=S,\ y_2(0)=1,\ y_4(0)=1,\ as\ \eta=0\\ y_2(\infty)=0,\ y_4(\infty)=0\ at\ \eta\to\infty \end{array}\right\}$

This problem is addressed by extending the results to the values 10^{-6} and then demonstrating that the results are numerically convergent.

3. Results and Discussion

The changed set of equations (7)-(9) was solved numerically based on the boundary circumstances (10) using the MATLAB in built solver bvp4c.

The impact of the magnetic parameter M on the temperature and velocity patterns are depicted in Fig. 29.1-29.2, respectively. It can be seen quite clearly in Fig. 29.1 that as the magnetic field factor grows, the fluid velocity drops. Furthermore, it can be seen quite clearly in Fig. 29.2 that as the *M* factor improves, the temperature rises. The retarding power, which is what fights the flow, likewise rises as M grows, and as a consequence, there is a diminution in velocity and a raise in thermal field. Figs. 29.3-29.4 provide a curve depicting the impact of the inertia factor *Fs* on the velocity and the temperature respectively. It is evident that axial velocity $f'(\eta)$ of the fluid would diminish as the inertia factor increases; however, the temperature is enhancing, and the rise is quite tiny. The resistance of the flow rises as the inertia factor enhances, which in turn causes the flow pattern in the porous media to slowly down, which in go causes the temperature to raise.

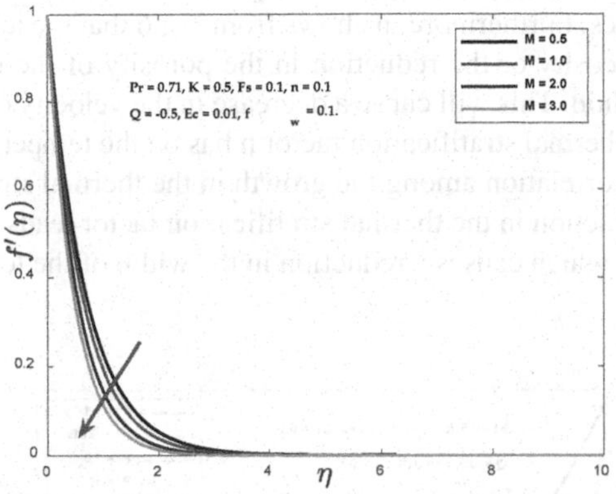

Fig. 29.1 M v/s Velocity.

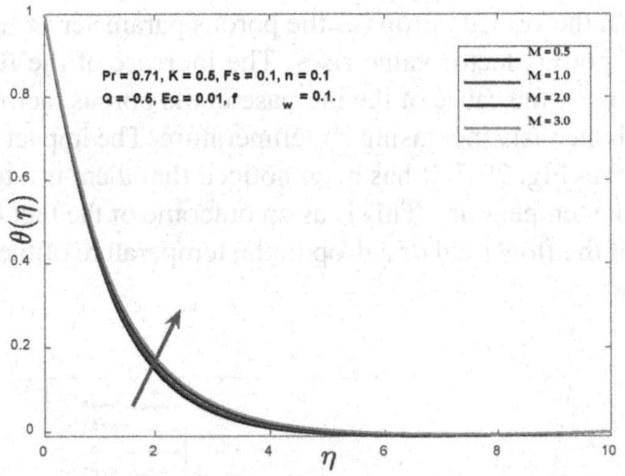

Fig. 29.2 M v/s Temperature.

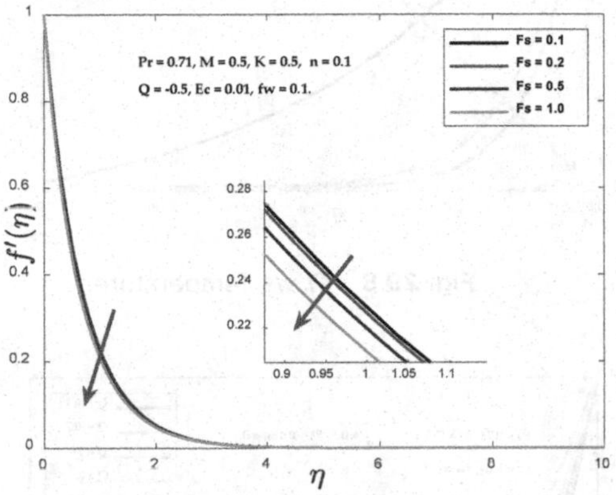

Fig. 29.3 *Fs* v/s Velocity.

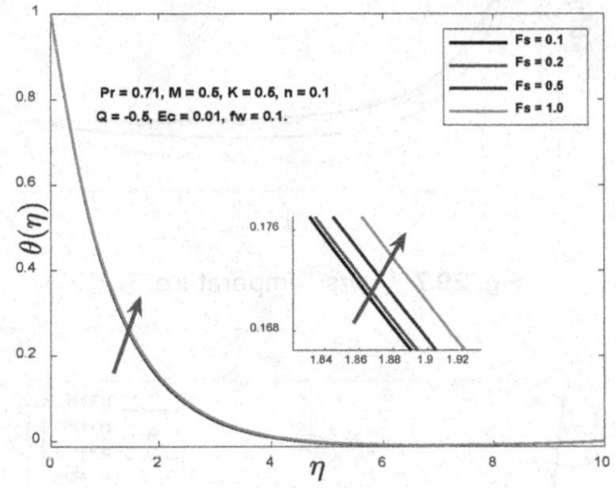

Fig. 29.4 *Fs* v/s Temperature.

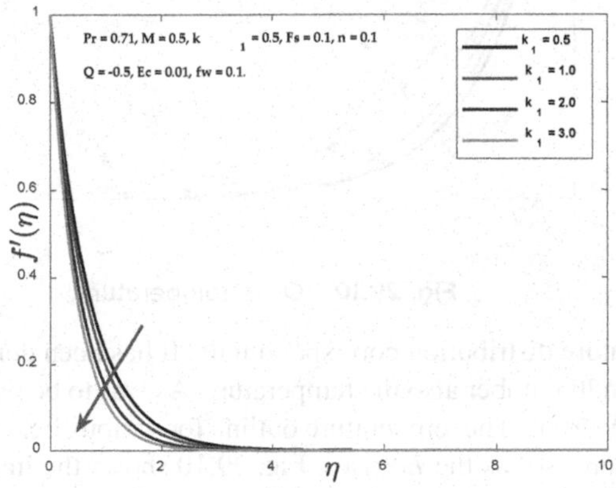

Fig. 29.5 k_1 v/s Velocity.

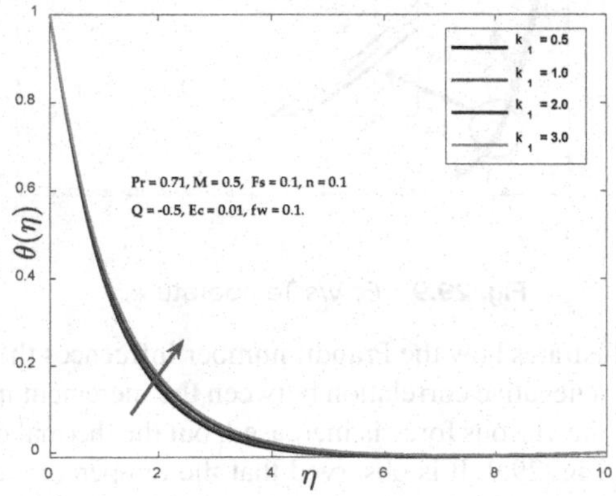

Fig. 29.6 k_1 v/s Velocity.

The influence of the porosity parameter on the temperature, and velocity of the fluid is seen in Fig. 29.5-29.6. This is evident of Fig. 29.5 that the velocity drops as the porous parameter value increases. Furthermore, it shows from Fig.6 that the temperature rises as the porous factor value rises. The increase of the fluid's viscosity or the reduction in the porosity of the permeable medium is likely the cause of the increase in the porous factor of the fluid. This will cause a decrease in the velocity of the fluid while simultaneously increasing its temperature. The impact that the thermal stratification factor n has on the temperature field is illustrated in Fig. 29.7. It has been noticed that there is a negative correlation among the growth in the thermal stratification factor and the temperature. This is as an outcome of the fact that a reduction in the thermal stratification factor leads in either a rise in the of the flow field or a drop in the temperature of the surface, which causes a reduction in the width of the temperature profile.

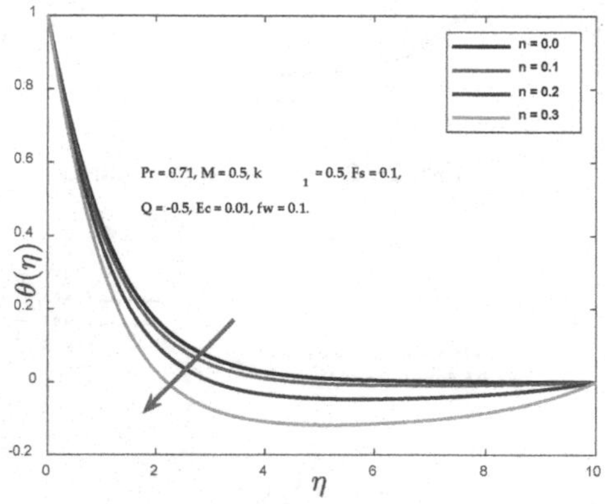

Fig. 29.7 *n* v/s Temperature.

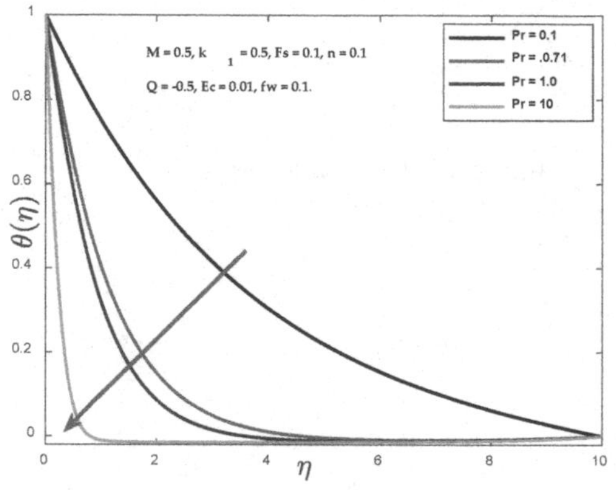

Fig. 29.8 *Pr* v/s Temperature.

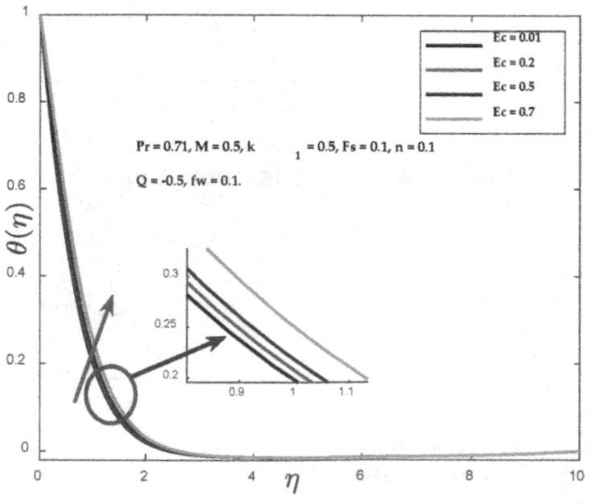

Fig. 29.9 *Ec* v/s Temperature.

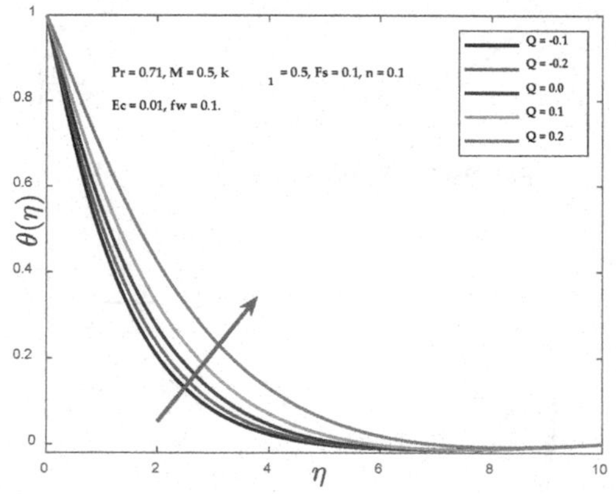

Fig. 29.10 *Q* v/s Temperature.

Fig. 29.8 illustrates how the Prandtl number influences the temperature distribution correspondingly. It has been demonstrated that there is a negative correlation between the increment in the Prandtl number also the temperature. As was to be predicted, as the Pr rises, the viscous force is increased, but the thermal diffusivity drops. The temperature outline for various Eckert numbers is shown in Fig. 29.9. It is observed that the temperature distribution rises as the *Ec* rises. Fig. 29.10 shows the impact of the heat absorption/generation factor on the temperature. As expected, rising temperatures are exhibited in response to rising values of the heat absorption/generation factor. In Fig. 29.11-29.12, the non-dimensional suction/blowing factor is shown to have a negative effect on fluid (velocity) outlines with respect to , assuming that all other variables remain constant velocity.). From Fig. 29.12, It is evident that as is increased, all values of temperature are enhances.

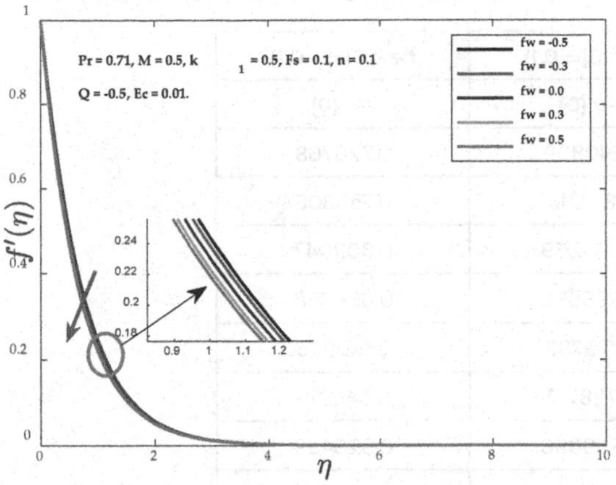

Fig. 29.11 *Pr* v/s velocity.

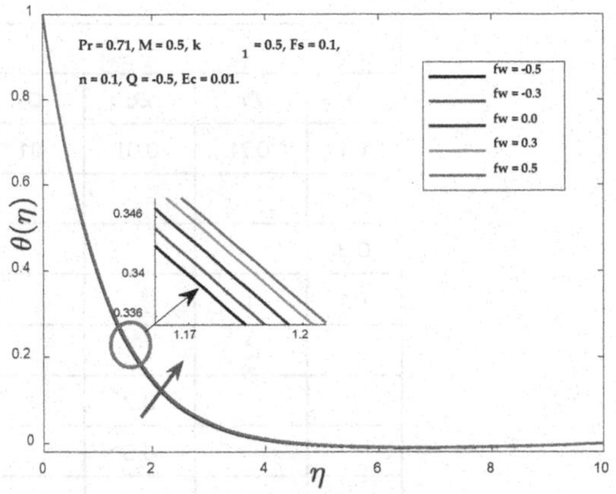

Fig. 29.12 *fw* v/s Temperature.

When we look at the Table 29.1, we see that the numerical schemes used in the most recent studies mean that the numerical results are accurate to a finite number of decimal places. Both the current and the prior limiting outcomes agree very well. The numerical computations of the $-f''(0)$ and $-\theta'(0)$ with suction and blowing case are present in the Table 29.2. Form the table with enhance values of M, Fs and k_1 the result in $-f''(0)$ increases and opposite observation is found in $-\theta'(0)$. The numerical results of the $-\theta'(0)$ with suction and blowing case are listed in the Table 29.3. Form the table with enhance values of n and Pr the result in $-\theta'(0)$ upsurges and opposite observation is found in $-\theta'(0)$ with increase of Ec and Q.

Table 29.1 Scales for comparison of $-\theta'(0)$ for numerous *Pr* values when $M = k_1 = Q = Ec = n = S = 0$.

Pr	Salleh et al[19] (CWT case)	Qasim et al [19]	Hassnien et al[20]	Present study
0.72	0.46317	0.46360	0.46325	0.463252
1	0.58198	0.58202	0.58198	0.581981
3	1.16522	1.16525	1.16525	1.165246
5	1.56806	1.56805	-	1.568055

Table 29.2 The numerical values of $-f''(0)$ and $-\theta'(0)$ with numerous values of *M*, *Fs* and k_1 when the further values are $Pr = 0.71$; $n = 0.1$; $Q = -0.5$; $Ec = 0.01$

			fw > 0(= 0.1)		*fw* < 0(= −0.1)	
M	*Fs*	k_1	$-f''(0)$	$-\theta'(0)$	$-f''(0)$	$-\theta'(0)$
0.5	0.1	0.5	1.487677	0.810809	1.388208	0.720768
1			1.652283	0.800240	1.552711	0.709773
1.5			1.801449	0.791529	1.701808	0.700773
0.5	0.2		1.509936	0.809883	1.410970	0.719777
	0.5		1.574908	0.807224	1.477312	0.716942
	0.1	1	1.652283	0.800240	1.552711	0.709773
		2	1.938852	0.784147	1.839161	0.693193

Table 29.3 The numerical values $-\theta'(0)$ with diverse values of n, Pr, Ec and Q when $M = 0.5$, $Fs = 0.1$, $k_1 = 0.5$.

				$fw > 0(= 0.1)$	$fw < 0(= -0.1)$
n	Pr	Ec	Q	$-\theta'(0)$	$-\theta'(0)$
0.1	0.71	0.01	-0.1	0.810809	0.720768
0.2				0.861713	0.753306
0.3				0.936269	0.802047
0.1	1			0.985656	0.859358
	10			3.779782	2.551946
		0.2		0.728717	0.646215
		0.5		0.599098	0.528499
			0.1	0.424089	0.286925
			0.2	0.297907	0.067999

4. Conclusions

This study's primary conclusion is as follows:

- The fluid's velocity drops but its temperature rises when the F_s, k_1, and M are increased.
- Low temperature profiles occur when the thermal stratification parameter is set to greater levels.
- The temperature profile rises as the Ec and Q values rise, but falls as the Pr rises. Also As f_w rises, it temperature increases mean while velocity falls.

REFERENCES

1. Nakayama, A. and Koyama, H. (1989): Similarity solutions for buoyancy-induced flows over non-isothermal curved surface in a thermally stratified porous medium, Appl. Sci. Research, 46:309–332.
2. Acharya, M., Singh, L. P. and Dash, G.C. (1999): Heat and mass transfer over an accelerating surface with heat source in the presence of suction and blowing, Int. Journal of Engineering Science, 37:189–211.
3. Chaudhary, M. A., Merkin, J. H. (1993): The effects of blowing and suction on free convection boundary layers on vertical surfaces with prescribed heat flux. Journal of Eng. Mathematics, 27(3), 265–292.
4. Muthucumaraswamy, R. (2002): Effects of a chemical reaction on a moving isothermal vertical surface with suction. Acta Mechanica, 155, 65–70.
5. Vajravelu, K., Hadjinicolaou, A. (1993): Heat transfer in a viscous fluid over a stretching sheet with viscous dissipation and internal heat generation. Int. Comm.in Heat and Mass Transfer, 20(3), 417–430.
6. Ali, M. E. (2006):The effect of variable viscosity on mixed convection heat transfer along a vertical moving surface. International Journal of Thermal Sciences, 45(1), 60–69.
7. B. Shankar Goud, G. Narender, E. Ranjit Kumar (2019): Stagnation point flow through a porous medium towards stretching surface in the presence of heat generation, International Journal of Engineering and Advanced Technology, 9(1), pp. 2646–2649.
. B. Shankar Goud (2020): Heat Generation/Absorption influence on steady stretched permeable surface on MHD flow of a micropolar fluid through a porous medium in the presence of variable suction/injection, International Journal of Thermofluids, vol.7-8, 100044.
9. B. Shankar Goud(2020):Thermal Radiation influences on MHD stagnation point stream over a stretching sheet with slip boundary conditions, Int. J. of Thermofluid Sci. and Tech., 7(2), Paper No. 070201.
10. Asogwa, Kanayo K, Shankr Goud, B (2022): Impact of velocity slip and heat source on tangent hyperbolic nanofluid flow over an electromagnetic surface with Soret effect and variable suction/injection, Part E: Journal of Process Mechanical Engineering, 1–13.
11. Yanala, Dharmendar Reddy Goud, B. Shankar (2022): MHD heat and mass transfer stagnation point Nanofluid flow along a stretching sheet influenced by thermal radiation, Journal of Thermal Analysis and Calorimetry, 147,11991–12003.
12. Chen, C., and Chen, C. (1990): Non-Darcian mixed convection along a vertical plate embedded in a porous medium. Applied Mathematical Modelling, 14(9), 482–488.

13. Elbashbeshy, E. (2003): The mixed convection along a vertical plate embedded in non-Darcian porous medium with suction and injection. Applied Mathematics and Computation, 136(1), 139–149.
14. Taklifi, A., and Aghanajafi, C. (2011): MHD non-Darcian flow through a non-isothermal vertical surface embedded in a porous medium with radiation. Meccanica, 47(4), 929–937.
15. A. Banshiwal and M. Goyal (2018): MHD non-darcian flow due to horizontal stretching sheet embedded in a porous medium with thermal stratification effects, J. of Naval Arc. and Marine Eng., 15(1), 65–73.
16. SallehMZ, Nazar R, Pop I (2010): Boundary layer flow and heat transfer over a stretching sheet with Newtonian heating. J Taiwan Inst Chem Eng 41: 651–655.
17. Muhammad Qasim, Ilyas Khan, Sharidan Shafie (2013): Heat Transfer in a Micropolar Fluid over a Stretching Sheet with Newtonian Heating, PLoS One. 2013; 8(4): e59393.
18. Hassanien, I. A., A. A. Abdullah, and R. S. R. Gorla (1998): Flow and heat Transfer in a power law fluid over a Nonisothermal Stretching Sheet, Math. Comput. Modell., 28, 105.

Note: All the figures and tables in this chapter were made by the authors.

Impending Inquisitions in Humanities and Sciences (ICIIHS-2022) – Dr. Mohan Varkolu et al. (eds)
© 2024 Taylor & Francis Group, London, ISBN 978-1-032-78829-6

Higher Overtone Vibrational Energies of Water Isotopologues using Lie Algebraic Model

30

P. Suneetha[1]
Department of Mathematics, V. R. Siddhartha Engineering College,
Vijayawada, India
Department of Mathematics, School of Science,
GITAM (Deemed to beUniversity), Hyderabad, India
J. Vijayasekhar[2]
Department of Mathematics, School of Science,
GITAM (Deemed to beUniversity), Hyderabad, India
B. M. Rao
Department of Physics, School of Science,
GITAM (Deemed to beUniversity), Hyderabad, India

ABSTRACT This investigation delves into the vibrational spectra of water isotopologues, namely Deuterium oxide (D_2O), and Water-^{18}O ($H_2{}^{18}O$), utilizing the Lie algebraic method to discern their fundamental and higher overtone frequencies. Central to this model is a Hamiltonian operator embodying the vibrational modes associated with stretching and bending motions, effectively depicted as an invariant and a Majorana operators.

KEYWORDS Lie algebraic model, Vibrational spectra, Water isotopologues

1. Introduction

Vibrational spectroscopy is a pivotal approach to unravelling the intricate structure of molecular systems and probing their intermolecular interactions. The realm of vibrational spectra bears significance for experimental and theoretical methodologies, serving as a bridge between observation and conceptualization.

The Dunham expansion and the potential approach, which have undergone extensive scrutiny and validation, are two well-established strategies that facilitate the characterization of molecular vibrations within this domain. The Dunham expansion is based on increasing energy levels by changing rovibrational quantum numbers. This gives a fuller picture of how molecules behave. On the other hand, the potential approach uses interatomic variables and potential coefficients to solve these vibrational problems by solving Schrodinger's equation in a complicated way. It is important to note that both methods are complicated and necessitate calibrating a wide range of parameters in order to produce accurate results.

Recognizing the limitations inherent in these two traditional avenues, Iachello introduced the innovative framework of the Lie algebraic model [Iachello et al. (1995)]. Iachello et al. (1991, 1992) show that the Hamiltonian operator can be elegantly

[1]vijayjaliparthi@gmail.com, [2]sprathip@gitam.in

DOI: 10.1201/9781003489436-30

described in terms of invariant operators. This describes the local and normal vibrational modes of the system. In recent years, Vijayasekhar and a group of collaborators have used the Lie algebraic model to calculate the vibrational frequencies of a wide range of polyatomic molecules [Vijayasekhar et al., 2022, 2023; Balla et al., 2021, 2022]. This model's application signifies a departure from the intricacies of conventional approaches, offering a more streamlined path to obtaining accurate vibrational spectra for complex molecular systems.

2. Lie Algebraic Model

For polyatomic molecules the Hamiltonian (molecular vibrations) form is [Iachello et al. (1995)]

$$h = e_0 + \sum_{i=1}^{m} k_1 C_i + \sum_{i<j}^{m} k_{ij} C_{ij} + \sum_{i<j}^{m} t_{ij} M_{ij} \quad (1)$$

For m stretching bonds i value changes from 1 to m. By spectroscopic data we can determine algebraic parameters k_i, k_{ij} and t_{ij}. Here C_i (uncoupled bonds) is an invariant operator with eigenvalues $-4(N_i v_i - v_i^2)$ and the operator C_{ij} (coupled bonds) with diagonal matrix elements,

$$\left\langle N_i, v_i; N_j, v_i \middle| C_{ij} \middle| N_i, v_i; N_j, v_i \right\rangle = 4\left[(v_i + v_j)^2 - (v_i + v_j)(N_i + N_j)\right]$$

Whereas the Majorana operator M_{ij} has both diagonal and non-diagonal matrix elements

$$\left\langle N_i, v_i; N_j, v_i \middle| M_{ij} \middle| N_i, v_i; N_j, v_i \right\rangle = (N_i v_j + N_j v_i - 2 v_i v_j)$$

$$\left\langle N_i, v_i + 1; N_j, v_j - 1 \middle| M_{ij} \middle| N_i, v_i; N_j, v_j \right\rangle = -\left[v_j (v_i + 1)(N_i - v_i)(N_j - v_j + 1)\right]^{1/2}$$

$$\left\langle N_i, v_i - 1; N_j, v_j + 1 \middle| M_{ij} \middle| N_i, v_i; N_j, v_j \right\rangle = -\left[v_i (v_j + 1)(N_j - v_j)(N_i - v_i + 1)\right]^{1/2}$$

$v_i (i = 1, 2, 3 ...)$ shows vibrational quantum numbers. Stretching bonds of molecule will be computed from the expression

$$N_i = \frac{\omega_e}{\omega_e x_e} - 1, i = 1, 2, ... \quad (2)$$

Where N_i $(i = 1, 2, 3...)$ is the Vibron number. In the above equation, the spectroscopic constants are ω_e and $\omega_e x_e$ [Irikura (2007)]. The single-oscillator fundamental mode's energy expression is attained by the first predict value for the parameter k_i, it is:

$$E(v = 1) = - 4k_i (N_i - 1). \quad ... (3)$$

Here k_{ij} may be taken as zero for first guess. The parameter t_{ij} will attained from the relation,

$$t_{ij} = \frac{|E_i - E_j|}{2N} \quad (4)$$

To get better results for the parameters k_i, k_{ij}, t_{ij} LSF (Least squares fitting) procedure is utilized. It starts from the values as given in the above equations (3) and (4).

3. Results

Table 30.1 Fitted parameters

Parameters	D_2O	$H_2{}^{18}O$
N (str.)	50	44
N (bend)	20	28
A_i(str.)	−13.073	−17.092
A_i(bend)	−11.831	−13.103
A_{ij} (str.)	−1.142	−1.911
A_{ij}(bend)	−2.923	−2.752
λ_{ij} (str.)	0.535	0.805
λ_{ij}(bend)	1.753	1.593

Table 30.2 Vibrational frequencies (cm^{-1})

Symmetry species (C_{2v}) point group	Mode	Vibrational frequencies			
		Experimental*		Calculated	
		D_2O	$H_2{}^{18}O$	D_2O	$H_2{}^{18}O$
Fundamental mode					
A_1 (sym. str.)	v_1	2671.46	3649.68	2670.032	3649.149
A_1 (bend)	v_2	1178.33	1588.27	1179.154	1587.075
B_1(anti str.)	v_3	2788.05	3741.58	2787.113	3741.852
First overtone					
A_1 (sym. str.)	$2v_1$	-	-	5276	7254
A_1 (bend)	$2v_2$	-	-	2289	3137
B_1(anti str.)	$2v_3$	-	-	5512	7408
Second overtone					
A_1 (sym. str.)	$3v_1$	-	-	7086	10872
A_1 (bend)	$3v_2$	-	-	3487	4709
B_1(anti str.)	$3v_3$	-	-	8346	11173
Third overtone					
A_1 (sym. str.)	$4v_1$	-	-	10610	14468
A_1 (bend)	$4v_2$	-	-	4689	6275
B_1(anti str.)	$4v_3$	-	-	11076	14873

*[Irikura, (2007), 36(2), Shimanouchi (1972)]

4. Conclusions

In this paper, we have utilized the Lie algebraic model to calculate the fundamental vibrational spectra of water isotopologues, specifically deuterium oxide (D_2O) and water-^{18}O ($H_2{}^{18}O$). Moreover, we extend our analysis to encompass vibrational frequencies up to the third overtone. Notably, the experimental data align harmoniously with the outcomes derived from our fundamental model, reinforcing the consistency between theoretical predictions and empirical observations.

5. Acknowledgments

Vijayasekhar gratefully acknowledges the Research Seed Grants (RSG) by Gandhi Institute of Technology and Management (GITAM), Deemed to be University, Hyderabad, India [Sanction Letter Ref: F.No. 2021/0114, dated 26-09-2022].

REFERENCES

1. F. Iachello, R.D. Levine, Algebraic Theory of Molecules, Oxford University Press, Oxford, 1995.
2. F. Iachello, S. Oss, R. Lemus, *J. Mol. Spectrosc.* 1991, 149 (1), 132–151.
3. F. Iachello, S. Oss, *J. Mol. Spectrosc.* 1992, 153, 225.
4. J. Vijayasekhar, M. Rao Balla, *Spectrochim. Acta A*, 2022, 264, 120289.
5. J. Vijayasekhar, *Ukr. J. Phys. Opt.* 2022, 23(3), 126.
6. J. Vijayasekhar, P. Suneetha, K. Lavanya, Ukr. J. Phys. Opt. 2023, 24(3), 193–199.
7. M. R. Balla, V. S. Jaliparthi, *Mol. Phys.* 2021, 115 (5), e1828634.
8. M. R. Balla, V. S. Jaliparthi, *Polycyclic Aromatic Compounds.* 2022, 42(7), 4684.
9. M. R. Balla, S. Venigalla, V. Jaliparthi, *Acta Phys. Pol. A.* 2021, 140 (2), 138.
10. K. K. Irikura, *J. Phys. Chem. Ref. Data.* 2007, 36(2), 389.
11. T. Shimanouchi, *Tables of Molecular Vibrational Frequencies*, National Bureau of Standards, 1972, Vol. I, 1–160.

Impending Inquisitions in Humanities and Sciences (ICIIHS-2022) – Dr. Mohan Varkolu et al. (eds)
© 2024 Taylor & Francis Group, London, ISBN 978-1-032-78829-6

Charaterization of Partial Lattices Via Measures

31

Y. V. Seshagiri Rao[1]
Associate Professor of Mathematics
Vignan Institute of Technology and Science, Deshmukhi, Hyderabad, Telangana, India

J. Pramada[2]
Associate Professor of Mathematics,
Keshav Memorial Institute of Technology, Narayanaguda, Hyderabad, Telangana, India

D. V. S. R. Anil Kumar[3]
Professor of Mathematics,
T.K.R. College of Engineering and Technology, Meerpet, Hyderabad, Telangana, India

ABSTRACT By defining outer measure (ω^*) on a set of partial lattices, we prove the class of ω^* measurable partial lattices is a sigma algebra and its restriction preserves a complete measure. The measure algebra defined on $L(\mathscr{B})$ is a countable subadditive whenever $\{M_i\}$ is a sequence of partial lattices in Q such that M is contained in join M_i. The extended real valued set function defined on a set H of all partial lattices is an outer measure and every member of the algebra of partial lattices (Q) is measurable with respect to ω^*.

KEYWORDS Partial lattice, Outer measure, Sigma algebra, Complete measure

1. Introduction

The sigma algebras, positive negative parts of signed lattice measures, different decompositions with respect to these lattice measures was studied by [1] Anil kumar et al. [4] defined partial lattice. In 2016 [1] defined the outer measure of a partial lattice, measure on Boolean lattice, studied the measurability of a partial lattices also verified various forms of Littlewoods's conditions with help of [5]. Later Y.V. [6] et al. established the definitions of locally-measurable lattice, complete-measure, saturated-lattice measure, and demonstrates that each lattice measure space can be incorporated into a complete-lattice measure space also set up an outcome that if ȷ is lattice sigma-finite measure then it is saturated. In section 2 we provide the preliminaries. In section 3, the class of ω^* measurable partial latticesin $\mathscr{H}$ is a sigma-algebra and restriction of ω^* preserves complete measure, and results that if $\{M_i\}$ is any sequences of partial lattices in Q such that M is contained in $\bigvee_{i=1}^{\infty} M_i$ and ω be a measure on an algebra $L(\mathscr{B})$ then ω is countably subadditive. If set $\mathscr{H}$ of all partial lattices then the extended real valued set function defined on $\mathscr{H}$ is an outer measure, every mumber of algebra of partial lattices (Q) is measurable with respect to ω^*

[1]yangalav@gmail.com, [2]pramadasksk@gmail.com, [3]dvsranilkumar@gmail.com

DOI: 10.1201/9781003489436-31

Notations

Let L be a lattice, H is a partial lattice, **P**(X) is the power set of X, $\mathscr{H}$ the set of all partial lattices of L, Q is an algebra of partial lattices, outer measure ω* and $L(\mathscr{B})$ is the class of ω* measurable partial lattices.

2. Preliminaries

Let L be a lattice contained **P**(X) such that if $G \in L$ then $G^c \in L$, a outer measure ω* is an extended real-valued set function defined on the set $\mathscr{H}$ of all partial lattices, H is a partial lattice contained L satisfying the following properties. Outer measure of empty set is zero, monotonicity and countable sub additive. For every partial lattice H, F is said to be measurable with reference to ω* if $\omega^*(H) = \omega^*(H \wedge F) + \omega^*(H \wedge F^c)$ where F is partial lattice and $F^c = X - F$. Also, ω* is regular, for any $\mathcal{E} > 0$ there is a partial lattice E of X and ω* measurable partial lattice M such that F is contained in M, $\omega^*(M) \leq \omega^*(M) + \mathcal{E}$.

Result2.1.[6] for any element of a partial lattice the order preserving valuation of an element is equal to its outer measure.

Theorem2.1.[3] If $(L, \mathscr{L}, \jmath)$ is a lattice measure space then we can find a complete measure space $(L, \mathscr{L}_0, \jmath_0)$ such that **(i)** $L \subseteq \mathscr{L}_0$ (ii) $E \in \mathscr{L}$ implies $\jmath(E) = \jmath_0(E)$ (iii) $E \in \mathscr{L}$ if and only if $E = A \vee B$, where $B \in \mathscr{L}$ and $A \subseteq C, C \in \mathscr{L}, \jmath(C) = 0$.

Result2.2.[6] Let Q be algebra of partial lattices and ɲ be a measure on an algebra $L(\mathscr{B})$. If $\{A_i\}$ is any sequence of partial lattices in Q such that $A \subseteq \bigvee_{i=1}^{\infty} A_i$ then $\jmath(A) \leq \sum_{i=1}^{\infty} \jmath(A_i)$.

Definition2.1.[5] Let X be any set, $L \subset \mathbf{P}(X)$ be a lattice there exists $E \in L$ implies $E^c \in L$, ℘ the set of all partial lattices of L. By an outer measure ɲ* we mean an extended real valued set function defined on the set ℘ of all partial lattices, $H \subset L$ satisfying the following properties.

(1) $\jmath^*(\phi) = 0$(2) $K_1 \subseteq K_2$ implies $\jmath^*(K_1) \leq \jmath^*(K_2)$ [monotonicity]

(3) $E \subseteq \bigvee_{i=1}^{\infty} E_i$ implies $\jmath^*(E) \leq \sum_{i=1}^{\infty} \jmath^*(E_i)$ [countable sub additive]

Definition2.2.[5] A lattice measure space $(X, L(\mathscr{B}), \jmath)$ is said to be complete if $L(\mathscr{B})$ contains all partial lattices of members of $L(\mathscr{B})$ of measure zero. That is, if $B \in L(\mathscr{B})$ and $\jmath(B) = 0$ and $A \subseteq B$ implies $A \in L(\mathscr{B})$.

Definition2.3. Todefine outer measure generated by ω

Write $\varpi^*(F) = \inf\{\sum_{i=1}^{\infty} \varpi(M_i) / \{M_i\}$ ranges over all sequence from Q such that $F \subseteq \bigvee_{n=1}^{\infty}\{M_i\}$ where ω be a measure on an algebra Q of partial lattices for any partial lattice F. Evidently $\omega^*(F) \geq 0 \ \forall F \in H$, H in $\mathscr{H}$ is ω* measurable if $\omega^*(F) \leq \omega^*(F \wedge H) + \omega^*(F \wedge H^C)$.

In this paper we start with a measure on an algebra Q of partial lattices and later we expand this to a measure defined on a sigma-algebra $L(\mathscr{B})$ containing Q. First, we define an outer measure on $\mathscr{H}$.

3. ω* Measurable Partial Lattice

Theorem3.1 The class of ω* measurable partial lattices in $\mathscr{H}$ are a sigma-algebra and restriction of ω* preserves complete measure.

Proof Claim: $L(\mathscr{B})$ is a sigma-algebra.

By definition of ω* measurable $\phi \in L(\mathscr{B})$ implies $L(\mathscr{B})$ is nonempty and also F^c is ω* measurable implies $F^c \in L(\mathscr{B})$.

Now to show that the join of two measurable partial lattices is measurable

For this let $F_1, F_2 \in L(\mathscr{B})$ implies $F_1 \vee F_2 \in L(\mathscr{B})$.

i.e., for any $F \in \mathscr{H}$ we have $\omega^*(K) = \omega^*[K \wedge (F_1 \vee F_2)] + \omega^*[K \wedge (F_1 \vee F_2)^c]$

Implies $\omega^*(K) \leq \omega^*[K \wedge (F_1 \vee F_2)] + \omega^*[K \wedge (F_1^c \wedge F_2^c)]$

Consider $\omega^*[K \wedge (F_1 \vee F_2)] + \omega^*[K \wedge (F_1^c \wedge F_2^c)] = \omega^*\{K \wedge [F_1 \vee (F_2 - F_2)]\} + \omega^*[K \wedge (F_1^c \wedge F_2^c)]$

$= \omega^*\{(K \wedge (F_1) \vee [K \wedge (F_2 \vee F_1^c)]\} + \omega^*[K \wedge (F_1^c \wedge F_2^c)] \leq \omega^*(K \wedge F_1) + \omega^*[K \wedge (F_2 \wedge F_1^c)] + \omega^*[K \wedge (F_1^c \wedge F_2^c)]$

$= \omega^*(K \wedge F_1) + \omega^*[(K \wedge F_1^c) \wedge F_2]\} + \omega^*[(K \wedge F_1^c) \wedge F_2^c] = \omega^*(K \wedge F_1) + \omega^*(K \wedge F_1^c)$ [since E_2 is ω* measurable] = ω*(K)

Therefore $\omega*(K)=\omega*[K\wedge(F_1\wedge F_2)]\}+\omega*[K\wedge(F_1\vee F_2)^c]$

Implies $F_1 \vee F_2$ is measurable. Implies $F_1 \vee F_2 \in L(\mathscr{B})$.

By induction the join of any finite number of measurable partial lattices is measurable. Therefore $L(\mathscr{B})$ is the algebra of partial Lattices. Let $\{F_i\}$ be a disjoint sequence of measurable partial lattices and write $F=\underset{i}{\vee}F_i$

Put $I_n=\overset{n}{\underset{i=1}{\vee}}F_i$. Clearly I_n is measurable $\forall n$. To show $\omega*(K)=\omega*(K\wedge I_n)+\omega*(K\wedge I_n^c)$

Since I_n is measurable $\omega*(K)=\omega*(K\wedge I_n)+\omega*(K\wedge I_n^c)$ but I_n is contained in F and F^c is contained in I_n^c

We have $\omega*(K)\geq\omega*(K\wedge I_n)+\omega*(K\wedge F^c)$. Since each F_n is measurable,

$$\omega*(K)\geq\omega*[(K\wedge I_n)\wedge F_n]+\omega*[(K\wedge F_n)\wedge F_n^c]+\omega*(K\wedge F^c)$$

Also, F_n contained in I_n, $I_n\wedge F_n=F_n$, $I_n\wedge F_n^c=I_{n-1}=\omega*(K\wedge F_n)+\omega*(K\wedge I_{n-1})+\omega*(K\wedge F^c)$

Since $\omega*(K\wedge I_{n-1})=\sum_{i=1}^{n-1}\omega*(K\wedge F_i)=\omega*(K\wedge F_n)+\omega*[K\wedge(\overset{n-1}{\underset{i=1}{\vee}})]+\omega*(K\wedge F^c)$

$$=\omega*(K\wedge F_n)+\sum_{i=1}^{n-1}\omega*(K\wedge F_i)=\omega*(K\wedge F^c)=\sum_{i=1}^{n}\omega*(K\wedge F_i)+\omega*(K\wedge F^c)$$

Therefore $\omega*(K)\geq\omega*(K\wedge F^c)+\sum_{i=1}^{n}\omega*(K\wedge F_i)\forall n$

Implies $\omega*(K)\geq\omega*(K\wedge F^c)+\sum_{i=1}^{\infty}\omega*(K\wedge F_i)$

Since $K\wedge F=\overset{\infty}{\underset{i=1}{\vee}}K\wedge F_i$ and $\omega*(K\wedge E)=\sum_{i=1}^{\infty}\omega*(K\wedge F_i)$

We have $\geq\omega*(K\wedge F^c)+\omega*(K\wedge F)$.

Therefore $\omega*(K)\geq\omega*(K\wedge F^c)+\omega*(K\wedge F)$

Hence F is measurable.

Since the join of any sequence of partial lattices in algebra Q can be replaced by the disjoint join of partial lattices in the algebra, it follows that $L(\mathscr{B})$ is a sigma-algebra.

Restriction of ω* preserves complete measure.

Let restriction of $\omega*$ be i.e., $=\omega*/L(B)$

By result 2.1 and [3] Theorem 3.1 (ii), we have $(F)=\omega*(F)\ \forall\ F\in L(B)$

Now $(\phi)=0$ implies $\omega*(\phi)=0$ and alsois a non – negative function defined on $L(B)$ which is the class of a $\omega*$ measurable partial lattices $L(B)$.To show is finitely-additive, let $F_1, F_2\in L(B)$ and $F_1\wedge F_2=\phi$

We prove this by induction, for $n=2$

Consider $(F_1 \vee F_2)=\omega*(F_1\vee F_2)$ (since F_1 is $\omega*$ measurable) $=\omega*[(F_1\vee F_2)\wedge F_1]+\omega*[(F_1\vee F_2)\wedge F_1^c]$

Clearly by absorption law and distributive law, $=\omega*(F_1)+\omega*(F_2)=(F_1)+(F_2)$

Therefore $(F_1\vee F_2)=(F_1)+(F_2)$

Assume for $n-1$, let $\{F_1, F_2, \ldots\ldots, F_n\}$ be a disjoint collection of partial lattices in $L(\mathscr{B})$. Evidently $\overset{n-1}{\underset{i=1}{\vee}}F_i$ and F_n are disjoint partial lattice in $L(\mathscr{B})$.

So $(\overset{n}{\underset{i=1}{\vee}}F_i)=[(\overset{n-1}{\underset{i=1}{\vee}}F_i)\vee F_n]=[(\overset{n-1}{\underset{i=1}{\vee}}F_i)]+(F_n)=\sum_{i=1}^{n-1}(F_i)+(F_n)=\sum_{i=1}^{n}(F_i)$

Hence this holds for n.

To show is measurable on $L(B)$. Let $\{F_i\}$ be a disjoint sequence of measurable partial lattices and $F = \bigvee_{i=1}^{\infty} F_i$

For each n we have $\bigvee_{i=1}^{n} F_i \subseteq F$. Implies $(F) \geq (\bigvee_{i=1}^{n} F_i) = \sum_{i=1}^{n} (F_i)$ $\forall$ n.

Implies $(F) \geq \sum_{i=1}^{\infty} (F_i)$ ------------ (1) But $(F) = \varpi^*(F) = \varpi^*(\bigvee_{i=1}^{\infty} F_i) \leq \sum_{i=1}^{\infty} \varpi^*(F_i) = \sum_{i=1}^{\infty} (F_i)$

Implies $(F) \leq \sum_{i=1}^{\infty} (F_i)$ -------------- (2)

From (1) and (2), $(F) = \sum_{i=1}^{\infty} (F_i)$. Therefore is measurable on $L(B)$.

To show is complete.

Let N be any measurable partial lattice such that $(N) = 0$ and $K \subseteq N$.

Now $K \subseteq N$, we have $\varpi^*(K) \leq \varpi^*(N) = (N) = 0$. Implies $\varpi^*(K) = 0$.

For S, $S \wedge K \subseteq K$, $S \wedge K^c \subseteq S$. Implies $\varpi^*(S \wedge K) \leq \varpi^*(K) = 0$.Implies $\varpi^*(S \wedge K) = 0$ and $\varpi^*(S \wedge K^c) \leq \varpi^*(S)$.

i.e., $\varpi^*(S) \geq \varpi^*(S \wedge K^c) + 0$ Implies $\varpi^*(S) \geq \varpi^*(S \wedge K^c) + \varpi^*(S \wedge K)$ $\forall$ S

Implies K is measurable. Implies $K \in L(B)$. So,$L(B)$ contains all partial lattices of measure zero.

Therefore is complete measure.

Hence the theorem

Result3.1 If $\{M_i\}$ is any sequence of partial lattices in Q such that M is contained in $\bigvee_{i=1}^{\infty} M_i$and ϖ be a measure on an algebra $L(B)$ then ϖ is countably subadditive.

Proof Take $N_n = M \wedge M_n \wedge \left(\bigvee_{i=1}^{n-1} M_i \right)^c$

Now $N_n \in Q$, $N_n \subseteq M_n$, also $N_n \subseteq M^c{}_m$ form< n,

For any $n > m$, $N_n \wedge N_m \subseteq M^c{}_m \wedge N_m \subseteq M^c{}_m \wedge M_m = \phi$.

Clearly $\{N_n\}$ is a sequence of disjoint partial lattices from M.

To show $M = \bigvee_n N_n$. By above clearly $N_n \subseteq M \wedge M_n$ $\forall$ n. Since $M \subseteq \bigvee_{i=1}^{\infty} M_i$.

Implies $M = M \wedge (\bigvee_{i=1}^{\infty} M_i) = \bigvee_{i=1}^{\infty} (M \wedge M_i)$. Implies $\bigvee_{n=1}^{\infty} N_n \subseteq \bigvee_{n=1}^{\infty} (M \wedge M_n) = M$.

Let $x \in M \subseteq \bigvee_n M_n$. Implies $x \in M_n$ for some n. For a smallest positive integer n such that $x \in M_n$

Implies $x \notin M_i \; \forall\, i < n$

Implies $x \notin (\bigvee_{i=1}^{n-1} M_i)$ Implies $x \in (M \wedge M_n) - (\bigvee_{i=1}^{n-1} M_i) = N_n$

Therefore $M \subseteq \bigvee_n N_n$

Therefore $M = \bigvee_n N_n$ and $\{N_n\}$ is a sequence of disjoint partial lattices in Q. Then by definition of measure on algebra we have

$\varpi(\bigvee_n N_n) = \sum_n \varpi(N_n)$

Implies $\varpi(M) = \sum_n \varpi(N_n) \leq \sum_n \varpi(M_n)$ [since $N_n \subseteq M_n$]

Observation3.1 If $M \in Q$ then $\varpi^*(M) = \varpi(M)$ where ϖ be a measure on Q

Proof Let $A \in Q$, clearly $M \subseteq M$ and $\varpi^*(M)$ is infimum we have $\varpi^*(M) \leq \varpi(M)$

By the result2.2, $\varpi(M) \leq \sum_{i=1}^{\infty} \varpi(M_i)$. Implies $\varpi(M) \leq \varpi^*(M)$ [By definition2.1]

Implies $\varpi(M) = \varpi^*(M)$

Result3.2 If set H of all partial lattices then the extended real valued set function defined on $\mathcal{H}$ is an outer measure

Proof Since $\phi \in L(B)$ clearly $\varpi^*(\phi) = 0$. Let us suppose $M \subseteq N$

Let $\{N_i\}$ where i = 1 to ∞ be a sequence of partial lattices in Q such that $N \subseteq \bigvee_{i=1}^{\infty} N_i$

Implies $M \subseteq N \subseteq \bigvee_{i=1}^{\infty} N_i$. Implies $M \subseteq \bigvee_{i=1}^{\infty} N_i$. Implies $\varpi^*(M) \leq \sum_{i=1}^{\infty} \varpi(N_i)$ [By definition2.1]

Implies $\varpi^*(M) \leq \operatorname{Inf}\{\sum_{i=1}^{\infty} \varpi(N_i) / N \subset \bigvee_{i=1}^{\infty} N_i$, where $\{N_i\}$ is a sequence of partial lattice in Q$\}$. Implies $\varpi^*(M) \leq \varpi^*(N)$. Therefore ϖ^* is monotone.

To show ϖ^* is countable sub-additive.

Let us suppose that $F \subseteq \bigvee_{i=1}^{\infty} F_i$

If $\varpi^*(F_i) = \infty$ for some i, then nothing to prove. Suppose $\varpi^*(F_i) < \infty$, $\forall$ i. Let $E > 0$.

Now $\omega^*(F_i)+$ is not a lower bound of the set $\{ \sum_{j=1}^{\infty} \omega(M_{ij}) \,/\, F_i \subseteq \vee_{j=1}^{\infty} M_{ij}$, where $\{M_{ij}\}$ is a sequence of partial lattices in Q$\}$.

Implies there exist $\{M_{ij}\}$ where j = 1 to ∞ of partial lattice such that $F_i \subseteq \vee_{j=1}^{\infty} M_{ij}$ and $\sum_{j=1}^{\infty} \omega(M_{ij}) < \omega^*(F_i)+$ .This is true for each i. Now $F \subseteq \vee_{i=1}^{\infty} F_i = \vee_{i=1}^{\infty} \vee_{j=1}^{\infty} M_{ij} = \vee_{i,j} M_{ij}$

So, by definition2.1, we have $\omega^*(F) \leq \sum_{i\,j} \omega(M_{ij}) = \sum_i \, [\sum_j \omega(M_{ij})] < \sum_i \, [\, \omega^*(F_i) + \,] = \sum_i \omega^*(F_i) + \sum_i$

$= \sum_i \omega^*(F_i) + \mathcal{E}$ [since $\sum_i \frac{1}{2^i} = 1$]. Since $\mathcal{E} > 0$ is arbitrary, we have $\omega^*(F) \leq \sum_{i=1}^{\infty} \omega^*(F_i)$.

Result3.3 Every member of algebra of partial lattices (Q) is measurable with respect to ω^*

Proof Let F be an arbitrary set.

Claim: $\omega^*(F) = \omega^*(F \wedge K) + \omega^*(F \wedge K^c)$. Clearly $\omega^*(F) \leq \omega^*(F \wedge K) + \omega^*(F \wedge K^c)$

Now it remains to show, $\omega^*(F) \geq \omega^*(F \wedge K) + \omega^*(F \wedge K^c)$

If $\omega^*(F) = \infty$ then there is nothing to prove

Suppose $\omega^*(F) < \infty$ and let $\mathcal{E} > 0$.

By definition of $\omega^*(F)$, $\omega^*(F) + \mathcal{E}$ is not lower bound of the set $\{ \sum_{i=1}^{\infty} \omega(K_i) \,/\, F \subseteq \vee_{i=1}^{\infty} K_i \}$.

Implies there exists a sequence $\{M_i\}$ of partial lattices in Q such that $F \subseteq \vee_{i=1}^{\infty} K_i$ and $\sum_{i=1}^{\infty} \omega(K_i) < \omega^*(F) + \mathcal{E}$.

For each i we can write $K_i = K_i \wedge X = K_i \wedge (K \vee K^c) = (K \wedge K_i) \vee (K_i \wedge K^c)$

Implies $\omega(K_i) = \omega(K \wedge K_i) + \omega(K_i \wedge K^c)$ $\forall$ i ------- (1) [By definition2.2]

Consider $F \wedge K \subseteq \vee_{i=1}^{\infty} (K \wedge K_i)$ and $F \wedge K^c \subseteq \vee_{i=1}^{\infty} (K_i \wedge K^c)$

Then $\omega^*(F \wedge K) \leq \sum_{i=1}^{\infty} \omega(K \wedge K_i)$ and $\omega^*(F \wedge K^c) \leq \sum_{i=1}^{\infty} \omega(K_i \wedge K^c)$ [By definition2.1]

Consider

$$\omega^*(F \wedge K) + \omega^*(F \wedge K^c) \leq \sum_{i=1}^{\infty} \omega(K \wedge K_i) + \sum_{i=1}^{\infty} \omega(K_i \wedge K^c) = \sum_{i=1}^{\infty} [\omega(K \wedge K_i) + \omega(K_i \wedge K^c)] = \sum_{i=1}^{\infty} \omega(K_i) < \omega^*(F) + \varepsilon$$

[from (1)].

Since $\varepsilon > 0$ is arbitrary, we have

$\omega^*(F) \geq \omega^*(F \wedge K) + \omega^*(F \wedge K^c)\ \forall E$

Implies K is measurable with respect to ω^*.

4. Conclusion

We proved that the class of ω^* measurable partial lattices in a sigma algebra and its restriction preserves a complete measure also the measure algebra defined on $L(\mathscr{B})$ is a countable subadditive whenever $\{M_i\}$ is a sequence of partial lattice in Q such that M is contained in join M_i we established results that the extended real valued set function defined on a set $\mathscr{H}$ of all partial lattices is an outer measure also every member of algebra of partial lattices (Q) is measurable with respect to ω^*. Finally, we conclude that the outer measure ω^* is called the outer measure induced by ω.

REFERENCES

1. Jordan Decomposition and its Uniqueness of Signed Lattice Measure, International Journal of Contemporary Mathematical Sciences (ISSN 1312-7586), 6(4):191–198, 2011.
2. Characterization of Outer Measure of Partial Lattices in a Countable Boolean Lattice, ARPN Journal of Engineering and Applied Sciences, Vol. 14, No. 4, February 2019, ISSN 1819-6608, Page No: 898–901.
3. The Characteristics of Partial Lattices and Irreducibility of Measurable Functions on Partial Lattice, International Journal of Scientific & Technology Research [IJSTR] Volume 8, Issue 09, September 2019 ISSN: 2277-8616, Page No: 639–641.
4. Gratzer.G, General Lattice Theory, Academic Press Inc., Volume 75, pages 380, Year 1978.
5. Royden. H.L., Real Analysis, 3rd ed., Macmillan Publishing, New York, Year 1981.
6. Y.V Seshagiri Rao, Thesis title "A Study of Partial Lattices on Countable Boolean Lattice Measures" Ph.D. awarded in 31th August 2019 at JNTUH.

Impending Inquisitions in Humanities and Sciences (ICIIHS-2022) – Dr. Mohan Varkolu et al. (eds)

Overtone Frequencies of Ozone using Vibrational Hamiltonian

32

K. Lavanya
Department of Mathematics, St. Francis College for Women, Begumpet, Hyderabad, India
Department of Mathematics, School of Science, GITAM (Deemed to be University), Hyderabad, India
J. Vijayasekhar*
Department of Mathematics, School of Science, GITAM (Deemed to be University), Hyderabad, India

ABSTRACT We employed the Hamiltonian operator based on the unitary Lie algebra to compute the combinational vibrational frequencies of ozone, extending up to the sixth overtone. A comparison between experimental findings and the fundamental vibrational spectra, as determined by the Lie algebraic model, was carried out. It has been affirmed that the outcomes derived from this approach exhibit a high level of accuracy, closely aligning with the experimental results.

KEYWORDS Vibrational frequencies, Lie algebraic method, Ozone, Vibrational Hamiltonian

1. Introduction

Various approaches exist, encompassing experimental techniques and theoretical approaches for determining vibrational frequencies in molecular systems. Notably, in 1981, Iachello et al. embarked on a pioneering endeavor where they harnessed the potential of the Hamiltonian operator formulated through Lie algebra techniques to unravel the vibrational spectra of molecules on a small size [Oss (1996), Iachello et al. (1990, 1995)]. This approach, grounded in the Morse potential function and the Schrödinger equation framework, offers insights into the intricate molecular dimensions characterizing diatomic entities. In the subsequent years following this pioneering investigation into molecular spectra, the applicability of this method was broadened to encompass molecules spanning medium to large sizes, signifying its adaptability and potential [Balla et al. (2021, 2022), Vijayasekhar et al. (2022, 2023)]. Alongside this theoretical trajectory, two well-established experimental methodologies have emerged for computing molecular vibrational spectra: the Dunham expansion and the Schrödinger equation approach. The first employs an expansion scheme that extends energy levels concerning rovibrational quantum numbers, furnishing a comprehensive perspective on molecular behavior. Conversely, the latter involves solving the Schrödinger equation to extrapolate the interatomic potential in terms of interatomic variables and potential coefficients, with parameter fitting being performed using available experimental data [Iachello et al. (1990)]. These experimental techniques entail a substantial demand for copious amounts of data to effectively calibrate the involved parameters - a demand that is not always reasonable due to practical limitations. In this context, the Lie algebraic approach emerges as a compelling alternative to mitigate this formidable challenge. Within the framework of Lie algebra, the Hamiltonian operator is elegantly expressed using algebraic operators and parameters, the specific values of which are contingent upon the nature of the molecule under investigation. This method thus serves as an innovative avenue to reconcile theoretical predictions with empirical observations while circumventing the substantial data requirements inherent in traditional approaches.

*Corresponding author: vijayjaliparthi@gmail.com

DOI: 10.1201/9781003489436-32

2. Vibrational Hamiltonian for the Triatomic Molecule, Ozone

The Lie algebraic technique, as elucidated by Iachello et al. (1995), Oss (1996), and Vijayasekhar (2018), illustrates the Hamiltonian operator through an assemblage of creation (and annihilation) operators, effectively delineating the system's local (uncoupled) and normal (coupled) modes. For a molecule comprising n interacting bonds, the Hamiltonian operator pertinent to the Lie algebraic method can be succinctly represented as follows:

$$h = e_0 + (a_1 l_1 + a_2 l_2 + \dots + a_n l_n) + (b_{12} l_{12} + b_{23} l_{23} + \dots + b_{pq} l_{pq}) + (c_{12} m_{12} + c_{23} m_{23} + \dots + c_{pq} m_{pq}),\ p < q \quad \dots (1)$$

Here, i vary between 1 and 2 for two stretched bonds (O-O and O-O) and a_i, b_{ij} and c_{ij} $(i, j = 1, 2, 3)$ are parameters, the values of which are calculated based on spectroscopic data. Where, l_i, l_{ij} are uncoupled and coupled bond invariant operators and provided by

$$\langle l_i \rangle = -4(u_i v_i - v_i^2) \quad \dots (2)$$

$$\langle u_i, v_i; u_j, v_j | l_{ij} | u_i, v_i; u_j, v_j \rangle = 4\left[(v_i + v_j)^2 - (v_i + v_j)(u_i + u_j) \right] \quad \dots (3)$$

and the Majorana operator, m_{ij} is used to describe local mode interactions in pairs. This contains diagonal and non-diagonal matrix elements,

$$\langle u_i, v_i; u_j, v_j | m_{ij} | u_i, v_i; u_j, v_j \rangle = (u_i v_j + u_j v_i - 2 v_i v_j)$$

$$\langle u_i, v_i + 1; u_j, v_j - 1 | m_{ij} | u_i, v_i; u_j, v_j \rangle = -\left[v_j (v_i + 1)(u_i - v_i)(u_j - v_j + 1) \right]^{1/2}$$

$$\langle u_i, v_i - 1; u_j, v_j + 1 | m_{ij} | u_i, v_i; u_j, v_j \rangle = -\left[v_i (v_j + 1)(u_j - v_j)(u_i - v_i + 1) \right]^{1/2} \quad \dots (4)$$

While v_2 indicates bending vibrations for Ozone, v_1 and v_3 stand for local stretching vibrations (o_3). Put $a_1 = a_2 = a$ and $u_1 = u_2 = u$ in equations (2), (3), and because two bonds (O-O) are equivalent (4). Except for u, which is dimension less, all parameters are in cm^{-1}.

The Vibron number u_i for stretching bonds of Ozone can be calculated using

$$u_i = \frac{\omega_e}{\omega_e x_e} - 1,\ i = 1, 2. \quad \dots (5)$$

The spectroscopic constants, ω_e and $\omega_e x_e$ for the O-O bond in the stretching mode are 1580.161 and 11.95127, respectively [Irikura (2007)].

The energy equation is used to calculate the initial value of the parameter a,

$$e(v = 1) = -4a(u - 1). \quad \dots (6)$$

Hence, $\bar{a}$ can be evaluated as,

$$\bar{a} = \frac{\bar{e}}{4(1 - u)} \quad \dots (7)$$

Where, $\bar{e}$ denotes the mean value of two distinct energies that are related to symmetric (e_1) and antisymmetric (e_2) combinations of local modes.

To calculate an initial value for c_{12}, using the relation,

$$c_{12} = \frac{|e_1 - e_2|}{2u} \quad \dots (8)$$

3. Results

Enclosed within the table provided below are the outcomes derived from employing the Lie algebraic model to compute the vibrational spectra of ozone:

Table 32.1 The vibrational spectra of Ozone (in cm^{-1})

$(V_1V_2V_3)$	Experimental [Shimanouchi (1972), Irikura (2007)]]	Lie algebraic model
(1 0 0) (symmetric)	1103	1102.9202
(0 0 1) (antisymmetric)	1042	1042.4187
(0 1 0) (bend)	701	702.3538
(2 0 0)		2126.8372
(0 0 2)		2243.7042
(0 2 0)		1389.9701
(1 0 1)		2122.9611
(3 0 0)		3201.0074
(0 0 3)		3087. 6529
(0 3 0)		2086.0054
(2 0 1)		3107.0731
(1 0 2)		3183.7327
(4 0 0)		4392.9006
(0 0 4)		4076.7107
(0 4 0)		2753.9120
(3 0 1)		4366.8826
(1 0 3)		4149.9301
(2 0 2)		4237.0171
(5 0 0)		5603.6623
(0 0 5)		5276.9428
(0 5 0)		3389.3371
(1 0 4)		5285.0062
(4 0 1)		5397.9326
(3 0 2)		5365.0626
(2 0 3)		5291.7724
(6 0 0)		6681.9065
(0 0 6)		6304.9073
(0 6 0)		4067.3108
(5 0 1)		6375.0587
(1 0 5)		6466.1045
(4 0 2)		6508.7752
(2 0 4)		6621.2274
(3 0 3)		6460.0059

$u = 131$, $a\,(str) = -1.752$, $a\,(bend) = 2.102$, $a_{ij}\,(str) = 1.074$, $a_{ij}\,(bend) = -0.836$, $c_{ij}\,(str) = 2.115$, $c_{ij}\,(bend) = 3.468$

4. Conclusion

Using the Lie algebraic method, the combined vibrational frequencies of ozone up to the sixth overtone have been calculated and are shown in Table 32.1. Additionally, the table provides insight into the symmetric and antisymmetric fundamental vibrational frequencies, offering a point of comparison against the available experimental data. This compilation of data is precious for individuals inclined towards empirical findings over theoretical constructs. Such practitioners, driven by a penchant for experimental results, may find an incentive to refine the combined oscillation energies within the domain of higher overtones.

5. Acknowledgments

Vijayasekhar gratefully acknowledges the Research Seed Grants (RSG) by Gandhi Institute of Technology and Management (GITAM), Deemed to be University, Hyderabad, India [Sanction Letter Ref: F.No. 2021/0114, dated 26-09-2022].

REFERENCES

1. S. Oss, *Adv. Chem. Phys.* 1996, 93, 455.
2. F. Iachello, R.D. Levine, Algebraic Theory of Molecules, Oxford University Press, Oxford, 1995.
3. F. Iachello, S. Oss, *J. Mol. Spectrosc.* 1990, 142(1), 85.
4. M. R. Balla, V. S. Jaliparthi, *Mol. Phys.* 2021, 115(5), e1828634.
5. M. R. Balla, S. Venigalla, V. Jaliparthi, *Acta Phys. Pol. A.* 2021, 140 (2), 138.
6. M. R. Balla, V. S. Jaliparthi, *Polycyclic Aromatic Compounds*. 2022, 42(7), 4684.
7. J. Vijayasekhar, M. Rao Balla, *Spectrochim. Acta A.* 2022, 264, 120289.
8. J. Vijayasekhar, *Ukr. J. Phys. Opt.* 2022, 23(3), 126.
9. J. Vijayasekhar, P. Suneetha, K. Lavanya, Ukr. J. Phys. Opt. 2023, 24(3), 193–199.
10. J. Vijayasekhar, Orient J Chem. 2018, 34(4), 2208.
11. K. K. Irikura, *J. Phys. Chem. Ref. Data.* 2007, 36(2), 389.
12. T. Shimanouchi, *Tables of Molecular Vibrational Frequencies*, National Bureau of Standards, 1972, Vol. I, 1–160.

Impending Inquisitions in Humanities and Sciences (ICIIHS-2022) – Dr. Mohan Varkolu et al. (eds)
© 2024 Taylor & Francis Group, London, ISBN 978-1-032-78829-6

Facial Emotion Recognition using Convolutional Neural Networks

33

Gopichand G
Assistant Professor Senior Grade-1, SCOPE, Vellore Institute of Technology, Vellore, Tamilnadu, India

I. Ravi Prakash Reddy
Professor, Department of Information Technology, GNITS, Hyderabad, India

Santhi H
Associate Professor, School of Computer Science and Engineering, Vellore Institute of Technology, Vellore, Tamilnadu, India

Vijaya Krishna Akula*
Assistant Professor, Department of Computer Science and Technology, GNITS, Hyderabad, India

ABSTRACT In this modern generation, where we are frequently facing factors which build on our high levels of stress, the human race is facing a big problem of deteriorating mental health. Moreover, in the current scenario of Covid-19, this issue has worsened because of fewer social interactions than ever before, increase in screen time and even lack of physical work. Estimating facial emotion has forever been a simple task for people, however, accomplishing a similar perfection with a computer is very difficult. With the new progression in computer vision and AI, it is feasible to distinguish feelings from pictures and videos. When we decided on the idea of research, our goal was to assist people having issues related to mental health and make them come out of the pain they are suffering from. The main idea behind our work was the motto that it acts as effective research, not only a comparative analysis. The execution was done using a web-app. In the emotion recognition and detection component, which is the main component of our research, Harr Cascades was utilized to find and detect the face of an individual in front of the webcam. Haar Cascades has been used for the detection of the facial region captured by the webcam. It works as follows: instead of applying all the six thousand features present in a frame, it groups the features into different stages and applies them one after another and not all the six thousand features at once. The next component would be the Deep Learning algorithm developed using CNN. OpenCV has been used for the entire development. Tensor Flow has been utilized to train the model that has been developed. The model works by running frame by frame in order to detect the emotion. The facial region that has been detected through the Haar Cascade Classifier is passed as input to the Convolutional Neural Network model that has been developed by us. The emotion having maximum softmax scores is given as the result on live webcam feed itself.

KEYWORDS Deep learning, Convolutional neural networks, Artificial Intelligence

1. Introduction

Emotions regularly intervene and work with connections among individuals. Thus, understanding emotions frequently carries setting to apparently odd or potentially complex social communication. Emotions can be perceived through a variety of means

*Corresponding author: vijayakrishnaakula2022@gnits.ac.in

DOI: 10.1201/9781003489436-33

like voice inflection, non-verbal communication, and more complex strategies such EEG. It is well known that emotions play an important part in a person's life. With vast variety of expressions or moments, the facial expression of a person expresses how they feel or in which mental state the person is currently. Facial expressions are the indispensable identifiers for human sentiments, since it relates to the feelings. The maximum of times (generally in 55% cases), the facial expression is a nonverbal method of passionate appearance, and it very well may be considered as substantial proof to reveal whether or not an individual is talking facts or not. The current methodologies fundamentally center around facial examination keeping foundation intact and henceforth developed a great deal of pointless and deceiving highlights that confound CNN preparing process. The current composition centers around five fundamental facial appearance classes revealed, which are displeasure/anger, sad/unhappy, smiling/happy, feared, and surprised/astonished Humans can show thousands of facial expressions during their interactions with others that vary in complexity, level of expressiveness and understanding. Emotion is often visible by subtle changes in one or several discrete features. Multiple surveys stated that verbal communication conveys thirty three percent of human communication, and nonverbal components convey approximately sixty six percent. Among the many intangibles, facial expressions are the most important conveyers of information in human communication. Thus, it is innate that face-to-face related researches have received a great deal of appreciation and attention over the past decades, not only in cognitive science, but also in computer-related applications and performances. To reach the optimal solution for emotion detection and recognition in detection from an environment, such as a webcam, both efficiency and accuracy must be achieved. This has many real time applications, such as recognition of facial expressions of online students, will allow teachers to tailor strategies and methods of teaching, provide students with more accurate feedback, and achieve excellent teaching standards, by providing how the students are perceiving the classes being taught in the online mode. The goal is to find the best solution for emotion recognition based on facial recognition in virtual, real-time learning environments. To reach this goal, it is mandatory and important to enhance the efficiency and accuracy of facial emotion recognition systems.

2. Motivation

In the course of last two decades, specialists have essentially progressed human facial emotion recognition with computer vision procedures. Since a long time, there have been numerous approaches to this issue, including utilizing (PHOG), AU aware facial features, boosted LBP descriptors, and RNNs. In this modern era, a person facing emotional distress has become a general aspect of life. But some cases might reach out of hand if not taken proper care and guidance.

Besides, we have a jump in the number of suicides in this decade, which had emotion conditions such as depression one of the main cause. Therefore, we have decided to research on a scenario which would not only help an individual understand his/her outer emotional state, but also his/her inner mental condition, which would help them to take preventive measures towards their psychological state before the situation would go out of hands.

3. Background of Research Work

Mental Health currently is being majorly affected due to the outbreak of Covid-19. This research will help the user to analyze their own emotions and help them find solutions if they are feeling stressed or depressed or angry. The research aims to provide two fronts to verify the user's emotions. One is to capture a live image of the user through a webcam and detect the emotion the user is feeling the most, through image processing. The other is to ask the user to give their response for a psychological quiz to detect their emotion. Its result will present the percentage of each emotion. Based on both the results, solutions will be provided to the user.

4. Proposed Approach

4.1 Various Components of Work

The main objective of the research is to detect Human Emotion, both internal and external. The research is divided into two modules. One module is based on detecting the user's emotion through his/her facial expressions implemented using image processing. The other module is based on detecting the emotion based on the responses given by the user for multiple questions asked through a quiz. Detecting emotion through image processing and quiz can further be divided into 4 components.

First, Haar Cascade classifier. It is used for the detection of the face in the frames that is being captured by the webcam. It works as follows: instead of applying all the six thousand features present in a frame, it groups the features into different stages and applies them one after another and not all the six thousand features at once. If any feature suggests that it is not part of the Face region, the algorithm will rule out that subregion as not-face. If the subregion passes all the stages, then it is classified as the Face region. Once the facial region is detected, then it runs through the deep learning algorithm.

Second is the deep learning algorithm that has been developed using Convolutional Neural Networks. Due to the recent fame and development, we have decided to use OpenCV for the entire development. Tensor Flow has been utilized to train the model that has been developed. The model works by running frame by frame in order to detect the emotion. The model works by capturing each frame after another to detect the emotion. The facial region detected through Haar Cascade Classifier is passed as input to CNN model that has been developed. The face region is sent through multiple layers of the developed network to train the system for emotion detection. Convolution is the first layer where features are extracted using a filter of chosen kernel size. Then it is sent through max pooling filter. Its function is to progressively reduce the spatial size of the representation to reduce the amount of parameters and computation in the network. Pooling layer operates on each feature map independently. Then it is sent through fully connected layers (FC) and softmax function are applied to classify an object with probabilistic values between 0 and 1. The emotion which has the maximum softmax score is showed as the result on live webcam feed itself.

Third is the Quiz. The user is presented with 24 questions which must be answered using a 5-point scale. Then the responses will go through an algorithm to give the percentage of each emotion: Stress, Happy, Anger, Sad, Disgust, and Fear. The result will signify what emotion the user is feeling. Fourth is the Solutions. The user is presented with solutions to help them overcome stress, depression and anger.

4.2 Proposed Mechanism

Convolutional neural Network (CNN) is the most famous method of analyzing pictures. CNN is unique in relation to a multi-flayer perceptron (MLP) as they have hidden layers, called convolutional layers. The proposed strategy depends on a two-level CNN system. Once the user lands on the website, they are provided with information about the research and how it works. The user can then begin analysis of their current emotion. First, we analyze the external emotion the user is feeling by capturing their live video through webcam. Each frame goes through Haar Cascade Classifier to detect the facial region. Once the face is detected, the sub-region goes through the deep learning algorithm based on CNN to predict the emotion the user is experiencing.

The result is based on the maximum softmax score given by the model. The accuracy of the model will depend on multiple factors. The user can see the result of external emotion detection on their live feed itself.

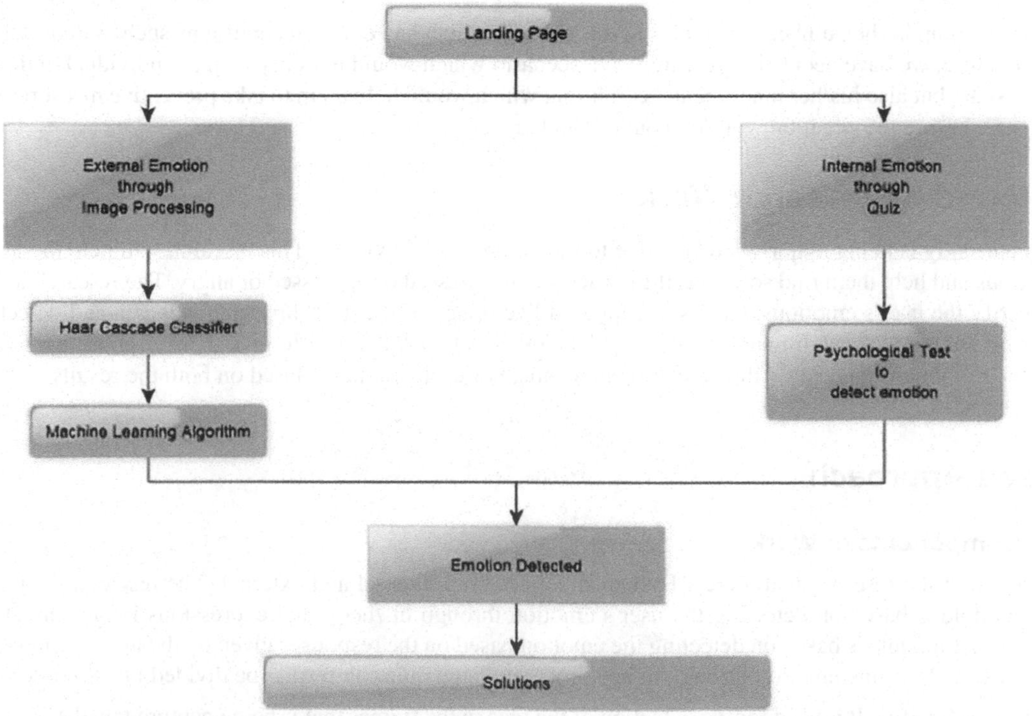

Fig. 33.1 Emotions recognition system using image processing and psychological test

Next, the user is asked to take a psychological test. He/she must answer all the questions using the 5 point scale. Once submitted, the result will showcase the percentage of each emotion: Stress, Happy, Anger, Sad, Disgust, and Fear, the user is feeling. Then, the results obtained through both the emotion detecting techniques will help the user understand their emotion. Solutions are given for low, moderate and high scores of stress and anger. The user can follow the solutions to overcome their current emotional state.

Haar Cascade classifier. It is used for the detection of the face in the frames that is being captured by the webcam. It works as follows: instead of applying all the six thousand features present in a frame, it groups the features into different stages and applies them one after another and not all the six thousand features at once. If any feature suggests that it is not part of the Face region, the algorithm will rule out that subregion as not-face. If the subregion passes all the stages, then it is classified as the Face region.

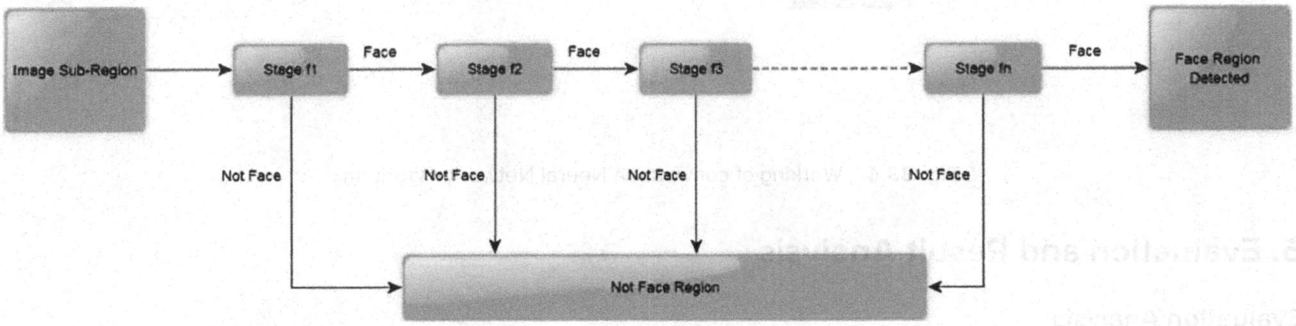

Fig. 33.2 Working of haar cascades classifier

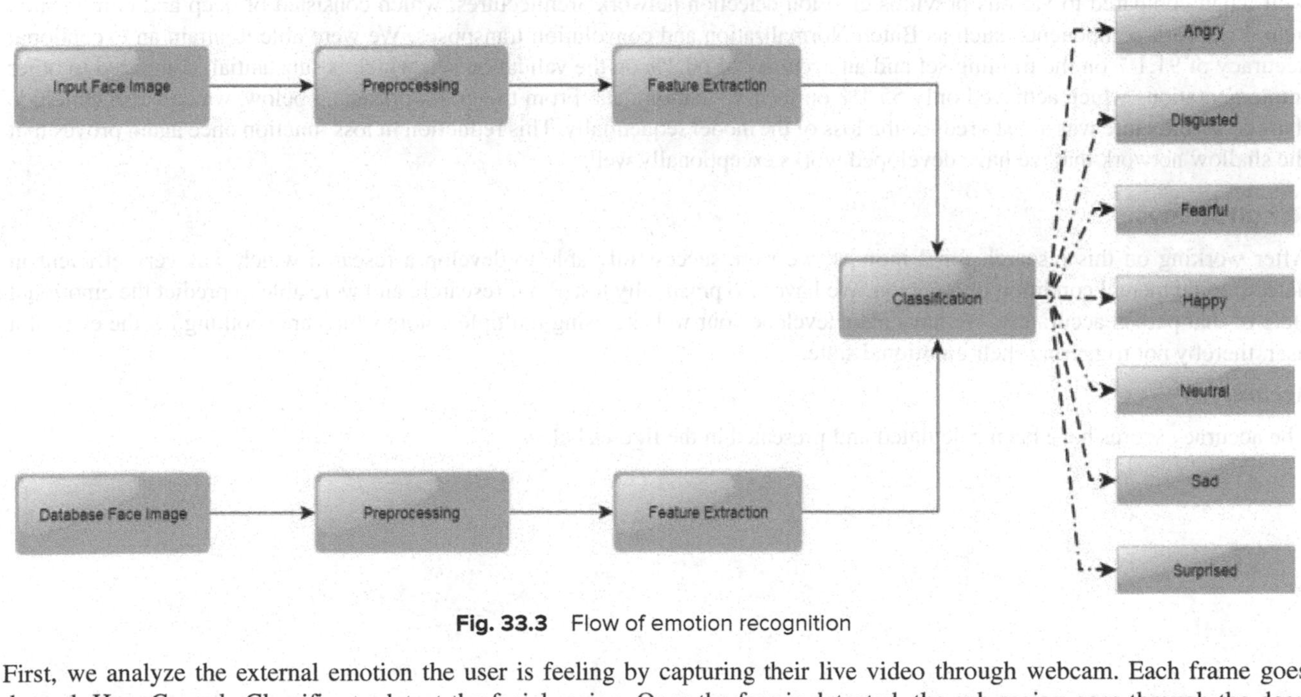

Fig. 33.3 Flow of emotion recognition

First, we analyze the external emotion the user is feeling by capturing their live video through webcam. Each frame goes through Haar Cascade Classifier to detect the facial region. Once the face is detected, the sub-region goes through the deep learning algorithm based on CNN to predict the emotion the user is experiencing. The result is based on the maximum softmax score given by the model. The accuracy of the model will depend on multiple factors. The user can see the result of external emotion detection on their live feed itself.

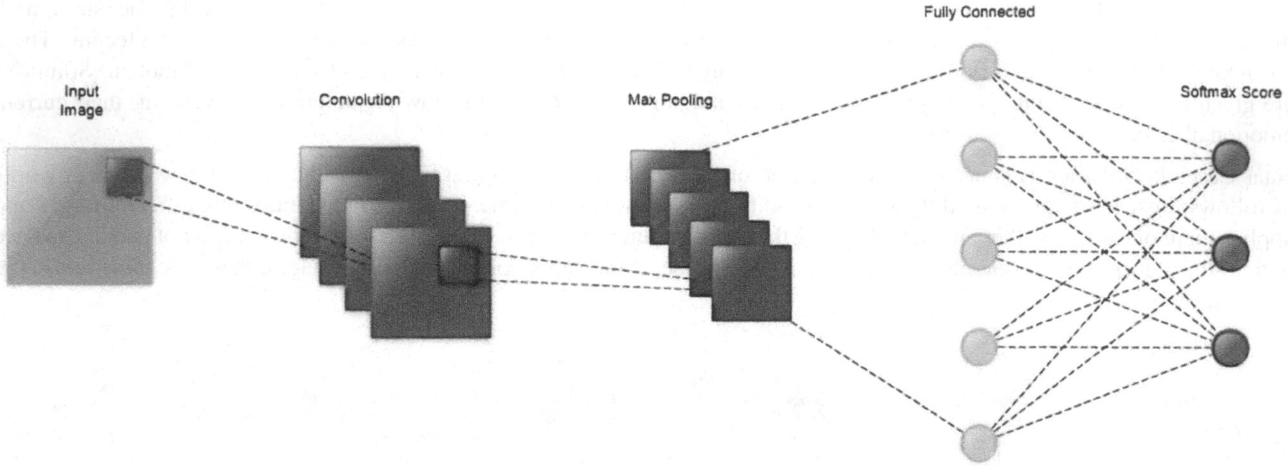

Fig. 33.4 Working of convolution Neural Network Algorithm

5. Evaluation and Result Analysis

Evaluation Analysis

The Neural Network architecture that we developed, though gives an assumption of a shallow network, works equivalently well when compared to various previous emotion detection network architectures, which consisted of deep and complicated neural network components, such as Batch Normalization and convolution transposes. We were able to attain an exceptional accuracy of 91.1% on the training set and an accuracy of 60.2% on the validation set, which is substantial, compared to other implementations which achieved only 55.1% on their validation data. From the graph presented below, we can also conclude that our architecture was able to reduce the loss of the model sequentially. This reduction in loss function once again proves that the shallow network that we have developed works exceptionally well.

Result Analysis

After working on this research for 2 months, we were successfully able to develop a research which was very efficient in detecting the mental condition of a person. We have also practically tested our research, and were able to predict the emotional state of that person accurately. We have also developed our website using multiple colors which are soothing for the eyes of a user, thereby not triggering their emotional state.

Accuracy:

The accuracy scores have been calculated and presented in the figures below:

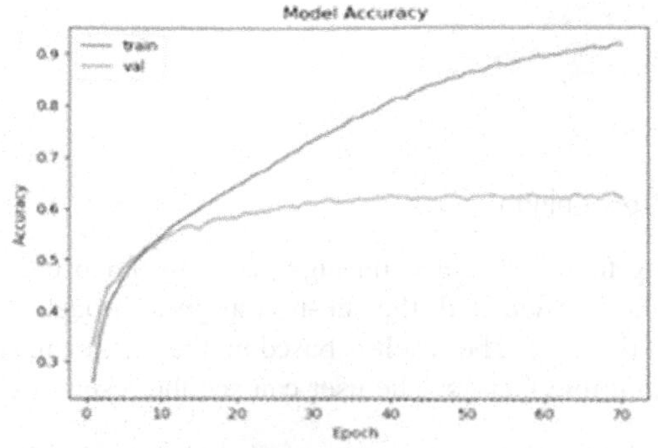

Fig. 33.5 Exceptional Model Accuracy of 91.1% on the training data and 60.2% on the validation data set

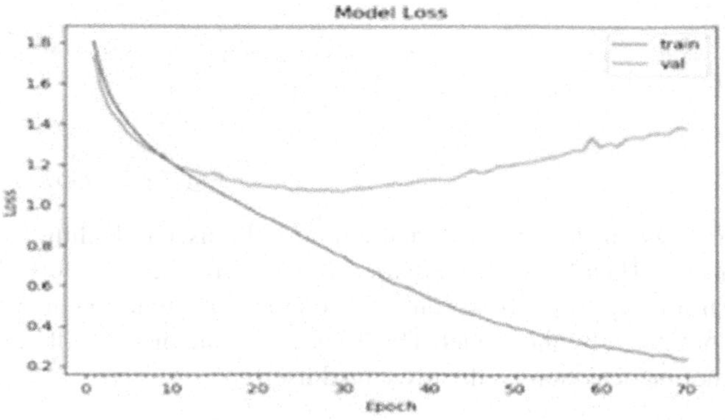

Fig. 33.6 Model Loss of both Trained and Validation data set

6. Conclusion

The objective of this research was to carry out real-time facial emotion recognition. Utilizing our exclusively prepared network with a face-detector given by OpenCV, we have been successful in completely developing an application. This research was created and tested and it successfully gave one's emotion both by image detection and Quiz. Image detection part is Realtime with on spot results on camera, while the quiz gives the percentage of what type of emotion one is feeling which includes Stress, Happy, Anger, Sad, Disgust, and Fear. While we accomplished an effective execution, significant improvements can be made by tending to a few key issues. Initially, a lot bigger dataset ought to be intended to work on the model's generalization. While we accomplished >60% exactness in laboratory and stable conditions (wonderful lighting, camera at eye level, subject confronting camera with an overstated expression), any deviation from that made the detection fall altogether. Specifically, any shadow regarding a matter's face would cause a mistaken identification as 'angry'. In addition, the webcam should be level to the subjects' countenances for precise classification. Expanding our self-created dataset to incorporate off-center faces might have tended to this issue. Heavier pre-processing of the information would have surely improved the accuracy of the system. As stated already, an especially difficult part of ongoing acknowledgment is choosing how to outline transition frames from neutral to fully understandable expressions of emotion. One feasible arrangement is to utilize a running average of the top classes reported by each frame which would lessen the issue of errors/noise brought about by dynamic expressions. Unfortunately, the slow frame-rate of our demonstration made this arrangement untenable. Regardless, our execution seemed to identify a person's emotional state reliably. Future executions that run at higher frame rates would require a running average.

REFERENCES

1. Reney, D., & Tripathi, N. (2015, April). An efficient method to face and emotion detection. In 2015 fifth international conference on communication systems and network technologies (pp. 493–497). IEEE.
2. De Silva, L. C., Miyasato, T., & Nakatsu, R. (1997, September). Facial emotion recognition using multi-modal information. In Proceedings of ICICS, 1997 International Conference on Information, Communications and Signal Processing. Theme: Trends in Information Systems Engineering and Wireless Multimedia Communications (Cat. (Vol. 1, pp. 397–401). IEEE.
3. Rajesh, K. M., & Naveenkumar, M. (2016, December). A robust method for face recognition and face emotion detection system using support vector machines. In 2016 International Conference on Electrical,Electronics, Communication, Computer and Optimization Techniques (ICEECCOT) (pp. 1–5). IEEE.
4. Kim, M. H., Joo, Y. H., & Park, J. B. (2005). Emotion detection algorithm using frontal face image. 제어로봇시스템학회:학술대회논문집, 2373–2378.
5. Dagar, D., Hudait, A., Tripathy, H. K., & Das, M. N. (2016, May). Automatic emotion detection model from facial expression. In 2016 International Conference on Advanced Communication Control and Computing Technologies (ICACCCT) (pp. 77–85). IEEE.
6. Maglogiannis, I., Vouyioukas, D., & Aggelopoulos, C. (2009). Face detection and recognition of natural human emotion using Markov random fields. Personal and Ubiquitous Computing, 13(1), 95–101.
7. Miyakoshi, Y., & Kato, S. (2011, March). Facial emotion detection considering partial occlusion of face using Bayesian network. In 2011 IEEE Symposium on Computers & Informatics (pp. 96–101). IEEE.
8. Chiranjeevi, P., Gopalakrishnan, V., & Moogi, P. (2015). Neutral face classification using personalized appearance models for fast and robust emotion detection. IEEE Transactions on Image Processing, 24(9), 2701–2711.
9. Yang, K., Wang, C., Sarsenbayeva, Z., Tag, B., Dingler, T., Wadley, G., & Goncalves, J. (2020). Benchmarking commercial emotion detection systems using realistic distortions of facial image datasets. The Visual Computer, 1–20.
10. Goldman, A. I., & Sripada, C. S. (2005). Simulationist models of face-based emotion recognition.Cognition, 94(3), 193–213.
11. Jaiswal, S., & Nandi, G. C. (2020). Robust real-time emotion detection system using cnn architecture.Neural Computing and Applications, 32(15), 11253–11262.
12. Bouhabba, E. M., Shafie, A. A., & Akmeliawati, R. (2011, May). Support vector machine for face emotion detection on real time basis. In 2011 4th International Conference on Mechatronics (ICOM) (pp. 1–6). IEEE.

Note: All the figures in this chapter were made by the authors.

Impending Inquisitions in Humanities and Sciences (ICIIHS-2022) – Dr. Mohan Varkolu et al. (eds)

Double Diffusive Analysis on Maxwell Fluid Flow through a Horizontal Porous Layer with the Effect of Soret and Heat Source

34

Pudari Chandra Mohan*
Mallareddy Institute of Engineering and Technology, Hyderabad, Telangana, India
Department of Mathematics, Koneru Lakshmaiah education foundation, Hyderabad, Telangana, India

Anjanna Matta
Department of Mathematics, Faculty of Science and Technology (IcfaiTech),
ICFAI Foundation for Higher Education, Hyderabad, Telangana, India

Suresh Kumar Y.
Department of Mathematics, Koneru Lakshmaiah education foundation,
Hyderabad, Telangana, India

ABSTRACT A study employed numerical techniques to explore how the flow of a Maxwell fluid is affected by the Soret effect. The research investigated the persistence of the flow within a porous layer positioned horizontally and exposed to heating and saltwater infiltration from its lower section. The Soret effect can significantly impact the behavior of the Maxwell fluid, and its specific manifestation depends on the composition of the mixture, temperature gradient, and material properties. Eigenvalue problems were solved to derive the wave number and critical Rayleigh number for various flow properties. This method enables the analysis of system behavior and prediction of potential instabilities or changes. The system was analyzed for linear stability to obtain a more comprehensive understanding of how the interplay between the heat source and the Soret effect influences the system. The shooting and Runge-Kutta methods were utilized to calculate the thermal and solutal gradient. The study examined the characteristics of velocity, concentration, and temperature under stable conditions, while considering the influence of the Soret effect and heat source.

KEYWORDS Maxwell fluid, Heat source, Soret effect, porous medium, Linear stability analysis

1. Introduction

Thermodiffusion in a horizontal porous medium is a subject of significant research activity due to its wide range of practical applications. In this paper, we have addressed the theory and numerical methodology that is frequently utilized in the Soret effect investigations, as well as a quick summay of the experimental and numerical work that has been done in this field. It is believed that numerical models will produce trustworthy results and lessen the need for expensive experiments because it is challenging to measure heat diffusion and mass diffusion coefficients precisely. Although various models have been put forth in the past, in some circumstances one may be better than others. Depending on the particular fluid mixture of interest, these models must be used when and how.

*chandu.shyamala@klh.edu.in

DOI: 10.1201/9781003489436-34

The Soret impact was considered during the investigation. An analytical solution that works for the Darcy model and for flows an enclosure with relatively large aspect ratios is presented in the study's first section. The study's second section involves a numerical analysis of all the conservation equations that support the analytical model.

Nield and Bejan's (2013) paper provides a thorough review of Hadley flow in porous media. Narayana et al. (2012) and Zhao et al. (2014) investigated the thermal instability in a porous medium for binary Maxwell fluid flow, both in linear and nonlinear scenarios, Chand, and Rana (2014) investigated the phenomenon of thermal convection in Maxwell fluids, specifically in the presence of porous media that exhibit saturation. Kim et al. in (2003), Chand (2013), and Akbar et al. (2013) employed linear theoretical analysis to investigate the impact of elastic parameters on the onset of convection in porous substances. The Maxwell fluid is analysed using a phenomenological model introduced by Alishayev (1974), which has become widely used for studying its constitutive equation. Which has been further employed by Tan and Matsuoka (2007).

Recently, there has been a growing interest in the study of pure thermal convection in Maxwell fluids that flow through porous media with saturation. Scholars, including Wang and Tan (2011) and Bishnoi and Goyal (2012), have made notable advancements in this area of research. A study was conducted by Michael and Yuriko (2009), which focused on the external flow in a planar direction of a uniform upper-converted Maxwell fluid. The phenomenon of double-diffusive convection in a Maxwell fluid containing a saturated porous layer has been explored by multiple researchers, including Sharma and Kumari (1993), Awad et al. (2010), Malashetty and Biradar (2011), Gaikward and Biradar (2013), and Gaikward and Dhanraj (2014). According to Hayat et al. (2008), earlier studies have focused on practical concerns such as the development of thermal energy storage systems, strategies for pollution control, and the properties of polymeric liquids. More recently, there have been several studies on Hadley flow, with Narayana et al. (2008) being among the contributors to this area.

In 2015, Anjanna and Narayana conducted research into the flow characteristics and heat transfer of a Maxwell fluid system with double-diffusive Hadley flow, which flows through a porous layer. Deepika et al. (2016) conducted a study to examine how flow and heat transfer behave in a porous medium when influenced by different factors, such as throughflow and concentration-based internal heat sources. One of their main objectives was to understand the effects of double-diffusive convection. An investigation was conducted by Anjanna et al. (2017) to explore the influence of several parameters, such as the internal heat source, on the flow behaviour and heat transfer of a horizontal porous layer in the context of double-diffusive Hadley-Prats flow. The objective of Anjanna's (2019) study was to examine how the stability of Hadley flow in a horizontal porous layer is affected by the presence of an internal heat source and non-uniform thermal gradient. The study aimed to understand how these factors affect the stability of the flow.

This study aims to evaluate how the Soret effect affects the flow in horizontal porous media under the influence of temperature and concentration gradients. The phenomenon of double-diffusive convection, which arises due to these gradients, has been extensively researched and is known to play a crucial role in various fields, such as the production of pharmaceuticals and high-quality crystals.

2. Mathematical Formulation

The system under consideration is a uniform and permeable layer of fluid-saturated material, bounded by impermeable planes located at $z = H/2$ and $z = -H/2$. The horizontal direction exhibits linear fluctuations in both temperature and solvent fields, while the vertical z-axis is oriented upward and runs in the opposite direction to that of gravity. The bounding planes at $z = H/2$ and $z = -H/2$ are maintained at the heated bottom wall and cold top wall. Similar solute concentrations top and bottom walls, respectively.

The temperature variance along the walls is denoted by ∇T, and a constant interior heat source Q is present. The imposed temperature gradient vector along the plates is represented by (βTx, βTy) and the concentration gradients are represented as (βCx, βCy). The flow of fluid in the horizontal porous material is described by Darcy's law.

The governing equations of the system are presented in a non-dimensional form.

$$\nabla^{*}.v^{*} \rightleftarrows = \rightleftarrows 0 \tag{1}$$

$$v^{*} \rightleftarrows = \rightleftarrows \rightleftarrows \frac{K}{\mu}\left(1+\gamma^{*}\frac{\partial}{\partial t^{*}}\right)\left(-\nabla^{*}P^{*} + \rightleftarrows \rho_{f}^{*}g\right) \rightleftarrows \tag{2}$$

$$(\rho c)_{m}\left(\frac{\partial T^{*}}{\partial t^{*}}\right) + (\rho C_{p})_{f}\, v^{*}.\nabla^{*}T^{*} = k_{m}\nabla^{*2}T^{*} + Q^{*} \tag{3}$$

$$\varphi\left(\frac{\partial c^*}{\partial t^*}\right)+v^*.\nabla^* C^* = D_m \nabla^{*2} C^* + \delta * \nabla^{*2} T^* \tag{4}$$

where

$$\rho_f^* = \rho_0[(1-\gamma_c(C^*-C_0)-\gamma_T(T^*-T_0)],$$

with the conditions at the plates are

$$w^*=0, C^* = C_0 - \frac{1}{2}(\pm\Delta C) - \beta_{c_x} x^* - \beta_{c_y} y^* T^* = T_0 - \frac{1}{2}(\pm\Delta T) - \beta_{T_x} x^* - \beta_{T_y} y^*, \quad z^* = \pm\frac{1}{2}H \tag{5}$$

The dimensional quantities are represented by the superscript '*'. Here, the Darcy velocity $v^* = (u^*, v^*, w^*)$, pressure P^*, stress relaxation characteristic time constant γ^*, concentration C^*, and temperature T^* are all present. The subscripts f and m denotes the fluid and porous media, respectively. Additionally, ρ, c and μ stands for the density, specific heat, and viscosity, respectively.

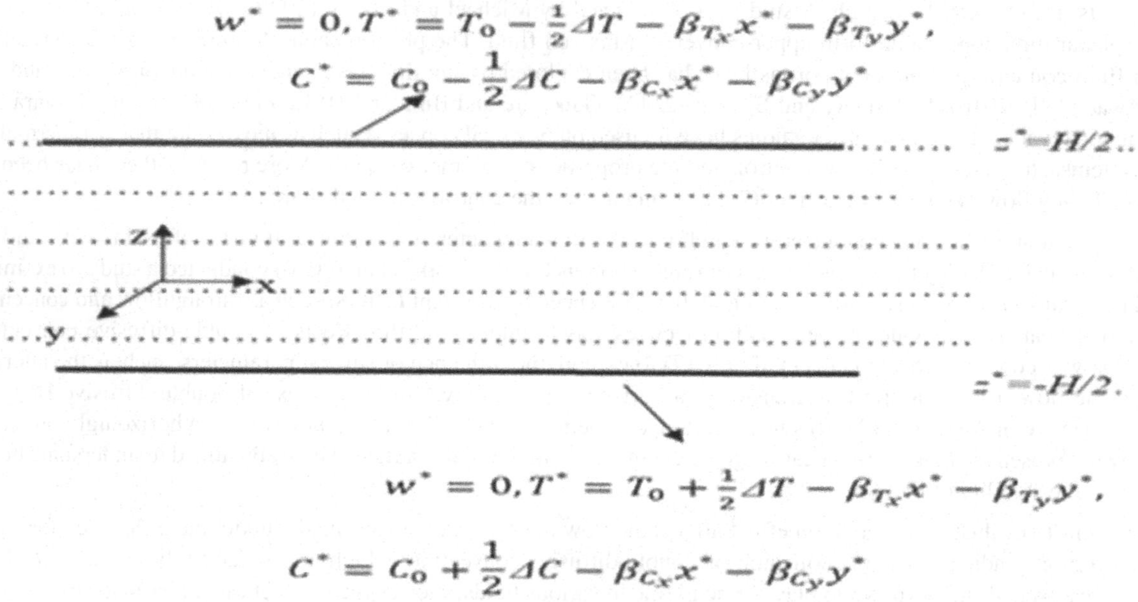

Fig. 34.1 The Physical System

We obtain the fundamental state solution of the physical system of equations, it is necessary to first solve them in non-dimensional form if they are initially in dimensional form. This involves non-dimensionalizing the variables used in the equations, such as D_m and k_m for concentration and thermal diffusivities, γ_C and γ_T for solutal and thermal expansion coefficients, and φ and k for porosity and permeability of the porous medium. By proposing non-dimensional variables for these parameters, accurate solutions can be obtained for the physical system equations.

$$(x,y,z)=\left(\frac{x^*}{H},\frac{y^*}{H},\frac{z^*}{H}\right), t=\frac{\alpha_m t^*}{AH^2}, v=\frac{Hv^*}{\alpha_m}, P=\frac{K(P^*+\rho_0 g z^*)}{\mu\alpha_m},$$

$$T=\frac{R_z(T^*-T_0)}{\Delta T}, C=\frac{S_z(C^*-C_0)}{\Delta C}, \gamma=\frac{\alpha_m \gamma^*}{AH^2}$$

$$\text{here } \alpha_m=\frac{k_m}{(\rho C_P)_f}, A=\frac{(\rho C)_m}{(\rho C_P)_f}, R_z=\frac{\rho_0 g \gamma_T K H \Delta T}{\mu\alpha_m}, S_z=\frac{\rho_0 g \gamma_C K H \Delta C}{\mu D_m}..$$

Along the z direction, solutal and thermal Rayleigh numbers are represented as S_z and R_z, respectively. Additionally, the horizontal temperature and concentration Rayleigh numbers are represented as

$$R_x=\frac{\rho_0 g \gamma_T K H^2 \beta_{T_x}}{\mu\alpha_m}, R_y=\frac{\rho_0 g \gamma_T K H^2 \beta_{T_y}}{\mu\alpha_m}, S_x=\frac{\rho_0 g \gamma_C K H^2 \beta_{C_x}}{\mu D_m}, S_y=\frac{\rho_0 g \gamma_C K H^2 \beta_{C_y}}{\mu D_m}..$$

The equations (1) to (4) reduced to non-dimensional variables form is given below:

$$\nabla.v = 0 \tag{6}$$

$$v = \left(1+\gamma\frac{\partial}{\partial t}\right)\left[-\nabla P + (T + \frac{1}{Le}C)k\right] \tag{7}$$

$$\frac{\partial T}{\partial t} + v.\nabla T = \nabla^2 T + QR_z \tag{8}$$

$$(\varphi/A)\frac{\partial C}{\partial t} + v.\nabla C = \frac{1}{Le}\nabla^2 C + \delta\nabla^2 T \tag{9}$$

And at plates the conditions become

$$w = 0, T = -\tfrac{1}{2}(\pm R_z) - R_x x - R_y y, C = -\tfrac{1}{2}(\pm S_z) - S_x x - S_y y \text{ at } z = \pm\tfrac{1}{2}. \tag{10}$$

The inclusion of concentration and temperature scaling has resulted in the appearance of S_z and R_z in the given bottom and top plate conditions. The temperature-to-solvent diffusivity ratio is known as the Lewis number, and it is represented by the following formula $Le = \frac{\alpha_m}{D_m}$. This demonstrates how the singular Rayleigh number may be written as $S_z = NLeR_z$, where N is the buoyancy ratio variable and is provided by $\frac{\gamma_C \Delta C}{\gamma_T \Delta T}$. $\delta = \frac{D^* C_0 \Delta C}{D \Delta T} D_m \Delta C$ which is the parameter describing the Soret effect, D^* is the Thermo-diffusion coefficient C_os the dimensional solute concentration at the cavity centre, ΔT is the Temperature difference, D_m is the mass diffusivity, ΔC is the concentration solute difference.

Steady-State Solution

The physical system equations (6) – (10) give a steady-state solution is given below.

$$T_s = \tilde{T}(z) - R_x x - R_y y, C_s = \tilde{C}(z) - S_x x - S_y y, \tag{11.1}$$

$$u_s = u(z), v_s = v(z), w_s = 0, P_s = P(x, y, z) \tag{11.2}$$

where
$$Du = R_x + \frac{S_x}{Le}, Dv = R_y + \frac{S_y}{Le} \tag{12.1}$$

$$D^2\tilde{T} = -uR_x - vR_y, \frac{1}{Le}D^2\tilde{C} = -uS_x - vS_y. \tag{12.2}$$

Where the differentiation with respect to z is denoted by D. Due to the absence of any resulting flow along the x and y directions are 0, then we have $\langle u\rangle = 0$, $\langle v\rangle = 0$, where $\langle\rangle$ represent the integration with respect to the variable z between the limits $z = -\frac{1}{2}$ to $\frac{1}{2}$. For the flow rate, concentration, and temperature in the medium, we arrive at the following solution.

$$u = (R_x + \frac{S_x}{Le})z, v = (R_y + \frac{S_y}{Le})z \tag{13.1}$$

$$\tilde{T} = -R_z z + \frac{1}{24}\lambda_1(z - 4z^3), \tilde{C} = -S_z z + \frac{1}{24}\lambda_2(z - 4z^3), \tag{13.2}$$

where λ_1 and λ_2 are given by

$$\lambda_1 = R_x^2 + R_y^2 + \frac{R_x S_x + R_y S_y}{Le} \qquad \lambda_2 = S_x^2 + S_y^2 + Le(R_x S_x + R_y S_y) \tag{14}$$

The fluid flow described by Eq. (13.1) is called the Hadley flow. Subsequently, we define the solutal and thermal Rayleigh number vectors by $\mathbf{S} = (S_x, S_y, S_z)$ and $\mathbf{R} = (R_x, R_y, R_z)$ respectively.

The Linear Stability Analysis

Take into consideration the perturbations of the form $T = T_s + \theta'$, $V = V_s + V'$, $P = P_s + p'$ and $C = C_s + c'$. The disturbance magnitude is represented by the prime symbol, and the subscript indicates the steady-state solution. The linearized perturbation equations taken neglected the nonlinear terms.

$$\nabla.V' = 0, \tag{15}$$

$$V' = \left(1+\gamma\frac{\partial}{\partial t}\right)\left[-\nabla p' + (\theta' + \frac{1}{Le}C')k\right], \tag{16}$$

$$\frac{\partial\theta'}{\partial t} + U\frac{\partial\theta'}{\partial x} + V\frac{\partial\theta'}{\partial y} - R_x u' - R_y v' + (D\tilde{T})w' = \nabla^2\theta', \tag{17}$$

$$(\varphi/A)\frac{\partial c'}{\partial t} + U\frac{\partial c'}{\partial x} + V\frac{\partial c'}{\partial y} - S_x u' - S_y v' + (D\tilde{C})w' = \frac{1}{Le}\nabla^2 c' + \delta\nabla^2\theta' \tag{18}$$

Choose the normal modes of the form.

$$[V', c', \theta', p'] = [V(z), c(z), \theta(z), p(z)]\exp\{(kx + ly - \sigma t)\} \tag{19}$$

and substituting in the above-linearized equations (15) - (18) and further removing p, u and v

$$(D^2 - \alpha^2)w + \alpha^2(1 - i\gamma\sigma)\left[\theta + \frac{1}{Le}c\right] = 0, \tag{20}$$

$$(D^2 - \alpha^2 + i\sigma - ikU - ilV)\theta + i\frac{2}{\alpha^2}(kR_x + lR_y)Dw - (D\tilde{T})w = 0, \tag{21}$$

$$\left(\frac{1}{Le}[D^2 - \alpha^2] + i(\varphi/A)\sigma - ikU - ilV\right)c + \delta[D^2 - \alpha^2) + i\frac{1}{\alpha^2}(lS_y + kS_x)Dw - (D\tilde{C})w = 0. \tag{22}$$

Where $D\tilde{T} = -R_z + \frac{1}{24}\lambda_1(1 - 12z^2)$ and $D\tilde{C} = -S_z + \frac{1}{24}\lambda_2(1 - 12z^2)$. λ_1 and λ_2 are given in equation (14). The boundary conditions are

$$W = 0, \theta = 0, C = 0 \text{ at } z = \pm\frac{1}{2}. \tag{23}$$

The impermeable nature of the plates and the absence of any temperature or concentration changes at the plates are represented by these conditions in equation (23). The aforementioned equations (20) through (23) represent an eigenvalue issue with R_z and Le, φ, A, R_x, R_y, S_x, S_y, S_z, σ, γ, k and l as parameters. The minimum value of R_z is the critical value because σ, k and l are different (with σ taking certain determined values). The total wave number $\alpha = \sqrt{k^2 + l^2}$ in the earlier example. In essence, wavenumber is to wavelength what frequency is to the time period. The frequency of a wave is the number of oscillations that occur at a point in unit time. The wave-number specifies the number of complete waves present at a given moment in a unit length. Their product is equal to the reciprocal of the wave speed. Wave number is used to determine the spatial frequency of a wave. In quantum mechanics, a wave number is used to calculate the canonical momentum by multiplying it to reduce Planck's constant. In wave optics, wave number is used to define wave scattering.

3. Results and Discussions

The study investigates the convection of viscoelastic fluid in a porous layer with the influence of heat source and the Soret effect, where the thermal and solutal fields exhibit linear variation along the walls. The study employs the Shooting Runga-Kutta method to compute the Rayleigh number R_z as the eigenvalue. The analysis primarily focuses on the linear assessment of thermal and solutal convection on viscoelastic fluid. The study's outcomes are compared with Neild's (1994) findings, which are presented in Table 34.1.

Table 34.1 R_x versus R_z for varying the Q values.

R_x	R_z at Q = 0		R_z at Q = 1
	Present	**Neild**	
0	39.4784	39.48	39.2368
10	42.0076	42.01	.41.7303
20	49.5486	49.56	49.1462
30	61.9566	62.28	61.2761
40	78.9671	79.24	77.7023
50	100.1163	100.9	97.5348
60	124.4724	126.4	117.8099

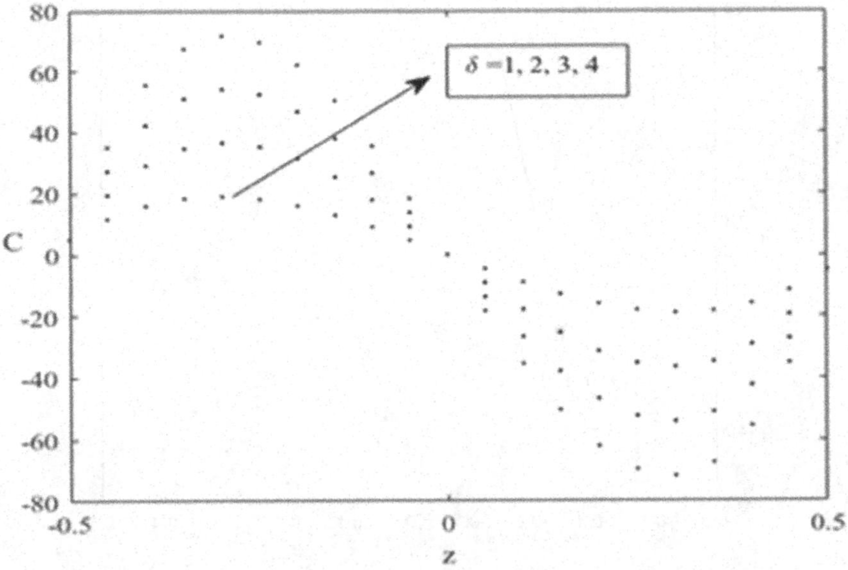

Fig. 34.2 Change of C with z for different at Q = 0 and $Rx = Ry = 10$

The Concentration profiles with respect to the Soret effect are shown in Fig. 34.2 at $Q = 0$ and $Rx = Ry = 10$. As increasing the Soret parameter the concentration is also increasing at the bottom plate and then decreases at the top plate. This means that over-stability is not affected by the presence of Soret effects in double-diffusive flow. The vertical boundary layer's thickness has an impact on the Soret parameter. When the heat and liquid buoyancy forces are interacting, the boundary thickness increases because of the Soret parameter.

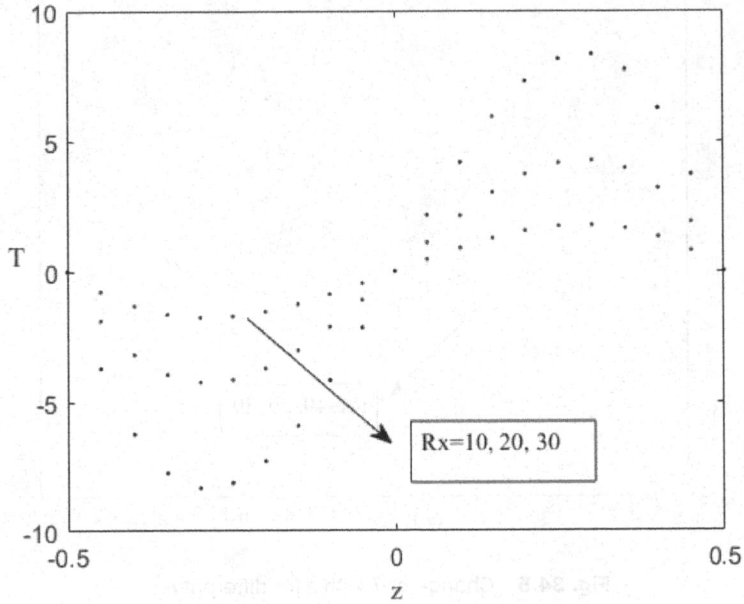

Fig. 34.3 Change of *T* with *z* for different *Rx* at Q = 0

The thermal profiles are asymmetric irrespective of *Rx* values seen in Fig. 34.3. Enhance in the horizontal thermal gradient (*Rx*) results causes to improve the temperature in the top the of horizontal porous layer. The thermal Rayleigh number (*Rz*), It was calculated for various arrangements of the flow's controlling factors. The concentration profiles for enhancement in the heat source results cause to improve in the temperature in the horizontal porous layer irrespective of gravity and mass flow shown in Fig. 34.4.

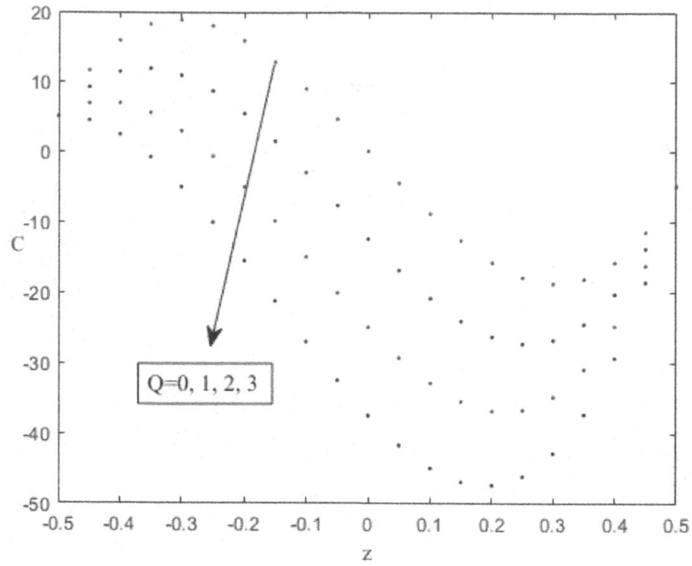

Fig. 34.4 Change of C with z for different Q and Rx = Ry = 10

By increasing the values of *Q*, the concentration profiles are decreased at the upper portion of the porous layer. As a result, the flow medium becomes unstable due to the heat source. As *Q* increases the Concentration profiles also increased near the lower plate and after the middle at the upper plate concentration is decreases in Fig. 34.4. It has been noted that when the heat source increases, the flow medium becomes significantly more unstable because of the increase in the world's temperature.

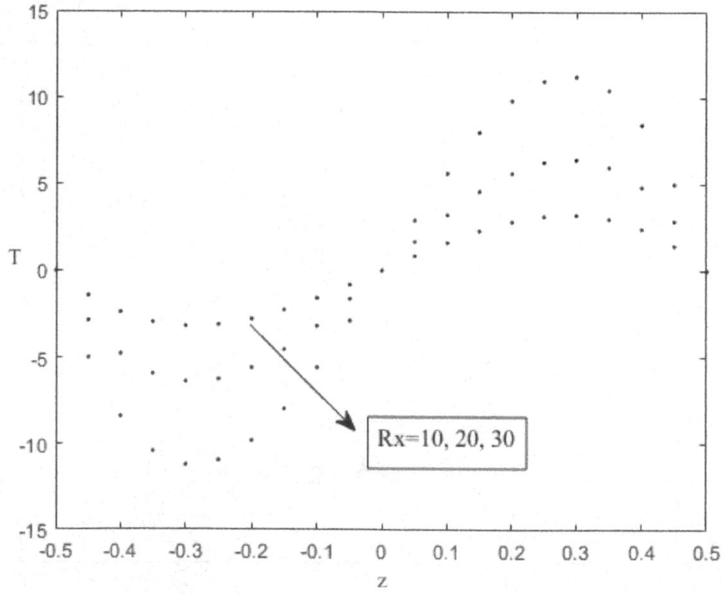

Fig. 34.5 Change of *T* with *z* for different *Rx*

Fig. 34.5 show that the temperature profile follows a similar pattern. As *Rx* increases, there is a non-linear increase in the thermal convection process, and the critical value of Rayleigh number (*Rz*) for oscillatory modes changes due to the presence of the stress relaxation parameter, exhibiting an oscillatory behaviour.

The velocity profiles with respect to concentration gradients (*Sx* and *Sy*) are shown in Fig. 34.6 and Fig. 34.7. As increasing the *Sx* velocity profiles also increases at the bottom to the top plate in Fig. 34.6. When the forces of thermal and solutal buoyancy are in opposition to one another, the boundary thickness decreased with the Soret parameter.

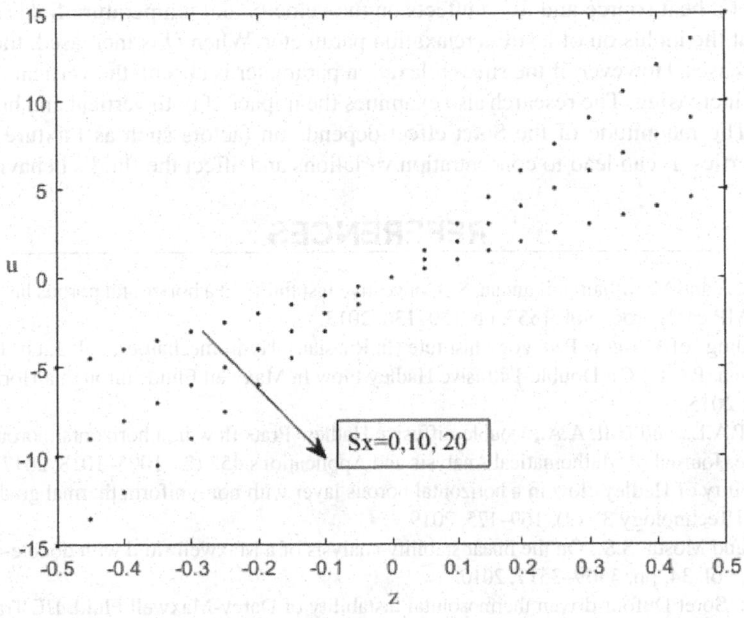

Fig. 34.6 Change of *u* with *z* for different *Sx*

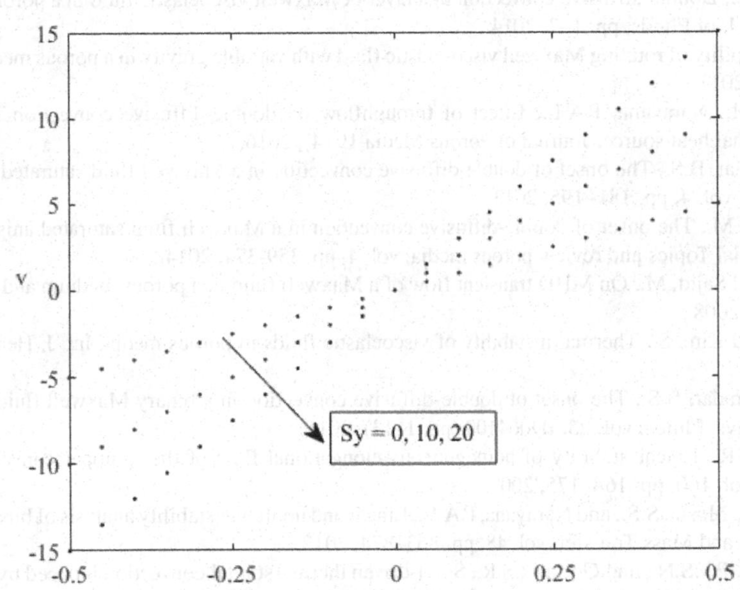

Fig. 34.7 Change of *v* with *z* for different *Sy*

The symmetric nature is also distributed. The velocity profiles move upward the porous layer in Fig. 34.6 and Fig. 34.7. The centre of the porous layer's velocity values improved as the concentration (*C*) increased seen in Fig. 34.6 and Fig. 34.7.

4. Conclusion

The focus of this study is on the horizontal direction of a layer of porous materials and how the convection of Maxwell fluid is affected by linear variations in concentration and temperature fields. The shooting Runge-Kutta method is utilized to analyse an eigenvalue problem involving the thermal Rayleigh number (*Rz*), which serves as the vertical component of the problem's eigenvalue. The study examines the temperature, concentration, and velocity profiles under different physical conditions of the flow field. The effects of a heat source and Soret effects on the velocity and temperature fields are investigated under two conditions, with and without the inclusion of a stress relaxation parameter. When Q is increased, the temperature at the centre of the porous body also increases. However, if the stress relaxation parameter is absent, the vertical thermal Rayleigh number's response is not consistently increasing. The research also examines the impact of both vertical and horizontal singular Rayleigh numbers in this situation. The magnitude of the Soret effect depends on factors such as mixture composition, temperature gradient, and material properties. It can lead to concentration variations and affect the fluid's behavior.

REFERENCES

1. Akbar, S.K., Nargund, A.L., and Maruthamanikandan, S., Convective instability in a horizontal porous layer saturated with a chemically reacting Maxwell fluid, AIP conf. proc., vol. 1557, pp. 130–136, 2013.
2. Alishayev, M.G., Proceedings of Moscow Pedagogy Institute (In Russian), Hydromechanics, vol. 3, pp. 166–174, 1974.
3. Anjanna, M., and Narayana, P.A.L., On Double-Diffusive Hadley Flow in Maxwell Fluids through a Horizontal Porous Layer, Journal of Porous Media 18 (12), 2015.
4. Anjanna, M., Narayana, P.A.L., and Hill, A.A., Double-diffusive Hadley–Prats flow in a horizontal porous layer with a concentration-based internal heat source, Journal of Mathematical Analysis and Applications 452 (2), 1005–1018, 2017.
5. Anjanna, M., On the stability of Hadley-flow in a horizontal porous layer with non-uniform thermal gradient and internal heat source, Microgravity Science and Technology 31 (2), 169–175, 2019.
6. Awed, F.G., Sibanda, P., and Mosta, S.S., On the linear stability analysis of a Maxwell fluid with double-diffusive convection, Applied Mathematical Modelling, vol. 34, pp. 3509–3517, 2010.
7. Bishnoi, J. and Goyal, N., Soret Dufour driven thermosolutal instability of Darcy-Maxwell Fluid, IJE Transactions A: Basics, vol. 25, pp. 367–377, 2012.
8. Chand, R., and Rana, G.C., Double diffusive convection in a layer of Maxwell viscoelastic fluid in a porous medium in the presence of Soret and Dufour effects, J. of Fluids, pp. 1–7, 2014.
9. Chand, R., Thermal instability of rotating Maxwell visco-elastic fluid with variable gravity in a porous medium, Journal of Indian math. Soc., vol. 80, pp. 23–31, 2013.
10. Deepika, N., Anjanna, M., Narayana, P.A.L., Effect of throughflow on double-diffusive convection in a porous medium with a concentration based internal heat source, Journal of Porous Media 19 (4), 2016.
11. Gaikward, S.N. and Biradar, B.S., The onset of double-diffusive convection in a Maxwell fluid saturated porous layer, Special Topics and review porous media, vol. 4, pp. 181–195, 2013
12. Gaikward, S.N., Dhanraj, M., The onset of double-diffusive convection in a Maxwell fluid saturated anisotropic porous layer with an internal heat source, Special Topics and review porous media, vol. 4, pp. 359-374, 2014.
13. Hayat, T., Fetecau, C., and Sajid, M., On MHD transient flow of a Maxwell fluid in a porous medium and rotating frame, Phys. Lett. A vol. 372, pp. 1639–1644, 2008.
14. Kim, M.C., Lee, S.B., and Kim, S., Thermal instability of viscoelastic fluids in porous media, Int. J. Heat Mass Transfer, vol. 46, pp. 5065–5072, 2003.
15. Malashetty, M. S. and Biradar, B.S., The onset of double-diffusive convection in a binary Maxwell fluid saturated porous layer with cross-diffusion effects, Phys. Fluids, vol. 23, ID064109 (pp. 1–13), 2011.
16. Michael, R., and Yuriko, R., Linear stability of homogeneous elongational flow of the upper convected Maxwell fluid, J. Non-Newtonian Fluid Mech, vol. 160, pp. 168–175, 2009.
17. Narayana, M., Sibanda, P., Mosta, S.S., and Narayana, P.A.L., Linear and nonlinear stability analysis of binary Maxwell fluid convection in a porous medium, Heat and Mass Transfer, vol. 48, pp. 863–874, 2012.
18. Narayana, P.A.L., Murthy, P.V.S.N., and Gorla, R.S.R., Soret-driven thermo-solutal convection induced by inclined thermal and solutal gradients in a shallow horizontal layer of a porous medium, J. Fluid Mech, vol. 612, pp. 1–19, 2008.
19. Nield, D.A. and Bejan, A., Convection in Porous Media, fourth ed., Springer-Verlag, New York, 2013.
20. Sharma, R. C. and Kumari, V., Thermosolutal convection in a Maxwell viscoelastic fluid in a porous medium. Czech. J. Phys., vol. 43, pp. 31–42, 1993.
21. Tan, W.C. and Matsuoka, T., Stability analysis of a Maxwell fluid in a porous medium heated from below, Phys. Lett. A., vol. 360, pp. 454–460, 2007.
22. Wang, S. and Tan, W.C., Stability analysis of Soret-driven double-diffusive convection of Maxwell fluid in a porous medium, Int. J. of Heat and Fluid Flow, vol. 32, pp. 88–94, 2011.
23. Zhao, M., Zhang, Q., and wang, S., Linear and nonlinear stability analysis of double-diffusive convection in a Maxwell fluid porous layer with internal heat source, J. Appl Math. Id 489279 pp. 1–12, 2014.
24. Neild, D., Convection in a porous medium with inclined temperature gradient: additional results. Int. J. Heat Mass Transfer, 37, pp. 3021–3025, 1994

Note: All the figures and table in this chapter were made by the authors.

Impending Inquisitions in Humanities and Sciences (ICIIHS-2022) – Dr. Mohan Varkolu et al. (eds)
© 2024 Taylor & Francis Group, London, ISBN 978-1-032-78829-6

Analysis of Fluid Flow in a Curved Channel Filled with a Brinkman-Darcy Porous Medium

35

R. Vijaya Sree[1]
Department of Mathematics, GITAM Deemed to be University, Hyderabad, India
Department of Mathematics, ACE Engineering College, Hyderabad, India

V. K. Narla*
Department of Mathematics, GITAM Deemed to be University, Hyderabad, India

ABSTRACT A laminar flow in a curved isotropic porous channel is studied between two impermeable walls. The flow inside the porous material is considered to be governed by a Brinkman-extended Darcy model. The flow is generated by an azimuthal pressure gradient that is fixed, and both the Brinkman and Darcy drag components have been taken into consideration. An analytical solution is obtained for the governing equations of the flow. The findings obtained indicate that the channel curvature and porous drag components included in the momentum equation have a considerable impact on the flow dynamics.

KEYWORDS Curved channel, Porous medium; Brinkman number, Isotropic permeability

1. Introduction

Many researchers are interested in investigating fluid flow through porous media, as well as describing and predicting flow features. Processes including cooling, drying, filtration and separation can all be considered as examples of engineering applications. Understanding the transport process in biological sciences necessitates a study of biological fluid movement. It may be encountered in biomedical operations involving reticulated foams that require blood flow to pass through the foam structure, such as during balloon angioplasty or stent implantation to lower the risk of stroke. Exploration and production of Hydrocarbon, geothermal power plants, groundwater flow, and glaciology transfer are just a few practical applications of geological porous media. There is much literature on fluid flow in a straight channel filled with porous medium but less on curved channels.

A theoretical analysis is done on the Dean problem in heterogeneous porous media of monotonic permeability variation vertically for a fully developed flow by S. Govender (2006). The streamlines are stated to be bunched closer to the inner wall for small curvature ratios and centrally for moderate and large curvature ratios. Using the Brinkman extended Darcy law, Avramenko et al. (2007) investigated a laminar flow in a curved channel filled with porous medium and having a rectangular cross-section. They incorporated Darcy and Brinkman drag elements in the momentum equation, and using the Fourier series, they developed an analytical solution for the velocity field. The velocity distribution was demonstrated to be dependent on the channel shape and the Darcy number. A helical pipe filled with fluid-saturated porous media was considered for the study by Cheng and Kuznetsov (2005), in which the flow was considered laminar. For analysing the momentum equation, the Brinkman and Forchheimer extensions of the Darcy law are considered along with flow inertia.

*Corresponding author: vnarla@gitam.edu
[1]rv.jayasree@gmail.com

DOI: 10.1201/9781003489436-35

In predicting a secondary flow, it was demonstrated that consideration of flow inertia is essential in a helical pipe. The impacts of the Darcy number, the Forchheimer coefficient, along with the curvature and torsion of the helical pipe are quantitatively explored on the axial flow velocity and secondary flow. Okechi and Asghar (2019) investigated the combined geometrical effects of curvature and grooves on viscous flows in a grooved curved channel in their work. The flow rate of a viscous flow in a grooved curved straight channel is affected by channel curvature, which differs from that of a viscous flow in a corrugated straight channel. Okechi and Asghar (2020) conducted a theoretical investigation on the viscous flow in a grooved curved channel filled with a porous medium. In order to describe the flow, a Darcy–Brinkman model is employed when the curved coordinates are taken into account. For corrugation amplitude, the effects of permeability of the medium, the channel curvature and the wall corrugations are examined using the approach of boundary perturbation. Kuznetsov and Avramenko (2007) examined laminar fluid flow in a channel created by two concentric cylindrical surfaces which is filled with porous medium and found the exact solution for the velocity profile. The solution to the momentum equation in cylindrical dimensions is provided using Bessel functions. The basic flow's stability is investigated using linear stability analysis. This demonstrates that the Dean number is affected by permeability of the medium and also the width of the channel. These findings might be helpful in the construction of bioseparators since Dean vortices provide a mechanism for better mass transfer. Narla et al. (2015) studied the flow of a Jeffrey nanofluid, undergoing peristalsis in a two-dimensional curved channel. Analysis of velocity distribution, temperature and concentration of nanoparticles is done for different parameters which govern the flow.

This paper investigates laminar flow in a curved channel saturated with porous material and confined by two impermeable walls. The fluid flow within the porous material is considered to be governed by a Brinkman extended Darcy model, which accounts for the presence of the Brinkman drag term. A steady azimuthal pressure gradient generates flow in the channel. Fluid flow is examined in relation to channel curvature and Darcy number. It should be noted that the channel curvature, viscosity ratio and porous drag factors substantially impact the flow dynamics.

2. Mathematical Formulation

Figure 35.1 shows a laminar flow in a curved channel saturated with porous media and confined by two impermeable walls. Consider a deformable curved channel with a diameter of $2a$ and a constant centre-line radius R.

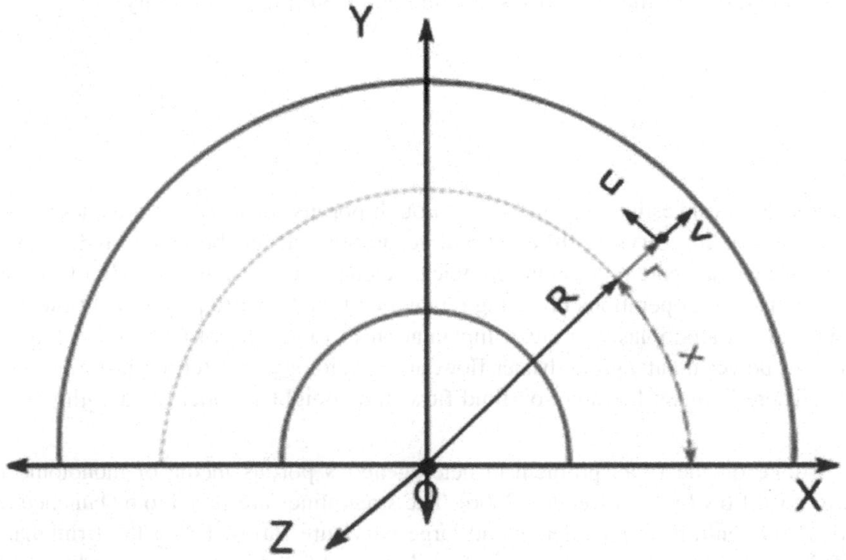

Fig. 35.1 Schematic view of a two-dimensional curved channel

The curvilinear coordinates (r, x, z) are defined with respect to rectangular Cartesian coordinates (X, Y, Z). The curved channel regime's geometry is represented in Fig. 35.1. The coordinate transformations' mathematical expressions are as follows:

$$X = (R+r)\cos\left(\frac{x}{R}\right), \quad Y = (R+r)\sin\left(\frac{x}{R}\right), \quad Z = z \tag{1}$$

The coordinate x is located along the curved channel's centre line, whereas r is normal to the x coordinate which is measured from the centre line. The flow is two-dimensional since there is no component in the z-direction. A steady azimuthal pressure gradient generates flow in the channel. The modified Brinkman extended Darcy model's corresponding mass conservation and momentum balance equations are given as:

$$\frac{R}{r+R}\frac{\partial u}{\partial x}+\frac{\partial v}{\partial r}+\frac{v}{r+R}=0 \tag{2}$$

$$\frac{R}{\rho(r+R)}\frac{\partial p}{\partial x}=\nu_{eff}\left[\left(\frac{R}{r+R}\right)^2\frac{\partial^2 u}{\partial x^2}+\frac{1}{r+R}\frac{\partial u}{\partial r}+\frac{\partial^2 u}{\partial r^2}-\frac{u}{(r+R)^2}+\frac{2R}{(r+R)^2}\frac{\partial v}{\partial x}\right]-\frac{\nu}{K}u-\frac{c_F}{\sqrt{K}}u^2, \tag{3}$$

$$\frac{1}{\rho}\frac{\partial p}{\partial r}=\nu_{eff}\left[\left(\frac{R}{r+R}\right)^2\frac{\partial^2 v}{\partial x^2}+\frac{1}{r+R}\frac{\partial v}{\partial r}+\frac{\partial^2 v}{\partial r^2}-\frac{v}{(r+R)^2}-\frac{2R}{(r+R)^2}\frac{\partial u}{\partial x}\right]-\frac{\nu}{K}v-\frac{c_F}{\sqrt{K}}v^2, \tag{4}$$

where x, r, and z are the cylindrical coordinates; u, v are the azimuthal and radial velocity components; p denotes the pressure, ν the kinematic viscosity of the fluid, ν_{eff} is the effective viscosity of the porous medium, ρ denotes the density, and K denotes the porous medium's permeability. The following are the supporting boundary conditions

$$u[\pm H]=0. \tag{5}$$

Let $\mathbf{q}=(u, v)$ be the velocity vector. $\mathbf{q}=(u(r), 0)$ is obtained by assuming a fully developed flow in the x-direction. The pressure gradient term $\partial p/\partial x$ becomes constant under this assumption, and we label it $-G$. As a result, the conservation of momentum balance is reduced to

$$\frac{R}{\rho(r+R)}G+\nu_{eff}\left[\frac{d^2u}{dr^2}+\frac{1}{r+R}\frac{du}{dr}-\frac{u}{(r+R)^2}\right]-\frac{\nu}{K}u-\frac{c_F}{\sqrt{K}}u^2=0(6) \tag{6}$$

The following non-dimensional variables are introduced:

$$r'=\frac{r}{H},\ u'=\frac{\mu u}{GH^2},\ M=\frac{\nu_{eff}}{\nu},\ \kappa=\frac{R}{H},\ Da=\frac{K}{H^2},\ F=\frac{c_F GH^3}{\rho\nu^2}, \tag{7}$$

where M represents the viscosity ratio, Da represents the Darcy number, F represents the Forchheimer number, and κ represents the curvature parameter. The corresponding nondimensionalization of equation (6) is given by (after dropping primes for simplicity)

$$M\left[\frac{d^2u}{dr^2}+\frac{1}{r+\kappa}\frac{du}{dr}-\frac{u}{(r+\kappa)^2}\right]-\frac{u}{Da}-\frac{F}{\sqrt{Da}}u^2=-\frac{\kappa}{r+\kappa}, \tag{8}$$

After nondimensionalization, the corresponding boundary conditions are:

$$u[\pm 1]=0. \tag{9}$$

2. Mathematical Solution

An analytical solution is obtained using the method of undetermined coefficients for the equation (8) when the Forchheimer number $F=0$ using the boundary conditions (9)

$$u=C_1 I_1\left[\frac{\kappa+r}{\sqrt{MDa}}\right]+C_2 K_1\left[\frac{\kappa+r}{\sqrt{MDa}}\right]+\frac{\kappa Da}{\kappa+r}, \tag{10}$$

where,

$$C_1 = \frac{Da\kappa\left((1-\kappa)K_1\left[\frac{\kappa-1}{\sqrt{DaM}}\right]+(\kappa+1)K_1\left[\frac{\kappa+1}{\sqrt{DaM}}\right]\right)}{(\kappa^2-1)\left(I_1\left[\frac{\kappa+1}{\sqrt{DaM}}\right]K_1\left[\frac{\kappa-1}{\sqrt{DaM}}\right]+I_1\left[\frac{1-\kappa}{\sqrt{DaM}}\right]K_1\left[\frac{\kappa+1}{\sqrt{DaM}}\right]\right)}$$

$$C_2 = \frac{Da\kappa\left((\kappa+1)I_1\left[\frac{\kappa+1}{\sqrt{DaM}}\right]+(\kappa-1)I_1\left[\frac{1-\kappa}{\sqrt{DaM}}\right]\right)}{(\kappa^2-1)\left(I_1\left[\frac{\kappa-1}{\sqrt{DaM}}\right]K_1\left[\frac{\kappa+1}{\sqrt{DaM}}\right]-I_1\left[\frac{\kappa+1}{\sqrt{DaM}}\right]K_1\left[\frac{\kappa-1}{\sqrt{DaM}}\right]\right)}$$

First order modified Bessel functions I_1 and K_2 are used here. In the curved channel, the mean velocity is obtained as

$$U = \sqrt{DaM}[-C_1 I_0\left[\frac{1-\kappa}{\sqrt{DaM}}\right]+C_1 I_0\left[\frac{\kappa+1}{\sqrt{DaM}}\right]+C_2 K_0\left[\frac{\kappa-1}{\sqrt{DaM}}\right] -C_2 K_0\left[\frac{\kappa+1}{\sqrt{DaM}}\right]]+Da\kappa\log\left[\frac{\kappa+1}{\kappa-1}\right]. \quad (11)$$

3. Results and Discussions

The velocity profile for various curvature parameters κ is depicted in Fig. 35.2. We see that for a fixed Da and M, as the curvature parameter κ increases, the velocity profile changes to a symmetric parabolic profile without any change at the centre line. The maximum velocity magnitude is observed near the inner wall of the curved channel. When going from a curved to a straight channel, an asymmetrical velocity becomes symmetrical. The velocity profile is depicted in Fig. 35.3 for various M. With the increase in M values for fixed Da and κ, the velocity is greatest at the lower wall, while the decrease of the velocity profile is less near the centre-line and shows substantial deviations towards the lower wall with an increasing M. The velocity profile for various Da levels is depicted in Fig. 35.4. We can observe that for fixed M and κ, the velocity profile is highest near the lower wall for low Da values and virtually a parabolic profile near the centre for higher Da values. The velocity profile for varied Darcy numbers for fixed M and κ in a straight channel is shown in Fig. 35.5. From this, we can observe that as the fluid moves from a curved to a straight channel, velocity is maximum near the centerline.

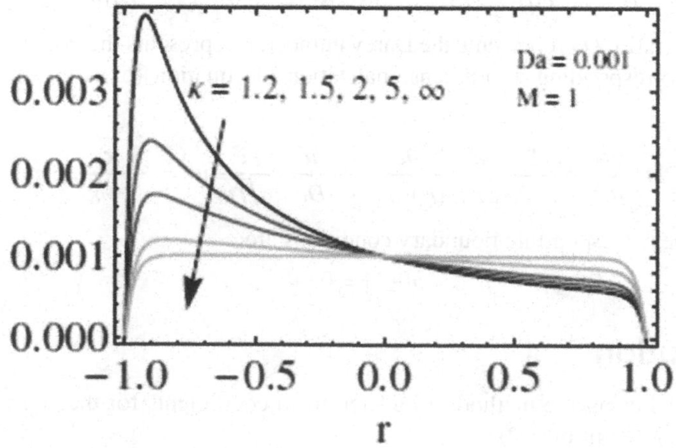

Fig. 35.2 Azimuthal velocity profiles for different curvature parameters κ.

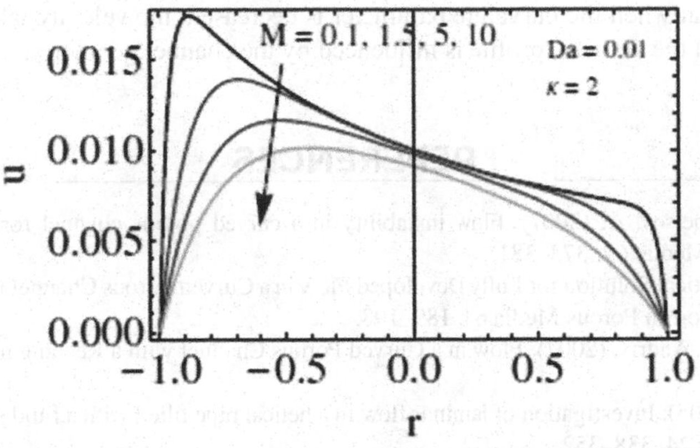

Fig. 35.3 Azimuthal velocity profiles for different viscosity ratios *M*.

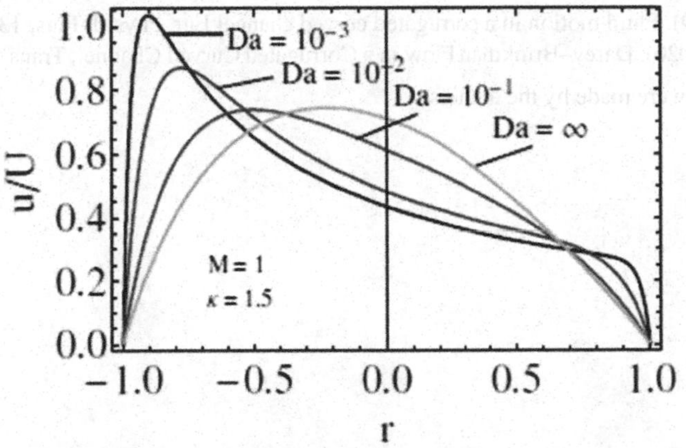

Fig. 35.4 Azimuthal velocity profiles for different Darcy numbers *Da* in a curved channel.

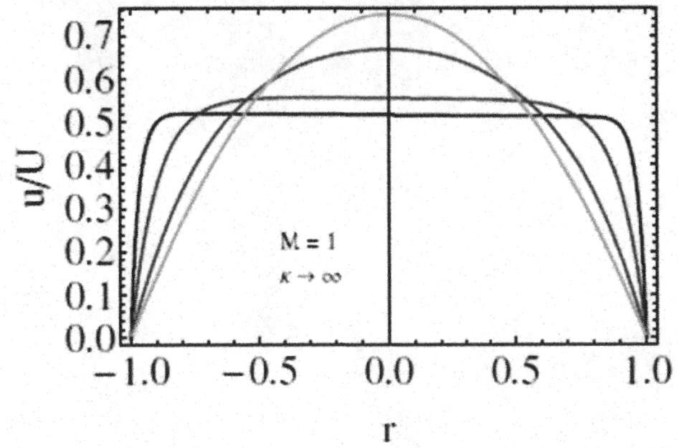

Fig. 35.5 Azimuthal velocity profiles for different Darcy numbers *Da* in a straight channel.

4. Conclusion

In a curved channel confined by two impermeable walls, we investigated laminar flow in a fluid-saturated porous material. In the momentum equation, we employed a Brinkman extended Darcy model. An analytical solution is obtained for a laminar velocity profile in a porous curved channel. Modified Bessel's functions are used to provide the solution to the momentum problem. It has been observed that on decreasing the Darcy number *Da*, the fullness of the velocity profile increases along the radial axis. It is also observed that when the curvature parameter is decreased, the velocity takes the maximum value near the lower wall. We can conclude that the velocity profile is influenced by the channel geometry, viscosity ratio and also the Darcy number.

REFERENCES

1. Avramenko, Andriy and Kuznetsov, A. (2007). Flow instability in a curved porous channel formed by two concentric cylindrical surfaces, Transport in Porous Media. 69: 373–381.
2. S. Govender, (2006). An Analytical Solution for Fully Developed Flow in a Curved Porous Channel for the Particular Case of Monotonic Permeability Variation, Transport in Porous Media 64, 189–198.
3. Kuznetsov, A. and Avramenko, Andriy, (2007). Flow in a Curved Porous Channel with a Rectangular Cross Section, Journal of Porous Media, 11, 241–246.
4. L. Cheng, A.V. Kuznetsov, (2005). Investigation of laminar flow in a helical pipe filled with a fluid saturated porous medium, European Journal of Mechanics B/Fluids 24, 338–352.
5. V.K. Narla, K.M. Prasad and J.V. Ramanamurthy, (2015). Peristaltic Transport of Jeffrey NanoFluid in Curved Channels, Procedia Engineering, 127, 869–876.
5. Okechi, N.F., Asghar, S. (2019). Fluid motion in a corrugated curved channel Eur. Phys. J. Plus, 134.
6. Okechi and Saleem Asghar (2020). Darcy–Brinkman Flow in a Corrugated Curved Channel, Transport in Porous Media, 135, 271–286.

Note: All the figures in this chapter were made by the authors.

Impending Inquisitions in Humanities and Sciences (ICIIHS-2022) – Dr. Mohan Varkolu et al. (eds)

Lie Group Analysis of Biomagnetic Nanofluid Flow Past a Stretching Sheet

36

K. Suresh Babu, V. Sugunamma
Department of Mathematics, Sri Venkateswara University, Tirupati, India

V. K. Narla*
Department of Mathematics, GITAM Deemed to be University, Hyderabad, India

ABSTRACT In this paper, bio-magnetic nanofluid flow with a magnetic dipole has been studied for fluid flow and heat transfer under boundary layer theory. Blood-based ferromagnetic nanofluid is considered in this model. Thermophysical properties are obtained by applying the Tiwari-Das and Maxwell-Garnett nanofluid models. The Lie group similiarity transformation is a powerful mathematical technique to transfor partial differential equations (PDEs) into simpler forms, including ordinary differential equations (ODEs) which are typically easier to solve. The simplified resultant equations are then solved numerically using the shooting method in Mathematica symbolic software. In this study, the graphs illustrate the influence of physical parameters such as ferromagnetic interaction parameter, Hartmann number, stretching parameter, and suction parameter on fluid flow and heat transfer characteristics.

KEYWORDS Stretching surface, Blood-Iron oxide (Fe_3O_4), Ferrofluid, Nanofluid, Magnetic Dipole-effect, Scaling group of transformation

1. Introduction

In modern times nanofluids have revolutionized the idea on heat transfer enhancement. This has been used in diverse technologies including turbulent flows. Nanofluids with a volume fraction of <5% of nanoparticles provide enhanced heat transfer. The magnetic nanofluid, in general is made up of three distinct types of materials (cobalt ferrite $CoFe_2O_4$, magnetite Fe_3O_4, and Mn–Zn ferrite Mn–$ZnFe_2O_4$) and non-magnetic (silver Ag, alumina Al_2O_3 and titania TiO_2) nano particles (Khan et al.[1]). The goal is to maximize thermal characteristics while minimizing concentration by dispersing and suspending nano particles in base fluids uniformly. The critical importance of nanoparticle mobility in nanofluids is studied by Jang and Choi [2] and shown results in their works. One of the most important mechanisms for controlling nano particle thermal activity in nanofluids is the molecular and nanoscale Brownian motion of nano particles. First, to investigate the heat transfer characteristics of micro fluids, Boungiorno [3] came up with several hypotheses. When it comes to increasing thermophysical qualities like thermal conductivity, viscosity and thermal diffusivity, nanofluids are a great tool to use.

Andersson and Valnes [4] original work focused on ferrofluid flow acrossa flat sheet using a magnetic dipole mode, which they found to be effective. The magnetic dipole impact on Maxwell liquid flow was examined by Kumar et al.[5]. Ferromagnetic fluid flow and magnetization were explored by Kumar et al.[6]. Researchers are increasingly focusing on the study of convection heat transfer in fluids. The heat transfer study that considers convective boundary limits is extremely important because of its

*Corresponding author: vnarla@gitam.edu

DOI: 10.1201/9781003489436-36

authority in operations that need high temperatures. Ovler [7] studied Lie group analysis for the algorithmic finding of precise solutions to partial differential equations and Lie has shown to be an effective method. Different fluid flow and heat transfer models have been tackled using the Lie Group analysis approach by several researchers (Ferdows et al.[8] and Rashidi et al.[9]).

A mathematical model is employed in this paper to investigate the flow of a blood-based ferrofluid over a stretched sheet in the presence of a magnetic dipole. It has been observed that the Lie group transformation approach is quite useful in solving highly non-linear differential equations. The computa-tional results have been obtained with the help of the fourth-order Runge-kutta with the shooting method. The detailed discussion has been made for variations of Bio-magnetic fluid velocity and temperature distribution. The current research focuses on the parameters that have the greatest effect on the ferromagnetic fluid's hydrodynamic properties at the interface.

2. Formulation in Mathematical Form

Investigating incompressible boundary layer flow and heat transfer over a stretched sheet in bio magnetic nanofluid. A flat sheet that is parallel to *x*-axis and perpendicular to the *y*-axis is considered as shown in Fig. 36.1. Let us assume the sheet gets stretched at a rate proportionate to the distance $u = cx$. The temperature of the stretched sheet is T_w, whereas that of the fluid far away from the sheet is T_∞, where $T_w < T_\infty$. A magnetic field strong enough to saturate the Bio-magnetic fluid is created when the magnetic dipole is placed underneath the sheet in the location designated *a*. Fig. 36.1 shows the flow arrangement schematically.

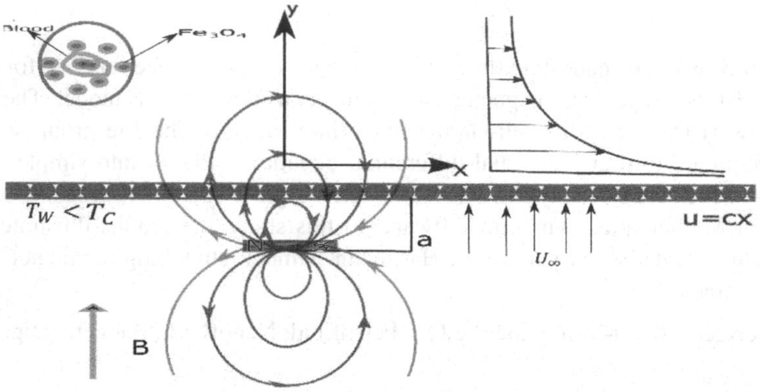

Fig. 36.1 An illustration of the flow arrangement

The following are the flow governing equations that regulate bio-magnetic nanofluid:

$$\frac{\partial u}{\partial x}+\frac{\partial v}{\partial y}=0 \tag{1}$$

$$\rho_{nf}\left(u\frac{\partial u}{\partial x}+v\frac{\partial v}{\partial y}\right)=\mu_{nf}\frac{\partial^2 u}{\partial y^2}+\mu_0 M\frac{\partial H}{\partial x}-\sigma_{nf}\frac{B_0^2}{x}u^2 \tag{2}$$

$$\left(\rho C_p\right)_{nf}\left(u\frac{\partial T}{\partial x}+v\frac{\partial T}{\partial y}\right)=\mu_0 T\frac{\partial M}{\partial T}\left(u\frac{\partial M}{\partial x}+v\frac{\partial M}{\partial y}\right)=k_{nf}\frac{\partial^2 T}{\partial y^2} \tag{3}$$

Associated boundary conditions are:

$$\left.\begin{array}{l} u=cx,\ v=v_w,\ T=T_w\ \ at\ y=0, \\ u\to 0,\ T\to T_\infty\ \ as\ \ y\to\infty. \end{array}\right\} \tag{4}$$

The axial and transient components of velocity are represented by *u* and *v*, the density is represented by ρ_{nf}, the viscous permeability is represented by μ_0, the magnetic strength is represented by B_0, and the electrical conductivity constant is represented by σ_{nf}. Magnetite nano particle concentration, nanofluid temperature and thermal conductivity are all denoted as ϕ, *K* and *T*, respectively.

The effective electrical conductivity σ_{nf} is given by

$$\frac{\sigma_{nf}}{\sigma_f} = \frac{(\sigma_s + 2\sigma_f) + 2(\sigma_s - \sigma_f)\phi}{(\sigma_s + 2\sigma_f) - (\sigma_s - \sigma_f)\phi}.$$

The electric conductivity of magnetic nanoparticles and base fluid is represented by σ_s and σ_{nf}, respectively. Following is a correlation of the nanofluid's effective density (ρ_{nf}) and capacitance $(\rho C_p)_{nf}$

$$\sigma_{nf} = (1-\phi)\sigma_f + \phi\sigma_s,$$
$$(\rho C_p)_{nf} = (1-\phi)(\rho C_p)_f + (\rho C_p)_s.$$

Using the Maxwell-Garnet model, the effective thermal conductivity K_{nf} of nanofluids may be estimated.

$$\frac{k_{nf}}{k_f} = \frac{(k_s + 2k_f) - 2(k_f - k_s)\phi}{(k_s + 2k_f) + 2(k_f - k_s)\phi}.$$

Where K_s and K_f signify the thermal conductivity of base fluid (blood) and nano particles (Iron oxide), respectively. Thermal and electrical properties of nano particle and fluid used in the present work are indicated in the Table 36.1.

Table 36.1 Nanofluid physical parameter values

Physical Properties	Blood(f)	Iron Oxide(Fe_3O_4)
$\rho(kgm^{-3})$	1053	5180
$K(Wm^{-1}K^{-1})$	0.492	9.7
$\sigma(Sm^{-1})$	0.8	0.74×10^6
$C_p(JKg^{-1}K^{-1})$	3594	670
Pr	21	

Magnetic Dipole

Under the influence of a magnetic field, the flow behavior of ferrofluid responds to the magnetic dipole, whose magnetic scalar potential is provided by the equation

$$\phi(x, y) = \frac{\alpha}{2\pi}\left[\frac{x}{x^2 + (y+a)^2}\right] \tag{5}$$

Because the gradient of the magnitude of H is directly proportional to the magnetic body force, Eqs. (2) and (3) we get from

$$H = \left[H_x^2 + H_y^2\right]^{\frac{1}{2}}, \tag{6}$$

$$\frac{\partial H}{\partial x} = \frac{\alpha}{2\pi}\left[\frac{-2x}{(y+a)^4}\right], \tag{7}$$

$$\frac{\partial H}{\partial x} = \frac{\alpha}{2\pi}\left[\frac{-2}{(y+a)^3} + \frac{4x^2}{(y+a)^5}\right]. \tag{8}$$

Lie Group Analysis

The following non-dimensional variables will now be discussed in further detail.

$$\left.\begin{aligned} x' = \frac{x}{L},\ y' = \frac{y}{L},\ u' = \frac{uL}{\vartheta_f},\ v' = \frac{vL}{\vartheta_f}, \\ T = T_\infty - \theta(T_\infty - T_w),\ M = k_f(T - T_\infty). \end{aligned}\right\} \tag{9}$$

Here, M be the magnetization with temperature T and k_f be the pyromagnetic coefficient. Moreover, the steam function ψ is carried out by

$$u = \frac{\partial \psi}{\partial y}, v = -\frac{\partial \psi}{\partial x}. \tag{10}$$

Equation (1) is automatically satisfied and using equations (9) and (10) in (2) and (3), we get the following set of equations

$$\left(\frac{\rho_{nf}}{\rho_f}\right)\left[\frac{\partial \psi}{\partial y'}\frac{\partial^2 \psi}{\partial y' \partial x'} - \frac{\partial \psi}{\partial x'}\frac{\partial^2 \psi}{\partial y'^2}\right] = \left(\frac{\mu_{nf}}{\mu_f}\right)\frac{\partial^3 \psi}{\partial y'^3} - \left(\frac{\sigma_{nf}}{\sigma_f}\right)\frac{H_a^2}{x'}\left(\frac{\partial \psi}{\partial y'}\right)^2 + \frac{2\beta\theta x'}{(y'+d)^4}, \tag{11}$$

$$\frac{(\rho C_p)_{nf}}{(\rho C_p)_f} Pr\left[\frac{\partial \psi}{\partial y'}\frac{\partial \theta}{\partial x'} - \frac{\partial \psi}{\partial x'}\frac{\partial \theta}{\partial y'}\right] = \mu_0 \vartheta_f (\in - \theta)\left[\frac{\partial \psi}{\partial y'}\frac{\partial H}{\partial x'} - \frac{\partial \psi}{\partial x'}\frac{\partial H}{\partial y'}\right] + \frac{k_{nf}}{k_f}\frac{\partial^2 \theta}{\partial y'^2}. \tag{12}$$

With the one-parameter scaling group of transformations, Lie group trans-formations are applied as

$$\Gamma : x^* = x' e^{\varepsilon\alpha_1}, y^* = y' e^{\varepsilon\alpha_2}, \psi^* = \psi e^{\varepsilon\alpha_3}, \theta^* = \theta e^{\varepsilon\alpha_4}, H^* = H e^{\varepsilon\alpha_5}. \tag{13}$$

Here ε the parameter of the group and an arbitrary real number α_i $(i = 1,2,3,...)$ can be used. Equations (9), (10) and boundary condition (11) are invariant under the point transformations (12). Using the transformation (12) in the equation (9), (10) we get

$$\left(\frac{\rho_{nf}}{\rho_f}\right) e^{-2\varepsilon\alpha_3 + 2\varepsilon\alpha_2 + \varepsilon\alpha_1}\left[\frac{\partial \psi^*}{\partial y^*}\frac{\partial^2 \psi^*}{\partial y^* \partial x^*} - \frac{\partial \psi^*}{\partial x^*}\frac{\partial^2 \psi^*}{\partial y^{*2}}\right] - \left(\frac{\mu_{nf}}{\mu_f}\right) e^{-\varepsilon\alpha_3 - 3\varepsilon\alpha_2}\frac{\partial^3 \psi^*}{\partial y^{*3}}$$

$$+\left(\frac{\sigma_{nf}}{\sigma_f}\right)\frac{e^{-2\varepsilon\alpha_3 + 2\varepsilon\alpha_2 + \varepsilon\alpha_1}}{qx^*}\left(\frac{\partial \psi^*}{\partial y^*}\right)^2 - e^{-\varepsilon\alpha_1 + 4\varepsilon\alpha_2 - \varepsilon\alpha_4}\frac{2\beta\theta^* x^*}{(y^* + d)^4} = 0, \tag{14}$$

$$\frac{(\rho C_p)_{nf}}{(\rho C_p)_f} Pr\, e^{-\varepsilon\alpha_3 - \varepsilon\alpha_4 + \varepsilon\alpha_2 + \varepsilon\alpha_1}\left[\frac{\partial \psi^*}{\partial y^*}\frac{\partial \theta^*}{\partial x^*} - \frac{\partial \psi^*}{\partial x^*}\frac{\partial \theta^*}{\partial y^*}\right] - \mu_0 \vartheta_f (\in - 0)$$

$$e^{-\varepsilon\alpha_3 - \varepsilon\alpha_5 + \varepsilon\alpha_2 + \varepsilon\alpha_1}\left[\frac{\partial \psi^*}{\partial y^*}\frac{\partial H^*}{\partial x^*} - \frac{\partial \psi^*}{\partial x^*}\frac{\partial H^*}{\partial y^*}\right] - \frac{k_{nf}}{k_f} e^{-\varepsilon\alpha_4 + 2\varepsilon\alpha_2}\frac{\partial^2 \theta}{\partial y'^2} = 0 \tag{15}$$

The invariant group (12) maintains the following relationships between parameters.

$$\alpha_1 + 2\alpha_2 - 2\alpha_3 = 3\alpha_2 - \alpha_3 = \alpha_1 + 2\alpha_2 - 2\alpha_3 = -\alpha_1 + 4\alpha_2 - 4\alpha_4, \tag{16}$$

$$\alpha_1 + \alpha_2 - \alpha_3 - \alpha_4 = \alpha_1 + \alpha_2 - \alpha_3 - \alpha_4 - \alpha_5 = \alpha_1 + \alpha_2 - \alpha_3 - \alpha_4 - \alpha_5 = 2\alpha_2 - \alpha_4. \tag{17}$$

Now we solve the resulting equation (16) and (17) we get

$$\alpha_2 = 0, \alpha_3 = \alpha_1, \alpha_4 = 0, \alpha_5 = 0 \tag{18}$$

thus, we have the following transformations of similarity

$$\eta = y, \psi = xf(\eta), \theta = \theta(\eta), \mathrm{H} = \mathrm{H} \tag{19}$$

The Similarity Equations are obtained by substituting (19) into (11) and (12).

$$\left(\frac{\mu_{nf}}{\mu_f}\right) f''' + \left(\frac{\rho_{nf}}{\rho_f}\right)\left[ff'' - f'^2\right] - \left(\frac{\sigma_{nf}}{\sigma_f}\right) f'^2 H_a^2 - \frac{2\beta\theta}{(\eta + d)^4} = 0, \tag{20}$$

$$\left(\frac{k_{nf}}{k_f}\right)\theta''' + \frac{(\rho C_p)_{nf}}{(\rho C_p)_f}\Pr(f\theta') + \lambda\beta(\varepsilon - \theta)\frac{2f}{(\eta + d)^3} = 0, \tag{21}$$

Along with the initial and boundary conditions

$$f(0) = S, f'(0) = 1, \theta(0) = 1 \text{ at } \eta = 0,$$
$$f'(\infty) \to 0, \theta(\infty) \to 0 \text{ at } \eta \to \infty. \tag{22}$$

Here Prandtl number (Pr) $= \dfrac{(\rho C_p)_f \vartheta_f}{K_f}$, ferromagnetic interaction parameter $\beta = \dfrac{\alpha\mu_0}{2\pi}\dfrac{\rho_f K_f (T_\infty - T_w)}{\mu_f^2}$,,

Hartmann number (Ha) $= \sqrt{\dfrac{\sigma_f \beta_0^2}{\rho_f}}$, suction parameter (S) $= \dfrac{-v_w L}{\vartheta_f}$, stretching parameter $\lambda = \dfrac{\mu_f^2 c}{\rho K_f (T_\infty - T_w)}$,

and dimensionless temperature parameter $(\epsilon) = \dfrac{T_\infty}{(T_\infty - T_w)}$.

Applied Numerical Analysis

Mathematica programming may now be used to solve the coupled nonlinear ordinary differential equations (20) and (21) with the boundary conditions (22) numerically. We consider $y_1 = \eta$, $f = y_2$, $f' = y_3$, $f'' = y_4$, $\theta = y_5$, $\theta' = y_6$. From (20) and (21), the simultaneous first-order ordinary differential equations are formed.

$$\begin{bmatrix} y_1' \\ y_2' \\ y_3' \\ y_4' \\ y_5' \\ y_6' \end{bmatrix} = \begin{bmatrix} 1 \\ y_3 \\ y_4 \\ \left(\dfrac{\mu_{nf}}{\mu_f}\right)\left[\left(\dfrac{\rho_{nf}}{\rho_f}\right)\left[y_2^2 - y_1 y_3\right] + \left(\dfrac{\sigma_{nf}}{\sigma_f}\right) y_2^2 H_a^2 + \dfrac{2\beta y_4}{(\eta+d)^4}\right] \\ y_6 \\ \left(\dfrac{k_{nf}}{k_f}\right)\left[-\left(\dfrac{\rho_{nf}}{\rho_f}\right)\Pr(y_1 y_5) - \lambda\beta(\varepsilon - y_4)\dfrac{2\beta y_1}{(\eta+d)^4}\right] \end{bmatrix} \tag{23}$$

along with the initial and boundary conditions

$$\begin{bmatrix} y_1 & y_2 & y_3 & y_4 & y_5 & y_6 \end{bmatrix}^T = \begin{bmatrix} 0 & S & 1 & u_1 & 1 & u_2 \end{bmatrix}^T \tag{24}$$

To solve the non-linear coupled system, the 4th-order Runge-Kutta method with a shooting approach is employed. $f'(\infty)$ and $\theta(\infty)$, the boundary conditions at infinity, limit the convenient beginning values of the cryptic initial conditions u_1 and u_2.

3. Results and Discussion

In this study, outcomes are offered for a Bio-magnetic nanofluid that is blood based iron oxide (Fe_3O_4). In this section, examination of the effects of different relevant parameters axial velocity and temperature. The velocity and temperature distributions are analyzed through graphs. The current work's default settings are considered $K_f = 0.492$, $K_s = 9.7$, $(C_p)_s = 670$, $(C_p)_f = 3594$, $\rho_s = 5180$, $\rho_f = 1053$, $\sigma_s = 0.74 \times 10^6$, $\sigma_f = 0.8$, $\Phi = 0.02$.

Velocity Profile

The effect of ferromagnetic interaction parameter β and Hartmann number Ha on velocity profile is illustrated in Fig. 36.2. Fig. 36.2(a) shows the relationship between velocity and the ferromagnetic interaction parameter β. Ferromagnetic effect is considered using the ferromagnetic interaction parameter and the dimensionless distance between the magnetic dipole's Center and the origin. Flow velocity is reduced when β increases in ferromagnetic liquid because of the increased viscosity caused by micron-sized ferrite particles floating in the fluid flow. Fig. 36.2(b) Influence of Hartmann-number(Ha) on velocity profile. The value of the Hartmann number rises, fluid mobility decreases. As a result, the velocity profile drops.

The non-linear stretching parameter's influence on the velocity profile is seen in Fig. 36.3. The thickness of the fluid's boundary wall grows when the stretching parameter is increased, and the fluid's mobility diminishes. As a result, the fluid's velocity reduces.

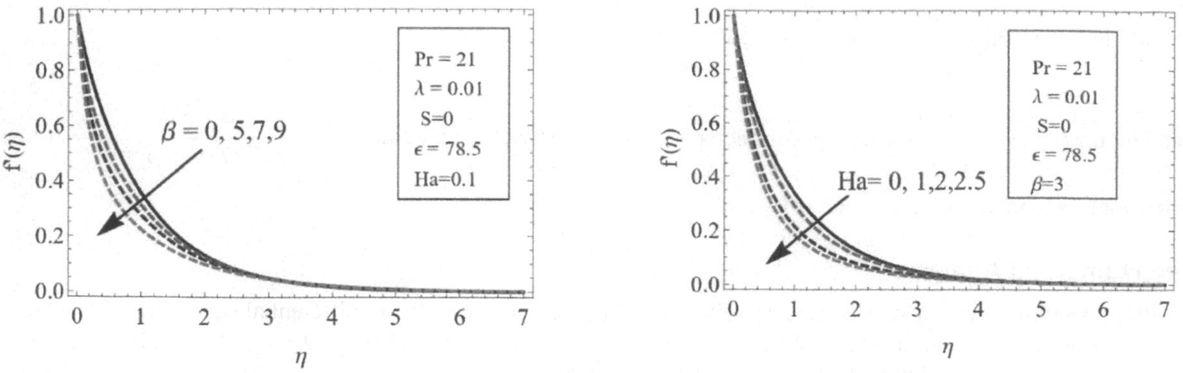

(a) (b)

Fig. 36.2 Velocity variation on ferromagnetic interaction parameter and Hartmann number

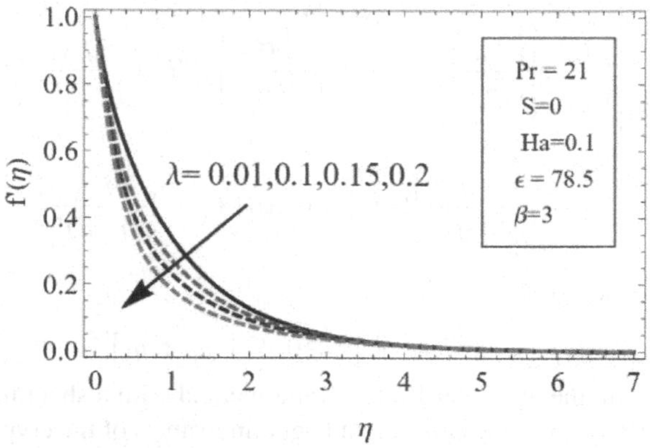

Fig. 36.3 Velocity on stretching parameter

Temperature Profile

The influence of ferromagnetic interaction parameter (β), Hartmann number (Ha), stretching parameter (λ) and suction parameter (S) on temperature is shown in Fig. 36.4. It can be shown in Fig. 36.4(a) that the temperature of the nanofluid rises with the increase in β. This is due to the reduction in the velocity of fluid and increases frictional heating involving within the fluid layers. Fig. 36.4(b) shows how a magnetic hydrodynamic parameter affects temperature distribution. It demonstrates that when the magnetic hydrodynamic parameter increases, so does the temperature. As seen in Fig. 36.4(c) the stretching parameter has a substantial impact on the temperature profile.

Fluid temperature rises as its surface stretches, increasing the thermal and momentum boundary layer thickness and slowing heat dissipation. Fig. 36.4(d) shows that when suction increases, the fluid temperature decreases. This is because the thickness of the thermal boundary layer reduces with increasing suction.

4. Conclusion

The fluid flow and heat transfer of a blood based nanofluid over a stretched sheet in the presence of a magnetic dipole is described in this article. The flow governing partial differential equations was transformed to a non-dimensional form and solved using the Lie group technique. Based on the computation results, we may take the following conclusions.

- Magnetic fluid flow and heat transfer are largely affected by the ferromagnetic interaction parameter. Increasing the ferromagnetic effect causes a reduction in velocity, which leads to a rise in temperature.
- An increasing parameter Ha (Hartmann number) and stretching parameter λ gives rise in temperature by a reduction in velocity.
- An increase in S (suction parameter) results in the depletion of the temperature profile.

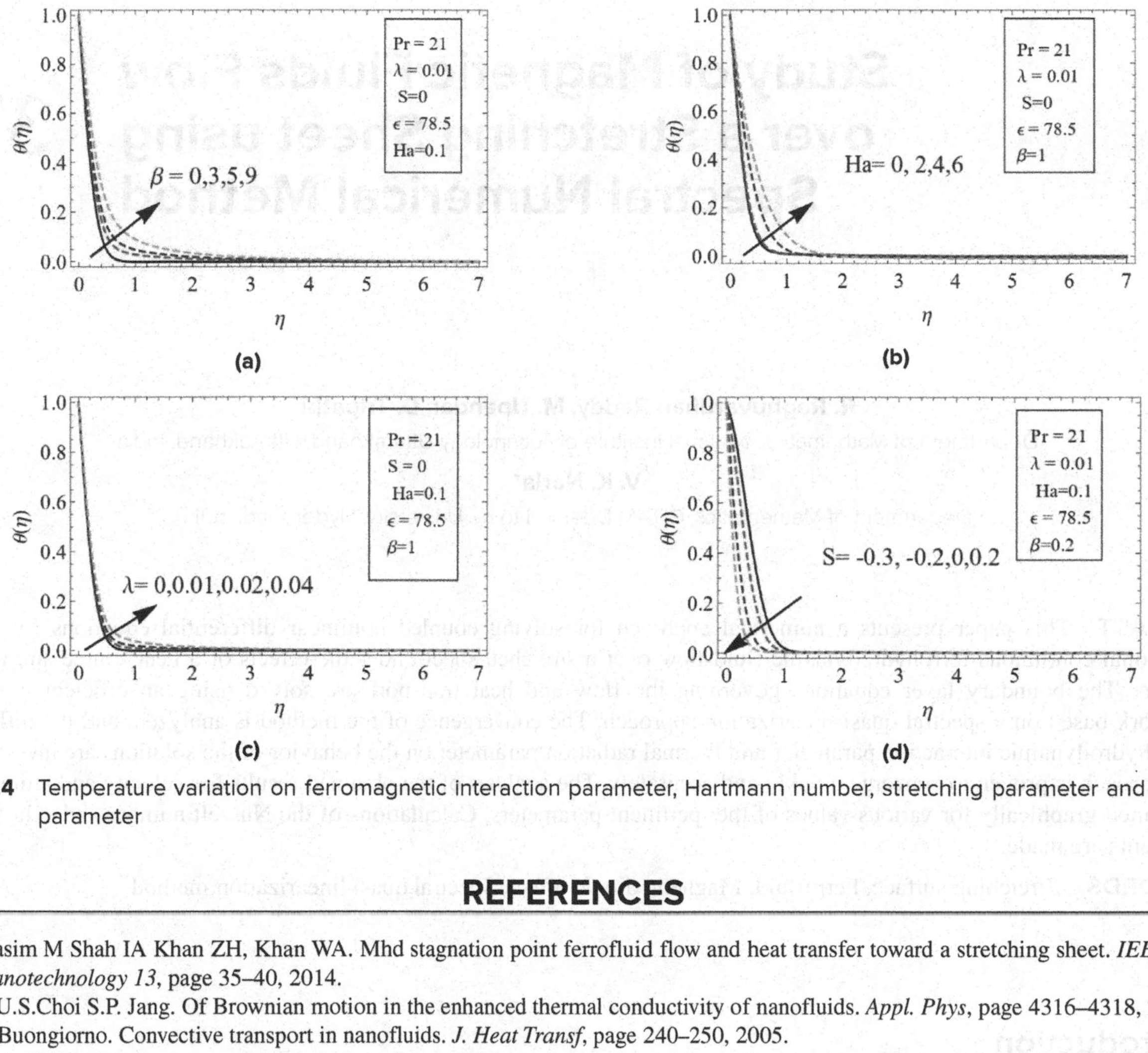

Fig. 36.4 Temperature variation on ferromagnetic interaction parameter, Hartmann number, stretching parameter and suction parameter

REFERENCES

1. Qasim M Shah IA Khan ZH, Khan WA. Mhd stagnation point ferrofluid flow and heat transfer toward a stretching sheet. *IEEE Trans Nanotechnology 13*, page 35–40, 2014.
2. S.U.S.Choi S.P. Jang. Of Brownian motion in the enhanced thermal conductivity of nanofluids. *Appl. Phys*, page 4316–4318, 2004.
3. J. Buongiorno. Convective transport in nanofluids. *J. Heat Transf*, page 240–250, 2005.
4. O.A. Valnes H.I. Andersson. Flow of a heated ferrofluid over a stretching sheet in the presence of a magnetic dipole. *Acta Mech*, page 39–47, 1998.
5. A.M. Jyothi B.C. Prasannakumara M. Ijaz Khan Y.M. Chu V. Kumar, J.K. Madhukesh. Analysis of single and multi-wall carbon nanotubes (swcnt/mwcnt) in the flow of Maxwell nanofluid with the impact of magnetic dipole. *Comput. Theor. Chem.*, 2021.
6. Abdullah M. Abusorrah Y.M. Mahrous Nidal H. Abu-HamdehAlibek Is-sakhovMohammadRahimi-Gorji B.C Prasannakumara R. Naveen Kumar,
7. R.J Punith Gowda. Impact of magnetic dipole on ferromagnetic hybrid nanofluid flow over a stretching cylinder. *Phys. Scr*, 2021.
8. P. Ovler. Applications of lie groups to differential equations. *Springer-Verlag*, 1993.
9. Md jashim Uddin M. Ferdows and A.A. Afifty.
10. M. Ferdows M.M Rashidi. E. Momoniat and A. Basiriparsa. Applications of lie groups to differential equations. *Math. Probl. Eng*, pages 1–24, 2014.

Note: All the figures and table in this chapter were made by the authors.

Impending Inquisitions in Humanities and Sciences (ICIIHS-2022) – Dr. Mohan Varkolu et al. (eds)

Study of Magnetic Fluids Flow over a Stretching Sheet using Spectral Numerical Method

37

R. Raghuvardhan Reddy, M. Upendar, D. Tripathi
Department of Mathematics, National Institute of Technology Uttarakhand, Uttarakhand, India

V. K. Narla*
Department of Mathematics, GITAM Deemed to be University, Hyderabad, India

ABSTRACT This paper presents a numerical approach for solving coupled nonlinear differential equations for a two-dimensional continuous ferrohydrodynamic fluid flow over a stretched sheet under the effects of a heat source and thermal radiation. The boundary layer equations governing the flow and heat transport are solved using an efficient numerical framework based on a spectral quasi-linearization approach. The convergence of the method is analyzed, and the influences of ferrohydrodynamic interaction parameter and thermal radiation parameter on the behavior of the solutions are investigated. This approach is proven to converge quickly and accurately. The analysis of the obtained results for velocity and temperature is presented graphically for various values of the pertinent parameters. Calculations of the Nusselt number and skin friction coefficients are made.

KEYWORDS Stretching surface, Ferrofluid, Magnetic dipole dffect, Spectral quasi linearization method

1. Introduction

It is possible to use the spectral quasi-linearization framework to solve the flow-governing differential equations by applying the right similarity transformations. The spectral quasi-linearization approach Bellman [1] yields numerical results with the necessary level of precision. There are few studies in the literature on fluid flow models that are solved by spectral approach (Motsa [2], Trefethen [3], Shahid et al. [4], Goqo et al. [5] and Shamshuddin et al. [6]). However, this approach was not utilized for the proposed model.

Ferrofluid contains suspended tiny particles of cobalt, magnetite, or iron. Such fluids are the compositing of small magnetic iron particles suspended in oil, most often kerosene, and treated with a surfactant to prevent the oleic acid from congealing are known as ferrofluids (Crane [7]). Ferrofluids are used in a variety of applications to minimize vibration, including computer hardware drives, various rotating shaft motors, and loudspeakers. Ferrofluid is a contrast agent that is utilized in magnetic resonance imaging in medicine. A growing number of researchers are seeing the potential for ferrofluid appli cations in the fast-developing fields of microelectromechanical systems and nanotechnology since the permanently magnetized particles have a spherical diameter of 10 nm (MEMS). Neuringer[8] and Rosensweig RE et al.[9] are studied saturated ferrofluids when heat and magnetic field gradients were coupled. Following this, numerous researchers were interested in the subject and conducted considerable investigations.

*Corresponding author: vnarla@gitam.edu

DOI: 10.1201/9781003489436-37

Nomenclature		Greek Symbols	
T	temperature(K)	α	magnetic field strength(A/m)
T_c	curie temperature(K)	β	ferromagnetic interaction parameter
T_w	Temperature at the surface(K)	ϵ	dimensionaless temperature parameter
$\tilde{u}$	fluid velocity(ms^{-1}) in $\tilde{x}$ direction(m)	η	similarity variable
v	fluid velocity(ms^{-1}) in $\tilde{y}$ direction(m)	λ	stretching parameter
C_f	skin friction coefficient	μ	dynamic viscosity($kgs^{-1}m^{-1}$)
C_p	specific heat ($Jkg^{-1}K^{-1}$)	μ_0	magnetic permeability
H	magnetic field(A/m)	ϕ	magnetic scalar potential
k	thermal conductivity($Wm^{-1}K^{-1}$)	ρ	density of fluid(kgm^{-3})
K^*	pyromagnetic coefficient	σ	Stefan-Boltzmann constant
M	magnetization(A/m)	σ_R	Rosseland mean absorption coefficient
Nr	thermal radiation parameters	ϑ	kinematic viscosity ($m^{-2}s^{-1}$)
Nu	Nusselt number		
Pr	Prandtl number		
qr	radiative flux		
Q_s	heat source parameter		
Re	Reynolds number		

Much emphasis has been paid to the study of heat transfer and flow over a stretched surface. This has widely been described in the literature see, (Khan [10] and Cortell [11]). The composition of the ambient liquid and the pace of stretching, determine the properties of products created through engineering procedures such as metallurgy and polymer extrusion. The magnetic field and radiation parameter were discovered to have a propensity to reduce the skin friction coefficient as for Andersson and Valnes [12]. Tzirtzi-lakis et al. [13] undertook experimental research on a ferrohydrodynamic fluid over a horizontal plate based on boundary layer principles in order to better understand the phenomenon. As for the result of Hayat et al. [14], radiation has a significant effect on controlling the heat transfer rate in the boundary layer flows. Because it involves heat transfer from electromagnetic waves that carry energy away from the generating object, thermal radiation is frequently included in engineering systems.

This study is organized as follows: in section 2, the problem is mathematically formulated for a 2D boundary layer flow of ferrohydrodynamic fluid over a stretched sheet with thermal radiation and heat source. Section 3 presents the spectral quasi-linearization method. Section 4 examines the numerical method for convergence. On the other hand, the papers end with a numerical result and discussion in section 5, and the summary of key findings is presented in section 6.

2. Mathematical Formulation

Consider the flow and heat transfer of a ferrofluid over a stretching sheet with a continuous two-dimensional incompressible boundary layer flow. The x-axis is parallel to the sheet and the y-axis is perpendicular to it as shown in Fig. 37.1. We assume that the sheet gets stretched at a rate proportionate to the distance $\tilde{u} = c\tilde{x}\tilde{T}_w$ is the temperature of the stretched sheet, and $\tilde{T}_c$ is the Curie temperature of the fluid far away from the sheet, where $\tilde{T}_w < \tilde{T}_c$. Assume that the magnetic dipole is positioned a below the sheet, resulting in a magnetic field strong enough to saturate the ferrofluid. Fig. 37.1 shows the flow arrangement schematically.

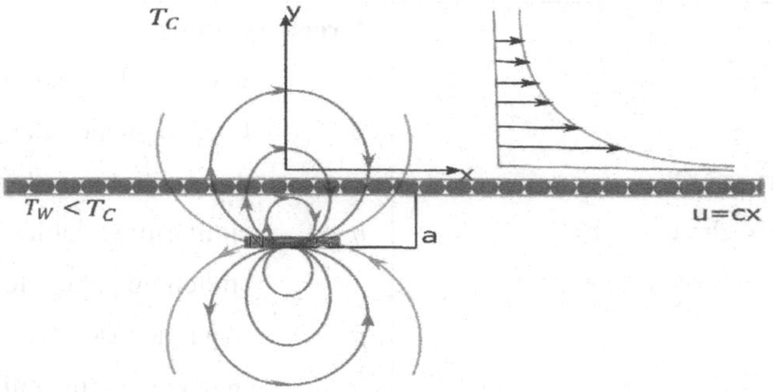

Fig. 37.1 A flow diagram is a visual depiction of the flow arrangement

The boundary layer flow governing by momentum and energy equations of ferrofluid are shown as

$$\frac{\partial \tilde{u}}{\partial \tilde{x}} + \frac{\partial \tilde{v}}{\partial \tilde{y}} = 0 \tag{1}$$

$$\rho\left[\tilde{u}\frac{\partial \tilde{u}}{\partial \tilde{x}} + \tilde{v}\frac{\partial \tilde{v}}{\partial \tilde{y}}\right] = \mu\frac{\partial^2 \tilde{u}}{\partial \tilde{y}^2} + \mu_0 M\frac{\partial H}{\partial \tilde{x}} \tag{2}$$

$$(\rho C_p)\left[\tilde{u}\frac{\partial \tilde{T}}{\partial \tilde{x}} + \tilde{v}\frac{\partial \tilde{T}}{\partial \tilde{y}}\right] = \mu_0 T\frac{\partial M}{\partial \tilde{T}}\left[\tilde{u}\frac{\partial H}{\partial \tilde{x}} + \tilde{v}\frac{\partial H}{\partial \tilde{y}}\right] = k\left[\frac{\partial^2 \tilde{T}}{\partial \tilde{y}^2}\right] + Q(\tilde{T}_c - \tilde{T}) - \frac{\partial q_r}{\partial \tilde{y}} \tag{3}$$

The boundary conditions associated with the flow are considered as

$$\left.\begin{aligned}&\tilde{u}(\tilde{x},o) = c\tilde{x}, \tilde{v}(\tilde{x},o) = 0, \tilde{T}(\tilde{x},o) = \tilde{T}_w \ at \ \tilde{y} = 0\\ &\tilde{u}(\tilde{x},\infty) \to 0, \tilde{T}(\tilde{x},\infty) = \tilde{T}_c \ at \ \tilde{y} \to \infty\end{aligned}\right\} \tag{4}$$

Magnetic Dipole

The flow of ferrofluid to the magnetic dipole is affected by the magnetic field. To calculate the magnetic dipole potential, we use the following formula

$$\phi(\tilde{x},\tilde{y}) = \frac{\alpha}{2\pi}\left[\frac{\tilde{x}}{\tilde{x}^2 + (\tilde{y}+a)^2}\right] \tag{5}$$

Using a linear relationship between the gradient of H and magnetic bodyforce, we arrive at this conclusion:

$$\frac{\partial H}{\partial \tilde{x}} = -\frac{\alpha \tilde{x}}{\pi(\tilde{y}+a)^4}. \tag{6}$$

$$\frac{\partial H}{\partial \tilde{x}} = \frac{\alpha}{\pi}\left[\frac{-1}{(\tilde{y}+a)^3} + \frac{2\tilde{x}^2}{(\tilde{y}+a)^5}\right], \tag{7}$$

A linear equation may be used to approximate the variation of magnetization M with temperature $\tilde{T}$ as $M = K^*(\tilde{T}_c - \tilde{T})$, and radiative heat flux is defined

$$q_r = \frac{-4}{3}\frac{\sigma}{\sigma_R}\frac{\partial \tilde{T}^4}{\partial \tilde{y}} \tag{8}$$

Taylor's series may be expanded around $\tilde{T}_c$ to determine $\tilde{T}^4$ as $\tilde{T}^4 \cong -3\tilde{T}_c^4 + 4T_c^3$.

Non-dimensional analysis

The following non-dimensional variables considered:

$$(\xi,\eta) = \frac{(\tilde{x},\tilde{y})}{\sqrt{v/c}}, (U,V) = \frac{(\tilde{u},\tilde{v})}{\sqrt{vc}}, \theta(\xi,\eta) = \frac{\tilde{T}_c - \tilde{T}}{\tilde{T}_c - \tilde{T}_w} = \theta_1(\eta) + \xi^2\theta_2(\eta) \tag{9}$$

The following similarity transformations are considered.

$$\eta = \tilde{y}, U = \xi f'(\eta), V = -f(\eta), \theta = \theta(\eta), H = H \tag{10}$$

The prime here signifies differentiation with regard to the η. Using (6) to (9) in the equations (2) and (3), the boundary value problem for the model expressed as:

$$f''' + ff'' - (f')^2 - \frac{2\beta\theta_1}{(\eta + d)^4} = 4 \tag{11}$$

$$(1+Nr)\theta_1'' + \Pr[f\theta_1' - Q_s\theta_1] + \lambda\beta(\theta_1 - \in)\left[\frac{2f}{(\eta+d)^3}\right] = 0 \tag{12}$$

$$(1+Nr)\theta_1'' + \Pr[2f\theta_2' - f\theta_2' + Q_s\theta_2] + \lambda\beta(\in - \theta_1)\left[\frac{2f'}{(\eta+d)^4} + \frac{4f}{(\eta+d)^5}\right] + \frac{2\lambda\beta f\theta_2}{(\eta+d)^3} = 0 \tag{13}$$

along with the initial and boundary conditions

$$\left.\begin{aligned} &f(0) = 0, f'(0) = 1, \theta_1(0) = 1, \theta_2(0) = 0, \\ &f'(\infty) = 0, \theta_1(\infty) = 0, \theta_2(\infty) = 0. \end{aligned}\right\} \tag{14}$$

Here $\Pr = \frac{\rho C_p \vartheta}{k}$, $\beta = \frac{\alpha\mu_0}{2\pi}\frac{\rho K^*(\tilde{T}_c - \tilde{T}_w)}{\mu^2}$, $\lambda = \frac{\mu^2 c}{\rho K^*(\tilde{T}_c - \tilde{T}_w)}$, $\in = \frac{\tilde{T}_c}{(\tilde{T}_c - \tilde{T}_w)}$, and $Nr = \frac{16}{3}\frac{\sigma(\tilde{T}_c - \tilde{T}_w)^3}{\sigma_R k}$

The physical quantities such as coefficient of skin friction (C_f) and Nusselt number (Nu) are defined as

$$-f''(0) + \frac{1}{2}C_f Re^{\frac{1}{2}}, -\theta_1'(0) = NuRe^{\frac{1}{2}}.$$

3. Spectral Quasi Linearization Method (SQLM)

The numerical solution of the nonlinearly coupled equations (11) to (13) according to the boundary conditions (14) has been achieved by the use of the spectral quasi-linearization technique. For the equations (11) to (13), the quasi-linearization approach is used, the resultant equations are as follows:

$$f_{r+1}''' + c_{1,r}f_{r+1}'' + c_{2,r}f_{r+1}' + c_{3,r}f_{r+1} + c_{4,r}(\theta_1)_{r+1} = r_1, \tag{15}$$

$$c_{5,r}f_{r+1} + (1+Nr)(\theta_1'')_{r+1} + c_{6,r}(\theta_1')_{r+1} + c_{7,r}(\theta_1)_{r+1} = r_2, \tag{16}$$

$$c_{8,r}f_{r+1}' + c_{9,r}f_{r+1} + c_{10,r}(\theta_1)_{r+1} + (1+Nr)(\theta_2'')_{r+1} + c_{11,r}(\theta_2')_{r+1} + c_{12,r}(\theta_2)_{r+1} = r_2 \tag{17}$$

and the boundary conditions take the form

$$\begin{aligned} &f_{r+1}(0) = f_{r+1}'(\infty) = 0, f_{r+1}'(0) = 1, (\theta_1)_{r+1}(0) = 1, \\ &(\theta_1)_{r+1}(\infty) = (\theta_2)_{r+1}(0) = 0, (\theta_2)_{r+1}(\infty) = 0. \end{aligned} \tag{18}$$

The coefficient are computed as

$$c_{1,r} = f_r, c_{2,r} = -2f_r', c_{3,r} = f_r'', c_{4,r} = \frac{-2\beta}{(\eta+d)^4},$$

$$c_{5,r} = \Pr\left(\theta_1'\right)_r - \frac{2\lambda\beta \in}{(\eta+d)^3} + \frac{2\lambda\beta(\theta_1)_r}{(\eta+d)^3}, c_{6,r} = \Pr f_r,$$

$$c_{7,r} = \frac{2\lambda\beta(f)_r}{(\eta+d)^3} - PrQ_s, c_{8,r} = 2Pr(\theta_2)_r + \frac{2\lambda\beta \in}{(\eta+d)^4} - \frac{2\lambda\beta(\theta_1)_r}{(\eta+d)^4},$$

$$c_{9,r} = Pr(\theta_2')_r + \frac{4\lambda\beta \in}{(\eta+d)^5} + \frac{2\lambda\beta(\theta_2)_r}{(\eta+d)^3} - \frac{4\lambda\beta(\theta_1)_r}{(\eta+d)^5}, c_{10,r} = \frac{-2\lambda\beta f_r'}{(\eta+d)^4} - \frac{4\lambda\beta f_r}{(\eta+d)^5},$$

$$c_{11,r} = Prf_r, c_{12,r} = -2Prf_r' - PrQ_s + \frac{2\lambda\beta f_r}{(\eta+d)^3},$$

$$r_1 = f_r f_r'' - f_r'^2, \quad r_2 = Prf_r(\theta_1')_r + \frac{2\lambda\beta(\theta_1)_r f_r}{(\eta+d)^3},$$

$$r_3 = -2Pr\, f_r'(\theta_2)_r + Prf_r(\theta_2')_r + \frac{2\lambda\beta(\theta_2)_r f_r}{(\eta+d)^3} - \frac{2\lambda\beta(\theta_1)_r f_r'}{(\eta+d)^4} - \frac{4\lambda\beta(\theta_1)_r f_r}{(\eta+d)^5} - \lambda R(f_r''')^2.$$

For $r = 1, 2, 3, \ldots$, it is possible to solve this system numerically in any way since it is a linear system of linked differential equations (15) to (18) with variable coefficients. The nodes in the interval $[-1, 1]$ are obtained by Gauss-Lobatto points using $z_i = \cos(\pi i/N)$, $i = 0, 1, 2, \ldots, N$. Spectral method works only in $z_i \epsilon [-1, 1]$ called computational domain (CD). This relation gives $N + 1$ collocation points in the interval $x_i \in [a,b]$,

$$\eta_i = x_i = \frac{a+b}{2} + \frac{(b-a)}{x} z_i$$

At the collocation nodes, the derivative of f_{r+1}, $(\theta_1)_{r+1}$ and $(\theta_2)_{r+1}$ are expressed as

$$\frac{\partial^p f_{r+1}}{\partial \eta^p} = \left(\frac{2}{L}\right)^p \sum_{i=0}^{N} D^p_{N,i} f_{r+1}(\eta_i) = \mathbf{D}^p F,$$

$$\frac{\partial^p (\theta_1)_{r+1}}{\partial \eta^p} = \left(\frac{2}{L}\right)^p \sum\nolimits_{i=0}^{N} D^p_{N,i} (\theta_1)_{r+1}(\eta_i) = \mathbf{D}^p \Theta_1, \tag{19}$$

$$\frac{\partial^p (\theta_2)_{r+1}}{\partial \eta^p} = \left(\frac{2}{L}\right)^p \sum_{i=0}^{N} D^p_{N,i} (\theta_2)_{r+1}(\eta_i) = \mathbf{D}^p \Theta_2.$$

Where **D** is the Chebyshev differentiation matrix scaled by $L/2$. In this case, the matrix **D** has order $(N + 1) \times (N + 1)$ with p derivative order. Substituting (19) into Eqs. (15) to (18), we obtain

$$\left[\mathbf{D}^3 + c_{1,r}\mathbf{D}^2 + c_{2,r}\mathbf{D} + c_{3,r} I\right] F_{r+1} + c_{4,r}(\Theta_1)_{r+1} = r_1, \tag{20}$$

$$c_{5,r} F_{r+1} + \left[(1+Nr)\mathbf{D}^2 + c_{6,r}\mathbf{D} + c_{7,r} I\right](\Theta_1)_{r+1} = r_2, \tag{21}$$

$$\left[c_{8,r}\mathbf{D} + c_{9,r} I\right] F_{r+1} + c_{10,r}(\Theta_1)_{r+1} + \left[(1+Nr)\mathbf{D}^2 + c_{11,r}\mathbf{D} + c_{12,r} I\right](\Theta_2)_{r+1} = r_3. \tag{22}$$

When the spectral approach is applied to the boundary conditions, the boundary conditions can be written as:

$$f_{r+1}(z_N) = 0, \sum\nolimits_{i=0}^{N} D_{N,1} f_{r+1}(z_N) = 1, (\theta_1)_{r+1}(z_N) = 1, (\theta_1)_{r+1}(z_N) = 0,$$

$$\sum\nolimits_{i=0}^{N} D^1_{0,i} f_{r+1}(z_0) = 0, (\theta_1)_{r+1}(z_0) = 0, (\theta_2)_{r+1}(z_0) = 0$$

The matrix representation of the system of equations (20-22) is as follows:

$$\begin{bmatrix} \mathbf{M}_{11} & \mathbf{M}_{12} & \mathbf{M}_{13} \\ \mathbf{M}_{21} & \mathbf{M}_{22} & \mathbf{M}_{23} \\ \mathbf{M}_{31} & \mathbf{M}_{32} & \mathbf{M}_{33} \end{bmatrix} \begin{bmatrix} F_{r+1} \\ (\Theta_1)_{r+1} \\ (\Theta_2)_{r+1} \end{bmatrix} = \begin{bmatrix} r_1 \\ r_2 \\ r_3 \end{bmatrix}. \tag{23}$$

Where

$$\mathbf{M}_{11} = \left[1 \ \operatorname{diag}(c_{1,r}) \ \operatorname{diag}(c_{2,r}) \ \operatorname{diag}(c_{3,r})\right]\left[\mathbf{D}^3 \ \mathbf{D}^2 \ \mathbf{D} \ I\right]^T,$$

$$\mathbf{M}_{12} = \operatorname{diag}(c_{4,r}), \mathbf{M}_{13} = 0, \mathbf{M}_{21} = \operatorname{diag}(c_{5,r}),$$

$$\mathbf{M}_{22} = \left[1 + Nr \ \operatorname{diag}(c_{6,r}) \ \operatorname{diag}(c_{7,r})\right]\left[\mathbf{D}^2 \ \mathbf{D} \ I\right]^T, \mathbf{M}_{22} = 0,$$

$$\mathbf{M}_{31} = \left[\operatorname{diag}(c_{8,r}) \ \operatorname{diag}(c_{9,r})\right]\left[\mathbf{D} \ I\right]^T, \mathbf{M}_{32} = \operatorname{diag}(c_{10,r}),$$

$$\mathbf{M}_{33} = \left[1 + Nr \quad \text{diag}(c_{11,r}) \quad \text{diag}(c_{12,r})\right]\left[\mathbf{D}^2 \quad \mathbf{D} \quad I\right]^T,$$

where $c_{i,r}$ is a matrix with $[c_{i,r}](N+1) \times (N+1)$.

Convergence Analysis

The SQLM is used to solve the coupled ordinary differential equations (11) to (13) with boundary conditions (14). The residual errors are generated to confirm the correctness of the numerical findings. The residual error is a measure of how close the numerical solution comes to matching the exact solution. The residual errors for equations (11) to (12) are hence represented as

$$Res(f) = f''' + ff'' - (f')^2 - \frac{2\beta\theta_1}{(\eta + d)^4}, \text{d}$$

$$Res(\theta_1) = (1 + Nr)\theta_1'' + \Pr\left[f\theta_1' - Q_s\theta_1\right] + \lambda\beta(\theta_1 - \epsilon)\left[\frac{2f}{(\eta + d)^3}\right]. \tag{24}$$

The infinity norms of (24) are defined as $\|\text{Res}(f)\|_\infty$, $\|\text{Res}(f)\|_\infty$. Several physical factors, such as the ferrohydrodynamic interaction parameter and the thermal radiation parameter, which were used to investigate how the residual norms are changed.

Table 37.1 Skin friction coefficient for each iteration with $Q_s = -0.5$.

Iterations	$-f''(0)$		
	$\beta = 0$	$\beta = 1$	$\beta = 2$
1	0.166666667	1.169273438	1.56356178
2	0.885148192	1.289924037	1.586312896
3	0.981932385	1.291499726	1.5863131
4	0.999061101	1.291499728	1.5863131
5	1.000470324	1.291499728	1.5863131
6	1.000483624	1.291499728	1.5863131
7	1.000483625	1.291499728	1.5863131
8	1.000483625	1.291499728	1.5863131

4. Results and Discussion

In this manuscript, Chebyshev spectral method is used to solve the problem of ferrofluid flow over a linear stretching sheet under isothermal boundary conditions with radiation. The nonlinear terms are linearized by a quasi-linearization framework. The influence of pertinent parameters such as ferrohydrodynamic interaction parameter β, thermal radiation parameter Nr, Prandtl Number Pr, and heat source Q_s on flow and heat transfer characteristics are analyzed and discussed.

Table 37.2 Nusselt number for each iteration with $Q_s = 0.5$.

Iterations	$-\theta_1'(0)$		
	$\beta = 0$	$\beta = 1$	$\beta = 2$
1	1.322875993	1.801750251	1.805675619
2	1.814255766	1.817731253	1.80250829
3	1.833300847	1.817444439	1.802508211
4	1.831912903	1.817444439	1.802508211
5	1.831669738	1.817444439	1.802508211
6	1.831667126	1.817444439	1.802508211
7	1.831667126	1.817444439	1.802508211
8	1.831667126	1.817444439	1.802508211

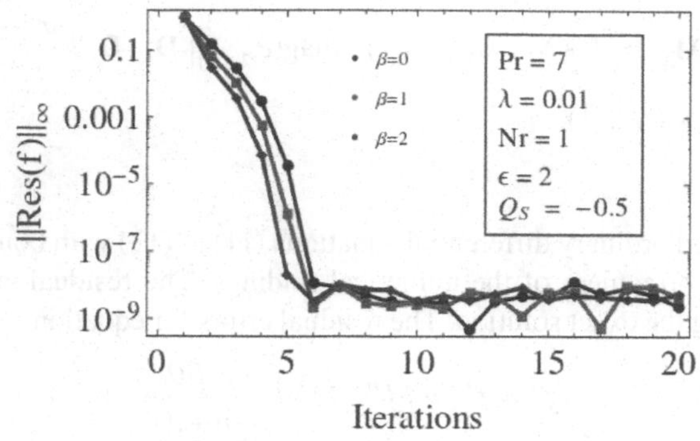

Fig. 37.2 Influence of iterations on for ferrohydrodynamic interaction parameter.

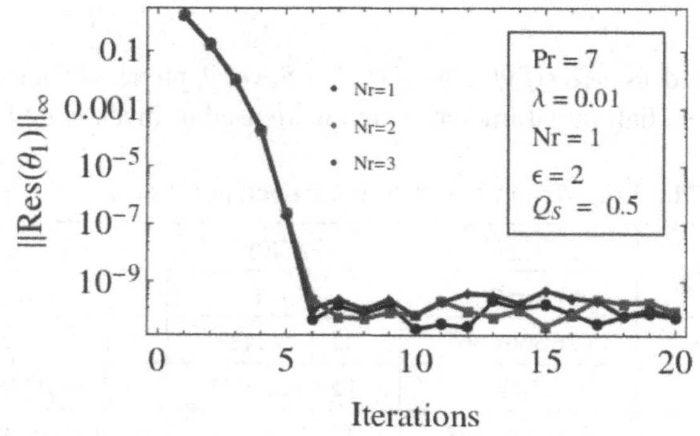

Fig. 37.3 Influence of iterations on for thermal radiation parameter.

Table 37.1 and Table 37.2 give a visual representation of how various ferrohydro dynamic interaction parameters affect the convergence of the solution to coefficient of skin friction and Nusselt number with the fixed values of $Pr = 7, \lambda = 0.01, Nr = 1$. It is observed from the table 37.1 that the convergence of tenth-order accuracy is achieved at the seventh iteration when $\beta = 0$ and the same accuracy is obtained at the third iteration when $\beta \geq 1$. It is also seen from the Table 37.2 that the Nusselt number achieved tenth-order accuracy at the sixth iteration when $\beta = 0$.

The effects of iterations on residual error are shown in Fig. 37.2 and 37.3. It is observed from the figures that the residual error of different β values decreases when the number of iterations increases. It is also noted that the accuracy of the method improves after the sixth iteration in both figures. The numerical method converges after the sixth iteration to a residual error norm lies between 10^{-7} and 10^{-10} as shown in Fig. 37.2. It can also be seen from Fig. 37.3 that the residual error of the spectral method converges at sixth iteration between 10^{-9} and 10^{-11}.

The parameter β represents the magnetic influences on the flow. If $\beta = 0$, the magnetic effect will vanish, causing the momentum and energy equations to decouple. An example of how β affects velocity and temperature is depicted in the Fig. 37.4. As demonstrated in Fig. 37.4, increasing β causes the velocity field to be decreased and the temperature to be rise.

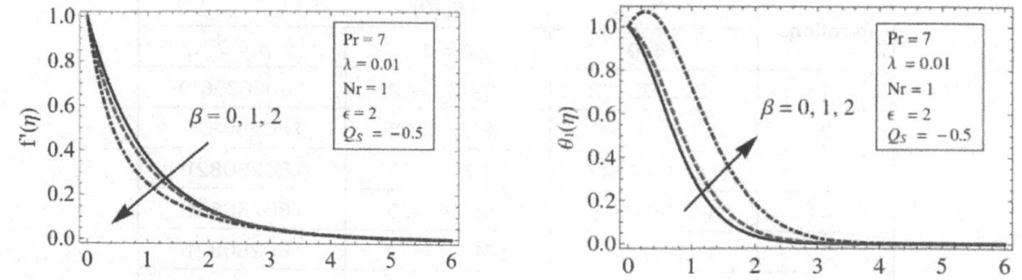

Fig. 37.4 Velocity and temperature variations with the ferrohydrodynamic interaction parameter

Fig. 37.5 depicts the influence of thermal radiation on the velocity and temperature profiles for a variety of *Nr* values. With an increase in the radiation parameter *Nr*, the velocity profile becomes more pronounced. Due to the action of thermal radiation, the thermal diffusivity of a medium is increased, resulting in a rise in the temperature profile of the medium.

Fig. 37.6 shows velocity and temperature curves as a function of *Pr*. This is because of the velocity boundary layer has a greater thickness when *Pr* increases and further reduces the temperature profile and the heat transfer. The influence of internal heat source production on velocity and temperature profile is shown in the Fig. 37.7. In the presence of a heat source, the velocity profile is increasing, while the temperature profile diminishes with heat source.

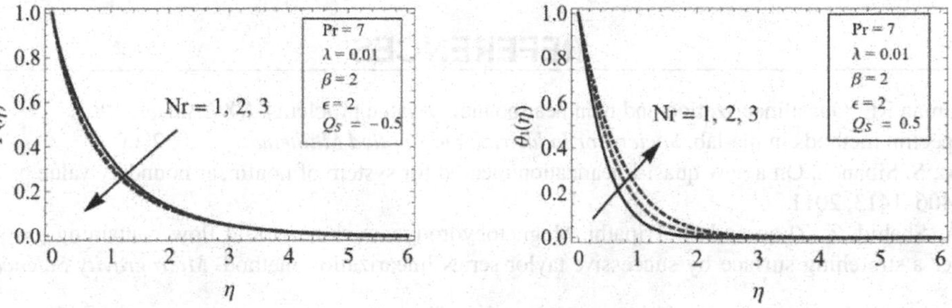

(a) (b)

Fig. 37.5 Velocity and Temperature variations with the thermal radiation parameter

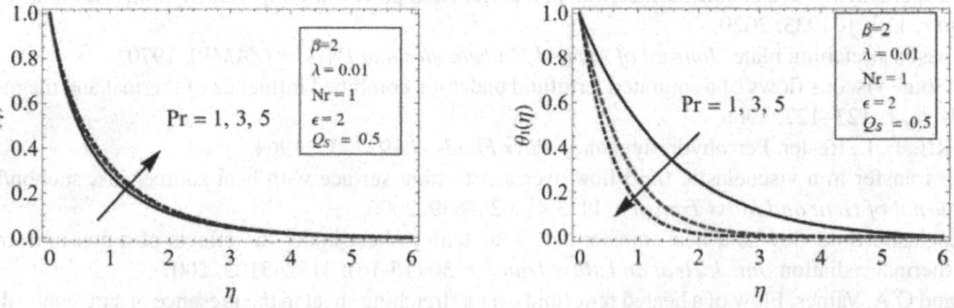

(a) (b)

Fig. 37.6 Velocity and Temperature variations with the Prandtl Number

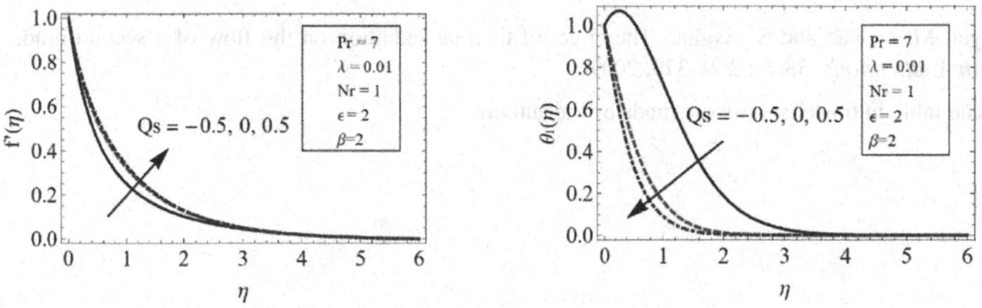

(a) (b)

Fig. 37.7 Velocity and temperature variations with the heat source

5. Conclusions

The numerical solution of the flow governing equations of ferrofluid flow across a stretched sheet that is subjected to a magnetic dipole was accomplished via the application of the spectral quasi-linearization approach. The results show the accuracy and efficiency of the spectral quasi-linearization method and its applicability to solve the flow governing nonlinear equations. The following are the significant results of the problem:

- As β increases, the axial velocity drops.
- The radiation parameter Nr increases, the velocity reduces little but the temperature increases.
- Prandtl numbers aid flow, minimizing heat transfer, hence they must be low for efficient cooling.
- As the heat source parameter increases, the velocity increase, and sink temperature changes more with than source.
- Residual error measures how closely numerical solution matches accu rate response.
- The skin friction coefficient and Nusselt number both converge in three iterations with a high ferro hydrodynamic interaction parameter.

REFERENCES

1. Kalaba RE. Bellman RE. Quasilinearization and nonlinear boundary-value problems. *NY:Elsevier*, 1965.
2. N. Trefethen. Spectral methods in matlab. *Society for Industrial and Applied Mathematics*, 10, 2000.
3. P. Shateyi Motsa, S. Sibanda. On a new quasi-linearization method for system of nonlinear boundary value problems. *Math. Methods Appl. Sci*, 34: 1406–1413, 2011.
4. M. M. Bhatti A. Shahid, Z. Zhou and D. Tripathi. Magnetohydrodynamics nanofluid flow containing gyrotactic microorganisms propagating over a stretching surface by successive taylor series linearization method. *Microgravity Science and Technology*, 30: 445–455, 2018.
5. S.P. Goqo, S.D. Oloniiju, H. Mondal, P. Sibanda, and S.S. Motsa. Entropy generation in mhd radiative viscous nanofluid flow over a porous wedge using the bivariate spectral quasi-linearization method. *Case Studies in Thermal Engineering*, 12: 774–788, 2018.
6. MD. Shamshuddin, S.O. Salawu, H.A. Ogunseye, and F. Mabood. Dissipative power-law fluid flow using spectral quasi-linearization method over an exponentially stretchable surface with hall current and power-law slip velocity. *International Communications in Heat and Mass Transfer*, 119: 104933, 2020.
7. L. Crane. Flow past a stretching plate. *Journal of Applied Mathematics and Physics (ZAMP)*, 1970.
8. J. L. Neuringer. Some viscous flows of a saturated ferrofluid under the combined influence of thermal and magnetic field gradients. *Int. J. Non linear Mech.*, 1: 123–127, 1966.
9. Jr. Rosensweig RE., E. L. Resler. Ferrohydrodynamics. *Phys Fluids*, 7:1927–37, 1964.
10. S.K. Khan. Heat transfer in a viscoelastic fluid flow over a stretching surface with heat source/sink, suction/blowing and radiation. *International Journal of Heat and Mass Transfer*, 49,(3-4): 628–639, 2006.
11. R. Cortell. Viscoelastic fluid flow and heat transfer over a stretching sheet under the effects of a non-uniform heat source, viscous dissipation and thermal radiation. *Int. J. Heat and Mass Transfer*, 50 (15-16): 3152–3162, 2007.
12. H.I. Andersson and O.A. Valnes. Flow of a heated ferrofluid over a stretching sheet in the presence of a magnetic dipole. *Acta Mechanica*, 128: 39–47, 1998.
13. N.G. Kafoussias E.E. Tzirtzilakis and A. Raptis. Numerical study of forced and free convective boundary layer flow of a magnetic fluid over a flat plate under the action of a localized magnetic field. *Zeitschrift frangewandte Mathematik und Physik ZAMP.*, 61(5): 929–947, 2010.
14. M. Sajid T. Hayat, M. Nawaz and S. Asghar. The effect of thermal radiation on the flow of a second grade fluid. *Computers and Mathematics with Applications*, 58(2): 369–379, 2009.

Note: All the figures and table in this chapter were made by the authors.

Impending Inquisitions in Humanities and Sciences (ICIIHS-2022) – Dr. Mohan Varkolu et al. (eds)

Facilitating Easy Navigation within a College Campus using Dijkstra's Algorithm: an Application of Graph Theory

38

K. Lavanya[1]
Assistant Professor, Department of Mathematics,
St. Francis College for Women, Begumpet, Hyderabad, India

K. Sri Sravanti[2]
Professor School of Architecture,
GITAM (Deemed to be University), Hyderabad, India

J. Vijayasekhar[3]
Associate Professor, Department of Mathematics,
School of Science, GITAM (Deemed to be University), Hyderabad, India

ABSTRACT In this paper, the shortest path and shortest distance are determined among the important spots on any college campus by applying Dijkstra's algorithm to an undirected simple weighted graph. Reaching destinations via the shortest routes is a necessity at all levels of daily life. It is the backbone of every navigation system. The buildings on campus, such as administrative, library, labs, and canteens, as well as the paths that connect them, are represented as a graph, with buildings as vertices and connecting paths as edges. An undirected simple weighted graph is obtained for this space network, with weights representing the distance between the buildings.

KEYWORDS Graph theory, Dijkstra's algorithm, Shortest distance

1. Introduction

In places like university campuses, gated communities, or companies like Google or Apple, where there are growing populations, it is becoming more difficult for people to go around on a daily basis by foot because of the larger and larger buildings. Everybody needs access to some basic facilities on a daily basis, such as canteens, banks, administrative offices, and libraries. It is therefore impossible to reach such common facilities via the quickest routes on a large campus. The art and science of designing structures, both built and unbuilt, is referred to as "Architecture" [Francis D.K. Ching and James F. Eckler John, 2012]. It is defined as the modification of our environment made possible by the use of forms, shapes, space, and light. According to French architect Le Corbusier, "Architecture is a masterly, correct, and magnificent dance of shapes beneath the light." Everyone has a distinct definition of architecture. Architecture is both a technical and an artistic and social field. It is the design of any constructed area, structure, or item, including furniture, items, and the fields of town planning, urban design, and landscape architecture. In order to combine data from many sources, Architects must act as mediators. Graphs can be used in real-world situations [V.K. Balakrishnan, 1997; Joel. Mott, 1985]. A transportation network is modelled using graphs, where nodes stand in for locations that deliver or receive goods and edges for the routes that connect them. The earliest and

[1]kalerilavanya@gamil.com, [2]skurri@gitam.edu, [3]vijayjaliparthi@gmail.com

DOI: 10.1201/9781003489436-38

well-known algorithm for single source shortest paths with non-negative edge weights in the graph theory is Dijkstra's algorithm [Adeel Javaid, 2013]. It is used in IP (Internet Protocol) routing to open the shortest path first (OSPF), designate file servers, and take robotic paths in Google Maps, social networking, telephone networks, and other applications. Recently, several researchers have been extensively using Dijkstra's algorithm to solve a variety of problems [Baoyi He, 2022; Wang Shu-Xi, 2012; A. Buzachis et al., 2022; Omoregbe et al., 2016; patel et al., 2014; Ojekudo et al., 2017]. In order to apply Dijkstra's method, we considered the college's buildings to be vertices, and the walking paths that connect them to be edges with weights that indicate how far apart they are from one another.

2. Methodology and Results

In this study, the shortest path and distance between blocks A, D, J, and H, as well as between the two canteens G and SHS on a college campus, are calculated. In these buildings, there are administrative offices, a number of departments, classrooms, and labs that staff and students use regularly. Staff and students can use the SD and SP determined to get to the canteens from these blocks. The paths linking each of these blocks and canteens on the college campus are measured in metres, and staff and students use them to get to their destinations. If there is a direct path and an edge drawn between the localities, they are considered neighbouring.

The following can be seen in the college campus plan.

A-block is next to blocks D and J, thus its vertex degree is 2; similarly, D-block is next to blocks A, H, and the canteen G, so its vertex degree is 3. Blocks D, J, and the SHS canteen are next to H-block, so H-block's vertex degree - deg(H) = 3. J-block has a vertex degree of 3 due to its proximity to blocks A, H, and canteen SHS. The SHS canteen has a vertex degree of 3 due to its proximity to blocks J, H, and canteen G. G canteen has a vertex degree of 2 due to its proximity to block D and canteen SHS.

The distances in metres between them are calculated according to their adjacency, and these numerical values are assigned to the edges as (A, D) = 178m, (A, J) = 37m, (D, H) = 284m, (D, G) = 451m, (H, J) = 69m, (H, SHS) = 49m, (J, SHS) = 51 m, and (G, SHS) = 300m.

With the help of this information, an undirected simple weighted graph with six vertices representing four blocks, two canteens, and eight edges is obtained, as shown in Fig. 38.1.

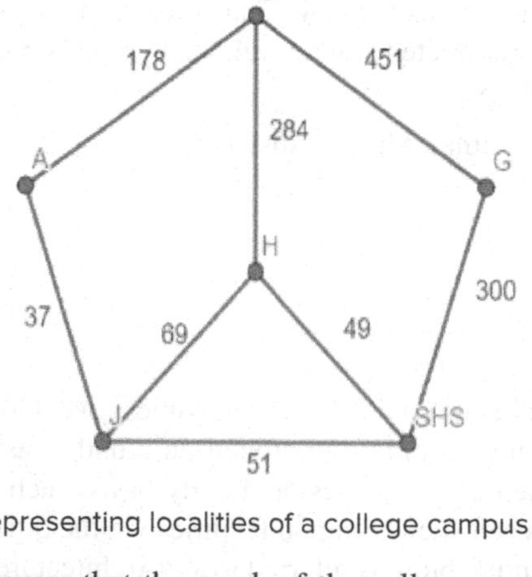

Fig. 38.1 Representing localities of a college campus into a graph

The graph theory fundamental theorem ensures that the graph of the college campus space network is a simple undirected weighted graph.

Theorem: If $V = \{v_1, v_2, \ldots v_n\}$ is the vertex set of an undirected graph G, then $\sum_{i=1}^{n} \deg(v_i) = 2|E|$.

The above said theorem [1] is applied for the graph shown in Fig. 38.1 as following.

Proof:

Let $V = \{A, D, J, H, G, SHS\}$ = number of vertices in the network

$$E = \{(A, D), (A, J), (D, H), (D, G), (H, J), (H, SHS), (J, SHS), (G, SHS)\}$$

Thus, $|V| = 6$, $|E| = 8$

Sum of degrees of all the vertices

$$\sum_{i=1}^{6} \deg(v_i) = \deg(A) + \deg(D) + \deg(H) + \deg(J) + \deg(SHS) + \deg(G).$$
$$= 2+3+3+3+3+2 = 16 = 2|E|.$$

SD and SP from fixed vertex A to other vertices are determined using Dijkstra's algorithm on an **undirected simple weighted** graph representing the college campus space network.

Theorem: Dijkstra's algorithm finds the shortest path (SP) and shortest distance (SD) from a fixed vertex v to any vertex i in the network if there is a path from v to i.

This theorem is applied to the graph in Fig. 38.1 representing the network of buildings in a college campus. In the graph A, D, J and H blocks and two canteens G, SHS in the college campus are vertices and the walking paths among them are edges with weights assigned to them. The weights indicate the distance between them.

Proof:

Let $V = \{A, D, J, H, G, SHS\}$ = number of vertices in the networks = source vertex = A, P = set of vertices with permanent labels. For vertex i, $L(i)$ indicates permanent label value and $L'(i)$ indicates tentative label value.

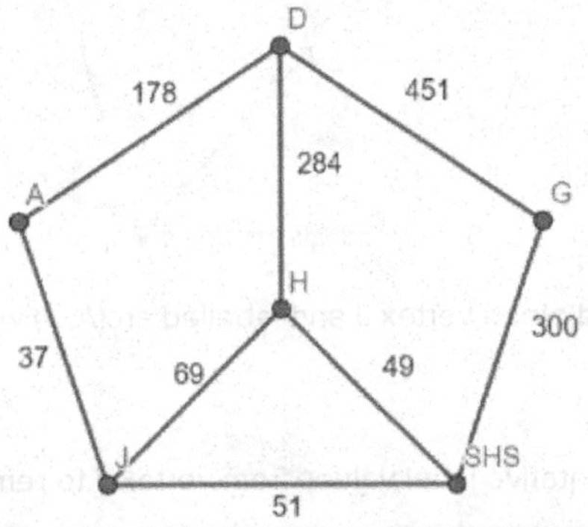

Fig. 38.2 Simple weighted graph of the space network

$$a(i, j) = \begin{cases} \text{weight of the arc from } i \text{ to } j \text{ if they are directly connected} \\ +\infty, \text{ if they is no direct connectivity between } i \text{ and } j \end{cases}$$

Each vertex i is assigned a label that is either permanent or tentative. The permanent label L(i) is the SD from A to i. The tentative label $L'(j)$ is an upper bound of L(i).

Initialization from the source vertex A.

P = {A} and L(A) = 0

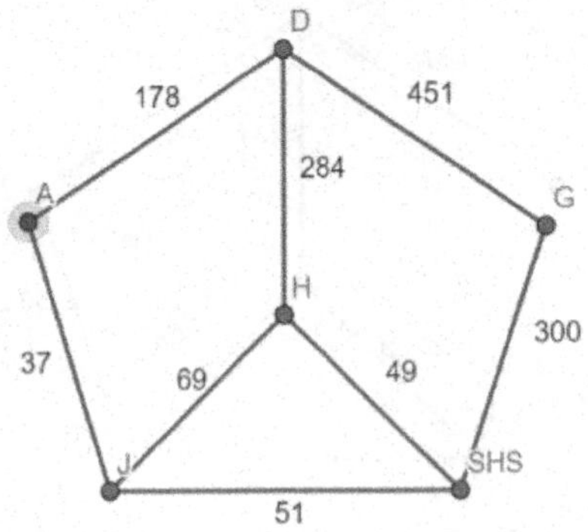

Fig. 38.3 Initialized from source vertex A indicating with red colour

Table 38.1 Tentative label values from vertex A to remaining vertices

$L'(D)$	178	Adjacent to A
$L'(J)$	37	Adjacent to A
$L'(H)$	∞	Non-Adjacent to A
$L'(SHS)$	∞	Non-Adjacent to A
$L'(G)$	∞	Non-Adjacent to A

From table $L'(J) = 37$ is shortest so adjoin J to P.

Therefore, $P = \{A, J\}$ and $L(J) = 37$ and arc (A, J) is labelled and indicated in red colour as shown in Fig. 38.4.

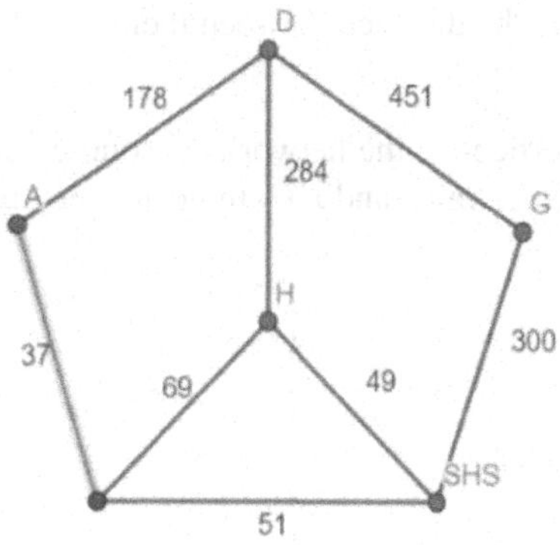

Fig. 38.4 Adjoined vertex J and labelled arc(A, J) with red colour

From vertex J

Table 38.2 Tentative label values from vertex J to remaining vertices

$L'(H) = Min\{L'(H), L(J) + \text{arc}(J, H)\} = Min\{\infty, 37 + 69\}$	106
$L'(SHS) = Min\{L'(SHS), L(J) + \text{arc}(J, SHS)\} = Min\{\infty, 37 + 51\}$	88
$L'(D) = Min\{L'(D), L(J) + \text{arc}(J, D)\} = Min\{178, 37 + \infty\}$	178
$L'(G) = Min\{L'(G), L(J) + \text{arc}(J, G)\} = Min\{\infty, 37 + \infty\}$	∞

From table $L'(SHS) = 88$ is shortest so adjoin SHS to P.

Therefore, $P = \{A, J, SHS\}$ and L(SHS) = 88 and arc (J, SHS) is labelled as shown in Fig. 38.5.

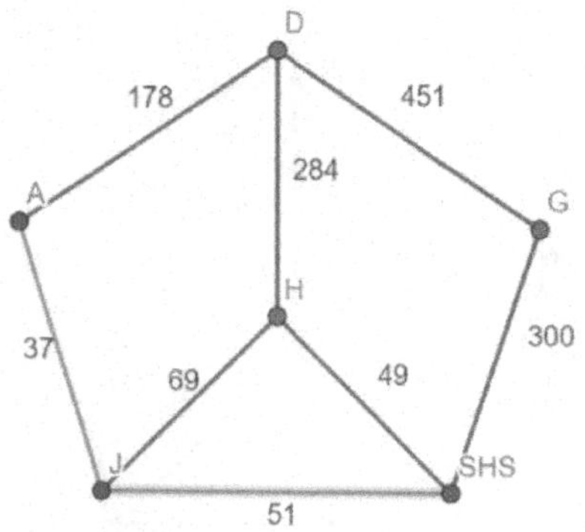

Fig. 38.5 Adjoined vertex SHS and labelled arc (J, SHS) with red colour

From vertex SHS

Table 38.3 Tentative label values from vertex SHS to remaining vertices

$L'(H) = Min\{L'(H), L(SHS) + arc(SHS, H)\} = Min\{106, 88 + 49\}$	106
$L'(G) = Min\{L'(G), L(SHS) + arc(SHS, G)\} = Min\{\infty, 88 + 300\}$	388
$L'(D) = Min\{L'(D), L(SHS) + arc(SHS, D)\} = Min\{178, 88 + \infty\}$	178

From table $L'(H) = 106$ is shortest. So, adjoin H to P. Therefore, $P = \{A, J, SHS, H\}$ and L(H) = 106 and arc (J, H) is labelled as shown in Fig. 38.6.

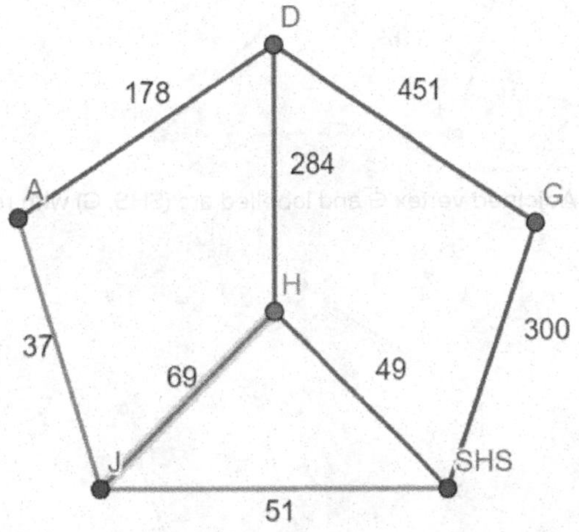

Fig. 38.6 Adjoined vertex H and labelled arc (J, H) with red colour

From vertex H

Table 38.4 Tentative label values from vertex H to remaining vertices

$L'(D) = Min\{L'(D), L(H) + arc(H, D)\} = Min\{178, 106 + 284\}$	178
$L'(G) = Min\{L'(G), L(H) + arc(H, G)\} = Min\{388, 106 + \infty\}$	388

From table $L'(D) = 178$ is shortest. So, adjoin D to P. Therefore $P = \{A, J, SHS, H, D\}$ and L(D) = 178 and arc (A, D) is labelled instead of arc(H, D) as shown in Fig. 38.7.

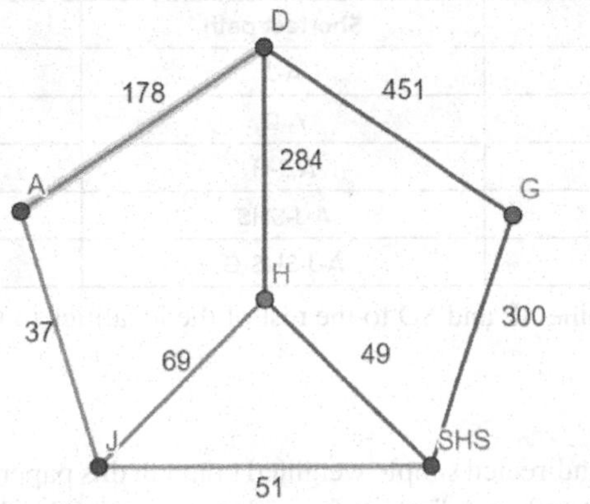

Fig. 38.7 Adjoined vertex D and labelled arc (A, D) with red colour

From vertex D

Later G is adjoined in P. Therefore, $P = \{A, J, SHS, H, D, G\}$ and L(G) = 388 and arc (SHS, G) is labelled as shown in Fig. 38.8

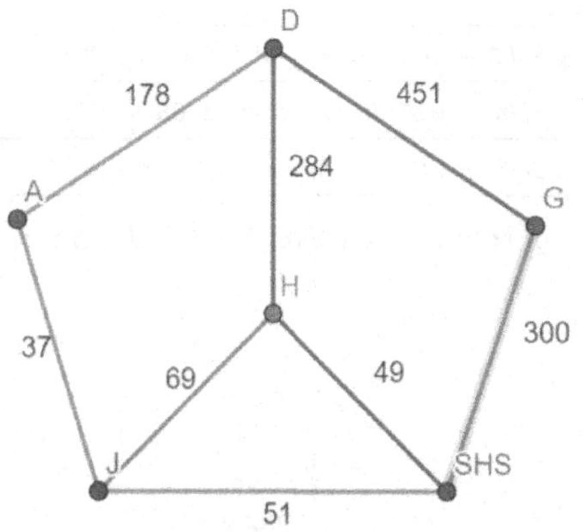

Fig. 36.8 Adjoined vertex G and labelled arc (SHS, G) with red colour

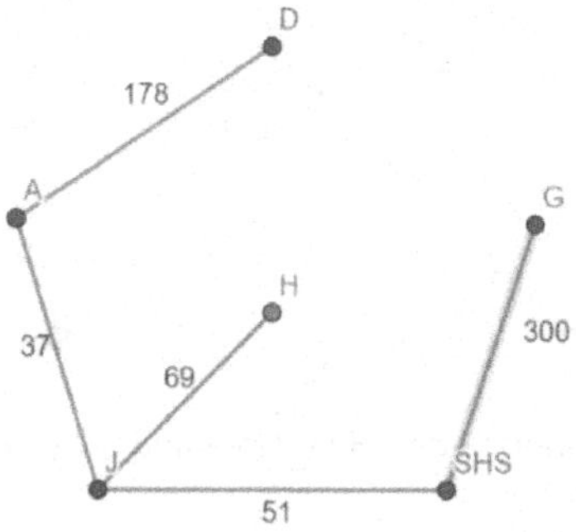

Fig. 38.9 SD & SP from vertex A to remaining vertices

Final result of SD and SP from fixed vertex A to all other vertices is determined as shown in Fig. 38.13.

$P = \{A, J, SHS, H, D, G\}$ Arc (A, J), arc (J, SHS), arc (J, H), Arc (A, D), arc (SHS, G) are labelled

Fig. 38.9 depicts the shortest path and shortest distance from locality A to the rest of the localities on the college campus using Dijkstra's algorithm.

Table 38.5 Shortest distance (SD) and shortest path (SP) from vertex A to remaining vertices

Locality	Shortest path	Shortest distance
J	A-J	37
D	A-D	178
H	A-J-H	106
SHS	A-J-SHS	88
G	A-J-SHS-G	388

Similar procedure is followed to determine SP and SD to the rest of the localities in the college campus.

3. Conclusions

The space network is converted into an undirected simple weighted graph in this paper. Since weights are non-negative numbers, the Dijkstra method is used. The shortest path and distance from block A to the other blocks and the two canteens on the college campus are calculated using Dijkstra's algorithm. To calculate the SP and SD among themselves, a similar process is used in the remaining places. According to their needs, college employees and students frequently visit the library, canteen, labs, and other vital locations. Finding the most direct routes between these locations enables them to travel quickly and comfortably. In the future, this application may be used to create navigational apps for locations like college campuses and gated communities.

REFERENCES

1. A. Buzachis, A. Celesti, A. Galletta, J. Wan and M. Fazio. (2022). Evaluating an Application Aware Distributed Dijkstra Shortest Path Algorithm in Hybrid Cloud/Edge Environments. IEEE Trans. Sustain. Comput. 7(2): 289–298.
2. Agarana, M, Omoregbe, N. and Ogunpeju, M. (2016). Application of Dijkstra Algorithm to Proposed Tramway of a Potential World Class University. *Applied Mathematics*. 7: 496–503.
3. Baoyi He. (2022). Application of Dijkstra algorithm in finding the shortest path. *J Phys Conf Ser*. 2181: 012005
4. Francis D.K. Ching and James F. EcklerJohn. (2012). Introduction to Architecture. Wiley & Sons Inc.
5. Javaid, M. A. (2013). Understanding Dijkstra's Algorithm . SSRN: http://dx.doi.org/10.2139/ssrn.2340905
6. Joel. Mott, Abraham Kandel Theodore P. Baker. (1985). Discrete Mathematics for computer scientists & Mathematics, Brady.
7. Ojekudo, Nathaniel Akpofure and Akpan, Nsikan Paul. (2017). An application of Dijkstra's Algorithm to shortest route problem. IOSR Journal of Mathematics. 13(3): 20–32.
8. Patel, V. and Bagar, C. (2014). A survey paper of Bellman-Ford algorithm and Dijkstra algorithm for finding shortest path in GIS application. International Journal of P2P Network Trends and Technology, 5: 1–4.
9. V.K. Balakrishnan. (1997). Schaum's Outline of Theory and Problems of Graph theory. McGraw Hill
10. Wang Shu-Xi. (2012). The Improved Dijkstra's Shortest Path Algorithm and Its Application. Procedia Eng. 29: 1186–1190.

Note: All the figures and tables in this chapter were made by the authors.

Impending Inquisitions in Humanities and Sciences (ICIIHS-2022) – Dr. Mohan Varkolu et al. (eds)

Application of Graph Theory to the Computation of Hydrocarbon Topological Indices: Ketone

39

Pranavi Jaina*
Department of Mathematics, School of Science,
GITAM (Deemed to be University), Hyderabad, India

Koppula Suresh
Department of Chemistry, School of Science,
GITAM (Deemed to be University), Hyderabad, India

J. Vijayasekhar
Department of Mathematics, School of Science,
GITAM (Deemed to be University), Hyderabad, India

ABSTRACT To explain the topological properties of the carbon atom skeleton or the graph of a hydrocarbon molecule with a hetero oxygen atom, numerous topological indices have been established. This paper suggests topological indices, such as the Wiener index, which is based on the degree and distance of a graph, to theoretically determine the boiling point of aliphatic compounds. In this study, the Wiener Index, a quantitative structure-activity relationship investigation, is applied to evaluate the aliphatic ketone (2,6-dimethylheptane-4-one).

KEYWORDS Topological indices, Wiener index, Aliphatic ketone

1. Introduction

There is a direct relationship between the quantity, type, and structural arrangement of the atoms in a molecule and its physical qualities, including its boiling point [1]. Aliphatic hydrocarbons share a set of defining characteristics, including an identical number and type of atoms, with any observed differences in physical attributes attributable only to shifts in the nature of the atom's structural interrelationships[2]. Assuming there is theoretical value, investigating how paraffin boiling points are affected by pure structural variation should be of interest. In this study, we will demonstrate that a molecular graph theory approach yields predictable and useful outcomes[3].

Physical and chemical properties of molecules are encoded in molecular graphs, which are based on organic molecular structures. As a result, a molecular graph is a basic graph in which vertices stand in for atoms and edges for chemical bonds[4]. Graph invariants known as topological indices of molecular graphs are employed to make predictions about chemical behaviour, such as the boiling and melting points of diverse compounds[5–8]. Wagner and Wang, for instance, provided an overview of key concepts and methods in chemical graph theory[9], drawing attention to topological indices such those based on distance, degrees, counts, etc. One of the earliest chemical topological indexes is the Wiener index, established by Harry Wiener in 1947[10]. He found that the molecular structure of paraffin was correlated with its boiling point.

*Corresponding author: jpranavi1006@gmail.com

DOI: 10.1201/9781003489436-39

The topological index of a molecule, known as the Wiener index, is calculated by adding the distances of the shortest pathways between all pairs of vertices in the chemical graph that stands in for the non-hydrogen atoms in the molecule.

Wiener index to show relationships between the topological structure of molecular graphs and the physical properties of organic molecules. It is one of the most common ways to describe the relationship between a chemical compound's property and its molecular graph.

To calculate a graph's Wiener index, or W(G), the total of the distances between every pair of vertices in the graph G.

$$W(G) = \frac{1}{2}\sum_{i=0}^{n}\sum_{j=1}^{n} d(v_i, v_j)$$

Where $d(v_i v_j)$ is the number of edges in a shortest path connecting the vertices v_i and v_j Wiener index of a complete graph K_n is

$$W(K_n) = \frac{n^2 - n}{2}$$

and path graph P_n,

$$W(P_n) = \frac{n^3 - n}{6}$$

Among all the trees on n vertices, the star $K_{1,n-1}$ has the lowest Wiener number and the path P_n has the largest Wiener number and hence

$$W(K_{1,n-1}) \le W(T) \le W(P_n)$$

2. Results and Discussion

In terms of carbon-carbon bonds, the path number w is defined as the sum of the distances between any two carbon atoms in the molecule. Calculation method in brief: Multiply the number of carbon atoms on one side of any bond by the number of carbon atoms on the other side; the sum of these values for all bonds is w

The gaseous stream emulating emissions from a leather company contains the volatile organic molecule aliphatic ketone 2,6-dimethyl heptane-4-one. Research into using a suspended-growth bioreactor (SGB) to biologically treat it has been undertaken.

Its associated graph G are shown in Fig. 39.1. Ten vertices make up the entire graph, and five of those vertices are pendent. We calculated the graph's Wiener Index for 2,6-dimethyl heptane-4-one and listed the results in Table 39.1 below.

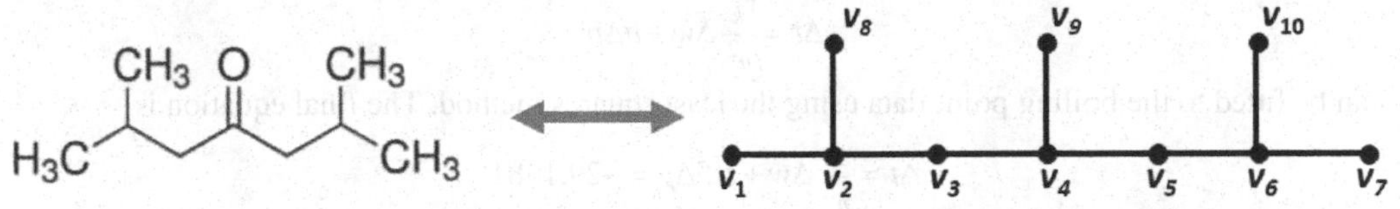

Fig. 39.1 The chemical structure and its graph of 2,6-dimethyl heptane-4-one

Source: Authors

Table 39.1 Calculation of wiener index of 2,6-dimethyl heptane-4-one

	1	2	3	4	5	6	7	8	9	10	W(G)
1	0	1	2	3	4	5	6	2	4	6	35
2	1	0	1	2	3	4	5	1	3	5	25
3	2	1	0	1	2	3	4	2	2	4	21
4	3	2	1	0	1	2	3	3	1	3	19
5	4	3	2	1	0	1	2	4	2	2	21

	1	2	3	4	5	6	7	8	9	10	W(G)
6	5	4	3	2	1	0	1	5	3	1	25
7	6	5	4	3	2	1	0	6	4	2	33
8	2	1	2	3	4	5	6	0	4	6	33
9	4	3	2	1	2	3	4	4	0	4	27
10	6	5	4	3	2	1	2	6	4	0	33

Source: Authors

$$w = \frac{35+25+21+19+21+25+33+33+27+33}{2} = 135$$

$$w_0 = \frac{1}{6}(n-1)n(n+1) = 120 \quad (n = \text{number of carbons})$$

$$\Delta w = w_0 - w = -15$$

Organic compounds boiling points, as well as all of their physical properties, are functionally determined by the number, type, and structural arrangement of the atoms in the molecule. Within an aliphatic group, both the number and type of atoms are constant, and variations in physical properties are caused solely by changes in structural interrelationships. The effect of pure structural variation on the boiling point of hydrocarbons is expected to be of theoretical interest. This paper will demonstrate that this approach can easily produce satisfactory results. The boiling points of the ketone are given by the linear formula

$$t_B = aw + bp + c$$

The number of pairs of carbon atoms separated by three carbon-carbon bonds is defined as the polarity number p.

$$C_1C_4,\ C_2C_5,\ C_3C_6,\ C_4C_7,\ C_4C_{10},\ C_4C_8,\ C_2C_9,\ C_9C_6$$

So, here p = 8

$$p_0 = n - 3 = 6, \Delta p = p_0 - p = -2$$

Let be the straight-chain ketone boiling point with structural variables w_0 and p_0, and let $\Delta t = t_0 - t_B$, $\Delta w = w_0 - w$, and $\Delta p = p_0 - p$. The expression then becomes for a ketone with structural variables w and p.

$$\Delta t = a\Delta w + b\Delta p$$

This has been expanded to include the entire hydrocarbon series. The following relationship was discovered to hold for a compound with n carbon atoms.

$$\Delta t = \frac{k}{n^2}\Delta w + b\Delta p$$

This equation can be fitted to the boiling point data using the least squares method. The final equation is

$$\Delta t = \frac{98}{n^2}\Delta w + 5.5\Delta p = -29.1481$$

Not only is the change in nomenclature important because of the ensuing simplification, but it also connects the boiling points of the branched isomers to the boiling points of the regular hydrocarbons, which have been considerably more intensively and accurately determined and correlated throughout the series. Egloff's equation[11] in particular.

$$t_0 = 745.42\log(n+4.4) - 689.4 = 150.76$$

reproduces the data to within their experimental limits.

$$\Delta t = t_0 - t_B = 179.91$$

The experimental data are reproduced by this equation with an average deviation of 0.5

3. Conclusion

In this paper, we provide the results of our investigation into various topological indices for polycyclic aromatic hydrocarbons that are based on degree. The Wiener index has been done accurately.

REFERENCES

1. Katritzky AR, Kuanar M, Slavov S, et al. Quantitative correlation of physical and chemical properties with chemical structure: Utility for prediction. Chem Rev. 2010;110:5714–5789.
2. Reddy MSB, Ponnamma D, Choudhary R, et al. A comparative review of natural and synthetic biopolymer composite scaffolds. Polymers (Basel). 2021;13.
3. Perdih A, Perdih B. On the structural interpretation of topological indices. Indian J Chem - Sect A Inorganic, Phys Theor Anal Chem. 2003;42:1219–1226.
4. Rouvray DH. The Rich Legacy of Half a Century of the Wiener Index. Topol Chem. 2002;16–37.
5. Stiel LI, Thodos G. The normal boiling points and critical constants of saturated aliphatic hydrocarbons. AIChE J. 1962;8:527–529.
6. Yu Y, Khalid A, Aamir M, et al. On Some Topological Indices of Metal-Organic Frameworks. Polycycl Aromat Compd. 2022;
7. Mondal S, De N, Pal A. Topological Indices of Some Chemical Structures Applied for the Treatment of COVID-19 Patients. Polycycl Aromat Compd. 2022;42:1220–1234.
8. Khalid A, Khan MA, Hussain M, et al. Topological Co-Indices of Molecular Structure of Porphyrazine Network. Polycycl Aromat Compd. 2022;
9. Suparyanto dan Rosad (2015). Predicting Anti HIV Activity of Quinolone Carboxylic Acids-Computation Approach Using Topological Indices. Suparyanto dan Rosad (2015. 2020;5:248–253.
10. Wiener H. Structural Determination of Paraffin Boiling Points. J Am Chem Soc. 1947;69:17–20.
11. David English W. Further Applications for Egloff's Boiling Point Equation. II. J Am Chem Soc. 1952;74:2927–2928.

Impending Inquisitions in Humanities and Sciences (ICIIHS-2022) – Dr. Mohan Varkolu et al. (eds)
© 2024 Taylor & Francis Group, London, ISBN 978-1-032-78829-6

Smart Agriculture and Automatic Irrigation System Using IoT

40

Roshan Raj A K, Mashetty Chakrapani, C. Vijay Krishna*, Kumar Datta, Ashrith Reddy, Prasanna Lakshmi Akella
Assistant Professor, Department of Electronics and communication Engineering, Koneru Lakshmaiah Education Foundation, Hyderabad, Telangana, India

ABSTRACT The goal of our project is to develop a smart agriculture and irrigation system that uses soil moisture sensor and a microcontroller to control the amount of water that is poured into the soil. In addition, we have included a facility that allows users to monitor their field from anywhere. This feature was enabled by the use of a WIFI integrated module that is used to send data to the cloud. In our case, the data accumulated by the smart system can be used to create a send-receive website for the users. The ability to monitor the moisture values of the soil is simplified depending on the user's needs. In addition, the OLED is provided that can be used to display the moisture content of the soil without an internet connection. However, this feature requires the user to manually visit the field and check the system.

KEYWORDS OLED display, NODEMCU ESP8266, Agriculture, Soil moisture, Irrigation, IoT

1. Introduction

Artificial irrigation is a process that involves the use of water in the field. There are various types of systems that are designed to provide the best possible service to the soil. They are usually replacing older equipment and delivering them to the soil more efficiently.

This technique is commonly used in combination with other methods to provide the best possible service to the soil. It eliminates the need for human intervention and allows the field irrigation system to be operated on its own. This project aims to develop a smart irrigation system. As such, in this project we have tried to:

1. Reduce the wastage of water in name of irrigation
2. Reduce manual efforts in irrigation of the fields
3. Increase the crop yielding of the field with a proper well- calculated way of irrigation rather than the commonly used prediction method

This report presents a system that will allow farmers to monitor and analyze the soil condition at any time.

It will allow them to make informed decisions regarding the best possible crop production and improve the efficiency of their operations. The system will also allow them to reduce their labor costs and water consumption. It will additionally help them make agriculture more profitable by providing them with the necessary tools and resources to improve their farming.

*Corresponding author: vijayjaliparthi@gmail.com

DOI: 10.1201/9781003489436-40

2. Methodology

Materials Used

A smart Agriculture & Automatic Irrigation system is composed of various crucial components are given as follows: -

- NODEMCU ESP8266
- Capacitive Soil Moisture Sensor
- Breadboard
- Connector wires/Jumper cables
- Laptop
- Single channel relay module
- Mini submersible pump
- OLED Display MODULE
- The software which we are used in the project are, ThingSpeak, Arduino IDE application for programming of NODEMCU.

NODEMCU ESP8266

The NODEMCU is an open-source device that can be used to create microcontroller packets. It is designed to be used in various operating systems. These packets are programmed to control the system's functions. In the example shown in Fig. 40.1, the smart agent is used to determine the system's working state and command it. Usually, NODEMCU boards are cheaper than other similar devices.

Fig. 40.1 Node MCU ESP8266

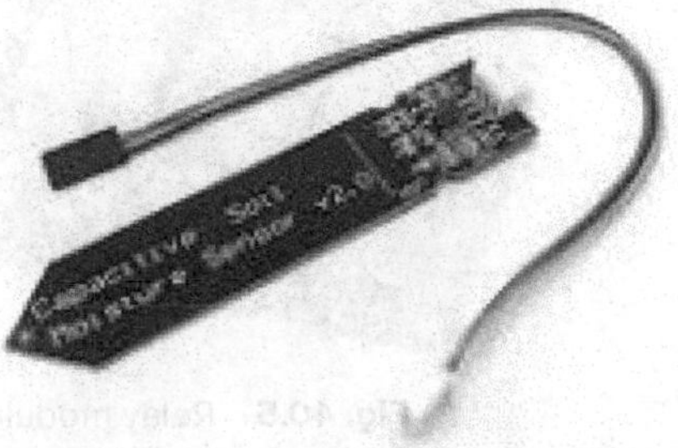

Fig. 40.2 Capacitive moisture sensor

Soil Moisture Sensor

The figure shown in Fig. 40.3 are from the Capacitive soil moisture sensor, which is designed to measure the moisture level of the soil. It performs two tests to determine the current's path to the dirt, and then inspects the barrier of the ground to check its moisture level. It's important to note that the water's proximity to the dirt makes it more inclined to guide the power, which means that the R(resistance) of the soil is lesser in wet soil. On the other hand, the dry soil has a better conductivity of intensity, which means that it can provide more surety than wet soil. The sensor is assembled on this property of intensity. There should be a point that supports the obstacle into voltage, this is done by making use of a circuit that shows us the inside of the sensor, which changes over the opposition into voltage.

Breadboard and Jumper Wires

A breadboard is a device that can be used to add a temporary model of a test circuit design. It can be made without any additional solid joint by connecting a single jump wire to a circuit board's header or a touch of the test circuit. A bounce wire, also known as a jumper, is an electrical wire that is typically used in the connections between a pair of wires. These wires are commonly referred to as "tinned" or "jumper wires." They are usually used in the interconnections of different components inside or outside a circuit board, without soldering them together.

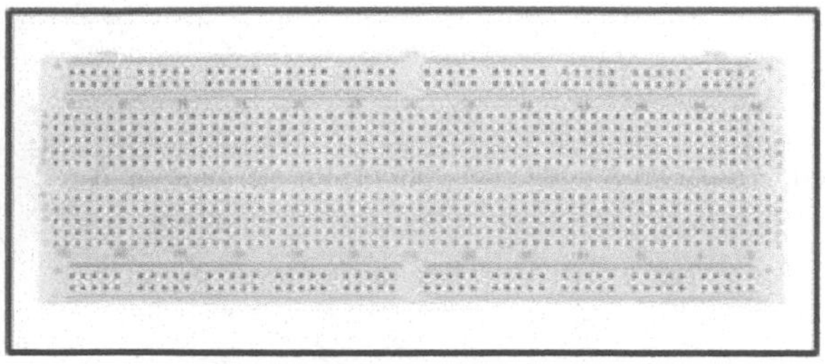

Fig. 40.3 Breadboard

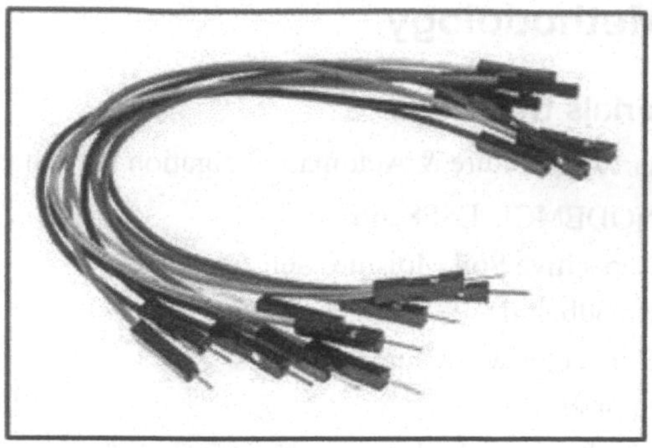

Fig. 40.4 Jumper wires

5V Four Channel Relay Module

5v relay module is perfect for single-chip microcontroller and family apparatus control applications. It has a triode drive that builds transfers curls and high impedance controller pins. This module's draws-down current maintains a strategic distance from the glitch and there are control supply and control pointer LEDs on the board. And some of all this device can turn off-on the motor by the command of NODEMCU

Fig. 40.5 Relay module

Fig. 40.6 Mini submersible pump

Mini Submersible Pump

A mini submersible pumps a water pump that is fully submerged in the water and pumps the water to the place wherever required. In our project, we are using this mini pump as a prototype or the replica of the Pump which is actually used in irrigating the fields.

LCD display Module

This is an I2C module that features a 128 x 64 dot matrix display. It's ideal for use when you need a small and ultra-small display. Compared to traditional LCDs, OLEDs offer a wide variety of advantages, such as high brightness, low power consumption, and a slim outline. It can be compatible with most 3.3V-5V microcontrollers.

Fig. 40.7 O LED Display

Fig. 40.8 ThingSpeak Site

Laptop

For coding the microcontroller (NodeMCU) and to view the stored and live data a laptop or any pc is required.

Thinkspeak Server

It is an application designed for the IoT. it grants us to create that type of application that collects the data from various sensors which are lined up with ThinkSpeak. At the ThingSpeak website, we can create the ThingSpeak Channel. A channel is the location in the server where the data is to be recorded. Each channel creates 8 fields by default for any kind of data, 3 location type fields, and 1 status type field. By creating our ThingSpeak Channel we can push the desired data to this channel, and let the ThingSpeak process our sensed data from the sensors, and allow the server to store the data and display it in final form of graphs or widgets.

3. Cost Estimation

The cost estimation of the project are as follows

Table 40.1

S.No.	Item Name	Specification	Unit	Cost/Piece (in Rs)	Total Cost (in Rs)
1	NodeMCU 8266	ESP8266 12E Board	1	500	500
2	Capacitive Soil Moisture Sensor	Capacitive Soil Moisture Sensor	1	150	150
3	Breadboard		1	150	150
4	Jumper Cables		30	1.5	45
5	Four Channel Relay Module	5v Switch	1	150	150
6	Mini Submersible Pump	5v-6v DC Pump	1	100	100
7	Dht 11 Sensor	Humidity & Temperature Sensor	1	70	70
8	Oled Display Module	0.96" I2C OLED Display	1	270	270
				Total	**1435/-**

3. Methodology

Connection Part

The connection part includes the connections of the circuit using a breadboard and other types of equipment through jump wires etc. This connection is done as follow-

NodeMCU 8266 A0 pin is connected with a capacitive soil moisture sensor. There are a total number of 3 pins in each capacitive soil moisture sensorthat are directly connected with the microcontroller i.e., NodeMCU 8266.

The way in which the pins are connected with each other is given below:

Soil Moisture Sensor:

- ESP8266 ⟶ Pin A0 ⟶ Pin AVOT
- Pin VCC ⟶ Pin 3V3
- Pin GND ⟶ GND

OLED Display:

- Pin SCK ⟶ Pin D1
- Pin SDA ⟶ Pin D2
- Pin GND⟶ Pin GND
- Pin VCC ⟶ Pin 3.3V

Relay Module:

- Pin VCC ⟶ Pin Vin
- Pin GND ⟶ Pin GND
- Pin IN ⟶ Pin D5

DC Motor:

- Pin Power ⟶ Pin Normally Open (Relay)
- Pin Ground ⟶ Pin Common (Relay)

DHT 11:

- Pin Power ⟶ Pin 3V3
- Pin Out ⟶ Pin GND
- Pin Ground ⟶ Pin D4

Data Transfer Part

Now the analog reading is taken from both the soil moisture sensor and sent to the NodeMCU 8266 board through Coding and jump wire the data is now shown in the OLED Screen, the Serial Monitor and the data is also sent to ThinkSpeak server through the pseudo code below:

```
#Include Libraries required for all the sensors
//Initializing pins for each senor to the appropriate pin connections on the NodeMCU 8266
Void Setup () {
//Initialize baud rate;
//Initialize the OLED values;
//Loop to Connect the module to Wi-Fi;
}
Void Loop () {
//Command to read data from Moisture sensor
soilMoisture=analogRead (Pin Value);
//Command to read temperature and humidity values;
Float temp = dht.readValue();
//Write multiple loops for condition checking and displaying data on the OLED display to the user;
//Writing a loop to check the moisture value and turn on or off the pump
If(moisture <30)
{
digitalWrite(pin,HIGH);
}
Else If(moisture <30)
{
digitalWrite(pin,LOW);
}
//Write a IF condition to push data onto the ThingSpeak server
}
```

Relay Part

The program senses when moisture sensor will detect the value of moisture content more than the desired value it will indicate the NodeMCU 8266 to switch the NO port (Normally Open) of the Relay to NC port (Normally closed). When the switch of the relay is at normally open it starts to surge the current through it and the current starts the motor is switched on and whenever the switch turns to (Normally Closed) the motor switches off without any external interference.

Powering the System

The system starts by powering the main unit with a 5-volt adapter or directly by battery. The water pump must be powered with a separate or external power source in case of higher hp of motors are used

Fig. 40.9 5 V Power adapter supply

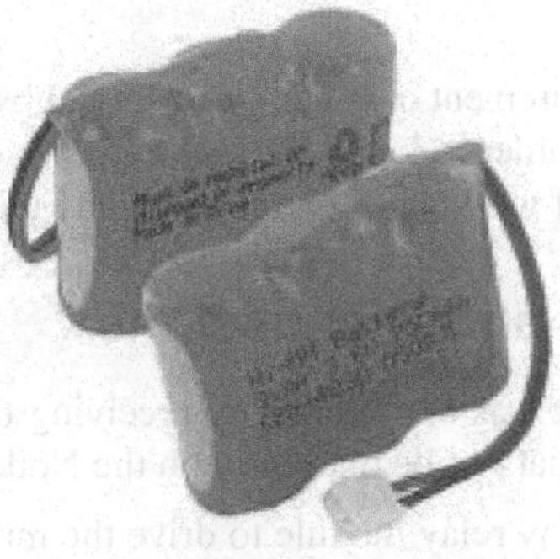

Fig. 40.10 External rechargeable batteries for motor

Initiation of the System

As the switch is turned on, the LEDs on the NodeMCU 8266 Blink, moisture sensor, and relay will also glow assuring the correct connection and also indicating the power is available to those devices.

This project is initiated by connecting to Wi-Fi. When the model turns on the Serial Monitor displays as **"Connecting Wi-Fi."**

As soon as the Wi-Fi is connected the system completes its initiation stage and initiates its process to detect the moisture content from the sensor.

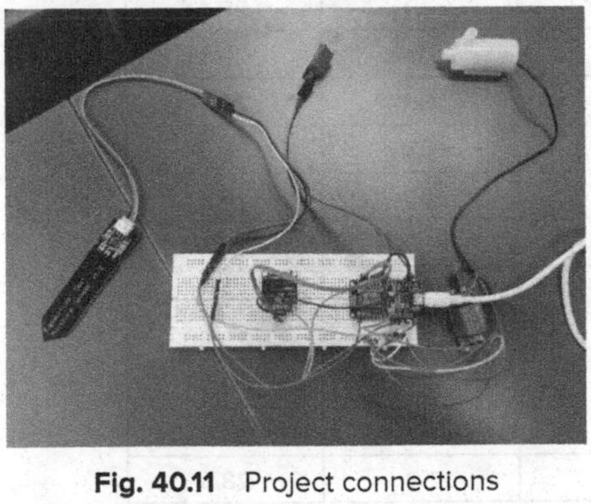

Fig. 40.11 Project connections

Fig. 40.12 Sensors application on the field

Output on the LCD Display

As the system is turned on the following output we will get in order

1. Shows the Moisture Percentage
2. The Air RH
3. The Air Temperature

Location of Application of Sensor on Field

The sensor will be deployed at any desirable [1] place which should be deeper than the pump throwing water once the water value goes below the soil level the sensor will detect the value and accordingly turn on or turn off the motor pump. Without any human intervention and the field will be prevented from being over irrigated.

Data Collection and its Monitoring on ThinkSpeak Webserver

The LCD displays very limited data. Our system is connected to the internet via a mobile phone's hotspot and the exact moisture content its level in percentage with respect to date and time is stored on the website and it can be monitored from around the on the ThinkSpeak web server. And the data can view in the form of Graphs, excel sheet format for better data viewing and its storage. And current moisture content value is also shown on the website separately from two different Sensors in the form of a percentage

The water requirement data can be observed by The Government before designing the Dam and other water storage facilities in that area in order to know the exact figures of water requirement in that area without actually visiting the field. These the methodology of working of Our System and this is how it is supposed to work.

5. Results and Discussions

The experiment was carried out by receiving the input data from the soil moisture sensor 1. The moisture sensor1 gives an analog output that can be read through the NodeMCU 8266 analog pin A0 and DHT 11 Data from Pin D5.

We are using a 5v relay module to drive the mini submersible motor pump. The data collected from various sensors are listed in the table given below. Table 40.2 contains the data from, humidity sensor, Soil moisture sensor Indicated as attributes as A1 which denotes the moisture value in percentage.

Table 40.2 Sample datasheet

DATE & TIME OF ENTRY	Entry_id	Soil Moisture	Air Humidity	Air Temp
2022-10-28 10:43:35 UTC	1	0	48	29.8
2022-10-28 10:44:09 UTC	2	0	49	29.8
2022-10-28 10:44:25 UTC	3	104	49	29.8
2022-10-28 10:44:41 UTC	4	37	49	29.4
2022-10-28 10:44:56 UTC	5	104	49	29.3
2022-10-28 14:35:32 UTC	6	1	53	28.5
2022-10-28 14:35:47 UTC	7	104	53	28.5
2022-10-28 14:36:02 UTC	8	104	53	28.5
2022-10-28 14:36:43 UTC	9	0	53	28.5
2022-10-28 14:36:58 UTC	10	0	53	28.5
2022-10-28 14:37:13 UTC	11	0	53	28.5
2022-10-28 14:38:03 UTC	12	-6	53	28.5
2022-10-28 14:38:41 UTC	13	-6	53	28.5
2022-10-28 14:38:56 UTC	14	2013	53	28.5
2022-10-28 14:39:12 UTC	15	0	53	28.5
2022-10-28 10:43:35 UTC	16	0	48	29.8
2022-10-28 10:44:09 UTC	17	0	49	29.8
2022-10-28 10:44:25 UTC	18	104	49	29.8
2022-10-28 10:44:41 UTC	19	37	49	29.4
2022-10-28 10:44:56 UTC	20	104	49	29.3
2022-10-28 14:35:32 UTC	21	1	53	28.5

Here the graph is plotted between moisture value with time. The graph of the moisture sensor is shown below: -

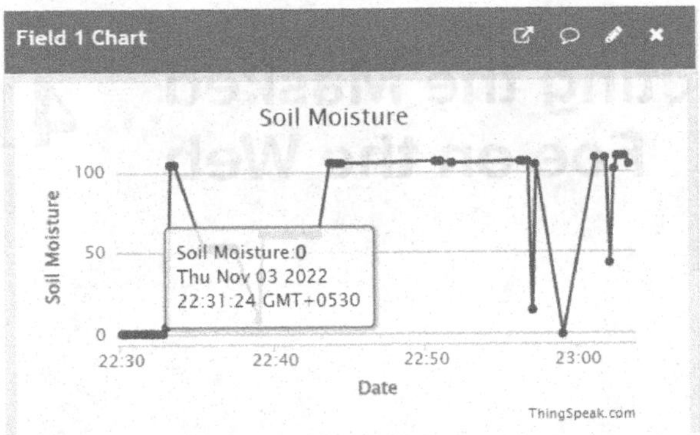

Fig. 40.13 This Graph display shows the data stats of the moisture content recorded by the sensor

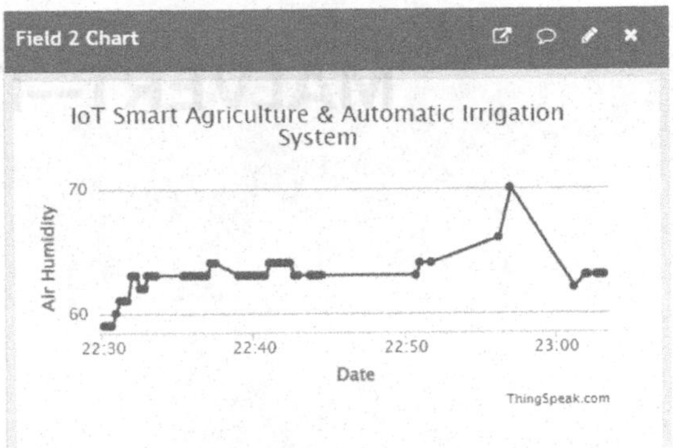

Fig. 40.14 This Graph display shows the data stats of the Air Humidity recorded by the sensor

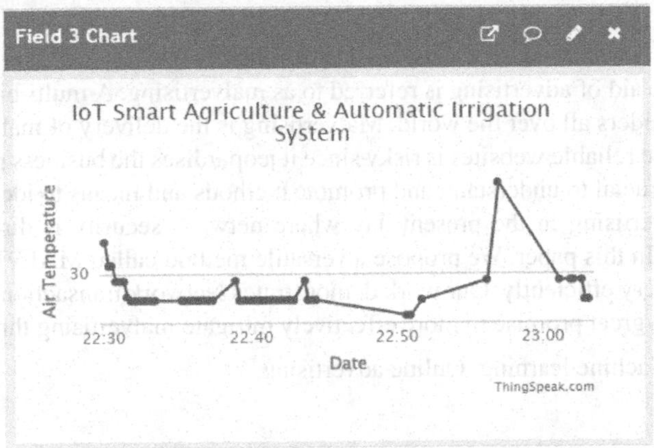

Fig. 40.15 This Graph display shows the data stats of the Temperature of Air recorded by the sensor

Fig. 40.16 OLED output

Hence, we can conclude that the above data are capable to interpret the aim of our project

REFERENCES

1. IJACT: International Journal of Advancements in Computing Technology, 2. Vol. 4, No. 5, pp. 83–90, 2012.
2. Xun-yi Ren, Lin-juan Chen, Hai-shan Wan, "Homomorphic Encryption and Its Security Application", JDCTA: International Journal of Digital Content Technology and its Applications, Vol. 6, No. 7, pp. 305–311, 2012.
3. Ken Cai." Internet of Things Technology Applied in Field Information Monitoring", Advances in Information Sciences and Service Sciences AISS, Vol.4, No.12, pp. 405–414, 2012.
4. Zhao Xing, Liao Guiping, Shi Xiaohui, Chen Cheng, and Li Wen. "Construction of agricultural service mode in IoT and cloud computing environment" [J]. Journal of Agricultural Mechanization Research, Vol. 4 pp. 142–147, 2012.
5. Serdaroglu, K. C., Onel, C., & Baydere, S. (2020, December). IoT Based Smart Plant Irrigation System with Enhanced learning. In 2020 IEEE Computing, Communications, and IoT Applications (ComComAp) (pp. 1–6). IEEE.
6. Pernapati, K. (2018, April). IoT based low-cost smart irrigation system. In 2018 Second International Conference on Inventive Communication and Computational Technologies (ICICCT) (pp. 1312–1315). IEEE.
7. Singh, P., & Saikia, S. (2016, December). Arduino-based smart irrigation using water flow sensor, soil moisture sensor, temperature sensor and ESP8266 Wi-Fi module. In 2016 IEEE Region 10 Humanitarian Technology Conference (R10-HTC) (pp. 1–4). IEEE.
8. Smart Irrigation System for Agriculture Using Internet of Things" Suresh, R.: http://dspace.library.iitkgp.ac.in/bitstream/123456789/29014/1/TH_2156.pdf

Note: All the figures and table in this chapter were made by the authors.

Impending Inquisitions in Humanities and Sciences (ICIIHS-2022) – Dr. Mohan Varkolu et al. (eds)
© 2024 Taylor & Francis Group, London, ISBN 978-1-032-78829-6

MALVERT—Detecting the Masked Foe on the Web

41

Buddiraju Pravallika, K. V. V. Satyanarayana, Praveen Tumuluru
Department of Computer Science and Engineering,
Koneru Lakshmaiah Education Foundation, Vaddeswaram, Guntur, India

ABSTRACT A harmful program that was developed with the aid of advertising is referred to as malvertising. A multi-billion dollar industry, online advertising helps fund web content providers all over the world. Malvertising is the delivery of malware in the form of an advertisement, and its appearance on otherwise reliable websites is risky since it jeopardises the business plans of reliable publishers and respectable online advertisers. It is crucial to understand and promote methods and means to identify and remove dangerous programs such as malware and Malvertising in the present day where network security is directly proportional to revenue in various sectors such as advertising. In this paper, We propose a versatile method called MALVERT. It works effectively against real-world malvertising activities very efficiently. Our work demonstrates Network transactions are analyzed using Machine learning and Neural Networks holds a great promise to more effectively mitigate malvertising threats.

KEYWORDS Malicious program, Malware, Malvertising, Machine learning, Online advertising.

1. Introduction

With the digital revolution and easy access to internet, people have become more curious about the happenings around the world and started consuming huge amounts of content on the internet. This digital consumption delicacy has been made an advantage of, by the ad agencies to generate huge revenues and to deliver ads based on user searches and recent activities. This has enabled crime-doers target unsuspecting victims with malicious advertisements on various websites, blogs, e-commerce stores or via emails etc. Also, in the recent times, with COVID-19 crisis raging around the world, related articles and advertisements have become the most popular traps. Software that is intended to compromise or harm a com- puter system without the owner's knowledge is referred to as malware. According to [1], the term" malware" is basically a catch-all term for all types of computer dangers. File infectors and stand-alone malware make up a straightforward classification of malware. Based on their specific actions, malware can also be categorised into worms, backdoors, trojans, rootkits, spyware, adware, etc. Since all current malware applications tend to have multiple polymorphic layers to avoid detection or to use side mechanisms to automatically update themselves to a newer version at short intervals of time in order to avoid detection by any antivirus software, malware detection through standard, signature-based methods is becoming increasingly challenging. For a demonstration of dynamic file analysis for malware detection using virtual environment emulation.

Malvertising is serving a malware (malicious program) through online advertisements with an aim to compromise user's computer system operations or data [2], [3].

DOI: 10.1201/9781003489436-41

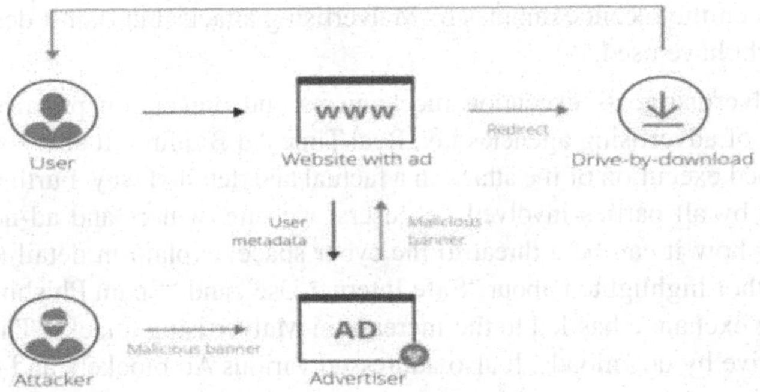

Fig. 41.1 Architecture diagram

Malvertising in the recent times has become a menace to the digital ecosystem. It is a malicious activity that leverages advertisements to deploy various forms of malware to the end user machines [4]. Malicious actors introduce harmful code onto reputable online ad networks. Users are frequently redirected to fraudulent websites by this code. When the web's income were analysed, it was found that the majority came from online adverts. Publisher websites, ad exchanges, ad servers, retargeting networks, and content delivery networks make up the intricate network that is the online advertising ecosystem. After a user clicks on an ad, numerous server redirections take place. Attackers take advantage of this intricacy to hide dangerous content where publishers and ad networks wouldn't normally look. So, this demands a huge focus and interest to understand the whole process behind the Malvertising attack and it is highly important that cyber security enthusiasts and experts focus on such critical areas and develop solutions to a better web and advertising hygiene. The general scenario of the malvertising is represented using the architecture diagram as shown in the Fig 41.1. The main motivation of the this study is 1. To identify ads with malicious content (malverts) and to understand which type of data exfiltration techniques are used. 2. To calculate Entropy values and understand whether the image file is malicious. 3. To examine the real vs malicious codes. 4. To inspect hidden malicious codes within WAV audio files. 5. To analyze the solutions using data mining, machine learning and deep analysis.

The remainder of this paper is structured as follows. Section 2 present the literature review on Malvertising. A detailed description of the proposed MALVERT is introduced in Section 3. The experimental settings and results interpretation are summarized in Section 4. Finally, in Section 5, conclusions are drawn.

2. Literature Review

Advertising is drawing consumers' attention towards a prod- uct or service a company provides, familiarize the audience with it and increase product reach. In order to hit the wide scale of audience, advertisements expanded their scope from televisions, magazines to web and mobile applications. Many world's top ranked companies are highly dependent on advertisements to market their products and to competitively impress the consumers. Users also rely enormously on these ads for product awareness and making purchase decisions. The exponential growth of the digital world has enabled the ads to reach every corner of the physical world. Due to the pandemic in the recent times, an increase in the online activities of consumers has given the companies a free hand to track the activities and push targeted ads to the audience. So, the efficiency of digital marketing has skyrocketed and hence the demand and use of digital advertising is also at an all time high. Taking advantage of this boom, the malefactors are extensively using digital advertising as a platform to inject malicious contents to compromise users' information/ system. For an advert to be hosted on a website, it has to undergo a great deal of background activity, commonly known as "Real Time Ad Bidding." Advertising businesses compete with one another in real-time auctions throughout this bidding process to have their advertisements shown on a certain webpage. Before they are shown on a webpage, online advertising may travel a long way through an ad network. As a result, a request for an advertisement to be displayed on a website may come from one firm, and the actual advertisement may come from a different company further down the chain. This makes it more difficult to determine and keep track of the sources of online advertisements. Online advertising has been extensively used by miscreants for malicious activities. We present the current research in Malvertising.

In [5] the author explains Malvertising and the efficient ways of identifying and protection methods. It presents the advanced forms of Malvertising that can attack the user and help to further analyze what types of advanced concepts may have been used by the criminal. The author highlight Drive-by-download and Drive by attacks mechanism and also mentioned how advanced malvertisements can install a malware on visitor's device. The paper also state how to identify which ads are real and which are potentially harmful. It also focus on the recent examples of Malvertising attacks but didn't describe which forms of advanced malvertising the malefactors might have used.

In [2] the author addresses Malvertising, its execution mechanisms and prevention practices. It provides the concepts of malvertising and the background of advertising agencies i.e., Real Time Ad Bidding. It also stated about malicious exploit kits and browser Plug-ins and described execution of the attack in a factual and detailed way. Further, the author provide the various preventive measures to be taken by all parties involved i.e., users, website owners and ad-networks. Next in [4] the author covers malvertising and explains how it can be a threat to the cyber space, explain in detail about adblockers and elucidates the risks of malvertising. The author highlighted about "Safe Internet Use" and "Spam Phishing" and also mentioned how the multi-party nature of the bidding exchange has led to the increase in Malvertising attacks. The paper also provides deceptive downloads, link hijacking and drive by downloads. It also addressed various Ad blockers and their working mechanism along with its effectiveness.

Later, in [5] explains how data mining offers solutions to Malvertising. It exploded the detail study and analyze of different data mining techniques to be implemented to detect and solve malware. Further, the author proposed solutions to malvertising such as scanning of online advertisement inventory. Utilize a service specifically created for website owners that manage ad inventory to secure advertisements. The service provider performs security audits on a regular basis to look for harmful ads. The author [6] covers raising malvertising risks and explains how machines learning can be used for advanced detection techniques.

3. Proposed System

In this work, we will consider a Research dataset to understand the patterns in data transfer in a network for detecting the malware existence. Here multiple concepts of data Analysis are used for understanding the data and form up the necessary steps of data pre-processing to get a clean and usable data. Then various machine learning algorithms and a Deep Learning model

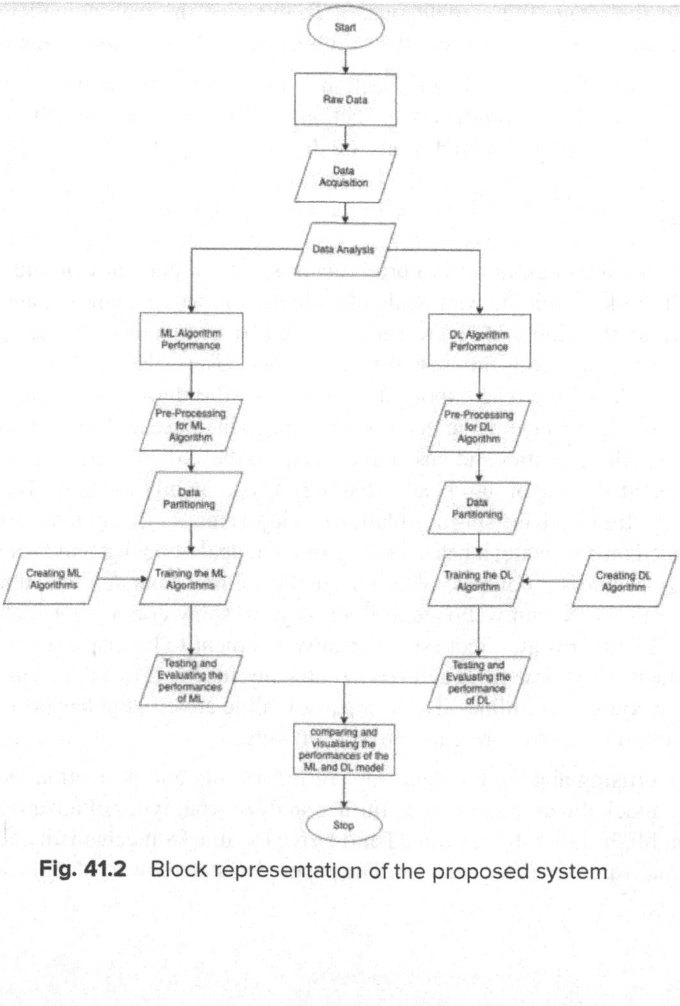

Fig. 41.2 Block representation of the proposed system

will be trained on the data and best one will be selected which will be used to understand the patterns in data and detect for presence of malicious content.

The block representation of the proposed system 2 helps us to understand the whole the proposed MALVERT in detail. Each stages is elaborated in details

Data Acquisition and analysis: In this stage the network transaction data is captured for the analysis.

Data pre-processing After acquiring the data from various source, pre-processing is done to clean and integrate the data. Application of Machine learning Models: After the pre- processing stage the various machine learning algorithms are applied to train the models.

Testing: Finally the trained models are tested to find the best model for detecting malvertising in the network transaction data.

	Android Malware		Training Set		Test Set	
Value	Count	Percent	Count	Percent	Count	Percent
0	199	50.00%	116	48.54%	83	52.20%
1	199	50.00%	123	51.46%	76	47.80%

A. Datasets

In our experiment, we used the Android Malware Genome Project's (MalGenome's) malicious dataset, which was published by a team of researchers [7]. The dataset includes actual Android malware from the real application. It has 49 different malware families. We have 1260 Android application packages (apks) altogether, and they contain trojans, spyware, adware, and other malicious software. Therefore, in line with our strategy, we must record the network traffic from each of these applications (apk files). Additionally, it occasionally occurs that the apk file does not produce any network traffic. In these situations, we choose to ignore them because they fall beyond the purview of our work and require the use of a separate detection method.

Supervised Machine Learning The supervised machine learning technique creates a model to make predictions using a known set of input data and responses by training a model to generate predictions for the response to incoming data based on evidence in the occurrence of hesitation. It is also possible to validate the developed model using a test dataset [17]. Machine learning under supervision can be divided into two main categories: Classification is the process through which an algorithm utilises data to forecast a category. Regression (When a value is used to predict a continuous measurement for an observation). The common classification algorithms used in the experiment are Support vector machines (SVM), Neural networks, Na¨ıve Bayes classifier, Decision trees, K-Nearest neighbors (k-NN). Apart we have also applied deep learning model to train the dataset.

B. Experimental Results

This section conducts a number of related experiments to gauge how well supervised machine learning algorithms per- form dependent on Android setting permissions for outcome evaluation. In the methodology section, we describe how we developed a model utilising a static analysis of the application configuration permissions, train and test datasets, evaluation standards, and the results of these algorithms' experiments. Table 41.1 provides a description of the dataset.

Table 41.1 Performance metrics with its formula

Metric	Formula
Recall	$\frac{TP}{TP+FN}$
Precision	$\frac{TP}{TP+FP}$

Metric	Formula
Specificity	$\frac{TP}{TP+FN}$
Sensitivity	$\frac{TN}{TN+FP}$
F-Measure (FM)	$\frac{2*Precision \times Recall}{Precision+Recall}$
ROC (AUC)	$\frac{Specificity+Sensitivity}{2}$
GM (GM)	$\sqrt{Specificity \times Sensitivity}$

C. Performance Measurement

A confusion matrix is a matrix which contains information about actual and predicted based on that the classification accuracy, commonly known as the correct classification rate, is one of the most basic measures for evaluating the effectiveness of predictive models. It calculates the proportion of correctly categorized samples in the total number of samples. Another metric is precision, which is determined by dividing the number of instances classified as faulty (TP + FP) by the number of cases accurately categorized as defective (TP + FP). Furthermore, recall is the proportion of instances accurately categorized as defective (TP) to the number of faulty examples (TP + FN). The F-score is a harmonic mean of accuracy and recall, and it has been utilized in numerous research. By computing trade-offs between TPR and FPR, ROC-AUC determines the area under the receiver operating characteristic (ROC) curve. We evaluated the performance using F-measure and AUC-ROC metrics for the proposed model I.

D. Technologies Used

The experiments are carried out using MATLAB (2017a) on a computer with the following specifications: System Model: MacBookPr011.5; Processor: Intel(R) Core(TM) i7-4870HQ CPU @ 2.506Hz (8 CPUs); CPU Speed: 2.SGHz 64-bit PC.

E. Result and Discussion

TSupport vector machines (SVM), neural networks (NN), naive bayes (NB), decision trees (DT), k-nearest neighbours (kNN), and convolution neural networks (CNN) were among the supervised machine learning methods used in our exper- iment. After applying these algorithms, the results got from the test are summarized in II.

Table 41.2 Result

Algorithms	Precision		Recal		Accuracy
	0	1	0	1	
CNN	0.96	0.96	0.97	0.96	0.96
KNN	0.81	0.84	0.89	0.89	0.86
DT	0.93	0.93	0.93	0.92	0.94
NN	0.87	0.9	0.85	0.89	0.89
SVM	0.91	0.93	0.92	0.94	0.93
NB	0.85	0.87	0.82	0.83	0.84

4. Conclusion

Malicious ads are omnipresent in modern web advertising, posing a severe risk to both legal businesses and web users. The answer to identifying fraudulent advertising is provided in this publication. We quantify the severity of the threat based on an extensive Web scan. According to prior studies, such attacks impacted a large number of publisher pages and key ad networks, including DoubleClick. Our analysis demonstrates that MALVERT combats actual malvertising operations very successfully. Our research shows how machine learning is used to evaluate network transactions, and neural networks hold significant promise for more effectively reducing the risks associated with malicious advertising.

REFERENCES

1. M. Yong Wong, M. Landen, M. Antonakakis, D. M. Blough, E. M. Redmiles, and M. Ahamad, "An inside look into the practice of malware analysis," in *Proceedings of the 2021 ACM SIGSAC Conference on Computer and Communications Security*, 2021, pp. 3053–3069.
2. A. K. Sood and R. J. Enbody, "Malvertising–exploiting web advertising," *Computer Fraud & Security*, vol. 2011, no. 4, pp. 11–16, 2011.
3. M. F. Ab Razak, N. B. Anuar, R. Salleh, and A. Firdaus, "The rise of "malware": Bibliometric analysis of malware study," *Journal of Network and Computer Applications*, vol. 75, pp. 58–76, 2016.
4. C. Dwyer and A. Kanguri, "Malvertising-a rising threat to the online ecosystem," *Journal of Information Systems Applied Research*, vol. 10, no. 3, p. 29, 2017.
5. S. Kumar, S. S. Rautaray, and M. Pandey, "Malvertising: A case study based on analysis of possible solutions," in 2017 *International Conference on Inventive Computing and Informatics (ICICI)*. IEEE, 2017, pp. 288– 291.
6. D. Gibert, C. Mateu, and J. Planes, "The rise of machine learning for detection and classification of malware: Research developments, trends and challenges," *Journal of Network and Computer Applications*, vol. 153, p. 102526, 2020.
7. Y. Zhou and X. Jiang, "Dissecting android malware: Characterization and evolution," in *2012 IEEE symposium on security and privacy*. IEEE, 2012, pp. 95–109.

Note: All the figures and tables in this chapter were made by the authors.

Impending Inquisitions in Humanities and Sciences (ICIIHS-2022) – Dr. Mohan Varkolu et al. (eds)

Computational Modeling of CNT's Nano Liquid Flow And Heat Transmission Over a Vertical Thin Film with Thermal Variation

42

S. V. Padma[1], M. P. Mallesh*
[1,1]*Research Scholar, Department of Engineering Mathematics,
Koneru Lakshmaiah Education Foundation Hyderabad Campus, Telangana, India
V. Rajesh[2]
Department of Mathematics, GITAM (Deemed to be University), Hyderabad Campus,
Rudraram Village, Patancheru (M), Medak, Telangana, India
O. D. Makinde[3]
Faculty of Military Science, Stellenbosch University, Private Bag X2, Saldanha 7395, South Africa

ABSTRACT We examine the augmentation of heat transfer of viscous incompressible time dependent natural convective flow of water based SWCNT's and MWCNT's nanoliquid past an exponentially accelerated vertical thin film with variable temperature, thermal radiation, viscous dissipation and heat generation. Using Galerkin finite element approach, the flow equations are solved. The alterations in velocity, temperature, skin friction coefficient (C_f) and Nusselt number (Nu) for diverse values of appropriate parameters such as Eckert Number (Ec), Volume fraction (δ), Thermal Grashof number (G), Radiation parameter (N), Heat generation parameter (Q), Acceleration parameter (a), and time (t) are examined in the form of graphs and tables. This study contributes an elaborate application in high temperature nanoliquid processing, chemical engineering coating operations, metallurgy, nano medicine material synthesis and space research as well as to enhance the cooling/heating performance of nuclear power plants and industrial systems.

KEYWORDS Carbon nanotubes (CNT's), Exponential accelerated vertical thin film, Free convection, Galerkin finite element method, Heat generation, Thermal radiation, Viscous dissipation, Water based nano liquid

1. Introduction

The technique of free transmission flows of nanoliquid over an exponentially accelerated vertical thin film has attracted many intellectuals owing to their wide applications in the field of chemical engineering and thermal engineering, geophysics, oceanography, drying processes and solidification of binary alloy. Besides this thin film mechanism can enhance the surface properties of solid surface like absorption, electric performance, diffusion, illumination and penetration etc. Rajesh et al. (2016) developed a numerical model of free convective flow past an up straight surface. Noor et al. (2018) investigated the behaviour of thin film and permeability which have impact on flow pattern when there is a stable temperature. Sravan et al. (2020) theoretically explored the impact of magnetism and viscous dissipation due to free convective nanoliquid flow over an accelerating up right plane. Recently, Anwar et al. (2021) explored the unsteady non linearly convective stream of thin film to analyze the features of heat and mass exchange.

*Corresponding author: malleshmardanpally@gmail.com,
[1]tumuluripadma@gmail.com, [2]v.rajesh.30@gmail.com, [3]makinded@gmail.com

DOI: 10.1201/9781003489436-42

CNT's nanoliquid acts as a capable fluid in the current development of science and technology. Among all the nano particles such as metals, metal oxides, carbides, nitrates or non-metals, CNT's have the high capability of heat conduction. Salma et al (2013), Salma et al (2014) experimentally studied the heat transfer characteristics of CNT's nanoliquid flows on a cone and plate. Analytical result of free convective flow of CNT's nanoliquid accompanied by magnetic field is developed by Ellahi et al (2015). Several authors have recently examined about CNT's nanoliquid flow on different geometries Shafiq (2020), Manjunatha (2021).

Free convection accompanied by heat absorption/generation, thermal radiation, viscous dissipation plays a significant role in the investigation of heat transfer characteristics in flowing liquids, geophysical flows and aerodynamics. Some of the pertinent studies on nanoliquid flows with heat generation on various physical models have been archived by Alsaedi et al (2012), Rajput (2017). Saleem et al (2019) analytically explored the non-Darcy 3D-flow of CNT's over a rotating stretchable disc. The time dependent flow of electrically conducting viscous liquid over an upright plane is analytically developed by Rajesh et al (2009). Imtiaz et al (2016) explored the convective flow of CNT's nanoliquid between rotating discs. Besides these Rajesh et al (2017), Sreedevi et al (2018), Feroz et al (2018), studied the impact of radiation on nanoliquid flows in different geometric models. Makinde et al (2018) numerically demonstrated the impact of radiation and heat source on two-dimensional boundary flow past a slithery stretching surface. Makinde et al (2018) numerically developed the mixed convection flow over an up straight surface with radiation and magnetic field. Further studies Rajesh and Chamkha (2014), Rajesh et al (2015), Promod et al (2020), on convective flows of nanoliquid with viscous dissipation on various physical models are acknowledged. Recently, Mallesh et al (2020) numerically scrutinized the influence of viscous dissipation on heat exchange of kerosene oil based nanoliquid over a vertical plane. Very recently, Prabhakar and Makinde (2022) numerically investigated the influence of Newtonian heating on hydromagnetic natural convective flow over an oscillating permeable surface.

2. Flow Analysis and Solution

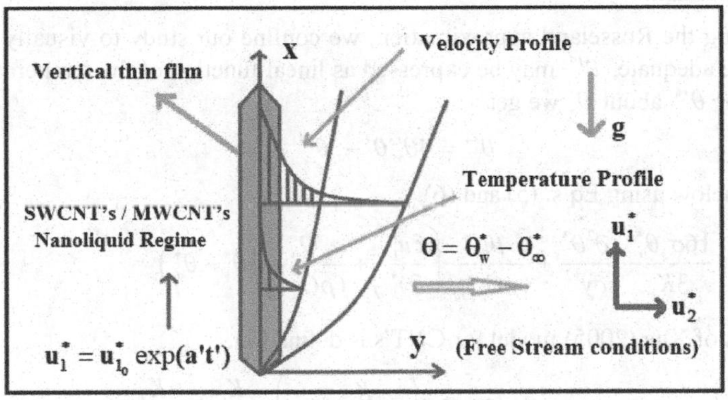

Fig. 42.1 Physical structure and coordinate system

In this simulation time dependent boundary layer flow of CNT's nanoliquid over an exponentially accelerated thin film in the presence of gravitational acceleration is considered, as shown in Fig. 42.1. The y-axis is taken normal to the thin film. Primarily thin film and liquid are kept stationary for $t' \leq 0$. At $t' > 0$, the film starts to move upwards with static velocity $u^*_{1_0}$. Temperature on the surface of thin film is elevated to $\theta^* = \theta^*_\infty$ and retained at the same level thereafter. In this work we presume Boussinesq approximation (Schlichting and Gerstein, (2001)) and Tiwari and Das (2007) nanoliquid model. In view of above presumptions, the following are the controlling equations.

$$\frac{\partial u_2^*}{\partial y^*} = 0 \tag{1}$$

$$\frac{\partial u_1^*}{\partial t'} + u_2^* \frac{\partial u_1^*}{\partial y^*} = \nu_{nf} \frac{\partial^2 u_1^*}{\partial y^{*2}} + \frac{(\rho B)_{nf}}{\rho_{nf}} g(\theta^* - \theta^*_\infty) \tag{2}$$

$$\frac{\partial \theta^*}{\partial t'} + u_2^* \frac{\partial \theta^*}{\partial y^*} = \frac{K_{nf}}{(\rho C_p)_{nf}} \frac{\partial^2 \theta^*}{\partial y^{*2}} + \frac{\mu_{nf}}{(\rho C_p)_{nf}} \left(\frac{\partial u_1^*}{\partial y^*}\right)^2 - \frac{1}{(\rho C_p)_{nf}} \frac{\partial q_r}{\partial y} + \frac{Q_0}{(\rho C_p)_{nf}} (\theta^* - \theta^*_\infty) \tag{3}$$

Table 42.1 Mechanical with physical valuables of CNT's nanoliquid

Model	ρ (Kg/m³)	C (Kg⁻¹/k⁻¹)	K (Wm⁻¹ K⁻¹)	β × 10⁻⁵ (K⁻¹)
Water	997.1	4179	0.613	21
SWCNT's	2600	425	6600	27
MWCNT's	1600	796	3000	44

Source: Sidra Aman et al (2017).

The frontier constraints as:

$$t' \le 0 : u_1^* = 0, \qquad \theta^* = \theta_\infty^* \qquad \text{for all } y^*$$

$$t' > 0 : u_1^* = u_{1_0}^* \exp(a't'), \qquad \theta^* = \theta_\infty^* + \left(\theta_w^* = \theta_\infty^*\right)At' \qquad \text{for } \quad y^* = 0 \tag{4}$$

$$u_1^* \to 0, \qquad \theta^* \to \theta_\infty^* \qquad \text{as} \qquad y^* \to \infty$$

Where $A = \dfrac{u_{1_0}^{*2}}{v_f}$

Rosseland approximation (Brewster, (1992)), is enacted to replicate one directional radiative heat flux q_r as,

$$q_r = -\frac{4\sigma_s}{3K_e} \cdot \frac{\partial \theta'^4}{\partial y} \tag{5}$$

Where K_e = mean absorption coefficient and

σ_s = Stefan-Boltzmann constant.

It is worth noting that by using the Rosseland approximation, we confine our study to visually thick nanoliquid. If thermal variations with in the flow are adequate, θ'^4 may be expressed as lineal function of the temperature, then by ignoring higher order terms in Taylor series for θ'^4 about θ_∞^* we get,

$$\theta'^4 = 4\theta_\infty'^3\theta' - 3\theta_\infty'^4 \tag{6}$$

Equation (3) can be taken as below using Eq's. (5) and (6)

$$\frac{\partial \theta^*}{\partial t'} + u_2^* \frac{\partial \theta^*}{\partial y^*} = \frac{1}{(\rho C_p)_{nf}}[K_{nf} + \frac{16\sigma_s \theta_\infty^{*3}}{3K_e}]\frac{\partial^2 \theta^*}{\partial y^{*2}} + \frac{\mu_{nf}}{(\rho C_p)_{nf}}\left(\frac{\partial u_1^*}{\partial y^*}\right) + \frac{Q_0}{(\rho C_p)_{nf}}\left(\theta^* - \theta_\infty^*\right) \tag{7}$$

Thermal conductivity in terms of Xue (2005) model for CNT's is defined as,

$$\frac{K_{nf}}{K_f} = \frac{(1-\delta) + 2\delta\left(\dfrac{K_{CNT}}{K_{CNT} - K_f}\right)\ln\dfrac{K_{CNT} + K_f}{2K_f}}{(1-\delta) + 2\delta\left(\dfrac{K_f}{K_{CNT} - K_f}\right)\ln\dfrac{K_{CNT} + K_f}{2K_f}}$$

The terms density ρ_{nf}, heat capacitance $(\rho C_p)_{nf}$ and thermal coefficient $(\rho\beta)_{nf}$ can be expressed as

$$\rho_{nf} = \delta\rho_{CNT} + (1-\delta)\rho_f$$

$$(\rho\beta)_{nf} = \delta(\rho\beta)_{CNT} + (1-\delta)(\rho\beta)_f \tag{8}$$

$$(\rho C_p)_{nf} = \delta(\rho C_p)_{CNT} + (1-\delta)(\rho C_p)_f$$

Dynamic viscosity with respect to Brinkman model of the nanoliquid, as:

$$\mu_{nf} = \frac{\mu_f}{(1-\delta)^{2.5}} \tag{9}$$

The relevant dimensionless specifications are as:

$$u_1=\frac{u_1^*}{u_{1_0}^*},\ u_2=\frac{u_2^*}{u_{1_0}^*},\ t=\frac{t'u_{1_0}^{*\,2}}{\nu_f},\ y=\frac{y^*u_{1_0}^*}{\nu_f}=\frac{y^*}{L_{ref}},\ \theta=\frac{\theta^*-\theta_\infty^*}{\theta_w^*-\theta_\infty^*},\ G=\frac{\beta_f\nu_f(\theta_w^*-\theta_\infty^*)}{\left(u_{1_0}^*\right)^3},\ E=\frac{\left(u_{1_0}^*\right)^2}{\left(C_p\right)_f\left(\theta_w^*-\theta_\infty^*\right)},\ Pr=\frac{\nu_f}{\alpha_f}=\frac{\left(\mu C_p\right)_f}{k_f}$$

$$N=\frac{k_f k_e}{4\sigma_s\theta_\infty^{*3}},\ Q=\frac{Q_0\nu_f}{(\rho C_p)_f u_{1_0}^{*2}},\ a=\frac{a'\nu_f}{u_{1_0}^{*\,2}} \tag{10}$$

Dimensionless flow control equations are:

$$\frac{\partial u_2}{\partial y}=0 \tag{11}$$

$$\frac{\partial u_1}{\partial t}+u_2\frac{\partial u_1}{\partial y}=G_1\frac{\partial^2 u_1}{\partial y^2}+G_2G\theta \tag{12}$$

$$\frac{\partial \theta}{\partial t}+u_2\frac{\partial \theta}{\partial y}=\left(G_3+\frac{4}{3N}G_4\right)\frac{1}{Pr}\frac{\partial^2\theta}{\partial y^2}+G_5\left(\frac{\partial u_1}{\partial y}\right)^2+G_4Q\theta \tag{13}$$

Where

$$G_1=\frac{1}{(1-\delta)^{2.5}}\cdot\frac{1}{(1-\delta)+\delta\left(\frac{\rho_{CNT}}{\rho_f}\right)},\ G_2=\frac{(1-\delta)+\delta\left(\frac{(\rho\beta)_{CNT}}{(\rho\beta)_f}\right)}{(1-\delta)+\delta\left(\frac{\rho_{CNT}}{\rho_f}\right)},\ G_3=\frac{K_{nf}}{K_f}\cdot\frac{1}{(1-\delta)+\delta\left(\frac{(\rho C_p)_{CNT}}{(\rho C_p)_f}\right)} \tag{14}$$

$$G_4=\frac{1}{(1-\delta)+\delta\left(\frac{(\rho C_p)_{CNT}}{(\rho C_p)_f}\right)},\ G_5=\frac{1}{(1-\delta)^{2.5}}\cdot\frac{1}{(1-\delta)+\delta\left(\frac{(\rho C_p)_{CNT}}{(\rho C_p)_f}\right)}$$

Dimensionless frontier constraints:

$$t\le 0:\ u_1=0\ \ \theta=0\ \text{ for all }\ y$$
$$t>0:\ u_1=\exp(at),\quad \theta=t\ \text{ at }\ y=0 \tag{15}$$
$$u_1\to 0,\quad \theta\to 0\ \text{ as }\ y\to\infty$$

3. Surface Engineering Properties

Skin friction and Nusselt number are two critical somatic specifications defined on the wall as mentioned below:

$$C_f=\frac{\tau_w}{\rho u_{1_0}^{*\,2}}\ \text{ and }\ Nu=\frac{q_w L_{ref}}{k_f\left(\theta_w^*-\theta_\infty^*\right)} \tag{16}$$

Here $\tau_w(skin-friction(shear\ stress))=\mu_{hnf}\left(\frac{\partial u_1^*}{\partial y^*}\right)_{y^*=0}$

$$q_w(heat\ flux(rate\ of\ heat\ transfer))=-\left(\frac{k_{nf}}{k_f}+\frac{4}{3N}\right)\left(\frac{\partial\theta^*}{\partial y^*}\right)_{y^*=0} \tag{17}$$

By applying the dimensionless variables, we get

$$C_f = \frac{1}{(1-\delta)^{2.5}}\left(\frac{\partial u_1}{\partial y}\right)_{y=0}, \quad Nu = -\left(\frac{k_{nf}}{k_f}+\frac{4}{3N}\right)\left(\frac{\partial \theta}{\partial y}\right)_{y=0} \tag{18}$$

Solution Procedure

A robust Galerkin Finite Element technique has ciphered the flow controlling Eq's (12) and (13) along with pertinent conditions (15). The elementary steps involved in this method are outlined by Rajesh and Chamkha (2014), Reddy (2006), and Bathe (1996). On employing Finite Element technique for Eq's (12) and (13) on the component (e) ($y_j \le y \le y_k$) as,

$$\int_{y_j}^{y_k} N^{*(e)\theta}\,[G_1\frac{\partial^2 u_1}{\partial y^2}+G_2G\theta-\frac{\partial u_1}{\partial t}]\,dy = 0 \tag{19}$$

On simplifying, we get

$$\frac{G_1}{l^{(e)}}\begin{bmatrix} u_{1_{i-1}} \\ u_{1_i} \\ u_{1_{i+1}} \end{bmatrix} + \frac{G_1 l^{(e)}}{6}\begin{bmatrix} 2 & 1 & 0 \\ 1 & 4 & 1 \\ 0 & 1 & 2 \end{bmatrix}\begin{bmatrix} \dot{u}_{1_{i-1}} \\ \dot{u}_{1_i} \\ \dot{u}_{1_{i+1}} \end{bmatrix} = \frac{l^{(e)}}{2}(G_2G\theta)\begin{bmatrix} 1 \\ 2 \\ 1 \end{bmatrix} \tag{20}$$

By setting the row members pertaining to the node to zero, we get from Eq. (20)., where $l^{(e)} = h$.,

$$\frac{G_1}{h}[-u_{1_{i-1}}+2u_{1_i}-u_{1_{i+1}}]+\frac{h}{6}[\dot{u}_{1_{i-1}}+4\dot{u}_{1_i}+\dot{u}_{1_{i+1}}]=\frac{G_2G\theta h}{2} \tag{21}$$

Using α – family of time marching schemes (Trapezoidal rule) we have.,

$$\dot{u}_{1_{i-1}}^{j+1}+4\dot{u}_{1_i}^{j+1}+\dot{u}_{1_{i+1}}^{j+1} = \frac{2}{k}[u_{1_{i-1}}^{j+1}-u_{1_{i-1}}^{j}+4u_{1_i}^{j+1}-4u_{1_i}^{j}+u_{1_{i+1}}^{j+1}-u_{1_{i-1}}^{j}] - [\dot{u}_{1_{i-1}}^{j+1}+4\dot{u}_{1_i}^{j+1}+\dot{u}_{1_{i+1}}^{j+1}] \tag{22}$$

On simplifying, we get

$$u_{1_{i-1}}^{j+1}\left(\frac{6G_1}{h^2}-\frac{2}{k}\right)+u_{1_i}^{j+1}\left(-\frac{12G_1}{h^2}-\frac{8}{k}\right)+u_{1_{i+1}}^{j+1}\left(\frac{6G_1}{h^2}-\frac{2}{k}\right)= u_{1_{i-1}}^{j}\left(-\frac{6G_1}{h^2}-\frac{2}{k}\right)+u_{1_i}^{j}\left(\frac{12G_1}{h^2}-\frac{8}{k}\right)+u_{1_{i+1}}^{j}\left(\frac{6G_1}{h^2}-\frac{2}{k}\right)-3G_2G\theta[\,T_i^j + T_i^{j+1}] \tag{23}$$

The system of Equations (23) is transformed into Matrix Tri-diagonal form. In addition, the above approach is also used for Eq. (13). Thomas algorithm (Carnahan et al (1969)) examines above equations for numerical solutions of velocity and temperature stencils.

Analysis of Grid Independence and convergence of Galerkin FEM

Grid independence work is accomplished to reduce the impact of mesh size on computing the output. To achieve precise findings, step sizes of time and space along the t and y-directions are chosen. A mesh evaluation with slightly updated values of the mesh distance in the t-direction and y-direction, k = 0.01 and h = 0.2, is performed in MATLAB to identify an ideal mesh system for the current simulation. No noteworthy changes were detected in the values of u_1 and θ. We achieved mesh independence for stable, convergent and consistent solutions which are shown in figures 3.

Comparative analysis of Numerical and Analytical Solution

In the absence of inertial term, viscous dissipation and heat generation terms Eq. (13) becomes,

$$\frac{\partial \theta}{\partial t} = \left(G_3+\frac{4G_4}{3N}\right)\frac{1}{P_r}\frac{\partial^2 \theta}{\partial y^2} \tag{24}$$

The analytical solution of Eq. (23), subject to frontier constraints (15) (when a=0) by applying Laplace transform method we get,

$$\theta = erfc\left(\frac{y\sqrt{p_r}}{2\sqrt{\left(G_3+\frac{4}{3N}G_4\right)t}}\right) \tag{25}$$

3. Result Analysis

In Figs 42.4–42.7 and 42.10–42.16, as G grows, temperature and velocity fields of SWCNT's and MWCNT's nanoliquid boosts up. The reason for this is that when Grashof number increases, the buoyancy force enhances the velocity field causing temperature enhancement, which leads Nu and values to increase as shown in Tables 42.2–42.5. It is seen in Figs 42.4–42.7 and 42.10–42.16, the temperature and velocity of both nanoliquid are boosting functions as Q enhances. Nu grows as the value of Q drops, which can be seen in Tables 42.2 and 42.5 whereas reverse trend is reflected for in Tables 42.3 and 42.4. Since Q impacts the more induced flow near the thin film, with the assistance of buoyancy force it generates a thermal state of the liquid. While the reverse effect of N is observed in Figs 42.4–42.6, 42.10–42.13 and Tables 42.2–42.5. The velocity profile of both CNT's nanoliquid increase as the volume fraction boot up, as seen in Figs 42.8 and 42.17 whereas the temperature profile also expands with enhancement in volume fraction as visualized in Figs 42.9 and 42.18. As volume fraction of nanoliquid grows, thermal conductivity enhances as a result thermal boundary thickness boosts up. Nu increases with the increment in volume fraction whereas C_f decreases with the decrement in volume fraction. These characteristics are connected to the influence of nano particle volume fraction on the calculation of nanoliquid characteristics, as indicated in Tables 42.2–42.5, and hence their strength or weakness in conducting and transmitting heat. It is seen in figures 4-6, 10-13 and 16 that velocity and temperature as boosting profiles of Ec and a for both CNT's nanoliquid. Since, Eckert number relates the kinetic energy to the activation energy of a fluid that characterize the amount of heat exchange caused due to friction between the fluid layers. A growth in velocity as well as temperature is therefore caused by an enhancement in the amount of heat dissipated by the viscous liquid. Tables 42.2 and 42.5, 42.3 and 42.4, illustrate an increase in C_f while a drop in Nu with the impact of Ec and a respectively.

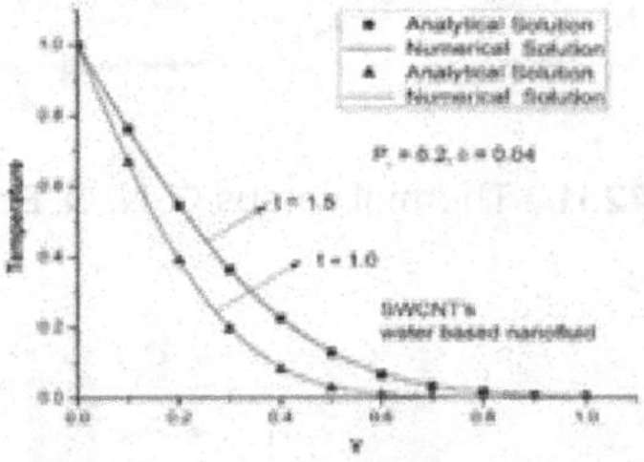

Fig. 42.2 Comparison of Numerical and Analytical Solution.

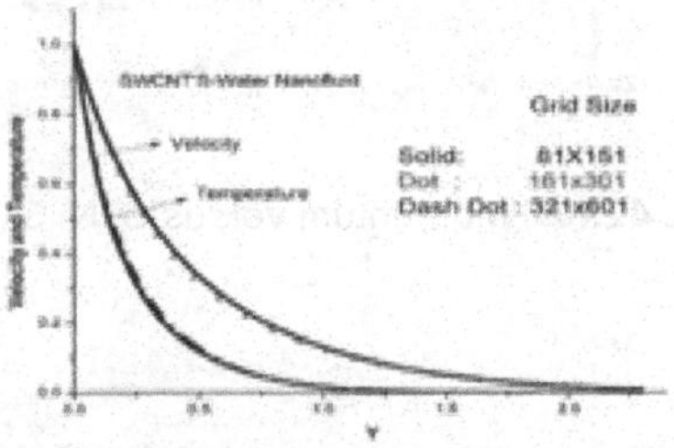

Fig. 42.3 Grid autonomy evaluation of momentum and thermal stencil.

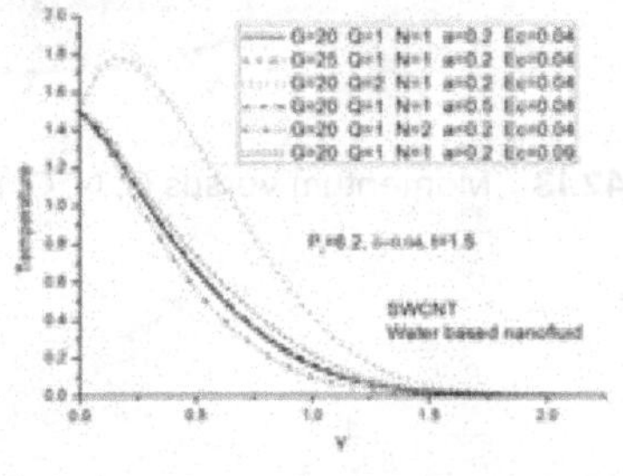

Fig. 42.4 Thermal versus G, Q, N, a, Ec.

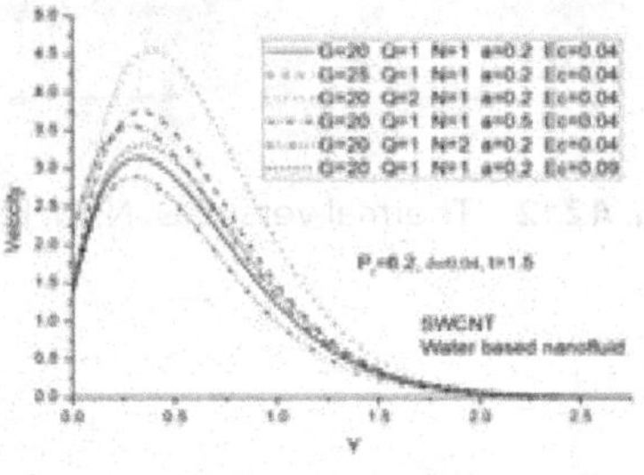

Fig. 42.5 Momentum versus G, Q, N, a, Ec.

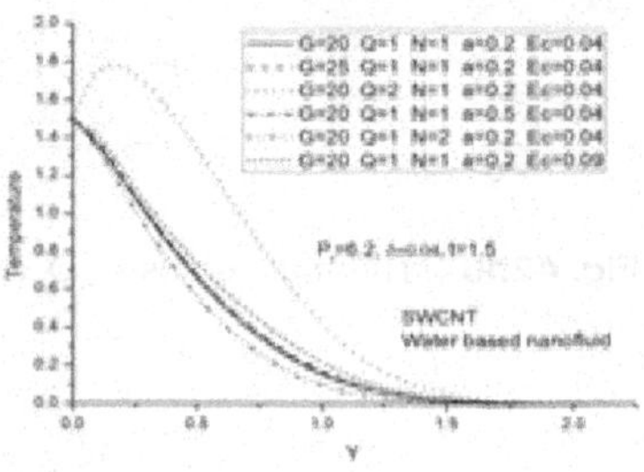

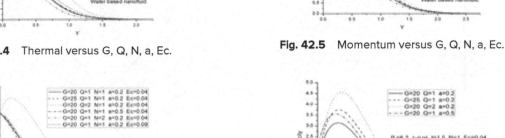

Fig. 42.6 Thermal versus G, Q, N, a, Ec.

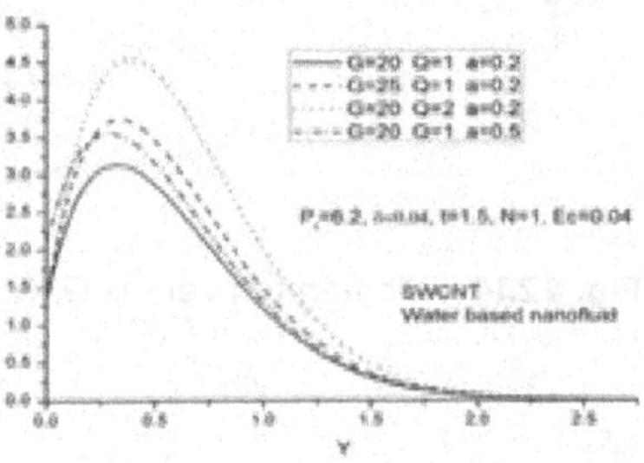

Fig. 42.7 Momentum versus G, Q, a.

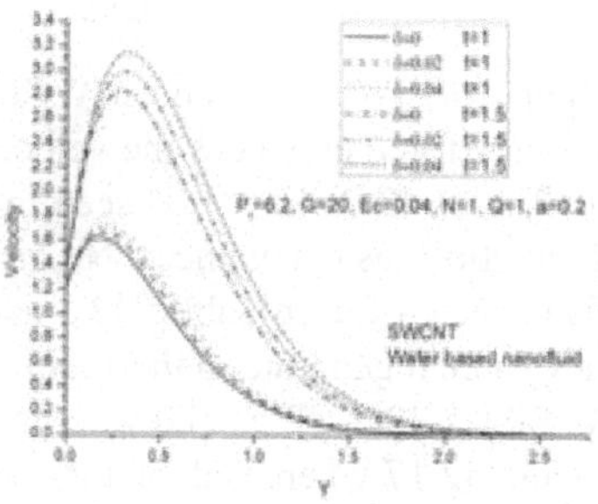

Fig. 42.8 Momentum versus δ, t.

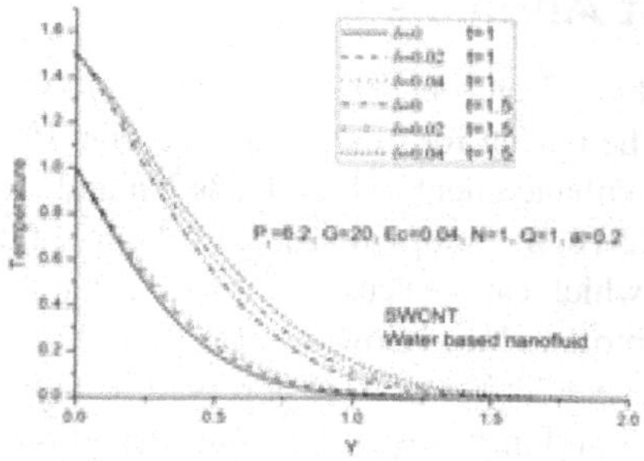

Fig. 42.9 Thermal versus δ, t.

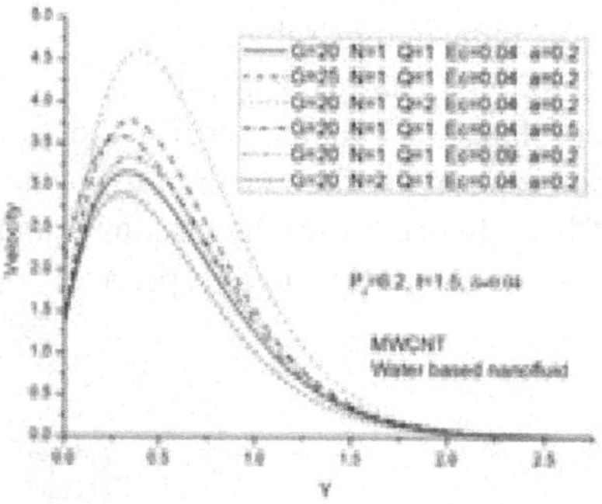

Fig. 42.10 Momentum versus G, N, Q, Ec, a.

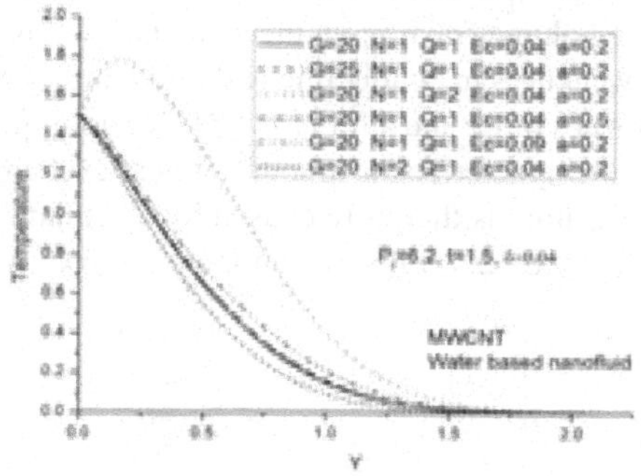

Fig. 42.11 Thermal versus G, N, Q, Ec, a.

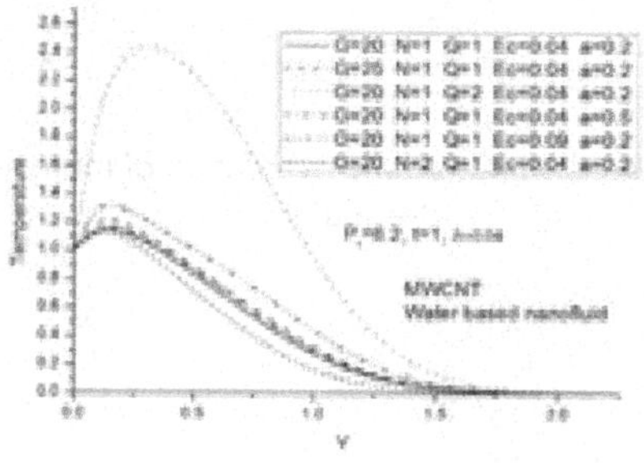

Fig. 42.12 Thermal versus G, N, Q, Ec, a.

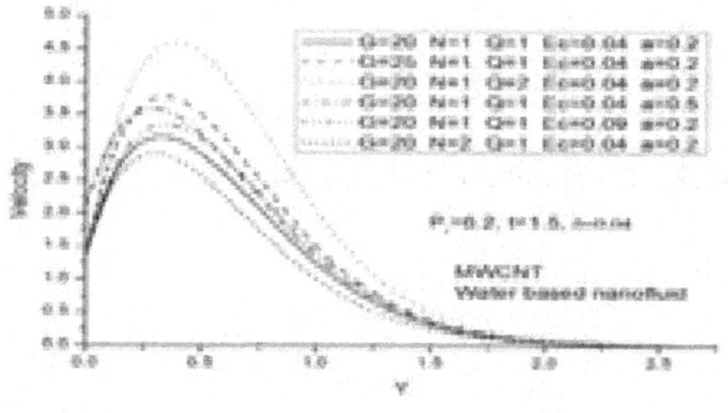

Fig. 42.13 Momentum versus G, N, Q, Ec, a.

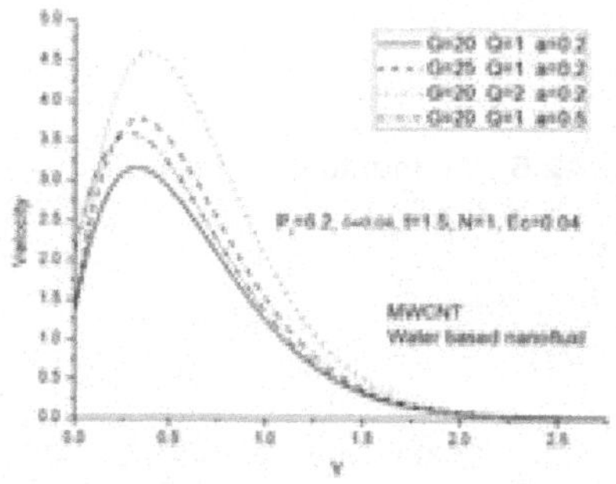

Fig. 42.14 Momentum versus G, Q, a.

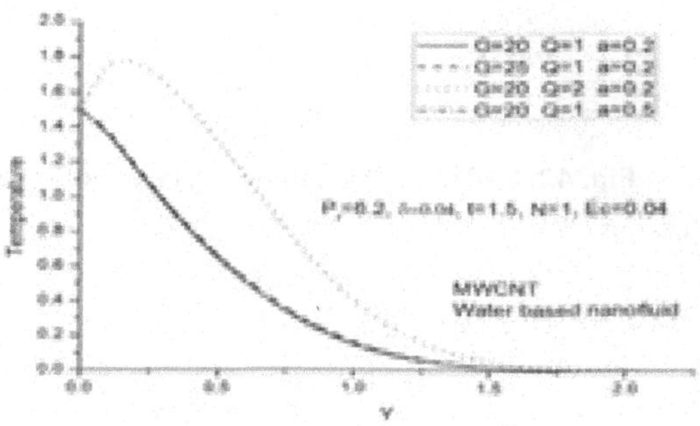

Fig. 42.15 Thermal versus G, Q, a.

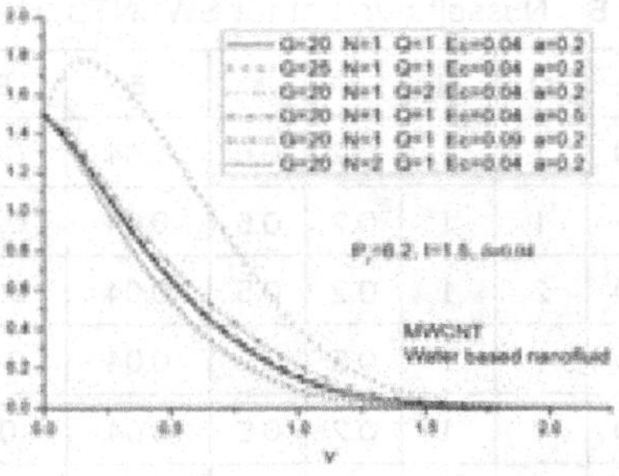

Fig. 42.16 Thermal versus G, N, Q, Ec, a.

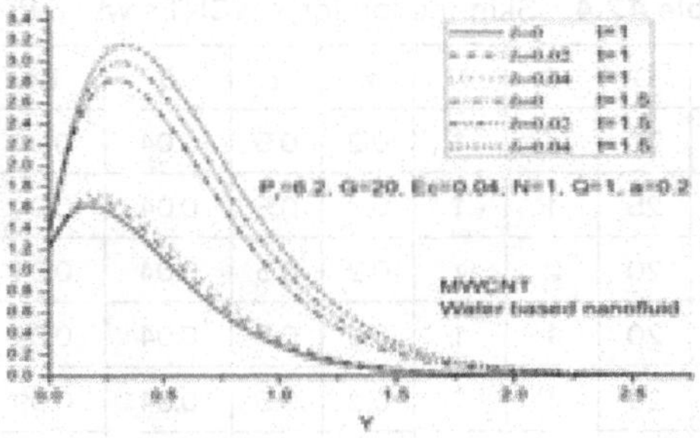

Fig. 42.17 Momentum versus δ, t.

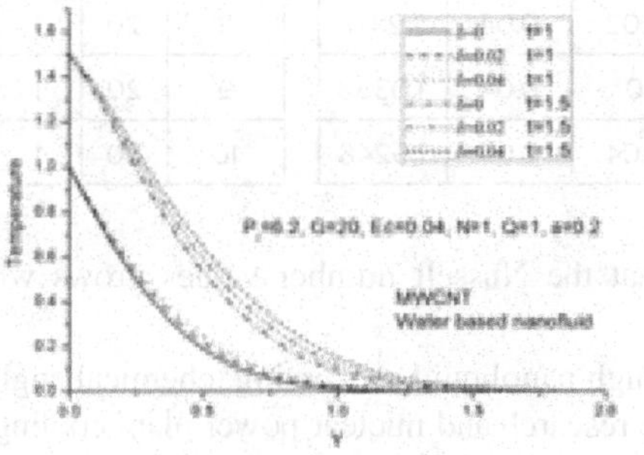

Fig. 42.18 Thermal versus δ, t.

Table 42.2 Nusselt number for SWCNT's when Pr = 6.2

S.No.	G	Q	N	a	t	δ	Ec	Cf
1	20	1	1	0.2	0.5	0.04	0.04	0.2367
2	25	1	1	0.2	0.5	0.04	0.04	0.5601
3	20	2	1	0.2	0.5	0.04	0.04	0.3519
4	20	1	1	0.5	0.5	0.04	0.04	0.0856
5	20	1	1	0.2	0.5	0.04	0.09	0.2863
6	20	1	1	0.2	1	0.04	0.04	3.0882
7	20	1	1	0.2	1.5	0.04	0.04	7.1934
8	20	1	1	0.2	0.5	0.02	0.04	0.1691
9	20	1	1	0.2	0.5	0	0.04	0.1029
10	20	1	2	0.2	0.5	0.04	0.04	0.1444

Table 42.3 Skin-friction for SWCNT's when Pr = 6.2

S.No.	G	Q	N	a	t	δ	Ec	Nu
1	20	1	1	0.2	0.5	0.04	0.04	1.4704
2	25	1	1	0.2	0.5	0.04	0.04	1.4724
3	20	2	1	0.2	0.5	0.04	0.04	1.0762
4	20	1	1	0.5	0.5	0.04	0.04	1.4543
5	20	1	1	0.2	0.5	0.04	0.09	1.4083
6	20	1	1	0.2	1	0.04	0.04	1.5388
7	20	1	1	0.2	1.5	0.04	0.04	0.7734
8	20	1	1	0.2	0.5	0.02	0.04	1.2639
9	20	1	1	0.2	0.5	0	0.04	1.0383
10	20	1	2	0.2	0.5	0.04	0.04	1.6650

4. Conclusions

1. With upsurge values in G, the velocity and temperature of CNT nanoliquid enhances. The Nusselt number and skin friction values rise with increment in G.
2. Similarly, current computation depicted the thermal resistance as an improving function with all of ascending in number of Heat generation (Q), Eckert number (Ec), acceleration (a), time(t) and volume fraction (δ) parameters.
3. In addition, the skin friction values drops while Nusselt number grows with the enhancement in volume fraction (δ).
4. The impact of radiation parameter, shows the velocity and temperature as declining functions for both CNT's nanoliquid.
5. The tables show that as Ec and a enhances, Nusselt number values grows while skin friction values drop.

Table 42.4 Skin-friction for MWCNT's when Pr = 6.2

S. No.	G	Q	N	a	t	δ	Ec	Nu
1	20	1	1	0.2	0.5	0.04	0.04	1.4300
2	25	1	1	0.2	0.5	0.04	0.04	1.4320
3	20	2	1	0.2	0.5	0.04	0.04	1.0477
4	20	1	1	0.5	0.5	0.04	0.04	1.4147
5	20	1	1	0.2	0.5	0.04	0.09	1.3709
6	20	1	1	0.2	1	0.04	0.04	1.4962
7	20	1	1	0.2	1.5	0.04	0.04	0.0754
8	20	1	1	0.2	0.5	0.02	0.04	1.2417
9	20	1	1	0.2	0.5	0	0.04	1.0383
10	20	1	2	0.2	0.5	0.04	0.04	1.6248

Table 42.5 Nusselt number for SWCNT's when Pr = 6.2

S.No.	G	Q	N	a	t	δ	Ec	Cf
1	20	1	1	0.2	0.5	0.04	0.04	0.2585
2	25	1	1	0.2	0.5	0.04	0.04	0.5823
3	20	2	1	0.2	0.5	0.04	0.04	0.3755
4	20	1	1	0.5	0.5	0.04	0.04	0.0575
5	20	1	1	0.2	0.5	0.04	0.9	0.4712
6	20	1	1	0.2	1	0.04	0.04	3.1159
7	20	1	1	0.2	1.5	0.04	0.04	7.2442
8	20	1	1	0.2	0.5	0.02	0.04	0.1792
9	20	1	1	0.2	0.5	0	0.04	0.1029
10	20	1	2	0.2	0.5	0.04	0.04	0.1623

6. It is illustrated from the tables that the Nusselt number values grows while the skin friction values drop with the enhancement in Ec and a.
7. This investigation is applicable to high nanoliquid processing, chemical engineering coating processes, metallurgy, nano medicine, material synthesis, space research and nuclear power plant cooling/heating performance.

5. Acknowledgements

The first author expresses her profound gratitude to KLEF for providing a research fellowship for the completion of the work.

REFERENCES

1. Rajesh, V., Chamkha, A., and Mallesh, M.P., "Nanofluid flow past an impulsively started vertical plate with variable surface temperature", *International Journal of Numerical Methods for Heat and Liquid Flow*, (2016), 26, 328–347.
2. Noor Saeed Khan., Saeed Islam., Taza Gul., Ilyas Khan., Waris Khan., Liaqat Ali., "Thin film flow of a second-grade fluid in porous medium past a stretching sheet with heat transfer", *Alexandria Engineering Journal*, (2018), 57, 1019–1031.
3. Sravan Kumara, T., Dinesh, A., Makinde, O.D., "Impact of Lorentz force and viscous dissipation on unsteady nanofluid convection flow over an exponentially moving vertical plate", *Mathematical models and computer simulations*, (2020), 12, 631–646.
4. Anwar Saeed., Poom Kumam., Saleem Nasir., TazaGul and Wiyada Kumam., "non-linear convective flow of the thin film nanofluid over an inclined stretching surface", *Scientific reports*, (2021), 11, 1–15.
5. Salma Halelfadl., Patrice Estelle., Bahadir Aladag., Nimeti Doner., Thierry Mare., "Viscosity of carbon nanotubes water-based nanofluids: Influence of concentration and temperature", *International Journal of thermal Sciences*, (2013), 71, 111–117.
6. Salma Halelfadl., Thierry Mare., Patrice Estelle., "Efficiency of carbon nanotubes water based nanofluids as coolants", *Experimental thermal and liquid science*, (2014), 53, 104–110.
7. Ellahi, R., Hassan, M., Zeeshan, A., "Study of natural convection MHD nanofluid by means of single and multi-walled carbon nanotubes suspended in a salt water solution", *IEEE transactions on Nanotechnology*, (2015), 14, 726–734.
8. Shafiq Ahmad., Sohail Nadeem., Noor Muhammad., Alibek Issakhov., "Radiative SWCNT and MWCNT nanofluid flow of Falkner-Skan problem with double stratification", *Physica A*, (2020), 547, 1–15.
9. Manjunatha, P.T., Punith Gowda, R.J., Naveen Kumar, R., Suresha, S., Deepak Umrao Sarwe., "Numerical simulation of carbon nanotubes nanofluid over vertically moving disc with rotation", *Partial differential equations in applied mathematics*, (2021), 4, 1–7.
10. Alsaedi, Awais M., and Hayat, T., "Effects of heat generation/absorption on stagnation point flow of nanofluid over a surface with convective boundary conditions", *Communications in Non-linear Science and Numerical Simulation*, (2012), 17, 4210–4223.
11. Rajput, U.S., and Gaurav Kumar., "Effect of heat absorption on MHD flow over a plate with variable wall temperature", *Journal of Applied Science and Engineering*, (2017), 20, 277–282.
12. Saleem Nasir., Zahir Shah., Saeed Islam., Waris Khan., Ebenezer Boynah., Muhammad Ayaz., Aurangzeb Khan., "Three dimensional Darcy-Forchheimer radiated flow of single and multi-walled carbon nanotubes over a rotating stretchable disc with convective heat generation and absorption", *AIP Advances*, (2019), 9, No 3, 1–10.

13. Rajesh, V and Vijaya Kumar Verma, S., "Radiation and mass transfer effects on MHD free convection flow past an exponentially accelerated vertical plate with variable temperature", *ARPN Journal of Engineering and Applied Sciences*, (2009), 4, 20–26.
14. Maria Imtiaz., Tasawar Hayat., Ahmed Alsaedi., Bashir Ahmad., "Convective flow of carbon nanotubes between rotating stretchable discs with thermal radiation effects", *International Journal of Heat and Mass Transfer*, (2016), 101, 948–957.
15. Rajesh, Vemula., Debnath, L., Chakrala, S., "Unsteady MHD free convective flow of nanofluid past and accelerated vertical plate with variable temperature and thermal radiation", *International Journal of Applied and Computational Mathematics*, (2017), 3, 1271–1287.
16. Sreedevi, Gandluru., Prasad Rao, D.R.V., and Makinde, O.D., "Hydromagnetic Oscillatory flow of a nanofluid with Hall effect and thermal radiation past vertical plate in a rotating porous medium", *Multi discipline modeling in materials and structures*, (2018), 14, 360–386.
17. Feroz Ahmed Soomro., Rizwan Ul Haq., Qasem, M., Mdallal, Al., Qiang Zhang., "Heat generation/absorption and non-linear radiation effects on stagnation point flow of nanofluid along a moving surface", *Results in physics*, (2018), 8, 404–414.
18. Makinde, O.D., Khan, Z.H., Ahmad R., Khan, W.A., "Numerical study of unsteady hydromagnetic radiating fluid flow past a slippery stretching sheet embedded in a porous medium", *Physics of Fluids*, (2018), 30, 1–7.
19. Makinde, O.D., Eegunjobi, A.S., Shuungula, O., Neossi-Nguetchue, S.N., "Hydromagnetic chemically reacting and radiating unsteady mixed convection Blasius flow past surface flat in a porous medium", *International Journal of Computing Science and Mathematics*, (2018), 9, 525–538.
20. Rajesh, V and Chamkha, A.J., "Unsteady convective flow past an exponentially accelerated infinite vertical porous plate with Newtonian heating and viscous dissipation", *International Journal of Numerical Methods for Heat and Liquid Flow*, (2014), 24, 1109–1123.
21. Rajesh, Vemula., Mallesh, M.P., Anwar Beg, O., "Transient MHD free convection flow and heat transfer nanofluid past an impulsively started vertical porous plate in the presence of viscous dissipation", *2nd International conference on nanomaterials and technologies*, (2015), 10, 80–89.
22. Promod Kumar, P., Bala Siddulu Malga, Sweta Matta., Lakshmi Appidi., "Finite element analysis for unsteady free convection and mass transfer flow of viscous fluid past an accelerated vertical porous plate with suction and chemical reaction", *AIP conference proceedings*, (2020), 2246, 1–8.
23. Mallesh, M.P., Rajesh, V., Kavitha, M., and Chamkha, Ali, J., "Study of time dependent free convective kerosene-nanofluid flow with viscous dissipation past a porous plate", *AIP conference proceedings*, (2020), 2246, 1–9.
24. Prabhakar Reddy, B., Makinde, O.D., "Newtonian heating effect on heat absorbing unsteady MHD radiating and chemically reacting free convection flow past an oscillating vertical porous plate", *International Journal of Applied Mechanics and Engineering*, (2022), 27, 168–187.
25. Schlichting, H., and Gerstein, K., Boundary layer theory, *Springer-Verlag*, 8th Edition, New York, USA, (2001).
26. Tiwari, R.K., and Das, M.K., *International Journal of Heat and Mass Transfer*, Vol 50, No 9-10, pp. 2002–2018, (2007).
27. Sidra Aman, Ilyas Khan., Zulkhibri Ismail, Mohd Zuki Salleh and Al-Mdallal, Qasem.M., "Heat transfer enhancement in free convection flow of CNTs maxwell nanofluids with four different types of molecular liquids", (2017), 7, 1–15.
28. Brewster, M.Q., *Thermal radiative transfer and properties*, New York: John Wiley and Sons, (1992).
29. Xue, Q., *Physica B: Condensed Matter*, (2005), 368, 302–307.
30. Rajesh, V., Chamkha, Ali.J., *International Journal of Numerical Methods for Heat and Fluid flow*, (2014), 24, 1109–1123.
31. Reddy, J.N., *An Introduction to the Finite Element Method*, Mc Graw-Hill, 3rd Edition, New York, USA, (2006).
32. Bathe, K.J., *Finite Element Procedures*, Prentice-Hall, Upper Saddle River, NJ, (1996).
33. Carnahan, B., Luther, H.A., and Wilkes, J.O., *Applied numerical methods*, John Wiley and Sons, New York, (1969).

Note: All the figures and tables in this chapter were made by the authors.

Impending Inquisitions in Humanities and Sciences (ICIIHS-2022) – Dr. Mohan Varkolu et al. (eds)

Deep Learning Methodologies Towards Leaf Disease Detection: A Review

43

Alampally Sreedevi[1]
Research scholar, Department of Computer Science and Engineering,
Koneru Lakshmaiah Education Foundation, Aziznagar, Moinabad Road, Hyderabad, Telangana, India

K. Srinivas[2]
Professor, Department of Computer Science and Engineering, Koneru Lakshmaiah Education Foundation, Aziznagar, Moinabad Road, Hyderabad, Telangana, India

ABSTRACT The main the purpose of this review article focusing on Deep Learning approaches to identify and confirm a plant leaf disease. Disease testimony at an early warning indication and how to avoid them before spreading into various plant sections is a challenge even for the expert's eye. So, it requires a sound system for detecting plant diseases in an early stage. Here, we discuss approaches Used to teach plant leaf recognition diseases by applying deep learning techniques and a summary of each method. Our analysis includes analysis of different approaches to allow upcoming research to learn the vast potential of Improvement of the plant's overall efficacy and perfection was done to make it possible to detect disease on the plant's leaves.

KEYWORDS Deep learning algorithms, Diagnosis, Plant, Leaf disease detection, Accuracy, CNN, Techniques, Analysis, Efficacy, Perfection

1. Introduction

58% of the people in India rely on agriculture for their livelihood. For the most part, Indian farming includes all types of crops, especially the most essential staples, such as wheat and rice. Contrary to popular belief, India's farmers gather other non-food crops, including tea, cotton, coffee, jute, and rubber (a brilliant fiber used to produce jute fabric and string). A wide variety of crops impact food crops, significantly damaging their economic value and jeopardizing the lives of farmers and the general food alike. Therefore, the primary areas of research for disease-causing diseases are diagnosing them and swiftly recovering from them. Progress on these issues has not stopped the challenges from frustrating India for years. If food crop yield crops are as high as one-fifth of the overall agricultural output, the complete yield losses cannot be discovered since various disease diseases keep them from making up the difference. The primary focus of the researcher is finding the condition of plants and their quick recovery.

It is critical for sustainable farming that images obtained by hyperspectral recognition be automatically recognized and handled. Such ideas lead to diseases that impact food crops, with considerable crop loss as a result. Recognized Plant disease infections are essential to plant development and provide opportunities for farmers to develop. It is considerably more complicated and expensive to continually monitor diseases affecting plant leaves in real-time. [1]

[1]sreedevi.a@klh.edu.in, [2]srirecw9@klh.edu.in

DOI: 10.1201/9781003489436-43

To counteract these difficulties, the researchers are using Image Processing Techniques to automate the disease diagnosis procedure. Researchers design and deploy deep learning algorithms to diagnose diseases in plants. As a result, this results in a reduction in labor effort, which increases productivity.

The next document is laid out as follows. First, the ideas are introduced. The second half of this report talks about Deep Learning and disease detection in plants. This third section addresses CNN and datasets that were used to do this study. A discussion of recent studies and research methods follows in the fourth part. The examination of several DL models is detailed in the fifth part. The conclusion of the paper is in the sixth part.

Deep Learning

DL subset of ML adds new characteristics, such as several hierarchical data representations and data alteration with various techniques. The capacity of DL to automatically extract qualities from a high level of data from raw data is its key benefit. Recent deep neural network usage has shown how end-to-end learning may be applied in various settings. Neural networks use complex computational mapping systems to translate inputs, such as pictures of sick plants, to outputs, such as crop diseases. A type of architecture with interconnected layers of nodes is a large neural network, passing through to the output layer while each one representing the input layer. The objective is to construct a deep network with a well-planned and exact mapping of input to output. A deep neural network is made by gradually modifying network parameters to enable the network to "learn" and develop in capability. Also great for addressing increasingly complex challenges because deep learning makes extensive use of complex models, allowing significant parallelization. When you have a deep body of existing data that illustrates the issue, apply these advanced machine learning models for classification or regression and enhance accuracy or reduce error. The key elements of deep learning include pooling, stride, padding, convolutions, encoding/decode, activation functions, and so on.

Feature engineering is significantly reduced with the implementation of Deep learning in image processing. Feature hand-engineering had a substantial influence on the total outcomes in previous classification challenges. Feature engineering is laborious and time-consuming when features must be changed due to changing issues or dataset conditions. Feature engineering, as a result, is a lengthy process that requires the knowledge of the user. [5]

For image categorization and image identification in agriculture and medicine, Deep Convolutional Neural Network (DCNN) is frequently crucial. DCNN requires an extensive training dataset, as well as substantial computer resources. Graphics processing units (GPUs) can help speed up the calculation requirements of high-performance computing. Therefore, one needs to have strong observation abilities to detect the typical signs of any plant disease. [6]

CNN

CNN is thought of as an important and somewhat mysterious unanticipated powerful learning method that applies 'filters executing convolution' in the image domain. We note that CNN is derived from the visual field through the retina to contrast CNN and neural networks. Starting with the nervous system, In order to empower natural eyesight and neural framework, CNN is necessary. CNN's design is rather complex. where excellent opportunities to exercise the brain's neurons are readily available. Order accuracy is outstanding [13]. The general utility of CNN in crop disease detection extends far and wide [14][15][16][17]. A class of machine learning known as CNN, which uses many images to improve detection accuracy, is capable of accomplishing this task. Some search engines, like Google, focus on producing features that focus on exact match searches. CNN, on the other hand, relies on recognizing targets In the realm of generalised identification, target detection [18], target segmentation [20] and object detection [21][22] are examples of applications of generalised identification.

The CNN [11][12] models are frequently used in image and video processing projects and other applications and domains.

Without explicitly naming them, CNN discovered that the filters they use are natural. Using these filters helps to remove the appropriate features from the input data while retaining the others.

CNN collects and assesses all of the spatial details of an image. The arrangement of pixels and the image's relationship between them are spatial characteristics. Cameras are essential because they help us identify the object correctly and where an object is located, and its relationship to other things in an image.

CNN uses parameter sharing as well. A feature map is combined to form the results of the filter applied to several input portions.

Dataset Description

Dele Lib, Yaser A. Nanehkarana, Defu Zhang, Junde Chen, together with 500 pictures of rice plant disease, use June Chen's custom AI plant Defu AI to diagnose several conditions rice plants. The photos have been saved in JPG format and organized into folders by theme. In the images below, lighting variety varies, as well as different backgrounds. Photoshop software is used to translate all the photos to the RGB model evenly, and then the scale of the images is calibrated. [23]

Over the course of this project, more than 2,200 pictures were used by S. S. Chouhan, S. Jain, and U. P. Singh to differentiate mango leaves. We have used the term "i.e." to indicate that 1070 of the 1130 pictures were photographs obtained in real-time (self-acquired), while the other 1130 images were part of the PlantVillage collection. The photos on this page fall into four categories. Images with mango leave suffering from the disease, without the condition, images are featuring many plants afflicted with the disease, and images in which the disease does not exist. Images are organized according to the category and labeled accordingly. Additionally, this assessment is based on real-time data collection. One thousand seventy photographs of mango tree leaves from SMVD University in Katra, J&K, India, were all snapped by hand. The collection contains both safe and infected leaf pictures. These models are less region than prior models. the findings reveal that According to the MCNN model, the proposed categorization would have higher accuracy. [24]

By taking pictures of diseased wheat at various resolutions, Wheat disorders have been identified by Mohsin Ahmad, Altaf Hussain, Imran Ahmad Mughal, and Haider Ali. Additionally, categories of wheat diseases like stem rust, yellow rust, and Powderly are included in the dataset. There were around 2,207 photos in each category. The classifier was trained by using a CNN. [25]

To identify Gathered from multiple sources, we utilized the Plant Village dataset, the Google website, and a worldwide AI challenge in which the prevalence of maize leaf disease at various times was studied to create an image collection of maize leaf disease. China's Hunan Academy of Agricultural Sciences assisted in this project. According to the diseases that damage maize leaves, this study separated up to 500 images into nine different groups. The images were taken from a variety of websites, including the Google and Plant Village websites. Before opening each image, you must first manually clip off the lesion's sub-image. Increasing the angle after the photo has been taken increases the overall number of shots. The total number of images expands as images are taken from diverse angles. Images are organized into specific categories based on the input of subject matter experts. [26]

These scientists have all used state-of-the-art technology. Six thousand eight hundred and ninety-four (6,794) images were brought together and formed the training collection for this dataset to find Among the most frequently diagnosed of tomato leaves is caused by three different organisms. Yellow leaf curl, early blight, and late blight are three of the most prevalent tomato leaf diseases. Preparation time involved acquiring training images, sorting and classifying them in advance, and putting them in a folder dedicated to the disease group, with the names functioning as labels in the experiment. [27]

Ashish is Mritunjay Nitish Gangwar, and Divya Tiwarim presented a detection model for finding potato leaf disease. The Plant village dataset that contains 2,152 photos of potato leaves was produced using the dataset that comprises 2,000 photographs of late blight and 152 images of healthy potato leaves. [29]

Nie et al. proposed using multi-task learning and attention model for identifying strawberry verticillium wilt. Researchers undertook the project to construct a large collection of 3531 distinct leaf categories, Each image is a leaf from one of the following six types: a a verticillium leaf, a petiole, and a leaf all in good health, a verticillium petiole, apple leaf, and eucalyptus leaf. To note additional information once the pictures have been collected, hand-annotations are used. [30]

A CNN for identifying sick jute leaves using a leaf image classification model was proposed by the authors Hasan, Hasan, K. M. Z., Ahamed, M. S., and Rakshit are M. Z. These researchers utilized two sets of jute leaf disease datasets to employ in their investigation, one that contains 350 disease-affected pictures and another that has 250 healthy images. They were able to increase the visual impact of the photos by rotating, resizing, and scaling them. [31]

The researchers tested their model by identifying Enhanced disease detection in maize plant leaves is done using a deep convolutional neural network. This study divided up to 500 images into nine different groups according to the illnesses that affect maize leaves. The images originated from a variety of sources, including the websites for Plant Village and Google. Eight categories deal with maize leaves that have been infected with an illness, and one for healthy ones. [32]

2. Literature Review

Yaser A. Nanehkarana, Dele Lib, Defu Zhang, and Junde Chen (2020) have worked on a deep learning technique for diagnosing and detecting rice plants diseases with excellent image processing and classification results. In Fig. 43.1, you can see the approach's framework. A novel DENS-INCEP deep learning architecture for rice disease detection is presented. The uppermost layers are erased since they were tied to the pre-trained DenseNet in the ImageNet dataset to create a new Softmax layer connected with the classifications. The first layer is compressed using a softmax classifier, and Using an inception module for transition learning, the bottom layer is enlarged. For both the public statistics dataset and the dataset with images of rice disease, it obtained an average prediction accuracy of 94%. The data demonstrate that the model performed admirably in the test. The information is intended to be used on mobile devices to analyse various plant disease information and provide recommendations comprehensively.

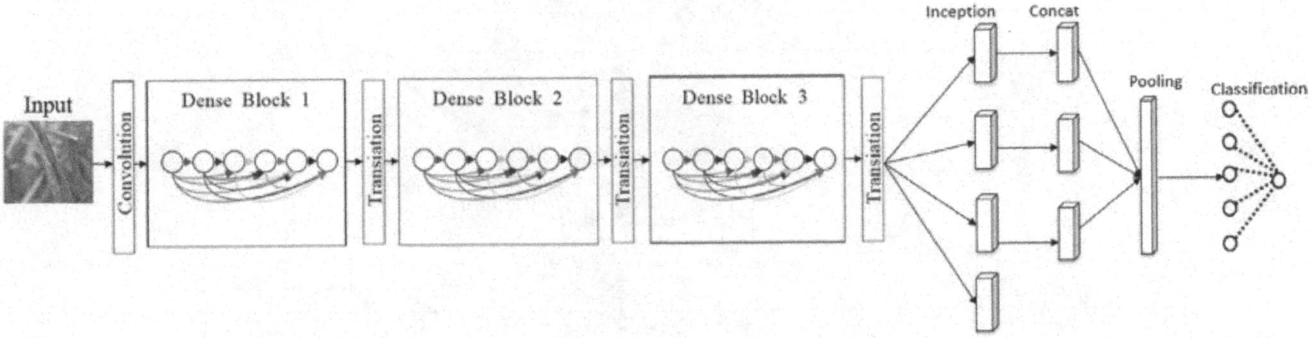

Fig. 43.1 DENS-INCEP architecture—a source of image [23]

U. P. Singh, S. S. Chouhan, and S. Jain propose classifying mango leaves using a multilayer convolutional neural network. In Fig. 43.2, we see the technique of the suggested system and an acceptable and effective approach for discovering the origin and symptoms of the disease. The MCNN method classifies mango leaves infected with anthracnose. According to the already accepted techniques, the suggested method's output is checked for accuracy and correct 97.13% of the time. It is an efficient and straightforward computational model. Among the authors' new works to be released are the following: Using Softmax activation rather than some other function results in an improved CNN that can better categorize different diseases. 2) The presence of discrepancies that arose when dealing with real-time data necessitated a countermeasure. While determining the severity of the disease, it is essential to consider other plants and the plants important to the financial sector. [24]

The picture classification project, which was directed by Altaf Hussain, Mohsin Ahmad, Imran Batch normalisation is used to assure that the model is not overtrained while preventing network overfitting. Combining the PRelu activation function with Adabound optimizer increases the total convergence speed and accuracy. See Fig. 43.3, which illustrates the model structure of this technique. Its accuracy was 84.54%. Farm equipment that helps preserve the wheat crop against diseases can also be regarded as a practical solution for farmers. In a proof-of-concept experiment, an experimental model found that deep models, particularly CNNs, performed better in classifying wheat diseases. Their future goal is to make their models less computationally intensive and lower the modeling depth on embedded devices like smartphones, Raspberry Pi, etc. In addition, understanding plant diseases via visualization is a frequent issue in deep learning, and thus can be used to visualise features [25].

A new technique for detecting maize leaf diseases was proposed by Wenzhuo Zhang, Mingjie LV, Guoxiong Zhou, Mingfang He, and Aibin Chen. In order to improve the appearance of maize in a complicated context, they created a framework for improving the image features of maize leavesThe fundamental Alexnet design is generated by the DMS-Robust Alexnet neural network. Dilated convolution and multi-scale convolution are used to enhance the DMS-Robust Alexnet's capacity to extract features. Batch normalisation is used to prevent network overfitting while guaranteeing that the model is not overtrained. Both the total convergence speed and accuracy are improved by combining the PRelu activation function and Adabound optimizer. The maize leaf disease feature improvement method has been confirmed in many tests to positively improve the DMS-Robust Alexnet functioning. This diagram illustrates the procedure proposed in the system in Fig. 43.4. [26] These methods show that the six diseases that affect maize leaves can be distinguished from healthy leaves. The characteristics of maize leaf diseases have already been improved. This model, known as the DMS-Robust Alexnet, helps detect and categorize pictures, with an accuracy of around 98.62%. The recommended method only requires you to eliminate traits, as proven by the study samples.

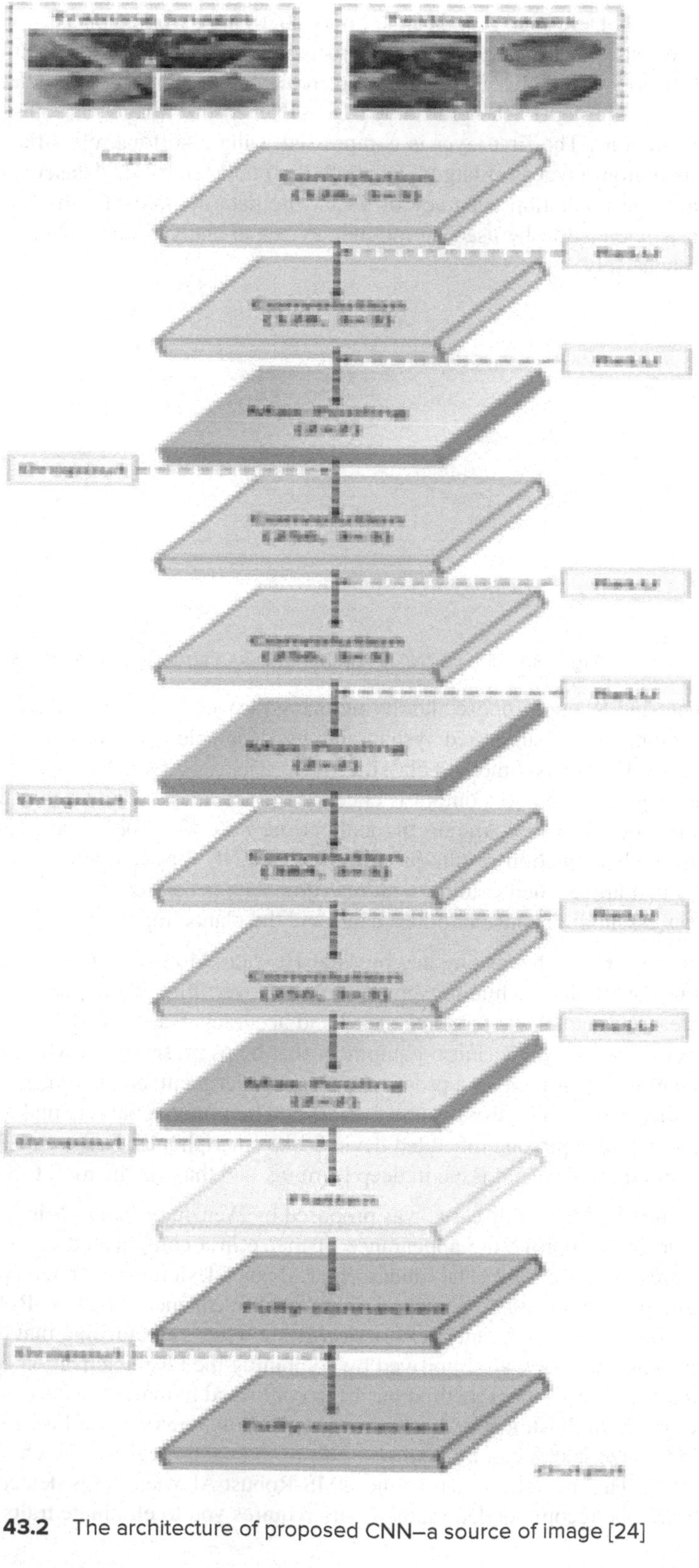

Fig. 43.2 The architecture of proposed CNN–a source of image [24]

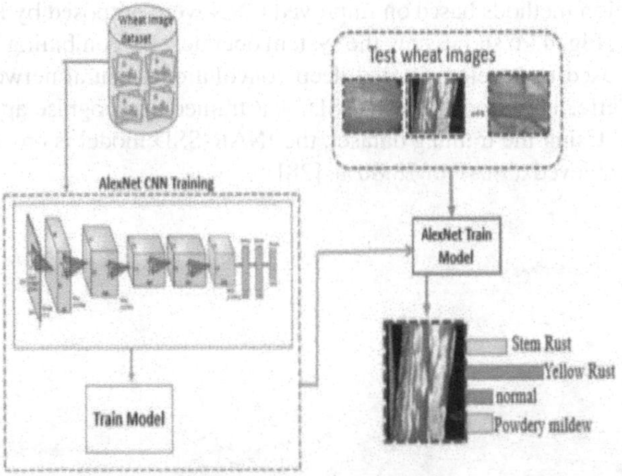

Fig. 43.3 Proposed deep learning methodology – a source of image [25]

Therefore, it is believed to be good in identifying maize leaf disease. [26] The systems approaches described here possess the ability to distinguish and identify healthy maize leaves from six unique maize leaf diseases. In the systems approaches, the use of individual characteristics is not necessary. This, in turn, is viewed as a worthwhile experience regarding recognising pictures of corn leaf disease. This technique will be effective for identifying future varieties of maize pests and diseases.

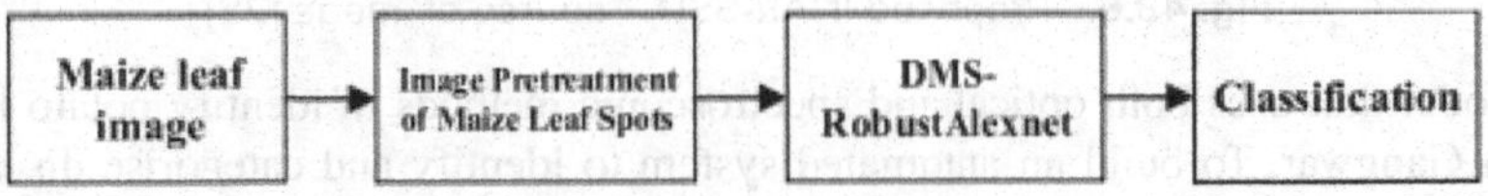

Fig. 43.4 DMS-Robust Alexnet framework frame work – source of image [26]

Yuequan Yang, Song Yu, Ding Jiang, Fudong Li, in collaboration with researchers from the Ding Jiang Institute of Plant Science and Technology, have developed a deep learning-based categorization method for issues with yellow leaf curl, late blight, and spot blight in tomato plants. Six 794 leaf images were of three distinct diseases and after data augmentation, the new Resnet-50 model was utilised to organise and categorise them. The Resnet-50 model will be utilised in this experiment, and instead of ReLU, a Leaky ReLU will be used to measure the influence of the Leaky ReLeaky Convolutional and ReLeaky. This is the model as illustrated. This approach tries to alleviate the problems caused by ReLU activation while simultaneously improving the capacity to gather more detailed information. In the training set, the recognition accuracy for the three diseases was 98.3%, but in the test set, it was only 98.0%. As contrast to that, the study demonstrated some progress; nonetheless, further refinements will be required in order to classify several diseases in one species.[27].

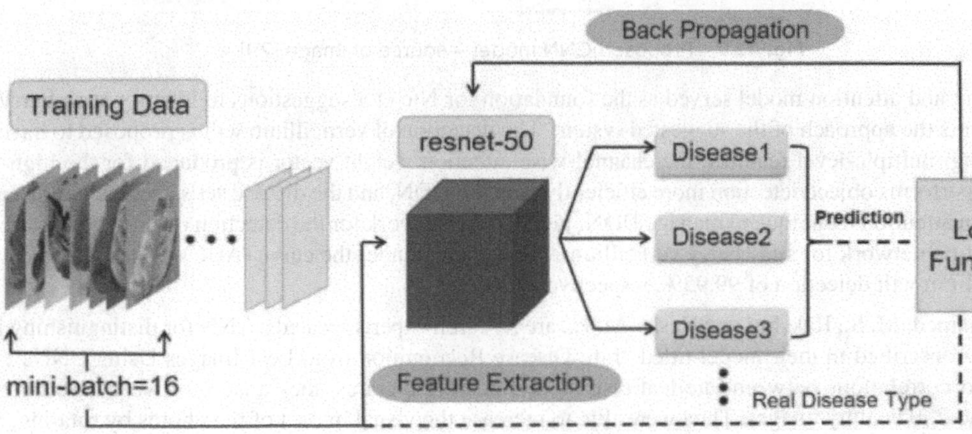

Fig. 43.5 Architecture of proposed CNN using Resnet-50 – source of image [27]

Apple leaf disease real-time detection methods based on improved CNN were proposed by Peng Jiang, Yuehan Chen, Bin Liu, Dorgjian He, and Chunquan Liang. Fig. 43.6 shows how the system operates. By combining The GoogLeNet Inception module with Rainbow concatenation was used to develop a novel deep convolutional neural network model. A deep learning-based GPU platform solution is called Caffe. This model, INAR-SSD, was trained to recognise apple leaf diseases using a dataset of 26,377 images of damaged leaves. Using the training dataset, the INAR-SSD model is created, and the testing dataset is used to evaluate it. Overall, the model achieved a mAP of 78.80%. [28]

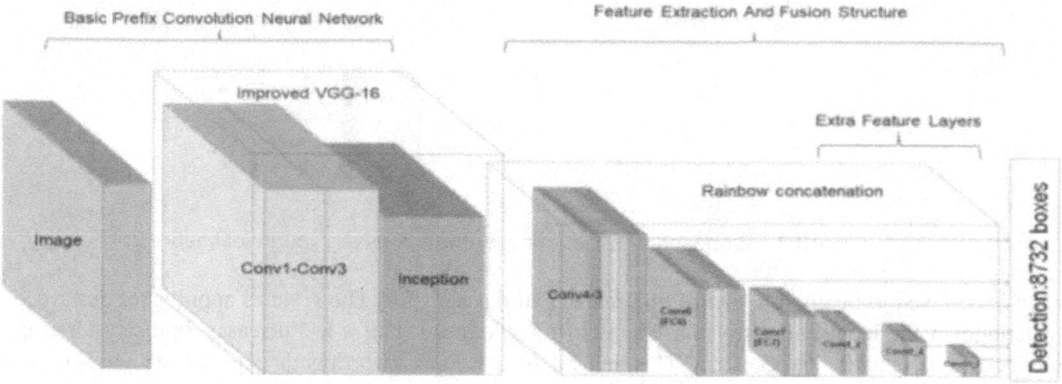

Fig. 43.6 Proposed INAR-SSD – source of image [28]

Ashish is Mritunjay A model that uses both optical and spectroscopic methods to identify potato leaf disease was proposed by Divya Tiwarim Nitish Gangwar. To build an automated system to identify and categorise disease in potato leaves, they employed the idea of transfer learning. Features were extracted from the input pictures with the help of VGG19, which were then passed through several classifiers such as logistic regression, neural networks, support vector machines, and k-nearest neighbours. This model is shown in Fig. 43.7. Researchers discovered that VGG19, using logistic regression, provided the most accurate state-of-the-art solution with a 97.7% classification accuracy. [29]

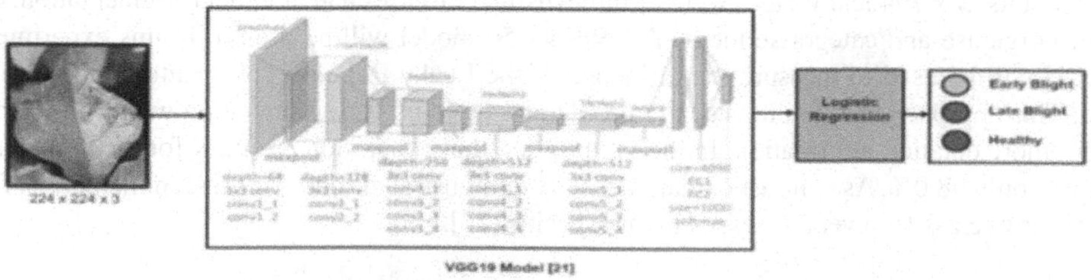

Fig. 43.7 Proposed CNN model – source of image [29]

A multi-task learning and attention model served as the foundation for Nie et alsuggestions to identify strawberry verticillium wilt. Fig. 43.8 presents the approach of the suggested system. The detection of verticillium wilt is proposed to have an attention mechanism. Utilizing multiple-level features, the channel-wise attention weight vector is produced for the high-level feature maps. The network performs object detection more efficiently than the DDN, and the disease set's accuracy of disease diagnosis is 100%. They use an attention technique to manage DDN, just like the network for the detection of strawberry verticillium wilt. The proposed detection network for strawberry verticillium wilt had area under the curve (AUC) values for object detection of 77.54% and verticillium wilt detection of 99.95%, respectively.[30].

Hasan, K. M. Z., Ahamed, M. S., Rakshit, and Hasan, M. Z., are research experts created a CNN for distinguishing jute diseased from leaf images, as described in their model titled "Jute Disease Recognition from Leaf Images Using CNNs". This is seen in Fig. 43.9. To find correlations between jute leaf disease and healthy pictures, they analysed two datasets—one with 350 sick photographs and 250 healthy images. They were able to increase the visual impact of the photos by rotating, resizing, and scaling them. At 80%, 86%, and 89% respectively, RF, KNN, and SVM can all be said to have achieved 80%, 86%, and 89% accuracy, and therefore the suggested DCCN model can provide a model with the best accuracy of 96%. [31].

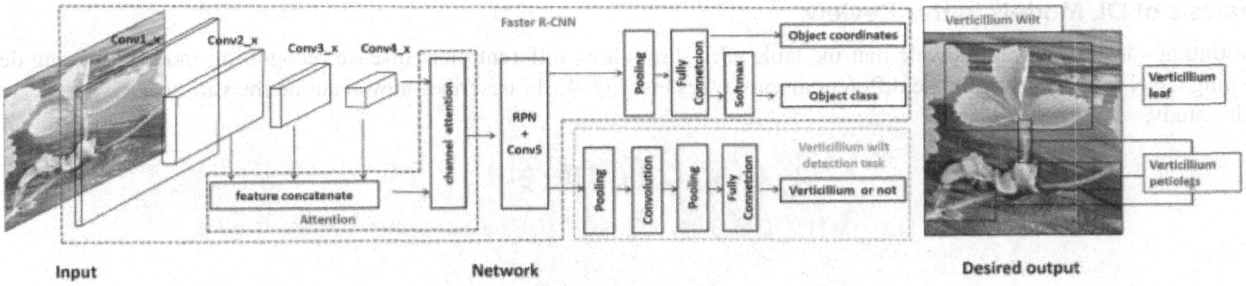

Fig. 43.8 Proposed SVWDN – source of image [30]

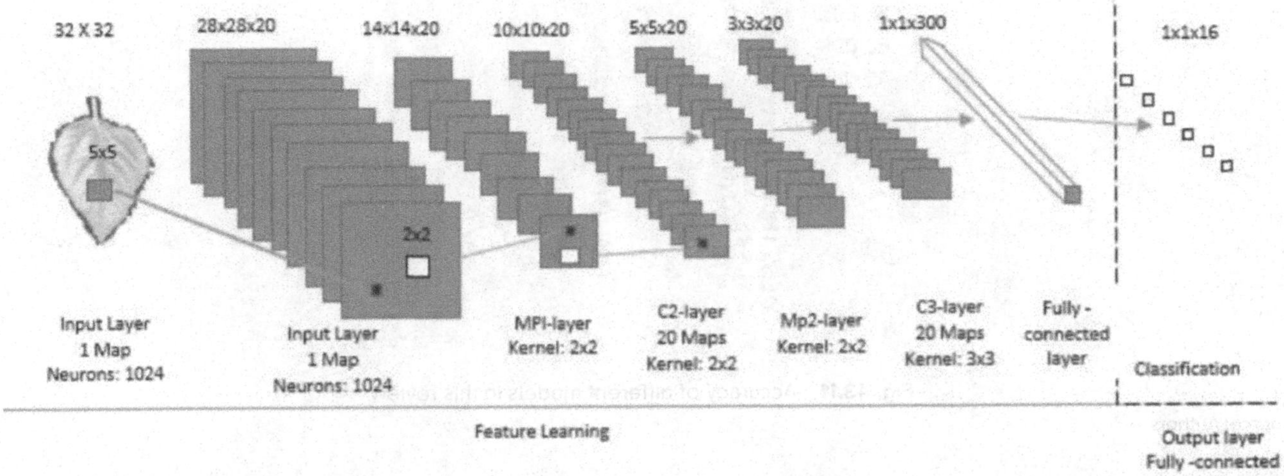

Fig. 43.9 Proposed DCCN – source of image [31]

Deep convolutional neural networks are used in a model created by Zhang, X., Qiao, Y., Meng, F., Fan, and Zhang to identify maize leaf diseases. Two enhanced deep convolutional neural network models, GoogLeNet and Cifar10, have 98.9% and 98.8% accuracy in identifying the nine varieties of maize leaves used in this study. These classification techniques enable the systems to collect samples from the training set with a wide range of variability, which are very resistant to variability in the test set (80 percent of the entire dataset used for training, 20 percent for testing). Model experiments have proven that diversifying pooling processes, adding a Relu function, and using dropout produce greater recognition accuracy. In Fig. 43.10, you can see an illustration of the CNN Model.

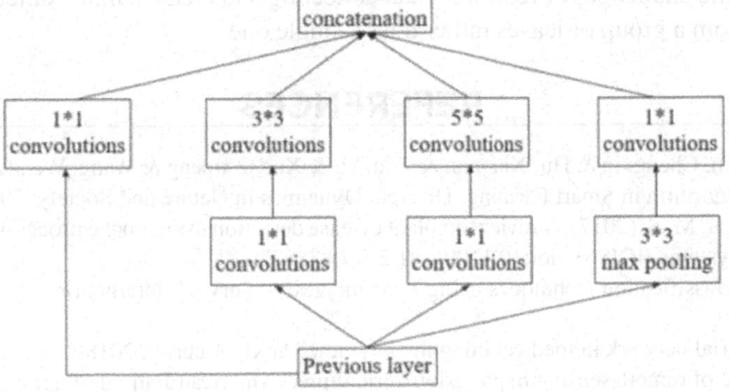

Fig. 43.10 Proposed Inception Module – source of image [32]

Analysis of DL Models in this Review

The findings in this survey indicate that the table 43.1 shows how different plant disease recognition models utilising deep learning CNN models discriminate different disease types and Fig. 43.11 describes how accurate the various CNN models are in this study.

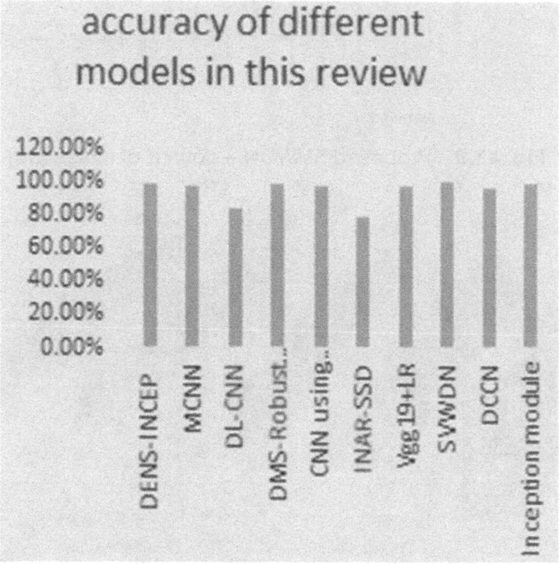

Fig. 43.11 Accuracy of different models in this review

Source: Authors

3. Conclusion

Despite significant shortcomings, this study reports that these disease detection techniques are precise and efficient enough to operate the leaf disease detection system. The CNN model with deep learning performs better according to our findings. Transferring a model is advantageous for training and generalisation; this is particularly beneficial for deep neural networks, which often need considerable fine-tuning to perform effectively. In other words, there is much room for improvement in this sector, given the results thus far.

4. Future Work

Future research will focus on the challenges of real-time data collecting and create a multi-object deep learning model that can even identify plant illnesses from a group of leaves rather than a single one.

REFERENCES

1. Guo, Yan & Zhang, Jin & Yin, Chengxin & Hu, Xiaonan & Zou, Yu & Xue, Zhipeng & Wang, Wei. (2020). Plant Disease Identification Based on Deep Learning Algorithm in Smart Farming. Discrete Dynamics in Nature and Society. 2020. 1–11. 10.1155/2020/2479172.
2. Bharate, A. A., & Shirdhonkar, M. S. (2017). A review on plant disease detection using image processing. 2017 International Conference on Intelligent Sustainable Systems (ICISS). doi:10.1109/iss1.2017.8389326.
3. G. G, "Plant Leaf Disease classification techniques using Leaf images: A Survey," International Journal of ChemTech Research, vol. 12, no. 5, pp. 212–216, 2019.
4. P. S. Y, "Generative adversarial network in medical imaging: a review," arXiv preprint, 2018.
5. Liaghat, "A review: The role of remote sensing in precision agriculture," American Journal of agricultural and biological sciences, vol. 5, no. 1, pp. 50–55, 2017.
6. D. H. Teke, "Saxena & Armstrong," in 6th International Conference on Recent Advances in Space Technologies (RAST), IEEE, Turkey, 2013.
7. S. J. Nikhil Shah, "Detection of Disease in Cotton Leaf using," IEEE, 2019.

Table 43.1 Recognising numerous plant diseases models making use of CNN models with deep learning in this review

	Ref. no	Agri Area	Year	Authors	Problem description	Data set used	Deep Learning model used	Outcome	Accuracy
Plant Disease Identification	[23]	Rice	2020	Chen, J., Zhang., Nanehkaran Y.A., & Li, D.	disease detection in rice plants	500 photos of rice plant disease from the Fujan Institute in Xiamen, China.	DENS-INCEP architecture	The overall accuracy of the system in detecting and classifying diseases was found to be satisfactory.	98.53%
	[24]	mango	2019	S. S. Chouhan, U. P. Singh, and S. Jain	Identifying the Types of Mango Leaves Affected by Anthracnose	1070 self-acquired images	Multilayer convolutional neural network (MCNN)	Using MCNN the authors were able to achieve computational accuracy.	97.13%
	[25]	wheat	2018	M. A. I. A. M. H. A. Altaf Hussain,	Automatic Disease Detection in Wheat Crop	Around 9000 images captured by camera devices	Deep Learning Convolutional Neural Network	Experimental model demonstrated that deep models and particularly CNNs surpass earlier classifications of wheat illnesses.	84.54%
	[26]	maize	2020	G. Z., M. H.,. A. C.,. W. Z.,. A. Y. H. MINGJIE LV	Maize Leaf Disease Identification	Plant village data set from the website	DMS-Robust Alexnet	The authors of this work were able to identify and Identify the six different maize leaf diseases from healthy leaves.	98.62%
	[27]	tomato	2020	D. Jiang, F. Li, Y. Yang, and S. Yu,	A Tomato Leaf Diseases Classification	6794 leaf pictures from AI-challenger	CNN using Resnet-50	This experiment has achieved a certain effect of single species	98.0%
	[28]	apple	2019	Peng Jiang, Yuehan chen, Bin Liu, Dorgjian he and chunquan liang	potato Leaf Disease detection model	2029 from laboratory and real apple field	INAR-SSD	This model is fully capable of real-time disease detection in apple leaves.	78.80%
	[29]	potato	2020	Divya Tiwari, Mritunjay Ashish, Nitish Gangwar	Using a Deep Learning Approach Based on Improved Convolutional Neural Networks, Apple Leaf Disease Real-Time Detection	2152 from platvillage dataset	Vgg19 + Logistic regression	This technique can help farmers identify infections early on and boost crop output.	97.7%
	[30]	Strawberry	2019	Nie, X., Xu, M., and Wang, L.	The Strawberry Verticillium Wilt Detection is based on Multi-Task Learning and Attention.	3531 images	SVWDN	Using SVWDN the authors were able to achieve computational accuracy	99.95%
	[31]	jute	2019	Hasan, M. Z., Hasan, K. M. Z., Ahamed, M. S., Rakshit, &	By applying CNN to classify leaf images, jute disease detection is possible.	600 images downloaded from internet	DCCN	CNN can assist farmers in taking precautions after cultivation.	96%
	[32]	maize	2018	Meng, F., Fan, C., & Zhang, Qiao, Y., Meng, & Zhang	Using enhanced deep convolutional neural networks, disease identification in maize leaves	500 images collected from platavillage dataset and google	Inception module	The overall accuracy of the system in detecting and classifying diseases was found to be satisfactory.	98.9%

Source: Authors

8. J. Sun, Y. Yang, X. He, and X. Wu, "Northern Maize Leaf Blight Detection Under Complex Field Environment Based on Deep Learning," in IEEE Access, vol. 8, pp. 33679–33688, 2020, doi: 10.1109/ACCESS.2020.2973658.
9. D. Z. A. N. a. D. L. Junde Chen, "Detection of rice plant diseases based on deep transfer learning," Wiley Online Library, 14 march 2020. [Online].
10. Kussul, N., Lavreniuk, M., Skakun, S., Shelestov, A., 2017. Deep learning classification of land cover and crop types using remote sensing data. IEEE Geosci. Remote Sens. Lett. 14 (5), 778–782. Kuwata, K., Shibasaki, R
11. Ning Li, Jinde Cao New synchronization criteria for memristor-based networks: Adaptive control and feedback control schemes Neural Networks, Volume 61, 2015, pp. 1–9.
12. Jürgen Schmidhuber, Deep learning in neural networks: An overview, Neural Networks, Volume 61, 2015, Pages 85–117, ISSN 0893-6080
13. Lu, Y., Yi, S., Zeng, N., Liu, Y., & Zhang, "Identification of rice diseases using deep convolutional neural networks", Neuro Computing, 267, 378–384, 2017.
14. Z. Lin, S. Mu, A. Shi, C. Pang, and X. Sun, "A novel method of maize leaf disease image identification based on a multichannel convolutional neural network", Trans. ASABE, vol. 61, no. 5, pp. 1461–1474, Oct. 2018.
15. G. Zhou, W. Zhang, A. Chen, M. He and X. Ma, "Rapid detection of rice disease based on FCM-KM and faster R-CNN fusion", IEEE Access, vol. 7, pp. 143190–143206, 2019.
16. Z. Limbo, H. Tian, G. Chunhui and M. Elhassan, "Real-time detection of Cole diseases and insect pests in wireless sensor networks", J. Intel. Fuzzy Syst., vol. 37, no. 3, pp. 3513–3524, Oct. 2019.
17. P. Jiang, Y. Chen, B. Liu, D. He, and C. Liang, "Real-time detection of apple leaf diseases using deep learning approach based on improved convolutional neural networks", IEEE Access, vol. 7, pp. 59069–59080, 2019.
18. X. Bai, Z. Cao, L. Zhao, J. Zhang, C. Lb, C. Li, et al., "Rice heading stage automatic observation by multi-classifier cascade-based rice spike detection method", Agri cult. Forest Meteorol., vol. 259, pp. 260–270, Sep. 2018.
19. K. P. Ferentinos, "Deep learning models for plant disease detection and diagnosis", Comput. Electron. Agricult., vol. 145, pp. 311–318, Feb. 2018.
20. M. A. Khan, T. Akram, M. Sharif, M. Awais, K. Javed, H. Ali, et al., "CCDF: Automatic system for segmentation and recognition of fruit crops diseases based on correlation coefficient and deep CNN features", Comput. Electron. Agricult., vol. 155, pp. 220–236, Dec. 2018.
21 Z. Lin, S. Mu, F. Huang, K. A. Mateen, M. Wang, W. Gao, et al., "A unified matrix-based convolutional neural network for fine-grained image classification of wheat leaf diseases", IEEE Access, vol. 7, pp. 11570–11590, 2019.
22. Chen, J., Zhang, D., Nanehkaran, Y. A., & Li, D. (2020). Detection of rice plant diseases based on deep transfer learning. Journal of the Science of Food and Agriculture. doi:10.1002/jsfa.10365 [Online]. Available: wileyonlinelibrary.com.
23. U. P. Singh, S. S. Chouhan, S. Jain, and S. Jain, "Multilayer Convolution Neural Network for the Classification of Mango Leaves Infected by Anthracnose Disease," in IEEE Access, vol. 7, pp. 43721–43729, 2019, doi: 10.1109/ACCESS.2019.2907383.
24. M. A. I. A. M. H. A. Altaf Hussain, "Automatic Disease Detection in Wheat Crop using Convolution Neural Network," in The 4th International Conference on Next Generation Computing 2018, 2018.
25. G. Z., M. H., A. C., W. Z., A. Y. H. MINGJIE LV, "Maize Leaf Disease Identification Based on Feature Enhancement and DMS-Robust Alexnet," IEEE Access, vol. 8, pp. 57952–57966, 2020.
26. D. Jiang, F. Li, Y. Yang, and S. Yu, "A Tomato Leaf Diseases Classification Method Based on Deep Learning," 2020 Chinese Control and Decision Conference (CCDC), Hefei, China, 2020, pp. 1446–1450, doi: 10.1109/CCDC49329.2020.9164457.
27. Jiang, P., Chen, Y., Liu, B., He, D., & Liang, C. (2019). Real-Time Detection of Apple Leaf Diseases Using Deep Learning Approach Based on Improved Convolutional Neural Networks. IEEE Access, 7, 59069–59080. doi:10.1109/access.2019.2914929.
28. Tiwari, Divyansh & Ashish, Mritunjay & Gangwar, Nitish & Sharma, Abhishek & Patel, Suhanshu & Bhardwaj, Suyash. (2020). Potato Leaf Diseases Detection Using Deep Learning. 461–466. 10.1109/ICICCS48265.2020.9121067.
29. Nie, X., Wang, L., Ding, H., & Xu, M. (2019). Strawberry Verticillium Wilt Detection Network Based on Multi-Task Learning and Attention. IEEE Access, 7, 170003–170011. doi:10.1109/access.2019.2954845
30. Hasan, M. Z., Ahamed, M. S., Rakshit, A., & Hasan, K. M. Z. (2019). Recognition of Jute Diseases by Leaf Image Classification using Convolutional Neural Network. 2019 10th International Conference on Computing, Communication and Networking Technologies (ICCCNT). doi:10.1109/icccnt45670.2019.8944907
31. Zhang, X., Qiao, Y., Meng, F., Fan, C., & Zhang, M. (2018). Identification of Maize Leaf Diseases Using Improved Deep Convolutional Neural Networks. IEEE Access, 6, 30370–30377. doi:10.1109/access.2018.2844405.

Impending Inquisitions in Humanities and Sciences (ICIIHS-2022) – Dr. Mohan Varkolu et al. (eds)

Modelling of Non-conducting Fluid Flow using Oldroyd-B Model with Nano-particles between Two Straight Channels in an Artery

44

Sunitha Rani Y.
Associate Professor, Department of Mathematics,
CMR Engineering College (Autonomous), Hyderabad
Kesava Reddy V.
Professor, Department of Mathematics, CMR Technical Campus, Hyderabad
Saroja Sandiri
Associate Professor, Department of Mathematics,
Malla Reddy Engineering College (Autonomous), Hyderabad
Kalyan Kumar Palaparthi*
Assistant Professor, Department of Mathematics,
KLEF Deemed to be University, Hyderabad

ABSTRACT In this paper we apply the perturbation technique and investigated some two-dimensional flows, assuming that the fluid to be inviscid and non-conducting together with Nano-particles, a partial differential equation is developed for the mass concentration of the Nano-particles is obtained by perturbation on the relaxation time λ_1, it is also discussed the distribution of the concentration of Nano-particles and found an exact solution in the case of non-conducting fluid in a wedge formed by two straight channels inclined at an angle.

KEYWORDS Harmonical, Nano-particles, Oldroyd-B model, Parallel plates, Viscous flow

1. Introduction

Flow of a Nano particles fluid through arteries is of common occurrence and hence merits thorough discussion. Take for example multiphase transport of Nano particles through an artery which is an important practical problem in drug targeting and in Non-surgical techniques. Also, there are very closely related problem as, for instance, steam flow along a pipe with water droplets. The problem of flow of a Nano particles viscous liquid through an artery under exponential pressure gradient was dealt with by sambasiva Rao (1969). Verma and Mathur (1973) solved the problem of flow through an artery under the influence of constant pressure gradient. The expression for the Nano particles velocity obtained by them does not satisfy the initial condition. Consequently, Dube and Sharma (1975), treating the same problem in a note, obtained expression for velocity of the fluid and Nano particles and gave the values of maximum velocity. Newal Kishore and Pandey (1977) considered the unsteady flow in a circular pipe due to a pressure gradient varying harmonically with respect to time. Rukmangadachari (1978) has recently discussed the Nano particles viscous fluid flow between oscillating coaxial circular cross section. Further by the same author the problem of the flow through coaxial circular cylinders caused by a pressure gradient varying exponentially

*Corresponding author: kalyan.palaparthi77@gmail.com

DOI: 10.1201/9781003489436-44

and also harmonically with respect to time has been considered in (1979). More recently, Ramana Prasad (1980) has dealt with an investigation of pulsating flow superposed on the unsteady laminar flow of a Nano particles viscous incompressible fluid through a circular tube under the influence of a periodic pressure gradient.

In this paper, we have investigated the problem of flow of a fluid with Nano particles through an artery, the fluid initially at rest, being set in motion due to sudden application of a pressure gradient. Using the Laplace transform we find the velocity fields of the fluid and Nano particles and also determine some physical quantities of interest viz., flux, drag etc on the Nano particles. Further, we evaluate the integral

$$\int_s \left(\mu\nabla^2 u - \frac{\partial p}{\partial z}\right) ds$$

over the cross-section s of the artery.

2. Governing Equations

The idealized incompressible elastic-viscous fluid considered here obeys the following constitutive equations by Oldroyd relating to the stress tensor S_{ik} and the rate of strain tensor $e_{ik} = \frac{1}{2}[u_{k,i} + u_{i,k}]$ (1)

$$S_{ik} = p_{ik} - pg_{ik} \tag{2}$$

$$\left(1+\lambda_1 \frac{\partial}{\partial t}\right) p^{ik} = 2\eta_0 \left(1+\lambda_2 \frac{\partial}{\partial t}\right) e^{ik} \tag{3}$$

Where u_i denotes the velocity vector, g_{ik} is the metric tensor, p_{ik} is the part of stress tensor related to change of shape of a material element, σ_e the electrical conductivity and p an isotropic pressure; η_0 coefficient of viscosity and λ_1, λ_2 are relaxation and retardation times respectively. The derivative $\frac{\partial}{\partial t}$ is the convected time-derivative.

The equations of motion of continuity are $p^{ik}, k = g_{ik}\, p, k + \sigma_e$ (4)

$$e^j_j = 0 \tag{5}$$

3. Mathematical Formulation

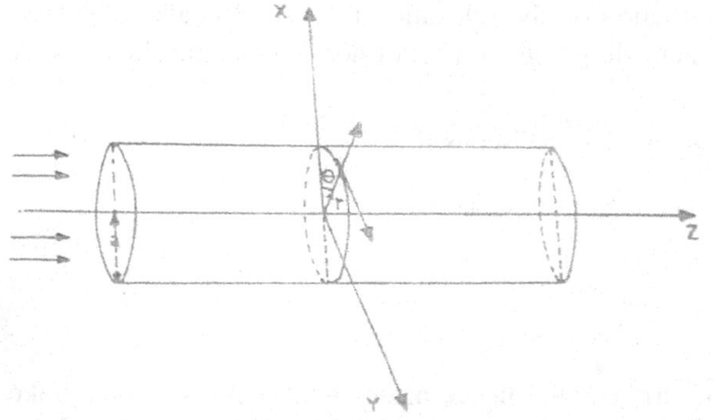

Fig. 44.1 Flow of Nano Particles in the direction of z in an artery

Consider unsteady, laminar flow of a Nano particles fluid through an long artery which assumes the shape of a circular cylinder of radius a. Consistent with the geometry of the problem we take the cylindrical polar coordinate system (z, r, θ) so that the z-axis coincides with the axis of the cylinder.

Let the flow be directed along the axis and be axi-symmetrical. Consequently only the z-components of velocity of the fluid and Nano particles denoted u and v by respectively will be different from zero.

Thus, we have $u = u(r, t)$ and $v = v(r, t)$.

Under these conditions, the equations for the Oldroyd-B model reduces to

$$\left(1+\lambda_1 \frac{\partial}{\partial t}\right)\frac{\partial u}{\partial t} = -\frac{1}{\rho}\left(1+\lambda_1 \frac{\partial}{\partial t}\right)\frac{\partial p}{\partial z} + \frac{\eta_0}{\rho}\left(1+\lambda_2 \frac{\partial}{\partial t}\right)\nabla^2 u + \frac{f}{\lambda_1}(v-u) \tag{6}$$

$$\left(1+\lambda_1 \frac{\partial}{\partial t}\right)\frac{\partial v}{\partial t} = \frac{f}{\lambda_1}(v-u) \tag{7}$$

$$\text{Where } \nabla^2 = \frac{\partial^2}{\partial t^2} + \frac{\partial}{r\partial r} \tag{8}$$

In the steady case $\frac{\partial u}{\partial t} - 0, \frac{\partial v}{\partial t} = 0$ and $v = u$ Suppose that the motion of fluid starts at $t = 0$ with $u = v = 0$ at $t = 0$.

This could be the case if $-\frac{1}{\rho}\left(1+\lambda_1 \frac{\partial}{\partial t}\right)\frac{\partial p}{\partial z} = H(t)$

where $H(t)$ is the Heaviside's unit step function satisfying the following function

$$H(t) = 0 \text{ for } t < 0$$
$$= 1 \text{ for } t \geq 0$$

Thus, the boundary and initial conditions are $u = 0$ at $r = a$ for all $t > 0$ (9)

$$u = 0 \text{ at } t = 0 \text{ for } 0 \leq r \leq a \tag{10}$$

and

$$-\frac{1}{\rho}\left(1+\lambda_1 \frac{\partial}{\partial t}\right)\frac{\partial p}{\partial z} = H(t) \tag{11}$$

Defining Laplace transforms of u and v by

$$\bar{u} = \int_0^\infty e^{-st} u \, dt \tag{12}$$

and

$$\bar{v} = \int_0^\infty e^{-st} v \, dt \tag{13}$$

where $Re\, s > 0$

we obtain the transformed equations of (6) and (7)

$$s\left(1+\lambda_1 \frac{\partial}{\partial t}\right)\bar{u} = \frac{1}{s} + \frac{\eta_0}{\rho}\left(1+\lambda_2 \frac{\partial}{\partial t}\right)\nabla^2\bar{u} + \frac{f}{\lambda_1}(\bar{v} - \bar{u}) \tag{14}$$

$$\bar{v} = \frac{1+\lambda_2}{1+\lambda_1}\bar{u} \tag{15}$$

Eliminating v from (14) and (15) we have

$$(\nabla^2 - \beta^2)\bar{U} = 0 \tag{16}$$

Wherc

$$\bar{U} = \bar{u} - \frac{\rho}{s\eta_0\beta^2} \tag{17}$$

$$\beta^2 = \frac{s\rho(1+f+s\lambda_2)}{\eta_0(1+s\lambda_1)} \tag{18}$$

Solving equations (15), (17),

ū may be obtained as

$$\bar{u} = \frac{\rho}{s\eta_0\beta^2} + AI_0(\beta r) + BK_0(\beta r) \tag{19}$$

Where A and B are arbitrary constants I_0 and K_0 are modified Bessel functions of the first and second kind respectively and or order zero.

Since the physical quantities should be finite at $r = 0$ we choose $B = 0$ since $K_0 \to \infty$ as $r \to 0$. Thus, using the boundary conditions (9) the solution obtained as

$$\bar{u} = \frac{\rho}{s\eta_0\beta^2}\left[1 - \frac{I_0(\beta r)}{I_0(\beta a)}\right] \tag{20}$$

$$\bar{v} = \frac{\rho}{s\eta_0\beta^2(1+s\lambda_1)}\left[1 - \frac{I_0(\beta r)}{I_0(\beta a)}\right] \tag{21}$$

By the inversion theorem $u = \frac{1}{2\pi i}\int_{c-i\infty}^{c+i\infty} e^{st}\bar{u}\, ds, t > 0$ (22)

$$= \frac{1}{2\pi i}\int_{c-i\infty}^{c+i\infty} e^{st} \frac{\rho}{\eta_0 s^2 \beta^2}\left[1 - \frac{I_0(\beta r)}{I_0(\beta a)}\right] ds \tag{23}$$

$$v = \frac{1}{2\pi i}\int_{c-i\infty}^{c+i\infty} e^{st}\bar{v}\, ds, t > 0 \tag{24}$$

$$= \frac{1}{2\pi i}\int_{c-i\infty}^{c+i\infty} e^{st} \frac{\rho}{s^2 \eta_0 \beta^2 (1+s\lambda_1)}\left[1 - \frac{I_0(\beta r)}{I_0(\beta a)}\right] ds \tag{25}$$

where c is greater than the real part of all the singularities of the integrand. The integrand in each case has a double pole at $s = 0$, simple poles at

$$s = -\frac{(1+f)}{\lambda_1} \text{ and } s = -p_{ni} \tag{26}$$

here

$$p_{ni} = \frac{1+f+\frac{\eta_0}{\rho}\lambda_1\alpha_n^2}{2\lambda_1} \pm \sqrt{\left(\frac{1+f+\frac{\eta_0}{\rho}\lambda_1\alpha_n^2}{2\lambda_1}\right)^2 - \frac{2\frac{\eta_0}{\rho}\alpha_n^2}{2\lambda_1}} \quad (n = 1, 2, 3, ...) \tag{27}$$

and i takes values 1 or 2 according as + or – sign is taken on the RHS of (27). Also, α_n are the roots of $J_0(\alpha_n a) = 0$, $J_n(x)$ being the Besel function of the first kind of order n.

Evaluating the residues at the poles, we finally obtain the velocity of the fluid and Nano particles respectively as

$$u = \frac{\rho(a^2 - r^2)}{4\eta_0} - \sum \frac{2\frac{\eta_0}{\rho}\alpha_n(1-p_{ni})^3 e^{-p_{ni}t}}{ap_{ni}^2(1+f-p_{ni}\lambda_1)[1+(1-p_{ni}\lambda_1)^2]} \frac{J_0(\alpha_n r)}{J_1(\alpha_n a)} \tag{28}$$

$$v = \frac{\rho(a^2 - r^2)}{4\eta_0} - \sum \frac{2\frac{\eta_0}{\rho}\alpha_n(1-p_{ni})^2 e^{-p_{ni}t}}{ap_{ni}^2(1+f-p_{ni}\lambda_1)[1+(1-p_{ni}\lambda_1)^2]} \frac{J_0(\alpha_n r)}{J_1(\alpha_n a)} \tag{29}$$

Here the first part of each of the expressions above represents the classical steady state solution for clean fluid flow.

Table 44.1

	Volume rates of flow	Drag	Velocities
Clean fluid	Q_c	D_c	u_c
Fluid with Nano particles	Q	D	u
Nano particles	Q^*	-	v

Numerical Results

In order to see the effect of Nano particles, numerical calculations have been carried out and the results are presented in Tables and Figures below. The variations of u_0 and u the clean and fluid with Nano particles velocities across the artery are shown for different values of f.

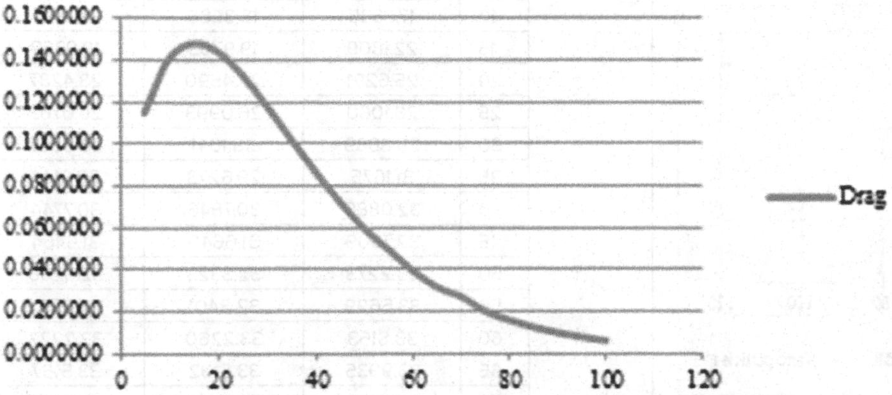

Fig. 44.2 Plot Showing the increase in the drag due to Nano particles for $f = 0.2$

Table 44.2

	t	Drag
time in sec	5	0.1147950
	10	0.1405115
	15	0.1479074
	20	0.1437833
	25	0.1324896
	30	0.1175108
	35	0.1015437
	40	0.0859674
	45	0.0716579
	50	0.0590058
	55	0.0488113
	60	0.0389159
	65	0.0312657
	70	0.0268873
	75	0.0198556
	80	0.0157183
	85	0.0123973
	90	0.0097465
	95	0.0076405
	100	0.0059745

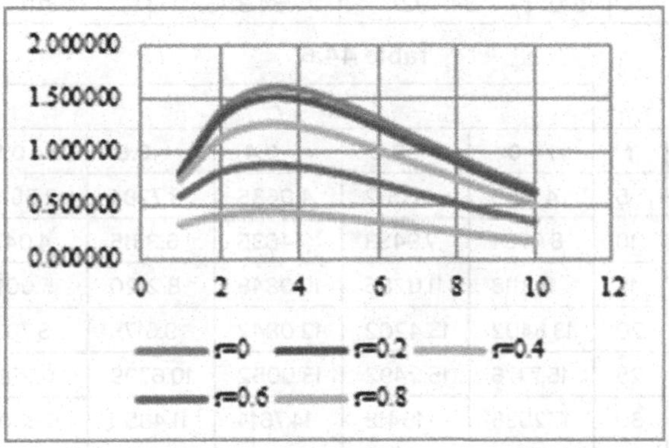

Fig. 44.3 Plot showing the decrease in the fluid velocity u due to nano particles for $f = 0.2$

Table 44.3

		r				
	t	$r = 0$	$r = 0.2$	$r = 0.4$	$r = 0.6$	$r = 0.8$
time in sec	5	0.813364	0.799922	0.743051	0.595805	0.333824
	10	1.382268	1.320261	1.131952	0.820080	0.421267
	15	1.595388	1.510042	1.264323	0.890449	0.446758
	20	1.591628	1.502724	1.249938	0.873328	0.435051
	25	1.477398	1.393901	1.157283	0.806770	0.400969
	30	1.313890	1.239390	1.028462	0.716508	0.355804
	35	1.135538	1.071091	0.888673	0.619008	0.307266
	40	0.961398	0.906821	0.752349	0.524024	0.260059
	45	0.801384	0.755887	0.627117	0.436791	0.216734
	50	0.659894	0.622428	0.516391	0.359669	0.178445

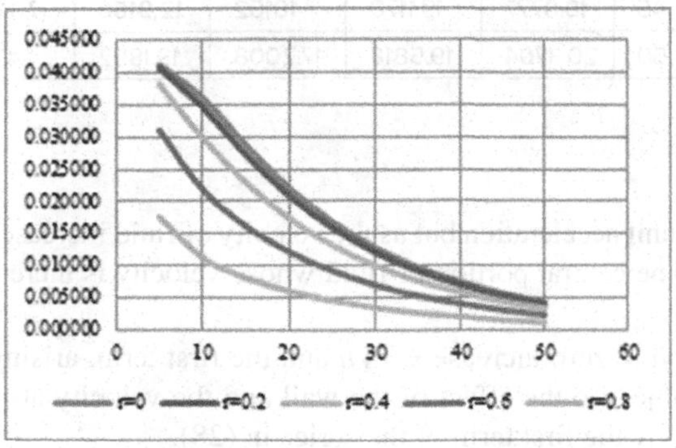

Fig. 44.4 Plot showing the relative velocity $u - v$, the lag in the motion of nano particles for $f = 0.2$

Table 44.4

		r				
	t	$r = 0$	$r = 0.2$	$r = 0.4$	$r = 0.6$	$r = 0.8$
time in sec	5	0.041262	0.040729	0.038277	0.031117	0.017631
	10	0.036153	0.034642	0.029947	0.021909	0.011288
	15	0.028758	0.027256	0.022902	0.016199	0.008129
	20	0.022161	0.020934	0.017436	0.012202	0.006072
	25	0.016913	0.015960	0.013257	0.009248	0.004590
	30	0.012869	0.012140	0.010076	0.007021	0.003482
	35	0.009783	0.009228	0.007657	0.005333	0.002645
	40	0.007435	0.007013	0.005818	0.004053	0.002009
	45	0.005650	0.005329	0.004421	0.003079	0.001527
	50	0.004294	0.004050	0.003360	0.002340	0.001160

Table 44.5

	t	Q_c = Clean Fluid	Q = Oldroyd - B Fluid	Q^* = Nano particles
time in sec	0	0.0035	0.0253	-0.1184
	5	10.5322	9.1289	9.0551
	10	17.3635	15.3586	15.3048
	15	22.1809	19.9763	19.9362
	20	25.6291	23.4590	23.4287
	25	28.1060	26.0993	26.0763
	30	29.8869	28.1041	28.0867
	35	31.1675	29.6273	29.6142
	40	32.0886	30.7846	30.7744
	45	32.7509	31.6640	31.6464
	50	33.2273	32.3323	32.3265
	55	33.5699	32.8401	32.8357
	60	33.8163	33.2260	33.2227
	65	33.9935	33.5192	33.5167
	70	34.1209	33.7421	33.7401
	75	34.2126	33.9114	33.9099
	80	34.2785	34.0401	34.0389
	85	34.3259	34.1378	34.1370
	90	34.3600	34.2121	34.2115
	95	34.3845	34.2686	34.2681
	100	34.4021	34.3115	34.3111

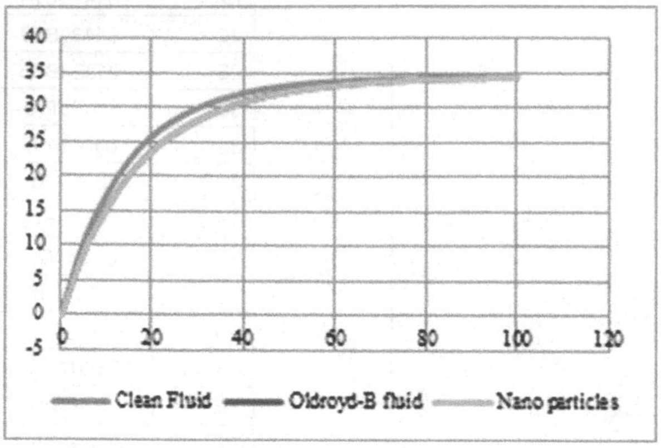

Fig. 44.5 Plot showing the volume rates of Q_c = Clean Fluid, Q = Oldroyd – B Fluid, Q^* = Nano particles

Table 44.6

		r				
	t	$r = 0$	$r = 0.2$	$r = 0.4$	$r = 0.6$	$r = 0.8$
time in sec	5	4.1655	4.1512	4.0635	3.7284	2.6547
	10	8.0734	7.9453	7.4635	6.3315	4.0471
	15	11.3118	11.0285	10.0848	8.2120	5.0013
	20	13.8402	13.4202	12.0842	9.6175	5.7031
	25	15.7776	15.2492	13.6052	10.6799	6.2310
	30	17.2536	16418	14.7614	11.4859	6.6309
	35	18.3761	17.7007	15.6401	12.0981	6.9344
	40	19.2293	18.5054	13078	12.5632	7.1650
	45	19.8777	19.1170	18152	12.9166	7.3402
	50	20.3704	19.5818	17.2008	13.1852	7.4734

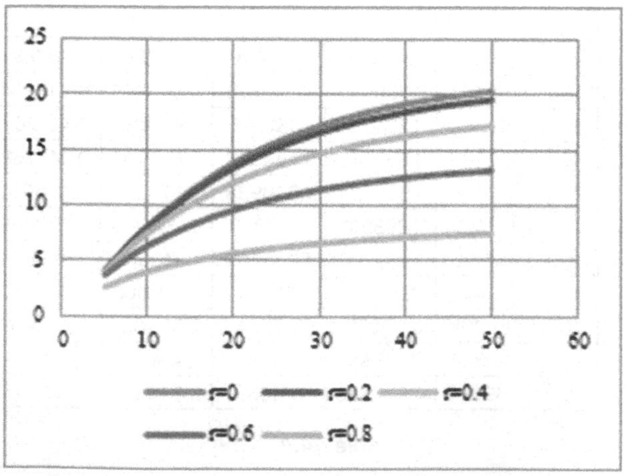

Fig. 44.6 Plot Showing the velocity profiles u of fluid

4. Conclusions

1. Initially the whole fluid is at rest and at $t = 0$ the whole fluid has unit acceleration but as the velocity of fluid increases the restraining influence of the wall spreads further into the fluid. The central portion of fluid whose velocity is increasing with t becomes narrower as t increases.
2. Also, the rapidity with which the terms of the series in (28) tend to zero increase with n and the first term, arising for $n = 1$, survives longest. As soon as all parts of the fluid are subject to the effect of the wall and the velocity at $r = 0$ tends to increase, the approach to the steady state is dominated by the first term of the series in (28).

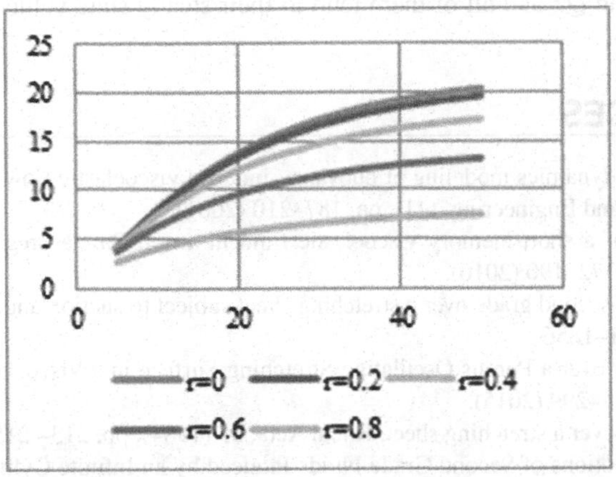

Fig. 44.7 Plot Showing the velocity profiles v of Nano particles

Table 44.7

	t	r = 0	r = 0.2	r = 0.4	r = 0.6	r = 0.8
time in sec	5	4.1243	4.1105	4.0253	3.6973	2.6371
	10	8.0373	7.9107	7.4335	3096	4.0358
	15	11.2830	11.0012	10.0619	8.1958	4.9932
	20	13.8180	13.3992	12.1016	9.6053	5.6971
	25	15.7607	15.2332	13.5919	10.6706	6.2264
	30	17.2407	16.6296	14.7513	11.4789	6.6274
	35	18.3663	17.6914	15.6324	12.0927	9318
	40	19.2219	18.4984	16.3020	12.5591	7.1630
	45	19.8721	19.1117	16.8108	12.9135	7.3387
	50	20.3661	19.5777	17.1974	13.1828	7.4722

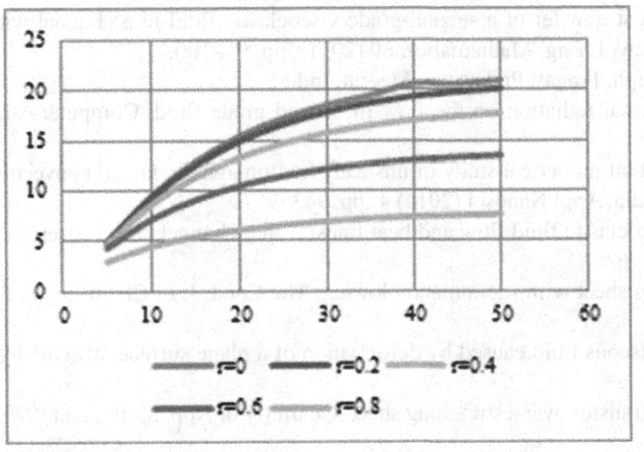

Fig. 44.8 Plot Showing the velocity profiles u_c of Clean fluid

Table 44.8

	t	r = 0	r = 0.2	r = 0.4	r = 0.6	r = 0.8
time in sec	5	4.9789	4.9511	4.8066	4.3242	2.9885
	10	9.4557	9.2656	8.5954	7.1516	4.4684
	15	12.9072	12.5385	11.3492	9.1025	5.4480
	20	15.4318	14.9229	13.3341	10.4908	6.1382
	25	17.255	16.6431	14.7625	11.4867	6.6320
	30	18.5675	17.8812	15.7898	12.2024	6.9867
	35	19.5116	18.7717	15287	12.7171	7.2417
	40	20.6791	19.4123	17.0601	13.0872	7.4251
	45	20.6791	19.8729	17.4423	13.3534	7.5570
	50	21.0303	20.2042	17.7172	13.5448	7.6518

3. It is also clear from Fig. 44.2 that curves of the Oldroyd-B fluid with Nano particles velocity are flatter than the corresponding curves of the clean fluid, which indicates that Nano particles delays the attainment of steady state by the fluid. Table 44.3 gives the decrease in fluid velocity due to the Nano particles and it may be noticed that maximum decrease occurs for $t = 15$ sec approximately.
4. In Fig. 44.2 it is shown the variation of the drag function with time for clean as well as Oldroyd-B fluid with Nano particles flow.
5. It is interesting to note that initially the Oldroyd-B fluid with Nano particles drag D is slightly greater than the clean fluid drag D_c.
6. The situation soon changes and thereafter D is less than D_c though both tend to their steady state value $\pi a^2 \rho$; of course, D trails behind D_c in this respect.
7. If we consider only the transient parts of D *and* D_c we note that D trails behind D is greater than D_c throughout. The transient parts in D dies down more slowly than that in D_c. It is noteworthy from Table 44.2 where values showing increase in drag due to Nano particles have been tabulated, that maximum increase occurs at $t = 15$ sec as in the case of the decrease in velocity.
8. In Table 44.4 the values of relative velocity $(u - v)$ showing the lag in the Nano particles particle motion in relation to fluid motion, are given.
9. Table 44.5 shows the value of volume rates of flown of clean fluid, Oldroyd-B fluid and Oldroyd-B fluid with Nano particles respectively denoted by Q_c, Q and Q^*.

10. It is clear that Q_c is greater than Q which is slightly greater than Q^* and all of them tend to their steady state value as t tends to ∞. Values of u, v and u_c are entered respectively.

REFERENCES

1. O. A. Bég, H.S. Takhar, G. Nath and M. Kumari, Computational fluid dynamics modeling of buoyancy-induced viscoelastic flow in a porous medium with magnetic field effects, Int. J. Applied Mechanics and Engineering, 6(1), pp. 187–210 (2001).
2. O. A. Bég, J. Zueco and S.K. Ghosh, Unsteady natural convection of a short-memory viscoelastic fluid in a non-Darcian regime: network simulation, Chemical Engineering Communications, 198, pp. 172–190 (2010).
3. R. Cortell, Flow and heat transfer of an electrically conducting fluid of second grade over a stretching sheet subject to suction and to a transverse magnetic field, Int. J. Heat Mass Transfer 2006; 49: pp. 1851–1856.
4. S. U. Khan, N. Ali, Z. Abbas, Hydromagnetic Flow and Heat Transfer over a Porous Oscillating Stretching Surface in a Viscoelastic Fluid with Porous Medium, PLOS ONE, DOI:10.1371/journal.pone.0144299 (2015).
5. K. R. Rajagopal, T. Y. Na and A. S. Gupta, Flow of a viscoelastic fluid over a stretching sheet. Rheol Acta, 23 (1984), pp. 213– 215.
6. C. Fetecau, C. Mocanu, Samiulhaq; C. Fetecau, Exact Solutions for Motions of Second Grade Fluids Induced by an Infinite Cylinder that Applies Arbitrary Shear Stresses to Fluid, Advanced Science Letters, 17, (2012), pp. 266–270.
7. M. Sajid, Z. Abbas, T. Javed and N. Ali, Boundary layer flow of an Oldroyd-B fluid in the region of stagnation point over a stretching sheet, Canadian J. of Phys., 88 (2010), pp. 635– 640.
8. J.C. Misra, G.C. Shit and S. Chandra, Hydromagnetic flow and heat transfer of a second-grade viscoelastic fluid in a channel with oscillatory stretching walls: application to the dynamics of blood flow. J. Eng. Mathematics, 69 (2011), pp. 91–100.
9. J.N. Kapur, et al., Non-Newtonian Fluid Flows – A survey Monograph, Pragati Prakashan, Meerut, India.
10. T. Hayat, M. Nawaz, M. Sajid and S. Asghar, The effects of thermal radiation on the flow of second grade fluid. Computers and Mathematics with Applications 58 (2009), pp. 369–379.
11. O. A. Beg, M. S. Khan, I. Karim, Md. M. Alam, M. Ferdows, Explicit numerical study of unsteady hydromagnetic mixed convective nanofluid flow from an exponentially stretching sheet in porous media, Appl Nanosci (2014) 4: pp. 943–957.
12. B. Raftari, K. Vajravelu, Homotopy analysis method for MHD viscoelastic fluid flow and heat transfer in a channel with a stretching wall, Commun Nonlinear Sci Numer Simulat 17 (2012) pp. 4149–4162.
13. P. S. Gupta, and A. S. Gupta, Heat and mass transfer on a stretching sheet with suction and blowing, The Cand. J. of Chem Engg., 55 (1977), 6, pp. 744–746.
14. K. B. Pavlov, Magnetohydrodynamics flow of an incompressible viscous fluid caused by deformation of a plane surface, Magnitnaya Gidrodinamika, 4 (1974), pp. 146–148.
15. Chakrabarti A and A. S. Gupta, Hydromagnetic flow heat and mass transfer over a stretching sheet. Quarterly of App. Math., 33 (1979), 1, pp. 73–78.

Impending Inquisitions in Humanities and Sciences (ICIIHS-2022) – Dr. Mohan Varkolu et al. (eds)
© 2024 Taylor & Francis Group, London, ISBN 978-1-032-78829-6

Classification of Tomato Leaf Disease Using Ensemble Learning

45

Alampally Sreedevi[1]
Research Scholar, Department of Computer Science and Engineering, Koneru Lakshmaiah Education Foundation, Aziznagar, Moinabad Road, Hyderabad, Telangana, India

K. Srinivas[2]
Professor, Department of Computer Science and Engineering, Koneru Lakshmaiah Education Foundation, Aziznagar, Moinabad Road, Hyderabad, Telangana, India

ABSTRACT Significant portions of the Indian economy is centred on agriculture. Despite a significant increase in population, agriculture continues to supply food for everyone. Unfortunately, pathogens can be foreseen at such an early stage in the crop cycle. The loss caused by crop diseases that reduce crop quality and output depends on the rapid and precise disease detection. Early disease detection helps to diagnose crop illnesses and prevents additional crop damage. Deep learning techniques have become the central objective to agricultural research. In order to categorise different forms of tomato diseases, we suggested an ensembled-DL-Model in this study. The most common tomato plant diseases, such as early blight, grey mould, serpentine miner, late blight, and caterpillar leaf damage. With the help of Inceptionv3 and Densenet201, top layers were adjusted. Densenet201 outperformed Inceptionv3 by 3%, and the proposed model, which uses data on tomato leaf diseases, provides the best single result for the test set with an accuracy of 86.52%. It illustrates the model performs better than competing models in terms of real-time performance and network complexity. I've observed that fine-tuning the entire pretrained model with fewer epochs is preferable because it requires less computation time and yields greater accuracy.

KEYWORDS Ensemble learning, InceptionV3, Densenet201, Real time dataset, Disease classification, Early leaf disease detection, DL-Model, Crop illness

1. Introduction

The agricultural sector is crucial for raising money and supplying the population's needs for food [1]. With an annual per capita consumption of around 20 kg, or 15% of all vegetable consumption, tomatoes are a significant vegetable crop in the globe[2]. The world's most widely grown vegetable crop is fresh tomatoes, producing more than 170 million tonnes annually. Tomatoes can grow on nearly any well-drained soil and go by the scientific name Solanum lycopersicum[3]. Similarly, 9 out of 10 farmers plant tomatoes on their properties. On their plots, many gardeners also cultivate tomatoes, so they can use them fresh in cooking and have a delightful lunch. The health, well-being, and financial stability of people can be impacted by numerous leaf diseases that can harm healthy tomato plants, resulting in decreased or poor-quality yields[4]. These leaf diseases drastically reduce the fruit output of tomato crops. Nevertheless, precisely monitoring the growth of their plants can occasionally prove challenging for farmers and gardeners. However, manual crop disease assessment is limited due to lower accuracy and limited accessibility of human resources. This approach has the difficult issue of being expensive, time-consuming, and ineffective [5]. Numerous studies have demonstrated that automated plant disease recognition can help solve this issue [6].

[1]sreedevi.a@klh.edu.in, [2]srirecw9@klh.edu.in

DOI: 10.1201/9781003489436-45

Because of its ability to image recognition and categorization, artificial intelligence and computer vision have experienced a significant increase. Deep learning and machine learning are two subsets of AI. Machine learning is a technique for training computers to perform tasks without explicit programming. Deep learning, a branch of machine learning that can be either supervised or unsupervised, is based on neural networks. Deep learning has recently grown in popularity as a result of the numerous computer vision and natural language processing applications it has. The different layers that the models integrate are referred to as "deep" in deep learning [8]. A class of deep learning models known as convolutional neural networks (CNNs) are widely usedin the classification of images due to their ability to learn features automatically and adaptively. Layers that are convolutional, pooling, and completely connected resemble the conventional CNN structure. [9]. Convolutional and pooling, the first two layers, are in charge of extracting features from the images, The extracted information are then used to classify the images using the fully connected layer. Examining an image to determine which class it belongs to is the process of categorising it[10]. In order to preserve the freshness of the leaves, our research focuses on recognising tomato leaf illnesses by using sophisticated image processing techniques. In this study, a machine-vision approach known as the Convolutional Neural Network (CNN) was employed.

The remaining portions of the research work is organised as follows: The "Literature survey" section gives a brief description of relevant literature's limitations. The suggested plant leaf disease diagnosis model is presented in the "proposed Method" section along with a quick overview of the various modules. The experimental details, analysis, and discussion of the Results can be found in the "Results and analysis" section. The "Conclusion" section provides a final overview of the investigation.

2. Literature Survey

Accurately detecting diseases is enhanced by deep learning for plant diseases. Due to a number of factors, including a lack of understanding about its practical application, the adoption rate of AI technologies in the agricultural sector is currently low. As a result, several studies have concentrated on identifying the primary AI technologies used in PA as well as their documented implementations. However, it might be difficult to recognise leaf diseases without underlying knowledge.

Using deep learning, an approach to recognise pests and diseases that impact tomatoes was proposed by Fuentes et al. To find features in photos of tomato leaves, they applied a region-based CNN approach. Durmuş and other. [12] On the publicly available PlantVillage dataset, CNN models from the AlexNet and SquezeeNet pre-trained frameworks were completely retrained to predict tomato pathogens and pests. Rangarajan et al. employed to determine 4 tomato pathogens, 500 healthy and unhealthy tomato leaves from the PlantVillage dataset were used[13]. They categorised the traits of the fully connected layer of the suggested CNN architecture using the Learning Vector Quantization (LVQ) algorithm. For the purpose of locating diseased locations, Brahimi et al. [14] created the saliency map approach. The accuracy of classification is increased by this kind of representation. Using a frame size of 13,262, pathogens impacting tomato crops have been identified using AlexNet and VGG 19 models. In order to attain 97.49% precision, the model is employed [15].

With a 95% accuracy rate, a transfer learning and CNN Model is utilised to find diseases that harm to crops [16]. In order to extract crucial disease classification features, Karthik et al. [17] To improve and upgrade the residual network, transfer learning was used, and a deep detection model structure for tomato leaf diseases was created. The original AlexNet and VGG16 networks' intricate topologies and numerous parameters restrict them from being employed in practical applications, even though transfer learning can yield higher recognition results. Increasing the dataset size is necessary, the author of [18] put forth a strategy for improving data using GANs. Xception, EfficientNet, AlexNet, DenseNet, ResNet, SEResNet, RestNet, and VGG16 were some of the Using CNN architectures, the dataset was trained).

The proposed model had a 98.70% accuracy rate. In order to assess the model's generalizability, only the Grape dataset was used. The Grape Dataset exclusively includes background-free leaf photos. This strategy will result in a decline in the model's performance on actual field images. To identify diverse diseases, the author [19] suggested a transfer learning approach.

The main contributions of our work are as follows:

- Deep learning models are developed and validated using real-time field images.
- Construct and evaluate deep learning models.
- The study also uses ensemble techniques, which aren't discussed in precise detail in more conventional articles on leaf disease prediction.
- The datasets have been analysed statistically and qualitatively in the study. Prior research has not specifically included pattern analysis of factors in regard to coronary heart disease prediction.

3. Proposed Method

This section outlines the proposed method for identifying tomato leaf disease. Using our proposed approach, we expect to be able to detect four tomato diseases: bacterial leaf steak, bacterial leaf blight, brown spot and sheath blight. The block diagram for the proposed work is shown in Fig. 45.1.

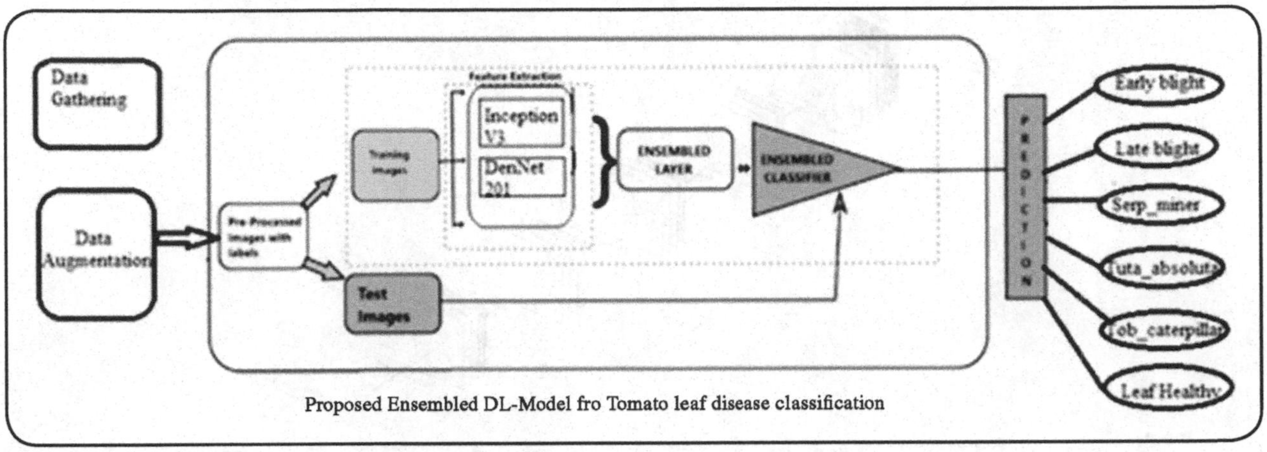

Proposed Ensembled DL-Model fro Tomato leaf disease classification

Fig. 45.1 Shows the proposed work's block diagram

Source: Authors

(i) **Image Acquisition:** As part of the data collection procedure involves gathering leaves from a variety of plants. The collection and processing of the images will be done in accordance with their format, size, and dimension. We visited farmer fields in Hyderabad's Kuntloor village, which is close to Hayat Nagar, to collect images of both healthy and sick tomato leaves in order to construct a real-time dataset. Tobacco caterpillar leaf damage, Leaf Gray mold, Leaf serpentine miner, leaf healthy, Tutaabsoluta leaf damage and Leaf Early Blight are among the classes in this dataset. There are 292 images in it.

(ii) **Transfer learning:** The process when a model created for one problem is used in some capacity to another problem that is comparable is known as "transfer learning," generally speaking. A neural network model is trained using the deep learning technique known as transfer learning on a problem that is comparable to the one being solved. A new model is then trained on the pertinent problem using one or more layers from the learned model. Reduced generalisation error and faster neural network model training time are two benefits of transfer learning. There are two options: starting the training process with the weights in the previously employed layers or changing it to address the new issue. Transfer learning is seen in this application as a particular kind of weight initialization method. We used our own collected dataset to retrain the Inception-V3 model and DenseNet201.

(a) **InceptionV3** - Top performers on ImageNet included Inception V3, which had accuracy ratings of 0.779 for top-1 and 0.937 for the top-5. InceptionV3 takes its name from the inception modules it employs, which are essentially miniature versions of the main model. The concept that you must choose what kind of convolution to make at each layer served as the basis for this research. What about a 3x3? or a 5x5? The concept is that you don't have to know if anything is preferable to do in advance. Additionally, this design enables the model to recover high abstraction features using larger convolutions while also recovering local features using smaller convolutions. More processing is required for the larger convolutions. In order to reduce the feature map's dimensionality, perform a 1x1 convolution, then perform the larger convolution after passing the resultant feature map through a ReLU. Because it will be utilised to decrease the dimensionality of its feature map, the 1x1 convolution is crucial. Dropout, Softmax Function, Concat Layer, and Fully Connected Layer. Weight sharing and sparse connectivity are used to characterise the convolution layer, which computes the neuronal output that is linked to the local regions that are currently present in the previous layer. It also shares the weights of the neurons under each particular feature map and is equivalent to kernels at the same layer. The feature outputs from Inception-V3 and the custom-generated segmented feature are both taken into a fully linked layer for the classification step.

(b) **DenseNet** - The recently released DCNN architecture known as DenseNet is among the best performers on ImageNet's 0.936 top-1 and 0.773 tip-5. Despite having fewer parameters, DenseNet performs on par with InceptionV3 in terms of performance (approximately 20 millions compare with approximate 23 millions of InceptionV3). In Fig. 45.3, one of the four dense blocks of DenseNet 201 is shown in broad terms. Every layer reads the state from the layer after it and updates the state of the layers before it. It affects the environment while sending important information at the same time. By concatenating characteristics rather than summing them as in ResNet, Dense Net architecture makes a distinct separation between information that is kept and contributed to the network.

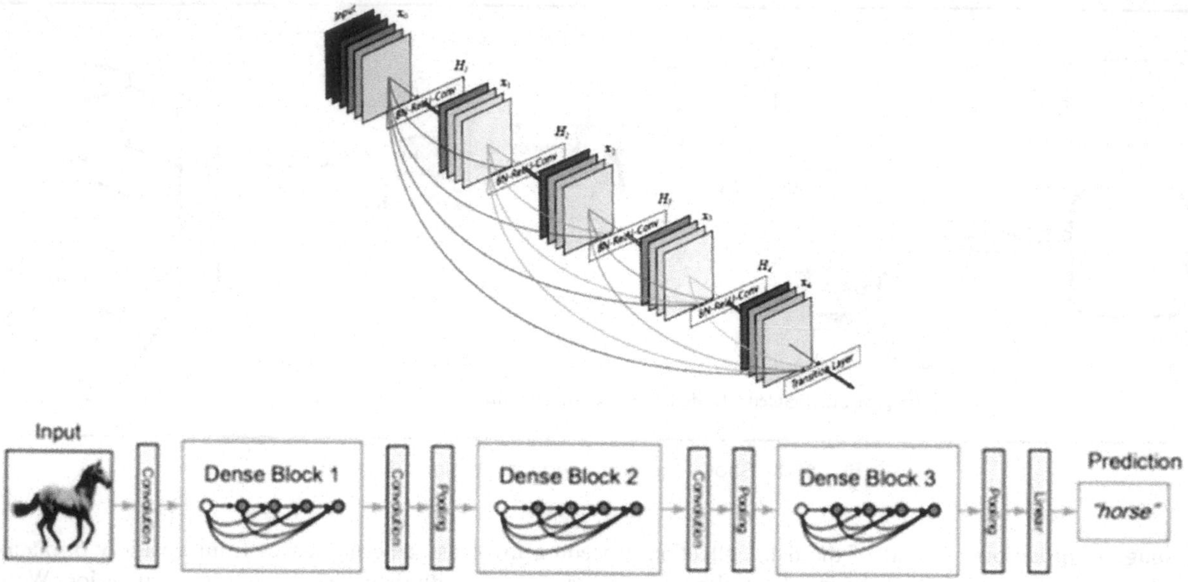

Fig. 45.2 Various DenseNet blocks and layers

Source: Original DenseNet paper

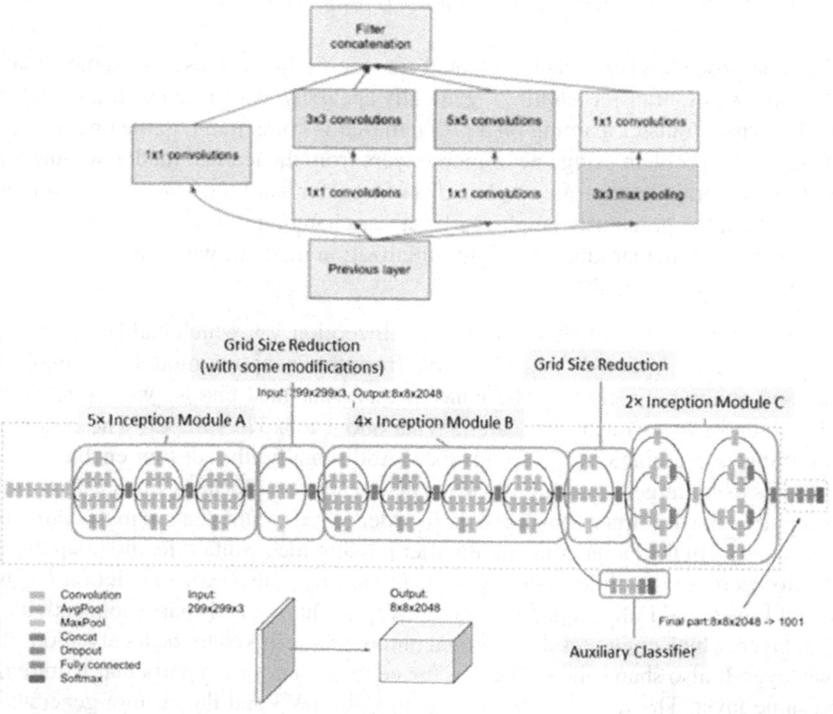

Fig. 45.3 Inception module and InceptionV3 with11 inception blocks [22]

4. Results and Analysis

Model	Validation	Test	Test loss
InceptionV3	80.92%	81.825%	0.4214
DenseNet201	83.69%	85.755%	0.3587
Ensemble(InceptionV3 and DenseNet201)	86.8%	86.52%	0.3115

Even though DenseNet201 has fewer parameters than Inception V3 in the table comparing the two models, it still produces the greatest results. With a test set holding 86.52% and a validation set holding 86.8%, the ensemble model of Inception V3 and DenseNet 201 produced the best results.

5. Conclusion

A difficult challenge is the detection of plant leaf diseases using image processing and machine learning. The current work provides an ensemble approach in order to identify plant leaf diseases.It employs transfer learning. Inception V3 and DenseNet201 ensemble models allowed me to develop a real-time dataset ensemble model that was 86.52% accurate on test set and 86.8% accurate on validation set. I found that using an ensemble model not only produces superior results but also speeds up convergence compared to using individual models. It is intended to reduce the network's complexity in subsequent research.

6. Future Work

The model can be enhanced in the future by extracting significant features using fusion approaches and by testing it with different dataset leaf samples.

REFERENCES

1. Pantazi, X.; Moshou, D.; Tamouridou, A. Automated leaf disease detection in different crop species through image features analysis and One Class Classifiers. Comput. Electron. Agric. 2018, 156, 96–104. [CrossRef].
2. Lalitpur, Dhapakhel, Tomato Plant Diseases Detection System Using Image Processing, 2018.
3. Valenzuela, M.F.M.; Restovi´c, F. Valorization of Tomato Waste for Energy Production. In Tomato Chemistry, Industrial Processing and Product Development; Royal Society of Chemistry: London, UK, 2019; pp. 245–258.
4. Zhu, X.K. Research on Tomato Disease Identification Based on Convolutional Neural Network; Beijing University of Technology: Beijing, China, 2020.
5. Kaur, M.; Bhatia, R. Development of an improved tomato leaf disease detection and classification method. In Proceedings of the 2019 IEEE Conference on Information and Communication Technology, Baghdad, Iraq, 15–16 April 2019; pp. 1–5.
6. Elnaggar, S.; Mohamed, A.M.; Bakeer, A.; Osman, T.A. Current status of bacterial wilt (Ralstonia solanacearum) disease in major tomato (Solanum lycopersicum L.) growing areas in Egypt. Arch. Agric. Environ. Sci. 2018, 3, 399–406. [CrossRef].
7. Wang, Y.; Zhang, H.; Liu, Q.; Zhang, Y. Image classification of tomato leaf diseases based on transfer learning. J. China Agric. Univ. 2019, 24, 124–130.
8. Saleem, M.H.; Potgieter, J.; Arif, K.M. Plant disease classification: A comparative evaluation of convolutional neural networks and deep learning optimizers. Plants 2020, 9, 1319. [CrossRef].
9. Guo, X.; Fan, T.; Shu, X. Tomato leaf diseases recognition based on improved multi-scale AlexNet. Trans. Chin. Soc. Agricult. Eng. 2019, 35, 162–169.
10. Zhou, C.; Zhou, S.; Xing, J.; Song, J. Tomato leaf disease identification by restructured deep residual dense network. IEEE Access 2021, 9, 28822–28831. [CrossRef].
11. Fuentes, A, Yoon, S, Kim, SC, Park, DS. 2017. A robust deeplearning-based detector for real-time tomato plant diseases and pests recognition. Sensors; 17(9): 2022.
12. Durmuş, H, Güneş, EO, Kırcı, M. Disease detection on the leaves of the tomato plants by using deep learning, proceedings of the 6th International Conference on Agro-Geoinformatics, Fairfax VA, USA, 2017, pp. 1–5. 13. Rangarajan, A.K., Purushothaman, R., Ramesh, A., 'Tomato crop disease classifcation using pre-trained deep learning algorithm, In: International Conference on Robotics and Smart Manufacturing (RoSMa2018)', Procedia Computer Science, pp. 1040–1047, 2018.
14. Brahimi, M., Arsenovic, M., Laraba, S., Sladojevic, S., Boukhalfa, K., Moussaoui, A., 2018. Deep learning for plant diseases: Detection and saliency map visualisation, in: Human and Machine Learning. Springer, pp. 93–117.

15. Rangarajan, A.K.; Purushothaman, R.; Ramesh, A. Tomato crop disease classification using pre-trained deep learning algorithm. Procedia Comput. Sci. 2018, 133, 1040–1047. [CrossRef]
16. Coulibaly, S.; Kamsu-Foguem, B.; Kamissoko, D.; Traore, D. Deep neural networks with transfer learning in millet crop images. Comput. Ind. 2019, 108, 115–120. [CrossRef]
17. Karthik, R.; Hariharan, M.; Anand, S.; Mathikshara, P.; Johnson, A.; Menaka, R. Attention embedded residual CNN for disease detection in tomato leaves. Appl. Soft Comput. 2020, 86, 105933.
18. Liu, B.; Tan, C.; Li, S.; He, J.; Wang, H. A data augmentation method based on generative adversarial networks for grape leaf disease identification. IEEE Access 2020, 8, 102188–102198. [CrossRef]
19. Kaya, A.; Keceli, A.S.; Catal, C.; Yalic, H.Y.; Temucin, H.; Tekinerdogan, B. Analysis of transfer learning for deep neural network based plant classification models. Comput. Electron. Agric. 2019, 158, 20–29. [CrossRef]

Impending Inquisitions in Humanities and Sciences (ICIIHS-2022) – Dr. Mohan Varkolu et al. (eds)
© 2024 Taylor & Francis Group, London, ISBN 978-1-032-78829-6

Identification of Most Important Nodes in Wireless Sensor Networks Using Centralities

46

Suneela Kallakunta[1]
Research Scholar, Department of ECE, GITAM (Deemed to be University), Visakhapatnam, India
Alluri Sreenivas[2]
Department of ECE, GITAM (Deemed to be University), Visakhapatnam, India

ABSTRACT This study undertook a comprehensive analysis to identify the most important nodes within a wireless sensor network of 30 nodes. This investigation involved a meticulous assessment through various centrality measures encompassing local (degree), global (betweenness, closeness, and Eigenvector), Katz, and subgraph centrality measures. As a result of this comprehensive evaluation, node ranks were assigned, spanning a scale of up to 10. This ranking system encapsulates the relative importance of nodes within the network hierarchy. This thorough exploration of node significance enriches the understanding of wireless sensor networks, contributing valuable insights for network optimization, design strategies, and informed decision-making.

KEYWORDS Graph theory, Centralities, Wireless Sensor Networks, Node Ranks

1. Introduction

Wireless sensor networks (WSNs) are rapidly gaining popularity. They can be utilized in various applications, such as military surveillance, geological monitoring (earthquakes, volcanoes, and tsunamis), agricultural control, etc. A wireless sensor network comprises a set of tiny sensor nodes that communicate with one another. Wireless communication and the miniaturization of electronics have made it possible to make wireless sensor networks. These networks comprise many small electronic devices that gather and send information. These sensors are called "nodes," and they have small amounts of memory, processing power, and energy storage. The nodes can process routes, transport data, and make "small" decisions. The base station(s) are particular nodes that might be permanent or movable. It uses the internet to connect wireless sensor networks to specified network domains [Mbiya et al., 2020]. Much recent research has presented methods for conserving energy and power consumption in wireless sensor networks and routing techniques to fit wireless sensor network limitations [Jain, 2016; Andrés Vázquez-Rodas et al., 2015]. Several research papers have been proposed to offer approaches to dealing with wireless sensor network challenges, such as node scheduling, energy harvesting, routing protocols, and node localization. Graph theory-based approaches are used to improve wireless sensor networks' network lifetime and energy usage. The centrality measure defines a node's importance and role in a network. It also aids in determining its significance and connectivity to nearby nodes [Gómez, 2019]. Centrality measures are used in wireless sensor networks to assess the importance and influence of individual nodes within the network. They help in the identification of essential nodes, the understanding of network dynamics, and the optimization of operations. Degree centrality, betweenness centrality, closeness centrality, eigenvector centrality, Katz centrality, and subgraph centrality are some of the most often used centrality measures in wireless sensor networks [Njotto, 2018]. Researchers and practitioners

[1]suneelakallakunta@gmail.com, [2]salluri@gitam.edu

DOI: 10.1201/9781003489436-46

can use centrality measures to analyze wireless sensor networks, identify important nodes, optimize routing and data collection, and improve network performance and efficiency.

2. Centrality Measures

2.1 Degree Centrality (DC)

Degree centrality is a way to measure how well a node can communicate with other nodes directly. Nodes with a high degree of centrality in WSNs have connections directly to many other nodes. These nodes are important hubs that help disseminate information and connect networks.

$$DC = e_i^T Ae$$

Where, A is the adjacency matrix of a network, e_i is the ith standard basis vector (i^{th} column of the identity matrix) and e is the vector of all entries one.

2.2 Betweenness Centrality (BC)

Betweenness centrality measures a node's importance in a network by calculating the number of times it appears in the shortest path between all pairs of nodes in a graph. Nodes with high betweenness centrality in WSNs are situated on many shortest paths between other nodes. These nodes play an essential role in information relay and network connectivity.

The betweenness centrality of a node v is given by

$$BC(v) = \sum_{i \neq j} \frac{\sigma_{ij}(v)}{\sigma_{ij}(v)}$$

Where $\sigma_{ij}(v)$ is the number of shortest paths from node i to node j passing through v and $\sigma_{ij}(v)$ is the number of shortest paths from node i to node j.

2.3 Closeness Centrality (CC)

Closeness centrality determines how quickly a node can reach all other nodes in the network. In WSNs, nodes with high closeness centrality have shorter average path lengths to other nodes, enabling them to disseminate information and coordinate network activities efficiently.

$$CC(i) = \frac{N-1}{e_i^T De}$$

Where N is the total number of nodes and D is the distance matrix.

2.4 Eigenvector Centrality (EVC)

Eigenvector centrality evaluates a node's influence based on its neighbors' influence. In WSNs, nodes with high eigenvector centrality are well connected and connected to other influential nodes. These nodes have a more significant impact on the overall network dynamics.

$$EVC = X_i = \frac{1}{\|AX_{i-1}\|} AX_{i-1}, i = 1,2,3, \ldots$$

Where X_0 is the unit column matrix.

2.5 Katz Centrality (KC)

It is a centrality measure that considers direct and indirect paths between nodes in a network. It provides node centrality scores based on the sum of contributions from immediate neighbors and neighbors at increasing distances. Katz centrality can be utilized to evaluate the significance and influence of sensor nodes in wireless sensor networks.

$$KC = (I - \alpha A)^{-1} e$$

Where I is an identity matrix of order n, α is called the attenuation factor. Here, $\alpha \in \left(0, \frac{1}{\lambda}\right)$, λ is principal eigenvalue of A.

2.6 Sub graph Centrality (SC)

Sub graph centrality is a measure of node centrality that considers the structural importance of subgraphs within a network. The sub graph centrality of node i in the network is given by

$$SC(i) = (e^A)_{ii}$$

3. Results

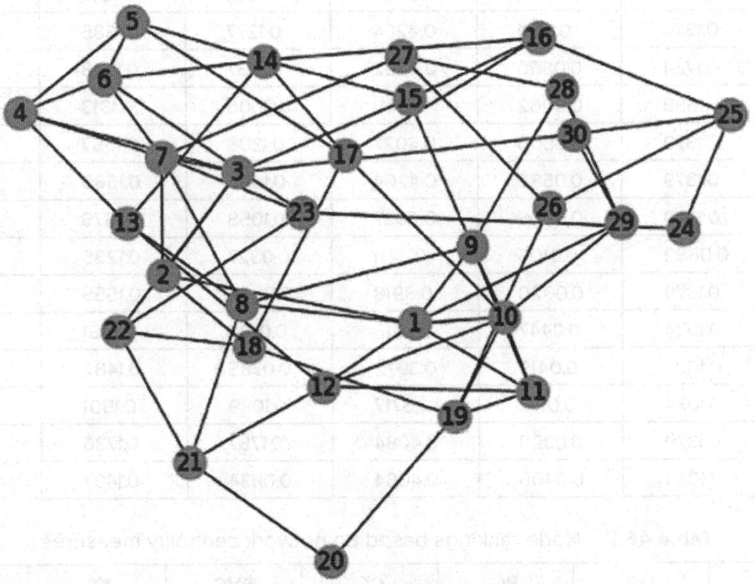

Fig. 46.1 Network with 30 nodes

Let us focus on the network depicted in the preceding diagram, which comprises 30 nodes. We focus on identifying the central node, employing a spectrum of centrality measures, including degree centrality, betweenness centrality, closeness centrality, eigenvector centrality, Katz centrality, and subgraph centrality. The results of these calculations, which determine the centrality values for each node within the network, are meticulously documented in Table 46.1. Moreover, to encapsulate the significance of each node concerning others, their respective rankings are compiled and presented in Table 46.2. This comprehensive analysis effectively characterizes the roles and importance of individual nodes within the network structure.

Table 46.1 Centrality values for network nodes

Node	DC	BC	CC	EVC	KC	SC
1	0.2413	0.0556	0.4531	0.3489	0.2369	23.0102
2	0.1379	0.0378	0.4531	0.1761	0.1798	7.8612
3	0.2068	0.0875	0.4754	0.1839	0.2053	11.3913
4	0.1724	0.0350	0.4202	0.1759	0.1935	11.4634
5	0.1379	0.0145	0.3866	0.1411	0.1744	8.9245
6	0.1724	0.0344	0.4084	0.1656	0.1927	11.7729
7	0.2413	0.1261	0.4833	0.2229	0.2248	15.6676
8	0.1724	0.0700	0.4603	0.2319	0.1982	10.8249
9	0.2068	0.0690	0.4531	0.2922	0.2164	16.6967
10	0.2413	0.0927	0.4677	0.3401	0.2367	22.6700
11	0.1724	0.0185	0.4202	0.2807	0.2035	16.7246

Node	DC	BC	CC	EVC	KC	SC
12	0.1724	0.0401	0.4142	0.2530	0.1981	13.3320
13	0.1379	0.0397	0.4202	0.1686	0.1756	7.1895
14	0.2068	0.0542	0.4393	0.1848	0.2041	12.8397
15	0.1379	0.0446	0.4461	0.1548	0.1757	7.1340
16	0.1379	0.0405	0.4084	0.1155	0.1689	7.0446
17	0.1724	0.0782	0.4461	0.1813	0.1915	9.7197
18	0.1379	0.0511	0.4264	0.1277	0.1685	5.7866
19	0.1724	0.0860	0.4202	0.1737	0.1848	8.7670
20	0.0689	0.0062	0.3411	0.0606	0.1313	2.8256
21	0.1379	0.0506	0.4027	0.1306	0.1657	5.7654
22	0.1379	0.0582	0.4264	0.1092	0.1669	5.6936
23	0.1379	0.0508	0.4027	0.1058	0.1675	6.5589
24	0.0689	0.0106	0.3411	0.0327	0.1285	2.6080
25	0.1379	0.0620	0.3918	0.0585	0.1555	4.4770
26	0.1034	0.0447	0.3918	0.0619	0.1451	3.6278
27	0.1034	0.0419	0.3972	0.0785	0.1482	3.9022
28	0.1034	0.0177	0.3717	0.1089	0.1501	4.5122
29	0.1379	0.0398	0.4084	0.1757	0.1736	7.9819
30	0.1034	0.0406	0.4084	0.0834	0.1497	3.9728

Table 46.2 Node rankings based on network centrality measures

Rank	DC	BC	CC	EVC	KC	SC
1	1, 7, 10	7	7	1	1	1
2		10	3	10	10	10
3		19	10	9	7	11
4	3, 9, 14	17	8	11	9	9
5		8	1, 2, 9	12	3	7
6		9		8	14	12
7	4, 6, 8,11	25		7	11	14
8		22	15, 17	14	8	6
9		1		3	12	4
10		14	14	17	4	3

4. Conclusions

In this paper, we have determined the central node using degree centrality, betweenness centrality, closeness centrality, eigenvector centrality, Katz centrality, and sub graph centrality with the help of Python for a 30-node network. In the future, we would like to extend it to any number of nodes. This research revolves around the meticulous exploration of centrality metrics within the context of a 30-node network. Our goal was to find the central node by using a variety of pivotal centrality measures, such as degree centrality, betweenness centrality, closeness centrality, eigenvector centrality, Katz centrality, and subgraph centrality. Notably, these calculations were conducted utilizing the programming capabilities of Python, offering a robust and versatile platform for network analysis. Through this endeavor, we gained insights into the intrinsic nature of the network and the roles different nodes play in shaping its dynamics. The application of these centrality measures contributed to a nuanced understanding of node significance and influence, which is essential for comprehending network behavior and optimizing its performance. Looking ahead, we aspire to broaden the scope of our research beyond the confines of a 30-node network. We aim to establish a methodology that seamlessly accommodates networks of varying sizes. By developing adaptable algorithms and methodologies, we aim to extend our insights to networks encompassing any number of nodes. This would enrich our understanding of network dynamics on a broader scale and provide valuable tools for network practitioners and researchers working with diverse network sizes and complexities.

REFERENCES

1. Gómez, S. (2019), Centrality in Networks: Finding the Most Important Nodes. In: Moscato, P., de Vries, N. (eds) Business and Consumer Analytics: New Ideas. Springer, Cham.
2. Njotto, L.L. (2018), Centrality Measures Based on Matrix Functions, Open Journal of Discrete Mathematics. 8: 79–115.
3. Mbiya S.M, Hancke G.P, Silva B. (2020), An Efficient Routing Algorithm for Wireless Sensor Networks based on Centrality Measures, Acta Polytechnica Hungarica. 17(1): 83–99.
4. Jain, A. (2016), Betweenness centrality-based connectivity aware routing algorithm for prolonging network lifetime in wireless sensor networks, *Wireless Networks*. 22: 1605–1624.
5. Andrés Vázquez-Rodas, Luis J. de la Cruz Llopis. (2015), A centrality-based topology control protocol for wireless mesh networks, Ad Hoc Networks, 24: 34–54.

Note: All the tables and figure in this chapter were made by the authors.

Impending Inquisitions in Humanities and Sciences (ICIIHS-2022) – Dr. Mohan Varkolu et al. (eds)

Air and Sound Pollution Tracking Using IoT and Machine Learning

47

R. Mamatha[1]
Research Scholar, Computer Science & Engineering, Koneru Lakshmaiah Education Foundation,
Deemed to be University, Hyderabad, Telangana, India
Assistant Professor, Computer Science & Engineering,
G. Narayanamma Institute of Technology & Science, Hyderabad, India

Aishwarya Koppula[2], K. Chandana[3]
Student, Computer Science & Engineering,
G. Narayanamma Institute of Technology & Science, Hyderabad, India

P. Lalitha Surya Kumari*
Professor, Computer Science & Engineering, Koneru Lakshmaiah Education Foundation,
Deemed to be University, Hyderabad, Telangana, India

ABSTRACT *Purpose:* Environmental pollution has gotten worse over the last century. With increasing population and industrialization, the level of air and noise pollution is increasing faster than ever before, posing serious threats to society.

Findings: This requires a surveillance system that enables a person to view and examine live air anomalies as well as sound contamination in a particular location. An effective system for keeping track of dangerous gas concentrations and dealing with the expanding issue of sound and air pollution is the Air and Sound Pollution Tracking System.

Methodology: To detect the presence of dangerous chemicals and compounds in the atmosphere, this model uses an air sensor and a sound sensor, and it feeds the information to the microcontroller continually. The model continuously monitors pollution levels and sends IoT reports of those levels to the web host. Remote sites transmit the captured data to the cloud over the Internet. A Machine Learning algorithm is used with the accounted information to model, predict, and monitor pollution levels, and the data is then analysed. Users can keep track of these predicted values using an app. If the model detects any air or noise pollutants above the threshold value, a notification is sent to the user, allowing them to take appropriate action to mitigate the problem.

KEYWORDS Air pollution, IoT, Machine learning, Noise pollution

1. Introduction

Ocean currents and fish migration enable the long-distance spread of marine pollutants. Unintentionally spilled radioactive material from a nuclear reactor can be swept up by winds and transported all over the world. One country's factory smoke enters another.

*Corresponding author: vlalithanagesh@gmail.com
[1]mamatha.racharla@gmail.com, [2]aishkoppula01@gmail.com, [3]chandanakondaveeti2000@gmail.com

DOI: 10.1201/9781003489436-47

Discharge of pollutants into the atmosphere that are harmful to both environment and the human health as a whole causes air pollutants. According to the World Health Organization (WHO), air pollution is a factor in almost 7 million deaths per year. Presently, 9 out of 10 people breathe air that exceeds WHO guideline limits. The majority of the effects of pollution are found in people who fall under low- and middle-income countries.

Clean Air Act gives the 1970-founded US Environmental Protection Agency (EPA) authority to protect public health by limiting the emissions of these harmful air pollutants. The primary source of air pollution is the burning of fossil fuels, such as coal, oil, and natural gas.

to do with noise pollution The only threat to the planet's living organisms is atmospheric pollution. One amongst the dangerous environmental risks to human health, according to reports of the World Health Organization (WHO), is noise pollution. According to estimates from the European Environment Agency (EEA), noise causes more than 72,000 hospital admissions and 16,600 premature deaths each year in just Europe. Along with us, animals are also negatively influenced by it. According to the reports given by National Park Service (NPS) of the United States, noise pollution unintentionally affects species and has a severe negative impact on the ecosystem. Noise pollution is not a label that applies to all noises, but is the term used to describe any obtrusive or unwanted sound that shows a negative effect on the health and over all well-being of people and other creatures. Noise pollution is a risk that cannot be seen. It exists on land and in the ocean even though it cannot be seen. The units used to measure sound volume are decibels. Noise levels of 85 dB or above may cause hearing damage. Loud rock concerts are one example of a sound source that exceeds this barrier, 90- to 115-decibel subway trains, and powerful lawnmowers (110 to 120 decibels). Any noise that is louder than 65 decibels (dB) is considered to constitute noise pollution by the WHO. Noise levels above 75 decibels (dB) are dangerous, while levels beyond 120 dB are uncomfortable. As a result, it is suggested that noise levels be kept below 65 dB during the day time and that exceeding 30 dB at night times which makes it difficult to sleep well. Noise pollution has an effect on millions of people every day. Hearing loss (NIHL) is the health issue associated with noise that occurs most frequently. Stress, sleeplessness, high blood pressure, and heart disease can all be brought on by loud noise. Not only children but people of all ages, can be affected by these health issues. Plentiful studies have demonstrated that children who live near congested roadways or airports experience stress in addition to other issues like memory, focus, and reading challenges. The healthiness and wellbeing of wildlife are also affected by noise pollution. Animals use sound for a variety of purposes, including locating food, warding off predators, and attracting mates. Because of noise pollution, they find it difficult to carry out these tasks, endangering their lives. Along with creatures on land, species that live in the ocean are increasingly impacted by noise pollution. The presence of ships, drilling platforms, seismic testing, and sonar equipment has caused the once-calm ocean environment to become noisy and chaotic. Noise pollution causes significant danger to whales and dolphins. Noise interferes with these sea creatures' ability to echolocate effectively. They communicate, forage, locate mates, and navigate using echolocation.

2. Literature Survey

Predictive learning, Clustering, frequent pattern mining, change detection, anomaly detection, and relationship mining were the six categories into which Atluri et al. subdivided their research. various spatiotemporal data mining techniques. This research largely focuses on mining common patterns utilizing association rule mining techniques that were once widely used. A Spatio-temporal association rule mining approach was put out by *Shaheen et al.* to find associations between different objects based on context. In this study, frequent item sets that are both positive and negative were identified using context variables and spatial-temporal series inputs. Using atomic class-association rule mining, *Shao et al.* presented ACAR, a

The supervised approach to predicting software issues. Also suggested was an algorithm for finding frequent association patterns in data produced by the internet of things and smart devices. Join-based algorithms, pattern growth algorithms, and tree-based search algorithms (such EclaT, TreeProjection, and others) were all categorized by Chee et al. as the three categories of frequent itemset mining techniques (such as Apriori, and DHP.) The FP-Growth method was modified to mine regular patterns by Antonelli et al. Growth with FP disadvantage exhibits poor performance on sparse datasets and has a convoluted data structure. Numerous tree-based data structures, such as node lists, node-set, and others, have been proposed in the past to address this. NegFIN, a data structure developed by *Aryabarzan* and colleagues, drastically decreased the overall execution time needed to mine frequent item collections. A temporal association rule mining approach based on trees was proposed by Wang et al. A tree data structure was utilized by Liang et al. to search for recurrent patterns. For data from moving flocks, Turdukulov et al. presented a frequent pattern discovery strategy. Eclat-close is a vertical miner algorithm that Szathmary devised. In addition, EclaT using a tree-based method was suggested by *Zhang et al.* These algorithms' main drawback is their high space complexity, which becomes particularly apparent for bigger transactions. Using the Apriori technique, *Qin et al.* projected the spatiotemporal effects of Particulate Matter in China. On the other hand, conventional algorithms like EClaT, FP-Growth, and Apriori were unable to adjust to spatiotemporal data contexts. Popular Spatio-Temporal association rule mining algorithms include Spatio-Temporal Apriori (STA). 2019 VOLUME 7 of 98922 An improvement on the Apriori method is STA. Frequent Pattern Mining on Time and Location-Aware Systems by *A. Aggarwal and D. Toshniwal* [1] Along with FPGrowth, 9 Eclat, and others, Air Quality Data is a well-known technique for mining association rules. There have also been a few Apriori variations proposed in the past that afterward gained a lot of traction. In uncertain datasets, Lin et al. also employed Apriori to find weighted frequent item groupings. Spatio-Temporal Apriori was used to repeatedly scan the database to provide candidate item sets. This required a lot of storage space and processing time. Second, there was a problem with the formation of an absurdly huge number of candidates when there were even fewer items. In order to extract association rules, the method advised significantly reducing the amount of queries to the transactional database. The overall execution time of the algorithm was reduced with this effort. The total execution time includes the duration of each database access as well as the total number of database accesses. Calendar map schemas (CMS), an additional approach for storing spatiotemporal data, were used in this study. The CMS-based strategy cuts down on database accesses, but it has the downside that recurrent CMS accesses could actually increase the algorithm's overall execution time. In the current work, hash keys were used instead of hash-based methods to prevent continually contacting CMS. Tree-based data structures, well known for their complex processing, were used, especially in earlier investigations. In our work, we combined CMS data structures with direct address hashing rather than employing tree-based data structures. We used hashing in this process since it takes less time. Among other hashing methods, direct address hashing is used to avoid collisions, which can increase time complexity. The shortcomings of each of these methods are thus discussed in this paper, and a hybrid of CMS-based and hash-based solutions is then proposed. The suggested design reduced the quantity of database accesses as well as the time required to access the database. There have also been previously described research that mine spatial and temporal connection principles. Park et al. mined association rules using hashing and a dynamic hash tree. ARMADA, an interval-based temporal association rule mining method, was proposed by Winarko and Roddick. Furthermore, it was recommended to use the STARminer extension to extract important patterns from the database by only considering Spatio-Temporal Association Rules with High Support and Confidence. The suggested design reduced the quantity of database accesses as well as the time required to access the database.

There have also been published a number of other research that mine spatial and temporal link ideas. Park et al. mined association rules using hashing and a dynamic hash tree. Winarko and Roddick proposed ARMADA, an interval-based temporal association rule mining technique. Furthermore, it was advised to only take into account Spatio-Temporal Association Rules with High Support and Confidence when using the STARminer extension to extract significant patterns from the database.

3. Proposed Methodology

The use of small, low-cost Air Quality Monitoring devices has undoubtedly increased, owing primarily to wireless data transfer. Therefore, we suggest a monitoring system that tracks the pollution levels in the air and noise. The air and sound sensors collect data from the environment in which they are installed and send it to a remote server via the internet. This data is retrieved from the cloud, delivered to users via an app, and fed into a machine-learning algorithm to predict pollution levels. This information is then used to derive correlations between commonly occurring pollutants. The current analysis of pollution levels is haphazard and conducted in various locations for various reasons. Create a system that continuously monitors environmental parameters, predicts values at regular intervals, and notifies users and relevant authorities to take the necessary actions to control the increased levels. The paper's objectives are as follows: (1) To develop a surveillance system that allows an individual to examine and evaluate the local sound pollution levels as well as live air characteristics; (2) To alert people in the event of detecting noise and air quality issues so that necessary measures can be taken to control the issue, and (3) To obtain correlations between frequently occurring pollutants and analyze the data. Requirements There are some requirements for our system to meet in order to achieve the best results. The requirements that we intend to include in the proposed system are as follows: (1) gas sensor MQ135, (2) Temperature and humidity sensor DHT11, (3) DEVKIT ESP32 V1, (4) Acoustic sensor2.

Approach The air and sound sensors in the suggested system gathers environmental data in real-time. The gadgets wirelessly communicate the data they have collected to a distant server for additional processing and analysis. The mobile user application receives the cloud data after it has been retrieved. The information is presented in a way that the typical person can read and understand. Finding the most accurate prediction and forecasting model for the rise or decline is the next stage of harmful air pollutants such as O_3, NO_2, SO_2, and CO according to WHO guidelines.

Two Machine Learning methods, Linear Regression and Random Forest have been used for time-series forecasting. The method is finished with an analysis of the data gathered to get knowledge about the relationships between the contaminants. For time-series forecasting, Linear Regression and Random Forest are two Machine Learning methods that have been used. The method is finished with an analysis of the data gathered to get knowledge about the relationships between the contaminants. The outcomes are shown on the hardware's display interface and are accessible via the cloud on any intelligent mobile device. IoT is primarily focused with using the OSI Layered Architecture to connect smart devices (embedded electronics devices) to the internet.

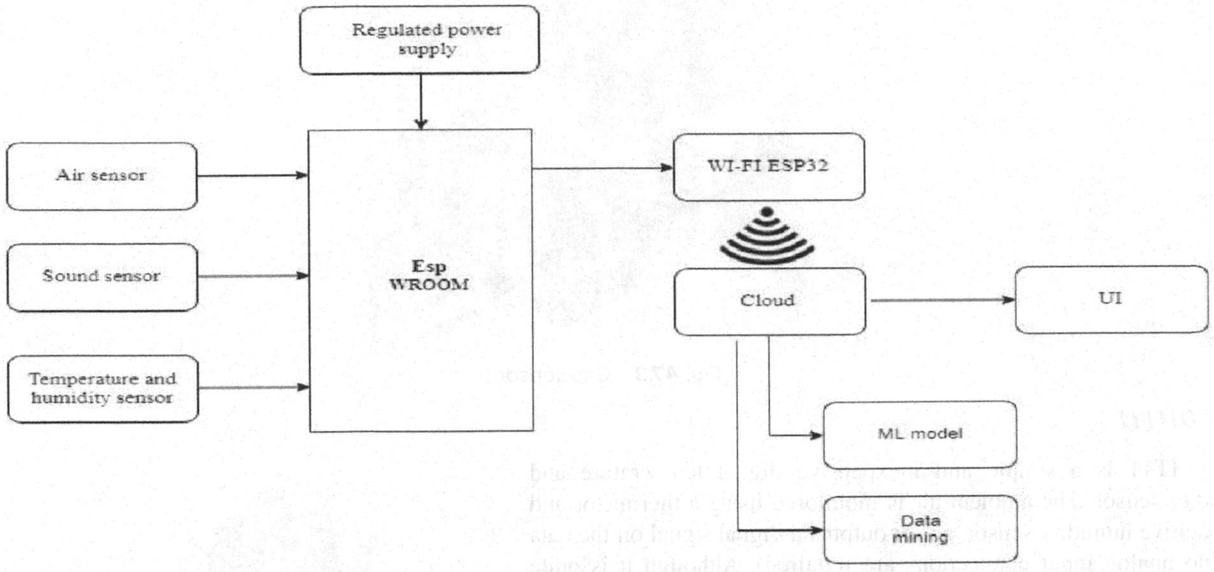

Fig. 47.1 Proposed system architecture

An international network of interconnected objects is created by the communication of all Air Sensors with a small embedded platform with network connectivity and internet access. Measurements of gas concentrations are made using the MQ135 Gas Sensor. To collect IoT data, this sensor data would be gathered and sent to the cloud. Through the user interface, the cloud's data is retrieved and presented to users in a way they can understand.

3.1 Components Used

3.1.1 ESP32

The Esp32 DevKit V1 is an evaluation board for the ESP-WROOM-32 module. The ESP32 is a single chip microcontroller that supports WiFi, Bluetooth, Low Power, and Ethernet serves as its focal point.

The ESP32 comes with an antenna, an RF balun, is a power amplifier, low-noise amplifiers, filters, and a power management module. On the printed circuit board, the total solution occupies the smallest amount of space. The board's TSMC 40nm low power 2.4 GHz dual-mode Wi-Fi and Bluetooth chips offer the finest RF and power performance.

They are suitable for a variety of applications and are safe, trustworthy, and flexible. The ESP32 module's internal flash memory is organized in a single flash area with 4096-byte-long pages. The flash begins at 0x00000, but certain locations are set aside for the Esp32 IDF SDK, and depending on whether BLE support is present, there are two distinct layouts.

Fig. 47.2 P32

The Esp32 DevKit v1 can be powered directly via either the "VIN" pin or the inbuilt USB Micro B connector. It chooses sources automatically. The device is powered by an external supply that ranges from 6 to 20 volts. If more than 12V is used, the voltage regulator may overheat and damage the device. The recommended voltage range is between seven and twelve volts.

3.1.2 Gas Sensor

The MQ135 gas sensor is made of sensitive SnO_2, which has a reduced conductivity in clean air. When the target flammable gas is present, the conductivity of the sensor rises along with the gas concentration. The MQ135 gas sensor is incredibly sensitive to ammonia, sulphur dioxide, and benzene steam in addition to smoke and other harmful gases.

Fig. 47.3 Gas sensor

3.1.3 DHT11

The DHT11 is a simple and inexpensive digital temperature and humidity sensor. The ambient air is monitored using a thermistor and a capacitive humidity sensor, and it outputs a digital signal on the data pin (no analog input connections are required). Although it is quite simple to operate, data collection requires careful timing. As a result, while using our library, sensor readings may be as much as two seconds old. This sensor's main notable limitation is that it can only deliver fresh data once every two seconds. This sensor is more compact and less expensive than the DHT22, but it also functions within a narrower temperature or humidity range, is less accurate, and is less precise. The 4.7K or 10K resistor that it comes with should be used.

Fig. 47.4 DTH11 Temperature and humidity sensor

3.1.4 Sound Sensor

The sound sensor module makes identifying sounds straightforward and is frequently used to gauge sound volume. Applications for this module include switching, monitoring, and security. For the convenience of usage, its precision may be simply modified.

Fig. 47.5 Sound sensor

The input to an amplifier, peak detector, and the buffer is a microphone. The sensor produces an output signal voltage in response to sound detection, which is transferred to a microcontroller for further processing.

4. Results and Discussion

Table 47.1 Excel files storing sensor readings

Pollution_ index
File Edit View Insert Format Data Tools Extensions Help Last edit was made 4 minutes ago by Chandana Kondaveeti

	A	B	C	D	E	F	G	H
278	3/25/2022	16:14:58	44	36.3	351	46.37	34	41
279	3/25/2022	16:15:06	44	36.3	348	45.82	34	41
280	3/25/2022	16:15:14	44	36.3	351	46.1	34	41
281	3/25/2022	16:15:23	44	36.3	356	45.64	34	40
282	3/25/2022	16:15:30	44	36.3	349	45.73	34	41
283	3/25/2022	16:15:38	44	36.3	348	45.64	34	41
284	3/25/2022	16:15:46	44	36.3	352	46.1	34	41
285	3/25/2022	16:15:54	44	36.3	351	46.19	34	41
286	3/25/2022	16:16:02	44	36.3	354	45.92	34	41
287	3/25/2022	16:16:10	44	36.3	354	46.19	34	41
288	3/25/2022	16:16:18	44	36.3	345	45.37	33	40
289	3/25/2022	16:16:26	44	36.3	352	46.92	34	42
290	3/25/2022	16:16:34	44	36.3	352	45.82	34	41
291	3/25/2022	16:16:43	44	36.3	348	46.55	34	41
292	3/25/2022	16:16:51	44	36.3	344	46.1	33	41
293	3/25/2022	16:16:59	44	36.3	348	46.01	33	41
294	3/25/2022	16:17:08	44	36.3	341	46.19	33	41
295	3/25/2022	16:17:15	44	36.3	347	46.55	33	41
296	3/25/2022	16:17:24	44	36.3	343	45.82	33	41
297	3/25/2022	16:17:31	44	36.3	348	46.37	33	41
298	3/25/2022	16:17:39	44	36.3	348	45.82	33	41
299	3/25/2022	16:17:47	44	36.3	348	46.46	34	41
300	3/25/2022	16:17:55	44	36.3	344	46.37	33	41
301	3/25/2022	16:18:03	44	36.3	340	46.92	34	42
302	3/25/2022	16:18:11	44	36.3	343	45.82	33	41
303	3/25/2022	16:18:18	44	36.3	343	45.82	33	41

Table 47.2 Predicting the pollution index

Pollution_ index
File Edit View Insert Format Data Tools Extensions Help Last edit was made seconds ago by Chandana Kondaveeti

A1 Date

	A	B	C	D	E	F	G	H	I
1541	3/25/2022	18:36:54	39	36.3	269	46.1	26	41	
1542									216.2809479
1543	3/25/2022	18:37:02	39	36.9	276	46.28	26	41	
1544									216.8771888
1545	3/25/2022	18:37:10	39	36.9	273	46.46	26	41	
1546									216.8771888
1547	3/25/2022	18:37:17	39	36.9	275	46.19	26	41	
1548									216.8771888
1549	3/25/2022	18:37:25	39	36.9	278	46.46	27	41	
1550									216.6122724

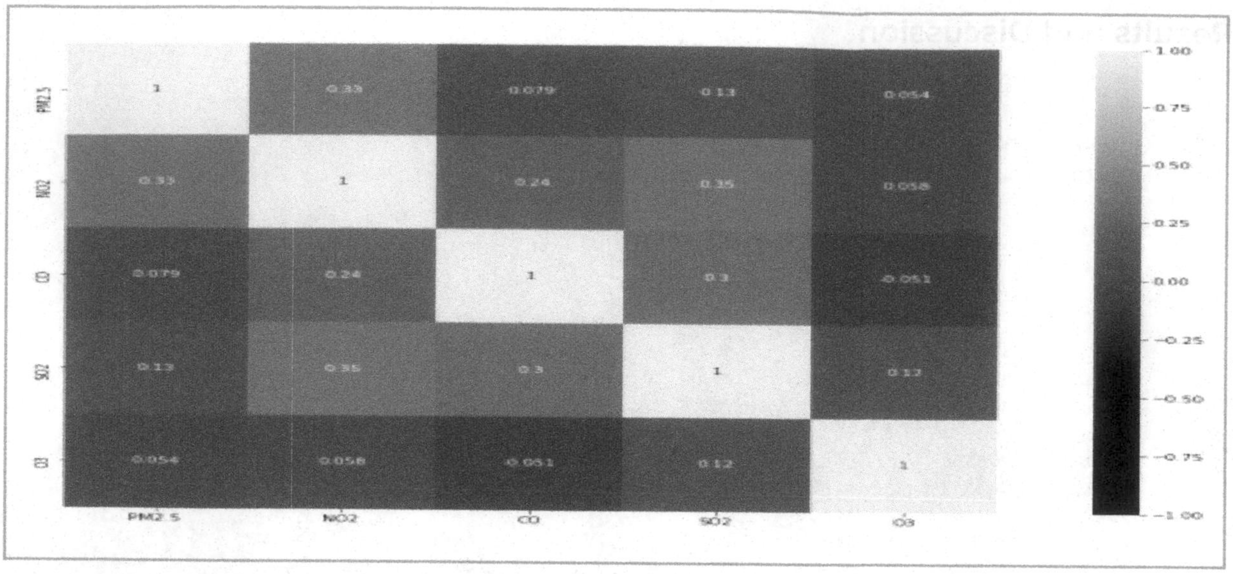

Fig. 47.6 Relationship matrix

The correlation matrix between five contaminants is shown in the graph above (O3, SO2, NO2, PM10, PM2.5)

As the correlation coefficient value approaches 0, the relationship between the two variables becomes less significant. The coefficient's sign indicates the relationship's direction; a + sign suggests a positive association, whereas a – sign indicates a negative relationship.

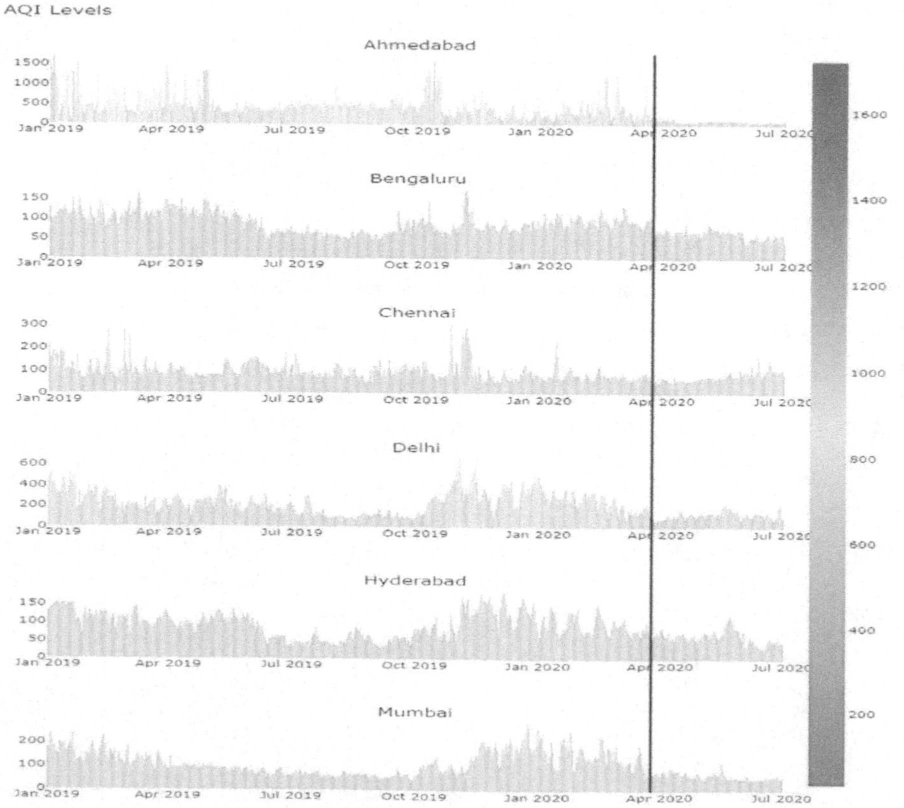

Fig. 47.7 AQI readings prior to and following COVID

The black vertical line denotes the date of the first lockdown in India.

The graph above shows how different pollutant levels changed between January 2019 and July 2020.

After March 25, 2020, it is obvious that all of the cities under consideration undergo a rapid decrease.

Pollutant levels in 2020 and 2019 are compared between January and April. This is to see if pollution levels in India have actually decreased or if they will remain low as summer approaches. A Time slider is located beneath the graph. Using a range slider, users can pan and zoom the X-axis while maintaining an overview of the chart. The level of pollution in India has been observed to decrease as summer approaches. The graphs above also support this. However, the drop in March 2020 is greater than in March 2019.

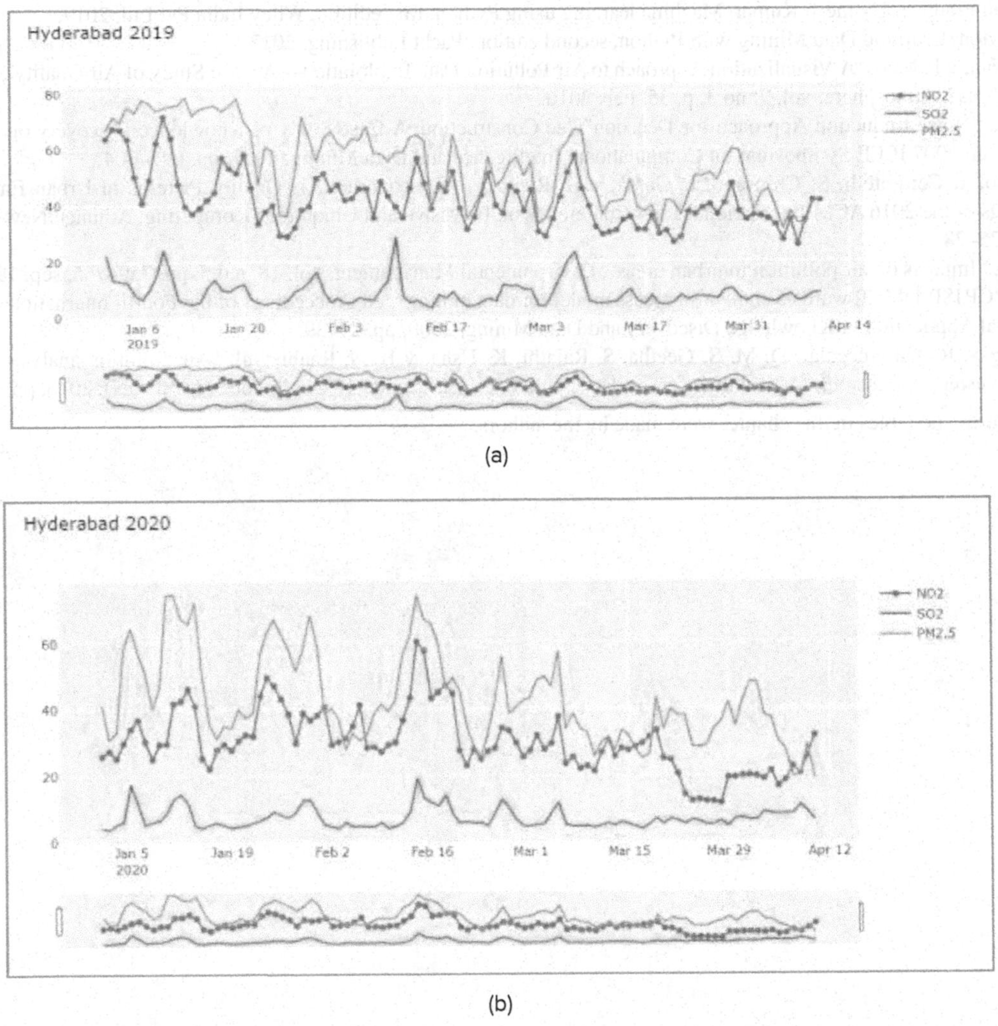

Fig. 47.8 (a) 2019 Pollutant concentration over period of time in HYD, (b) Pollutant concentration in 2020 overtime in HYD

5. Conclusion

The current state of the environment is deteriorating faster than ever before. Increased technological advancements, peak industrial development, scientific discoveries and inventions, and increased vehicle use all contribute to environmental degradation. We are unaware of the rising levels of pollution because we lack a system for tracking them. We have a tendency to disregard pollution-reduction measures. An efficient tracking system that takes real-time values from the environment is required. This project provides a system that tracks data, predicts future values based on past data, and alerts users of a specific area if the parameter values exceed the threshold values. It also obtains and analyses data on correlations between commonly occurring pollutants. This allows environmentalists to take the necessary steps right away.

REFERENCES

1. A. Aggarwal and D. Toshniwal, "*Frequent pattern mining on time and location aware air quality data*", IEEE Access, vol. 7, pp. 98921–98933, 2019.
2. Diana Arce, Juan-FernandoLima, Marcos Orellana, John Ortega, "*Discovering behavioral patterns among air pollutants: A data mining approach*".
3. Fabiana Martins Clemente, Ales Popovic, Sara Silva, and Leonardo Vanneschi,"*A Machine Learning Approach to Predict Air Quality in California*", Volume 2020, Article ID 8049504
4. K. Cornelius, N. K. Kumar, S. Pradhan, P. Patel and N. Vinay, "*An Efficient Tracking System for Air and Sound Pollution using IoT*", 2020 6th International Conference on Advanced Computing and Communication Systems (ICACCS), pp. 22–25, 2020.
5. Manaranjan Pradhan, U dinesh Kumar, Machine learning using Python, first edition, Wiley India Pvt Ltd, 2019.
6. Robert Layton, Learning Data Mining with Python, second edition, Packt Publishing, 2017.
7. H. Li, H. Fan, y F. Mao, "A Visualization Approach to Air Pollution Data Exploration—A Case Study of Air Quality Index (PM2.5) in Beijing, China", Atmosphere, vol. 7, no 3, p. 35, Feb. 2016.
8. K. Fukuda, "Noise Reduction Approach for Decision Tree Construction: A Case Study of Knowledge Discovery on Climate and Air Pollution", en 2007 IEEE Symposium on Computational Intelligence and Data Mining, 2007, pp. 697–704.
9. L. Cagliero, T. Cerquitelli, S. Chiusano, P. Garza, y G. Ricupero, "Discovering Air Quality Patterns in Urban Environments", en Proceedings of the 2016 ACM International Joint Conference on Pervasive and Ubiquitous Computing: Adjunct, New York, NY, USA, 2016, pp. 25–28.
10. E. Wagner, "Impacts on air pollution in urban areas", Environmental Management, vol. 18, n.o 5, pp. 759–765, sep. 1994.
11. R. Wirth, "CRISP-DM: Towards a standard process model for data mining", en Proceedings of the Fourth International Conference on the Practical Application of Knowledge Discovery and Data Mining, 2000, pp. 29–39.
12. G. R. Kingsy, R. Manimegalai, D. M. S. Geetha, S. Rajathi, K. Usha, y B. N. Raabiathul, "Air pollution analysis using enhanced K-Means clustering algorithm for real time sensor data", en 2016 IEEE Region 10 Conference (TENCON), 2016, pp. 1945–1949.

Note: All the figures and tables in this chapter were made by the authors.

Impending Inquisitions in Humanities and Sciences (ICIIHS-2022) – Dr. Mohan Varkolu et al. (eds)

Exploiting Inherent Linguistic Information to Detect Aspects and Associated Sentiments of Online Course Reviews

48

Aparna Bulusu*
Research Scholar, Dept of Computer Science Engineering, K L University, Hyderabad

D. S. Rao
Professor, Dept of Computer Science Engineering, K L University, Hyderabad

ABSTRACT Online Courses and distance based education have gained a lot of traction over the last few years with huge numbers of learners enrolling in these courses. Course feedback provided by learners is a great source of information regarding the quality of course content, teaching expertise of instructors, difficulty levels of assignments and evaluation models. NLP techniques make it possible to carry out fine grained aspect detection and sentiment analysis through use of Machine Learning classifiers. This paper attempts to utilize inherent linguistic information within natural language in order to determine possible aspect terms and employs a lexicon and rule-based algorithm called VADER to detect sentiment polarities. Knowledge gained about multiple course aspects can be of great help to course and instructional designers as well as prospective learners of MOOC platforms
KEYWORDS NLP, Aspcct detection, VADER , Lexicon, Course reviews

1. Introduction

Sentiment Analysis is a common NLP task used for understanding user feedback regarding products, services, political agendas etc. across domains. While it was initially construed as a simple task of figuring out the sentiment of a complete document, the field of sentiment analysis has now expanded to figuring out sentiments at a very detailed and fine grained level. Advances in Machine and Deep Learning fields and availability of robust open source packages in the NLP arena are making it possible to carry out allied tasks like aspect extraction and sentiment detection with good accuracy. Product reviews and social media posts have been the most common targets for SA tasks, however Sentiment Analysis is now being applied across multiple domains to gain insights into user expectations and behavior.

Although MOOC courses were available from many years, the Covid pandemic has led to a drastic increase in the number of enrolled users. Multiple course formats are now available through various online educational platforms like Coursera, EdX etc and some courses have millions of users pursuing a course concurrently. In this scenario, student feedback about these courses are a rich source of information that can be utilized by students, instructors and MOOC platform providers as well for various purposes.

Multiple algorithms and machine learning classifiers are available to process NLP data and carry out sentiment analysis. The paper attempts to identify aspects from course reviews and detect their sentiments using a specific sentiment analyzer called Vader.

*Corresponding author: aparnabulusu79@gmail.com

DOI: 10.1201/9781003489436-48

2. Literature Review

2.1 Prevalence of MOOC Courses and Importance of Student Feedback

MOOC Course offerings as well as the number of students being enrolled in MOOC's has been going up over the years. Learners are seeking out MOOC courses due to easy accessibility, availability of multiple course options, enhancing skills and advancing in careers and sometimes for the pure joy of learning new things as per Christensen and Gayle. Jara, Magdalena et al. reported that Collecting student feedback is a very important aspect of higher education in both offline and online modes. Collecting and analyzing feedback is very critical in maintaining academic quality in making enhancements to course content and delivery structure. Course feedback of online courses generates huge amounts of text data which reflects the feelings and opinions of learners regarding course content, instructors and other factors as per Liu, Sannyuya, et al.

MOOCs have radicalised the elearning and distance education space. Student enrolment and retention ratios are important aspects for MOOC platform providers and instructional designers. Adamopoulos observed from analysing student feedback of these courses that the most important factors are the platform through which the course is being offered, course discipline/ stream, difficulty level of the course, time required per week, course pace, assessment and evaluation model and availability of course certificate.

2.2 Sentiment Analysis—Types and Techniques

According to Bo and Lee, Sentiment analysis or Opinion mining/ Polarity detection is the use of automated techniques to detect the strength and polarity of users feelings/opinions about a variety of issues the internet and world wide web have made it possible for users to express their opinions through a variety of platforms and these reviews are a very rich source of information that can be mined for various purposes.

Sentiment analysis can be carried out at multiple levels of granularity as reported by McDonald, Ryan, et al. Initial research was focused on detecting sentiment of a large quantity of text, also referred to as document level SA. Sentence level SA focuses on detecting sentiments at each individual sentence level and it is possible that a single document might contain multiple sentences with different polarities. In depth fine grained analysis is now being carried out since a single review might contain a mix of sentiments regarding multiple features or aspects.

In terms of techniques used, machine learning classifiers have been used extensively to carry out SA tasks. Naive Bayes classifiers, Decision trees, Random Forests, Support Vector MAchines and Maximum Entropy classifiers have all been reported to provide good accuracy levels in detecting sentiments. Neethu et al report that Deep learning networks are also being used extensively since Natural language processing can be performed efficiently with use of RNN's and LSTM's. Auto encoders, Denoising Auto encoders, Convolutional Neural networks, Bi directional Recurrent neural networks, Long Short Term Memory networks, attention mechanisms and memory networks yield state of the art results in various SA tasks as per Madhukar and Ramesh.

2.3 Utilising Language Features

Efficacy of NLP tasks depends on the effective representation of words through various techniques. Common embeddings that are preferred for SA tasks include Continuous Bag of Words (CBOW), One hot encoding, and more richer representations like Glove and Fasttext as stated by Yu. Current research is focused on using transformers and bi directional representation with algorithms like BERT to achieve SOTA results on a multitude of NLP tasks.

This paper attempts to make use of inherent relationships and order among words based on syntactic and grammatical information in order to identify aspect terms from course reviews posted by students on online MOOC platforms. Vader sentiment analyzer is then implemented on extracted aspects to detect sentiment polarity. VADER (Valence Aware Dictionary for sEntiment Reasoning) makes use of a gold standard sentiment lexicon that was specifically created to understand and analyze twitter feeds and similar posts along with a set of generalized rules to consider grammatical and syntactic considerations to help with detecting sentiments (Hutto, Clayton, and Eric Gilbert). This rule based approach of VADER sentiment analysis engine has been shown to yield very high accuracy in multiple domains.

3. Methodology and Algorithm

3.1 Sequence of Steps Followed

i. Gather course reviews
ii. Carry out cleaning and preprocessing
iii. Use Spacy Module to determine POS tags and dependency parse trees
iv. Utilise various rules to determine which terms are part of aspect sentiment pairs
v. Use VADER algorithm on aspect term lists generated in previous step
vi. Convert polarity reported by VADER to a number score
vii. Compare the score returned by VADER with the numerical star rating available along with reviews on course websites
viii. Difference between these scores helps determine accuracy of this technique in determining sentiment polarity of aspect based terms in course reviews.

3.2 Process Flow

Course reviews were extracted using web scraping tools from Coursera Platform. Data cleaning and preprocessing were then carried out to retain text of review and a number rating. SpaCy, an open source software library is used to perform dependency parsing on the reviews to extract aspect terms of interest.

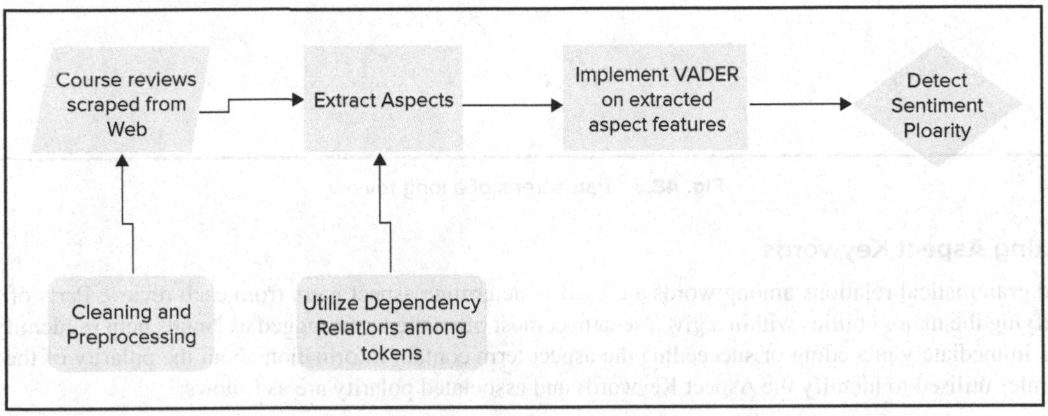

Fig. 48.1 Process flow to identify aspects and detect sentiments using VADER

For example, Consider the following review:

"The course is good, instructor did a great job, however the assignments were unclear and time consuming."

We can identify three different aspects being referred to in this single review, namely: Course, Instructor and Assignment. The first two aspects have a positive sentiment associated with them while the third aspect can be considered neutral.

Performing dependency parsing allows us to identify terms that can be considered aspects depending on the context with which those terms are being used specifically identifying elements like adjective modifiers, adverb modifiers, open causal complements, negations and compound words leads to identification of aspect terms within the reviews, The following images show the output of dependency parsing using spaCy modules on short and long reviews and how this helps in identifying aspect terms:

3.3 Generating Parse Trees

Consider the following Review: **"A very Good Course"**

The dependency parsing of this sentence yields the following parse tree (Fig. 48.2) along with sufficient grammatical information regarding the terms involved.

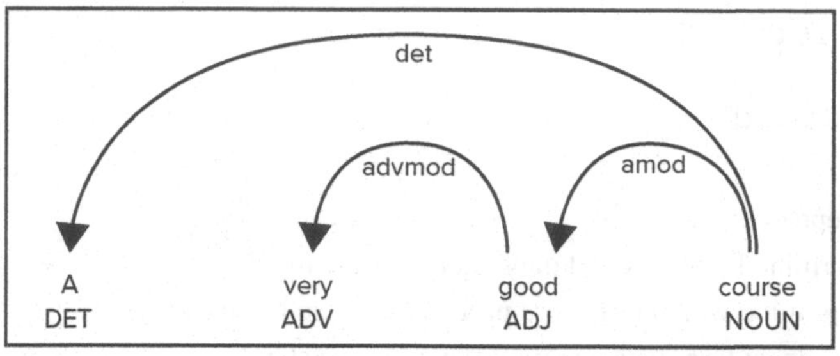

Fig. 48.2 Identifying grammatical relations through dependency parsing

These trees can also be used with very long sentences as shown through parse tree depicted in Fig. 48.3. Consider the Review:

txt = "A very good course with basic knowledge for data science. However, it is a little bit challenging if you are fresh to this area. "

The Parse trees contain information regarding the POS tag associated with each word as well as grammatical information regarding relationships among all these terms.

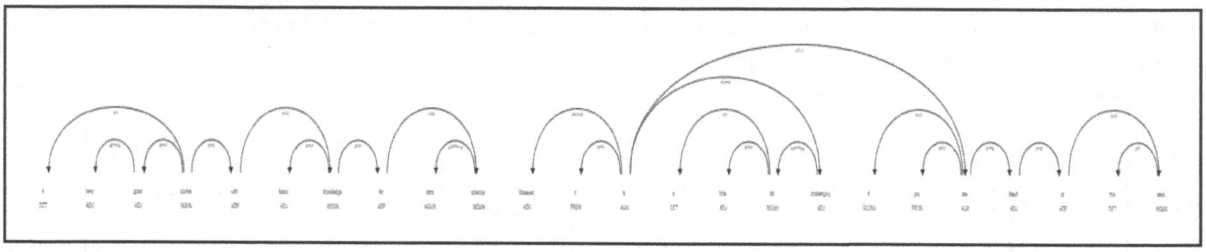

Fig. 48.3 Parse trees of a long review

3.4 Extracting Aspect Keywords

POS tags and grammatical relations among words are used to determine aspect pairs from each review. Parts of Speech tags help in identifying the major entities within a given sentence, most often the terms tagged as Nouns help in identifying aspects and the terms immediately preceding or succeeding the aspect term contain information about the polarity of the said aspect. Some of the rules utilised to identify the Aspect Keywords and associated polarity are as follows:

- Compound Nouns are identified where POS tag of a Token is Noun and is preceded by a token with tag 'Compound'
- Similarly aspect pairs can be identified by tokens with 'Noun' tag preceded by 'Amod' and 'ADJ' tags OR
- Tokens with 'VERB' tag preceded by 'ADV' tags

Review	rating	compound_nouns	aspect_keywords
Some of the details of the hyperparameters I w...	4	[]	[]
Fantastic course! Wish my university offered ...	5	[(TensorBoard inputs, inputs)]	[only suggestion, Fantastic course]
It explains many problems you might meet when ...	5	[(ml theory, theory), (theory course, course)]	[many problems, actually implementing, other t...
I understand the DL hyper parameters to good e...	3	[(DL parameters, parameters), (hyper parameter...	[good extent]
Andrew does an amazing job of explaining the i...	4	[]	[subtle manner, intricate level, amazing job, ...

Fig. 48.4 Extraction of Aspect Keywords

3.5 Running VADER on top of Extracted Words

Vader sentiment analyzer was then run on top of the extracted aspects to detect the associated sentiments. *VADER* stands for **V**alence **A**ware **D**ictionary for s**E**ntiment **R**easoning and uses 5 Heuristics: Punctuation, Capitalization, Degree Modifiers, Polarity shift due to conjunctions and Polarity negation. The following example illustrates how VADER algorithm uses these rules to determine a sentiment score.

i. Punctuation, namely the exclamation point (!), amplifies the intensity without changing the semantic focus. For instance, the intensity of "The course is awesome!!!!" is higher than that of "The course is awesome."

ii. Without changing the semantic orientation, capitalization—more precisely, using ALL-CAPS to emphasise a sentiment-relevant word when there are other non-capitalized terms nearby—increases the magnitude of the sentiment intensity. "The test SUCKS." is a stronger statement than "The test sucks".

iii. Degree modifiers affect sentiment intensity by either raising or decreasing the intensity. They are also known as intensifiers, booster words, or degree adverbs. For example, "The professor taught extremely well" has more intensity than "The professor taught well,"

iv. Polarity shifting: The contrastive conjunction "but" denotes a change in sentiment polarity, with the text's subsequent sentiment taking precedence. For instance, "The course is fine, but the assignments are tough." has opposing sentiments, with the second half determining the final score.

v. Polarity Negation Capture: In Approximately 90% of instances, negation reverses the polarity of the text by looking at the continuous sequence of 3 items that comes before a sentiment-filled lexical characteristic. The line "The course isn't really that good." is an example of a negated sentence.

Sentiment polarity ['pos' or 'neg' for positive or negative] reported by VADER were converted into numerical values and compared with ratings already included with reviews to check for accuracy.

4. Results

A small dataset of hundred plus reviews was used. A few reviews where aspects couldn't be identified returned a NaN result and were discarded. Comparison of Vader analysis with existing ratings showed that an accuracy of 70% was realised. Small size of the dataset could be a potential reason for the low accuracy levels reported.

```
accuracy
0.7083333333333334
f1 score
0.8292682926829269
recall
0.9855072463768116
precision
0.7157894736842105
```

Fig. 48.5 Results: accuracy and other metrics

5. Conclusion

Grammatical knowledge inherent within Natural language can be exploited effectively to identify fine grained aspects from given text. Results show that there is scope to improve the methodology by working with a larger dataset and by including more comprehensive lexical relationships in the aspect term detection stage. Future studies can focus on identifying negative aspects automatically from a large number of course reviews and this information can be of immense help for instructors and course designers.

REFERENCES

1. Adamopoulos, Panagiotis. "What makes a great MOOC? An interdisciplinary analysis of student retention in online courses." (2013).
2. Bonta, Venkateswarlu, and Nandhini Kumaresh and N. Janardhan. "A comprehensive study on lexicon based approaches for sentiment analysis." *Asian Journal of Computer Science and Technology* 8.S2 (2019): 1-6.
3. Christensen, Gayle, et al. "The MOOC phenomenon: Who takes massive open online courses and why?." *Available at SSRN 2350964* (2013).

4. Hutto, Clayton, and Eric Gilbert. "Vader: A parsimonious rule-based model for sentiment analysis of social media text." *Proceedings of the international AAAI conference on web and social media*. Vol. 8. No. 1. 2014.
5. Jara, Magdalena, and Harvey Mellar. "Quality enhancement for e-learning courses: The role of student feedback." *Computers & Education* 54.3 (2010): 709–714.
6. Liu, Sannyuya, et al. "Unfolding sentimental and behavioral tendencies of learners' concerned topics from course reviews in a MOOC." *Journal of Educational Computing Research* 57.3 (2019): 670–696.
7. Madhukar Rao, G., and D. Ramesh. "Supervised learning techniques for big data: a survey. IJCTA." Int Sci Press 9 (2016): 3811–3891.
8. McDonald, Ryan, et al. "Structured models for fine-to-coarse sentiment analysis." *Proceedings of the 45th annual meeting of the association of computational linguistics*. 2007.
9. Neethu, M. S., and R. Rajasree. "Sentiment analysis in twitter using machine learning techniques." *2013 fourth international conference on computing, communications and networking technologies (ICCCNT)*. IEEE, 2013.
10. Pang, Bo, and Lillian Lee. "Opinion mining and sentiment analysis." *Foundations and Trends® in information retrieval* 2.1–2 (2008): 1-135.
11. Yu, Liang-Chih, et al. "Refining word embeddings for sentiment analysis." *Proceedings of the 2017 conference on empirical methods in natural language processing*. 2017.

Note: All the figures in this chapter were made by the authors.

Impending Inquisitions in Humanities and Sciences (ICIIHS-2022) – Dr. Mohan Varkolu et al. (eds)

Analytical Models for Materialized View Maintenance Methods

49

Sasidhar K*, Prathipati Ratna Kumar, Nandula Anuradha, Ashish Arun Kumar
Department of Computer Science and Engineering,
Koneru Lakshmaiah Education Foundation, KL Deemed to be University,
Hyderabad, Telangana, India
Nanduri Naga Raju
Department of Management Studies, Dadi Institute of Engineering and Technology,
Visakhapatnam, Andhra Pradesh, India

ABSTRACT This manuscript explores the usage of information warehouses for data integration, decision support, and research purposes. Organizations employ these warehouses for asset planning, information management, and determination. Essential tools like OLAP extract, coordinate, and store relevant data from distributed sources, processing them within the warehouse. The focus of this theory lies in addressing the challenges associated with realizing a Data Warehouse. Calculation and maintenance are crucial for achieving benefits such as efficient access, reliable performance, and improved information availability. However, the large number of information sources, increasing size of each source, and substantial information content pose difficulties. To overcome these challenges, we propose simplified methods for supporting information warehousing. Our observations indicate that materialized view support is more manageable in homogeneous information sources than in heterogeneous ones. Through multiple tests, we analyzed the number of updates within a given information collection from various sources. We established connections between suggested systems and multiple information sources, effectively demonstrating their efficacy. Materialized view maintenance reviews play a significant role in supporting views within data warehousing. Separate systems for materialized view support were also examined. By evaluating view maintenance using a test system, we found that pre-processed information in materialized views enhances query execution. Additionally, a combined information update aids in materialized view assistance. During the recreation of materialized view maintenance, we discovered that storing more information at the information distribution center reduces support costs.

KEYWORDS OLAP, Data warehouse, Query execution, Information distribution

1. Introduction

While data is invaluable, it may not be easy to organize and display it in an appropriate way. An organization's ability to arrange the release of massive data sets today puts it to the test. As the variety and sophistication of these apps expand, IT personnel face increasing challenges in providing support. Database systems often contain much of the information that an application needs. Data gathered is constantly isolated, for the most part as per divisions, groups or even land region. Associations need to relocate the data and access information from assorted endeavor business application within determined time periods. The information frequently required to the clients is put away as appeared sees in the data warehouse. A view is a determined connection

*sasidhar@klh.edu.in

DOI: 10.1201/9781003489436-49

characterized as far as base relations. A view in this way characterizes a capacity from a lot of base tables to a determined table; this capacity is ordinarily recomputed each time the view is referenced.

2. Comparison with Existing Methods

The counting algorithm and rundown delta scheme are compared to the suggested technique for materialized view maintenance. The foundation for this is the prior model. We compared two datasets, each containing 100,000 entries from the "emp paper publish" connection and 500,000 tuples from the "emp_rschr" relation. The time and effort required to compute the modifications and the time and effort required to update the materialized view are factors in the total cost of materializing and sustaining a view. Location: Data warehouse location. The findings for the three methods are shown in Table 49.1 and Table 49.2 in terms of maintenance time for inserting and updating tuples, respectively. This data shows that our suggested materialized view maintenance approach outperforms competing methods. This is because incorporating the necessary adjustments does not require extra computational time.

Table 49.1 Illustrates a cost analysis of maintaining a materialized view once an insert is made (in terms of both time and money)

Number of records	Delta table method (Second)	Counting algorithm (Second)	Proposed View maintenance method (Second)
1x10^3	19.04	22.06	16.02
2x10^3	25	28.9	19.08
3x10^3	30.8	37.08	25.04
4x10^3	36.4	44.04	30.5
5x10^3	42.08	49.4	35.05

Table 49.2 Consider the time and money spent on upkeep vs the updates that are generated

Number of records	Delta table method (Second)	Counting algorithm (Second)	Proposed View maintenance method (Second)
1x10^3	14	18.02	11.06
2x10^3	20.05	29.07	17.04
3x10^3	29	36.08	24.02
4x10^3	36.02	43.07	32.07
5x10^3	48.04	57.08	43.08

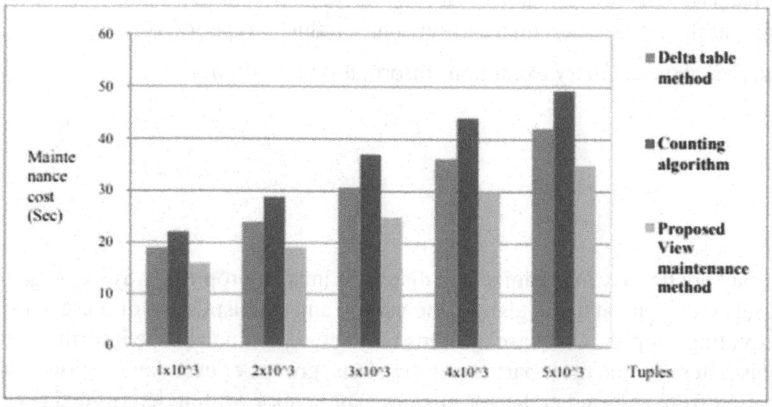

Fig. 49.1 Cost-benefit analysis of maintenance interventions that result in insertion

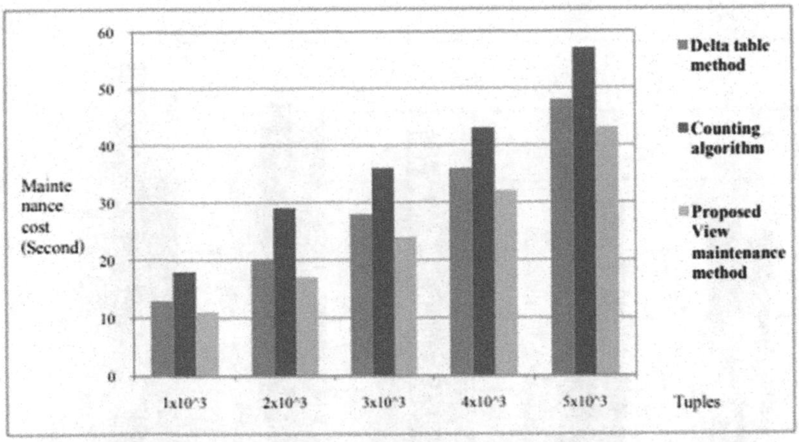

Fig. 49.2 Analysis of upkeep costs caused by updates

The findings that were obtained from analyzing and contrasting the three ways for adding and updating tuples indicate that the materialized view maintenance strategy that was presented is superior to the other two approaches. This advantage might be attributed to the lack of extra calculation time needed at the data warehouse location in order to incorporate changes into the final materialized view. We can reduce the amount of time spent on the view maintenance process while also improving its overall performance by synchronizing the propagate and refresh procedures.

In order to further evaluate the proposed materialized view maintenance technique, we conducted a comparison by varying the size of the "emp_rschr" relation while keeping the delta relation fixed at 5000 tuples. Table 49.3 and Table 49.4 display the results of the three techniques for tuple insertion and updating, respectively. These tables illustrate the elapsed time variation as the base relation size changes, while the delta relation size remains constant. From the analysis, it is observed that the propagating time remains relatively constant regardless of the increase in the base relation size. This is because the propagation technique is not dependent on the base relation size. Furthermore, it can be observed from the tables that the proposed materialized view maintenance strategy outperforms the other two techniques, indicating its superiority in terms of performance.

Table 49.3 Create a time and money analysis of view maintenance when the base relation size is changed to generate inserts

Number of records	Delta table method (Second)	Counting algorithm (Second)	Proposed View maintenance method (Second)
5x10^4	31.7	34	27.01
10x10^4	37.05	44.03	33.06
15x10^4	43.8	47.02	38
20x10^4	48.8	51.01	43.8
25x10^4	52.8	57.02	48.04

Table 49.4 Shows a comparison of the amount of time and money required to maintain a materialized view while making updates of varied base relation sizes

Number of records	Delta table method (no. 2)	Counting algorithm (no. 2)	Proposed View maintenance method (no. 2)
5x10^4	35.02	34.06	29.2
10x10^4	40.05	40.09	35.13
15x10^4	47.02	45	41.09
20x10^4	51.05	51	46.08
25x10^4	56.28	59.5	51.8

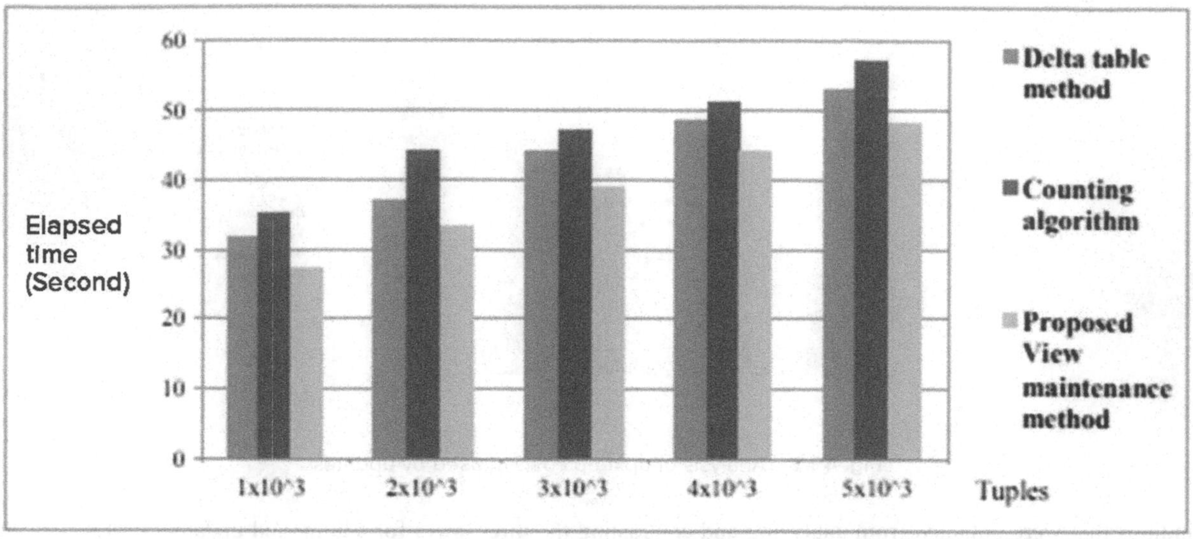

Fig. 49.3 A cost comparison of several source relation insertion producing alterations

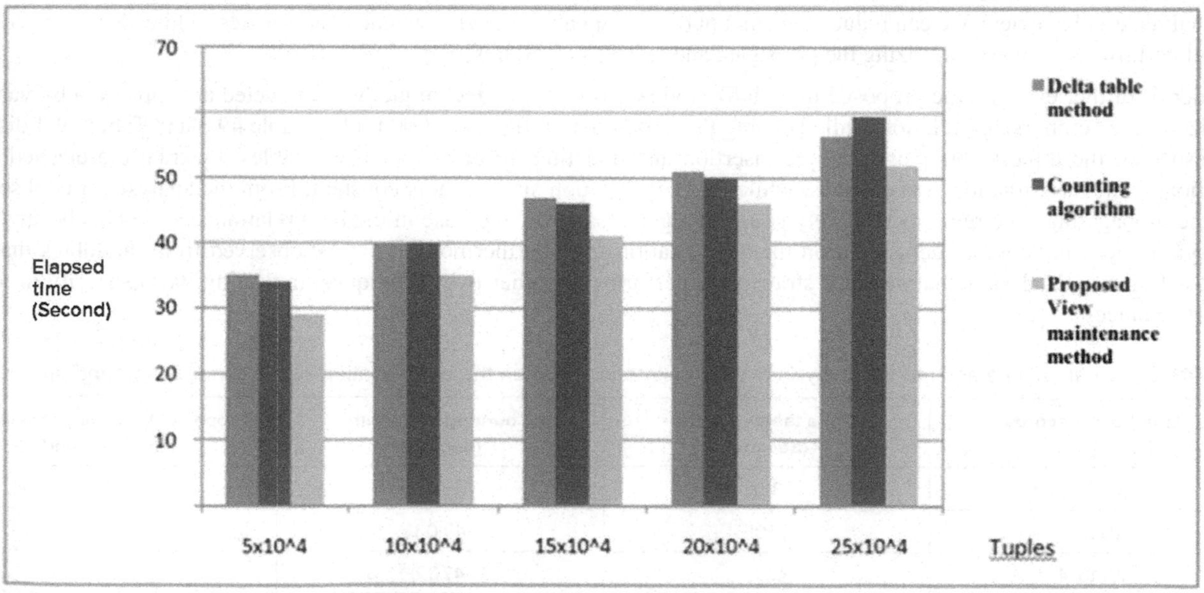

Fig. 49.4 Source-relation-dependent cost comparisons of update-initiated adjustments

In addition, we performed a performance analysis to compare the efficiency of the suggested materialized view maintenance strategy across different database sizes. Our results demonstrate that the experiment employing the proposed technique exhibits a significantly reduced maintenance time for the materialized view in the data warehouse (as illustrated in Table 49.5), when compared to both the tree-based strategy and the counting algorithm.

Fig. 49.5 illustrates the execution time as a function of the database size. It is evident from the graph that the proposed view maintenance technique significantly reduces the execution time, resulting in faster query execution and minimizing the overall query response time. This highlights the efficiency and effectiveness of the proposed technique in optimizing the performance of the system.

Table 49.5 Performance comparisons: Effect of database size (kb) on execution time (second)

Database Size (kb)	Tree Based Method (Second)	Counting Algorithm (Second)	Proposed View maintenance method (Second)
0.5	3.5	3.8	2.1
1	4.1	4.5	3.2
1.5	4.8	5.5	4.3
2	5.8	18.5	4.8
2.5	6.7	31.3	5.6
3	8.8	40.4	6.9

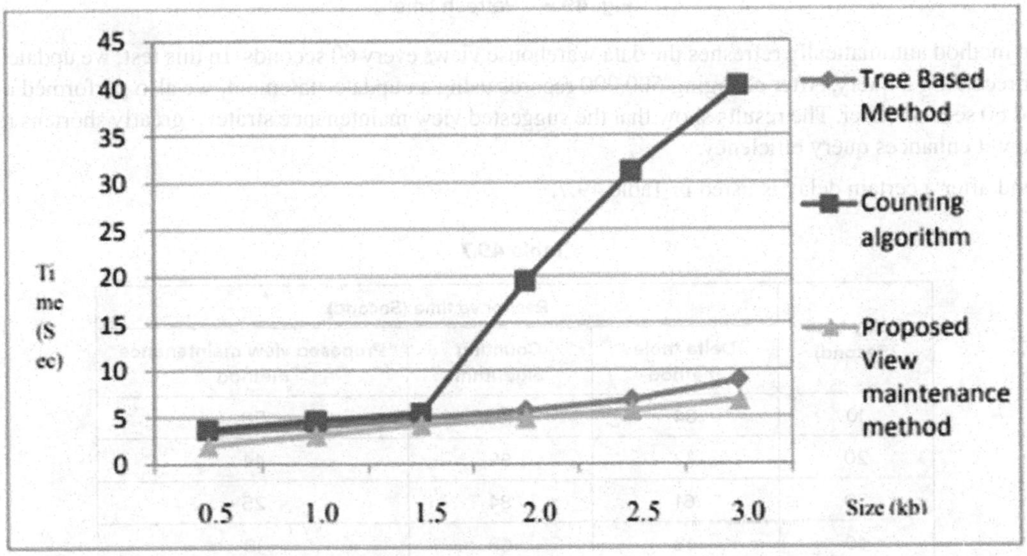

Fig. 49.5 Effect of database size (kb) on execution time (second)

Two major conclusions may be drawn from the results of the tests performed. To begin, our suggested incremental maintenance method has a reasonably constant execution time independent of the size of the database. Second, in most circumstances, incremental maintenance has a shorter execution time than re-computation does. We find that our suggested incremental view maintenance technique beats re-computation and displays efficiency in handling such circumstances, which is especially important given the normally enormous size of data warehouses.

To conduct a comprehensive evaluation of the data warehouse view maintenance framework's performance, we carried out an supplementary experiment centered around the query response time as a key parameter. We assessed the refresh time by executing queries with varying record counts: 1×10^5, 5×10^5, 10×10^5, 15×10^5, and 20×10^5, obtained from the data sources. On average, the data warehouse was refreshed with 5×10^5 records.

Table 49.6 Refresh time (Second)

No. of Records	Refresh time (Sec.)
1×10^5	30
5×10^5	67
10×10^5	94
15×10^5	127
20×10^5	162

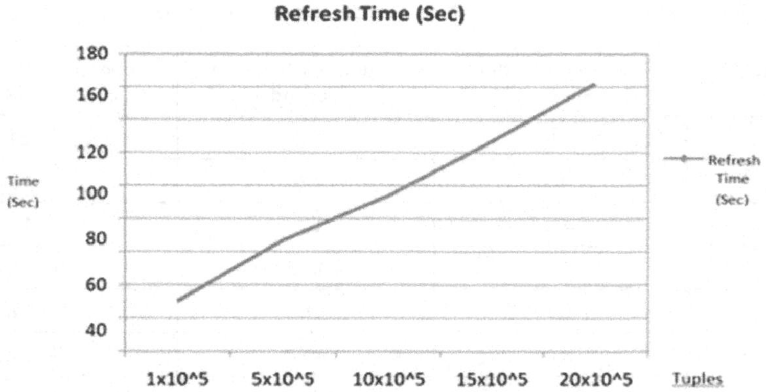

Fig. 49.6 Refresh time

The suggested method automatically refreshes the data warehouse views every 60 seconds. In this test, we updated the database instantly after receiving a query. After changing 500,000 records using an update statement, we also performed queries 10, 20, 30, 40, 50, and 60 seconds later. The results show that the suggested view maintenance strategy greatly shortens reaction times. This shows how it enhances query efficiency.

Time to respond after a certain delay is listed in Table 49.7.

Table 49.7

Delay time (Second)	Response time (Second)		
	Delta table method	Counting algorithm	Proposed view maintenance method
10	84	102	58
20	72	95	44
30	61	84	25
40	48	68	18
50	38	51	14
60	18	37	04

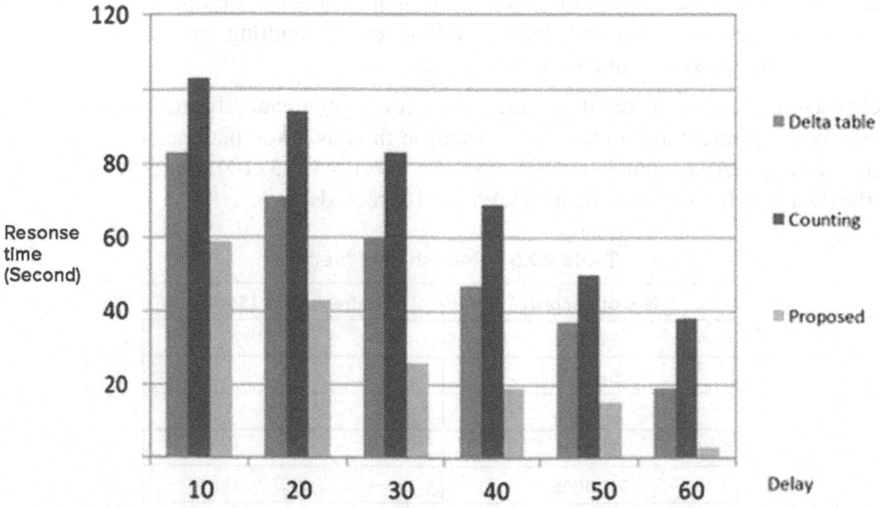

Fig. 49.7 Response time (Second) after delay

By analyzing the query response time in different scenarios (as depicted in Figure 7), we can observe that the framework requires less time to respond if a query is sent after a sufficient time interval. This is because the view is maintained before the user's query arrives. This eliminates the need for ongoing maintenance and only requires computation time for the query result. For instance, when a query is sent 10 seconds after the update, the framework requires 68 seconds with the delta table technique. It requires 83 seconds using the counting algorithm and only 59 seconds with the proposed view maintenance technique. In the case of a 60-second delay, the response time using the delta table technique was 19 seconds, which is six times higher than the alternative view maintenance strategy. Similarly, it was 12 times higher than the proposed technique. According to these results, the proposed view maintenance technique significantly improves response time compared to other methods.

If we consider a large retail store where 100,000 records are being added to the data warehouse every working day, and assume that there are already an average of 2,000,000 records in the data warehouse, this would imply that during a span of 12 hours (business hours), approximately 100,000 records would be added to the warehouse views.

(a) Proposed view maintenance method For 1 hour average number of records are = 100000/12 = 8333.33 records. For 1 minute average number of records are = 8333.33/60 = 138.89 records. For 1 second average number of records are = 138.89/60 = 2.31 records.

(b) For Delta table method For 1 hour average number of records are = 10444.16 records. For 1 minute average number of records are = 194.07records. For 1 second average number of records are = 4.51 records.

(c) For Counting algorithm For 1 hour average number of records are = 12083.32 records. For 1 minute average number of records are = 234.59 records. For 1 second average number of records are = 5.74 records

A query run before view maintenance reveals that the proposed view maintenance method outperforms the delta table method and the counting algorithm in terms of accessible data. In the proposed view maintenance method, the user would only miss out on 138.89 new records, which represents 0.0139% of the total records. It is considered a negligible percentage. With this method, the error ratio is extremely low at 0.0139%. The delta table method, however, would result in missing 194.07 new records, equivalent to 4.63% of the total records. This results in an error ratio of 3.63%. The user would also miss 234.59 new records if the counting algorithm were used, representing 4.06% of the total records. This approach has a 5.021% error ratio. It is evident from these findings that the proposed view maintenance technique performs better in terms of data accessibility. In addition, it offers greater time efficiency than the delta table method and counting algorithm. It takes 25.33% less time than the delta table method and 45% less time than the counting algorithm to perform the proposed method. Comparatively to delta table and counting algorithms, the proposed view maintenance method yields superior results regarding data accessibility and efficiency.

3. Conclusion

Information warehouses are used for data integration, decision support, and research. Organizations use it for asset planning, determining, and information management. OLAP and other fundamental leadership tools extract, coordinate, and store relevant data from a distributed information source and process in an information warehouse. In this theory, we have focused on issues that facilitate Data Warehouse materialisation. To get realised prospective benefits like professional access, reliable performance, and high information accessibility, calculation and maintenance are essential. Due to the large number of information sources, growing size of each datum source, and high information content, these two needs are problematic. We found simplified information warehousing see support methods. We observed that homogenous information sources make materialised view support easier than heterogeneous ones. Multiple situations test the number of updates in a given informative collection from various sources. We linked the suggested systems to several information sources and proved their success. A materialised view maintenance review covers most of the view support in data warehousing. Materialized view support systems are examined separately. We examined view maintenance using test system. Pre-processed information in the materialised view improves query execution. A blended information update provides materialise see assistance. During materialised sight maintenance recreation, we found that if more information is stored at the information. distribution centre reduces support cost.

REFERENCES

1. All Mabhouh, A., & Ahmad, A. (2010). Identifying Quality FActors within Data Warehouse. Second International Conference on Computer Research and Development (pp. 65–72). Kuala Lumpur: IEEE.
2. He, B., Wang, R., Chen, Y., Lelescu, A., & Rhodes, J. (2007). BlwTL: A Business Information Warehouse Toolkit and LAnguage for Warehousing Simplification and Automation. SIGMOD (pp. 1041–1052). Beijing: ACM.

3. Sorensen, J. O. (2007). Even Straight Forward Data Warehouses are Complicated. DOLAP (pp. 103–104). Portugal: ACM.
4. Sheta, O. E., & Eldeen, A. N. (2013). The Technology of using a Data Warehouse to support Decision Making in Health Care. International Journal of Datbase Management, 75–86.
5. Karayannidis, N., Tsois, A., Sellis, T., Pieringer, R., Markl, V., Ramsak, F., et al. (2012). PRocessing Star Queries on Hierarchichally-Clustered Fact Tables. 28th VLDB Conference. Hong Chong.
6. Guo, Y., Tang, S., Tong, Y., & Yang, D. (2006). Triple Driven Data Modeling Methodology in Data Warehousing: A Case Study. DOLAP. Virginia: ACM.
7. Subramanian, A., Douglas, L. S., & Nelson, A. C. (2006). Strategic Planning for Data Warehousing in the Public Sector. 29th Annual Hawaii International Conference on System Sciences, 54–61.
8. Armstrong, R., & Diego, S. (2007). Data Warehousing: Dealing with the Growing Pains. 13th International Conference on Data Engineering.
9. Saeki, S., Bhalla, S., & Hasegawa, M. (2012). Parallel Generation of Base Relation Snapshots for Materialized View Maintenanace in Data Warehouse Environment. Interantional Conference on Parallel Processing Workshops. IEEE Computer Society.
10. BAssiliades, N., Vlahavas, I., Ahmad, K., & Houstis, E. N. (IEEE Transactions on Knowledge and Data Engineering). InterBAse-KB: Integrating a Knowledge Base System with a Multi database System for Data Warehousing. IEEE.
11. Paim, F. R., & Castro, J. F. (2013). DWARF: An Approach for Requirements Definition and MAnagement for Data Warehouse Systems. IEEE International Requirements Engineering Conference. IEEE Computer Society.
12. Casati, F., Casrellanos, M., Dayal, U., & Salazar, N. (2012). A Generic Solution for warehousing Business Process Data. International Workshop on very Large Data Warehouses.
13. Armstrong, R. (2007). Data Warehosuing Dealing With the Growing Pains. 199–205.
14. Edjlali, Edjlali, R., Feinberg, D., Beyer, M. A., & Adrian, M. (2012). The State of Data Warehousing in 2012. Gartner.
15. Farooq, F., & Sarwar, S. M. (2010). Real Time Data Warehousing for Business Intelligence. FIT. Islamabad: ACM.
16. Friedrich, J. R. (2005). Meta-Data Version and Configuration Management in Multi-Vendor Environments. SIGMOD (pp. 799–804). Maryland: ACM.
17. Abdelouarit, A. E., Merouani, M. E., & Medouri, A. (2014). The bitmap index advantages on the data warehouses. American Academic and Scholarly Research Journal, 376–382.
18. Rohman, K. M., & Parent, M. (2011). Gaining Insight from Data Warehouse: The Competence Maturity Model. 34th Hawaii International Conference on System Sciences (pp. 1–10). Hawaii: IEEE.
19. Gomes, P., Farinha, J., & Trigueiros, J. M. (2006). A Data Quality Metamodel Extension to CWM. 4th Asia PAcific Conference on Conceptual Modelling (pp. 17–25). Victoria: Australian Computer Society.
20. Bernardino, J., & Madeira, H. (2011). Experimental Evaluation of a New Distributed Partitioning Technique. Interantional Database Engineering and Applications Symposium. Grenoble.

Note: All the figures and tables in this chapter were made by the authors.

Impending Inquisitions in Humanities and Sciences (ICIIHS-2022) – Dr. Mohan Varkolu et al. (eds)
© 2024 Taylor & Francis Group, London, ISBN 978-1-032-78829-6

Blockchain Consensus: Analogical Comparison and Challenges

50

Rashi Saxena[1]
Research Scholar, Department of Computer Science and Engineering,
Koneru Lakshmaiah Education Foundation, Hyderabad

P. Lalitha Surya Kumari[2]
Professor, Department of Computer Science and Engineering,
Koneru Lakshmaiah Education Foundation, Hyderabad

ABSTRACT Blockchain is a distributed ledger that has attracted widespread interest in various domains. So many sectors have started to integrate blockchain technology into their applications and services. To comprehend the significance and applicability of blockchain, it is essential to grasp its major components, functional properties, and architecture. As a distributed ledger, a blockchain network's peer nodes require a consensus mechanism to assure its correct operation. There are several consensus mechanisms in the present paper, each with their own unique performance and security properties. One consensus algorithm cannot fulfil the needs of all applications. Existing consensus algorithms must be technically analyzed to determine their benefits, drawbacks, and potential uses. We have found and investigated characteristics that are connected with the efficiency and safety of blockchain consensus. Regarding these factors, the consensus mechanisms are reviewed and contrasted. This paper focuses on a research gap in the design and evaluation of efficient consensus methods. This paper will help developers and academics analyze and construct new modified consensus algorithms.

KEYWORDS Blockchain, Consensus algorithms, Cybersecurity, Distributed systems

1. Introduction

Blockchain technology emerged to overcome the risks and inefficiencies in business transactions. It has revolutionized the structure of industries and businesses [1] [2] [3] [4], [5], [6]. When we say "blockchain," we mean a shared ledger that is spread out across the nodes of a business network. Transaction information is stored in blocks that are linked to each other to form a chain. [49]. It makes it easier for a company to keep track of its assets. An asset may be anything of value, including things that are physical and things that are not tangible [47]. The automobile, the land, the home, and the cash are all examples of physical assets. Intellectual assets, such as copyrights, properties, patents, or brands, are examples of intangible assets. Other examples include patents. Transactions are bundled together and stored in a block in a manner determined by the block size that is specified. By utilizing a timestamp, Block verifies the order in which transactions took place. It is simpler to validate newly added blocks if each block stores a hash of the prior block. Because of this, the overall blockchain's integrity is bolstered because no malicious or corrupted blocks can be inserted in between two legitimate blocks [50]. This property is known as immutability because it ensures that existing blocks cannot be changed in any way. Every node in the network is responsible for updating the ledger whenever a transaction takes place. Every complete node has a copy of the whole ledger in its storage (the blockchain). The architecture of blockchain is shown in Fig. 50.1.

[1]rashisaxena.cse@klh.edu.in, [2]vlalithanagesh@gmail.com

DOI: 10.1201/9781003489436-50

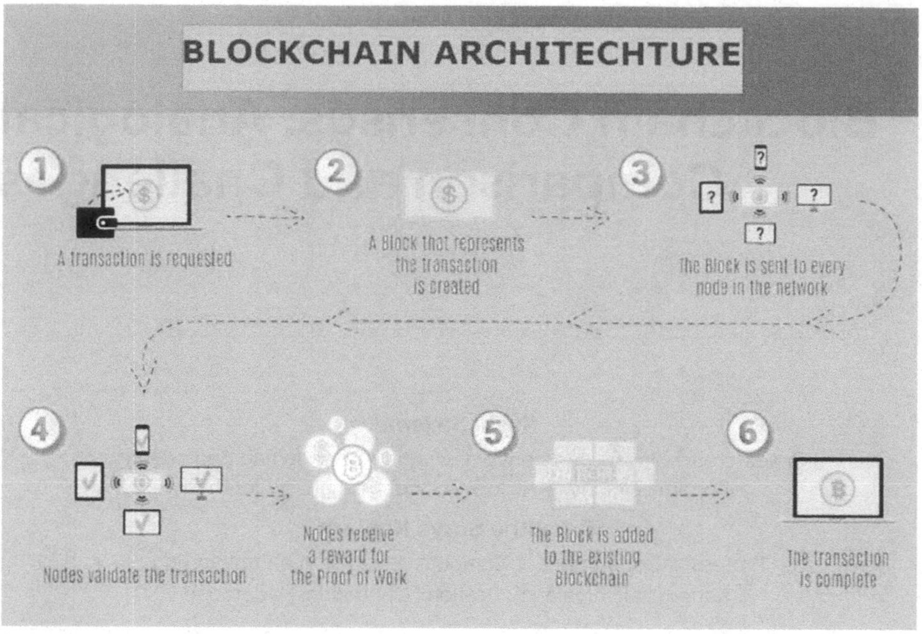

Fig. 50.1 Blockchain architecture

Source: Authors

As compared to traditional distributed databases, blockchain provides significant advantages such as reduced cost and time, no reliance on a third party, and security of assets [7] and [8]. Each computer in a distributed database system maintains its own copy of the database and uses an intermediary to process transactions. Intermediaries not only add costs but also delay the implementation of agreements.

This setup is insecure because the integrity of a single intermediary might be attacked, causing widespread damage to the firm. Blockchain transactions are safe and verifiable. These problems have been resolved in the blockchain's architecture. The network nodes reach a collective agreement on the veracity of the data, bypassing the necessity for third parties. [9] [10]. When all the nodes agree on the same version of the truth, there is no way for anybody to alter the data, making the consensus process a crucial part of the blockchain's design. In a decentralized network, all the nodes need to agree on the blockchain's current state. This makes it more difficult for an attacker to insert a compromised block into the blockchain. When developing a blockchain system, the choice of consensus algorithm is critical. And while developing a consensus algorithm, it is important to keep in mind a few key factors.

This paper discusses the consensus mechanisms in distributed systems and their significance. Various consensus algorithms designed both for distributed environments in general and specifically for blockchain are reviewed. Parameters relevant to the consensus algorithm are identified and discussed, which will aid the developers in evaluating and designing a new consensus algorithm. The paper is organized as follows: Section 2 reviews the work done related to the design and comparison of consensus algorithms in blockchain. Section 3 discusses the consensus mechanism in distributed systems. Section 4 includes a comparative analysis of some recently designed consensus algorithms. Section 5 presents the research opportunities and concludes the paper.

2. Collateral Work

2.1 Secure Sharding Algorithm for Open Blockchain [11]

For permissionless blockchains, ELASTICO is presented as a scalable agreement algorithm. The mining process is very computational; therefore, the transaction rate grows linearly with the amount of work put in. Simply put, increased computational resources allow for the processing of more transaction blocks. This method can withstand competition against opponents with only a quarter of the available computing power. The primary concept is to partition a system into manageable pieces called committees. This approach is parallelized so that different committees handle different sets of transactions. The agreement criterion is implemented in a probabilistic manner, setting this approach apart from the traditional Byzantine algorithm. Each trustworthy procedure verifies the accuracy of the agreed-upon value by comparing it to a constraint function. Only solutions that work within the constraints of the constraint function will be considered.

Byzantine consensus is used by every committee. The last committee brings everyone's input together. The strength of the defined security algorithm is demonstrated by validating it against a set of security properties. Up to 1600 nodes are used in these EC2 experiments. Bitcoin's open-source code is supplemented with new instructions for implementing this method [11].

2.2 SCP: A Computationally Scalable Byzantine Consensus Algorithm for Blockchain [12]

SCP is a new Byzantine-based blockchain consensus algorithm. The meaning of "computationally scalable" is that the algorithm can change bandwidth consumption by changing factors like PoW difficulty. Increasing computational power will boost throughput. Miners with less processing power can vote in the SCP. Transactions are uniformly distributed among data blocks, decentralizing the network. No central certificate is required, eliminating single-point failure. SCoin uses the SCP algorithm. The Merkel tree prevents double spending. Checking if a transaction has one output prevents double spending locally. SCP scales linearly with CPU capacity without quadratic bandwidth growth. Experiments are done with 80, 40, 20, and 10 cores on Amazon EC2 [12].

2.3 Leader-free Byzantine Consensus Algorithm [13]

In this paper, we examine the Byzantine consensus problem for asynchronous systems. An asynchronous system that transitions to synchronous mode is called "partial synchronous." Designing a leaderless, deterministic algorithm for a partially synchronous system is difficult. But in rounds of a leader-free algorithm, all the nodes talk to each other. To reach agreement in a partially synchronous system, this study proposes a leader-free procedure. We generalize the leaderless consensus procedure from synchronous systems to semi-synchronous ones. This work considers an interactive consistency problem-based consensus technique for synchronous systems. An element of a process is a group of values that is computed as part of that process. When done properly, both sets of values result from the processes. Two new features are added to this algorithm's repertoire. There is the option of employing a parametrized consensus process, wherein disagreements based on varying fault models are tolerated. Strong validation, another approach presented in this research [13], aims to make the consensus ideal.

2.4 On the Security and Performance of Proof-of-Work Blockchain [14]

To evaluate the security and effectiveness of blockchains powered by the PoW consensus in relation to numerous network and consensus related characteristics, a quantitative methodology is provided. Most of the blockchains that are currently in existence use PoW consensus. Regarding security evaluation, the PoW variants have not garnered much attention in the literature. There has not been much research done on how PoW blockchain performance and security relate to one another. To simulate the network and consensus mechanism of a blockchain, a simulator was used in this paper to study this relationship. The effects of changing the block size and block interval on selfish mining and double spending are investigated. It has been discovered that the blockchain's high block reward will reduce the likelihood of double spending attacks. Security is not penalized with a block size of 1 MB and a block interval period of 1 minute [14].

2.5 Blockchain Consensus: An Analysis of Proof-of-Work and its Applications [15]

An in-depth analysis of the proof of work was performed by the writers. With the changing difficulty of the problems, we examine the average mining time, the number of invalid blocks, and the length of forks. The amount of time it takes to mine additional blocks rises in tandem with the problem's complexity. The many uses of blockchain technology are also discussed. The proof-of-work (PoW) consensus technique is extensively implemented in blockchain implementations. Cryptocurrency Bitcoin was the first to do so. With this algorithm, every node casts a vote by independently computing the solution to a proof of work, and together they use their computing capacity to generate new blocks. Bitcoin uses a hash-based proof of work to determine a nonce value. The assumption of this algorithm's security is that no single node has more than 50% of the network's total processing power. If this occurs, the node that builds the longest chain gains control of the whole network. An effective voting mechanism is required to guarantee consensus among network nodes. A Nakamoto consensus selects a leader by lottery, with the node that draws the winning ticket becoming the network's new coordinator. Bitcoin functions like a lottery, where the first node to solve a riddle is awarded a prize. The block is subsequently broadcast by the leader, and the other nodes vote to implicitly accept it. Modern Byzantine fault tolerance (BFT) algorithms use explicit voting, but the tried-and-true Byzantine fault tolerance (BFT) procedures are also used to get all the nodes to agree [15].

2.6 The Blockchain Consensus Layer and BFT [16]

Our study conducted research into the applicability of Byzantine fault tolerance algorithms in addition to blockchain techniques. The concept of blockchain is broken down into its component parts and examined from a layered perspective. People have also talked about a solution that uses both the Nakamoto algorithm and the BFT [16].

2.7 Byzantine Consensus for Consortium Blockchain [17]

Here is a technique for reaching consensus that does not rely on signatures, leaders, or randomness. By using this approach, the multi valued-based Byzantine consensus problem can be simplified into a binary Byzantine consensus problem. A value is decided upon in a period that is constant when binary instances of consensus are conducted in parallel. Theoretically, the recommended design is shown to be the most effective option. [17]

2.8 Implicit Consensus: Blockchain with Unbounded Throughput [18]

Each node would operate its own blockchain under the suggested implicit consensus architecture. Unrestricted throughput is the main advantage of this setup. The BFT termination property is swapped out for a self-interested one. All transactions may not necessarily reach consensus. The message complexity of this technique is linear as a result. There is a proposal to use a different kind of block called checkpoint blocks, which do not require consensus like transaction blocks do. Theoretical examinations of performance and other factors are conducted. [18]

2.9 Design and Implementation of a Proof-of-Stake Consensus Algorithm for Blockchain [19]

This research explores PoS implementation on the blockchain. Bitcoin's PoW is a hot topic. It is used to pick a block signer based on miners' work on challenging arithmetic problems. PoS is introduced to reduce energy waste. Nodes with higher stakes might add more blocks to the network. A fresh block signer is chosen depending on a miner's stake. This reduces the power needed to mine a block compared to PoW's brute-force approach. Ouroboros and Casper are PoS consensus protocols. Ouroboros randomly chooses a stakeholder using a secure coin-flipping algorithm and time slot synchronization. Casper guarantees less about how much an enemy need to cause disruption. [19]

2.10 PPCoin: Peer-to-Peer Cryptocurrency with Proof-of-Stake [20]

This study addresses proof-of-stake methods that decrease energy usage without affecting blockchain network security. Cryptocurrency peer-to-peer design: PPcoin has no energy dependence. Coin age is considered when designing proof-of-stake algorithms. Coin age is how long a currency is retained. Unlike PoW, a block's originator is determined by its wealth (stake). Block mining or generation does not pay. It is cheaper since miners do not have to race to do the math. Node hash targets are not fixed depends on coinage. Highest coin age eaten on the blockchain wins. This technique simplifies double spending attacks by requiring only a specific amount of currency to introduce a falsified block to a blockchain. [20]

2.11 Proof-of-Trust Consensus Algorithm for Enhancing Accountability in Crowdsourcing Services [21]

Proposed crowdsourcing consensus algorithm Node trust determines transaction validators. Shamir's secret-sharing technique and RAFT leader election are employed. Traditional BFT scalability issues are resolved. Four phases comprise the consensus algorithm. Phase 1 focuses on Raft leader election in the consortium. Next, voters choose transaction validators. A collection of nodes (chosen in phase 2) validates transactions in phase 3. The last phase ties validated transactions to the blockchain. The suggested technique can manage node failures such that $x3y + 1$, where x is the number of network nodes and y is the number of byzantine nodes, The consensus process creates four nodes. Consortium leader, gateway, normal ledger management, and validator nodes the suggested algorithm's agreement, validity, performance, fairness, liveliness, and scalability are evaluated. Discussions include attack scenarios. In hybrid consortium architecture, consortium and validator group sizes remain the same despite rapid node growth. This improves BFT's scalability. One machine simulates experiments. PoT is compared to ripple, consortium, and shared consensus. Concurrent transactions are also well studied. PoT is more accurate, scalable, and performant than other consensus solutions. [21]

2.12 RMBC: Randomized Mesh Blockchain Using a DBFT Consensus Algorithm [22]

An algorithm for reaching consensus based on a diversity of opinions is developed and explained here. The primary focus of this method was to solve the two issues that plagued the PoW and BFT algorithms, respectively. There is an overhead cost associated with POW. When using BFT, the prevention of invalid transactions becomes impossible when the number of malevolent entities within a network increase by a significant amount. The procedure for reaching a consensus and agreeing on terms has two stages. The BFT algorithm is used in the first step of the process. The second phase is the process of grouping relevant departments together. After that, an individual from each department is chosen to serve as the verifier. If the first agreement phase and the second agreement phase are the same, then the transaction is confirmed. Experiments are not utilized to their full potential to validate the proposed algorithm architecture. Here are the results of a quick quantitative comparison of PoW solutions and BFT solutions.[22]

2.13 PoPF: A Consensus Algorithm for JCLedger [23]

Using a concept known as JointCloud, developers can personalize cloud-based services to their needs. JCLedger is a solution for JointCloud that is built on blockchain technology. PoPF is the name of the consensus method that has been suggested for JCLedger, and this paper details it (Proof of Participation and Fees). Because Proof of Work requires a lot of extra computational power, it is challenging to have it implemented by JCLedger. As a result, an approach for reaching consensus that uses significantly less computer power has been presented. The cost paid by the participant and the number of times the participant appeared as an accountant are taken into consideration during the selection process for the candidates for mining. The authors have performed an experimental analysis and simulation to evaluate the algorithm's performance in relation to the distribution of accountants. Experiments do not involve the accuracy and performance-related components of anything being tested. [23]

2.14 The Ripple Algorithm: A Consensus Algorithm [24]

Ripple is presented as a consensus technique that makes use of mutually trusted subnetworks across the whole network. To keep the network running smoothly, the protocol is run at regular intervals of a few seconds. When everyone agrees, the books are closed. If there is no fork, all nodes in the network should have the same copy of the most recent closed ledger. This is a circular protocol. Each node starts the consensus process by officially announcing a candidate list that includes all the legitimate transactions it has seen up until that point. Nodes then aggregate their candidate sets and vote on whether all transactions were executed correctly. There is a comparison between the total number of votes and a certain threshold, which determines whether the transaction is validated. Then, in the last round, a transaction is approved only if 80% or more of the network's nodes approve it. When all transactions that meet this condition are checked and written down, a new closed ledger is made.[24]

2.15 Proof of Vote: A High-Performance Consensus Algorithm Based on the Vote Mechanism and Consortium Blockchain [25]

The Proof of Vote (POV) consensus algorithm is proposed as a more effective alternative to the Proof of Work (POW) technique. It uses a voting system to confirm that the blocks are legitimate. The paradigm of a consortium network defines four roles. We have the commissioner position, the butler candidate position, the butler position, and the regular user position. Overall, the proposed method has demonstrated the lowest power consumption. [25]

2.16 Proof of Work [26]

Miners participating in proof-of-work are rewarded for being the first to solve a mathematical puzzle. The entire strength of the blockchain network determines the difficulty of a problem. To mine, miners need to solve a complex mathematical puzzle, which requires a lot of computing power. A major drawback of this technique is the additional processing time it requires. [26]

Qualitative Comparisons [27] compares PoS, PoW, BFT, PoET, and Federated BFT using qualitative characteristics. The study [28] compared PoX and hybrid consensus methods qualitatively. In [29], five consensus algorithms are examined for Byzantine and crash fault tolerance, verification speed, throughput, and scalability. [30] Compared qualitatively, some consensus algorithms Recent algorithms are not compared. Giang et al. [43] compared vote-based and proof-based blockchain consensus techniques. The authors briefly reviewed how these two consensuses technique's function. Agreement, node joining, number of executing nodes, decentralization, trust, node identities, and security risks are reviewed. [44] compares PoW-based solutions to BFT replication-based alternatives. Node identity management, consensus finality, scalability, performance, power consumption, attacker tolerance, network synchrony, and correctness proofs are compared at an important level. [45] reviews consensus techniques for their robustness. We discuss permissioned consensus protocols. Elli et al. [46] addressed permissioned blockchain sharding. Permitted blockchain reduces enemy power. reviewing the design's confidentiality. Sharding for permissioned asset management.

The aforementioned bodies of literature have certain gaps in their coverage. Neither the speed nor the safety of the suggested algorithms have been thoroughly evaluated. When assessing the merits of blockchain consensus solutions, performance and security are two crucial factors to consider. To evaluate the agreed-upon solutions, it is necessary to consider these facets. Studies that have analyzed blockchain consensus algorithms have often neglected to account for crucial factors relating to the protocols' performance and safety. The comparison is too superficial to be useful in determining whether a consensus solution is mature. Most analyses that compare algorithms only do so qualitatively, without any quantitative measures of performance. To add insult to injury, not all modern consensus methods are included in the evaluation. In this work, we have isolated certain crucial factors for the efficiency and safety of a consensus method. The consensus methods that have been presented lately are compared and examined considering the determined parameters. This study will aid programmers in making the best choice when it comes to the algorithm design of their blockchain application thanks to the information it provides. The lack of studies evaluating consensus methods will also be brought to light.

3. Consensus in Distributed Blockchain

The concept of consensus is fundamental to distributed systems in general and is not unique to blockchain technology. It is applicable to circumstances in which the same state of a data item must be maintained by numerous processes or nodes simultaneously. There are two primary categories of blockchains: permission-less and permissioned blockchains. Nodes in a blockchain that do not require permission to access are anonymous. A new transaction block that has been tampered with can be inserted, which can result in a fork. A fork takes place whenever a legitimate transaction fails to match up with an invalid one. The basic objective of a consensus algorithm is to bring about agreement among the network's nodes in such a way that they all concur on a single value as being the correct one. In a blockchain that requires permission to access, the nodes are not anonymous and are instead treated as known entities.

Despite this, reaching consensus remains essential to solve the problem because the nodes cannot be trusted. The architecture of blockchain is illustrated in Fig. 50.2. In situations where the nodes of a network are broken or they communicate in an unreliable manner, reaching a consensus is often something that should be considered. In distributed systems, consensus can be reached regarding the various failures that can occur. In addition, when creating a consensus system, the type of communication model that is used, such as synchronous or asynchronous, is also taken into consideration. There are distinct kinds of failures, such as crash failure, transitory failure, omission failure, security failure, software failure, byzantine failure, temporal failure, and environmental perturbations.

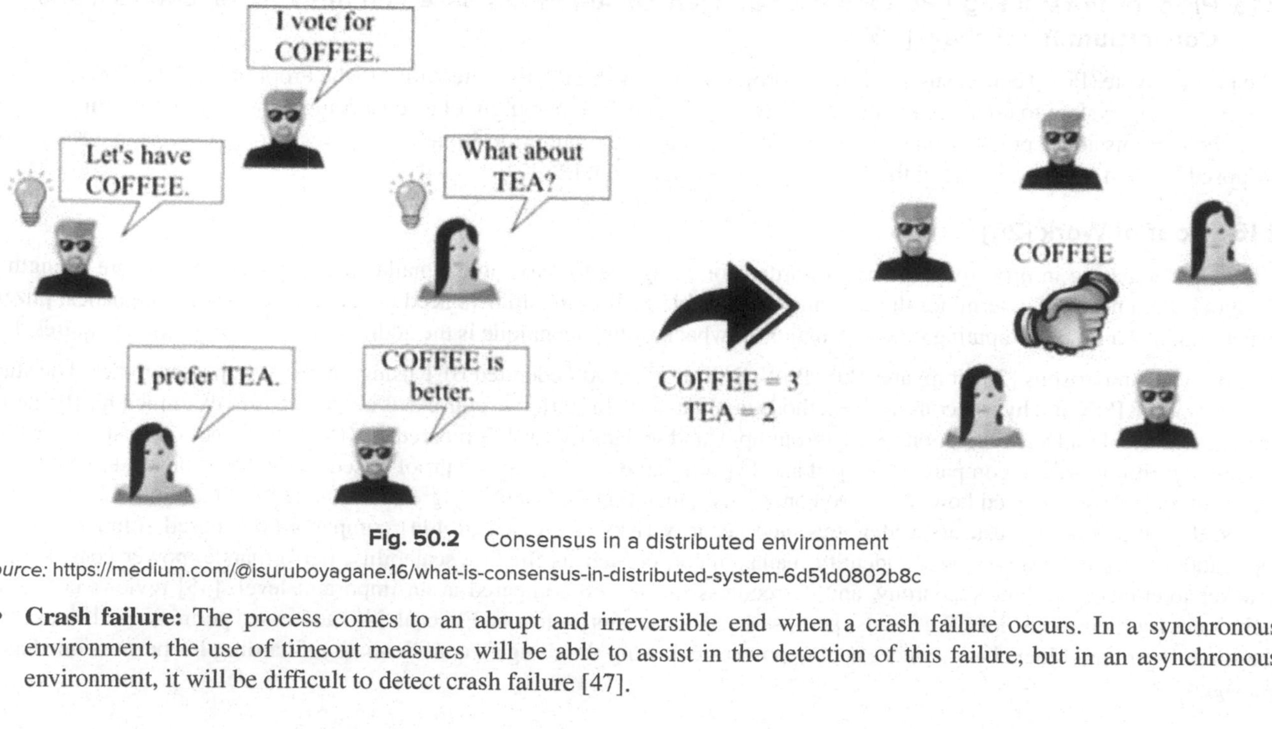

Fig. 50.2 Consensus in a distributed environment

Source: https://medium.com/@isuruboyagane.16/what-is-consensus-in-distributed-system-6d51d0802b8c

- **Crash failure:** The process comes to an abrupt and irreversible end when a crash failure occurs. In a synchronous environment, the use of timeout measures will be able to assist in the detection of this failure, but in an asynchronous environment, it will be difficult to detect crash failure [47].

- **Transient failure:** These flaws are permanent and nondeterministic. It may be a hardware issue caused by poor batteries or a power surge. When it comes to software, these mistakes could be bugs in internal routines that are not found during testing.
- **Omission failure:** These failures are caused by problems with the transmission, such as a full buffer, a collision, or a broken transmitter.
- **Security failures:** Failures in security are caused by security attacks and impersonation. This could cause data corruption.
- **Software failure:** Design and modelling errors are the causes of software failures. Other failures, including crashes or omissions, may result from this type of failure.
- **Byzantine failure:** Byzantine failure is when all system members experience different symptoms. It prevents members from agreeing. These failures confound and stress the system. One observer may see a failed server while another sees it working. The server is not failing since the observers have conflicting information [51].
- **Temporal failure:** Temporal failures happen because of missing the deadline. In other words, accurate results might be produced, but they would be too late to be of any service. In real-time systems, this category is extremely important [51].
- **Environmental perturbations:** This kind of failure takes place when the solution is not made adaptive to the shifting conditions of the environment. If the environment were to change, the outcome might no longer be accurate [31], [47].

Blockchains operate a decentralized ledger that is not overseen by any one entity. There is a chance that malicious actors will be incentivized heavily to try to introduce flaws. So, it is important for blockchain to consider the Byzantine problem and its solutions. A Byzantine fault is one in which every component in the system experiences the problem in a unique way. It makes it harder for everyone to agree on anything. Because of this, the system becomes more vulnerable to failures and less able to recover from them. A server, for instance, may appear broken to one person but fully operational to another. Due to the lack of unanimity amongst the observers, the server cannot be marked down. The term "Byzantine" originates from the "general's difficulty" of the Byzantine Empire. It is a negotiation conundrum involving a group of generals who are besieging a city. The generals need to settle on an assault strategy for the city. Each commander will have his or her own preference for whether to retreat or advance. It is imperative that all generals come to the same conclusion. When one of a group of generals betrays the others, complications arise. The betrayal involves selective voting for a less-than-optimal plan to sow discord. Say, for instance, that there are eleven generals, one of whom is a traitor while the other ten are completely trustworthy. The traitor will communicate with both groups, sending a vote of retreat to those who favor it and an attack vote to those who favor it if there is a tie. In such a situation, half of the army will retire while the other half advances. A consensus problem is especially challenging to overcome in asynchronous networks. To address this, several randomized consensus solutions have been offered in the literature. These methods modify the three required properties of Byzantine consensus—agreement, termination, and validity—to make them optional. Both the Ben-Or consensus algorithm and the Rabin consensus algorithm are well-known examples of randomized approaches. Consensus algorithms based on randomness come in a variety of forms, including Monte Carlo and Las Vegas. Another type of consensus algorithm deals with an elected leader deciding on a probability for itself. The loosened validity characteristic allows for additional categorization of the Monte Carlo consensus. One consequence of using a randomized approach to reach consensus is that it compromises the distributed system's security. Randomized [32, 33, 34], deterministic [35, 36], Monte Carlo [37, 38], Las Vegas [39], leader-based [40], and leader-free [13] algorithms are the most common types of consensus algorithms. Additionally, synchronous, partial synchronous, and asynchronous models of communication are used to classify each method. Guarantees for agreement, validity, and termination can be obtained using randomized consensus, but only up to a certain probability. Deterministic solutions are those for which no probabilities are involved. To achieve consensus, the Monte Carlo technique is executed independently on each process or node using a set of ranging data. To arrive at a global consensus value, values from multiple processes are averaged together. Each round of consensus in Las Vegas also has a likelihood. The clocks in Monte Carlo are different from those in Las Vegas. The clock in Monte Carlo can be accurately predicted, unlike in Las Vegas, where it is more of a crapshoot. Termination of consensus is also possible with leader-based consensus techniques. In addition, we present a leaderless consensus algorithm. Fig. 50.3 gives a detailed look at consensus techniques in distributed systems, and Fig. 50.2 shows how their generic architecture works.

In addition to the classification of consensus algorithms in distributed environments, the classification may also be done in relation to blockchain. Consensus algorithms, which are unique to blockchain technology, can be divided into two distinct groups: proof-based consensus and vote-based consensus. Proof-based consensus requires the node that wants to join the network to demonstrate that it is better qualified than the other nodes to conduct the appending task.

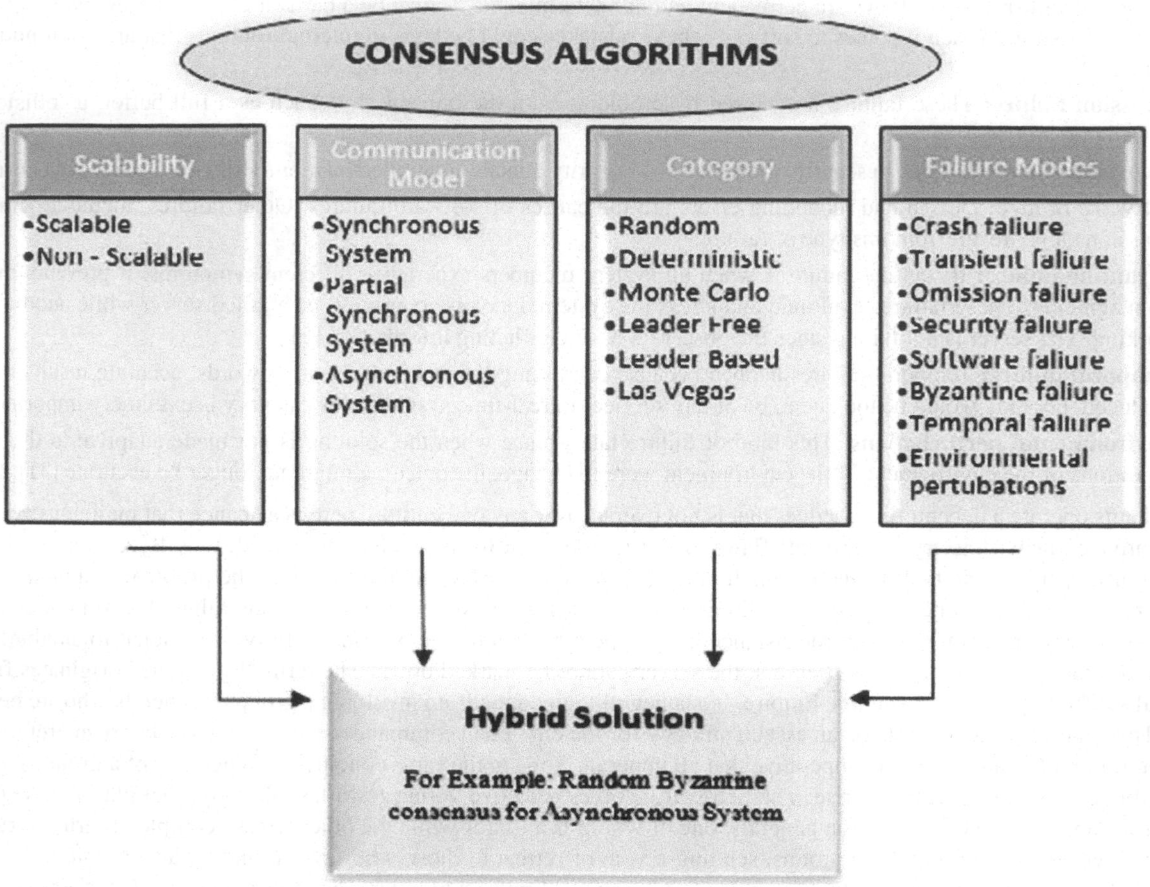

Fig. 50.3 Detailing of consensus algorithms

Source: Authors

When a vote is used to reach consensus, it is necessary for each node in the network to communicate and share the outcome of each new transaction block that it verifies with the other nodes in the network. This is what is known as "vote-based consensus. After considering the results of the votes cast by many people, the ultimate choice is made.

4. Analogical Comparison of Consensus Algorithms

This section examines the parameters of the blockchain consensus algorithm. Comparing blockchain consensus algorithms requires comparing blockchain type, transaction rate, scalability, adversary tolerance model, experimental setup, latency, throughput, bandwidth, communication model, communication complexity, attacks, energy consumption, mining, consensus category, and consensus finality. ELASTICO [11], leader-free Byzantine consensus [17], implicit consensus [18], blockchain with unbounded throughput [18], proof of trust [21], DBFT consensus [22], PoPF [23], Ripple [24], proof of vote [25], and proof of work [26] are compared. Table I–III compares consensus algorithms. Parameters and consensus algorithm comparisons are shown below.

- **Blockchain Type:** Public, private, and consortium blockchains exist [41]. Blockchain type determines consensus algorithm membership control. When analyzing consensus methods, check what kind of membership is assumed. The business application determines the blockchain type.
- **Scalability:** Scalability is necessary in today's big data environment. Scalability is gained when more nodes execute transaction blocks. Provenance and ELASTICO are scalable. PoW and implicit consensus are not scalable. Other compared methods are not scalable yet.

- **Adversary Tolerance Model:** The adversary tolerance model determines the fraction of the blockchain network that can sustain failure or attack without influencing consensus. The blockchain consensus algorithm has an adversary model threshold. better enemy threshold value. ELASTICO has superior enemy control.
- **Performance related parameters:** Some existing consensus methods have not been evaluated for performance. Theoretically, utilizing correctness proofs Along with this, a quantitative examination of various consensus algorithms' performance and security is needed. Each consensus algorithm must focus on latency, throughput, and bandwidth. Other algorithms are not evaluated for these performance features except for ELASTICO.
- **Complexity and communication model:** The sender waits for a response in synchronous communication. In asynchronous communication, the sender does not wait for the recipient's answer. PoW, PoT, ripple, and implicit consensus are options for real-time applications. If an application expects more read operations, choose the synchronous model for immediate response. ELASTICO and leader-free consensus [17] assume synchronous communication. Leader free consensus [17] has lower communication costs than ELASTICO and PoT. The rest of algorithms' communication costs are not analyzed.
- **Attacks:** Ripple is vulnerable to Sybil attacks, in which a single attacker controls several network nodes by generating different IP addresses, virtual machines, and user identities. Leader-free consensus protects PoT from attack. The algorithms are not assessed for security attacks in a blockchain network. Review algorithmic security attacks.
- **Energy consumption:** how much energy the blockchain network's hardware infrastructure uses. All consensus algorithms' energy consumption is not tested.
- **Mining and consensus:** This category define how blockchain mining works. It is tied to verification. Proof-based consensus works well in large networks. Vote-based consensus requires fewer nodes. ELASTICO, PoW, PoPF, and implicit consensus are best for large networks. Other consensus procedures (used in comparison) would fit better.
- **Consensus finality:** The transaction is concluded and cannot be reversed. It is crucial when creating a blockchain consensus protocol. Consensus finality is probabilistic or absolute. As a brick moves further down the chain, its risk of reversing decreases. A transaction is finalized instantly after being added to the blockchain. ELASTICO, PoPF, and implicit consensus can attain absolute consensus finality.

In summary, several features exist to evaluate a consensus algorithm with respect to its working, performance, and security related aspects.

5. Conclusion & Future Work

Many applications and companies are transitioning toward solutions that are based on blockchains because of the current trend toward blockchain technology. As a result, it became essential to conduct a thorough investigation of blockchain technology to determine its capabilities and level of safety. Recently, there has been some study done towards analyzing the many consensus algorithms that are already used in blockchain and suggesting a new one. This work has just begun. In this article, we have gone to great lengths to discuss the consensus processes, including their categories and the significance they have in a distributed setting. Both in a broad sense for a distributed system and more particularly for the blockchain, consensus procedures are being considered. We looked at several newly suggested consensus algorithms and compared them regarding the number of factors that may have a substantial effect on the algorithm. The issues of performance and safety are covered by the metrics that have been established for comparison. Following that, a discussion is had on the algorithms, along with each of the detected parameters. In addition to this, there are several additional factors that should also be given careful consideration. Network architecture (such as a completely linked graph), transaction rate, consistency reached by the consensus solution, concurrency check, transaction verification time, and complexity of rounds are all factors that are taken into consideration (if the consensus algorithm involved multiple rounds or phases). These factors may be combined for a deeper comparison in future research. The comparative perspective that has been offered in this research has brought to light the factors that, regarding certain new algorithms, need more study and investigation. It is possible to conduct a comprehensive comparison, both qualitatively and quantitatively, that would fill in the gaps in the comparison that has been described in this study. Experiments need to be done in a cluster to get a good idea of the pros and cons of the different consensus algorithms in terms of what big data needs.

REFERENCES

1. Christidis, K., & Devetsikiotis, M. (2016). Blockchains and Smart Contracts for the Internet of Things. IEEE Access, 4, 2292–2303. https://doi.org/10.1109/access.2016.2566339
2. Zhao, J. L., Fan, S., & Yan, J. (2016, December). Overview of business innovations and research opportunities in blockchain and introduction to the special issue. Financial Innovation, 2(1). https://doi.org/10.1186/s40854-016-0049-2
3. M. Crosby, P. Pattanayak, S. Verma, and V. Kalyanaraman, "Blockchain technology: Beyond bitcoin," Applied Innovation, vol. 2, pp. 6–10, 2016.
4. Yermack, D. (2017, January 10). Corporate Governance and Blockchains. Review of Finance, rfw074. https://doi.org/10.1093/rof/rfw074
5. MacDonald, T. J., Allen, D. W. E., & Potts, J. (2016). Blockchains and the Boundaries of Self-Organized Economies: Predictions for the Future of Banking. SSRN Electronic Journal. https://doi.org/10.2139/ssrn.2749514
6. Pilkington, M. (n.d.). Blockchain technology: principles and applications. Research Handbook on Digital Transformations, 225–253. https://doi.org/10.4337/9781784717766.00019
7. T. McConaghy, "Blockchain, throughput, and big data," Bitcoin Startups Berlin, October, vol. 28, 2014.
8. Dan Brandon, "The Blockchain: The Future of Business Information Systems," International Journal of the Academic Business World, vol. 10, no. 2, pp. 33–40, 2016.
9. Mattila et al., "The blockchain phenomenon: The disruptive potential of distributed consensus architectures," The Research Institute of the Finnish Economy, Tech. Rep., 2016.
10. Watanabe, H., Fujimura, S., Nakadaira, A., Miyazaki, Y., Akutsu, A. and Kishigami, J. J. (2015). Blockchain contract: A complete consensus using blockchain. 2015 IEEE 4th Global Conference on Consumer Electronics (GCCE). https://doi.org/10.1109/gcce.2015.7398721
11. Luu, L., Narayanan, V., Zheng, C., Baweja, K., Gilbert, S. and Saxena, P. (2016). A secure sharding protocol for open blockchains. Proceedings of the 2016 ACM SIGSAC Conference on Computer and Communications Security. https://doi.org/10.1145/2976749.2978389
12. Luu, L., Narayanan, V., Zheng, C., Baweja, K., Gilbert, S. and Saxena, P., "SCP: A computationally-scalable byzantine consensus protocol for blockchains," IACR Cryptology ePrint Archive, vol. 2015, p. 1168, 2015.
13. Borran, F. and Schiper, A. (2010). A leader-free Byzantine consensus algorithm. Distributed Computing and Networking, 67–78. https://doi.org/10.1007/978-3-642-11322-2_11
14. Gervais, A., Karame, G. O., Wüst, K., Glykantzis, V., Ritzdorf, H. and Capkun, S. (2016). On the security and performance of proof of work blockchains. Proceedings of the 2016 ACM SIGSAC Conference on Computer and Communications Security. https://doi.org/10.1145/2976749.2978341
15. Porat, A. Pratap, P. Shah, and V. Adkar, "Blockchain Consensus: An Analysis of Proof-of-Work and its Applications"
16. Abraham, D. Malkhi et al., "The blockchain consensus layer and BFT," Bulletin of EATCS, 3, no. 123, 2017.
17. Crain, V. Gramoli, M. Larrea, and M. Raynal, "(leader/randomization/signature)-free byzantine consensus for consortium blockchains," arXiv preprint arXiv:1702.03068, 2017.
18. Ren, K. Cong, J. Pouwelse, and Z. Erkin, "Implicit Consensus: Blockchain with Unbounded Throughput," arXiv preprint arXiv:1705.11046, 2017.
19. Garcia Ribera, "Design and implementation of a proof-of-stake consensus algorithm for blockchain," B.S. thesis, Universitat Politècnica de Catalunya, 2018.
20. King and S. Nadal, "Ppcoin: Peer-to-peer Cryptocurrency with Proof-of-Stake," self-published paper, August, vol. 19, 2012.
21. Zou, J., Ye, B., Qu, L., Wang, Y., Orgun, M. A. and Li, L. (2019). A proof-of-trust consensus protocol for enhancing accountability in crowdsourcing services. IEEE Transactions on Services Computing, 12(3), 429–445. https://doi.org/10.1109/tsc.2018.2823705
22. Jeon, S., Doh, I. and Chae, K. (2018). RMBC: Randomized mesh blockchain using DBFT consensus algorithm. 2018 International Conference on Information Networking (ICOIN). https://doi.org/10.1109/icoin.2018.8343211
23. Fu, X., Wang, H., Shi, P. and Mi, H. (2018). PoPF: A consensus algorithm for JCLedger. 2018 IEEE Symposium on Service-Oriented System Engineering (SOSE). https://doi.org/10.1109/sose.2018.00034
24. Schwartz, N. Youngs, A. Britto et al., "The ripple protocol consensus algorithm," Ripple Labs Inc White Paper, vol. 5, 2014.
25. Li, K., Li, H., Hou, H., Li, K., & Chen, Y. (2017). Proof of vote: A high-performance consensus protocol based on Vote Mechanism and Consortium blockchain. 2017 IEEE 19th International Conference on High Performance Computing and Communications; IEEE 15th International Conference on Smart City; IEEE 3rd International Conference on Data Science and Systems (HPCC/SmartCity/DSS). https://doi.org/10.1109/hpcc-smartcity-dss.2017.61
26. Nakamoto, "Bitcoin: A Peer-to-Peer Electronic Cash System," 2008.
27. Baliga, A. (2017). Understanding blockchain consensus models. Persistent, 4(1), 14.
28. Wang, W., Hoang, D. T., Hu, P., Xiong, Z., Niyato, D., Wang, P., Wen, Y., & Kim, D. I. (2019). A Survey on Consensus Mechanisms and Mining Strategy Management in Blockchain Networks. IEEE Access, 7, 22328–22370. https://doi.org/10.1109/access.2019.2896108

29. Mingxiao, D., Xiaofeng, M., Zhe, Z., Xiangwei, W., & Qijun, C. (2017, October). A review on consensus algorithm of blockchain. In 2017 IEEE international conference on systems, man, and cybernetics (SMC) (pp. 2567–2572). IEEE.
30. Chalaemwongwan, N., & Kurutach, W. (2018, January). Notice of Violation of IEEE Publication Principles: State of the art and challenges facing consensus protocols on blockchain. In 2018 International Conference on Information Networking (ICOIN) (pp. 957–962). IEEE.
31. "Homepage.divms.uiowa.edu," [online] http://homepage.divms.uiowa.edu/ghosh/16612.week10.pdf
32. Toueg, S. (1984). Randomized asynchronous Byzantine agreements. Cornell Univ., Dep. of Computer Science.
33. Bracha, G., & Rachman, O. (n.d.). Randomized consensus in expected o (n2log N) operations. Distributed Algorithms, 143–150. https://doi.org/10.1007/bfb0022443
34. Aspnes, J. (2003). Randomized protocols for asynchronous consensus. Distributed Computing, 16(2-3), 165–175. https://doi.org/10.1007/s00446-002-0081-5
35. Fischer, M. J., Lynch, N. A. and Paterson, M. S. (1985). Impossibility of distributed consensus with one faulty process. Journal of the ACM, 32(2), 374–382. https://doi.org/10.1145/3149.214121
36. Dolev, D., Dwork, C. and Stockmeyer, L. (1987). On the minimal synchronism needed for distributed consensus. Journal of the ACM, 34(1), 77–97. https://doi.org/10.1145/7531.7533
37. L. Scott et al., "Comparing consensus Monte Carlo strategies for distributed Bayesian computation," Brazilian Journal of Probability and Statistics, vol. 31, no. 4, pp. 668–685.
38. Scott, S. L., Blocker, A. W., Bonassi, F. V., Chipman, H. A., George, E. I., & McCulloch, R. E. (2016). Bayes and big data: The consensus Monte Carlo algorithm. International Journal of Management Science and Engineering Management, 11(2), 78–88.
39. Hideaki Ishii, & Tempo, R. (2008). Las Vegas randomized algorithms in Distributed Consensus problems. 2008 American Control Conference. https://doi.org/10.1109/acc.2008.4586880
40. Mostefaoui, A., & Raynal, M. (2001). Leader-based consensus. Parallel Processing Letters, 11(01), 95–107.
41. "Different types of blockchains in the market and why we need them," [online] https://coinsutra.com/different-types-blockchains/.
42. "Digiconomist: Cryptocurrency Fraud and Risk Mitigation" [online] https://digiconomist.net/
43. Nguyen, G. T., & Kim, K. (2018). A survey about consensus algorithms used in blockchain. Journal of Information processing systems, 14(1), 101–128.
44. Vukolić, M. (2016). The quest for scalable blockchain fabric: Proof-of-work vs. BFT Replication. Open Problems in Network Security, 112–125. https://doi.org/10.1007/978-3-319-39028-4_9
45. Cachin, C. (2017). Blockchains and consensus protocols: Snake oil warning. 2017 13th European Dependable Computing Conference (EDCC). https://doi.org/10.1109/edcc.2017.36
46. Androulaki, E., Cachin, C., De Caro, A., & Kokoris-Kogias, E. (2018). Channels: Horizontal scaling and confidentiality on permissioned blockchains. Computer Security, 111–131. https://doi.org/10.1007/978-3-319-99073-6_6
47. Chaudhry, N., & Yousaf, M. M. (2018). Consensus algorithms in Blockchain: Comparative Analysis, challenges, and opportunities. 2018 12th International Conference on Open Source Systems and Technologies (ICOSST). https://doi.org/10.1109/icosst.2018.8632190
48. https://medium.com/@isuruboyagane.16/what-is-consensus-in-distributed-system-6d51d0802b8c)
49. Sharma and D. Jain, "Consensus Algorithms in Blockchain Technology: A Survey," 2019 10th International Conference on Computing, Communication, and Networking Technologies (ICCCNT), 2019, pp. 1–7, doi: 10.1109/ICCCNT45670.2019.8944509.
50. Naz, S., & Lee, S. U.-J. (2020). Why the new consensus mechanism is needed in blockchain technology? 2020 Second International Conference on Blockchain Computing and Applications (BCCA). https://doi.org/10.1109/bcca50787.2020.9274461
51. George, J. T. (2022). Introducing blockchain applications: understand and develop blockchain applications through distributed systems. Apress.

Impending Inquisitions in Humanities and Sciences (ICIIHS-2022) – Dr. Mohan Varkolu et al. (eds)

Mining of Cascading Spatio-Temporal Frequent Patterns from Massive Data Sets

51

M. Vasavi*
Research Scholar, SRM Institute of Science and Technology, Kattankulanthur, India

A. Murugan
Associate Professor, SRM Institute of Science and Technology, Kattankulanthur, India

K. Venkatesh Sharma
Professor, CVR College of Engineering, Hyderabad, India

ABSTRACT In this research, we aim to propose a novel algorithm for frequent pattern mining in spatiotemporal datasets. Our approach will incorporate filtering techniques, known as CSTPM (Cascading Spatiotemporal Pattern Mining), as well as methods without filtering such as K-Means, Affinity Propagation-Graph based algorithm, Univariate Autoregressive Model, Vector Auto Regression Models, and Moving Average Model for Time Series Forecasting. The primary objective is to effectively reduce the number of non-prevalent patterns and optimize the time required to complete the mining process. These suggestions are based on the findings derived from various research endeavors conducted in the field. By incorporating filters and advanced algorithms, we anticipate that our proposed approach will enhance the efficiency and effectiveness of frequent pattern mining in spatiotemporal datasets. The goal is to uncover patterns that occur frequently and have significant relevance in the given context, while minimizing the computational overhead and improving the overall mining process.

KEYWORDS K-Means, Spatiotemporal, Time series, Data set

1. Introduction

Spatio-temporal data is often used to describe events or locations over time. It encompasses data that involves changing architectures over time and/or events moving across fixed geometries. Spatio-temporal data finds applications in various domains such as road traffic, environmental monitoring, weather prediction, and more. There are several pattern mining techniques specifically designed for spatio-temporal data:

1. Sequential Pattern Mining: This technique considers both space and time dimensions, but only takes into account time in a forward direction. It focuses on finding patterns that occur in a sequential order.
2. Co-occurrence Pattern Mining: Co-occurrence pattern mining takes into account both space and time in all directions. It aims to discover patterns where events or locations co-occur in both space and time.
3. Cascaded Pattern Mining: Cascaded spatiotemporal patterns (CSTP) are event categories that are partially ordered, with occurrences happening at the same or different times. This technique aims to identify patterns that exhibit a cascading effect.

*vm4200@srmist.edu.in

DOI: 10.1201/9781003489436-51

The research motivation behind studying spatio-temporal patterns lies in understanding how event types relate to the cardinality of exponential candidate patterns. It involves addressing the computational complexity of ST neighborhood enumeration to evaluate the interest measure, while also considering the challenges associated with statistical interpretation.

2. Literature Survey

2.1 A Related Study on STDM

Other survey articles have assessed related content from various disciplines because STDM has been demonstrated for years. Computing principles are used in both climate science and remote information [1] and [2]. The work of ST data prediction data mining is highlighted in [3], which also presents STDM statistics from a computational perspective. [5, 6] provides trajectory data. Introduces ST data clustering. The essay details STDM issues and methods [7].

This article summarizes and applies characteristics.

2.2 Characteristics of Spatio-temporal Data

Applications include ST correlation, partial ordering, ST clustering, spatial aggregation, ST association, ST resolution, and ST auto regression.

3. Research Objectives

The objective of this research is to enhance the following areas:

- Developing new interest measures for cascading spatio-temporal patterns that are statistically interpretable and computationally efficient.
- Proposing innovative pattern mining algorithms to capture both trivial and complete patterns.
- Conducting performance evaluations of the pattern mining algorithms using real and synthetic datasets.
- Performing performance analysis using algebraic cost models.
- Validating the applicability of the proposed model by considering various pattern families.

The research will be conducted in several phases:

Phase 1 (CSA): In this phase, a comprehensive current state-of-the-art or survey paper will be conducted, with the following outcomes:

- Reviewing the current state of the art in frequent pattern mining for spatio-temporal patterns.
- Examining recent advancements in research on interest measures and filtering techniques.

Phase 2: This phase aims to demonstrate the effectiveness of various methods, resulting in the following outcomes:

- Presenting a novel pre-processing technique for normalizing spatio-temporal datasets using the proposed filtering technique to minimize noise in prevalent patterns.
- Investigating significant types of pattern mining methods.
- Assessing the applicability of interest measures on prevalent patterns.
- Demonstrating an improved pattern mining method that yields higher prevalent patterns.
- Analyzing the time complexity and accuracy of the methods.

Phase 3: This phase focuses on exploring the applicability of the CSTP Miner or K-Mean method, leading to the following outcomes:

- Evaluating the effectiveness of the CSTP Miner on prevalent patterns using the defined interest measures.
- Analyzing the time complexity and accuracy of the method.

Phase 4: This phase aims to showcase the application of neighborhood analysis, resulting in the following outcomes:

- Presenting a novel proposed algorithm named Affinity Propagation-Graph based algorithm, along with the Univariate Auto-Regressive model, Vector Auto-Regression models, and Moving Average Model for Time Series Forecasting framework for prevalent patterns.

- Demonstrating improvements in accuracy achieved through the proposed techniques.
- Analyzing the overall time complexity of the framework.

4. Proposed Methodology

Researchers seek frequent pattern mining strategies to improve the spatio temporal domain. For spatial pattern prediction and computational scalability, spatio temporal research approaches oppose prevailing patterns. This section covers frequent pattern mining and interest measure and computational term frameworks. Autoregressive models, frequent pattern recognition with interest measures, and graph-based methods are used in this research review. Thus, to attain current patterns and cost reduction, employ other ways. This study also introduces a modified pattern recognition algorithm using interest measure for spatial temporal patterns with lower time complexity.

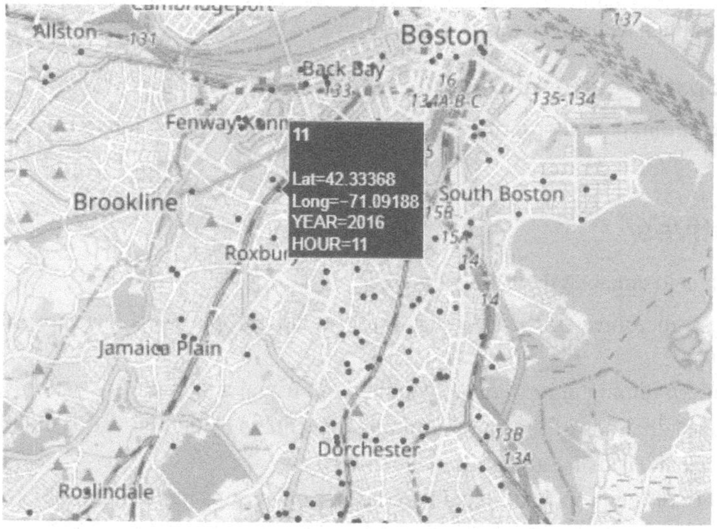

Fig. 51.1 Testing on Spatio data set

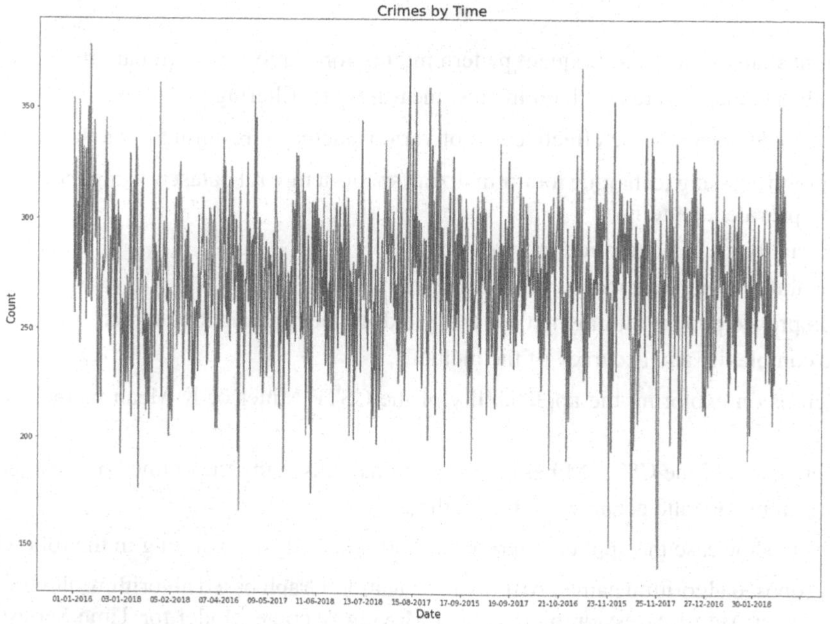

Fig. 51.2 Testing on temporal data set

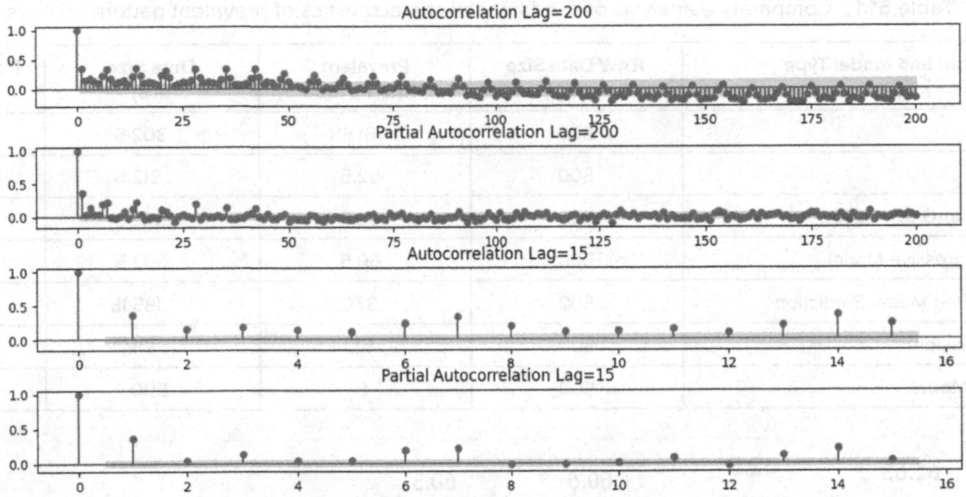

Fig. 51.3 Testing on temporal data set with autocorrelation, partial autocorrelation

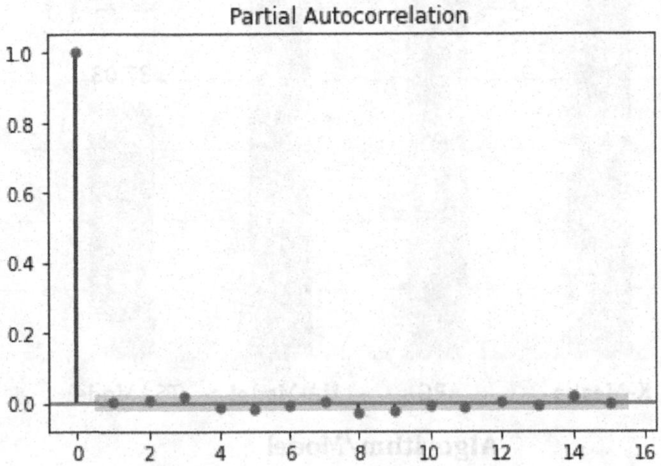

Fig. 51.4 P-value for autocorrelation

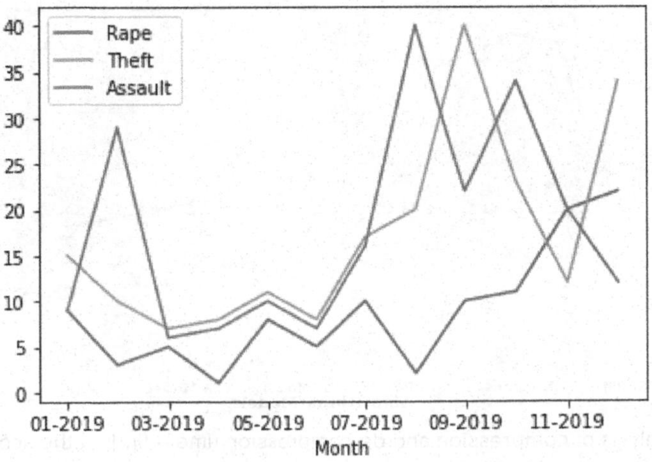

Fig. 51.5 Testing on temporal data set (events and months)

Table 51.1 Comparative analysis on fundamental characteristics of prevalent pattern analysis

Algorithm and Model Type	RAW Data Size (KB)	Prevalent Patterns ratio	Data Size (KB)	Patterns (CSTPM) Time (Sec)
CSTPM	500	61.65	302.5	6.12
K-Means	500	62.5	312.5	6.5
Affinity Propagation-Graph based	500	60.6	303	5.6
Univariate autoregressive Model	500	60.5	302.5	5.3
Time-series AutoReg Model Prediction	500	37.03	185.15	3.2
Vector Auto regression	500	19.74	98.7	2.8
Moving Average Model	500	0	500	0

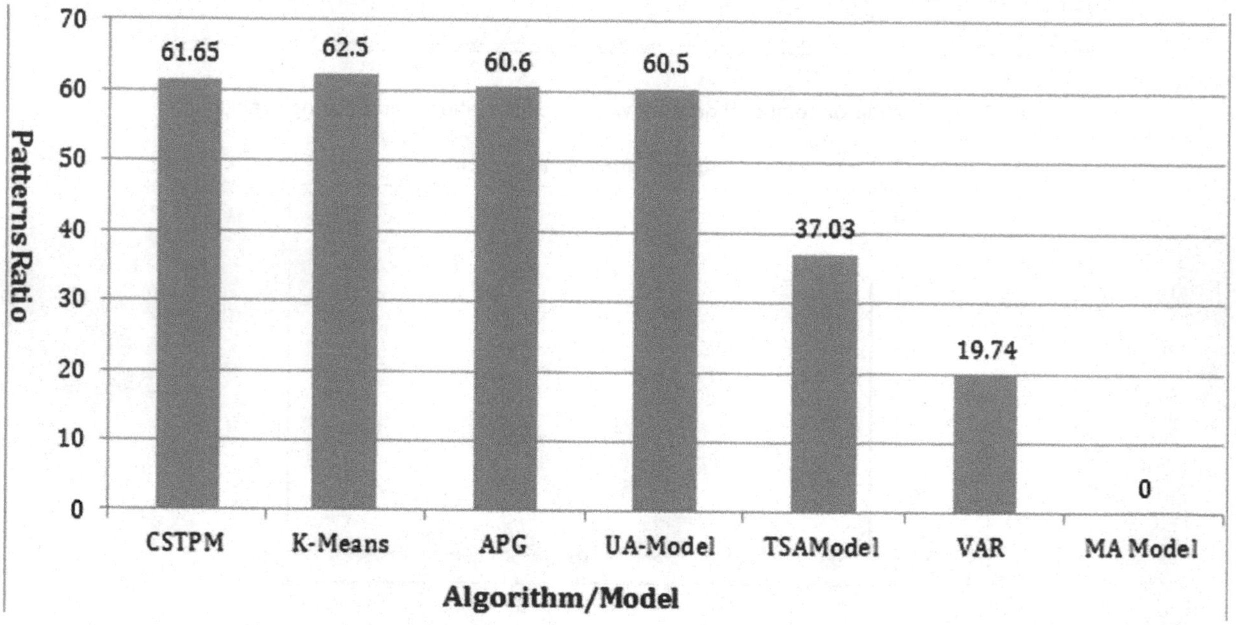

Fig. 51.6 Visual analysis of patterns ratio – finding the scope for improvements

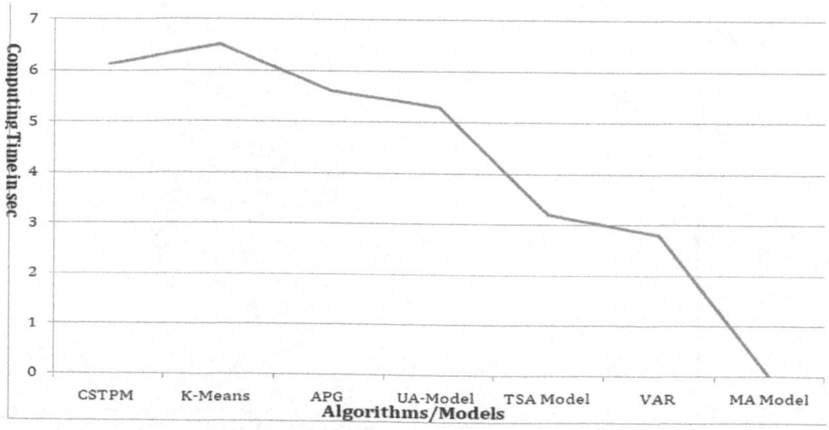

Fig. 51.7 Visual analysis of compression and de-compression time – finding the scope for improvements

Table 51.2 Analysis of patterns with filters in CPI, CPR values

Pattern From dataset	Prevalent pattern	CPI, CPR values	Non prevalent pattern	CPI,CPR values	Computed for cpi (size=1)	CPI,CPR values
Location (neighborhood analysis)	500	2.09,1.92	300	1.02,2.45	CPI(1)..CPI(7)	1.02,1.92
Location (without neighborhood analysis)	500	1.03,1.01	210	2.3,3.46	CPI(1)..CPI(7)	1.03,1.01
Time (neighborhood analysis)	500	3.01,2.35	320	1.45,2.86	CPI(1)...CPI(7)	1.45,2.35
Time (with neighborhood analysis)	500	1.21,2.25	145	1.5,2.34	CPI(1)...CPI(7)	1.21,2.25
Moving average (threshold value)	500	1.35,2.0	126	1.29,2.67	CPI(1)...CPI(7)	1.29,2.0

Calculating CPI:

Step1: Directed_neighbor_relation_R= 7 Threshold_CPI= 0.5

Step2: CPR=Total_instances_Participating_event/Total_instances_dataset

Step3: CPI=(df2.index+[(math.floor(CPR))])/Total_instances_dataset CPI=min(CPI)

Step4: Applying CPI in CSPTM

Step5: Patterns with given threshold value by CPI is compared for each iteration

CSTPM:

Step1: flag=1,size=1 Input:-Directed_neighbor_relation_R= 7,Threshold_CPI= 0.5

Step2: for size in range(1,Directed_neighbor_relation_R):

```
        flag=1
                CPI_instances_applied = [Threshold_CPI/size]
                CPI_value = filter(CSTPM, CPI_instances_applied)
        flag=0
                considering the CPI_value for size draw the graph
```

Step3: Interest measure CPI is compared for each iteration and patterns of similar CPI value is listed

Step4: Change the Directed_neighbor_relation_R value to get next location events

Table 51.3 Comparison of algorithms

TKDE STP-CSTPM Transaction of Knowledge and Data Engineering ST partitioning-based CSTPM	SIAM NL-CSTPM A simple nested loop-based CSTPM
Spatial partitioning based on join algorithm	Time ordering to limit join candidates and performs a nested loop determine spatial neighbours
Filters are used to get $O(n^2)$ computation for CPI evaluation for each pattern decreases computational cost	Filters are not used for CPI evaluation for each pattern increases computational cost
CPI evaluation, cost model has asymptotically best performance	CPI evaluation, cost model has asymptotically less performance
CPI threshold 0.5 for synthetic data sets and 0.15 for real data sets. Temporal neighbourhood sizes set at 15.53 miles and 1.04 days respectively for synthetic datasets, 0.31 miles and 10 days for the real data set.	CPI threshold 0.5 for synthetic data sets and 0.15 for real data sets. Temporal neighbourhood sizes set at 15.53 miles and 1.04 days respectively for synthetic datasets, 0.31 miles and 10 days for the real data set.
Synthetic data sets with sizes ranging from 5000 to 26000 instances . Real data set from 5000 to 33000 instances.	Synthetic data sets with sizes ranging from 5000 to 26000 instances . Real data set from 5000 to 33000 instances.
Performance based on clumpiness 4 is factor 2 (instances)	Performance based on clumpiness 25 is factor 10 (instances)

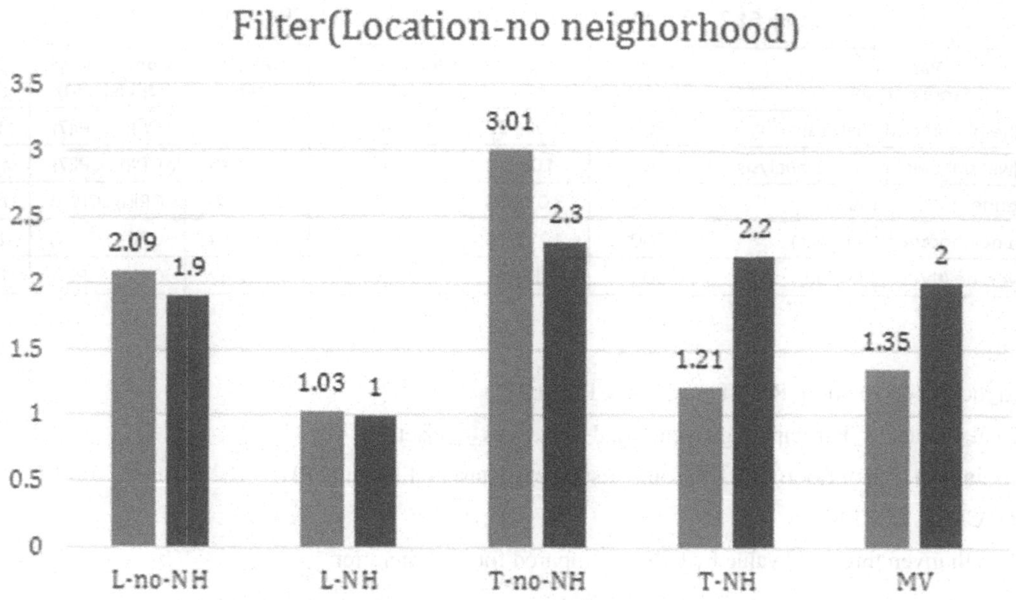

Fig. 51.8 Testing on prevalent patterns with CPI value

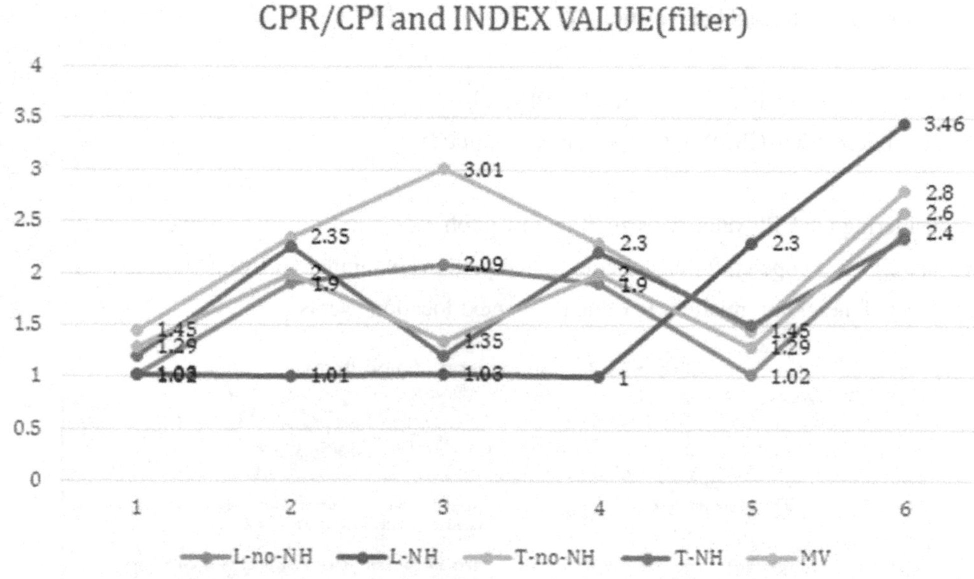

Fig. 51.9 Testing on prevalent patterns with CPR, CPI and index value

5. Methodology

The growth of data sets and retrieval patterns relies on spatio-temporal information, as it enhances the representation and understanding of processes and events, particularly for prevalent patterns.

- Pattern complexity plays a crucial role because non-prevalent patterns can hinder event recognition within a time and location-based data collection.
- Consequently, traditional pattern recognition methods are not suitable for addressing this problem.
- Therefore, phase II of this research aims to tackle the challenge of achieving optimal computational complexity for spatial data sets, preserving temporal and spatial information in patterns.

The analysis included the following aspects using CSTP Miner and Graph Generated techniques for sizes 1 to 7:

- Illustration of Miner for each size (1 to 7).
- Utilization of Miner based on topographic neighborhood relations.
- Exploration of inter-dependent events and their locations.
- Incorporation of topological terminology for each event occurrence within different sizes.
- Implementation of the Affinity Propagation-Graph based algorithm.

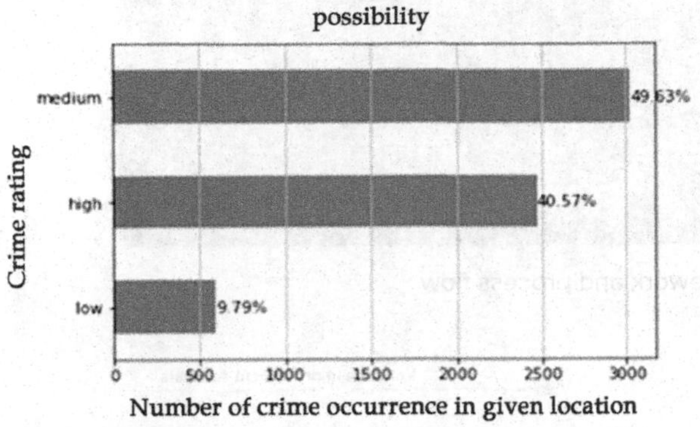

Fig. 51.10 Prediction of event rate at given location.

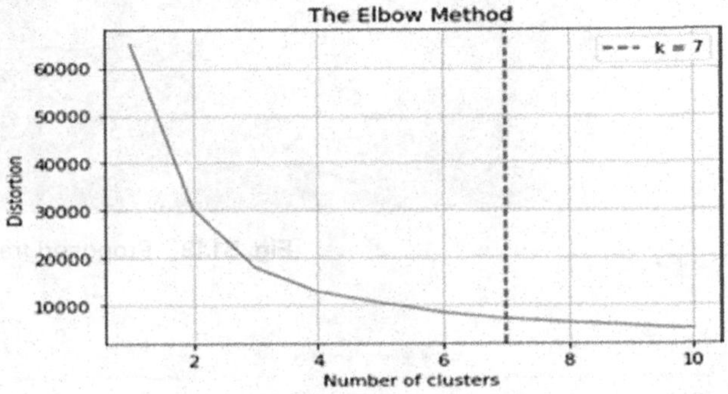

Fig. 51.11 K-Means algorithm

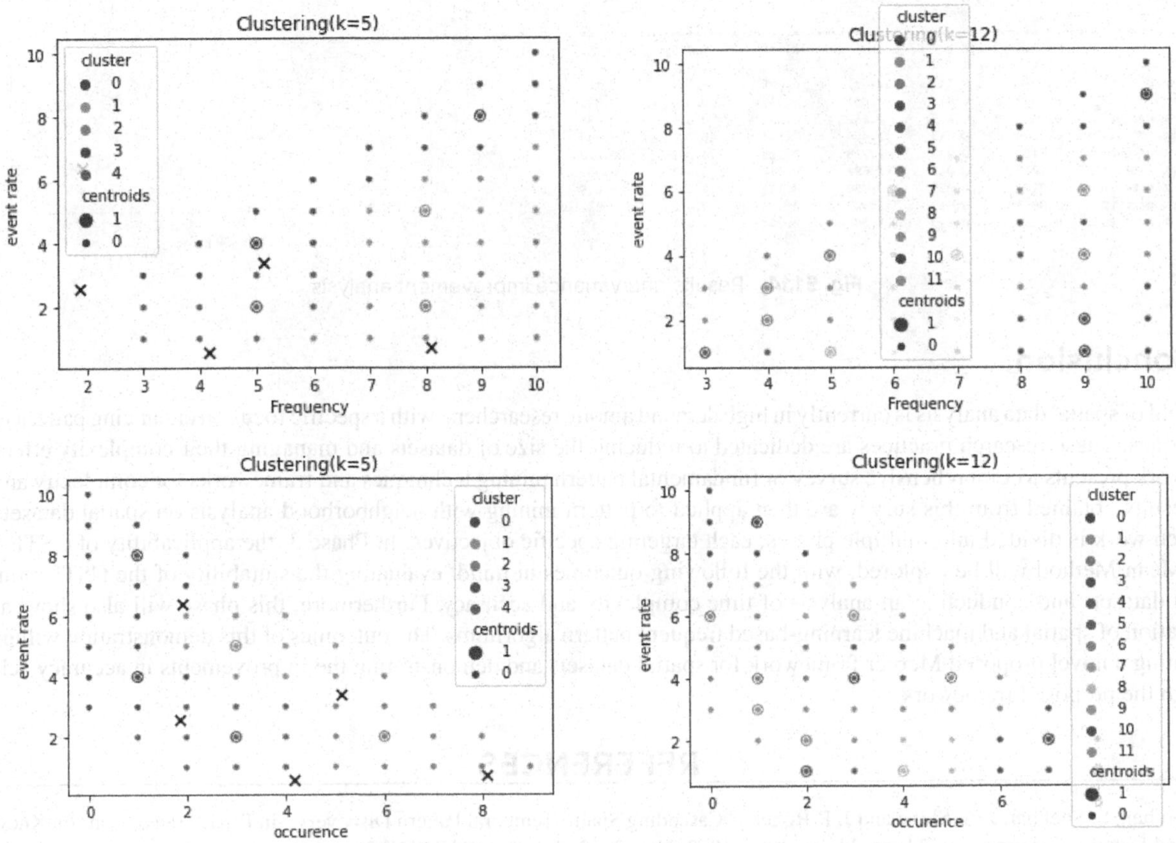

Fig. 51.12 K-Means algorithm

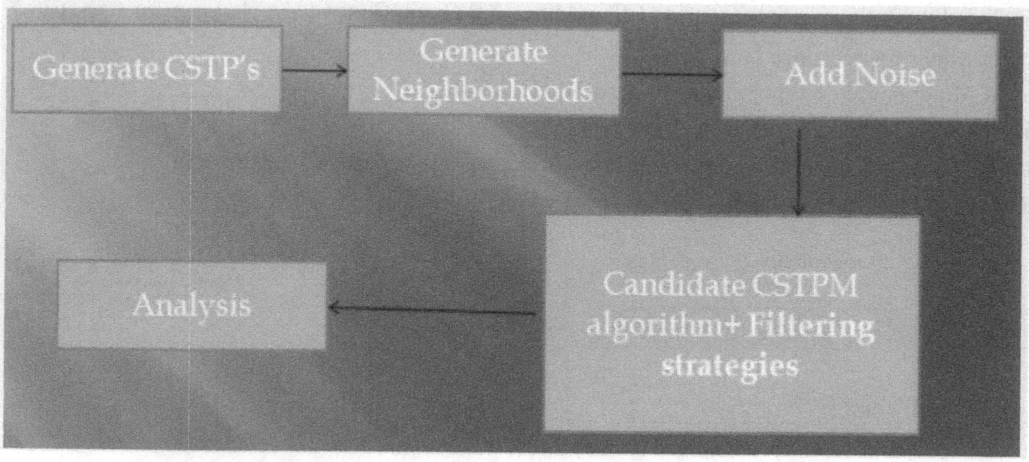

Fig. 51.13 Proposed framework and process flow

AutoReg Model Results

Dep. Variable:	event rate	No. Observations:	5995
Model:	AutoReg(8)	Log Likelihood	-13505.936
Method:	Conditional MLE	S.D. of innovations	2.309
Date:	Fri, 29 Jul 2022	AIC	1.677
Time:	13:18:04	BIC	1.688
Sample:	8	HQIC	1.681
	5995		

	coef	std err	z	P>|z|	[0.025	0.975]
intercept	3.9080	0.141	27.811	0.000	3.633	4.183
event rate.L1	-0.0007	0.013	-0.052	0.959	-0.026	0.025
event rate.L2	0.0110	0.013	0.852	0.394	-0.014	0.036
event rate.L3	0.0189	0.013	1.460	0.144	-0.006	0.044
event rate.L4	-0.0157	0.013	-1.213	0.225	-0.041	0.010
event rate.L5	-0.0218	0.013	-1.691	0.091	-0.047	0.003
event rate.L6	-0.0101	0.013	-0.780	0.435	-0.035	0.015
event rate.L7	0.0063	0.013	0.487	0.626	-0.019	0.032
event rate.L8	-0.0257	0.013	-1.988	0.047	-0.051	-0.000

Roots

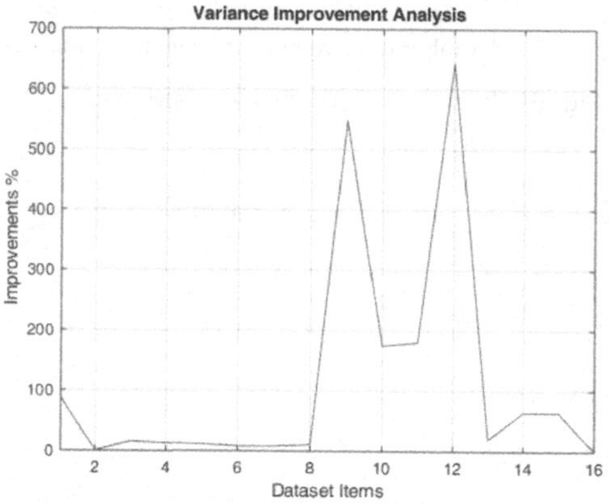

Fig. 51.14 Results and variance improvement analysis

6. Conclusion

The field of spatial data analysis is currently in high demand among researchers, with a specific focus on enhancing pattern mining techniques. These research practices are dedicated to reducing the size of datasets and managing their complexity effectively. This work presents a comprehensive survey of fundamental pattern mining techniques and frameworks for complexity analysis. The results obtained from this survey are then applied to pattern mining with neighborhood analysis on spatial datasets. The research work is divided into multiple phases, each targeting specific objectives. In Phase 3, the applicability of CSTP Miner or K-Mean Method will be explored, with the following outcomes in mind: evaluating the suitability of the CSTP Miner for spatial datasets and conducting an analysis of time complexity and accuracy. Furthermore, this phase will also showcase the application of spatial and machine learning-based frequent pattern algorithms. The outcomes of this demonstration will include presenting a novel proposed Merger framework for spatial datasets and demonstrating the improvements in accuracy achieved through the proposed framework.

REFERENCES

1. Mohan, S. Shekhar, J. A. Shine and J. P. Rogers, "Cascading Spatio-Temporal Pattern Discovery," in IEEE Transactions on Knowledge and Data Engineering, vol. 24, no. 11, pp. 1977–1992, Nov. 2012, doi: 10.1109/TKDE.2011.146.
2. C. Yu, W. Ding, M. Morabito and P. Chen, "Hierarchical Spatio-Temporal Pattern Discovery and Predictive Modeling," in IEEE Transactions on Knowledge and Data Engineering, vol. 28, no. 4, pp. 979–993, 1 April 2016, doi: 10.1109/TKDE.2015.2507570.

3. X. He, L. Wang, Y. Liu and L. Han, "Node correlation causal inference of spatio-temporal association data based on Bayesian method," 2019 IEEE 4th Advanced Information Technology, Electronic and Automation Control Conference (IAEAC), Chengdu, China, 2019, pp. 408–412, doi: 10.1109/IAEAC47372.2019.8997973.
4. L. Kesheng, N. Yikun, L. Zihan and D. Bin, "Data Mining and Feature Analysis of College Students' Campus Network Behavior," 2020 5th IEEE International Conference on Big Data Analytics (ICBDA), Xiamen, China, 2020, pp. 231–237, doi: 10.1109/ICBDA49040.2020.9101257.
5. Y. Chen, Q. Fu and J. Zhu, "Finding Next High-Quality Passenger Based on Spatio-Temporal Big Data," 2020 IEEE 5th International Conference on Cloud Computing and Big Data Analytics (ICCCBDA), Chengdu, China, 2020, pp. 447–452, doi: 10.1109/ICCCBDA49378.2020.9095695.
6. P. Zhou and F. Yuan, "Study on Numerical Simulation Model of Spatio-Temporal Evolution in Dynamic Gob Areas," 2020 International Conference on Intelligent Transportation, Big Data & Smart City (ICITBS), Vientiane, Laos, 2020, pp. 1056–1060, doi: 10.1109/ICITBS49701.2020.00233.
7. J. Xu, C. Guan, H. Dai, D. Yang, L. Xu and J. Kai, "Incentive Mechanisms for Spatio-Temporal Tasks in Mobile Crowdsensing," 2019 IEEE 16th International Conference on Mobile Ad Hoc and Sensor Systems (MASS), Monterey, CA, USA, 2019, pp. 55–63, doi: 10.1109/MASS.2019.00016.
8. A Case Study on Spatio-Temporal Data Mining of Urban Social Management Events Based on Ontology Semantic Analysis by Shaohua Wang, Xianxiong Liu, Haiyin Wang, and Qingwu Hu. Published: 19 June 2018.
9. S. G. Teo, J. Cao and V. C. S. Lee, "DAG: A General Model for Privacy-Preserving Data Mining: (Extended Abstract)," 2020 IEEE 36th International Conference on Data Engineering (ICDE), Dallas, TX, USA, 2020, pp. 2018–2019, doi: 10.1109/ICDE48307.2020.00228.
10. X. Qiu and S. Ji, "Research On Mine Data Sharing Scheme Based On Blockchain," 2020 International Conference on Computer Engineering and Application (ICCEA), Guangzhou, China, 2020, pp. 154–157, doi: 10.1109/ICCEA50009.2020.00040.
11. B. S. Panda and R. Kumar Adhikari, "A Method for Classification of Missing Values using Data Mining Techniques," 2020 International Conference on Computer Science, Engineering and Applications (ICCSEA), Gunupur, India, 2020, pp. 1–5, doi: 10.1109/ICCSEA49143.2020.9132935.
12. The big data analysis and mining of people's livelihood appeal based on tim series modeling and algorithm May 2020 DOI: 10.1109/HPBDIS49115.2020.9130588Conference: 2020 International Conference on High Performance Big Data and Intelligent Systems (HPBD&IS)

Note: All the figures and tables in this chapter were made by the authors.

Impending Inquisitions in Humanities and Sciences (ICIIHS-2022) – Dr. Mohan Varkolu et al. (eds)
© 2024 Taylor & Francis Group, London, ISBN 978-1-032-78829-6

Mathematical Model for Herschel-Bulkley Fluid Flow through an Elastic Tube with Multiple Constructions

52

K. Maruthi Prasad
Department of Mathematics, School of Science, GITAM (Deemed to be University), Hyderabad, Telangana, India
Sreekala Bheema*
Research Scholar, Department of Mathematics, School of Science, GITAM (Deemed to be University)
Department of Basic Science and Humanities, BVRIT Hyderabad College of Engineering for Women, Nizampet Road, Rajiv Gandhi Nagar, Bachupally, Hyderabad

ABSTRACT A computational model for the movement of Herschel-Bulkley fluid with an pliable tube was developed that takes consideration of two stenosis and multiple non-uniform cross sections. Closed form solutions have been obtained, assuming that the stenosis is minor. These flow variables have been examined as a result of the influence of various parameters, the flow resistance expressions and shear strain at the wall have been obtained. By the methods of Rubinow & Keller and Mazumdar the flux variation is calculated. Herschel-Bulkley fluid in an elastic tube for the impact of numerous pertinent parameters on change of flux along the tube radius has been examined. It has been found that the fluid flux increases with the elastic parameters and exit pressure, but not with inlet pressure. Additionally, it has been observed that the power law index, wall shear stress, and stenosis height all increase the velocity of the fluid. Also a comparison of the Rubinow & Keller and Mazumdar elasticity models is taken into consideration.

KEYWORDS Multiple constructions, Herschel-Bulkley fluid, Rubinow & Keller model, Mazumdar model, Elastic tube, Fluid flux

1. Introduction

The greatest cause of death for both men and women worldwide today is cardiovascular disease, which is mostly brought on by coronary artery illness [Young 1968]. The narrowing or occlusion of coronary arteries, typically brought on by stenosis, is known as coronary artery disease. Stenosis is an unnatural and abnormal growth that impairs normal blood flow. By restricting the actual blood vessel and its function, the stenosis might reduce the amount of blood that reaches the heart. Due to limited blood flow, the heart is deprived of vital nutrients and oxygen that it need to function. If the blood supply to the heart is significantly more than the blood supply, a heart stroke or attack could happen [Boyd 1961].

Many researchers have taken into account the experimental and theoretical studies on blood flow over arteries with minor stenosis [Young DF and Tsai 1973, Macdonald D A 1979, Chaturani P and PonnalagaruSwami 1986], which are more beneficial for the diagnosis of cardiovascular illnesses and in the design of health devices used in dialysis and other procedures.

*bheemasreekala@gmail.com

DOI: 10.1201/9781003489436-52

By considering blood as a Newtonian fluid, Young [1979] and Misra et al. [2011] examined the flow characteristics in an artery wall that was stenotic. Although the blood is a suspension of cells, it exhibits non-Newtonian behavior at low shear rates in small arteries [Charm S.E. and Kurland G (1965), Huckaba.C.E. and Hahn A.N, (1968)]. The Newtonian performance may be true in superior arteries. Blood flows in big vessels, where Newtonian behavior is anticipated, and small vessels, where non-Newtonian effects are observed, in various ways [Buchanana et al (2000), Mandal et al (2005), Ismail (2007), Maruthi Prasad and Radhakrishnamacharya (2007)]. In small channels, blood behaves less like a power law or Bingham fluid and more like a Herschel-Bulkley fluid [Chaturani and Samy (1985)]. The peristaltic movement of H-B fluid in an inclined tube was investigated by [Vajravelu et al (2005)].

The link between flow and pressure gradient is explained by the Poiseuille law, among other significant contributions to the study of fluid dynamics. According to Poisuille's law, the flow of a viscous, incompressible fluid over a rigid tube is a constant function of the pressure difference between the tube's ends. However, the pressure flow relation is never linear in the mammal vascular beds. The flexible characteristics of blood arteries and their relative high dispensability have been connected to the non-linearity.

The majority of the vascular system is known to be composed of a multifaceted network of branching pliable tubes. Young [1808] was one of the first significant figures to diagnose the significance of pliability for the beat surge coming to the heart. Following, Rubinow & Keller [1972] and others proved how radius of the tube may be estimated by balancing the tension in the wall with the transmural pressure differential under the premise that the Poiseuille rule is locally valid. This equilibrium provides a method by which blood vessels can change dimensions and, as a result, their conductivity in response to pressure differences. It can be expressed by a simple relation. With this assertion, Burton [1951] is in agreement.

Experimental research were carried out by physiologists including Fry [1968], Brecher [1952], and Rodbard [1955] that helped to understand the arterial flow to the blood and the heart movement via thepulmonic arrangement in qualitative terms. The stable flow over a vascular scheme can be compared to the flow through an elastic tube, according to [Brecher (1956), Banister & Torrance (1960), and Permutt et al. (1962)]. This tube is raised to as a starling resistor since it was developed as a laboratory instrument to raise the vascular system is resistance [Knowlton & Starling (1912)]. Due to physiology's requirements, a measurable description of flow via pliable tubes is necessary.

By taking Herschel-Bulkley fluid, Vajraveylu et al. [2011] studied the scenario of introducing a cylinder into an pliable tube to perceive variations in the blood flow pattern. By using lubrication examination to simulate the distortion of the tube wall, Takaghi and Balmforth [2011] calculated the pushing efficiency. An experimental finding on non-Newtonian flow properties in inflatable elastic tubes was presented by Nahar et al. in 2013. Sochi [2014] used the concept of cylindric-shaped elastic tubes to derive analytical formulas for the Newtonian flow characteristics.

As a result, investigating the impact of a non-Newtonian fluid in the presence of numerous stenosis may aid in improving knowledge of the function of fluid dynamics in the expansion and development of arterial diseases.

These findings served as inspiration for the development of a mathematical explanation for the flow of Herschel-Bulkley fluid along an pliable tube with numerous non-uniform cross sections and binary stenosis. Closing form solutions have been found, presuming a slight stenosis. The influences of numerous parameters on these flow variables have been researched, and the flow resistance expressions and shear strain at the wall have been obtained. By using the Rubinow & Keller and Mazumdar methods, the flux variation is determined. Graphs are used to calculate and explain the impact of a variety of important factors on the flow change along the tube radius for Herschel-Bulkley fluid in a flexible tube.

2. Mathematical Formulation

Consider the steady flow of Herschel-Bulkley fluid through an pliable tube with two stenosis of radius and length . Centre line of the tube coincides with axis using cylindrical polar coordinate system . The stenosis should be minimal and develop axially symmetrical. The tube's radius is taken as (MARUTHI PRASAD AND RADHAKRISHNAMACHARYA [2012])

$$h(z)=\frac{R(z)}{R_0}=\begin{cases} 1 & 0\le z\le d_1 \\ 1-\frac{\delta_1}{2R_0}\left[1+\cos\frac{2\pi}{L_1}\left(z-d_1-\frac{L_1}{2}\right)\right] & d_1\le z\le d_1+L_1 \\ 1 & d_1+L_1\le z\le B_1-\frac{L_2}{2} \\ 1-\frac{\delta_2}{R_0}\left[1+\cos\frac{2\pi}{L_2}(z-B_1)\right] & B_1-\frac{L_2}{2}\le z\le B_1 \\ \frac{R^*(z)}{R_0}-\frac{\delta_2}{2R_0}\left[1+\cos\frac{2\pi}{L_2}(z-B_1)\right] & B_1\le z\le B_1+\frac{L_2}{2} \\ \frac{R^*(z)}{R_0} & B_1+\frac{L_2}{2}\le z\le B \end{cases} \tag{2.1}$$

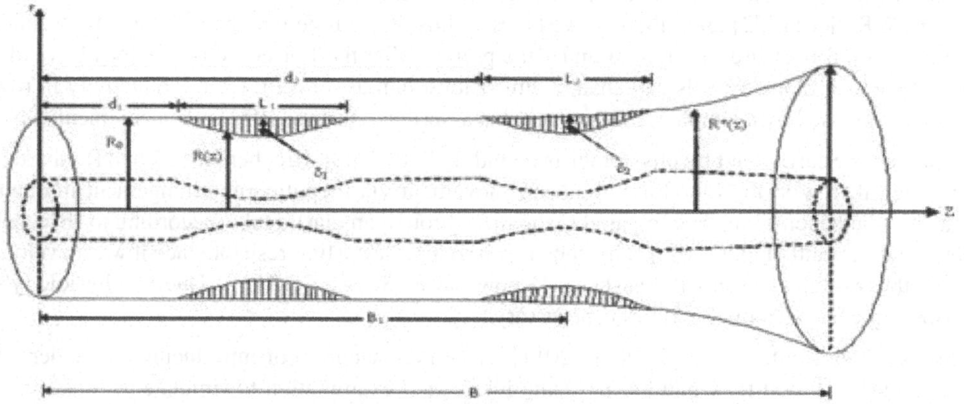

Fig. 52.1 The physical model of geometry

$\boldsymbol{R_0}$ is radius of the tube without stenosis whereas $\boldsymbol{R(z)}$ is radius of the stenotic tube

According to (VAJRAVELU et al. [2011]), the basic momentum governing equation is

$$\frac{1}{r}\frac{\partial}{\partial r}(r\tau_{rz})=-\frac{\partial p}{\partial z} \tag{2.2}$$

where r_{rz} H. B. fluid is shear stress and is determined by

$$\tau_{rz}=\left(\frac{\partial u}{\partial r}\right)+\tau_0,\qquad \tau_{rz}>\tau_0 \tag{2.3}$$

$$\frac{\partial u}{\partial r}=0,\quad \tau_{rz}>\tau_0 \tag{2.4}$$

Here p is the pressure, u is the velocity and τ_0 is yield stress. If shear stress is lesser than yield stress, it should be note that core region forms as a plug i.e., $\tau_{rz}<\tau_0$, and In such area, the velocity gradient vanishes according to equation (2.4). However, as mentioned above for $\tau_{rz}>\tau_0$, this fluid model is what we employ.

The boundary conditions are

τ_{rz} is finite at $r=0$ (2.5)

$u=0$ at $r=h(z)$ (2.6)

2.1 Solution of the Problem

The following are non-dimensional quantities used to solve the equations (2.2) and (2.3) under boundary conditions (2.5) and (2.6)

$$\bar{r}=\frac{r}{h_0},\bar{h}=\frac{h}{h_0},\bar{z}=\frac{z}{L},\bar{u}=\frac{u}{U},\bar{P}=\frac{h_0^{n+1}}{L\mu U^n},\bar{\tau}_0=\frac{\tau_0}{\mu\left(\frac{U}{h_0}\right)^n},\bar{\tau}_{rz}=\frac{\tau_{rz}}{\mu\left(\frac{U}{h_0}\right)^n}$$

where $P=-\frac{\partial p}{\partial z}$, length of the tube L, average velocity U, h_0 is radius of the tube in absence of elasticity and radius of plug flow region. The governing equations become after dropping the bars

$$\frac{1}{r}\frac{\partial}{\partial r}(r\tau_{rz})=-\frac{\partial p}{\partial z}=P \tag{2.7}$$

Boundary conditions

$$\tau_{rz} \text{ is finite at } r=0 \tag{2.8}$$

$$u=0 \text{ at } r=h(z) \tag{2.9}$$

The velocity is determined as the result of solving eq. (2.7) under boundary conditions (2.8) and (2.9)

$$u=\frac{2}{P\left(\frac{1}{n}+1\right)}\left[\left(\frac{P}{2}h-\tau_0\right)^{\frac{1}{n}+1}-\left(\frac{P}{2}r-\tau_0\right)^{\frac{1}{n}+1}\right] \tag{2.10}$$

Using boundary condition (2.4), upper limit for the plug flow region (i.e. region between $r=0$ and $r=r_0$ for which $\tau_{rz}<\tau_0$)) is

$$r_0=\frac{2\tau_0}{P} \tag{2.11}$$

Using condition $\tau_{rz}=\tau_h$ at $r=h$ (Bird et al. [2007]), we obtain

$$P=\frac{2\tau_0}{h} \tag{2.12}$$

Hence $\frac{\tau_0}{\tau_h}=\frac{r_0}{h}=\tau,\ for\ 0<\tau<1$ (2.13)

Using the relation (2.13) by taking $r=r_0$ in equation (2.10), we obtain the plug flow velocity is

$$u_p=\left(\frac{p}{2}\right)^k\frac{h^{k+1}}{k+1}(1-\tau)^{k+1}\quad for\ 0\le r\le r_0 \tag{2.14}$$

Flux Q through any cross-section is given by

$$Q=\left[\int_0^{\tau_0}u_p r\,dr+\int_{\tau_0}^{h}ur\,dr\right] \tag{2.15}$$

On integrating,

$$Q=\frac{h^{k+1}}{k+1}\left(\frac{p}{2}\right)^k(1-\tau)^{k+1}\left[1-\frac{2(1-r)(k+2+\tau)}{(k+2)(k+3)}\right] \tag{2.16}$$

Flux Q for a tube with a variable radius h is given by the equation (2.16). We assume Poisseuille rule for H-B fluid flow in an pliable tube and examine its significances in the following section using the fact that difference arises because of the pliability of the tube wall.

2.2 Theoretical Determination of Flux

The steady flow Q of a rigid H-B fluid with viscosity in an elastic tube of length L and radius h is now theoretically calculated. Assume the fluid pressures enters the tube at p_1 and exits at p_2, with p_0 being the pressure outside the tube. The pressure in the fluid at z falls from $p(0) = p_1$ to $p(L) = p_2$ if z represents the length of tube from the inlet end. Tube may expand or contract as a result of pressure variance between the interior and outside of the tube, and because the wall of the tube is elastic, this may cause a change in the cross-sectional shape. So the pressure differential will affect conductivity of the tube at z. We take $\sigma_1 = \sigma_1[p(z) - p_0]$ into consideration as a function of $p(z) - p_0$. This conductivity is taken to be same as that of tube with a uniform cross section matches at z. Suppose that relationship between Q and the pressure gradient is

$$Q = \sigma_1(p - p_0)\left(-\frac{\partial p}{\partial z}\right)^{\frac{1}{n}} \tag{2.17}$$

Where $\sigma_1(p - p_0) = F\, h^{\frac{1}{n}+3}$

And
$$F = \frac{(1-\tau)^{k+1}}{(k+1)2^k}\left[1 - \frac{2(1-\tau)(k+\tau+2)}{(k+2)(k+3)}\right] \tag{2.18}$$

Note that equation (2.18) simplifies according to Rubinow & Keller [1972] for Newtonian fluid flow in an pliable tube when $n = 1$ and $\tau = 0$. Using the inlet condition $p(0) = p_1$ and integrating equation (2.18) with respect to z at $z = 0$, we obtain

$$Q^n z = \int_{p(z)-p_0}^{p_1-p_0} \left(\sigma_1(p')\right)^n dp' \tag{2.19}$$

In which $p' = p(z) - p_0$. In terms of Q and z and this equation implicitly determines $p(z)$.

In order to determine Q, we assign $z = 1$ and $p(1) = p_2$ in equation (2.19)

$$Q^n = \int_{p(z)-p_0}^{p_1-p_0} \left(\sigma_1(p')\right)^n dp' \tag{2.20}$$

In the current scenario, radius h is a function of $p - p_0$, i.e. $h = h(p - p_0)$. Equation (2.20) can be written as follows

$$Q^n = F^n \int_{p_2-p_0}^{p_1-p_0} h^{3n+1} \tag{2.21}$$

Equation (2.21) can also be resolved if the form of the function $h(p - p_0)$ is known. If the hoop tension or stress $T(h)$ is known as a function of h in the tube wall, the equilibrium condition determines $h(p - p_0)$.

$$\frac{T(h)}{h} = p - p_0 \tag{2.22}$$

3. Application to Flow through a Stenotic Artery

Considering the wall elasticity of a blood vessel with stenosis and utilizing the pressure differential $p(z) - p_0$, it is believed that the conduct σ_1 is the same as that of a uniform arterial tube having cross section z.

We can identify the arterial flow using two alternative methods.

4. Rubinow and Keller Method

The constant pressure-volume relationship of a 4 cm long segment of the human external iliac artery, which is subsequently transformed into a tension versus length curve, is used to calculate the flow through an artery. Rubinow & Keller [1972] produced the following equation using least squares technique:

$$T(h) = t_1\,(h - 1) + t_2(h - 1)^5 \tag{2.23}$$

Where $t_1 = 13$, $t_2 = 300$

On simplification, we obtain by substituting (2.23) into (2.22)

$$dp' = \left[\frac{t_1}{h^2} + t_2 \left(4h^3 - 15h^2 + 20h - 10 + \frac{1}{h^2} \right) \right] dh \tag{2.24}$$

Using (2.24), (2.21) can be written as

$$Q^n = F^n \int_{p_2 - p_0}^{p_1 - p_0} h^{3n+1} \left[\frac{t_1}{h^2} + t_2 \left(4h^3 - 15h^2 + 20h - 10 + \frac{1}{h^2} \right) \right] dh \tag{2.25}$$

On further simplification, we get

$$Q = F(g(h_1) - g(h_2)) \tag{2.26}$$

Where

$$g(h) = \left[t_1 \frac{h^{3n}}{3n} + t_2 \left(\frac{4h^{3n+5}}{3n+5} - \frac{15h^{3n+4}}{3n+4} + \frac{20h^{3n+3}}{3n+3} - \frac{10h^{3n+2}}{3n+2} + \frac{h^{3n}}{3n} \right) \right]^{\frac{1}{n}}$$

$$h_1 = h(p_1 - p_0)$$

$$h_2 = h(p_2 - p_0) \tag{2.27}$$

We observe that by solving equation (2.22) with $p = p_1$ and $p = p_2$, respectively, we can find h_1 and h_2. We see that the outcomes from Rubinow & Keller [1972] for the flow of Newtonian fluid (when $n = 1$, $\tau = 0$) in an pliable tube are reduced by equation (2.27).

4.1 Mazumdar Method

According Mazumdar [1992], the relationship between tension and expression can be expressed as follows:

$$T(h) = A(e^{kh} - e^{k}) \tag{2.28}$$

Substituting Eq. (2.28) in Eq. (2.22) at $A = 0.007435$ and $k = 5.2625$, we get

$$p' = p - p_0 = A\left[\frac{e^{Kh}}{h} - \frac{e^{K}}{h} \right] \tag{2.29}$$

$$dp' = A\left[e^{Kh} \left(\frac{K}{h} - \frac{1}{h^2} \right) + \frac{e^{K}}{h^2} \right] dh \tag{2.30}$$

Eq. (2.30) into Eq. (2.21) yields the flux

$$Q^n = AF^n \int_{p_2 - p_0}^{p_1 - p_0} h^{3n+1} \left[e^{Kh} \left(\frac{K}{h} - \frac{1}{h^2} \right) + \frac{e^{K}}{h^2} \right] dh \tag{2.31}$$

We observe that h_1 and h_2 are found by finding a solution to (2.22) with $p = p_1$ and $p = p_2$ correspondingly. The flow rate for Herschel-Bulkley's model is assumed in an pliable tube using the previous integral (2.31), which was evaluated numerically.

5. Results and Discussion

Mathematica software was used to compute the numerical impacts of different parameters on flux, and the results are represented graphically in figures 2 to 17 by taking $\frac{R^*}{R_0} = EXp[\beta B^2 (Z - B)^2]$.

Figures 52.2 to 52.9 show variations in volume flow rate using the Rubinow and Keller method for various parameters.

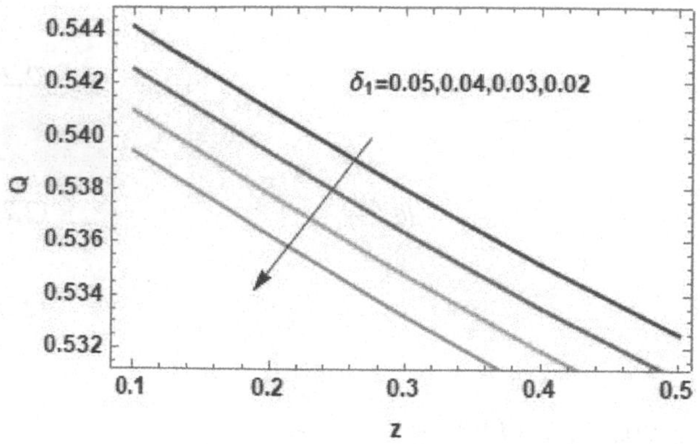

Fig. 52.2 Variation of flux Q vs. z with height of the stenosis δ_1 for $\delta_2 = 0.02$, $\tau = 0.2$, $n = 2$, $t_1 = 13$, $t_2 = 300$, $P_2 - P_0 = 0.2$, $P_1 - P_0 = 1$ (Rubinow and Keller)

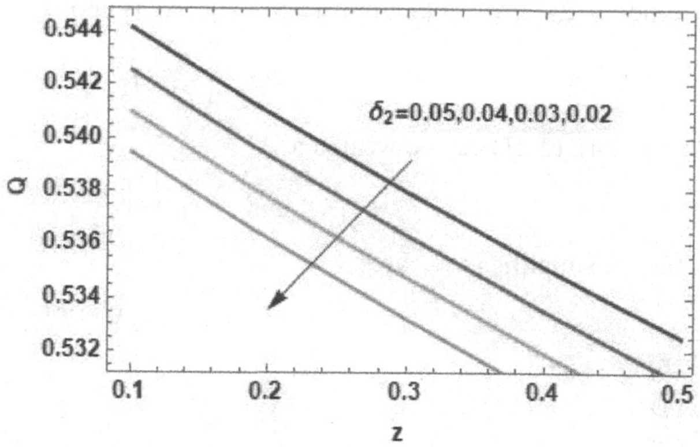

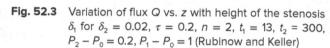

Fig. 52.3 Variation of flux Q vs. z with height of the stenosis δ_1 for $\delta_2 = 0.02$, $\tau = 0.2$, $n = 2$, $t_1 = 13$, $t_2 = 300$, $P_2 - P_0 = 0.2$, $P_1 - P_0 = 1$ (Rubinow and Keller)

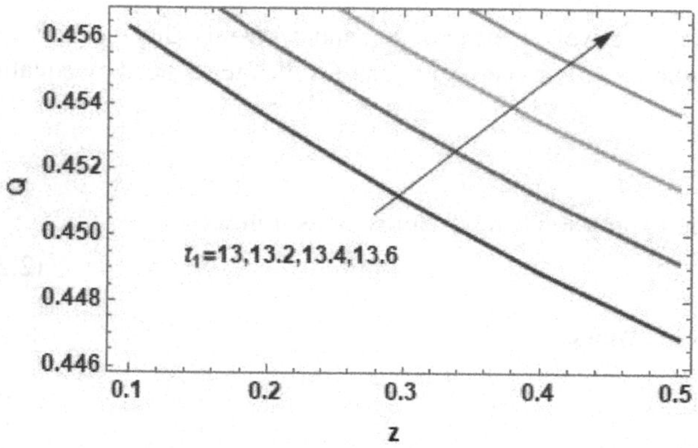

Fig. 52.4 Variation of flux Q vs. z with elastic parameter t_1 for $t_2 = 300$, $\tau = 0.2$, $n = 3$, $\delta_1 = 0.02$, $\delta_2 = 0.02$, $P_2 - P_0 = 0.2$, $P_1 - P_0 = 1$ (Rubinow and Keller)

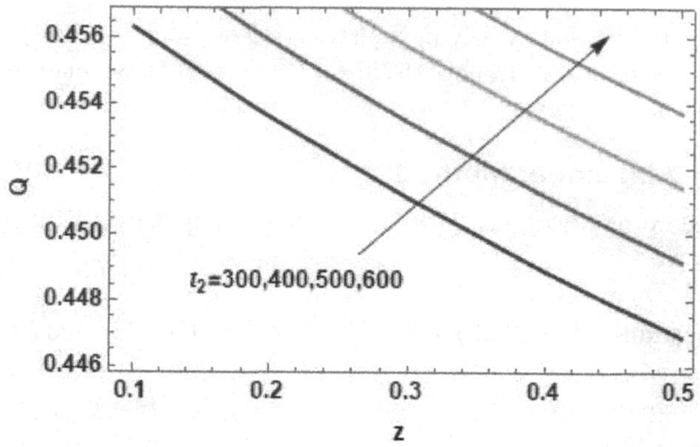

Fig. 52.5 Variation of flux Q vs. z with elastic parameter t_1 for $t_2 = 13$, $\tau = 0.2$, $n = 3$, $\delta_1 = 0.02$, $\delta_2 = 0.02$, $P_2 - P_0 = 0.2$, $P_1 - P_0 = 1$ (Rubinow and Keller)

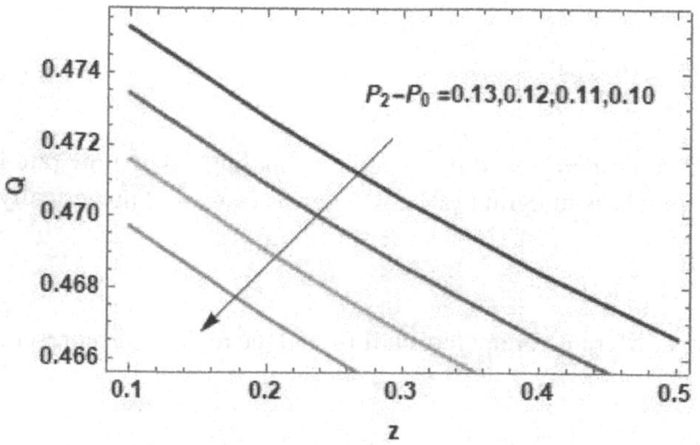

Fig. 52.6 Variation of flux Q vs. z with inlet pressure $P_2 - P_0$ for $P_1 - P_0 = 1$, $t_1 = 13$, $t_2 = 300$, $\tau = 0.2$, $n = 3$, $\delta_1 = 0.02$, $\delta_2 = 0.02$ (Rubinow and Keller)

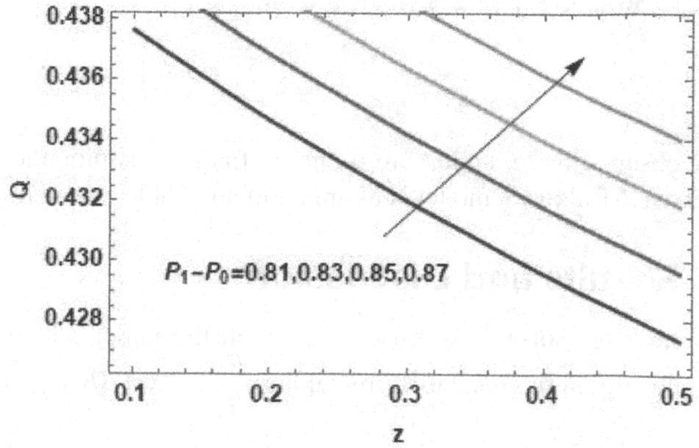

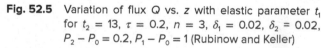

Fig. 52.7 Variation of flux Q vs. z with outlet pressure $P_2 - P_0$ for $P_1 - P_0 = 1$, $t_1 = 13$, $t_2 = 300$, $\tau = 0.2$, $n = 3$, $\delta_1 = 0.02$, $\delta_2 = 0.02$ (Rubinow and Keller)

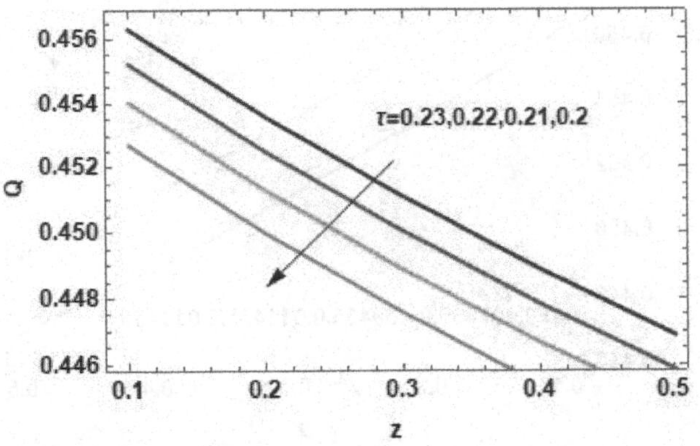

Fig. 52.8 Variation of flux Q vs. z with yield stress τ for $t_1 = 13$, $t_2 = 300$, $\tau = 0.2$, $n = 3$, $\delta_1 = 0.02$, $\delta_2 = 0.02$, $P_2 - P_0$, $P_1 - P_0 = 1$ (Rubinow and Keller)

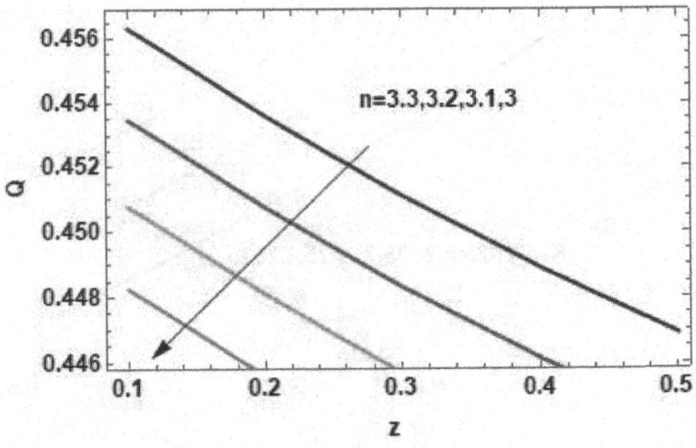

Fig. 52.9 Variation of flux Q vs. z with power law-index n for $t_1 = 13$, $t_2 = 300$, $\tau = 0.2$, $n = 3$, $\delta_1 = 0.02$, $\delta_2 = 0.02$, $P_2 - P_0$, $P_1 - P_0 = 1$ (Rubinow and Keller)

Figures 52.2 and 52.3 show that the flux is decreasing as the height of the stenosis, shown by δ_1 and δ_1, increases.

Figures 52.4 and 52.5 show the performances of the flux Q versus the z-axis as a function changes in the elastic parameters, particularly t_1 and t_2. The other elastic parameter, t_2, is fixed, an increase in the elastic parameter t_1 results are rise in the flow Q. The other elastic parameter, t_2, also follows same trend. Fig. 52.5 illustrates this graphically.

Figures 52.6 and 52.7 demonstrate, respectively, the flux profiles Q with pressure difference changes. When the outlet pressure $(p_2 - p_0)$ is increased, it has the effect of decreasing the flow for a stable value of the inlet pressure $(p_1 - p_0)$, which causes Q to fall. As a result, Q decreases with rising outlet pressure. When we fix the outlet pressure $(p_2 - p_0)$ and vary the input pressure, the behavior is the exact reverse (see Fig. 52.7).

Fig. 52.8 shows graphically how the change of flux Q with the z-axis is determined for numerous yield stress values. Fig. 52.8 shows that the flux Q depends on yield stress and that it falls with increasing yield stress τ for a certain z-axis and power-law index n.

Fig. 52.9 depicts the power-law index behavior on flux Q with thez-axis. We observed from the graphical representation that the flux Q drops with rising values of the power-law index n, but increases as the z-axis increases.

Figures 52.10 to 52.17 illustrate how the Mazumdar method changes the volume flow rate for various parameters.

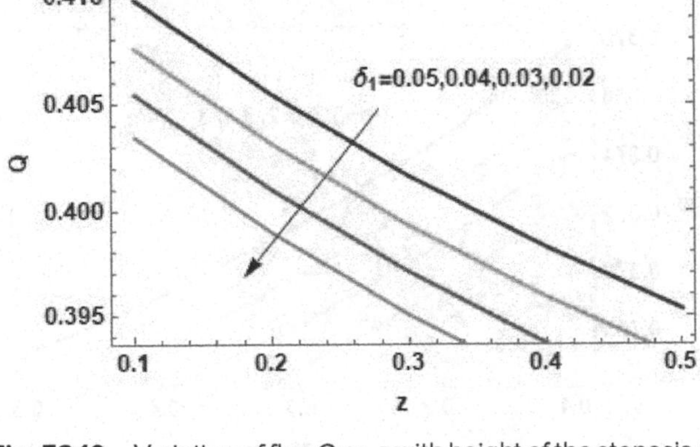

Fig. 52.10 Variation of flux Q vs. z with height of the stenosis δ_1 for $\delta_2 = 0.02, \tau = 0.2, n = 2, K = 5.2625, A = 0.007435$, $P_2 - P_0 = 0.2$, $P_1 - P_0 = 1$ (Mazumdar)

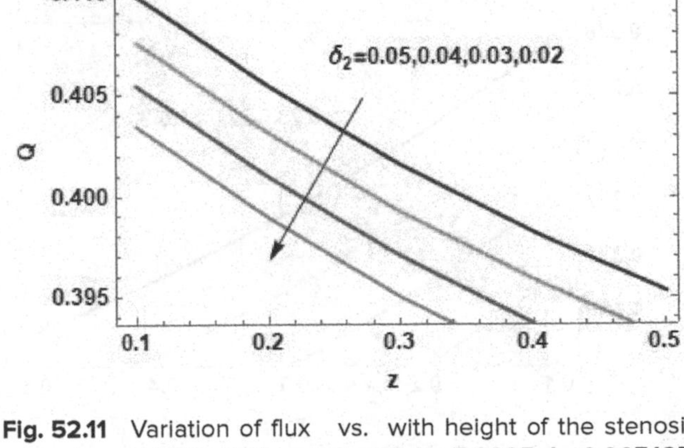

Fig. 52.11 Variation of flux vs. with height of the stenosis δ_2 for $\delta_1 = 0.02, \tau = 0.2, n = 2, K = 5.2625, A = 0.007435$, $P_2 - P_0 = 0.2$, $P_1 - P_0 = 1$ (Mazumdar)

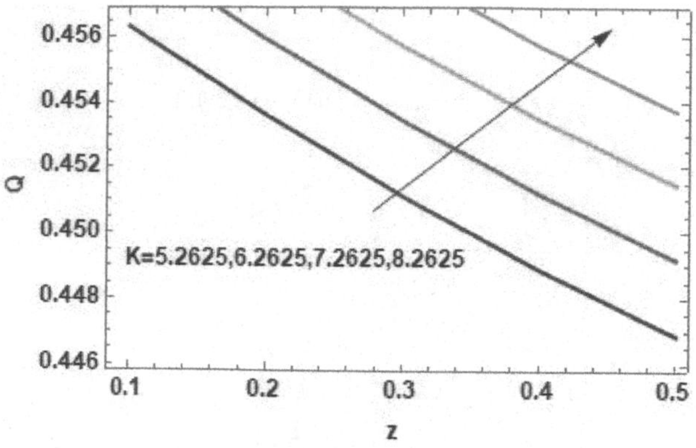

Fig. 52.12 Variation of flux Q vs. z with elastic parameter K for $A = 0.007435, \tau = 0.2, n = 3, \delta_1 = 0.02, \delta_2 = 0.02, P_2 - P_0 = 0.2, P_1 - P_0 = 1$ (Mazumdar)

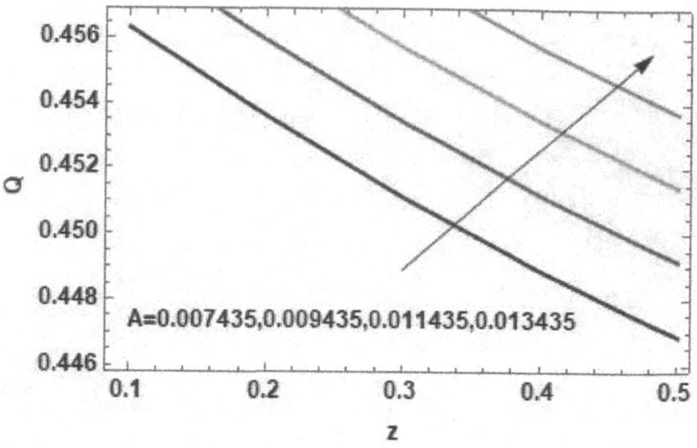

Fig. 52.13 Variation of flux Q vs. z with elastic parameter A for $K = 5.2625, \tau = 0.2, n = 3, \delta_1 = 0.02, \delta_2 = 0.02, P_2 - P_0 = 0.2, P_1 - P_0 = 1$ (Mazumdar)

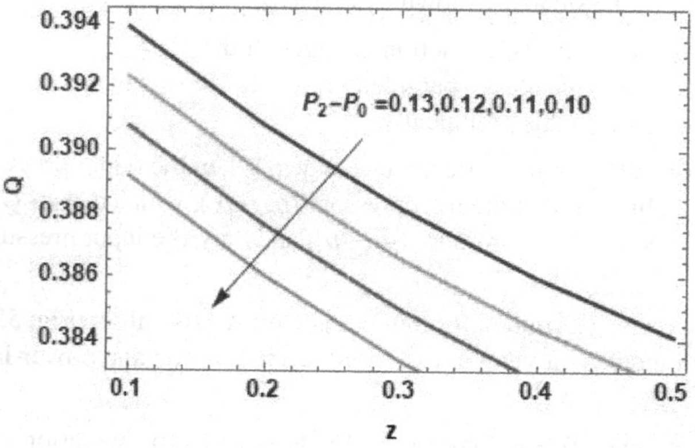

Fig. 52.14 Variation of flux Q vs. z with inlet pressure $P_2 - P_0$ for $P_1 - P_0 = 1, K = 5.2625$, for $A = 0.007435, \tau = 0.2, n = 3, \delta_1 = 0.02, \delta_2 = 0.02$ (Mazumdar)

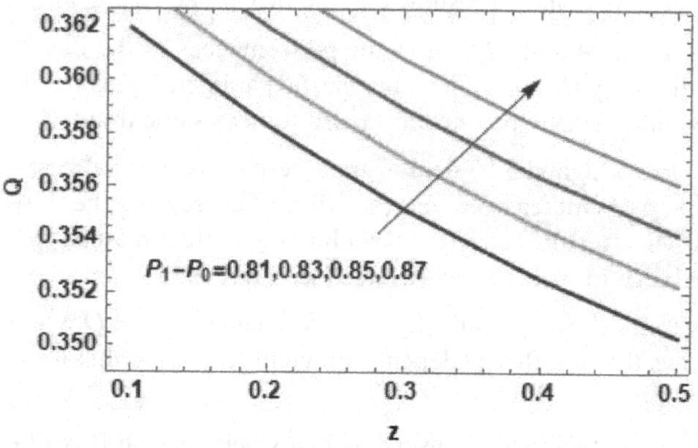

Fig. 52.15 Variation of flux Q vs. z with outlet pressure $P_1 - P_0$ for $P_2 - P_0 = 0.1, K = 5.2625$, for $A = 0.007435, \tau = 0.2, n = 3, \delta_1 = 0.02, \delta_2 = 0.02$ (Mazumdar)

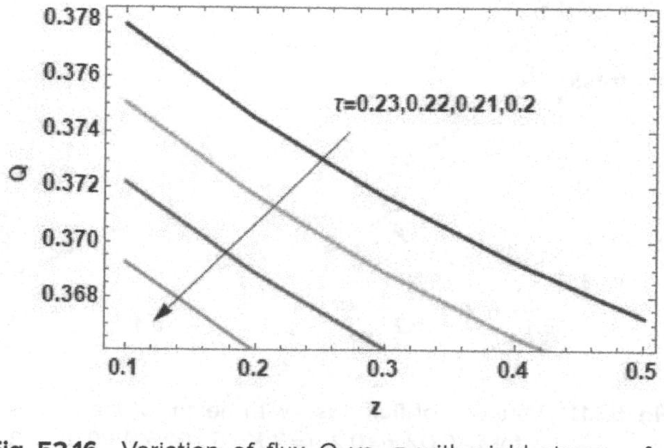

Fig. 52.16 Variation of flux Q vs. z with yield stress τ for $K = 5.2625, A = 0.007435, n = 3, \delta_1 = 0.02, \delta_2 = 0.02, P_2 - P_0 = 0.2, P_1 - P_0 = 1$ (Mazumdar)

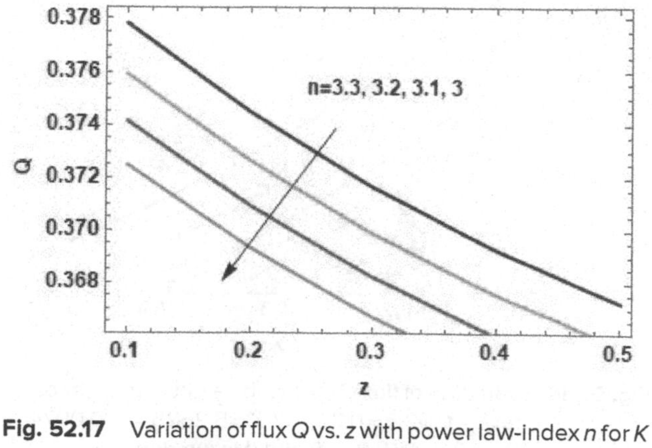

Fig. 52.17 Variation of flux Q vs. z with power law-index n for $K = 5.2625, A = 0.007435, \tau = 0.2, \delta_1 = 0.02, \delta_2 = 0.02, P_2 - P_0 = 0.2, P_1 - P_0 = 1$ (Mazumdar)

Figures 52.10–52.11 show that the flux is decreasing as the heights of the stenosis δ_1 and δ_2 increases.

The behaviors of flux Q vs z-axis with variations in the elastic parameters, especially K and A can be seen in figures 52.12 and 52.13. The other elastic parameter A is fixed, an increase of elastic parameter K results in an upsurge in the flow Q. The other elastic parameter, A, also follows same trend. Fig. 52.13 depicts this graphically.

Figures 52.14 and 52.15 display flux profiles Q with the pressure difference changes, respectively. The influence of increasing values of exit pressure $(p_1 - p_0)$ causes the flow Q to drop for a fixed value of inlet pressure $(p_2 - p_0)$, and as a result, Q decreases as the exit pressure rises. When the inlet pressure is varied while the output pressure $(p_2 - p_0)$ is fixed, behavior is the opposite (see Fig. 52.15).

For various yield stress τ values, the change of flux Q with the z-axis is calculated and graphically shown in Fig. 52.16. Fig. 52.16 shows that the flux Q depends on yield stress and decreases with increasing yield stress τ for a certain z-axis and power-law index n.

Fig. 52.17 depicts the power-law index behavior on flux Q with the z-axis. We observed from the graphical depiction that the flux Q grows with rising values of the z-axis but drops with rising values of the power-law index n.

Both the Rubinow & Keller and Mazumdar methods demonstrate these results.

6. Conclusions

The results were studied for various values of the pertinent parameters, including radius, stenosis and the stress of shear stability of tube wall. Among the interesting findings are:

1.As the tube radius is increases, the flow increases.

- 2. The rising values of power-law index (n) and yield stress , the flow rate decreases.
- 3. As the elastic parameter values increase, the flow rate rises.
- 4. The flux and outlet pressures have different impacts on each other.
- 5. As the stenosis parameters are decreased, the flow rate increases.

REFERENCES

1. Young, D.F., Effects of Time Dependent Stenosis on Flow through a Tube. *Journal of Engineering and Industrial Transactions, ASME, 90, 248-254,1968.*
2. Boyd, H. and Anderson, L., A Method for Reinsertion of the Distal Biceps Brachii Tendon. The Journal of Bone & Joint Surgery, 43, 1041-1043, 1961.
3. Young, D.F. and Tsai, F.Y., Flow Characteristics in Models of Arterial Stenosis—Steady Flow. Journal of Biomechanics, 6, 395-411, 1973.
4. MacDonald, D.A., On Study Flow through Modelled Vascular Stenoses. *Journal of Biomechanics, 12, 13-20, 1979.*
5. P. Chaturani and R. P. Swamy, "Pulsatile Flow of Casson's Fluid through Stenosed Arteries with Applications to Blood Flow". *Biorheology, Vol. 23, No. 5, pp. 499–511, 1986.*
6. Young, D.F., Fluid mechanics of arterial stenosis. *Journal of Biomechanical Engineering, Vol. 101, pp. 151–175, 1979.*
7. Mishra, R.K. and Singh, A., Spring Theory as an Approach to the Unification of Fields. *International Journal of Research and Review in Applied Sciences, 8, 1–7, 2011.*
8. S E Charm, G S Kurland, Static method for determining blood yield stress. *Nature, 216(5120), 1121–1123, 1967.*
9. C E Huckaba, A W Hahn, A generalized approach to the modeling of arterial blood flow. *The bulletin of mathwmatical biophysics, 30(4), 645–662, 1968.*
10. Buchanan, P. C., Lindstrom, D. J., Mittlefehldt, D. W., Koeberl, C. and Reimold, W. U., The South African polymict eucrite Macibini. *Meteoritics and Planetary Science, 35, 1321–1331, 2000.*
10. Mandal, S., Guptan, P., Owusu-Ansah, E., Banerjee, U., Mitochondrial regulation of cell cycle progression during development as revealed by the tenured mutation revealed in Drosophila. *Flybase, 9(6), 843–854, 2005.*
12. Ismail, K.M., Evaluation of Cysteine as Environmentally Friendly Corrosion Inhibitor for Copper in Neutral and Acidic Chloride Solutions. *Electrochimica Acta, 52, 7811–7819, 2007.*
13. K. Maruthi Prasad and G. Radhakrishnamacharya, Peristaltic transport of a Herschel-Bulkley fluid in a channel in the presence of magnetic field of low intensity. *International Journal of Computational Intelligence Research and Applications, 1, 71–81, 2007.*

14. P Chaturani, R P Samy, A study of non-Newtonian aspects of blood flow through stenosed arteries and its applications in arterial diseases. Biorheology, Vol. 22(6), pp. 521–531, 1985.
15. K. Vajravelu, S. Sreenadh, V. Ramesh Babu, Peristaltic transport of a Herschel-Bulkley fluid in an inclined tube. *International Journal of Non-linear Mechanics, Vol. 40(1), pp. 83–90, 2005.*
16. T. Young, Philos. T. R. Soc. Lond. 98, 164, 1808.
17. S I Rubinow, J B Keller, Flow of a viscous fluid through an elastic tube with applications to blood flow. *J Theor Biol., 35(2), 299–313, 1972.*
18. Mazumdar, N.J. Biofluid Mechanics, World Scientific, Chapter.5, Singapore. 1992.
19. A.C. Burton, On the physical equilibrium of small blood vessels. *Am J Physiol, 1951.*
20. D. L. Fry, Comput. Biomed. Res. 2, 111, 1968.
21. G. A. Brecher, Mechanism of venous flow under different degrees of aspiration. *Am J Physiol., 169(2), 423–433, 1952.*
22. Simon Rodbard and Walter Stone, Pressor Mechanisms induced by intracranial Compression. *Circulation, 12(5), 883–890, 1955.*
23. G. A. Brecher, Experimental evidence of ventricular diastolic suction. *Circulation Research, 4(5), 513–518, 1956.*
24. J. Banister, R. W. Torrance, The Effects of the Tracheal Pressure upon flow: Pressure relations in the Vascular bed of isolated Lungs. *Quarterly Journal of Experimental Physiology and Cognate Medical Sciences, 45(4), 352–367, 1960.*
25. S. Permutt, B. Bromberger-Barnea, H. N. Bane, Alveolar Pressure, Pulmonary Venous Pressure, and the Vascular Waterfall. *Research Articles, 19(4), 239–260, 1962.*
26. F. P. Knowlton, E. H. Starling, The influence of variations in temperature and blood-pressure on the performance of the isolated mammalian heart. *The journal of Physiology, 44(3), 206–219, 1912.*
27. K. Vajravelu, K. V. Prasad, Jinho Lee, Changhoon Lee, I. Pop, Robert A. Van GorderConvective heat transfer in the flow of viscous Ag-water and Cu-water nanofluids over a stretching surface. *International Journal of Thermal Sciences, 50(5), 843–851, 2011.*
28. D. Takagi and N. J. Balmforth, Peristaltic pumping of viscous fluid in an elastic tube. *J. Fluid Mech., vol. 672, pp. 196–218, 2011.*
29. M. Hasanuzzaman, k. Nahar, Md. Mahabub Alam, Rajib Roychowdhury and Masayuki Fujita, Physiological, Biochemical, and Molecular Mechanisms of Heat Stress Tolerance in Plants. *International Journal of Molecular Sciences, 14(5), 9643–9684, 2013.*
30. Taha Sochi, The flow of Newtonian and power law fluids in elastic tubes. *International Journal of Non-linear mechanics, Vol. 67, pp. 245–250, 2014.*
31. K. Maruthi Prasad, G. Radhakrishnamacharya, Effect of multiple stenoses on Herschel–Bulkley fluid through a tube with non-uniform cross-section. *International e-journal of engineering mathematics, Theory and Application, 1, 69–76, 2007.*
32. K. Vajravelu, S. Sreenadh, P. Devaki, K. V. Prasad, Mathematical model for a Herschel-Bulkley fluid flow in an elastic tube, *Central European Journal of Physics, 9(5), 1357–1365, 2011.*
33. R. B. Bird, W. E. Stewart, E. N. Lightfoot, Transport Phenomena, *2nd ed. (Wiley, New York, 2007), 11.*

Note: All the figures in this chapter were made by the authors.

Impending Inquisitions in Humanities and Sciences (ICIIHS-2022) – Dr. Mohan Varkolu et al. (eds)
© 2024 Taylor & Francis Group, London, ISBN 978-1-032-78829-6

On Common Coupled Fixed Point for ψ-Contractive in Partially Ordered Bipolar-Metric Spaces Using Compatible Mappings

53

G. N. V. Kishore[1]
Department of Engineering Mathematics and Humanities,
SRKR Engineering College, Bhimavaram, Andhra Pradesh, India.

B. V. Appa Rao[2]
Department of Engineering Mathematics, Koneru Lakshmaiah Education Foundation,
Vaddeswaram, Guntur, Andhra Pradesh, India

B. Srinuvasa Rao[3]
Department of Mathematics, Dr. B. R. Ambedkar University,
Srikakulam, Andhra Pradesh, India

Y. Phani*
Department of Mathematics, Shri Vishnu Engineering College for Women(A),
Vishnupur, Bhimavaram, AP, India

G. Suri Babu[4]
WET Centre, Department of Civil Engineering, SRKR Engineering College,
Bhimavaram, Andhra Pradesh, India

ABSTRACT In this paper, we proved some coupled fixed point theorems for Jungck type mappings by using ψ-contraction condition for covariant mappings in bipolar metric spaces. Also we give some examples which supports our results.

KEYWORDS Covariant mapping, Jungck type mappings, Coupled fixed point, Bipolar metric spaces

1. Introduction and Preliminaries

Very recently, in 2016 Mutlu and G¨*u*rdal [21] introduced the notion of Bipolar metric spaces, which is one of generalizations metric spaces. Also they investigated some fixed point and coupled fixed point results on this space,see ([21],[22]).

In this paper, we proved some coupled fixed point theorems for multivalued maps . Also we provide examples, which supports our main results. First we recall some basic definitions and results.

Definition 1.1. ([21]) *Let A and B be a two non-empty sets. Suppose that $d : A \times B \to [0, +\infty)$ be a mapping satisfying the following properties:*

(B_0) *If $d(a,b) = 0$ then $a = b$ for all $(a,b) \in A \times B$,*

*Correspondence Author: phaniyedlapalli23@gmail.com
[1]kishore.apr2@gmail.com, gnvkishore@srkrec.ac.in, [2]bvardr2010@kluniversity.in, [3]srinivasabagathi@gmail.com, [4]gsuribabu8@gmail.com
2000 Mathematics Subject Classification. 54H25, 47H10, 54E50.

DOI: 10.1201/9781003489436-53

(B_1) If $a = b$ then $d(a,b) = 0$, for all $(a,b) \in A \times B$,
(B_2) If $d(a,b) = d(b,a)$, for all $a,b \in A \cap B$.
(B_3) If $d(a_1,b_2) \leq d(a_1,b_1) + d(a_2,b_1) + d(a_2,b_2)$ for all $a_1,a_2 \in A$, $b_1,b_2 \in B$.

Then the mapping d is called a Bipolar-metric on the pair (A,B) and the triple (A,B,d) is called a Bipolar-metric space.

Definition 1.2. ([21]) *Assume (A_1, B_1) and (A_2, B_2) as two pairs of sets and a function as $F : A_1 \cup B_1 \rightrightarrows A_2 \cup B_2$ is said to be a covariant map. If $F(A_1) \subseteq A_2$ and $F(B_1) \subseteq B_2$, and denote this with $F : (A_1,B_1) \rightrightarrows (A_2,B_2)$. And the mapping $F : A_1 \cup B_1 \rightrightarrows A_2 \cup B_2$ is said to be a contravariant map. If $F(A_1) \subseteq B_2$ and $F(B_1) \subseteq A_2$, and write $F : (A_1,B_1) \leftrightarrows (A_2,B_2)$. In particular, if d_1 and d_2 are bipolar metric on (A_1,B_1) and (A_2,B_2), respectively, we some time use the notation $F : (A_1,B_1,d_1) \rightrightarrows (A_2,B_2,d_2)$ and $F : (A_1,B_1,d_1) \leftrightarrows (A_2,B_2,d_2)$.*

Definition 1.3. ([21]) *Assume (A,B,d) be a bipolar metric space. A point $v \in A \cup B$ is termed as a left point if $v \in A$, a right point if $v \in B$ and a central point if both. Similarly, a sequence $\{a_n\}$ on the set A and a sequence $\{b_n\}$ on the set B are called a left and right sequence respectively. In a bipolar metric space, sequence is the simple term for a left or right sequence. A sequence $\{v_n\}$ as considered convergent to a point v, if and only if $\{v_n\}$ is a left sequence, v is a right point and $\lim_{n\to+\infty} d(v_n, v) = 0$; or $\{v_n\}$ is a right sequence, v is a left point and $\lim_{n\to+\infty} d(v, v_n) = 0$. A bisequence $(\{a_n\},\{b_n\})$ on (A, B, d) is sequence on the set $A \times B$. If the sequence $\{a_n\}$ and $\{b_n\}$ are convergent, then the bisequence $(\{a_n\},\{b_n\})$ is said to be convergent. $(\{a_n\},\{b_n\})$ is Cauchy sequence, if $\lim_{n,m\to+\infty} d(a_n, b_m) = 0$. In a bipolar metric space, every convergent Cauchy bisequence is biconvergent. A bipolar metric space is called complete, if every Cauchy bisequence is convergent, hence biconvergent.*

2. Main Results

Theorem 2.1. *Let $(A, B, \leq)$ be a partially ordered set and there exist a bipolar metric d on (A, B) such that (A, B, d) is complete bipolar metric space. Let $F : (A^2, B^2) \rightrightarrows (A, B)$ and $g : (A,B) \rightrightarrows (A, B)$ be two covariant maps. Suppose F has g-mixed monotone property and there exists $\psi \in \Psi$ and*

$$\alpha : (A^2 \cup B^2) \times (A^2 \cup B^2) \to [0. + \infty)$$

$$\alpha\left((g(a), g(b)), (g(p), g(q))\right) d\left(F(a, b), F(p, q)\right) \leq \psi\left(\frac{d(g(a), g(p)) + d(g(b), g(q))}{2}\right) \tag{1}$$

for all a, $b \in A$ and p, $q \in B$ with $g(a) \geq g(p)$ and $g(b) \leq g(q)$. Suppose also that

(2.1.1) *F and g are α- admissible*

(2.1.2) *There exist $a_0,b_0 \in A \cup B$ such that*

$\alpha((g(a_0), g(b_0)),(F(a_0,b_0), F(b_0, a_0))) \geq 1$ *and* $\alpha((g(b_0), g(a_0)), (F(b_0, a_0), F(a_0, b_0))) \geq 1$

(2.1.3) *$F(A^2 \cup B^2) \subseteq g(A \cup B)$ and g is continuous.*

(2.1.4) *$\{F,g\}$ is compatible in $A \cup B$.*

If there exist a_0, $b_0 \in A \cup B$ such that $g(a_0) \leq F(a_0, b_0)$ and $g(b_0) \geq F(b_0, a_0)$. Then F and g has coupled fixed point that is there exist $a,b \in A \cup B$ such that

$F(a, b) = g(a) = a$ and $F(b, a) = g(b) = b$

Proof. Let a_0, $b_0 \in A$ and p_0, $q_0 \in B$ be such that

$\alpha((g(a_0), g(b_0)), (F(a_0, b_0), F(b_0, a_0))) \geq 1$, $\alpha((g(b_0), g(a_0)), (F(b_0, a_0), F(a_0, b_0))) \geq 1$ and

$\alpha((g(p_0), g(q_0)), (F(p_0, q_0), F(q_0, p_0))) \geq 1$, $\alpha((g(q_0), g(p_0)), (F(q_0, p_0), F(p_0, q_0))) \geq 1$ and

$g(a_0) \leq F(a_0, b_0) = g(a_1)$, $g(b_0) \geq F(b_0, a_0) = g(b_1)$ and $g(p_0) \leq F(p_0, q_0) = g(p_1)$, $g(q_0) \geq F(q_0, p_0) = g(q_1)$.

Let $a_2,b_2 \in A$ and $p_2,q_2 \in B$ be such that $F(a_1, b_1) = g(a_2)$, $F(b_1,a_1) = g(b_2)$ and $F(p_1, q_1) = g(p_2)$, $F(q_1, p_1) = g(q_2)$

Continuing this process, we can construct two bisequences $(\{a_n\},\{p_n\})$ and $(\{b_n\},\{q_n\})$ in (A,B) as follows, for all $n \geq 0$

$F(a_n, b_n) = g(a_{n+1})$, $F(b_n, a_n) = g(b_{n+1})$ and $F(p_n, q_n) = g(p_{n+1})$, $F(q_n, p_n) = g(q_{n+1})$

Now we will show that $g(a_n) \leq g(a_{n+1})$, $g(b_n) \geq g(b_{n+1})$ and $g(p_n) \leq g(p_{n+1})$, $g(q_n) \geq g(q_{n+1})$ for all $n \geq 0$ (2)

For $n = 0$, since, $g(a_0) \leq F(a_0, b_0) = g(a_1)$, $g(b_0) \geq F(b_0, a_0) = g(b_1)$ and $g(p_0) \leq F(p_0, q_0) = g(p_1)$, $g(q_0) \geq F(q_0, p_0) = g(q_1)$.
That is we have $g(a_0) \leq g(a_1)$, $g(b_0) \geq g(b_1)$ and $g(p_0) \leq g(p_1)$, $g(q_0) \geq g(q_1)$. Thus (2) holds for $n = 0$.
Now suppose that (2) holds for some fixed $n \geq 0$.
Then, since, $g(a_n) \leq g(a_{n+1})$, $g(b_n) \geq g(b_{n+1})$ and $g(p_n) \leq g(p_{n+1})$, $g(q_n) \geq g(q_{n+1})$
Therefore, by g-mixed monotone property of F, we have

$$g(a_{n+2}) = F(a_{n+1}, b_{n+1}) \geq F(a_n, b_{n+1}) \geq F(a_n, b_n) = g(a_{n+1}),$$
$$g(b_{n+2}) = F(b_{n+1}, a_{n+1}) \leq F(b_n, a_{n+1}) \leq F(b_n, a_n) = g(b_{n+1})$$

and

$$g(p_{n+2}) = F(p_{n+1}, q_{n+1}) \geq F(p_n, q_{n+1}) \geq F(p_n, q_n) = g(p_{n+1}),$$
$$g(q_{n+2}) = F(q_{n+1}, p_{n+1}) \leq F(q_n, p_{n+1}) \leq F(q_n, p_n) = g(q_{n+1})$$

From above, we conclude that

$$g(a_{n+1}) \leq g(a_{n+2}),\ g(b_{n+1}) \geq g(b_{n+2}) \text{ and } g(p_{n+1}) \leq g(p_{n+2}),\ g(q_{n+1}) \geq g(q_{n+2})$$

Thus, by mathematical induction, we conclude that (2) holds for all $n \geq 0$. Since, F and g are α-admissible, we have

$$\alpha((g(a_0),g(b_0)),(g(a_1),g(b_1))) = \alpha((g(a_0),g(b_0)),(F(a_0,b_0),F(b_0,a_0))) \geq 1 \text{ implies}$$
$$\alpha((F(a_0,b_0),F(b_0,a_0)),(F(a_1,b_1),F(b_1,a_1))) = \alpha((g(a_1),g(b_1)),(g(a_2),g(b_2))) \geq 1$$

and

$$\alpha((g(p_0),g(q_0)),(g(p_1),g(q_1))) = \alpha((g(p_0),g(q_0)),(F(p_0,q_0),F(q_0,p_0))) \geq 1 \text{ implies}$$
$$\alpha((F(p_0,q_0),F(q_0,p_0)),(F(p_1,q_1),F(q_1,p_1))) = \alpha((g(p_1),g(q_1)),(g(p_2),g(q_2))) \geq 1$$

Thus by mathematical induction, we have

$$\alpha((g(a_n),g(b_n)),\ (g(a_{n+1}),\ g(b_{n+1}))) \geq 1,\ \alpha((g(p_n),\ g(q_n)),\ (g(p_{n+1}),\ g(q_{n+1}))) \geq 1 \tag{3}$$

Similarly, for all $n \in N$ we have

$$\alpha((g(b_n),g(a_n)),(g(b_{n+1}),g(a_{n+1}))) \geq 1, \alpha((g(q_n),g(p_n)),(g(q_{n+1}),g(p_{n+1}))) \geq 1 \tag{4}$$

From (1) and conditions (2.1.1) and (2.1.2) of hypothesis, we get

$$\begin{aligned} d(g(a_n), g(p_{n+1})) &= d\left(F(a_{n-1}, b_{n-1}), F(p_n, q_n)\right) \\ &\leq \alpha\left((g(a_{n-1}), g(b_{n-1})), (g(p_n), g(q_n))\right) d\left(F(a_{n-1}, b_{n-1}), F(p_n, q_n)\right) \\ &\leq \psi\left(\frac{d(g(a_{n-1}), g(p_n)) + d(g(b_{n-1}), g(q_n))}{2}\right) \end{aligned} \tag{5}$$

and

$$\begin{aligned} d(g(b_n), g(q_{n+1})) &= d\left(F(b_{n-1}, a_{n-1}), F(q_n, p_n)\right) \\ &\leq \alpha\left((g(b_{n-1}), g(a_{n-1})), (g(q_n), g(p_n))\right) d\left(F(b_{n-1}, a_{n-1}), F(q_n, p_n)\right) \\ &\leq \psi\left(\frac{d(g(b_{n-1}), g(q_n)) + d(g(a_{n-1}), g(p_n))}{2}\right) \end{aligned} \tag{6}$$

On adding (5) and (6), we get

$$\frac{d(g(a_n), g(p_{n+1})) + d(g(b_n), g(q_{n+1}))}{2} \leq \psi\left(\frac{d(g(a_{n-1}), g(p_n)) + d(g(b_{n-1}), g(q_n))}{2}\right) \tag{7}$$

Repeating the above process, we get

$$\frac{d(g(a_n), g(p_{n+1})) + d(g(b_n), g(q_{n+1}))}{2} \leq \psi^n\left(\frac{d(g(a_0), g(p_1)) + d(g(b_0), g(q_1))}{2}\right) \tag{8}$$

On the other hand

$$\begin{aligned} d(g(a_{n+1}), g(p_n)) &= d\left(F(a_n, b_n), F(p_{n-1}, q_{n-1})\right) \\ &\leq \alpha\left((g(a_n), g(b_n)), (g(p_{n-1}), g(q_{n-1}))\right) d\left(F(a_n, b_n), F(p_{n-1}, q_{n-1})\right) \\ &\leq \psi\left(\frac{d(g(a_n), g(p_{n-1})) + d(g(b_n), g(q_{n-1}))}{2}\right) \end{aligned} \quad (9)$$

and

$$\begin{aligned} d(g(b_{n+1}), g(q_n)) &= d\left(F(b_n, a_n), F(q_{n-1}, p_{n-1})\right) \\ &\leq \alpha\left((g(b_n), g(a_n)), (g(q_{n-1}), g(p_{n-1}))\right) d\left(F(b_n, a_n), F(q_{n-1}, p_{n-1})\right) \\ &\leq \psi\left(\frac{d(g(b_n), g(q_{n-1})) + d(g(a_n), g(p_{n-1}))}{2}\right) \end{aligned} \quad (10)$$

On adding (9) and (10), we get

$$\frac{d(g(a_{n+1}), g(p_n)) + d(g(b_{n+1}), g(q_n))}{2} \leq \psi\left(\frac{d(g(a_n), g(p_{n-1})) + d(g(b_n), g(q_{n-1}))}{2}\right) \quad (11)$$

Repeating the above process, we get

$$\frac{d(g(a_{n+1}), g(p_n)) + d(g(b_{n+1}), g(q_n))}{2} \leq \psi^n\left(\frac{d(g(a_1), g(p_0)) + d(g(b_1), g(q_0))}{2}\right) \quad (12)$$

Moreover,

$$\begin{aligned} d(g(a_n), g(p_n)) &= d\left(F(a_{n-1}, b_{n-1}), F(p_{n-1}, q_{n-1})\right) \\ &\leq \alpha\left((g(a_{n-1}), g(b_{n-1})), (g(p_{n-1}), g(q_{n-1}))\right) d\left(F(a_{n-1}, b_{n-1}), F(p_{n-1}, q_{n-1})\right) \\ &\leq \psi\left(\frac{d(g(a_{n-1}), g(p_{n-1})) + d(g(b_{n-1}), g(q_{n-1}))}{2}\right) \end{aligned} \quad (13)$$

and

$$\begin{aligned} d(g(b_n), g(q_n)) &= d\left(F(b_{n-1}, a_{n-1}), F(q_{n-1}, p_{n-1})\right) \\ &\leq \alpha\left((g(b_{n-1}), g(a_{n-1})), (g(q_{n-1}), g(p_{n-1}))\right) d\left(F(b_{n-1}, a_{n-1}), F(q_{n-1}, p_{n-1})\right) \\ &\leq \psi\left(\frac{d(g(b_{n-1}), g(q_{n-1})) + d(g(a_{n-1}), g(p_{n-1}))}{2}\right) \end{aligned} \quad (14)$$

On adding (13) and (14), we get

$$\frac{d(g(a_n), g(p_n)) + d(g(b_n), g(q_n))}{2} \leq \psi\left(\frac{d(g(a_{n-1}), g(p_{n-1})) + d(g(b_{n-1}), g(q_{n-1}))}{2}\right) \quad (15)$$

Repeating the above process, we get

$$\frac{d(g(a_n), g(p_n)) + d(g(b_n), g(q_n))}{2} \leq \psi^n\left(\frac{d(g(a_0), g(p_0)) + d(g(b_0), g(q_0))}{2}\right) \quad (16)$$

For an arbitrary $\epsilon > 0$, there exists $n(\epsilon) \in N$ such that

$$\sum_{n \geq n(\epsilon)} \psi^n\left(\frac{d(g(a_0), g(p_1)) + d(g(b_0), g(q_1))}{2}\right) + \sum_{n \geq n(\epsilon)} \psi^n\left(\frac{d(g(a_0), g(p_0)) + d(g(b_0), g(q_0))}{2}\right) < \frac{\epsilon}{2}$$

and

$$\sum_{n \geq n(\epsilon)} \psi^n\left(\frac{d(g(a_1), g(p_0)) + d(g(b_1), g(q_0))}{2}\right) + \sum_{n \geq n(\epsilon)} \psi^n\left(\frac{d(g(a_0), g(p_0)) + d(g(b_0), g(q_0))}{2}\right) < \frac{\epsilon}{2}$$

Using the property (B_4), for each $n;m \in N$ be such that $n(\epsilon) < n < m$. Then, from (8), (12) and (16), we have

$$\begin{aligned}
&\frac{d(g(a_n),g(p_m))+d(g(b_n),g(q_m))}{2}\\
\leq\ & \frac{(d(g(a_n),g(p_{n+1}))+d(g(b_n),g(q_{n+1})))}{2}+\frac{(d(g(a_{n+1}),g(p_{n+1}))+d(g(b_{n+1}),g(q_{n+1})))}{2}\\
&+\cdots+\frac{(d(g(a_{m-1}),g(p_{m-1}))+d(g(b_{m-1}),g(q_{m-1})))}{2}\\
&+\frac{(d(g(a_{m-1}),g(p_m))+d(g(b_{m-1}),g(q_m)))}{2}\\
\leq\ & \psi^n\left(\frac{d(g(a_0),g(p_1))+d(g(b_0),g(q_1))}{2}\right)+\psi^{n+1}\left(\frac{d(g(a_0),g(p_0))+d(g(b_0),g(q_0))}{2}\right)\\
&+\cdots+\psi^{m-1}\left(\frac{d(g(a_0),g(p_0))+d(g(b_0),g(q_0))}{2}\right)+\psi^{m-1}\left(\frac{d(g(a_0),g(p_1))+d(g(b_0),g(q_1))}{2}\right)\\
\leq\ & \sum_{k=n}^{m-1}\psi^k\left(\frac{d(g(a_0),g(p_1))+d(g(b_0),g(q_1))}{2}\right)+\sum_{k=n+1}^{m-1}\psi^k\left(\frac{d(g(a_0),g(p_0))+d(g(b_0),g(q_0))}{2}\right)\\
\leq\ & \sum_{n\geq n(\epsilon)}\psi^n\left(\frac{d(g(a_0),g(p_1))+d(g(b_0),g(q_1))}{2}\right)+\sum_{n\geq n(\epsilon)}\psi^n\left(\frac{d(g(a_0),g(p_0))+d(g(b_0),g(q_0))}{2}\right)\\
<\ & \frac{\epsilon}{2}
\end{aligned}\tag{17}$$

and

$$\begin{aligned}
&\frac{d(g(a_m),g(p_n))+d(g(b_m),g(q_n))}{2}\\
\leq\ & \frac{(d(g(a_m),g(p_{m-1}))+d(g(b_m),g(q_{m-1})))}{2}+\frac{(d(g(a_{m-1}),g(p_{m-1}))+d(g(b_{m-1}),g(q_{m-1})))}{2}\\
&+\cdots+\frac{(d(g(a_{n+1}),g(p_{n+1}))+d(g(b_{n+1}),g(q_{n+1})))}{2}\\
&+\frac{(d(g(a_{n+1}),g(p_n))+d(g(b_{n+1}),g(q_n)))}{2}\\
\leq\ & \psi^{m-1}\left(\frac{d(g(a_1),g(p_0))+d(g(b_1),g(q_0))}{2}\right)+\psi^{m-1}\left(\frac{d(g(a_0),g(p_0))+d(g(b_0),g(q_0))}{2}\right)\\
&+\cdots+\psi^{n+1}\left(\frac{d(g(a_0),g(p_0))+d(g(b_0),g(q_0))}{2}\right)+\psi^n\left(\frac{d(g(a_1),g(p_0))+d(g(b_1),g(q_0))}{2}\right)\\
\leq\ & \sum_{k=n}^{m-1}\psi^k\left(\frac{d(g(a_1),g(p_0))+d(g(b_1),g(q_0))}{2}\right)+\sum_{k=n+1}^{m-1}\psi^k\left(\frac{d(g(a_0),g(p_0))+d(g(b_0),g(q_0))}{2}\right)\\
\leq\ & \sum_{n\geq n(\epsilon)}\psi^n\left(\frac{d(g(a_1),g(p_0))+d(g(b_1),g(q_0))}{2}\right)+\sum_{n\geq n(\epsilon)}\psi^n\left(\frac{d(g(a_0),g(p_0))+d(g(b_0),g(q_0))}{2}\right)\\
<\ & \frac{\epsilon}{2}
\end{aligned}\tag{18}$$

From (17) and (18), we have

$$d(g(a_n), g(p_m)) + d(g(b_n), g(q_m)) < \epsilon,\ d(g(a_m), g(p_n)) + d(g(b_m), g(q_n)) < \epsilon,$$

Hence, $(g(a_n), g(p_n))$ and $(g(b_n), g(q_n))$ are Cauchy bisequences in (A, B, d). Since, (A, B, d) is complete, therefore, $(g(a_n), g(p_n))$ and $(g(b_n), g(q_n))$ are convergent in (A,B,d). There exists, $a,b \in A$ and $p,q \in B$ such that

$$\lim_{n\to\infty} F(a_n,b_n) = \lim_{n\to\infty} g(a_{n+1}) = p, \quad \lim_{n\to\infty} F(b_n,a_n) = \lim_{n\to\infty} g(b_{n+1}) = q$$

$$\lim_{n\to\infty} F(p_n,q_n) = \lim_{n\to\infty} g(p_{n+1}) = a, \quad \lim_{n\to\infty} F(q_n,p_n) = \lim_{n\to\infty} g(q_{n+1}) = b$$

Since g is continuous, we have

$$\lim_{n\to\infty} g\,(F(a_n,b_n)) = \lim_{n\to\infty} g^2(a_{n+1}) = g(p),\ \lim_{n\to\infty} g\,(F(b_n,a_n)) = \lim_{n\to\infty} g^2(b_{n+1}) = g(q)$$

$$\lim_{n\to\infty} g\,(F(p_n,q_n)) = \lim_{n\to\infty} g^2(p_{n+1}) = g(a),\ \lim_{n\to\infty} g\,(F(q_n,p_n)) = \lim_{n\to\infty} g^2(q_{n+1}) = g(b)$$

Since, $\{F, g\}$ is compatible mapping; therefore, we have

$$\lim_{n\to\infty} F\,(g(a_n),g(b_n)) = g(p),\ \lim_{n\to\infty} F\,(g(b_n),g(a_n)) = g(q) \tag{19}$$

and

$$\lim_{n\to\infty} F\,(g(p_n),\ g(q_n)) = g(a),\ \lim_{n\to\infty} F\,(g(q_n),\ g(p_n)) = g(b) \tag{20}$$

Now we will prove $g(a) = p$, $g(b) = q$ and $g(p) = a$, $g(q) = b$.

Now consider,

$$\begin{aligned} d(g^2(a_{n+1}), g(p_{n+1})) &= d(F(ga_n, gb_n), F(p_n, q_n)) \\ &\le \alpha\left((g^2(a_n), g^2(b_n)), (g(p_n), g(q_n))\right) d(F(ga_n, gb_n), F(p_n, q_n)) \\ &\le \psi\left(\frac{d(g^2(a_n), g(p_n)) + d(g^2(b_n), g(q_n))}{2}\right) \end{aligned} \tag{21}$$

and

$$\begin{aligned} d(g^2(b_{n+1}), g(q_{n+1})) &= d(F(gb_n, ga_n), F(q_n, p_n)) \\ &\le \alpha\left((g^2(b_n), g^2(a_n)), (g(q_n), g(p_n))\right) d(F(gb_n, ga_n), F(q_n, p_n)) \\ &\le \psi\left(\frac{d(g^2(b_n), g(q_n)) + d(g^2(a_n), g(p_n))}{2}\right) \end{aligned} \tag{22}$$

Combining (21) and (22)

$$\begin{aligned} \frac{d(g^2(a_{n+1}),g(p_{n+1}))+d(g^2(b_{n+1}),g(q_{n+1}))}{2} &\le \psi\left(\frac{d(g^2(a_n),g(p_n))+d(g^2(b_n),g(q_n))}{2}\right) \\ &< \frac{d(g^2(a_n),g(p_n))+d(g^2(b_n),g(q_n))}{2} \end{aligned}$$

Letting $n \to \infty$, we get

$$\frac{d(g(p), a) + d(g(q), b)}{2} < \frac{d(g(p), a) + d(g(q), b)}{2}$$

which yields that $d(g(p), a) + d(g(q), b) = 0$ implies $g(p) = a$, $g(q) = b$. Similarly, we can prove that $g(a) = p$, $g(b) = q$. Now we will show that $F(a,b) = p$, $F(b,a) = q$ and $F(p,q) = a, F(q,p) = b$ For all $n \ge 0$, by using (1) and property (B_4), since g is continuous, we have

$$\begin{aligned} d\left(F(a,b),p\right) &\le d\left(F(a,b), g^2(p_{n+1})\right) + d\left(g^2(a_{n+1}), g^2(p_{n+1})\right) + d\left(g^2(a_{n+1}), p\right) \\ &= d\left(F(a,b), F(g(p_n), g(q_n))\right) + d\left(g^2(a_{n+1}), g^2(p_{n+1})\right) + d\left(g^2(a_{n+1}), g(p)\right) \\ &\le \alpha\left((g(a), g(b)), (g^2(p_n), g^2(q_n))\right) d\left(F(a,b), F(g(p_n), g(q_n))\right) \\ &\quad + d\left(g^2(a_{n+1}), g^2(p_{n+1})\right) + d\left(g^2(a_{n+1}), g(p)\right) \\ &\le \psi\left(\frac{d(g(a),g^2(p_n))+d(g(b),g^2(q_n))}{2}\right) \\ &\quad + d\left(g^2(a_{n+1}), g^2(p_{n+1})\right) + d\left(g^2(a_{n+1}), g(p)\right) \to 0 \text{ as } n \to \infty. \end{aligned}$$

Therefore, $d(F(a,b),p) = 0$ implies $F(a,b) = p$. Similarly, we can show $F(b,a) = q$ and $F(p,q) = a, F(q,p) = b$.

Therefore,

$$F(a,b) = p = g(a),\ F(b,a) = q = g(b) \text{ and } F(p,q) = a = g(p),\ F(q,p) = b = g(q).$$

Hence, we have proved that F and g has coupled coincidence point.

On the other hand,

$$\begin{aligned} d(a,p) &= d\left(\lim_{n\to\infty} g(p_n), \lim_{n\to\infty} g(a_n)\right) \\ &= \lim_{n\to\infty} d(g(a_n), g(p_n)) = 0 \end{aligned}$$

and

$$\begin{aligned} d(b,q) &= d\left(\lim_{n\to\infty} g(q_n), \lim_{n\to\infty} g(b_n)\right) \\ &= \lim_{n\to\infty} d(g(b_n), g(q_n)) = 0 \end{aligned}$$

Therefore,

$a = p$ and $b = q$ and hence $F(a,b) = g(a) = a$ and $F(b,a) = g(b) = b$ Now we will prove the uniqueness, we begin by taking $(a^*,b^*) \in A^2 \cup B^2$ be another coupled fixed point of F and g. If $(a^*,b^*) \in A^2$, then we have

$$\begin{aligned} d(a^*,a) = d(g(a^*),g(a)) &= d(F(a^*,b^*),F(a,b)) \\ &\leq \alpha((g(a^*),g(b^*),(g(a),g(b)))\, d(F(a^*,b^*),F(a,b)) \\ &\leq \psi\left(\tfrac{d(g(a^*),g(a))+d(g(b^*),g(b))}{2}\right) \\ &< \tfrac{d(g(a^*),g(a))+d(g(b^*),g(b))}{2} \\ &< \tfrac{d(a^*,a)+d(b^*,b)}{2} \end{aligned}$$

and

$$\begin{aligned} d(b^*,b) = d(g(b^*),g(b)) &= d(F(b^*,a^*),F(b,a)) \\ &\leq \alpha((g(b^*),g(a^*),(g(b),g(a)))\, d(F(b^*,a^*),F(b,a)) \\ &\leq \psi\left(\tfrac{d(g(b^*),g(b))+d(g(a^*),g(a))}{2}\right) \\ &< \tfrac{d(g(b^*),g(b))+d(g(a^*),g(a))}{2} \\ &< \tfrac{d(b^*,b)+d(a^*,a)}{2} \end{aligned}$$

Therefore,

$$\frac{d(a^*,a)+d(b^*,b)}{2} < \frac{d(a^*,a)+d(b^*,b)}{2}$$

which yields that $d(a^*,a) + d(b^*,b) = 0$ implies $d(a^*,a) = 0, d(b^*,b) = 0$ that is $a = a^*$ and $b = b^*$. Hence (a,b) is unique coupled fixed point of F and g.

Theorem 2.2. *Let $(A, B, \leq)$ be a partially ordered set and there exist a bipolar metric d on (A, B) such that (A, B, d) is complete bipolar metric space. Let $F : (A^2, B^2) \Rightarrow (A, B)$ be a covariant map. Suppose F has mixed monotone property and there exists $\psi \in \Psi$ and $\alpha : (A^2 \cup B^2) \times (A^2 \cup B^2) \to [0. +\infty)$*

$$\alpha((a,b),(p,q))\, d(F(a,b),F(p,q)) \leq \psi\left(\frac{d(a,p)+d(b,q)}{2}\right) \tag{23}$$

for all $a,b \in A$ and $p,q \in B$ with $a \geq p$ and $b \leq q$. Suppose also that

(2.1.1) F is α-admissible

(2.1.2) There exist $a_0,b_0 \in A \cup B$ such that

$$\alpha((a_0,b_0)), (F(a_0,b_0), F(b_0,a_0))) \geq 1 \text{ and } \alpha((b_0,a_0), (F(b_0,a_0), F(a_0,b_0))) \geq 1$$

If there exist $a_0,b_0 \in A \cup B$ such that $a_0 \leq F(a_0,b_0)$ and $b_0 \geq F(b_0,a_0)$. Then F have coupled fixed point that is there exist $a,b \in A \cup B$ such that $F(a,b) = a$ and $F(b,a) = b$

Example 2.3. *Let $A = \{U_m(R)/U_m(R)$ is upper triangular matrices over $R\}$ and $B = \{L_m(R)/L_m(R)$ is lower triangular matrices over $R\}$ with the bipolar metric $d(P,Q) = \sum_{i,j=1}^{m} |p_{ij} - q_{ij}|$, for all $P = (p_{ij})_{m\times m} \in U_m(R)$ and*

$Q = (q_{ij})_{m\times m} \in L_m(R)$. *On the set (A, B), consider the following relation:*

$$(P, Q) \in A^2 \cup B^2, P \preceq Q \Leftrightarrow p_{ij} \leq q_{ij}$$

where $\leq$ *is usual ordering. Then clearly, (A,B,d) is a complete bipolar metric space and (A, B,* $\preceq$*) is a partially ordered set. Let* $F : (A^2, B^2) \rightrightarrows (A, B)$ *be defined as*

$$F(P,Q) = \left(\frac{p_{ij} + q_{ij}}{5}\right)_{m\times m} + \frac{2}{5}(I_{ij})_{m\times m} \; for \; all \; (P = (p_{ij})_{m\times m}, Q = (q_{ij})_{m\times m}) \in A^2 \cup B^2$$

and $g : A \cup B \to A \cup B$ *by* $g(P) = (p_{ij})_{m\times m}$ *and let* $\psi : [0, 1) \to [0, 1)$ *by* $\psi(t) = \frac{2t}{5}$ *for* $t \in [0,1)$

Let the bisequence (P_n, Q_n) *in (A, B) such that* $\lim_{n\to\infty} F(P_n, Q_n) = \lim_{n\to\infty} g(P_n) = X$

and $\lim_{n\to\infty} F(Q_n, P_n) = \lim_{n\to\infty} g(Q_n) = Y$. *Then obviously,* $X = Y = \frac{2}{5}(I_{ij})_{m\times m}$

Now for all $n \geq 0$, $g(P_n) = (p_{nij})_{m\times m}$ *and* $g(Q_n) = (q_{nij})_{m\times m}$ *and*

$$F(P_n, Q_n) = \left(\frac{p_{n_{ij}} + q_{n_{ij}}}{5}\right)_{m\times m} + \frac{2}{5}(I_{ij})_{m\times m}, \; F(Q_n, P_n) = \left(\frac{q_{n_{ij}} + p_{n_{ij}}}{5}\right)_{m\times m} + \frac{2}{5}(I_{ij})_{m\times m}$$

Then its follows that $\lim_{n\to\infty} d(g(F(P_n,Q_n)), F(g(P_n),g(Q_n))) = 0$, $\lim_{n\to\infty} d(g(F(Q_n,P_n)), F(g(Q_n),g(P_n))) = 0$.

Hence, the mappings F and g are compatible in $A \cup B$.

Consider the mapping $\alpha : (A^2 \cup B^2) \times (A^2 \cup B^2) \to [0, \infty)$ *be such that*

$$\alpha((g(P), g(Q)), (g(R), g(S))) = \begin{cases} (I_{ij})_{m\times m} \; if \; P \succeq Q, R \preceq S \\ (O_{ij})_{m\times m} \; Otherwise \end{cases}$$

Then obviously, F has the g-mixed monotone property, also there exist $P_0 = (O_{ij})_{m\times m}$ *and* $Q_0 = (I_{ij})_{m\times m}$ *such that*

$$F((O_{ij})_{m\times m}, (I_{ij})_{m\times m}) = (\frac{O_{ij} + I_{ij}}{5})_{m\times m} + \frac{2}{5}(I_{ij})_{m\times m} \succeq (O_{ij})_{m\times m}$$

and

$$F((I_{ij})_{m\times m}, (O_{ij})_{m\times m}) = (\frac{O_{ij} + I_{ij}}{5})_{m\times m} + \frac{2}{5}(I_{ij})_{m\times m} \preceq (I_{ij})_{m\times m}$$

Taking $(P = (p_{ij})_m \times_m, Q = (q_{ij})_m \times_m), (R = (r_{ij})_m \times_m, S = (s_{ij})_m \times_m) \in A^2 \cup B^2$ *with* $P \geq R$, $Q \preceq S$ *that is* $p_{ij} \geq r_{ij}$, $q_{ij} \leq s_{ij}$, *we have*

$$\begin{aligned} d(F(P,Q), F(R,S)) &= d(\tfrac{p_{ij}+q_{ij}}{5} + \tfrac{2}{5}(I_{ij}), \tfrac{r_{ij}+s_{ij}}{5} + \tfrac{2}{5}(I_{ij})) = \tfrac{1}{5}\sum_{i,j=1}^{m} |(p_{ij} + q_{ij}) - (r_{ij} + s_{ij})| \\ &\leq \tfrac{1}{5}\left(\sum_{i,j=1}^{m} |p_{ij} - r_{ij}| + \sum_{i,j=1}^{m} |q_{ij} - s_{ij}|\right) \\ &\leq \tfrac{1}{5}(d(g(P), g(R)) + d(g(Q), g(S))) \\ &\leq \tfrac{2}{5}\left(\tfrac{d(g(P),g(R)) + d(g(Q),g(S))}{2}\right) \end{aligned}$$

Therefore, all the conditions of Theorem (2.1) holds for $\frac{2t}{5}$ *for all* $t > 0$ *and we see that* $F(A^2 \cup B^2) \subseteq g(A \cup B)$ *and also verified that F has g-monotone property and* $((\frac{2}{5}I_{ij})_{m\times m}, (\frac{2}{5}I_{ij})_{m\times m})$ *is the coupled fixed point of F and g.*

REFERENCES

1. S. Banach, *Sur les operations dans les ensembles abstraits etleur applications aux equations integrales*, Fund. Math. 3..133-181(1922).
2. M. Turinici, *Abstract Comparison Principles and Multivariable Gronwall-Bellman Inequalities*, Journal of Mathematical Analysis and Applications, 117, 100-127(1986). http://dx.doi.org/10.1016/0022-247X(86)90251-9.
3. Ran, A.C.M. and Reurings, M.C.B, *A Fixed Point Theorem in Partially Ordered Sets and Some Applications to Matrix Equations*, Proceedings of the American Mathematical Society, 132, 1435-1443 (2004). http://dx.doi.org/10.1090/S0002-9939-03-07220-4.
4. Nieto, J.J. and Rodr´iguez-L´opez, *Contractive Mapping Theorems in Partially Ordered Sets and Applications to Ordinary Differential Equations*, Order, 22, 223-239(2005). http://dx.doi.org/10.1007/s11083-005-9018-5.

5. Nieto, J.J. and Rodr´iguez-L´opez, *Existence and Uniqueness of Fixed Point in Partially Ordered Sets and Applications to Ordinary Differential Equations*, Acta Mathematica Sinica, English Series, 23, 2205(2007). http://dx.doi.org/10.1007/s10114-005-0769-0.
6. T. G. Bhaskar, V. Lakshmikantham, *Fixed point theorems in partially ordered metric spaces and applications*, Nonlinear Anal. MA65(2006)1379-1393 .http://dx.doi.org/ 10.1016/j.na.2005.10.017.
7. D. Guo, and V. Lakshmikantham, *Coupled Fixed Points of Nonlinear Operators with Applications*, Nonlinear Analysis, 11, 623-632(1987). http://dx.doi.org/10.1016/0362546X(87)90077-0.
8. V. Lakshmikantam, Ljubomir *C´iri´c*, *Coupled fixed point theorems for nonlinear contractions in partially oredered metric spaces*, Nonlinear Analysis 70(2009)43414349 .http://dx.doi.org/ 10.1016/j.na.2008.09.020.
9. B. S.Choudhury, and A. Kundu, *A Coupled Coincidence Point Result in Partially Orderedmetric Spaces for Compatible Mappings*, Nonlinear Analysis, 73, 2(2010). http://dx.doi.org/10.1016/j.na.2010.06.025.
10. B. Samet, C. Vetro, and P. Vetro, *Fixed Point Theorems for $\alpha - \psi$*-Contractive Type Mappings, Nonlinear Analysis, 75, 2154-2165(2012). http://dx.doi.org/10.1016/j.na.2011.10.014.
11. W.Shatanawi, B.Samet, and M.Abbas,*Coupled Fixed Point Theorems for Mixed Monotone Mappings in Ordered Partial Metric Spaces*, Mathematical and Computer Modelling, 55 , 680-687(2012).
12. T. Abdeljawad, *Coupled Fixed Point Theorems for Partially Contractive Type Mappings*, Fixed Point Theory and Applications, 2012, 148(2012).
13. H. Aydi, B.Samet, and C.Vetro, *Coupled Fixed Point Results in Cone Metric Spaces for WCompatible Mappings*, Fixed Point Theory and Applications, 2011, 27(2011).
14. B.S.Choudhury, K. Das, and P. Das, *Coupled Coincidence Point Results for Compatible Mappings in Partially Ordered Fuzzy Metric Spaces*, Fuzzy Sets and Systems, 222, 84-97(2012).
15. A. Amini-Harandi, *Coupled and Tripled Fixed Point Theory in Partially Ordered Metric Spaces with Application to Initial Value Problem*, Mathematical and Computer Modelling, 57, 2343-2348(2013).
16. G.Jungck, *Compatible Mappings and Common Fixed Points*, International Journal of Mathematics and Mathematical Sciences, 9, 771-779(1996).
17. E.Karapinar, and B.Samet, *Generalized α–ψ*-Contractive Type Mappings and Related Fixed Point Theorems with Applications, Abstract and Applied Analysis, 2012, Article ID: 793486.
18. N.V.Luong, and N.X.Thuan, *Coupled Fixed Points in Partially Ordered Metric Spaces and Application*, Nonlinear Analysis: Theory, Methods and Applications, 74, 983-992(2011). http://dx.doi.org/10.1016/j.na.2010.09.055.
19. H.K.Nashine, B.Samet, and C.Vetro, *Coupled Coincidence Points for Compatible Mappings Satisfying Mixed Monotone Property*, The Journal of Nonlinear Science and Applications, 5 , 104-114(2012).
20. H.K. Nashine, Z. Kadelburg, and S.RadenoviC´, *Coupled Common Fixed Point Theorems for W^*-Compatible Mappings in Ordered Cone Metric Spaces*, Applied Mathematics and Computation, 218, 5422-5432(2012).http: //dx.doi.org/10.1016/j.amc. 2011.11.029.
21. Ali Mutlu, Utku G¨urdal, *Bipolar metric spaces and some fixed point theorems*, J. Nonlinear Sci. Appl. 9(9), 5362-5373, 2016.
22. Ali Mutlu, K¨ubra O¨zkan, Utku G¨urdal, *Coupled fixed point theorems on bipolar metric spaces*, European journal of pure and applied mathematics. Vol. 10, No. 4, 2017, 655-667.
23. M. Mursaleen, S.A.Mohiuddine, and R.P.Agarwal, *Coupled Fixed Point Theorems for α–ϕContractive Type Mappings in Partially Ordered Metric Spaces*, Fixed Point Theory and Applications, 2012, 228.
24. Preeti, Sanjay Kumar, *Coupled Fixed Point for* (α,ψ)*-Contractive in Partially Ordered Metric Spaces Using Compatible Mappings*,Applied Mathematics, 2015, 6 , 1380-1388 Published Online July 2015 in SciRes. http://www.scirp.org/journal/am, http://dx.doi.org/10.4236/am.2015.68130.

Impending Inquisitions in Humanities and Sciences (ICIIHS-2022) – Dr. Mohan Varkolu et al. (eds)

On General Solutions of Polytropes of Lane-Emden Equation in (r, P) Variables: A Review

54

Nagaraju Vuppala*
Research Scholar, Dept. of Mathematics, JNTU, Hyderabad
Asst. Professor, Dept. of Mathematics, CMRCET (Autonomous), Hyderabad

M. N. Rajashekar
Professor, Dept. of Mathematics, JNTU, Hyderabad

ABSTRACT Nonlinear singular differential equations are challenging to solve, and they frequently have no accurate solution. It is only logical to search for the presence of the analytical solution and numerical solution since the exact solution does not exist. In this work we focused on both sides of nonlinear singular boundary value problems (SBVPs) and discussed various analytical and numerical approaches that have been developed to handle a class of nonlinear singular differential equations. Solution of Lane – Emden equation have generally been considered in (ξ, θ) variables. It is shown that r and P are the only suitable variables for the study of the structure of polytrophic configuration near the origin in the form of a series for radius $r < 1$ of the Lane-Emden equation.

KEYWORDS Boundary value problem, Nonlinear singular differential equations, Lane-Emden equation, Polytropes

1. Introduction

The Poisson's equation for the gravitational potential of a self-gravitating, spherically symmetric, and polytropic fluid at hydrostatic equilibrium is known as the Lane-Emden equation in astrophysics. Analysis of the diffusive movement and chemical reaction of species inside a porous catalyst particle is one of the key areas in which the Lane-Emden equation is applied. The most practicable catalyst geometry was chosen by Aris and Metha & Aris, who were among the first to address this problem as a boundary value problem by assuming power-law kinetics for a single chemical reaction occurring inside a spherical catalyst particle.

2. Mathematical Formulation

In this paper we proceed with the assumption that in steady state of gravitational equilibrium and the density distribution such that the mean density p(r), interior to a given point r inside the star does not increase outward from the centre. Since we have a symmetrical distribution of a matter, the total pressure P, the density p and M(r), the mass enclosed inside r, are given by

$$\mu(r)=\int_0^r \rho.4\pi r^2, d\mu(r)=\rho.4\pi r^2 dr \tag{1}$$

*Corresponding aurhor: nagrajuvuppala@gmail.com

DOI: 10.1201/9781003489436-54

$$\rho(r) = \frac{\mu(r)}{\frac{4}{3}\pi r^3} \tag{2}$$

Consider an infinitesimal cylinder at a distance r from the centre, the height dr and of unit cross sectional at right angles. Let P be the pressure at r and let the increment in P as we go from r to r+ be dP

Therefore the force – dP acting on the element must be counter acted by gravitational attractive force between M(r) and dr.

Thus, for equilibrium, we should have

$$\frac{dp}{dr} = -\frac{GM(r)}{r^2}\rho \tag{3}$$

From (3) and (1) we have our fundamental equation of equilibrium

$$\frac{1}{r^2}\frac{d}{dr}\left(\frac{r^2}{\rho}\frac{dp}{dr}\right) = -4\pi G\rho \tag{4}$$

Here we study the equilibrium configuration in which a relation of the kind

$$\mathrm{P} = \mathrm{k}\,\rho^{1+\frac{1}{n}} \tag{5}$$

Equation (5) holds, where k and n are constants.

According to our assumption, we can write

$$\rho = \lambda\theta^n,\ p = k\lambda^{1+\frac{1}{n}}\theta^{n+1} \tag{6}$$

Where λ is for the present an arbitrary constant. From (6) and (4) we have

$$\left[\frac{(n+1)k}{4\pi G}\lambda^{\frac{1}{n}-1}\right]\frac{1}{r^2}\frac{d}{dr}\left(r^2\frac{d\theta}{dr}\right) = -\theta^n \tag{7}$$

With the introduction of the dimensionless variable ξ, this is defined by

$$r = \alpha\xi\ ,\alpha = \left[\frac{(n+1)k}{4\pi G}\lambda^{\frac{1}{n}-1}\right]^{\frac{1}{2}} \tag{8}$$

Equation (7) becomes

$$\frac{1}{\xi^2}\frac{d}{d\xi}\left(\xi^2\frac{d\theta}{d\xi}\right) = -\theta^n \tag{9}$$

Equation (9) is the Lane-Emden equation of index n. This equation governs the density distribution in any region where the relation (6) is valid. The region of validity of (6) need not of course extend throughout the entire mass. We shall consider first only complete polytrope, that is, equilibrium configurations in which a relation of the kind (6) is valid for the entire mass. In that case we can choose λ to be equal to the central density ρ_c, we must seek a solution of (9) which satisfies the boundary conditions

$$\theta = 1, \frac{d\theta}{d\xi} = 0 \text{ at } \xi = 0 \tag{10}$$

Equation (4) which governs the Equilibrium configuration contains two dependent variables P and ρ. A pressure density relation therefore is necessary to eliminate one dependent variable from the equations as given in (6)

$P = k\rho^{1+\frac{1}{n}},\ \rho = \lambda\theta^n$ The resulting equation (9) can be put in a suitable form.

From equations (4) and (9), it is clear that the equation (9) can be relevant only for such values of n for which d P and ρ remain functions θ and $d\theta$, simultaneously, For n = 0 $\rho = \lambda \neq f(\theta)$ although dP remains a function of θ. for n = –1, dP is not a function of θ although ρ remains a function of θ. Thus we see that for n = 0 and –1 only other value of dP and ρ is a function of θ

and not both simultaneously. Moreover, transformations in (8) are not defined for n = 0 and n = −1. Thus we see that neither transformations connecting dependant variables P and ρ, nor transformations connecting independent variables r and ξ are relevant. Hence equation (9) must remain relevant for n = 0 and n = −1.

Further the pressure – density relation $k\rho^{1+\frac{1}{n}}$ for n = 0 and n = −1 gives ρ =1 and P = k respectively, i.e., for these two values of n there is no functional relation between P and ρ, but for n tending to zero and minus one, P and ρ still remain related . Hence n = 0 and n = −1 in $P = k\rho^{n+1}$, should be regarded as limiting cases and we should consider equations for n tending to zero and minus one and not for n = 0 and n = −1. Hence if we eliminate one dependent variable from (4) with the help of $P = \rho^{1+\frac{1}{n}}$, then the resulting equation can be significant only, if the process of elimination involves some mathematical operation over the pressure-density relation, because otherwise solutions for n=0 and n= -1 in the resulting equation will correspond to n = 0 and n = −1 in $P = \rho^{1+\frac{1}{n}}$ and not to n tending to zero and minus one. of course , even if the process of elimination of a dependent variable does not involve any mathematical operation over the pressure – density relation, the resulting equation shall relevant for further transformation. Elimination of ρ from (4) with the help of $P = k\rho^{1+\frac{1}{n}}$ leads to

$$\frac{1}{r^2}\frac{d}{dr}(r^2\rho^{-\frac{n}{n+1}}\frac{dP}{dr}) = -4\pi G k^{\frac{-2n}{n+1}}\rho^{\frac{n}{n+1}} \tag{11}$$

Equation (11) is the Lane – Emden equation in terms of P and r.

3. General Solutions of Polytropes near the Origin

Solution of Lane – Emden equation have generally been considered in (ξ, θ) variables. During 1966 Shambunath has shown that r and P are the only suitable variables for the study of the structure of polytropic configuration near the origin as the variables ξ and θ are not defined at the origin. In this paper author gives a general solution of (11) in the form of a series for r < 1.

Equations (11) can be written in the symbolic form as

$$P_2 - \frac{n}{n+1}\frac{P_1^2}{P} + \frac{2P_1}{r} = kP^{\frac{2n}{n+1}} \tag{12}$$

Where $k = -4\pi G k^{\frac{-2n}{n+1}}$

Let us assume a solution of (12) near the origin, i.e., r < 1 in the form of a series

$$P = P_0 + cr^2 + dr^4 + \tag{13}$$

Where r = 0 gives $P = P_0$ and $\frac{dP}{dr} = 0$ giving boundary conditions at the origin and consequently series contains only terms of even powers in r. By substituting in (13) and (14), we get

$$P\left(2c + 12d^3 + ...\right) - \frac{n}{n+1}\left(2cr + 4dr^3 + ...\right)^2 + \frac{2}{r}\left(2cr + 4dr^3 + ...\right) = k\left(P_0 + cr^2 + dr^4 + ...\right) \tag{14}$$

$$P = P_0 + \frac{k}{6}P_0^{\frac{2n}{n+1}}r^2 + \frac{k^2}{45}\frac{n}{n+1}P_0^{\frac{3n-1}{n+1}}r^4 + ... \tag{15}$$

Where K is negative, the terms in (A) are alternatively positive and negative. The expression (A) gives the general solution for all positive values of n including n = 0. For the verification of the general solution (A) of equation (12) we proceed to find the solution of (12) for n = 0,1,2,3,4 and 5 and verify the results derived therefrom.

3.1 Case 1: For n = 0

Equation (12) becomes. for n = 0

$$\frac{d}{dr}\left(r^2\frac{dP}{dr}\right) = r^2 k \tag{16}$$

By integrating (14) twice, we get

$$P = \frac{k}{6}r^2 - \frac{c_1}{r} + c_2 \tag{17}$$

c_1, c_2 are constants of integration.

Equation (14) has singularity at origin and $c_1 = 0$ we have

$$P = C_2 + \frac{k}{6}r^2 \tag{18}$$

Where $P = P_0$ for $r = 0$ at the origin, hence $C_2 = P_0$, thus we have

$$P = P_0 + \frac{k}{6}r^2 \tag{19}$$

Above result as given in (16) is also obtained on putting n = 0 in the general solution (15)

3.2 Case (ii) for n = 1

Equation (12) for n = 1 becomes

$$P_2 + \frac{2P_1}{r} - \frac{1}{2}\frac{P_1^2}{P} = kP \tag{20}$$

Since P = f taking the solution of the differential equation (19) by expansion as a Taylor's series,we have

$$P = P_0 + r\left(\frac{dP}{dr}\right)_{r=0} + \frac{r^2}{L^2}\left(\frac{d_2P}{dr^2}\right)_{r=0} + \frac{r^3}{L^3}\left(\frac{d_3P}{dr^3}\right)_{r=0} + \ldots\ldots \tag{21}$$

We proceed to determine the value of P_1, P_2, P_3 etc in (21)

As $r = 0$, $P_1 = 0$ and $\lim_{r\to 0}\frac{P_1}{r} \to P_2$

Substituting the values of P_1 and $\frac{P_0}{r}$ in (19) we get $P_2 = k\frac{P_0}{3}$

By successive differentiations of (19) and substituting of P_1, P_2 as $r \to 0$ we get $P_3 = 0, P_4 = \frac{4K^2}{15}P_0$ etc.,

Hence the first terms in Taylor's series (18) give

$$P = P_0\left(1 + \frac{k}{6}r^2 + \frac{k^2}{90}r^4\right) \tag{22}$$

Which is exactly the same as will be obtained from the general solution (15) on putting n = 1.

Similarly, solutions for other values of n can be obtained from the general solution (15)

3.3 Case (iii) n = 5

Equation (12) on putting n = 5 becomes

$$P_2 - \frac{5}{6}\frac{P_1^2}{P} + \frac{2P_1}{r} = KP^{\frac{5}{3}} \tag{23}$$

We proceed to determine the values of P_1, P_2 which as $r \to 0, P_1 = 0, P_2 = \frac{k}{3}P_0^{\frac{5}{3}}$

By differentiating (22) twice and substituting the values of P_1, P_2 etc. we get P_3 as $r \to 0, P_4 = \frac{4}{9}k^2 P_0^{\frac{7}{3}}$ Substituting the values in (20) we have

$$P = P_0 + \frac{k}{6} P_0^{\frac{5}{3}} r^2 + \frac{K^2}{54} P_0^{\frac{7}{3}} r^4 + \dots\dots\dots\dots \tag{24}$$

Again equation (23) will be the same as will be obtained from the general solution for n = 5.

We thus see that the equation

$$\frac{1}{r^2}\frac{d}{dr}\left(r^2 P^{\frac{-n}{n+1}} \frac{dP}{dr}\right) = kP^{\frac{n}{n+1}} \tag{25}$$

Governing the equilibrium of gaseous configuration, can be studied in the neighbourhood of the centre for all positive values of n, including $n = 0$. Thus by taking a sufficient number of terms in the series solution (12) we can calculate the values of P for $r < 1$ to any required degree of accuracy. By taking finite number of terms, the series solution (12) can be used to study the structure of polytrophic configuration for any value of n.

4. Conclusion

We have identified the integral manifold on which boundary value problem solutions must necessarily reside. This will be carried out in the appropriate blown-up and warped coordinates. We also offer our thoughts on the numerical implementation.

REFERENCES

1. Dunninger, D.R.; Kurtz, J.C. Existence of Solutions for Some Nonlinear Singular Boundary Value Problems. J. Math. Anal. Appl. 1986, 115, 396–405.
2. Lin, H.S. Oxygen Diffusion in a Spherical Cell with Nonlinear Oxygen Uptake Kinetics. J. Theor. Biol. 1976, 60, 449–457.
3. McElwain, D.L.S. A Re-examination of Oxygen Diffusion in a Spherical Cell with Michaelis—Menten Oxygen Uptake Kinetics. J. Theor. Biol. 1978, 71, 255–263.
4. Anderson, N.; Arthurs, A.M. Analytical Bounding Functions for Diffusion Problems with Michaelis–Menten Kinetics. Bull. Math. Biol. 1985, 47, 145–153.
5. Duggan, R.C.; Goodman, A.M. Pointwise Bounds for Nonlinear Heat Conduction Model of the Human Head. Bull. Math. Biol. 1986, 48, 229–236.
6. Flesch, U. The Distribution of Heat Sources in the Human Head: A Theoretical Consideration. J. Theor. Biol. 1975, 54, 285–287.
7. Gray, B.F. The Distribution of Heat Sources in the Human Head—Theoretical Considerations. J. Theor. Biol. 1980, 82, 473–476.
8. Anderson, N.; Arthurs, A.M. Complementary Extremum Principles for a Nonlinear Model of Heat Conduction in the Human Head. Bull. Math. Biol. 1981, 43, 341–346.
9. Keller, J.B. Electrohydrodynamics I. The Equilibrium of a Charged Gas in a Containor. J. Rational Mech. Anal. 1956, 5, 715–724.
10. Chamber, P.L. On the Solution of the Poisson-Boltzmann Equation with Application to the Theory of Thermal Explosions. J. Chem. Phys. 1952, 20, 1795–1797.
11. Kazutaka, N.; Toshiyuki, T.; Shin, S. A modified arrhenius equation. Chem. Phys. Lett. 1989, 160, 295–298.
12. Verma, A.K.; Tiwari, D. On Some Computational Aspects of Hermite wavelets on a Class of SBVPs Arising in Exothermic Reactions. arXiv 2019, arXiv:1911.00495.
13. Chandrashekhar, S. An Introduction to the Study of Stellar Structure; Dover: New York, NY, USA, 1939.
14. Escudero, C.; Hakl, R.; Peral, I.; Torres, P.J. Existence and nonexistence Results for a Singular Boundary Value Problem Arising in the Theory of Epitaxial Growth. Math. Meth. Appl. Sci. 2014, 37, 793—807.
15. Rachůnková, I.; Koch, O.; Pulverer, G.; Weinmxuxller, E. On a Singular Bundary Value Problem Arising in the Theory of Shallow Membrane Caps. J. Math. Anal. Appl. 2007, 332, 523–541.
16. Flockerzi, D.; Sundmacher, K. On Coupled Lane–Emden Equations Arising in Dusty Fluid Models. J. Phys. Conf. Ser. 2011, 268, 012006.
17. Agarwal, R.P.; Hodis, S.; O'Regan, D. 500 Examples and Problems of Applied Differential Equations; Springer: Berlin, Germany, 2019; p. 388.
18. Ciarlet, P.G.; Natterer, F.; Varga, R.S. Numerical Methods of Higher-Order Accuracy for Singular Nonlinear Boundary Value Problems. Numer. Math. 1970, 15, 87–99.
19. Jamet, P. On the Convergence of Finite-Difference Apprxoimations to One-Dimensional Singular Boundary Value Problems. Numer. Math. 1970, 14, 355–378.
19. Russell, R.D.; Shampine, L.F. Numerical Methods for Singular Boundary Value Problems. SIAM J. Numer. Anal. 1975, 12, 13–36.

20. Shampine, L.F. Boundary Value Problems for Ordinary Differential Equations. II. Patch Bases and Monotone Methods. SIAM J. Numer. Anal. 1969, 6, 414–431. Mathematics 2020, 8, 1045 42 of 50
21. Chawla, M.M.; Katti, C.P. Finite Difference Methods and Their Convergence for a Class of Singular Two Point Boundary Value Problems. Numer. Math. 1982, 39, 341–350.
22. Chawla, M.M. A Fourth Order Finite-Difference Method Based on Uniform Mesh for Singular Two-Point Boundary Value Problems. J. Comp. Appl. Math. 1987, 17, 359–364.
23. Chawla, M.M.; Katti, C.P. A Finite-Difference Method for a Class of Singular Two Point Boundary Value Problems. IMA J. Numer. Anal. 1984, 4, 457–466.
24. Chawla, M.M.; Katti, C.P. A Unifrom Mesh Finite Difference Method for a Class of Singular Two-Point Boundary Value Problems. SIAM J. Numer. Anal. 1985, 22, 561–565.
25. Aris R 1965 Introduction to the analysis of chemical reactors (Englewood Cliffs, N.J.: Prentice-Hall)
26. Metha B N and Aris A 1971 J. of Mathematical Analysis and Applications 36 611

Impending Inquisitions in Humanities and Sciences (ICIIHS-2022) – Dr. Mohan Varkolu et al. (eds)
© 2024 Taylor & Francis Group, London, ISBN 978-1-032-78829-6

Some Existence of Unique Fixed Solutions in Partially Ordered S_b-Metric Spaces

55

G. N. V. Kishore[1]
Department of Engineering Mathematics and Humanities,
SRKR Engineering College, Bhimavaram, Andhra Pradesh, India

B. V. Appa Rao[2]
Department of Engineering Mathematics, Koneru Lakshmaiah Education Foundation,
Vaddeswaram, Guntur, Andhra Pradesh, India

Y. Phani*
Department of Mathematics, Shri Vishnu Engineering College for Women(A),
Vishnupur, Bhimavaram, AP, India

G. Suri Babu[3]
WET Centre, Department of Civil Engineering, SRKR Engineering College,
Bhimavaram, Andhra Pradesh, India

ABSTRACT In this paper we give some fixed point theorems in partially ordered complete S_b-metric space by using generalized contractive conditions. We also furnish an example which supports our main result.

KEYWORDS S_b-metric space, w-Compatible pairs, S_b - Completeness.

1. Introduction

Recently Sedghi et al. [8] defined S_b-metric spaces using the concept of S-metric spaces. [9].

The aim of this paper is to prove some unique fixed point theorems for generalized contractive conditions in complete S_b-metric spaces. Throughout this paper R, R⁺ and N denote the set of all real numbers, non-negative real numbers and positive integers respectively.

First we recall some definitions, lemmas and examples.

Definition 1.1. ([9]) *Let X be a non-empty set. A S–metric on X is a function*

$S : X^3 \to [0,+\infty)$ *that satisfies the following conditions for each* $x,y,z,a \in X$,

$(S1) : 0 < S(x, y, z)$ *for all* $x, y, z \in X$ *with* $x \neq y \neq z$,

$(S2) : S(x, y, z) = 0 \Leftrightarrow x = y = z$,

*Correspondence Author: phaniyedlapalli23@gmail.com
[1]kishore.apr2@gmail.com, gnvkishore@srkrec.ac.in; [2]bvardr2010@kluniversity.in; [3]gsuribabu8@gmail.com
2000 Mathematics Subject Classification. 54H25, 47H10, 54E50.

DOI: 10.1201/9781003489436-55

$(S3)$: *$S(x, y, z) \leq S(x, x, a) + S(y, y, a) + S(z, z, a)$ for all $x, y, z, a \in X$.*

Then the pair (X,S) is called a S-metric space.

Definition 1.2. *([8]) Let X be a non-empty set and $b \geq 1$ be given real number. Suppose that a mapping $S_b : X^3 \to [0, \infty)$ be a function satisfying the following properties :*

(S_b1) $0 < S_b(x, y, z)$ for all $x, y, z \in X$ with $x \neq y \neq z$,

(S_b2) $S_b(x, y, z) = 0 \Leftrightarrow x = y = z$,

(S_b3) $S_b(x, y, z) \leq b(S_b(x, x, a) + S_b(y, y, a) + S_b(z, z, a))$ for all $x, y, z, a \in X$.

Then the function S_b is called a S_b-metric on X and the pair (X,S) is called a S_b-metric space.

Remark 1.3. *([8])It should be noted that, the class of S_b-metric spaces is effectively larger than that of S-metric spaces. Indeed each S-metric space is a S_b-metric space with $b = 1$.*

Following example shows that a S_b-metric on X need not be a S-metric on X.

Example 1.4. ([8]) *Let (X, S) be a S-metric space, and $S_*(x, y, z) = S(x, y, z)^p$, where $p > 1$ is a real number. Note that S_* is a S_b-metric with $b = 2^{2(p-1)}$. Also, (X, S_*) is not necessarily a S-metric space.*

Definition 1.5. ([8]) *Let (X, S) be a S_b-metric space. Then, for $x \in X$, $r > 0$ we defined the open ball $B_S(x, r)$ and closed ball $B_S[x,r]$ with center x and radius r as follows respectively:*

$$B_S(x, r) = \{y \in X : S(y, y, x) < r\},$$

$$B_S[x, r] = \{y \in X : S(y, y, x) \leq r\}.$$

Lemma 1.6. ([8]) *In a S_b-metric space, we have*

$$S(x, x, y) \leq bS(y, y, x)$$

and

$$S(y, y, x) \leq bS(x, x, y)$$

Lemma 1.7. ([8]) *In a S_b-metric space, we have*

$$S(x, x, z) \leq 2bS(x, x, y) + b^2S(y, y, z)$$

Definition 1.8. ([8]) *If (X,S) be a S_b-metric space. A sequence $\{x_n\}$ in X is said to be:*

1. *S_b-Cauchy sequence if, for each $\epsilon > 0$, there exists $n_0 \in N$ such that $S(x_n, x_n, x_m) < \epsilon$ for each $m, n \geq n_0$.*
2. *S_b-convergent to a point $x \in X$ if, for each $\epsilon > 0$, there exists a positive integer n_0 such that $S(x_n, x_n, x) < \epsilon$ or $S(x, x, x_n) < \epsilon$ for all $n \geq n_0$ and we denote by $\lim_{n\to\infty} x_n = x$.*

Definition 1.9. ([8]) *A S_b-metric space (X, S) is called complete if every S_bCauchy sequence is S_b-convergent in X.*

Lemma 1.10. ([8]) *If (X, S) be a S_b-metric space with $b \geq 1$ and suppose that $\{x_n\}$ is a S_b-convergent to x, then we have*

(i) $\frac{1}{2b}S(y,x,x) \leq \lim_{n\to\infty} \inf S(y,y,x_n) \leq \lim_{n\to\infty} \sup S(y,y,x_n) \leq 2bS(y,y,x)$ *and*

(ii) $\frac{1}{b^2}S(x,x,y) \leq \lim_{n\to\infty} \inf S(x_n,x_n,y) \leq \lim_{n\to\infty} \sup S(x_n,x_n,y) \leq b^2S(x,x,y)$ *for all $y \in X$*

In particular, if $x = y$, then we have $\lim_{n\to\infty} S(x_n,x_n,y) = 0$.

Now we prove our main results.

2. Main Results

Definition 2.1. *Let $(X, S_b, \preceq)$ be a partially ordered complete S_b-metric space which is also regular, let $f : X \to X$ be mapping, we say that f is satisfy (ψ, ϕ) contraction if there exists $\psi, \phi : [0,\infty) \to [0,\infty)$ are such that*

(2.1.1) *f is non-decreasing*

(2.1.2) ψ is continuous and monotonically non - decreasing and ϕ *is lower semi continuous*

(2.1.3) $\psi(t) = 0 = \phi(t)$ *if and only if* $t = 0$

(2.1.4) $\psi(\alpha t) = \alpha\, t$ and $\phi(t) > 0$ for $t > 0$

(2.1.5) $\psi\left(S_b\left(fx, fx, fy\right)\right) \leq \frac{1}{4b^4}\psi\left(M_f^i\left(x,y\right)\right) - \phi\left(M_f^i\left(x,y\right)\right)$, $\forall\, x, y \in X$, $x \preceq y$, $i = 3,4,5$ *and*

$$M_f^5\left(x,y\right) = \max\left\{\ S_b(x,x,y), S_b(x,x,fx), S_b(y,y,fy), S_b(x,x,fy), S_b(y,y,fx)\ \right\}.$$

$$M_f^4\left(x,y\right) = \max\left\{\ S_b(x,x,y), S_b(x,x,fx), S_b(y,y,fy), \frac{1}{4b^4}\left[S_b(x,x,fy) + S_b(y,y,fx)\right]\ \right\}.$$

$$M_f^3\left(x,y\right) = \max\left\{\ S_b(x,x,y), \frac{1}{4b^4}\left[S_b(x,x,fx) + S_b(y,y,fy)\right], \frac{1}{4b^4}\left[S_b(x,x,fy) + S_b(y,y,fx)\right]\ \right\}.$$

Definition 2.2. *Suppose* $(X, \preceq)$ *is partially ordered set, and* f *is a mapping of* X *into itself. We say that* f *is non - decreasing if for every* $x, y \in X$,

$$x \preceq y \text{ implies that } fx \preceq fy. \tag{2.1}$$

Theorem 2.3. *Let* $(X, S_b, \preceq)$ *be an ordered complete* S_b *metric space, which is also regular and let* $f : X \to X$ *be satisfies* (ψ, ϕ) *- contraction with* $i = 5$. *If there exists* $x_0 \in X$ *with* $x_0 \preceq fx_0$, *then* f *has unique fixed point in* X.

Proof. Since f is a mapping from X into X, there exists a sequence $\{x_n\}$ in X such that

$$x_{n+1} = f x_n,\ n = 0, 1, 2, 3, \cdots$$

Case(i): If $x_n = x_{n+1}$, then x_n is fixed point of f.

Case(ii): Suppose $x_n \neq x_{n+1}\ \forall\, n$.

Since $x_0 \preceq fx_0 = x_1$ and f is increasing, it follows that

$$x_0 \preceq fx_0 \preceq f^2x_0 \preceq f^3x_0 \preceq \cdots \preceq f^nx_0 \preceq f^{n+1}x_0 \preceq \cdots$$

Now

$$\begin{aligned}\psi\left(S_b\left(fx_0, fx_0, f^2x_0\right)\right) &= \psi\left(S_b\left(fx_0, fx_0, fx_1\right)\right)\\ &\leq \frac{1}{4b^4}\psi\left(M_f^5\left(x_0, x_1\right)\right) - \phi\left(M_f^5\left(x_0, x_1\right)\right),\end{aligned}$$

where

$$\begin{aligned}M_f^5\left(x_0, x_1\right) &= \max\left\{\begin{array}{l} S_b\left(x_0, x_0, x_1\right), S_b\left(x_0, x_0, fx_0\right), S_b\left(x_1, x_1, fx_1\right)\\ S_b\left(x_0, x_0, f^2x_0\right), S_b\left(fx_0, fx_0, fx_0\right)\end{array}\right\}\\ &= \max\left\{\ S_b\left(x_0, x_0, fx_0\right), S_b\left(fx_0, fx_0, f^2x_0\right), S_b\left(x_0, x_0, f^2x_0\right)\ \right\}.\end{aligned}$$

Therefore

$$\begin{aligned}\psi\left(S_b\left(fx_0, fx_0, f^2x_0\right)\right) &\leq \frac{1}{4b^4}\psi\left(\max\left\{\ S_b\left(x_0, x_0, fx_0\right), S_b\left(fx_0, fx_0, f^2x_0\right), S_b\left(x_0, x_0, f^2x_0\right)\ \right\}\right)\\ &\quad -\phi\left(\max\left\{\ S_b\left(x_0, x_0, fx_0\right), S_b\left(fx_0, fx_0, f^2x_0\right), S_b\left(x_0, x_0, f^2x_0\right)\ \right\}\right)\\ &\leq \frac{1}{4b^4}\psi\left(\max\left\{\ S_b\left(x_0, x_0, fx_0\right), S_b\left(fx_0, fx_0, f^2x_0\right), S_b\left(x_0, x_0, f^2x_0\right)\ \right\}\right).\end{aligned}$$

By the definition of ψ, we have that

$$S_b\left(fx_0, fx_0, f^2x_0\right) \leq \max\left\{\begin{array}{l}\frac{1}{4b^4}\ S_b\left(x_0, x_0, fx_0\right),\\[2mm] \frac{1}{4b^4}\ S_b\left(fx_0, fx_0, f^2x_0\right),\\[2mm] \frac{1}{4b^4}\ S_b\left(x_0, x_0, f^2x_0\right)\end{array}\right\}. \tag{2.2}$$

But

$$\begin{aligned}\frac{1}{4b^4}S_b\left(x_0, x_0, f^2x_0\right) &\leq \frac{1}{4b^4}\left[2bS_b\left(x_0, x_0, fx_0\right) + b^2S_b\left(fx_0, fx_0, f^2x_0\right)\right]\\ &\leq \max\left\{\frac{1}{2b}S_b\left(x_0, x_0, fx_0\right), \frac{1}{2}S_b\left(fx_0, fx_0, f^2x_0\right)\right\}.\end{aligned}$$

From (2.2), we have that

$$S_b\left(fx_0, fx_0, f^2x_0\right) \leq \max\left\{\frac{1}{2b}S_b\left(x_0, x_0, fx_0\right), \frac{1}{2}S_b\left(fx_0, fx_0, f^2x_0\right)\right\}.$$

If $\frac{1}{2}S_b\left(fx_0, fx_0, f^2x_0\right)$ is maximum, we get a contradiction. Hence

$$S_b\left(fx_0, fx_0, f^2x_0\right) \leq \frac{1}{2b}S_b\left(x_0, x_0, fx_0\right) \tag{2.3}$$

Also

$$\begin{aligned}\psi\left(S_b\left(f^2x_0, f^2x_0, f^3x_0\right)\right) &= \psi\left(S_b\left(fx_1, fx_1, fx_2\right)\right)\\ &\leq \tfrac{1}{4b^4}\psi\left(M_f^5\left(x_1, x_2\right)\right) - \phi\left(M_f^4\left(x_1, x_2\right)\right),\end{aligned}$$

where

$$\begin{aligned}M_f^5\left(x_1, x_2\right) &= \max\left\{\begin{array}{l} S_b\left(fx_0, fx_0, f^2x_0\right), S_b\left(fx_0, fx_0, f^2x_0\right), S_b\left(f^2x_0, f^2x_0, f^3x_0\right)\\ S_b\left(fx_0, fx_0, f^3x_0\right), S_b\left(f^2x_0, f^2x_0, f^2x_0\right)\end{array}\right\}\\ &= \max\left\{ S_b\left(fx_0, fx_0, f^2x_0\right), S_b\left(f^2x_0, f^2x_0, f^3x_0\right), S_b\left(fx_0, fx_0, f^3x_0\right) \right\}.\end{aligned}$$

Therefore

$$\begin{aligned}\psi\left(S_b\left(f^2x_0, f^2x_0, f^3x_0\right)\right) &\leq \tfrac{1}{4b^4}\psi\left(\max\left\{\begin{array}{l} S_b\left(fx_0, fx_0, f^2x_0\right), S_b\left(f^2x_0, f^2x_0, f^3x_0\right),\\ S_b\left(fx_0, fx_0, f^3x_0\right)\end{array}\right\}\right)\\ &\quad -\phi\left(\max\left\{\begin{array}{l} S_b\left(fx_0, fx_0, f^2x_0\right), S_b\left(f^2x_0, f^2x_0, f^3x_0\right),\\ S_b\left(fx_0, fx_0, f^3x_0\right)\end{array}\right\}\right)\\ &\leq \tfrac{1}{4b^4}\psi\left(\max\left\{\begin{array}{l} S_b\left(fx_0, fx_0, f^2x_0\right), S_b\left(f^2x_0, f^2x_0, f^3x_0\right),\\ S_b\left(fx_0, fx_0, f^3x_0\right)\end{array}\right\}\right).\end{aligned}$$

By the definition of ψ, we have that

$$S_b\left(f^2x_0, f^2x_0, f^3x_0\right) \leq \max\left\{\begin{array}{l} \frac{1}{4b^4}\, S_b\left(fx_0, fx_0, f^2x_0\right),\\ \frac{1}{4b^4}\, S_b\left(f^2x_0, f^2x_0, f^3x_0\right),\\ \frac{1}{4b^4}\, S_b\left(fx_0, fx_0, f^3x_0\right)\end{array}\right\}. \tag{2.4}$$

But

$$\begin{aligned}\frac{1}{4b^4}S_b\left(fx_0, fx_0, f^3x_0\right) &\leq \frac{1}{4b^4}\left[2bS_b\left(fx_0, fx_0, f^2x_0\right) + b^2S_b\left(f^2x_0, f^2x_0, f^3x_0\right)\right]\\ &\leq \max\left\{\frac{1}{2b}S_b\left(fx_0, fx_0, f^2x_0\right), \frac{1}{2}S_b\left(f^2x_0, f^2x_0, f^3x_0\right)\right\}.\end{aligned}$$

From (2.4), we have that

$$S_b\left(f^2x_0, f^2x_0, f^3x_0\right) \leq \max\left\{\frac{1}{2b}S_b\left(fx_0, fx_0, f^2x_0\right), \frac{1}{2}S_b\left(f^2x_0, f^2x_0, f^3x_0\right)\right\}.$$

If $\frac{1}{2}S_b\left(f^2x_0, f^2x_0, f^3x_0\right)$ is maximum, we get a contradiction. Hence

$$\begin{aligned}S_b\left(f^2x_0, f^2x_0, f^3x_0\right) &\leq \frac{1}{2b}S_b\left(fx_0, fx_0, f^2x_0\right)\\ &\leq \frac{1}{(2b)^2}S_b\left(x_0, x_0, fx_0\right).\end{aligned}$$

Continuing this process, we can conclude that

$$\begin{aligned} S_b\left(f^n x_0, f^n x_0, f^{n+1} x_0\right) & \leq \frac{1}{(2b)^n} S_b\left(x_0, x_0, f x_0\right) \\ & \rightarrow 0 \text{ as } n \rightarrow \infty \end{aligned}$$

That is

$$\lim_{n \rightarrow \infty} S_b\left(f^n x_0, f^n x_0, f^{n+1} x_0\right)=0. \tag{2.5}$$

Now we prove that $\{f^n x_0\}$ is Cauchy sequence in (X,S_b). On contrary we suppose that $\{f^n x_0\}$ is not Cauchy. Then there exist $\epsilon > 0$ and monotonically increasing sequence of natural numbers $\{m_k\}$ and $\{n_k\}$ such that $n_k > m_k$.

$$S_b\left(f^{m_k} x_0, f^{m_k} x_0, f^{n_k} x_0\right) \geq \epsilon \tag{2.6}$$

and

$$S_b\left(f^{m_k} x_0, f^{m_k} x_0, f^{n_k-1} x_0\right)<\epsilon. \tag{2.7}$$

From (2.6) and (2.7), we have

$$\begin{aligned} \epsilon & \leq S_b\left(f^{m_k} x_0, f^{m_k} x_0, f^{n_k} x_0\right) \\ & \leq 2b S_b\left(f^{m_k} x_0, f^{m_k} x_0, f^{m_k+1} x_0\right)+b^2 S_b\left(f^{m_k+1} x_0, f^{m_k+1} x_0, f^{n_k} x_0\right). \end{aligned}$$

Letting $k \rightarrow \infty$ and apply ψ on both sides, we have that

$$\begin{aligned} \psi\left(\frac{\epsilon}{b^2}\right) & \leq \lim_{k \rightarrow \infty} \psi\left(S_b\left(f^{m_k+1} x_0, f^{m_k+1} x_0, f^{n_k} x_0\right)\right) \\ & =\lim_{k \rightarrow \infty} \psi\left(S_b\left(x_{m_k+1}, x_{m_k+1}, x_{n_k}\right)\right) \\ & =\lim_{k \rightarrow \infty} \psi\left(S_b\left(f x_{m_k}, f x_{m_k}, f x_{n_k-1}\right)\right) \\ & \leq \frac{1}{4b^4} \lim_{k \rightarrow \infty} \psi\left(M_f^5\left(x_{m_k}, x_{n_k-1}\right)\right)-\lim_{k \rightarrow \infty} \phi\left(M_f^5\left(x_{m_k}, x_{n_k-1}\right)\right) \end{aligned} \tag{2.8}$$

where

$$\begin{aligned} & \tfrac{1}{4b^4} \lim_{k \rightarrow \infty} M_f^5\left(x_{m_k}, x_{n_k-1}\right) \\ & =\tfrac{1}{4b^4} \lim_{k \rightarrow \infty} \max \left\{\begin{array}{c} S_b\left(f^{m_k} x_0, f^{m_k} x_0, f^{n_k-1} x_0\right), S_b\left(f^{m_k} x_0, f^{m_k} x_0, f^{m_k+1} x_0\right), \\ S_b\left(f^{n_k-1} x_0, f^{n_k-1} x_0, f^{n_k} x_0\right), S_b\left(f^{m_k} x_0, f^{m_k} x_0, f^{n_k} x_0\right), \\ S_b\left(f^{n_k-1} x_0, f^{n_k-1} x_0, f^{m_k+1} x_0\right) \end{array}\right\} \\ & <\lim_{k \rightarrow \infty} \max \left\{\begin{array}{c} \tfrac{\epsilon}{4b^4}, 0, 0, \tfrac{1}{4b^4} S_b\left(f^{m_k} x_0, f^{m_k} x_0, f^{n_k} x_0\right), \\ \tfrac{1}{4b^4} S_b\left(f^{n_k-1} x_0, f^{n_k-1} x_0, f^{m_k+1} x_0\right) \end{array}\right\} \\ & =\lim_{k \rightarrow \infty} \max \left\{\begin{array}{c} \tfrac{\epsilon}{4b^4}, \tfrac{1}{4b^4} S_b\left(f^{m_k} x_0, f^{m_k} x_0, f^{n_k} x_0\right), \\ \tfrac{1}{4b^4} S_b\left(f^{n_k-1} x_0, f^{n_k-1} x_0, f^{m_k+1} x_0\right) \end{array}\right\}. \end{aligned}$$

But

$$\begin{aligned} \lim_{k \rightarrow \infty} \tfrac{1}{4b^4} S_b\left(f^{m_k} x_0, f^{m_k} x_0, f^{n_k} x_0\right) & \leq \lim_{k \rightarrow \infty} \tfrac{1}{4b^4}\left[\begin{array}{c} 2b\, S_b\left(f^{m_k} x_0, f^{m_k} x_0, f^{n_k-1} x_0\right) \\ +b^2 S_b\left(f^{n_k-1} x_0, f^{n_k-1} x_0, f^{n_k} x_0\right) \end{array}\right] \\ & <\tfrac{\epsilon}{2b^3}. \end{aligned}$$

Also

$$\lim_{k\to\infty} \tfrac{1}{4b^4} S_b\left(f^{n_k-1}x_0, f^{n_k-1}x_0, f^{m_k+1}x_0\right) \le \lim_{k\to\infty} \tfrac{1}{4b^4}\left[\begin{array}{c} 2bS_b\left(f^{n_k-1}x_0, f^{n_k-1}x_0, f^{m_k}x_0\right) \\ +b^2 S_b\left(f^{m_k}x_0, f^{m_k}x_0, f^{m_k+1}x_0\right)\end{array}\right]$$
$$< \tfrac{\epsilon}{2b^2}.$$

Therefore

$$\frac{1}{4b^4}\lim_{k\to\infty} M_f^5\left(x_{m_k}, x_{n_k-1}\right) \le \max\left\{\tfrac{\epsilon}{4b^4}, \tfrac{\epsilon}{2b^3}, \tfrac{\epsilon}{2b^2}\right\} = \frac{\epsilon}{2b^2}.$$

From (2.8), by the definition of ψ, we have that

$$\frac{\epsilon}{b^2} \le \frac{\epsilon}{2b^2}.$$

It follows that

$$1 \le \frac{1}{2}.$$

It is a contradiction. Hence $\{f^n x_0\}$ is Cauchy sequence in complete regular S_b metric space $(X, S_b, \preceq)$. By completeness of (X, S_b), it follows that the sequence $\{f^n x_0\}$ is converges to z in (X, S_b). Thus

$$\lim_{k\to\infty} f^n x_0 = \alpha = \lim_{k\to\infty} f^{n+1}x_0.$$

Since x_n, $\alpha \in X$ and X is regular, it follows that either $x_n \preceq \alpha$ or $\alpha \preceq x_n$.

Now we have to prove that α is fixed point of f.

Suppose $f\alpha \neq \alpha$,

From (2.1.5), by the definition of ψ and Lemma (1.10), we have that

$$\psi\left(\frac{1}{2b}S_b(f\alpha, f\alpha, \alpha)\right) \le \lim_{n\to\infty}\inf \psi\left(S_b\left(f\alpha, f\alpha, f^{n+1}x_0\right)\right)$$
$$\le \frac{1}{4b^4}\lim_{n\to\infty}\inf \psi\left(M_f^5(\alpha, x_n)\right) - \lim_{n\to\infty}\inf \phi\left(M_f^5(\alpha, x_n)\right). \tag{2.9}$$

Here

$$\lim_{n\to\infty}\inf M_f^5(\alpha, x_n) = \lim_{n\to\infty}\inf \max\left\{\begin{array}{c} S_b(\alpha,\alpha,x_n), S_b(\alpha,\alpha,f\alpha), S_b(x_n,x_n,fx_n), \\ S_b(\alpha,\alpha,fx_n), S_b(x_n,x_n,f\alpha)\end{array}\right\}$$
$$\le \lim_{n\to\infty}\sup \max\left\{0, S_b(\alpha,\alpha,f\alpha), 0, 0, S_b(x_n,x_n,f\alpha)\right\}$$
$$\le \max\left\{bS_b(f\alpha,f\alpha,\alpha), b^3 S_b(f\alpha,f\alpha,\alpha)\right\}$$
$$= b^3 S_b(f\alpha,f\alpha,\alpha).$$

Hence from (2.9), we have that

$$\psi\left(\frac{1}{2b}S_b(f\alpha, f\alpha, \alpha)\right) \le \frac{1}{4b^4}\psi\left(b^3 S_b(\alpha,\alpha,f\alpha)\right) - \lim_{n\to\infty}\inf \phi\left(M_f^5(\alpha, x_n)\right)$$
$$\le \frac{1}{4b^4}\psi\left(b^3\, S_b(f\alpha, f\alpha, \alpha)\right).$$

By the definition of ψ, we have that

$$1 \leq \frac{1}{2}.$$

It is a contradiction. So that α is fixed point of f.

Suppose α^* is another fixed point of f such that $\alpha \neq \alpha^*$

Consider

$$\begin{aligned}\psi\left(S_b\left(\alpha,\alpha,\alpha^*\right)\right) &\leq \frac{1}{4b^4}\,\psi\left(M_f^4\left(\alpha,\alpha^*\right)\right)-\phi\left(M_f^4\left(\alpha,\alpha^*\right)\right)\\ &= \frac{1}{4b^4}\,\psi\left(\max\{S_b\left(\alpha,\alpha,\alpha^*\right),S_b\left(\alpha^*,\alpha^*,\alpha\right)\}\right)-\phi\left(\max\{S_b\left(\alpha,\alpha,\alpha^*\right),S_b\left(\alpha^*,\alpha^*,\alpha\right)\}\right)\\ &\leq \frac{1}{4b^4}\,\psi\left(bS_b\left(\alpha,\alpha,\alpha^*\right)\right).\end{aligned}$$

By the definition of ψ, we have that

$$b^3 \leq \frac{1}{4}.$$

It is a contradiction.

Hence α is unique fixed point of f in (X,S_b).

Example 2.4. *Let* $X = [0,1]$ *and* $S : X \times X \times X \to \mathbb{R}^+$ *by* $S_b(x, y, z) = (|y + z - 2x| + |y - z|)^2$ *and* $\preceq$ *by* $a \preceq b \iff a \leq b$*, then* $(X, S_b, \preceq)$ *is complete ordered* S_b*-metric space with* $b = 4$*. Define* $f : X \to X$ *by* $f(x) = \frac{x}{32\sqrt{2}}$*. Also define* $\psi, \phi : \mathbb{R}^+ \to \mathbb{R}^+$ *by* $\psi(t) = t$ *and* $\phi(t) = \frac{t}{8b^4}$.

$$\begin{aligned}\psi\left(S_b(fx,fx,fy)\right) &= \left(|fx+fy-2fx|+|fx-fy|\right)^2\\ &= \left(2\left|\frac{x}{32\sqrt{2}}-\frac{y}{32\sqrt{2}}\right|\right)^2\\ &= \frac{1}{8b^4}S_b(x,x,y)\\ &\leq \frac{1}{8b^4}M_f^5(x,y)\\ &\leq \frac{1}{4b^4}\psi\left(M_f^5(x,y)\right)-\phi\left(M_f^5(x,y)\right),\end{aligned}$$

where

$$M_f^5(x,y)=\max\left\{\ S_b(x,x,y),S_b(x,x,fx),S_b(y,y,fy),S_b(x,x,fy),S_b(y,y,fx).\ \right\}$$

Hence from Theorem 2.3, 0 is unique fixed point of f.

Theorem 2.5. *Let* $(X, S_b, \preceq)$ *be an ordered complete* S_b *metric space and let* $f : X \to X$ *be satisfies* (ψ, ϕ) *- contraction with* $i = 3$ *or* 4*. If there exists* $x_0 \in X$ *with* $x_0 \preceq fx_0$*. Then* f *has unique fixed point in* X.

Proof. If we replace $M_f^3(x,y)$ or $M_f^4(x,y)$ in place of $M_f^5(x,y)$, the rest of proof follows from Theorem 2.3.

Theorem 2.6. *Let* $(X, S_b, \preceq)$ *be an ordered complete* S_b *metric space and let* $f: X \to X$ *be satisfies*

$$S_b\left(fx,fx,fy\right)\leq \frac{1}{4b^4}M_f^i\left(x,y\right)-\varphi\left(M_f^i\left(x,y\right)\right),$$

where $\varphi : [0,\infty) \to [0,\infty)$ *and* $i = 3,4,5$*. If there exists* $x_0 \in X$ *with* $x_0 \preceq fx_0$*. Then* f *has unique fixed point in* X.

Proof. The proof follows from Theorem 2.3 and Theorem 2.5 by taking $\psi(t) = t$ and $\phi(t) = \varphi(t)$.

Theorem 2.7. *Let* $(X, S_b, \preceq)$ *be an ordered complete* S_b *metric space and let* $f: X \to X$ *be satisfies*

$$S_b(fx, fx, fy) \leq \lambda M_f^i(x, y),$$

where $\lambda \in = \left[0, \frac{1}{4b^4}\right)$ *and* $i = 3, 4, 5$*. If there exists* $x_0 \in X$ *with* $x_0 \preceq fx_0$*. Then* f *has unique fixed point in* X.

REFERENCES

1. Abbas M and Ali khan M and Randenovic S, *Common coupled fixed point theorems in cone metric spaces for w-compatible mappings*. Appl.Math. Comput., 217, (2010), 195–202.
2. Banach S, *Theorie des Operations lineaires*, Manograic Mathematic Zne, Warsaw, Poland, 1932.
3. Czerwik S, *Contraction mapping in b-metric spaces*, Acta Mathematica et Informatica Universitatis Ostraviensis, 1 (1993), 5–11.
4. Lakshmikantham V and Ciric Lj, *Coupled fixed point theorems for nonlinear contractions in partially ordered metric spaces*. Nonlinear Analysis. Theory, Methods and Applications, 70(12),(2009),4341-4349.
5. Mustafa Z, Sims B, *A new approach to generalized metric spaces*. J Nonlinear Convex Anal, 7(2), (2006), 289–297.
6. Kishore G.N.V, Rao K. P. R and Hima Bindu V.M.L, *Suzuki type unique common fixed point theorem in partial metric spaces by using (C): condition with rational expressions*, Afr. Mat. (2016), DOI 10.1007/s13370-017-0484- x
7. Sedghi S, Altun I, Shobe N and Salahshour M, *Some properties of S-metric space and fixed point results*, Kyung pook Math. J., 54(2014), 113–122.
8. Sedghi S, Gholidahneh, Dosenovic T, Esfahani J and Radenovic S, *Common fixed point of four maps in S_b-metric spaces*, Journal of Linear and Topological Algebra, Vol.5(2), (2016), 93–104.
9. Sedghi S, Shobe N and Aliouche A, *A generalization of fixed point theorem in S- metric spaces*. Mat. Vesnik, 64 (2012), 258–266.
10. Sedghi S, Shobe N and T.Dosenovic, *Fixed point results in S-metric spaces*, Nonlinear Functional Analysis and Applications, 20(1), (2015), 55–67.

Impending Inquisitions in Humanities and Sciences (ICIIHS-2022) – Dr. Mohan Varkolu et al. (eds)

Spectroscopic and Thermal Properties of VO^{2+} Doped Barium Molybdenum Tellurite Glass System

56

V. Vamsipriya
Department of Physics, Osmania University, Hyderabad, Telangana, India
J. Bhemarajam
Department of Humanities and Sciences, Vardhaman College of Engineering, Shamshabad, Hyderabad, Telangana, India
R. V. Neeraja, M. Prasad*
Department of Physics, Osmania University, Hyderabad, Telangana, India

ABSTRACT The glass samples were synthesized with composition $69TeO_2$-(30-x)MoO_3-xBaO: $1V_2O_5$ (where x = 0, 7.5, 15, 22.5, 30) by conventional melt-quenching technique. Density, molar volume and Oxygen packing density (OPD) were determined. It is observed that density and molar volume both are increasing when MoO_3 is replaced by BaO. Infrared (IR), Raman spectroscopic techniques were employed for structural characterization of the samples. Structural transformations in the glass network is revealed by Raman and IR spectra which showed that with the addition of BaO in the glass system, transformation of TeO_4 (tbp) units to TeO_{3+1} units and then to TeO_3 (tp) units takes place. While the coordination number of Mo changes from +4 to +6 that corresponds to the decrease in the total number of NBO with increasing content of BaO in the glass matrix. Differential scanning calorimetry (DSC) is used to understand the effect of BaO and MoO_3 content on thermal properties of tellurium glass. UV-VIS spectroscopic technique is employed for determining direct, indirect band gaps, Urbach energies, refractive index, molar refractivity, metallization criterion and optical basicity of the TMB samples. It is found that band gap energies increased and refractive index values decreased with increasing in the concentration of BaO. Spin Hamiltonian parameters are also evaluated by Electron Paramagnetic Resonance studies which followed the relation $g_{\parallel} < g_{\perp} < g_e$ that reveal that VO^{2+} exists in octahedral site with tetragonal distortion with C_{4v} symmetry.

KEYWORDS TML glasses, XRD, FTIR spectroscopy, Raman; DSC; UV-VIS spectroscopy; Spin Hamiltonian

1. Introduction

Tellurium glasses have gained interest in the recent years due to their technological importance because of their unique properties. Tellurite glasses have large linear and non-linear refractive indices properties, high transmission range, low melting temperatures, low energy phonons, high refractive index, high rare earth (RE) ions soluability and high chemical stability [1-4]. These properties have made these glasses potential materials for laser hosts and optical switching devices. Tellurium oxide donot have the ability to form the glass under normal conditions because Te-O bond is strongly covalent so it does not allow the requisite amount of distortion to form glass. It forms glass when a modifier oxide such as alkali, alkaline earth or both or other glass formers are added. The structure of these glasses is based on tellurium crystalline α and β forms.

*Corresponding author: prasad5336@yahoo.co.in

DOI: 10.1201/9781003489436-56

Based on various investigations done on the glass formation and structure of Tellurium glasses it is reported that there are two basic structural units in tellurite glasses. They are trigonal bi-pyramidal (tbp) TeO_4 and trigonal pyramidal (tp) TeO_3 with an intermediate TeO_{3+1} Polyhedron. There are four oxygen atoms coordinated with one atom of tellurium in TeO_4 units which has equatorial oxygen unoccupied. There are two bridging oxygens and a non-bridging oxygen (NBO) in TeO_3 unit. This NBO forms double bond with tellurium atom as Te=O. Addition of modifiers such as BaO into the glass network breaks the random network structure by increasing the concentration of TeO_3 and TeO_{3+1} units [5, 6].

Molybdenum oxide MoO_3 also forms stable glasses when doped with different oxides. Molybdenum oxide can exist in multiple valence states like +3, +4, +5 or +6 which help it increase its glass network formation tendency [7]. Attention of recent research has been drawn on tellurite glasses with MoO_3 due to their properties that include semiconducting, transport, memory, threshold switching and non-linear optical properties which make these glasses have applications in making electrode materials in lithium ion batteries, electrochromic devices, sensors, super capacitors etc. It is known that the mixture of two glass formers has an influence on conductivity and thermal stability [8].

When alkali and alkaline earth metal oxides are added to the tellurium oxide glasses cause modifications in structural, thermal and Optical properties and enhances chemical durability. It is also known from the studies that BaO because of its high atomic number enhances absorption of x-rays and gamma rays that increase the radiation shielding properties of tellurite glasses [9]. Barium tellurite glasses gained interest because it can produce zero stress lead free optical glasses and non-toxicity compared to lead based glasses [10][11]. The aim of the work presented in this paper is to synthesize $69TeO_2$-(30-x) MoO_3-xBaO glass system doped with 1 mol % of V_2O_5 and to investigate the effect of replacing MoO_3 with BaO on structural, optical and thermal characteristics of the prepared TMB glass samples.

2. Experimental Studies and Characterization Techniques

$69TeO_2$-(30-x) MoO_3-xBaO where x takes the values 0, 7.5, 15, 22.5, 30 with 1 mol% V_2O_5 samples were synthesized using conventional melt-quench technique. The list of composition is given in Table 56.1. Starting chemicals are taken in required proportions and are powdered into a fine mixture in a mortar and pestle and is taken in a porcelain crucible. The mixture is then melted at about 800°C for 30 min in an electrical furnace. The molten mixture is stirred regularly to get homogeneous melt and then quenched by pouring melts on the stainless-steel preheated plate and is pressed with a pre-heated steel rod at room temperature. The sample is then annealed at 250°C for 6hrs to remove thermal stress and strains. XRD patterns were taken for prepared samples with Pan Analytical X'pert ProX-ray diffractometer at room temperature at scanning rate of 2deg/min and with 2Θ varied between 10 to 80°. The density values (ρ gm/cc) of the synthesized glasses are determined by Archimedes principle by immersing in Xylene. From density values of the glass samples V_m and OPD were computed. FT-IR transmittance spectroscopic studies were performed by means of FTIR 8400 Shimadzu spectrometer for a wavelength ranging from 4000-250 cm^{-1} at the room temperature of TMB glass samples using KBr pellets which are pressed in vacuum at 22 tons load for 2 minutes. This procedure gave transparent pellet with about 2wt% of glass. Raman spectra was measured for TMB glass samples in the back scattering geometry by using Ocean optics RSI 2001 S Raman spectrometer in the range 100-1500 cm^{-1}. The excitation light source is a solid-state diode laser with the constant output of 500 mW at a wavelength of 514 nm. The DSC curves were obtained for the TMB glasses by DSC instrument, SDT Q600. The glass transition (Tg) temperatures were measured by placing approximately 15mg of the prepared sample in a Nitrogen environment with the rate of flow 15 Psi in an open platinum pan heated at a rate 10°C/min upto 450°C. UV-VIS spectra were recorded in the wavelength ranging from 200 nm to 800 nm by a Camspec M-350 double beam UV-Vis Spectrophotometer at room temperature. Electron Paramagnetic Resonance measurements for all the TMB glasses were performed at X-band frequencies in the range 9.1 to 9.6 GHz on a JOEL FX1X series EPR spectrometer employing 100 kHz modulation field at room temperature.

Table 56.1 Composition and code of the samples are shown below

Sample code	Composition
TMB1	$69TeO_2$-$30MoO_3$-$1V_2O_5$
TMB2	$69TeO_2$-$22.5MoO_3$-7.5BaO-$1V_2O_5$
TMB3	$69TeO_2$-$15MoO_3$-15BaO-$1V_2O_5$
TMB4	$69TeO_2$-$7.5MoO_3$-22.5BaO-$1V_2O_5$
TMB5	$69TeO_2$-30BaO-$1V_2O_5$

3. Results and Discussion

3.1 XRD

Fig. 56.1 presents XRD spectra of $69TeO_2$-(30-x) MoO_3-xBaO-$1V_2O_5$ glasses which did not demonstrate any definite and sharp peaks confirming the glasses formed are amorphous in nature.

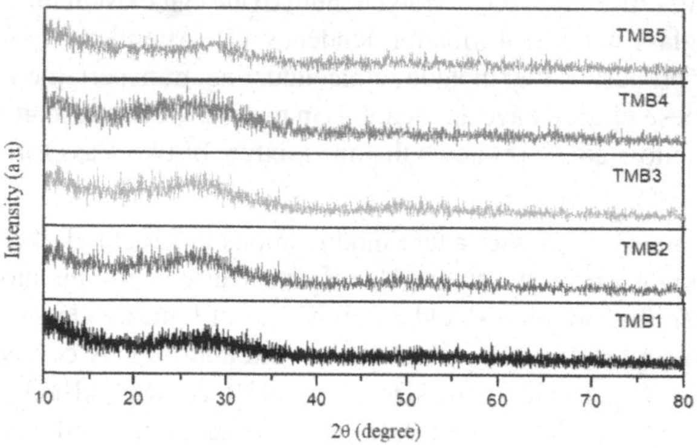

Fig. 56.1 XRD of TMB series

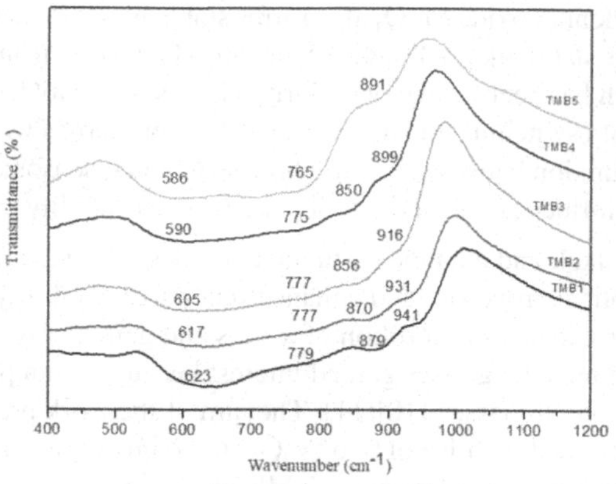

Fig. 56.2 FTIR spectra of TMB series

3.2 Physical Properties

Physical parameters of TMB glasses doped with vanadium ion were computed by Archimedes principle. The physical parameters of the prepared system are calculated using Xylene as an immersion liquid by the formula given in equation (1)

$$\rho = \frac{a}{a-b} X \rho_X \tag{1}$$

"Here a, b are weights of samples in air and xylene liquid and density of immersed liquid is ρ x = 0.865 gm/cm^3. Using the density, molar volume (V_m) of TMB glasses were computed by the equation (2).

$$\text{Molar volume } V_m = \frac{\sum M}{\rho} \tag{2}$$

Here 'M' represents molecular weight and ρ is the density.

OPD values of samples were determined using the values of density using equation (3).

$$\text{Oxygen Packing Density (OPD)} = \left(\frac{\rho}{M}\right) X O_n \tag{3}$$

Here o_n is the no. of oxygens per unit formula. ρ, V_m and OPD values are summarized in Table 56.2. One of the important variables to understand the structural modifications in the glass network is Density parameter. The abrupt variation in the values of density is observed when there is a structural change, change in coordination number, geometric configuration, cross linking density and structural packing in the glasses. From measurements it is found that the values of densities of the samples are increasing from 5.08 to 5.40 gm/cm^3 with replacement of MoO_3 by BaO. This is because low molecular weight MoO_3 is replaced with high molecular weight BaO.

With increase in the densities and molar volumes of TMB series also showed an increase from 30.48 to 31.20 (cc/mol) which indicates an increase in free volume in the sample at an expense of MoO_3 with BaO. It is predicted and found for many glass systems that the molar volume and density varies inversely but in the present work it is found different. This increase in the molar volume with increase in BaO content is may be because of that when a higher atomic radius BaO (0.135 nm) is replacing MoO_3 (0.06 nm) which has lower atomic radius and due to an increase in interatomic spacing between the atoms or decrease in bond length that may be because of increase in the number of NBOs[12]. Oxygen packing density values also showed a decrease from TMB1 to TMB5 from 76.43 to 54.59 (mol/l) with decreasing concentration of MoO_3 and increase in BaO content in the prepared TMB samples. Decrease in OPD values shows a decrement in the number of oxygen in structure.

Table 56.2 Physical parameters of TMB series

Sample code	Density ρ(g/cc)	Molar volume V_m (cc/mol)	OPD (mol/l)
TMB1	5.09	30.48	76.43
TMB2	5.21	30.55	71.36
TMB3	5.34	30.84	65.81
TMB4	5.39	31.02	60.60
TMB5	5.40	31.20	54.59

3.3 FTIR Spectroscopy

FTIR spectra for TMB glass samples are presented in Fig. 56.2 and band positions and band assignments are shown in Tables 56.3 and 56.4. From the spectra it is found that prepared samples exhibit four vibrational modes which are from 644-594 cm^{-1}; 779-765 cm^{-1}; 879-850 cm^{-1}; 941-916 cm^{-1}.

Table 56.3 FTIR band positions of TMB series

Sample Code	Band1 (cm^{-1})	Band2 (cm^{-1})	Band3 (cm^{-1})	Band4 (cm^{-1})
TMB1	623	779	879	941
TMB2	617	777	870	931
TMB3	605	777	856	916
TMB4	590	775	850	899
TMB5	586	765	---	891

The stretching vibrational band of the Te-O bonds of deformed TeO4 structural units in trigonal bipyramids (tbp) structural units is exhibited around 586-623 cm^{-1}[13]. Second band at 779 cm^{-1} was attributed to stretching TeO_3 unit vibrations of Te-O in trigonal pyramids(tp) with NBOs inside TMB glass structure [14]10, 20, 30, 40, 50 and 60 mol% were synthesized. The X-ray diffraction patterns (XRD. Modifier ions in glass network destroy the structure and create non-bridging oxygens (NBO). This gradually transforms TeO_4 units to TeO_3 units. The band at 623 cm^{-1} shifted to 586 cm^{-1} i.e., this band shifted from higher wavenumber side to lower wavenumber side and from spectra it was observed that the intensity of the band at 630 cm^{-1} is decreasing as the content of BaO is increasing which may be because of decrease in number of TeO_4 units in the glass system and it is shifting towards the lower wavenumber side when MoO_3 content is decreasing with increase in BaO content. This may be due lighter molecule is replaced by a heavier molecule. So the frequency of vibration is reduced and the band showed a shift towards lower wavenumber. The band at 891 cm^{-1} is ascribed to stretching vibrations of MoO_4[15][16] and shoulder at 941 cm^{-1} is ascribed to the stretching vibrations of Mo=O related bonds of distorted MoO_6 unit [17] organometallic, bioinorganic, and coordination compounds. From fundamental theories of vibrational spectroscopy to applications in a variety of compound types, this has been extensively updated. New topics include the theoretical calculations of vibrational frequencies (DFT method.This band disappeared when MoO_3 content decreases below 15 mol% and when it is replaced by BaO.

Table 56.4 FTIR band assignments of TMB series

Wavenumber (cm^{-1})	Band assignments
586-623	Stretching vibrations of Te-O bonds in TeO_4 tbp
765-779	Stretching vibrations of Te-O bond in TeO_3 units
850-879	stretching vibrations of MoO_4 units
891-941	stretching vibrations of Mo=O bonds of deformed MoO_6 units

3.4 Raman Spectroscopy

Raman spectra for the present samples were recorded and shown in Fig. 56.3 and the positions of bands and the assignments are tabulated in Tables 56.5 and 56.6 respectively.

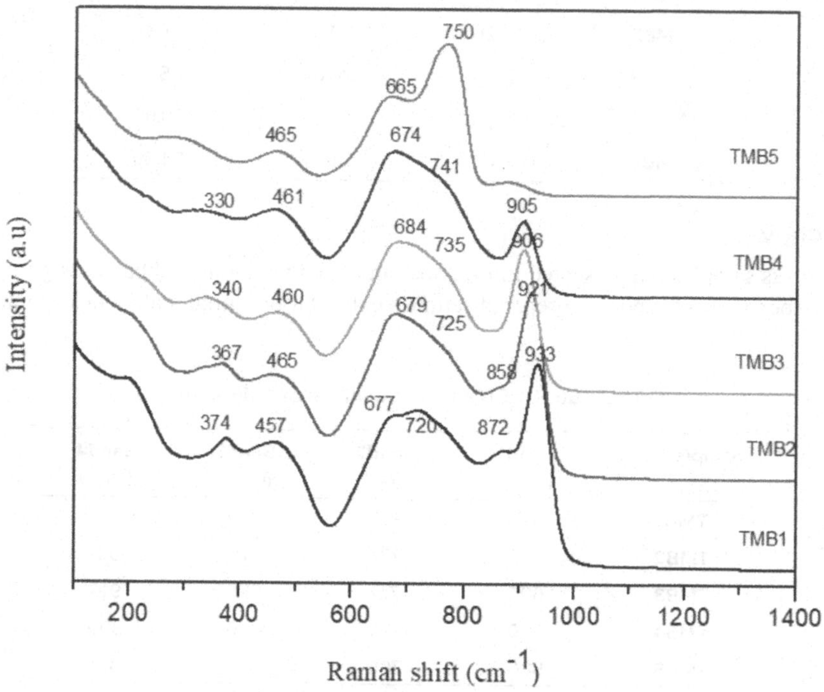

Fig. 56.3 Raman spectra of TMB series

Table 56.5 Raman band positions of TMB series

Sample Code	Band1 (cm^{-1})	Band2 (cm^{-1})	Band3 (cm^{-1})	Band4 (cm^{-1})	Band5 (cm^{-1})	Band6 (cm^{-1})
TMB1	374	457	677	720	872	933
TMB2	367	465	679	725	858	921
TMB3	340	460	684	735	---	906
TMB4	330	461	674	741	---	905
TMB5	---	465	665	750	---	---

The Raman peak at 374 cm^{-1} is due to the bending vibrations of Mo-O-Mo linkages. Band at 457 cm^{-1} is attributed to stretching vibrational mode of Te-O-Te bond linkages that are created from the corner shared from TeO_3, TeO_{3+1} as well as TeO_4 structural units [18]. The band at 677 cm^{-1} is due to stretching vibrations of TeO_4 structural units in glass system [19]acting on raphe cell body autoreceptors, also mediates inhibitory postsynaptic potentials as a result of release from collaterals from neighboring raphe neurons. This may involve a ligand (5-HT which represents the concentration of TeO_4. The same is also seen in FTIR spectra. The peak at 720 cm^{-1} is a Raman mode of Te-O-(NBO) of TeO_3 units. This was also observed in FTIR spectroscopy. The intensity of this band increased when BaO content is increased which gives evidence that addition of BaO breaks the Te-O bond, forming TeO_3 units. The band 872 cm^{-1} is ascribed to stretching vibrations of Mo-O-Mo linkages in MoO_4 and MoO_6 units[20]. This band is shown only in TMB1 and TMB2 samples where the MoO_3 content is more than 20 mol% and intensity of this band disappeared when MoO_3 content is less and is replaced by BaO. Band at 933 cm^{-1} is due to Mo-O-Te vibrations that are formed due to the rearrangement of NBOs in Te_{3+1} units and breaking of Mo-O-Mo bonds[21].

Table 56.6 Raman band assignments of TMB series

Wavenumber (cm^{-1})	Assignments
330-374	bending vibrations of Mo-O-Mo linkages
457-465	symmetrical stretching of Te-O-Te linkages
665-677	stretching vibrations of TeO_4 structural units
720-750	stretching vibrations of TeO_3 units
850-870	stretching vibrations of MoO_4 units
905-933	stretching vibration of Mo-O bonds of MoO_4 or MoO_6 units

3.5 Differential Scanning Calorimetry

Fig. 56.4 shows the DSC thermograms of the prepared TMB glass samples and Table 56.7 shows the T_g, onset crystalline temperatures (T_o) of the samples. The thermal stability given by $\Delta T = T_o - T_g$, is also presented. It is known that the transition temperatures depends majorly on composition of glass and as well on cross link density and bond enthalpy values. It is observed that DSC curves for the present series shows an endothermic peak that corresponds to glass transition temperature. From the observed values it was observed that transition temperatures (T_g) of the glass series TMB increased from 311°C to 350°C which indicates the rigidity of the glass samples. Transition temperature of presented samples increased with increasing content of BaO. As the concentration of MoO_3 is reducing and is replaced by BaO transition temperatures increased. This may be due to increase in the crosslink density and average force constant. The bond strength of BaO is 138 KJ/mol and the bond strength of MoO_3 is 386 KJ/mol. Crystalline onset temperature (T_o) shows an increase from TMB1 to TMB3 and also shows a slight lower value for TMB5 sample. This indicates that the crystalline onset temperature increased as MoO_3 is replaced with BaO. Thermal stability (ΔT) value of TMB glass system decreased with an increasing BaO content[22].

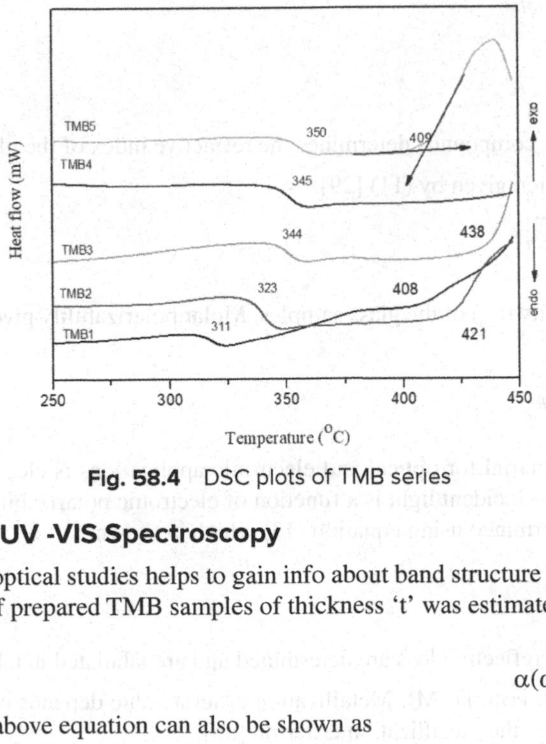

Fig. 58.4 DSC plots of TMB series

Table 56.7 DSC data of TMB series

Sample code	T_g(°C)	T_o(°C)	ΔT(°C)
TMB1	311	408	97
TMB2	323	421	98
TMB3	344	438	94
TMB4	345	--	--
TMB5	350	409	59

3.6 UV -VIS Spectroscopy

The optical studies helps to gain info about band structure and energy gap of a material [23]. The coefficient optical absorption (α) of prepared TMB samples of thickness 't' was estimated using the equation (4) or (5) [24].

$$\alpha(\omega) = \frac{\ln\frac{I_o}{I}}{t} \tag{4}$$

The above equation can also be shown as

$$\alpha(\omega) = \frac{2.303A}{t} \tag{5}$$

here absorbance is represented by A and sample thickness by t.

In amorphous materials $\alpha(\omega)$ depends exponentially on photon energy which is expressed by the equation (6) [25]

$$\alpha(\omega) = B\exp\left(\frac{hv}{\Delta E}\right) \tag{6}$$

where ΔE represents band tail width of localized state and B represents a constant.

values are calculated using Mott and Davis relation [26]given by equation (7)

$$\alpha hv = B(hv - E_{opt})^n \tag{7}$$

where B represents a constant, $h\nu$ is energy of incident photon and value of n depends on the transition type. For indirect transition n takes the value 2 and for direct transition it takes the value ½ [27] for two optical transition in k-space. Direct transition may occur between lowest minimum of the conduction band and highest maximum of valence band lie in different k space regions and indirect transition occurs with transition from top of valence band to bottom of conduction band. In case of indirect band gap the absorption coefficient increases relatively slowly with the increase of the photon energy till the photon energy crosses direct energy gap and then the absorption coefficient increases rapidly with further increase in incident photons. It is well known from the studies that amorphous materials fit for indirect band gap equation that is given by the equation (8)

$$(\alpha h\nu)^{1/2} = B(h\nu - E_{opt}) \tag{8}$$

From the linear region of the Tauc's plot shown in Fig. 56.6, indirect band gap energy was determined. Equation (9) is used to determine direct band gap energies. It is determined from the linear regions of Tauc's plots given by Fig. 56.7.

$$(\alpha h\nu)^2 = B(h\nu - E_{opt}) \tag{9}$$

Both indirect and direct optical band gap energy values are evaluated and are given in table 56.8.

Urbach energies are determined from the reciprocal values of slopes of linear regions of plot between ln (α) verses $h\nu$ that is shown in Fig. 56.8.

Using the relation given by the equation (10) proposed by Dimitrov and Sakka [28]refractive index (n) is determined as

$$\frac{n^2-1}{n^2+2} = 1-\left(\frac{E_{opt}}{20}\right)^{1/2} \tag{10}$$

Where E_{opt} represents indirect band gap.

The interaction of light with the electrons of atoms of glass network compounds determines the refractive index of the glass.

The estimation of R_m of prepared glasses is done by using the equation given by (11) [29].

$$R_m = V_m\left[1-\sqrt{\frac{E_{opt}}{20}}\right] \tag{11}$$

Lorentz-Lorentz equation (12) is used to calculate Molar polarizability (α_m) of the glass samples. Molar polarizability gives the no. of electrons recognized with an applied electric field.

$$\alpha_m = V_m\left[\frac{3}{4\pi N_A}\right]R_m \tag{12}$$

One of the important parameters that make the glass a potential material for optical and electronic applications is electronic polarizability. The non-linear response exhibited by the glasses to the incident light is a function of electronic polarizability of elements that constitute the glass matrix. Reflection loss (R_L) is determined using equation (13) which is shown.

$$R_L = \frac{V_M}{R_M} \tag{13}$$

Here V_M is molar volume and R_M is molar refractivity. The values of reflection loss are determined and are tabulated in table 8.

The nature of material formed can be characterized by metallization criteria (M). Metallization criteria value depends on the values of reflection loss. Following equation (14) is used to determine the metallization criterion [30].

$$M = 1-\frac{V_M}{R_M} \tag{14}$$

Optical parameters are determined and tabulated in Table 56.8. It is found that cut-off wavelength values showed non-linear variation. Cut-off wavelengths showed increment from TMB1 to TMB5 whose values vary from 541nm to 621nm. An absorption band at around 600 nm is exhibited by the prepared samples that correspond to the $^2B_{2g} \rightarrow ^2E_g$ transitions of vanadium ion [31]. The optical energies for both direct and indirect transitions found to decrease from TMB1 to TMB5. The decrease in the value of E_{opt} for both direct and indirect transitions is attributed to the presence of Non-bridging oxygens in TMB network resulting in increase in disorderness. Urbach energy helps to characterize the degree of disorder in the glass system. Its value showed an increase. The increase in its energy values indicates an increment in fragility of glasses which is the increase in TeO_3 concentration that changed the glass structure which indicates the formation of defects that causes decrease in band gap value. In the present work as the MoO_3 is replaced with BaO it is found that refractive index values showed an increase from 2.955 to 2.992 from TMB1 to TMB5.

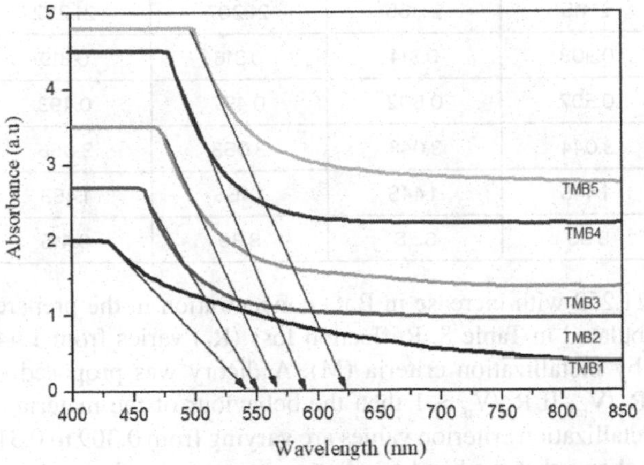

Fig. 56.5 Optical absorption spectra of TMB series

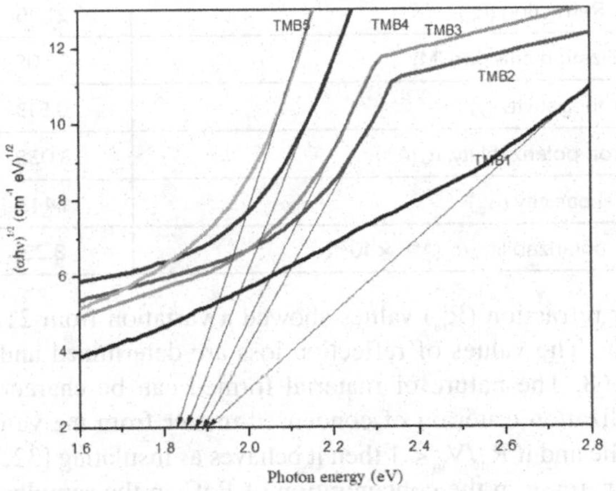

Fig. 56.6 Tauc's plots to evaluate indirect band gap energies of TMB glass series

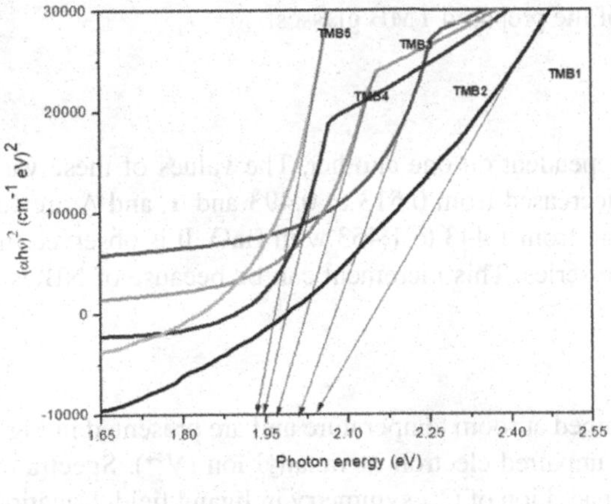

Fig. 56.7 Tauc's plots to evaluate direct band gap energies of TMB glass series

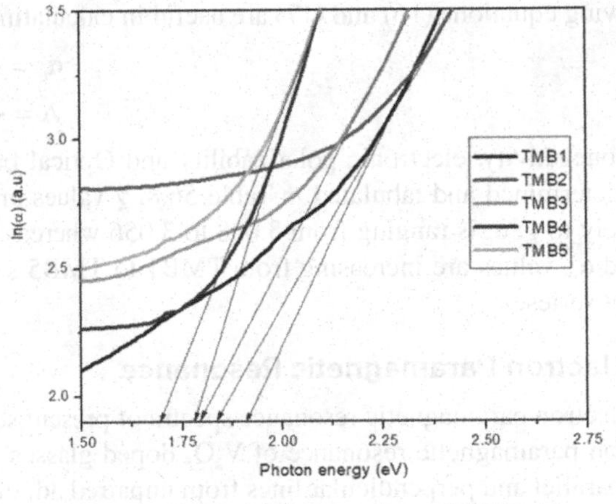

Fig. 56.8 Urbach energy plots of TMB series

Table 56.8 Optical data of TMB series

Sample code	TMB1	TMB2	TMB3	TMB4	TMB5
Cut-off wavelength (nm)	541	550	565	589	621
E_{opt} indirect band gap (eV)	1.91	1.89	1.87	1.85	1.83
E_{opt} direct band gap (eV)	2.04	2.01	1.97	1.94	1.93
Urbach energy (ΔE)	0.48	0.49	0.51	0.52	0.53
Refractive index (n)	2.957	2.964	2.975	2.984	2.992
Reflection Loss R_L	1.447	1.446	1.457	1.463	1.468
Molar Refraction (R_m)	21.06	21.115	21.165	21.207	21.252
Metallization criterion (M)	0.309	0.308	0.314	0.316	0.319
Electronegativity (χ)	0.513	0.507	0.502	0.497	0.493
Electron polarizability, α_e $(A^o)^3$	3.038	3.044	3.048	3.053	3.056
Optical basicity (Λ_{th})	1.443	1.446	1.449	1.466	1.453
Molar polarizability, α_m $(A^o)^3 \times 10^{24}$	8.33	8.36	8.38	8.39	8.415

Molar refraction (R_m) values showed a variation from 21.06 to 21.252 with increase in BaO concentration in the prepared sample. The values of reflection loss are determined and are tabulated in Table 8. Reflection loss (R_L) varies from 1.447 to 1.468. The nature of material formed can be characterized by metallization criteria (M). A theory was proposed on metallization criterion of condensed matter from the values of R_m/V_m. If $R_m/V_m > 1$ then the behaviour of the material is metallic and if $R_m/V_m < 1$ then it behaves as insulating [32, 33]. Metallization criterion values are varying from 0.309 to 0.319 with increase in the concentration of BaO in the samples which shows that the band gap between valence and conduction band is increasing by decreasing the width of both the bands. From the values of M it is concluded that the behaviour of material is metallic. Electronegativity (χ) of the TMB series is determined by using the equation given by (15) [34].

$$\chi = 0.2688\, E_g \tag{15}$$

Following equations (16) and (17) are useful in calculating α_e and Λ of the prepared TMB glasses.

$$\alpha_e = -0.9\chi + 3.5 \tag{16}$$

$$\Lambda = -0.5\chi + 1.7 \tag{17}$$

Electronegativity, electronic polarizability and Optical basicity are dependent on one another. The values of these variables were determined and tabulated in Table 56.8. χ values are found to decreased from 0.513 to 0.493 and α_e and Λ are varying inversely to χ. α_e is ranging from 3.038 to 3.056 whereas Λ is changing from 1.443 to 1.453 with BaO. It is observed that the R_m and α_m values are increasing from TMB1 to TMB5 samples in the series. This increment can be because of NBOs in the present series.

3.7 Electron Paramagnetic Resonance

The Electron paramagnetic resonance spectra of present series are obtained at room temperature and are presented in Fig. 56.9. Electron paramagnetic resonance of V_2O_5 doped glasses arise due to unpaired electron in vanadyl ion (V^{4+}). Spectra exhibit eight parallel and perpendicular lines from unpaired $3d_1$ electron of vanadyl ion of C_{4V} symmetry in ligand field. Equation (18) gives spin-Hamiltonian which is employed in the analysis [35, 36].

$$H = \beta\{g_\perp H_z S_z + g_\parallel (H_x S_x + H_y S_y)\} + A_\parallel I_z S_z + A_\perp (S_x I_x + S_y I_y) \tag{18}$$

where H_X, H_y and H_Z are the components of magnetic field. Here spin component of electron are given by S_x, S_y, S_z and the nucleus are given by I_x, I_y, I_z. The Spin-Hamiltonian parameters of both parallel and perpendicular lines are specified by the following equations (19) and (20) respectively as

$$H_\parallel(m) = H_\parallel(0) - mA_\parallel \left\{\frac{63}{4} - m^2\right\} \frac{A_\perp^2}{2H_\parallel(0)} \tag{19}$$

$$H_{\perp}(m) = H_{\perp}(0) - mA_{\perp}\left\{\frac{63}{4} - m^2\right\}\frac{A_{\perp}^2 + A_{\parallel}^2}{4H_{\perp}(0)} \quad (20)$$

$g_{\parallel}$ and $g_{\perp}$ depends upon local site symmetry of the field in the glass network. Spin-Hamiltonian parameter values of the VO^{2+} ion are calculated for different samples of TMB glass system are reported in Table 56.9.

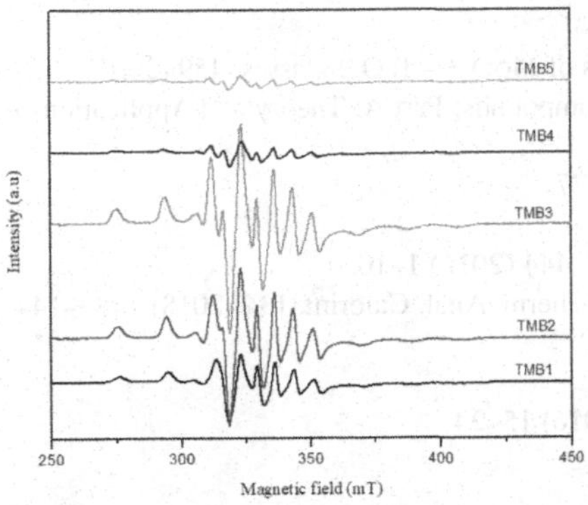

Fig. 56.9 EPR spectra of TMB series

Table 56.9 Spin Hamiltonian parameters of TMB series

Sample code	$g_{\parallel}$	$g_{\perp}$	$A_{\parallel} \times 10^{-4}$ (cm^{-1})	$A_{\perp} \times 10^{-4}$ (cm^{-1})	$\Delta g_{\parallel}/\Delta g_{\perp}$
TMB1	1.901	1.941	197	58	1.652
TMB2	1.922	1.989	195	60	6.037
TMB3	1.935	1.998	194	54	15.651
TMB4	1.912	1.985	193	51	5.219
TMB5	1.965	1.997	185	42	7.037

It is observed that spin Hamiltonian parameter values $g_{\parallel}$, $g_{\perp}$, $A_{\parallel}$ and $A_{\perp}$ showed that the glass network is influencing local symmetries of the vanadium ions. It is clear from $g_{\parallel}$, $g_{\perp}$, $A_{\parallel}$, $A_{\perp}$ and $\Delta g_{\parallel}/\Delta g_{\perp}$ values that the vanadium is present distorted octahedral symmetry. In the present glass system we have obtained that $g_{\parallel} < g_{\perp} < g_e$ and $|A_{\parallel}| > |A_{\perp}|$ which confirms that the vanadium occupied octahedral positions with tetragonal compression [37]. The value of $\Delta g_{\parallel}/\Delta g_{\perp}$ gives the degree of distortion and is observed that it varies nonlinearly in the present glasses.

4. Conclusion

The TMB glass system with composition $69TeO_2$-(30-x)MoO_3-xBaO with 1 mol% V_2O_5 is synthesized by conventional melt-quenching technique and amorphous character of TMB glasses are proven by X-ray diffraction spectra. The physical parameters were calculated and were established that the density increased and molar volume values also increased and OPD decreased when MoO_3 is replaced by BaO. FT-IR and Raman spectroscopy reveals that the glass system consists of TeO_4, TeO_3 and MoO_4 and MoO_6 units. It also shows that Mo coordination changes from 4 to 6. From DSC studies it is found that the glass transition temperature of the present samples increases with increased content of BaO. Thermal stability is observed to be high for TMB2 sample. Optical properties showed that direct and indirect band gap values decreased from TMB1 to TMB5 sample ranging from 2.00-1.95 eV for direct band gap and from 1.91 to 1.85 for indirect band gap energies. Urbach energy values were also determined. Refractive index of the glass series increased from 2.957 to 2.984. Molar refraction, Reflection loss (R_L), Metallization criterion (M), polarizability and electronegativity were also determined. From EPR studies Spin Hamiltonian parameters are also evaluated which reveals that VO^{2+} exists in octahedral site with tetragonal distortion.

REFERENCES

1. El. Mallawany R, The optical properties of Tellurite glasses, J. Appl. Phys. (1992)1774-4.
2. M.R. Zaki, D. Hamani, M. Dutreilh-Colas, J.R. Duclère, O. Masson, P. Thomas, J. Non. Cryst. Solids. 484 (2018) 139–148.
3. G.W. Brady, J. Chem. Phys. 27 (1957) 300–303.
4. S. Badamasi, Y.A. Tanko, Sci. World J. 13 (2018) 95–99.
5. M. Lesniak, R. Szal, B. Srarzyk, M. Gajek, M. Kochanowicz, J. Zmojda, P. Dorosz, M. Sitarz, D. Dorosz, J. of Thermal Analysis and Calorimetry 138, 4295-4302 (2019).
6. Mc Laughlin JC, Tagg SL, Zwanziger JW, J. Phys Chem B, 105 (2001) 67–75.
7. I.J. Badovinac, I.K. Piltaver, I. Šariá, R. Peter, M. Petravić, Mater. Tehnol. 51 (2017) 617–621. https://doi.org/10.17222/mit.2016.199.
8. S.V. Pershina, A.A. Raskovalov, B.D. Antonov, T.V. Yaroslavtseva, O. G. Reznitskikh, Ya.V. Baklanova, E.D. Pletneva, J. Non Cryst. Solids 430 (2015) 64–72.

9. A. Azuraida, M.K. Halimah, M. Ishak, L. Hasnimulyati, S.I. Ahmad, Chalcogenide Lett. 17 (2020) 187–196.
10. M. Guignard, J.W. Zwanziger, J. Non. Cryst. Solids. 353 (2007) 1662–1664.
11. A. Kaur, A. Khanna, L.I. Aleksandrov, J. Non. Cryst. Solids. 476 (2017) 67–74.
12. E.S. Nurbaisyatul, K. Azman, H. Azhan, W.A.W. Razali, A. Noranizah, S. Hashim, Y.S.M. Alajerami, Opt. Spectrosc. (English Transl. Opt. i Spektrosk. 116 (2014) 413–417.
13. S.S. Al-rawi, A.H. Jassim, H.A. Al-hilli, FTIR Spectra of Molybdenum Tellurite Glasses, 2 (2006) 23–25.
14. N. Elkhoshkhany, H.M. Mohamed, E. Sayed Yousef, Results Phys. 13 (2019) 102222.
15. D. Souri, Middle-East J. Sci. Res. 5 (2010) 44–48.
16. M. Pal, K. Hirota, Y. Tsujigami, H. Sakata, Structural and electrical properties of MoO 3 – TeO 2 glasses, 459 (2001).
17. K. Nakamoto, Infrared and Raman Spectra of Inorganic and Coordination Compounds: Part A: Theory and Applications in Inorganic Chemistry: Sixth Edition, 2008.
18. C. Duverger, M. Bouazaoui, S. Turell, J. Non-Cryst. Solids 220 (1997) 169–177.
19. Swapna, G. Upender, M. Prasad, J. Taibah Univ. Sci. 11 (2017) 583–592.
20. A. Kaur, A. Khanna, F. González, C. Pesquera, B. Chen, J. Non. Cryst. Solids. 444 (2016) 1–10.
21. J.L. Gomes, A. Gonçalves, A. Somer, J. V. Gunha, G.K. Cruz, A. Novatski, J. Therm. Anal. Calorim. 134 (2018) 1439–1445.
22. E.R. Shaaban, Y.B. Saddeek, M. Abdel Rafea, Philos. Mag. 89 (2009) 27–39.
23. A.A. Higazy, A. Hussein, Radiat. Eff. Defects Solids. 133 (1995) 225–235.
24. V. Uma, K. Maheshvaran, K. Marimuthu, G. Muralidharan, J. Lumin. 176 (2016) 15–24.
25. M.A. Hassan, C.A. Hogarth, J. Mater. Sci. 23 (1988) 2500.
26. E.A. Davis, N.F. Mott, Phil. Mag. 22 (1970) 903.
27. S.H. Alazoumi, S.A. Aziz, R. El-Mallawany, U.S. ad Aliyu, H.M. Kamari, M.H.M.M. Zaid, K.A. Matori, A. Ushah, Results Phys. 9 (2018) 1371–1376.
28. V. Dimitrov, S. Sakka, J. Appl. Phys. 79 (1996) 1736–1740.
29. T. Honma, R. Sato, Y. Benino, T. Komatsu, V. Dimitrov, J. Non-Cryst. Solids 272 (2000) 1.
30. K. Pradeesh, J. X. Oton, V.K. Agotiya, M. Raghavendra, G. Vijaya Prakash, Opt. Mater. 31 (2008) 155.
31. G. Keerti Marita, J. Santhan Kumar, Sandhya Cole, International J. of Physics and Research, 4 (2013) 17.
32. MN. Azlan, MK. Halimah, SZ. Shafinas, WM. Daud, P Chalcogenide lett. 11 (2014) 319.
33. V. Dimitrov, T. Komatsu, J. Univ. Chem. Tech. Metallur. 45 (2010) 219.
34. X. Zhao, X. Wang, H. Lin, Z. Wang, Phys. B Condens. Matter 403 (2008) 2450–2460.
35. F. Mohd Fudzi, H.M. Kamari, A. Abd Latif, A. Muhammad Noorazlan, J. Nanomater. 2017 (2017). https://doi.org/10.1155/2017/4150802.
36. G. Hecht, T. S. Johntson, J. Chem. Phys. 46 (1967) 23.
37. M. Sugathri, K. Jyothi, P.M. Rao, B. Apparao, EPR and Optical Absorption studies on Vanadium doped Glasses, 10 (2017) 267–273.

Note: All the figures and tables in this chapter were made by the authors.

Impending Inquisitions in Humanities and Sciences (ICIIHS-2022) – Dr. Mohan Varkolu et al. (eds)

A Unique Common Coupled Fixed Point Theorem for Rational Contraction in Partial–*b*–Metric Spaces

57

G. N. V. Kishore[1]
Department of Engineering Mathematics and Humanities,
SRKR Engineering College, Bhimavaram, Andhra Pradesh, India

Y. Phani*
Department of Mathematics, Shri Vishnu Engineering College for Women(A),
Vishnupur, Bhimavaram, AP, India

B. V. Appa Rao[2]
Department of Engineering Mathematics, Koneru Lakshmaiah Education Foundation,
Vaddeswaram, Guntur, Andhra Pradesh, India

G. Suri Babu[3]
WET Centre, Department of Civil Engineering, SRKR Engineering College,
Bhimavaram, Andhra Pradesh, India

ABSTRACT In this paper we proved a common coupled fixed point theorem for Jungck type maps and using rational contraction condition in partial–*b*–metric space.

KEYWORDS Partial *b*–metric space, Weakly compatible mapping, Rational contraction, Partial metric space

1. Introduction

In 1993, Czerwik [2],[3] extended results related to the *b*-metric spaces. In 1994, Matthews [4] introduced the concept of partial metric space in which the self-distance of any point of space may not be zero. In 2013, Shukla [6] generalized both the concept of *b*-metric and partial metric spaces by introducing the partial *b*-metric spaces.

Bhaskar and Lakshmikantham [7] introduced the notion of a coupled fixed point and proved some coupled fixed point theorems for mixed monotone mappings in ordered metric spaces.

In this paper we proved a common coupled fixed point theorem for Jungck type maps in partial–*b*–metric spaces.

2. Preliminaries

In order to obtain our results we need to consider the followings.

Definition 2.1. ([1]) *Let X be a nonempty set and let $s \geq 1$ be a given real number. A function $d : X \times X \to [0,\infty)$ is called a $b-$ metric if for all $x,y,z \in X$ the following conditions are satisfied:*

*Correspondence Author: phaniyedlapalli23@gmail.com
[1]kishore.apr2@gmail.com, gnvkishore@srkrec.ac.in, [2]bvardr2010@kluniversity.in, [3]gsuribabu8@gmail.com

DOI: 10.1201/9781003489436-57

$d(x,y) = 0$ *if and only if* $x = y$;

$d(x,y) = d(y,x)$;

$d(x,y) \leq s(d(x,z) + d(z,y))$.

The pair (X, d) is called a b–metric space. The number $s \geq 1$ *is called the coefficient of (X,d).*

Definition 2.2. ([4]) *Let X be a nonempty set. A function* $P : X \times X \to [0,\infty)$ *is called a partial metric if if for all* $x,y,z \in X$ *the following conditions are satisfied:*

(i) $x = y$ *if and only if* $P(x, x) = P(x, y) = P(y, y)$;

(ii) $P(x, x) \leq P(x, y)$;

(iii) $P(x, y) = P(y, x)$;

(iv) $P(x, y) \leq P(x, z) + P(z, y) - P(z, z)$.

The pair (X, P) is called a partial metric space.

Remark 2.3. *It is clear that the partial metric space need not be a metric spaces, since in a b - metric space if* $x = y$, *then* $d(x, x) = d(x, y) = d(y, y) = 0$. *But in a partial metric space if* $x = y$ *then* $P(x, x) = P(x, y) = P(y, y)$ *may not be equal zero. Therefore the partial metric space may not be a b-metric space.*

On the other hand, Shukla ([6]) introduced the notion of a partial b-metric space as follows:

Definition 2.4. ([6]) *Let X be a nonempty set and* $s \geq 1$ *be a given real number. A function* $P_b : X \times X \to [0, \infty)$ *is called a partial b-metric if if for all x, y,* $z \in X$ *the following conditions are satisfied:*

(P_b1) $x = y$ *if and only if* $P_b(x, x) = P_b(x, y) = P_b(y, y)$;

(P_b2) $P_b(x, x) \leq P_b(x, y)$; (P_b3) $P_b(x, y) = P_b(y, x)$;

(P_b4) $P_b(x, y) \leq s(P_b(x, z) + P_b(z, y) - P_b(z, z))$.

The pair (X, P_b*) is called a partial b–metric space. The number* $s \geq 1$ *is called the coefficient of (X,* P_b*).*

Remark 2.5. *The class of partial b-metric space (X,* P_b*) is effectively larger than the class of partial metric space, since a partial metric space is a special case of a partial b-metric space (X,* P_b*) when* $s = 1$. *Also, the class of partial b- metric space (X,* P_b*) is effectively larger than the class of b-metric space, since a b- metric space is a special case of a partial b–metric space (X,* P_b*) when the self - distance* $P(x;x) = 0$. *The following examples shows that a partial b– metric on X need not be a partial metric, nor a b-metric on X see also* ([5], [6]) [5, 6].

Example 2.6. ([6]) *Let* $X = [0,1)$. *Define a function* $P_b : X \times X \to [0,\infty)$ *such that* $P_b(x;y) = [\max\{x,y\}]^2 + |x - y|^2$, *for all* $x,y \in X$. *Then* (X,P_b) *is called a partial b–metric space with the coefficient* $s = 2 > 1$. *But,* P_b *is not a b-metric nor a partial metric on X.*

Definition 2.7. ([5]) *Every partial b-metric* P_b *defines a b-metric* d_{Pb}, *where* $d_{Pb}(x,y) = 2P_b(x,y) - P_b(x,y) - P_b(y,y)$, *for all* $x,y \in X$

Definition 2.8. ([5]) *A sequence* $\{x_n\}$ *in a partial b-metric space* (X,P_b) *is said to be:*

(i) P_b*-convergent to a point* $x \in X$ *if* $\lim_{n\to\infty} P_b(x, x_n) = P_b(x, x)$

(ii) *a* P_b*-Cauchy sequence if* $\lim_{n,m\to\infty} P_b(x_n, x_m)$ *exists and is finite;*

(iii) *A partial b-metric space* (X,P_b) *is said to be* P_b*-complete if every* P_b*Cauchy sequence* $\{x_n\}$ *in X is* P_b *converges to a point* $x \in X$ *such that*

$$\lim_{n,m\to\infty} P_b(x_n, x_m) = \lim_{n\to\infty} P_b(x, x_n) = P_b(x, x)$$

Lemma 2.9. ([5]) *A sequence* $\{x_n\}$ *is a* P_b*-Cauchy sequence in a partial b- metric space* (X,P_b) *if and only if it is a b-Cauchy sequence in the b-metric space* (X,d_{Pb}).

Lemma 2.10. ([5]) *A partial b-metric space* (X,P_b) *is* P_b*-complete if and only if the b-metric space* (X,d_{Pb}) *is b-complete. Moreover* $\lim_{n,m\to\infty} d_{Pb}(x_n, x_m) = 0$ *if and only if*

$$\lim_{m\to\infty} P_b(x_m,x) = \lim_{m\to\infty} P_b(x_n,x) = P_b(x,x)$$

Definition 2.11. ([7]) *Let X be a nonempty set. An element* $(x,y) \in X \times X$ *is called a coupled fixed point of the mapping* $F : X \times X \rightarrow X$ *if* $x = F(x, y)$ *and* $y = F(y, x)$.

Now we prove our main result.

3. Main Results

Theorem 3.1. *Let* (X, P_b) *Partial b- metric space, let* $R : X \times X \rightarrow X$ *and* $A : X \rightarrow X$ *be mappings satisfying the following Let f,* $T : (A, B) \rightrightarrows (A,B)$ *be a α-admissible mappings on complete bipolar metric space (A,B,d) such that* $f(A \cup B) \subseteq T(A \cup B)$*. Suppose that the following conditions are hold:*

(3.1.1) $s.P_b\,(R(x, y), R(u, v)) \leq k.\max \left\{ \begin{array}{c} P_b(Ax, Au), P_b(Ay, Av) \\ \frac{P_b(Ax,R(x,y)),P_b(Au,R(u,v))}{1+P_b(Ax,Au)}, \frac{P_b(Ay,R(y,x)),P_b(Av,R(v,u))}{1+P_b(Ay,Av)} \end{array} \right\}$

(3.1.2) $R(X \times X) \subseteq A(X)$ *and* $A(X)$ *is complete subspace of X*

(3.1.3) *(R, A) is weakly compatible pair.*

Then R and A have a unique common coupled fixed point in X.

Proof. Let x_0, y_0 be arbitrary points in X. From (3.1.2), There exist sequences $\{x_n\}$, $\{y_n\}$, $\{u_n\}$ and $\{v_n\}$ in X such that

$$R(x_n, y_n) = Ax_{n+1} = u_n \quad R(y_n, x_n) = Ay_{n+1} = v_n \text{ for all } n \geq 0.$$

For simplification we denote $R_n = \max\{P_b(u_n,u_{n+1}),P_b(v_n,v_{n+1})\}$.

Case(i): Suppose that $u_n = u_{n+1}$ and $v_n = v_{n+1}$ for some n

Claim: $un+1 = un+2$ and $vn+1 = vn+2$

Suppose $u_{n+1} \neq u_{n+2}$ and $v_{n+1} \neq v_{n+2}$. From (3.1.1) consider

$$\begin{aligned} P_b\,(u_{n+1},u_{n+2}) &\leq s.P_b\,(R\,(x_{n+1},y_{n+1}),R\,(x_{n+2},y_{n+2})) \\ &\leq k\max \left\{ \begin{array}{c} P_b\,(u_n,u_{n+1}),P_b\,(v_n,v_{n+1}) \\ \frac{P_b\,(u_n,u_{n+1}),P_b\,(u_{n+1},u_{n+2})}{1+P_b\,(u_n,u_{n+1})}, \frac{P_b\,(v_n,v_{n+1}),P_b\,(v_{n+1},v_{n+2})}{1+P_b\,(v_n,v_{n+1})} \end{array} \right\} \\ &\leq k.\max\{P_b\,(u_{n+1},u_{n+2}),P_b\,(v_{n+1},v_{n+2})\}. \end{aligned}$$

Similarly,

$$P_b\,(v_{n+1}, v_{n+2}) \leq k.\max\{P_b\,(u_{n+1}, u_{n+2}), Pb\,(v_{n+1}, v_{n+2})\}$$

It follows that

$$\begin{aligned} \max\{P_b\,(u_{n+1}, u_{n+2}), P_b\,(v_{n+1}, v_{n+2})\} &\leq k.\max\{P_b\,(u_{n+1}, u_{n+2}), P_b(v_{n+1}, v_{n+2})\} \\ &< \max\{P_b\,(u_{n+1}, u_{n+2}), P_b\,(v_{n+1}, v_{n+2})\}, \end{aligned}$$

is contradiction. Hence $u_{n+1} = u_{n+2}$ and $v_{n+1} = v_{n+2}$.

Continuing this way we can conclude that $u_n = u_{n+l}$ and $v_n = v_{n+l}$ for all l.

It is clear that $\{u_n\}$ and $\{v_n\}$ are Cauchy sequence in (X, P_b).

Case (ii): Suppose that $u_n \neq u_{n+1}$ and $v_n \neq v_{n+1}$ for all $n \geq 0$.

From (3.1.1), we have that

$$\begin{aligned} P_b\,(u_n,u_{n+1}) &\leq s.P_b\,(R\,(x_n,y_n),R\,(x_{n+1},y_{n+1})) \\ &\leq k\max \left\{ \begin{array}{c} P_b\,(u_{n-1},u_n),P_b\,(v_{n-1},v_n) \\ \frac{P_b\,(u_{n-1},u_n),P_b\,(u_n,u_{n+1})}{1+P_b\,(u_{n-1},u_n)}, \frac{P_b\,(v_{n-1},v_n),P_b\,(v_n,v_{n+1})}{1+P_b\,(v_{n-1},v_n)} \end{array} \right\} \\ &\leq k.\max\{P_b\,(u_{n-1},u_n),P_b\,(v_{n-1},v_n),P_b\,(u_n,u_{n+1}),P_b\,(v_n,v_{n+1})\}. \\ &\leq k.\max\{R_{n-1},R_n\}. \end{aligned}$$

Similarly,

$$P_b(v_n, v_{n+1}) \leq k.\max\{R_{n-1}, R_n\}.$$

Thus $R_n \leq k.\max\{R_{n-1}, R_n\}$.

If R_n is maximum, we have $R_n \leq k.R_n < R_n$. Which is a contradiction. Hence R_{n-1} is maximum. Thus

$$\begin{aligned} R_n &\leq k.R_{n-1} \\ &\leq R_{n-1} \end{aligned} \tag{1}$$

Therefore, $\{R_n\}$ is decreasing sequence and converges to $t \geq 0$. Suppose $t > 0$, letting $n \to \infty$ in (1),we have that $t \leq k.t < t$, is a contradiction . Hence $t = 0$.

Thus $\lim_{n\to\infty} R_n = 0$. It follows that

$$\lim_{n\to\infty} P_b(u_n, u_{n+1}) = 0 = \lim_{n\to\infty} P_b(v_n, v_{n+1}). \tag{2}$$

From (2) and (P_b2), we have that

$$\lim_{n\to\infty} P_b(u_n, u_n) = 0 = \lim_{n\to\infty} P_b(v_n, v_n). \tag{3}$$

From definition of d_{Pb}, (2) and (3), we have that

$$\lim_{n\to\infty} d_b(u_n, u_{n+1}) = 0 = \lim_{n\to\infty} d_b(v_n, v_{n+1}). \tag{4}$$

Now we prove that $\{u_n\}$ and $\{v_n\}$ are Cauchy sequence in Partial –b-metric space (X,P_b). If sufficient to prove that $\{u_n\}$ and $\{v_n\}$ are Cauchy sequence in b- metric space (X,d_{Pb})

On contrary suppose that either $\{u_n\}$ and $\{v_n\}$ are not Cauchy sequence. This gives that $d_{Pb}(u_n,u_m) \nrightarrow 0$ and $d_{Pb}(v_n,v_m) \nrightarrow 0$ as $n, m \to\infty$. Consequently, $\max\{d_{Pb}(u_n,u_m),d_{Pb}(v_n,v_m)\} \nrightarrow 0$ as $n,m \to\infty$.

Then there exist an $\epsilon > 0$ and monotonically increases sequences of natural numbers $\{m_k\}$, $\{n_k\}$ such that $n_k > m_k > k$

$$\max\{d_{P_b}(u_{n_k}, u_{m_k}), d_{P_b}(v_{n_k}, v_{m_k})\} \geq \tag{5}$$

and

$$\max\{d_{P_b}(u_{n_k-1}, u_{m_k}), d_{P_b}(v_{n_k-1}, v_{m_k})\} < \epsilon. \tag{6}$$

From (5) and (6), we have that

$$\begin{aligned} \epsilon &\leq \max\{d_{P_b}(u_{n_k}, u_{m_k}), d_{P_b}(v_{n_k}, v_{m_k})\} \\ &\leq s.\max\{d_{P_b}(u_{m_k}, u_{n_k-1}), d_{P_b}(v_{m_k}, v_{n_k-1})\} + s.\max\{d_{P_b}(u_{n_k-1}, u_{n_k}), d_{P_b}(v_{n_k-1}, v_{n_k})\} \\ &< s.\epsilon + s.\max\{d_{P_b}(u_{n_k-1}, u_{n_k}), d_{P_b}(v_{n_k-1}, v_{n_k})\} \end{aligned}$$

Taking upper limit as $k \to\infty$ and from (2), we have that

$$\epsilon \leq \lim_{k\to\infty} \sup \max\{d_{P_b}(u_{n_k}, u_{m_k}), d_{P_b}(v_{n_k}, v_{m_k})\} \leq s.\epsilon. \tag{7}$$

Also

$$\begin{aligned} \epsilon &\leq \max\{d_{P_b}(u_{n_k}, u_{m_k}), d_{P_b}(v_{n_k}, v_{m_k})\} \\ &\leq s.\max\{d_{P_b}(u_{m_k}, u_{n_k+1}), d_{P_b}(v_{m_k}, v_{n_k+1})\} + s.\max\{d_{P_b}(u_{n_k+1}, u_{n_k}), d_{P_b}(v_{n_k+1}, v_{n_k})\} \end{aligned}$$

Taking upper limit as $k \to \infty$ and from (4), we have that

$$\frac{\epsilon}{s} \leq \lim_{k\to\infty} \sup \max\{d_{P_b}(u_{m_k}, u_{n_k+1}), d_{P_b}(v_{m_k}, v_{n_k+1})\}. \tag{8}$$

On other hand

$$\begin{aligned} &\max\{d_{P_b}(u_{m_k}, u_{n_k+1}), d_{P_b}(v_{m_k}, v_{n_k+1})\} \\ \leq\ & s.\max\{d_{P_b}(u_{m_k}, u_{n_k}), d_{P_b}(v_{m_k+1}, v_{n_k})\} + s.\max\{d_{P_b}(u_{n_k}, u_{n_k+1}), d_{P_b}(v_{n_k}, v_{n_k+1})\} \end{aligned}$$

Taking upper limit as $k \to \infty$ and from (4), we have that

$$\lim_{k\to\infty} \sup \max \{d_{P_b}(u_{m_k}, u_{n_k+1}), d_{P_b}(v_{m_k}, v_{n_k+1})\} \le \epsilon s^2 \tag{9}$$

also, from (5), we have that

$$\begin{aligned}
\epsilon &\le \max \{d_{P_b}(u_{n_k}, u_{m_k}), d_{P_b}(v_{n_k}, v_{m_k})\} \\
&\le s.\max \{d_{P_b}(u_{m_k}, u_{m_k+1}), d_{P_b}(v_{m_k}, v_{m_k+1})\} + s.\max \{d_{P_b}(u_{n_k+1}, u_{n_k}), d_{P_b}(v_{n_k+1}, v_{n_k})\} \\
&\le \left\{ \begin{array}{c} s.\max \{d_{P_b}(u_{m_k}, u_{m_k+1}), d_{P_b}(v_{m_k}, v_{m_k+1})\} \\ +s^2.\max \{d_{P_b}(u_{m_k+1}, u_{n_k+2}), d_{P_b}(v_{m_k+1}, v_{n_k+2})\} \\ +s^2.\max \{d_{P_b}(u_{n_k+2}, u_{n_k}), d_{P_b}(v_{n_k+2}, v_{n_k})\} \end{array} \right\} \\
&\le \left\{ \begin{array}{c} s.\max \{d_{P_b}(u_{m_k}, u_{m_k+1}), d_{P_b}(v_{m_k}, v_{m_k+1})\} \\ +s^2.\max \{d_{P_b}(u_{n_k+1}, u_{n_k+2}), d_{P_b}(v_{n_k+1}, v_{n_k+2})\} \\ +s^3.\max \{d_{P_b}(u_{n_k+2}, u_{n_k+1}), d_{P_b}(v_{n_k+2}, v_{n_k+1})\} \\ +s^3.\max \{d_{P_b}(u_{n_k+1}, u_{n_k}), d_{P_b}(v_{n_k+1}, v_{n_k})\} \end{array} \right\}
\end{aligned}$$

Taking upper limit as $k \to \infty$ and from (4), we have that

$$\frac{\epsilon}{s^3} \le \lim_{k\to\infty} \sup \max \{d_{P_b}(u_{m_k+1}, u_{n_k+2}), d_{P_b}(v_{m_k+1}, v_{n_k+2})\}.$$

On other hand

$$\begin{aligned}
&\max \{d_{P_b}(u_{m_k+1}, u_{n_k+2}), d_{P_b}(v_{m_k+1}, v_{n_k+2})\} \\
&\le s.\max \{d_{P_b}(u_{m_k+1}, u_{m_k}), d_{P_b}(v_{m_k+1}, v_{m_k})\} + s.\max \{d_{P_b}(u_{m_k}, u_{n_k+2}), d_{P_b}(v_{m_k}, v_{n_k+2})\} \\
&\le \left\{ \begin{array}{c} s.\max \{d_{P_b}(u_{m_k+1}, u_{m_k}), d_{P_b}(v_{m_k+1}, v_{m_k})\} \\ +s^2.\max \{d_{P_b}(u_{m_k}, u_{n_k}), d_{P_b}(v_{m_k}, v_{n_k})\} \\ +s^3.\max \{d_{P_b}(u_{n_k}, u_{n_k+1}), d_{P_b}(v_{n_k}, v_{n_k+1})\} \\ +s^3.\max \{d_{P_b}(u_{n_k+1}, u_{n_k+2}), d_{P_b}(v_{n_k+1}, v_{n_k+2})\} \end{array} \right\}
\end{aligned}$$

Taking upper limit as $k \to \infty$ and from (4), (7) we have that

$$\lim_{k\to\infty} \sup \max \{d_{P_b}(u_{m_k}, u_{n_k+1}), d_{P_b}(v_{m_k}, v_{n_k+1})\} \le \epsilon s^2 \tag{10}$$

Now

$$\begin{aligned}
P_b(u_{m_k+1}, u_{n_k+2}) &\le s.P_b(u_{m_k+1}, u_{n_k+2}) \\
&\le s.P_b(R(x_{m_k+1}, y_{m_k+1}), R(x_{n_k+2}, y_{n_k+2})) \\
&\le k.max \left\{ \begin{array}{c} P_b(u_{m_k}, u_{m_k+1}), P_b(v_{m_k}, v_{m_k+1}) \\ \frac{P_b(u_{m_k}, u_{m_k+1}), P_b(u_{n_k+1}, u_{n_k+2})}{1+P_b(u_{m_k}, u_{n_k+1})}, \frac{P_b(v_{m_k}, v_{m_k+1}), P_b(v_{n_k+1}, v_{n_k+2})}{1+P_b(v_{m_k}, v_{n_k+1})} \end{array} \right\}.
\end{aligned}$$

Similarly,

$$P_b(v_{m_k+1}, v_{n_k+2}) \le k.max \left\{ \begin{array}{c} P_b(u_{m_k}, u_{m_k+1}), P_b(v_{m_k}, v_{m_k+1}) \\ \frac{P_b(u_{m_k}, u_{m_k+1}), P_b(u_{n_k+1}, u_{n_k+2})}{1+P_b(u_{m_k}, u_{n_k+1})}, \frac{P_b(v_{m_k}, v_{m_k+1}), P_b(v_{n_k+1}, v_{n_k+2})}{1+P_b(v_{m_k}, v_{n_k+1})} \end{array} \right\}.$$

Thus

$$\max \left\{ \begin{array}{c} P_b(u_{m_k+1}, u_{n_k+2}), \\ P_b(v_{m_k+1}, v_{n_k+2}) \end{array} \right\} \le k.max \left\{ \begin{array}{c} P_b(u_{m_k}, u_{m_k+1}), P_b(v_{m_k}, v_{m_k+1}) \\ \frac{P_b(u_{m_k}, u_{m_k+1}), P_b(u_{n_k+1}, u_{n_k+2})}{1+P_b(u_{m_k}, u_{n_k+1})}, \\ \frac{P_b(v_{m_k}, v_{m_k+1}), P_b(v_{n_k+1}, v_{n_k+2})}{1+P_b(v_{m_k}, v_{n_k+1})} \end{array} \right\}.$$

Taking upper limit as $k \to \infty$ and from (2), (9) and (10) we have that

$$\epsilon.s^3 \le k\epsilon s^2$$

Sub Case (i): If $s = 1$

$$\epsilon \leq k.\epsilon < \epsilon$$

is contradiction. Sub Case (ii): If $s > 1$

$$\epsilon.s^3 \leq k\epsilon s^2$$

It follows that $s \leq k < 1$, is contradiction. Hence, $\{u_n\}$ and $\{v_n\}$ are Cauchy sequence in (X, d_b).

Suppose $A(x)$ is complete subspace of X. Then $\{Ax_{n+1}\}$ and $\{Ay_{n+1}\}$ are converges to a and b in $(A(X),d_{Pb})$, thus $d_{Pb}(Ax_{n+1},a) = 0 = d_{Pb}(Ay_{n+1},b)$ for some $a = Ax$ and $b = Ay$. we have that

$$P_b(a,a) = \lim_{n,m\to\infty} P_b(Ax_n, Ax_m) = \lim_{n\to\infty} P_b(Ax_n, a) = \lim_{n\to\infty} P_b(Ax_{n+1}, a) = 0. \tag{11}$$

and

$$P_b(b,b) = \lim_{n,m\to\infty} P_b(Ay_n, Ay_m) = \lim_{n\to\infty} P_b(Ay_n, b) = \lim_{n\to\infty} P_b(Ay_{n+1}, b) = 0. \tag{12}$$

Now we claim that $R(x, y) = a$ and $R(y, x) = b$ From (3.1.1), we have that

$$\begin{aligned} & P_b\left(R(x,y), R(x_n,y_n)\right) \\ \leq\ & s.P_b\left(R(x,y), R(x_n,y_n)\right) \\ \leq\ & \max\left\{ \begin{array}{c} P_b\left(a, Ax_n\right), P_b\left(b, Ay_n\right) \\ \frac{P_b(a,R(x,y)).P_b(Ax_n,Ax_{n+1})}{1+P_b(a,Ax_n)}, \frac{P_b(b,R(y,x)).P_b(Ay_n,Ay_{n+1})}{1+P_b(b,Ay_n)} \end{array} \right\} \to 0 \text{ as } n \to \infty. \end{aligned}$$

It follows that $R(x, y) = a = Ax$.

Similarly, $R(y, x) = b = Ay$.

Since (R, A) is weakly compatible pair, We have $R(a, b) = Aa$ and $R(b, a) = Ab$.

From (3.1.1), we have that

$$\begin{aligned} & P_b\left(R(a,b), R(x_n,y_n)\right) \\ \leq\ & s.P_b\left(R(a,b), R(x_n,y_n)\right) \\ \leq\ & \max\left\{ \begin{array}{c} P_b\left(Aa, Ax_n\right), P_b\left(Ab, Ay_n\right) \\ \frac{P_b(Aa,R(a,b)).P_b(Ax_n,Ax_{n+1})}{1+P_b(a,Ax_n)}, \frac{P_b(b,R(b,a)).P_b(Ay_n,Ay_{n+1})}{1+P_b(b,Ay_n)} \end{array} \right\}. \end{aligned}$$

Letting $n \to \infty$, we have that

$$P_b(R(a, b), a) \leq k. \max\{P_b(R(a, b), a), P_b(R(b, a), b)\}$$

Similarly we can also prove that

$$P_b(R(b, a), b) \leq k. \max\{P_b(R(a, b), a), P_b(R(b, a),b)\}$$

Thus

$$\max\{P_b(R(a, b), a), P_b(R(b, a), b)\} \leq k. \max\{P_b(R(a, b), a), P_b(R(b, a), b)\}$$

It follows that $R(a, b) = a = Aa$ and $R(b, a) = b = Ab$. Therefore (a, b) is common coupled fixed point of R and A for uniqueness let us suppose (w, z) be another common coupled fixed point of R and A. such that $a \neq w$ and $b \neq z$. Now from (3.1.1), we have that

$$\begin{aligned} & P_b(a,z) = P_b\left(R(a,b), R(z,w)\right) \\ \leq\ & s.P_b\left(R(a,b), R(z,w)\right) \\ \leq\ & \max\left\{ \begin{array}{c} P_b\left(a, z\right), P_b\left(b, w\right) \\ \frac{P_b(a,a).P_b(b,b)}{1+P_b(a,b)}, \frac{P_b(z,z).P_b(w,w)}{1+P_b(z,w)} \end{array} \right\}. \\ \leq\ & k.\max\left\{P_b(a,z), P_b(b,w)\right\}. \end{aligned}$$

Similarly,

$$P_b(b, w) \leq k. \max\{P_b(a, z), P_b(b, w)\}$$

Therefore,

$$\max\{P_b(a, z), P_b(b, w)\} \leq k. \max\{P_b(a, z), P_b(b, w)\} < \max\{P_b(a, z), P_b(b, w)\}$$

It is a contradiction. Hence (a, b) is unique common coupled fixed point of R and A.

Now we prove that $a = b$ From (3.1.1), we have

$$\begin{aligned} P_b(a,b) &= P_b\left(R(a,b), R(b,a)\right) \\ &\leq s.P_b\left(R(a,b), R(b,a)\right) \\ &\leq \max\left\{\begin{array}{c} P_b(a,b), P_b(b,a) \\ \frac{P_b(a,a).P_b(b,b)}{1+P_b(a,b)}, \frac{P_b(a,a).P_b(b,b)}{1+P_b(a,b)} \end{array}\right\}. \\ &\leq k.P_b(a,b). \end{aligned}$$

It follows that $a = b$. Hence (a, a) is unique common coupled fixed point of R and A.

Acknowledgments: The authors are very thanks to the reviewers and editors for valuable comments, remarks and suggestions for improving the content of the paper.

REFERENCES

1. H. Aydi, M. Bota, E. Karapinar, S. Mitrovic, *A fixed point theorem for set valued quasicontractions in b-metric spaces*, Fixed Point Theory and its Applications, 2012, 2012:88, 8 pp.
2. S. Czerwik, *Nonlinear set-valued contraction mappings in b-metric spaces*, Atti Sem. Mat. Fis. Univ. Modena. 46 (1998), 263 - 276.
3. S. Czerwik, *Contraction mappings in b-metric spaces*, Acta Math. Inform. Univ. Osrav., 1 (1993), 5-11.
4. S. G. Matthews, *Partal metric topology*, Proc. 8th Summer Conference on General Topology and Applications, Ann. N.Y. Acad. Sci., 728 (1994), 183-197.
5. Z. Mustafa, J. R. Roshan, V. Parvaneh, Z. Kadelburg, *Some common fixed point result in ordered partial b-metric spaces*, Journal of Inequalities and Applications, (2013), 2013:562.
6. S. Shukla, *Partial b-metric spaces and fixed point theorems*, Mediterranean Journal of Mathematics, doi:101007/s00009-013-0327-4, (2013).
7. T. G. Bhaskar, V. Lakshmikantham, *Fixed point theorems in partially ordered metric spaces and applications*, Nonlinear Anal. 65 (2006) 1379–1393.

Impending Inquisitions in Humanities and Sciences (ICIIHS-2022) – Dr. Mohan Varkolu et al. (eds)
© 2024 Taylor & Francis Group, London, ISBN 978-1-032-78829-6

Higher Overtone Vibrational Frequencies of Functional Erythrocytes using U(2) Lie Algebraic Model

58

P. Suneetha[1]
Department of Mathematics,
V. R. Siddhartha Engineering College, Vijayawada, India
J. Vijayasekhar[2]
Department of Mathematics,
School of Science, GITAM (Deemed to beUniversity), Hyderabad, India

ABSTRACT In this study, a significant insight has been garnered by applying the U(2) Lie algebraic model. Specifically, the investigation focuses on the computation of Raman spectra about the first and second overtone vibrational frequencies within the context of oxygenated and deoxygenated functional erythrocytes. Within this framework, the Hamiltonian operator associated with the Lie algebraic model delineates vibrational modes containing C-H and C-C stretching.

KEYWORDS Lie algebraic model, Hamiltonian operator, Metalloporphyrins, Functional erythrocytes

1. Introduction

The Lie algebraic model [Iachello et al (1995), Oss (1996)] is a mathematical tool applied to analyze molecular vibrations from a symmetry perspective and interpret experimental molecular physics observations. The Lie algebraic model identifies the molecular point group, which describes the symmetry operations that leave the molecule invariant. The symmetry operations set encompasses rotations, reflections, inversions, and other transformations. The molecular point group is characterized by a mathematical group that is accompanied by a corresponding Lie algebra. With the help of the Lie algebraic model, symmetry-adapted basis functions can be made that have transformation properties that match the irreducible representations of the molecular point group. The use of these basis functions enables the analysis of vibrational modes. The selection rules for vibrational-rotational couplings can be determined by studying the Lie algebraic properties. The Hamiltonian operation, which is associated with the Lie algebraic model, will provide the vibrational spectra of the molecule [Balla et al (2021, 2022). Infrared and Raman spectroscopy are commonly used methods for obtaining the vibrational spectra of metalloporphyrins [Bayden et al (2007)]. Metalloporphyrins are coordination complexes consisting of a porphyrin ligand and a central metal ion. As a result of the presence of both metal-ligand and ligand-ligand vibrations, the vibrational spectra of these compounds have been interesting. In metalloporphyrins, the central metal ion can be transition metals like iron, cobalt, nickel, copper, or zinc. Red blood cells, or erythrocytes, are specialized cells in the bloodstream that transport oxygen to tissues and remove carbon dioxide. Heme, a metalloporphyrin, is one of the primary components of erythrocytes. It contains a metal ion (iron) coordinated by nitrogen atoms from the porphyrin ligand, positioned at the center of the porphyrin ring. In 2008, Karumuri et al. used the Lie algebraic model for the fundamental stretching vibrations of medium-sized molecules to study the vibrational spectra of nickel octaethyl porphyrins, then extended their work to nickel tetraphenyl porphyrin, copper octaethyl porphyrin, magnesium

[1]sprathip@gitam.in, [2]vijayjaliparthi@gmail.com

DOI: 10.1201/9781003489436-58

octaethyl porphyrin, copper tetramesityl porphyrin, and the resonance Raman spectra of oxygenated and deoxygenated functional erythrocytes of D4h molecular point group [S.R. Karumuri et al (2008, 2009, 2010, 2011, 2012, 2013, 2015, 2016)]. These studies clearly show that calculations are limited to fundamental mode vibrations only. Hence, this paper presents the first and second overtone vibrational spectra of oxygenated and deoxygenated functional erythrocytes.

2. Lie Algebraic Model

The Hamiltonian operator [Vijayasekhar et al (2022, 2023)] associated with the Lie algebraic model for the n vibrations of metalloporphyrins is

$$H = E_0 + \sum_{i=1}^{n} A_i C_i + \sum_{i<j}^{n} A_{ij} C_{ij} \sum_{i<j}^{n} \left(m_{ij}^1 + m_{ij}^2 \right) \lambda_{ij} M_{ij}, \quad (1)$$

with adjacent and opposite interaction coefficients are given by

$$m_{ij}^1 = \begin{cases} 1, (i,j) = (1,2), (2,3), (3,4), (4,5), \ldots \\ 0, \; otherwise \end{cases}$$

$$m_{ij}^2 = \begin{cases} 1, (i,j) = (1,3), (2,4), (3,5), (4,6), \ldots \\ 0, \; otherwise \end{cases}$$

Here, C_i, C_{ij}, M_{ij} are operators, and A_i, A_{ij}, λ_{ij} are algebraic parameters, which are determined from the expressions

$$\langle C_i \rangle = -4(N_i v_i - v_i^2) \quad \ldots (2)$$

$$\left\langle N_i, v_i; N_j, v_j \left| C_{ij} \right| N_i, v_i; N_j, v_j \right\rangle = 4(v_i + v_j)(v_i + v_j - N_i - N_j) \quad \ldots (3)$$

$$\left.\begin{aligned} &\left\langle N_i, v_i; N_j, v_j \left| M_{ij} \right| N_i, v_i; N_j, v_j \right\rangle = v_i N_j + v_j N_i - 2 v_i v_j \\ &\left\langle N_i, v_i + 1; N_j, v_j - 1 \left| M_{ij} \right| N_i, v_i; N_j, v_j \right\rangle = -[v_j (v_i + 1)(N_i - v_i)(N_j - v_j + 1]^{1/2} \\ &\left\langle N_i, v_i - 1; N_j, v_j + 1 \left| M_{ij} \right| N_i, v_i; N_j, v_j \right\rangle = -[v_i (v_j + 1)(N_j - v_j)(N_i - v_i + 1]^{1/2} \end{aligned}\right\} \quad \ldots (4)$$

$$\langle C_i \rangle = -4(N-1), \langle C_{ij} \rangle = -4(2N-1), \langle M_{ij} \rangle = \begin{cases} -N(i \neq j) \\ N(i = j) \end{cases} \quad \ldots (5)$$

$$E = -4A_i(N-1), \lambda_{ij} = \frac{|E_a - E_b|}{2N}, \lambda'_{ij} = \frac{|E_a - E_b|}{4N} \quad \ldots (6)$$

Where v_i and v_j are the vibrational quantum numbers of bonds i and j, respectively, and N indicates the vibron number. E_a and E_b are the energies of two local modes.

3. Results

Table 58.1 shows how we used the Lie algebraic model with the algebraic parameters that fit the system [Karumuri et al., 2011] to figure out the Raman spectra for oxygenated and deoxygenated erythrocytes. The ensuing outcomes, detailing the frequencies of the first and second overtones, are presented in Table 58.2.

Table 58.1 Values of the parameters (in cm^{-1}, N is dimensionless)

Parameters	oxygenated			deoxygenated		
	C_a–C_m	C_b–C_b	C_m–H	C_a–C_m	C_b–C_b	C_m–H
A_i	−2.2103	−2.2805	−9.6825	−2.1230	−1.9289	−10.5320
A_{ij}	−1.0172	1.0151	2.5610	-1.1002	0.5234	2.4950
λ_{ij}	0.0369	0.0297	0.2581	0.0594	0.0345	0.2576
λ_{ij}'	0.1073	0.1029	0.0981	0.0203	0.1302	0.0934
N	140	140	44	140	140	44

Table 58.2 First and second overtone vibrational frequencies of erythrocytes (cm^{-1})

Vibrational mode	Local coordinates	Symmetry species	Lie algebraic model	
			I overtone	II overtone
ν_{10}	(C_a-C_m) a-str	B_{1g}	29765.82	403.24
ν_{37}	(C_a-C_m) a-str	E_{1u}	2765.07	4340.11
ν_{28}	(C_a-C_m) str	B_{2g}	2387.43	3865.68
ν_{19}	(C_b-C_b)	B_{1g}	2735.30	4103.13
ν_{11}	(C_b-C_b)	B_{1g}	2689.88	4137.42
ν_{21}	(C_m-H)	B_{1g}	2107.52	3449.40
ν_{42}	(C_m-H)	E_{1u}	2018.21	3264.69
ν_{13}	(C_m-H)	B_{1g}	1987.73	3120.01
$\nu_5 + \nu_{18}$	(C_m-H)	B_{1g}	2065.63	3238.55

4. Conclusion

The present investigation reveals the vibrational frequencies corresponding to the first and second overtones within the Raman spectra of both oxygenated and deoxygenated erythrocytes. Specifically, the study delves into the vibrational behaviors encompassing the asymmetric stretching mode of C_a-C_m, the stretching mode of C_b-C_b, and the in-plane bending mode of C_m-H. The implications of this research extend towards fostering further exploration into higher overtone frequencies of biomolecules that have hitherto remained relatively unexplored.

5. Acknowledgments

Vijayasekhar gratefully acknowledges the Research Seed Grants (RSG) by Gandhi Institute of Technology and Management (GITAM), Deemed to be University, Hyderabad, India [Sanction Letter Ref: F.No. 2021/0114, dated 26-09-2022].

REFERENCES

1. Iachello, F., Levine, R. D., Algebraic theory of molecules, Oxford: Oxford University Press, 1995.
2. Oss, S., Adv. Chem. Phys., 1996, 93: 455.
3. Bayden R. Wood, Peter Caspers, Gerwin J. Pupples, Anal. Bioanal. Chem., 2007, 387: 1691
4. S. R. Karumuri, N.K. Sarkar, J. Choudhury, R. Bhattacharjee, Mol. Phys., 2008, 106(14): 1733–1737.
5. S. R. Karumuri, N.K. Sarkar, J. Choudhury, R. Bhattacharjee, J. Mol. Spectrosc., 2009, 255(2): 183–188.
6. S. R. Karumuri, J. Mol. Spectrosc., 2010, 259(2): 86–92.
7. S. R. Karumuri, Chem. Phys. Lett., 2010, 27(10): 103301(1-4).
8. S.R. Karumuri, J. Vijayasekhar, V. Sreeram., V.U.M. Rao, M. V.B. Rao, J. Mol. Spectrosc., 2011, 269: 119.
9. S. R. Karumuri, Indian J. Phys., 2012, 86(12): 1147–1153.
10. S. R. Karumuri et al., Chin. Phy. B., 2013, 22(9): 090304.
11. S.R. Karumuri, Eur. Phys. J., 2015, 69, 281.
12. S.R. Karumuri, K. G. Sravani, Mol. Phys. 2016, 114 (5): 643–649.
13. M. R. Balla, V. S. Jaliparthi, Mol. Phys. 2021, 115 (5), e1828634.
14. M. R. Balla, V. S. Jaliparthi, Polycyclic Aromatic Compounds. 2022, 42(7), 4684.
15. J. Vijayasekhar, M. Rao Balla, Spectrochim. Acta A, 2022, 264, 120289.
16. J. Vijayasekhar, Ukr. J. Phys. Opt. 2022, 23(3), 126.
17. J. Vijayasekhar, P. Suneetha, K. Lavanya, Ukr. J. Phys. Opt. 2023, 24(3), 193–199.

Note: All the tables in this chapter were made by the authors.

Impending Inquisitions in Humanities and Sciences (ICIIHS-2022) – Dr. Mohan Varkolu et al. (eds)

C-Class Function and 3-Tupled Unique Common Fixed Point Result in Partial Metric Spaces

59

G. N. V. Kishore[1]
Department of Engineering Mathematics and Humanities,
SRKR Engineering College, Bhimavaram, Andhra Pradesh, India

A. Kiran Kumar[2]
Department of Engineering Mathematics and Humanities,
SRKR Engineering College, Bhimavaram, Andhra Pradesh, India

B. V. Appa Rao[3]
Department of Engineering Mathematics, Koneru Lakshmaiah Education Foundation,
Vaddeswaram, Guntur, Andhra Pradesh, India

Y. Phani*
Department of Mathematics, Shri Vishnu Engineering College for Women (A),
Vishnupur, Bhimavaram, AP, India

G. Suri Babu[4]
WET Centre, Department of Civil Engineering, SRKR Engineering College,
Bhimavaram, Andhra Pradesh, India

ABSTRACT In this paper, we obtain a unique common 3-tupled fixed point theorem along with C - class function in partial metric spaces and also furnish an example to support our main result.

KEYWORDS Partial metric, Weakly compatible maps, 3-Tupled fixed point, Complete space, C-Class function

1. Introduction and Mathematical Preliminaries

The notion of a partial metric space was introduced by Matthews [17] as a part of the study of denotational semantics of data flow networks. In fact, it is widely recognized that partial metric spaces play an important role in constructing models in the theory of computation and domain theory in computer science (see e.g. [24, 27, 28, 8, 14, 20, 15, 22, 25]).

Matthews [17, 18], Oltra and Valero [19] and Romaguera [23] and Altun et al. [2] proved some fixed point theorems in partial metric spaces for a single map (see also [1, 9, 10, 11, 12, 13, 3, 4, 5, 26, 20]).

Recently, the concept of coupled fixed points was introduced by Bhaskar and Lakshmikantham [7]. In [4], Aydi proved some coupled fixed point theorems for the mappings satisfying contractive conditions in partial metric spaces.

*Correspondence Author: phaniyedlapalli23@gmail.com
[1]kishore.apr2@gmail.com, gnvkishore@srkrec.ac.in, [2]kirankumarappana@gmail.com, [3]bvardr2010@kluniversity.in, [4]gsuribabu8@gmail.com
2000 Mathematics Subject Classification. 54H25, 47H10, 54E50

DOI: 10.1201/9781003489436-59

In this paper, obtain a unique common 3-tupled fixed point theorem for four maps satisfying a contractive condition along with C-class function in partial metric spaces.

First we recall some basic definitions and lemmas which play crucial role in the theory of partial metric spaces.

Definition 1.1. *(See [17, 18]) A partial metric on a nonempty set X is a function $p : X \times X \to R^+$ such that for all $x,y,z \in X$:*

(p_1) $x = y \Leftrightarrow p(x, x) = p(x, y) = p(y, y)$, (p_2) $p(x, x) \leq p(x, y)$, $p(y, y) \leq p(x, y)$,

(p_3) $p(x, y) = p(y, x)$,

(p_4) $p(x, y) \leq p(x, z) + p(z, y) - p(z, z)$.

The pair (X, p) is called a partial metric space (PMS).

If p is a partial metric on X, then the function $p^s : X \times X \to R^+$ given by

$$p^s(x, y) = 2p(x, y) - p(x, x) - p(y, y) \tag{1.1}$$

is a metric on X.

Example 1.2. *(See e.g. [18, 10, 11, 1]) Consider $X = [0, \infty)$ with $p(x,y) = \max\{x, y\}$. Then (X, p) is a partial metric space. It is clear that p is not a (usual) metric. Note that in this case $p^s(x, y) = |x - y|$.*

Example 1.3. *(See [9]) Let $X = \{[a, b] : a, b, \in R, a \leq b\}$ and define $p([a, b], [c, d]) = \max\{b, d\} - \min\{a, c\}$. Then (X, p) is a partial metric space.*

In a partial metric space (X, p) , it is clear that

(i) if $x = y$, then $p(x,y)$ may not be zero,
(ii) If $p(x,y) = 0$ then $x = y$,
(iii) If $x \neq y$, then $p(x,y) > 0$.

Each partial metric p on X generates a T_0 topology τ_p on X which has as a base the family of open p-balls $\{B_p(x, \varepsilon), x \in X, \varepsilon > 0\}$, where $B_p(x, \varepsilon) = \{y \in X : p(x, y) < p(x, x) + \varepsilon\}$ for all $x \in X$ and $\varepsilon > 0$.

We now state some basic topological notions (such as convergence, completeness, continuity) on partial metric spaces (see e.g. [17, 18, 2, 1, 10, 13].)

Definition 1.4.

1. *A sequence $\{x_n\}$ in the PMS (X, p) converges to the limit x if and only if $p(x, x) = \lim_{n\to\infty} p(x, x_n)$.*
2. *A seque nce $\{x_n\}$ in the PMS (X, p) is called a Cauchy sequence if $\lim_{n,m\to\infty} p(x_n, x_m)$ exists and is finite.*
3. *A PMS (X, p) is called complete if every Cauchy sequence $\{x_n\}$ in X converges with respect to τ_p, to a point $x \in X$ such that $p(x,x) = \lim_{n,m\to\infty} p(x_n, x_m)$.*
4. *A mapping $F : X \to X$ is said to be continuous at $x_0 \in X$ if for every $\epsilon > 0$, there exists $\delta > 0$ such that $F(B_p(x_0, \delta)) \subseteq B_p(Fx_0, \epsilon)$.*

Remark 1.5. [1,10,11]

1. *If $\{x_n\}$ is converges to z in (X, p), then $\lim_{n\to\infty} p(x_n, y) \leq p(z, y)$ for all $y \in X$.*
2. *If $\{x_n\}$ is converges to z in (X, p) with $p(z, z) = 0$, then $\lim_{n\to\infty} p(x_n, y) = p(z, y)$ for all $y \in X$.*

The following lemma is one of the basic results in PMS([17, 18, 1, 2, 10, 13]).

Lemma 1.6.

1. *A sequence $\{x_n\}$ is a Cauchy sequence in the PMS (X,p) if and only if it is a Cauchy sequence in the metric space (X, p^s).*
2. *A PMS (X, p) is complete if and only if the metric space (X, p^s) is complete. Moreover*

$$\lim_{n\to\infty} p^s(x,x_n) = 0 \Leftrightarrow p(x,x) = \lim_{n\to\infty} p(x,x_n) = \lim_{n,m\to\infty} p(x_n,x_m) \tag{1.2}$$

Definition 1.7. [7]. *An element $(x,y) \in X \times X$ is called a coupled fixed point of mapping $F : X \times X \to X$ if $x = F(x, y)$ and $y = F(y, x)$.*

In 2013, K. P. R. Rao et. al. [[21]] gives the following definitions.

Definition 1.8 ([21]). *An element $(x,y,z) \in X \times X \times X$ is called a 3-tupled fixed point of the mapping $T : X \times X \times X \to X$ if $x = T(x, y, z)$, $y = T(y, z, x)$ and $z = T(z, y, x)$.*

Definition 1.9 ([21]). *An element $(x,y,z) \in X \times X \times X$ is called (i) a 3-tupled coincident point of mappings $T : X \times X \times X \to X$ and $f : X \to X$ if $fx = T(x, y, z)$, $fy = T(y, z, x)$ and $fz = T(z, x, y)$;*

[[21]] (ii) a common 3-tupled fixed point of mappings $T : X \times X \times X \to X$ and $f : X \to X$ if $x = fx = T(x, y, z)$, $y = fy = T(y, z, x)$ and $z = fz = T(z, x, y)$.

Definition 1.10 ([21]). *The mappings $T : X \times X \times X \to X$ and $f : X \to X$ are called w - compatible if $f(T(x, y, z)) = T(fx, fy, fz)$, $f(T(y, z, x)) = T(fy, fz, fx)$ and $f(T(z, x, y)) = T(fz, fx, fy)$ whenever $fx = T(x, y, z)$, $fy = T(y, z, x)$ and $fz = T(z, x, y)$.*

In 2014 the concept of *C*-class functions(see Definition 1.11) was introduced by A.H. Ansari in [30] that is pivotal result in fixed point theory ,for example see numbers (1),(2),(9) and (15) from Example 1.12.

Definition 1.11. [30]*A mapping $F : [0, \infty)^2 \to R$ is called C-class function if it is continuous and satisfies following axioms:*

1. $F(s, t) \le s$;
2. $F(s,t) = s$ *implies that either* $s = 0$ *or* $t = 0$; *for all* $s, t \in [0, \infty)$.

Note for some F we have that $F(0, 0) = 0$.

We denote *C*-class functions as C.

Example 1.12. [30]*The following functions $F : [0, \infty)^2 \to R$ are elements of C, for all $s, t \in [0, \infty)$:*

1. $F(s, t) = s - t$, $F(s, t) = s \Rightarrow t = 0$;
2. $F(s, t) = ms$, $0 < m < 1$, $F(s, t) = s \Rightarrow s = 0$;
3. $\frac{s}{(1+t)^r}$ $r \in (0, \infty)$, $F(s, t) = s \Rightarrow s = 0$ or $t = 0$;
4. $F(s, t) = \log(t + a^s)/(1 + t)$, $a > 1$, $F(s, t) = s \Rightarrow s = 0$ *or* $t = 0$;
5. $F(s, t) = \ln(1 + a^s)/2$, $a > e$, $F(s,1) = s \Rightarrow s = 0$;
6. $F(s, t) = (s + l)^{(1/(1+t)r)} - l$, $l > 1$, $r \in (0,\infty)$, $F(s,t) = s \Rightarrow t = 0$;
7. $F(s, t) = s\log_{t+a} a$, $a > 1$, $F(s,t) = s \Rightarrow s = 0$ *or* $t = 0$;
8. $s - \left(\frac{1+s}{2+s}\right)\left(\frac{t}{1+t}\right)$, $F(s, t) = s \Rightarrow t = 0$;
9. $F(s, t) = s\beta(s)$, $\beta : [0,\infty) \to [0,1)$,*and is continuous*, $F(s,t) = s \Rightarrow s = 0$;
10. $s - \frac{t}{k+t}$, $F(s, t) = s \Rightarrow t = 0$;
11. $F(s, t) = s - \varphi(s)$,$F(s,t) = s \Rightarrow s = 0$,*here* $\varphi : [0,\infty) \to [0,\infty)$ *is a continuous function such that* $\varphi(t) = 0 \Leftrightarrow t = 0$;
12. $F(s, t) = sh(s,t)$,$F(s,t) = s \Rightarrow s = 0$,*here* $h : [0,\infty) \times [0,\infty) \to [0,\infty)$*is a continuous function such that* $h(t,s) < 1$ *for all* $t, s > 0$;
13. $F(s, t) = s - \left(\frac{2+t}{1+t}\right)t$, $F(s, t) = s \Rightarrow t = 0$.
14. $F(s, t) = \sqrt[n]{\ln(1 + s^n)}$, $F(s, t) = s \Rightarrow s = 0$.
15. $F(s, t) = \phi(s)$, $F(s,t) = s \Rightarrow s = 0$,here $\phi : [0,\infty) \to [0,\infty)$ is a continuous function such that $\phi(0) = 0$, and $\phi(t) < t$ for $t > 0$,
16. $F(s, t) = \frac{s}{(1+ s)^r}$; $r \in (0, \infty)$, $F(s, t) = s \Rightarrow s = 0$;

Definition 1.13 ([31]). *A function $\psi : [0,\infty) \to [0,\infty)$ is called an altering distance function if the following properties are satisfied:*

(i) ψ is non-decreasing and continuous,

(ii) $\psi(t) = 0$ *if and only if* $t = 0$.

Definition 1.14. *An ultra altering distance function is a continuous, nondecreasing mapping $\varphi : [0,\infty) \to [0,\infty)$ such that $\varphi(t) > 0$,$t > 0$ and $\varphi(0) \ge 0$*

Remark 1.15. *We denote Φ_u set ultra altering distance functions*

Now we prove our main result.

2. Main Result

Let Ψ denotes the set of all functions $\psi : [0,\infty) \to [0,\infty)$ satisfying

(φ_1) ψ is non-decreasing and continuous,

(φ_2) $\psi(t) = 0$ if and only if $t = 0$,

(φ_3) $\psi(s + t) < \psi(s) + \psi(t)$ for $s,t > 0$,

Theorem 2.1. *Let (X,p) be a partial metric space and let $S,T : X \times X \times X \to X$ and $f,g : X \to X$ be mappings satisfying*

(i) $\psi(p(S(x, y, z), T(u, v, w))) \leq \frac{1}{3}F\left(\begin{array}{c}\psi\left(p(fx,gu)+p(fy,gv)+p(fz,gw)\right),\\ \varphi\left(\frac{p(fx,gu)+p(fy,gv)+p(fz,gw)}{3}\right)\end{array}\right)$

$\forall x, y, z, u, v, w \in X$ *where* $\psi \in \Psi$, $\varphi \in \Phi_u$, $F \in C$,

(ii) $S(X \times X \times X) \subseteq g(X), T(X \times X \times X) \subseteq f(X)$,

(iii) *either $f(X)$ or $g(X)$ is a complete subspace of X,*

(iv) *the pairs (f,S) and (g,T) are w - compatible.*

Then S and f or T and g have 3-tupled coincidence point in $X \times X \times X$.

or

S, T, f and g have a unique common 3-tupled fixed point in $X \times X \times X$ of the form (α, α, α).

Proof. Let x_0, y_0, z_0 be arbitrary points in X. From *(ii)*, there exist sequences $\{x_n\}$, $\{y_n\}$, $\{z_n\}$, $\{u_n\}$, $\{v_n\}$ and $\{w_n\}$ in X such that

$$\begin{aligned}
u_{2n} &= gx_{2n+1} = S(x_{2n}, y_{2n}, z_{2n}),\\
v_{2n} &= gy_{2n+1} = S(y_{2n}, z_{2n}, x_{2n}),\\
w_{2n} &= gz_{2n+1} = S(z_{2n}, x_{2n}, y_{2n}),\\
u_{2n+1} &= fx_{2n+2} = T(x_{2n+1}, y_{2n+1}, z_{2n+1}),\\
v_{2n+1} &= fy_{2n+2} = T(y_{2n+1}, z_{2n+1}, x_{2n+1}),\\
w_{2n+1} &= fz_{2n+2} = T(z_{2n+1}, x_{2n+1}, y_{2n+1}), \quad n = 0, 1, 2, \cdots
\end{aligned} \tag{2.1}$$

Case(a): Suppose $u_m = u_{m+1}, v_m = v_{m+1}$ and $w_m = w_{m+1}$ for some m.

If $m = 0$. Then T and g have a 3 - tupled coincedence point in $X \times X \times X$.

If $m = 1$. Then S and f have a 3 - tupled coincedence point in $X \times X \times X$ and so on.

Thus S and f or T and g have 3 - tupled coincidence point in $X \times X \times X$.

Case(b):Assume $u_n \neq u_{n+1}$ or $v_n \neq v_{n+1}$ or $w_n \neq w_{n+1}$ for all n. Then by (i), we have

$$\begin{aligned}
\psi(p(u_2,u_1)) &= \psi(p(S(x_2,y_2,z_2),T(x_1,y_1,z_1)))\\
&\leq \tfrac{1}{3}F(\psi\left(p(fx_2,gx_1)+p(fy_2,gy_1)+p(fz_2,gz_1)\right), \varphi\left(\tfrac{p(fx_2,gx_1)+p(fy_2,gy_1)+p(fz_2,gz_1)}{3}\right))\\
&= \tfrac{1}{3}F(\psi\left(p(u_1,u_0)+p(v_1,v_0)+p(w_1,w_0)\right), \varphi\left(\tfrac{p(u_1,u_0)+p(v_1,v_0)+p(w_1,w_0)}{3}\right)).
\end{aligned}$$

$$\begin{aligned}
\psi(p(v_2,v_1)) &= \psi(p(S(y_2,z_2,x_2),T(y_1,z_1,x_1)))\\
&\leq \tfrac{1}{3}F(\psi\left(p(fy_2,gy_1)+p(fz_2,gz_1)+p(fx_2,gx_1)\right), \varphi\left(\tfrac{p(fy_2,gy_1)+p(fz_2,gz_1)+p(fx_2,gx_1)}{3}\right))\\
&= \tfrac{1}{3}F(\psi\left(p(u_1,u_0)+p(v_1,v_0)+p(w_1,w_0)\right), \varphi\left(\tfrac{p(u_1,u_0)+p(v_1,v_0)+p(w_1,w_0)}{3}\right)).
\end{aligned}$$

And

$$\begin{aligned}
\psi(p(w_2,w_1)) &= \psi(p(S(z_2,x_2,y_2),T(z_1,x_1,y_1)))\\
&\leq \tfrac{1}{3}F(\psi\left(p(fz_2,gz_1)+p(fx_2,gx_1)+p(fy_2,gy_1)\right), \varphi\left(\tfrac{p(fz_2,gz_1)+p(fx_2,gx_1)+p(fy_2,gy_1)}{3}\right))\\
&= \tfrac{1}{3}F(\psi\left(p(u_1,u_0)+p(v_1,v_0)+p(w_1,w_0)\right), \varphi\left(\tfrac{p(u_1,u_0)+p(v_1,v_0)+p(w_1,w_0)}{3}\right)).
\end{aligned}$$

so,

$$\begin{aligned}
\psi(p(u_2,u_1)+p(v_2,v_1)+p(w_2,w_1)) \leq\ & F(\psi\left(p(u_1,u_0)+p(v_1,v_0)+p(w_1,w_0)\right),\\
& \varphi\left(\tfrac{p(u_1,u_0)+p(v_1,v_0)+p(w_1,w_0)}{3}\right)).
\end{aligned}$$

Similarly

$$\psi(p(u_3,u_2)+p(v_3,v_2)+p(w_3,w_2)) \leq F(\psi(p(u_2,u_1)+p(v_2,v_1)+p(w_2,w_1)), \varphi\left(\frac{p(u_2,u_1)+p(v_2,v_1)+p(w_2,w_1)}{3}\right)).$$

Hence by induction, we have

$$\begin{aligned}\psi(p(u_n,u_{n+1})+p(v_n,v_{n+1})+p(w_n,w_{n+1})) &\leq F(\psi(p(u_n,u_{n-1})+p(v_n,v_{n-1})+p(w_n,w_{n-1}), \\ &\quad \varphi\left(\frac{p(u_n,u_{n-1})+p(v_n,v_{n-1})+p(w_n,w_{n-1})}{3}\right) \\ &\leq \psi(p(u_n,u_{n-1})+p(v_n,v_{n-1})+p(w_n,w_{n-1}).\end{aligned}$$

Put $\lambda_{n+1} = p(u_n, u_{n+1}) + p(v_n, v_{n+1}) + p(w_n, w_{n+1})$. Then $\{\lambda_n\}$ is a non - increasing sequence of real numbers and must converge to $\lambda \geq 0$.

Letting $n \to \infty$ in (2.2), we have that

$$\begin{aligned}\psi(\lambda) &= \psi\left(\lim_{n\to\infty}\lambda_{n+1}\right) \leq F(\lim_{n\to\infty}\psi(p(u_n,u_{n-1})+p(v_n,v_{n-1})+p(w_n,w_{n-1}), \\ &\quad \lim_{n\to\infty}\varphi\left(\frac{p(u_n,u_{n-1})+p(v_n,v_{n-1})+p(w_n,w_{n-1})}{3}\right) \\ &= F(\psi(\lambda),\varphi\left(\frac{\lambda}{3}\right)),.\end{aligned}$$

By the definition of C- class function we have that $\lambda = 0$.

Thus

$$\lim_{n\to\infty}[p(u_n,u_{n+1}) + p(v_n,v_{n+1}) + p(w_n,w_{n+1})] = 0.$$

It follows that

$$\lim_{n\to\infty} p(u_n, u_{n+1}) = \lim_{n\to\infty} p(v_n, v_{n+1}) = \lim_{n\to\infty} p(w_n, w_{n+1}) = 0. \quad (2.3)$$

From (p_2), we have

$$\lim_{n\to\infty} p(u_n, u_n) = \lim_{n\to\infty} p(v_n, v_n) = \lim_{n\to\infty} p(w_n, w_n) = 0. \quad (2.4)$$

From definition of p^s, (2.3) and (2.4), we have

$$\lim_{n\to\infty} p^s(u_n, u_{n+1}) = \lim_{n\to\infty} p^s(v_n, v_{n+1}) = \lim_{n\to\infty} p^s(w_n, w_{n+1}) = 0. \quad (2.5)$$

Now we prove that $\{u_{2n}\},\{v_{2n}\}$ and $\{w_{2n}\}$ are Cauchy sequences.

On contrary, suppose that $\{u_{2n}\}$ or $\{v_{2n}\}$ or $\{w_{2n}\}$ is not Cauchy.

Then there exists an $\epsilon > 0$ and monotone increasing sequences of natural numbers $\{2m_k\}$ and $\{2n_k\}$ such that $n_k > m_k$,

$$p^s(u_2m_k, u_2n_k) + p^s(v_2m_k, v_2n_k) + p^s(w_2m_k, w_2n_k) \geq \epsilon \quad (2.6)$$

and

$$p^s(u_2m_k, u_2n_{k-2}) + p^s(v_2m_k v_2n_{k-2}) + p^s(w_2m, w_2n_{k-2}) < \epsilon. \quad (2.7)$$

From (2.6) and (2.7), we have

$$\begin{aligned}\epsilon &\leq p^s(u_{2m_k},u_{2n_k})+p^s(v_{2m_k},v_{2n_k})+p^s(w_{2m_k},w_{2n_k}) \\ &\leq p^s(u_{2m_k},u_{2n_k-2})+p^s(v_{2m_k},v_{2n_k-2})+p^s(w_{2m_k},w_{2n_k-2}) \\ &\quad +p^s(u_{2n_k-2},u_{2n_k-1})+p^s(v_{2n_k-2},v_{2n_k-1})+p^s(w_{2n_k-2},w_{2n_k-1}) \\ &\quad +p^s(u_{2n_k-1},u_{2n_k})+p^s(v_{2n_k-1},v_{2n_k})+p^s(w_{2n_k-1},w_{2n_k}) \\ &< \epsilon+p^s(u_{2n_k-2},u_{2n_k-1})+p^s(v_{2n_k-2},v_{2n_k-1})+p^s(w_{2n_k-2},w_{2n_k-1}) \\ &\quad +p^s(u_{2n_k-1},u_{2n_k})+p^s(v_{2n_k-1},v_{2n_k})+p^s(w_{2n_k-1},w_{2n_k}).\end{aligned}$$

Letting $k \to \infty$ and using (2.5), we have

$$\lim_{k\to\infty}[p^s(u_{2m_k},u_{2n_k})+p^s(v_{2m_k},v_{2n_k})+p^s(w_{2m_k},w_{2n_k})] = \epsilon. \quad (2.8)$$

By definition of p^s and (2.4), we have

$$\lim_{k\to\infty}[p(u_{2m_k},u_{2n_k})+p(v_{2m_k},v_{2n_k})+p(w_{2m_k},w_{2n_k})]=\frac{\epsilon}{2}. \tag{2.9}$$

And,

$$\begin{aligned}\epsilon &\leq p^s(u_{2m_k},u_{2n_k})+p^s(v_{2m_k},v_{2n_k})+p^s(w_{2m_k},w_{2n_k})\\ &\leq p^s(u_{2m_k},u_{2m_k-1})+p^s(v_{2m_k},v_{2m_k-1})+p^s(w_{2m_k},w_{2m_k-1})\\ &\quad +p^s(u_{2m_k-1},u_{2n_k})+p^s(v_{2m_k-1},v_{2n_k})+p^s(w_{2m_k-1},w_{2n_k}).\end{aligned}$$

Letting $k\to\infty$ and using (2.5), we have

$$\epsilon\leq\lim_{k\to\infty}[p^s(u_{2m_k-1},u_{2n_k})+p^s(v_{2m_k-1},v_{2n_k})+p^s(w_{2m_k-1},w_{2n_k})]. \tag{2.10}$$

Also

$$\begin{aligned}&p^s(u_{2m_k-1},u_{2n_k})+p^s(v_{2m_k-1},v_{2n_k})+p^s(w_{2m_k-1},w_{2n_k})\\ &\quad\leq p^s(u_{2m_k-1},u_{2m_k})+p^s(v_{2m_k-1},v_{2m_k})+p^s(w_{2m_k-1},w_{2m_k})\\ &\quad\quad +p^s(u_{2m_k},u_{2n_k})+p^s(v_{2m_k},v_{2n_k})+p^s(w_{2m_k},w_{2n_k})\end{aligned}$$

Letting $k\to\infty$, we have

$$\lim_{k\to\infty}[p^s(u_{2m_k-1},u_{2n_k})+p^s(v_{2m_k-1},v_{2n_k})+p^s(w_{2m_k-1},w_{2n_k})]\leq\epsilon. \tag{2.11}$$

From (2.10) and (2.11) we have

$$\lim_{k\to\infty}[p^s(u_{2m_k-1},u_{2n_k})+p^s(v_{2m_k-1},v_{2n_k})+p^s(w_{2m_k-1},w_{2n_k})]=\epsilon. \tag{2.12}$$

By definition of ps and (2.4), we have

$$\lim_{k\to\infty}[p(u_{2m_k-1},u_{2n_k})+p(v_{2m_k-1},v_{2n_k})+p(w_{2m_k-1},w_{2n_k})]=\frac{\epsilon}{2}. \tag{2.13}$$

On other hand we have

$$\begin{aligned}&p^s(u_{2m_k},u_{2n_k})+p^s(v_{2m_k},v_{2n_k})+p^s(w_{2m_k},w_{2n_k})\\ &\quad\leq p^s(u_{2m_k},u_{2n_k+1})+p^s(v_{2m_k},v_{2n_k+1})+p^s(w_{2m_k},w_{2n_k+1})\\ &\quad\quad +p^s(u_{2n_k+1},u_{2n_k})+p^s(v_{2n_k+1},v_{2n_k})+p^s(w_{2n_k+1},w_{2n_k}).\end{aligned}$$

Letting $k\to\infty$ and using (2.4), (2.5) and (2.8), we have

$$\begin{aligned}\epsilon &\leq\lim_{k\to\infty}[p^s(u_{2m_k},u_{2n_k+1})+p^s(v_{2m_k},v_{2n_k+1})+p^s(w_{2m_k},w_{2n_k+1})]+0\\ &=2\lim_{k\to\infty}[p(u_{2m_k},u_{2n_k+1})+p(v_{2m_k},v_{2n_k+1})+p(w_{2m_k},w_{2n_k+1})].\end{aligned}$$

Thus,

$$\frac{\epsilon}{2}\leq\lim_{k\to\infty}[p(u_{2m_k},u_{2n_k+1})+p(v_{2m_k},v_{2n_k+1})+p(w_{2m_k},w_{2n_k+1})]. \tag{2.14}$$

Now

$$\begin{aligned}\psi(p(u_{2m_k},u_{2n_k+1})) &=\psi(p(S(x_{2m_k},y_{2m_k},z_{2m_k}),T(x_{2n_k+1},y_{2n_k+1},z_{2n_k+1})))\\ &\leq\tfrac{1}{3}F(\psi\,(p(u_{2m_k-1},u_{2n_k})+p(v_{2m_k-1},v_{2n_k})+p(w_{2m_k-1},w_{2n_k}))\,,\\ &\quad\varphi\left(\tfrac{p(u_{2m_k-1},u_{2n_k})+p(v_{2m_k-1},v_{2n_k})+p(w_{2m_k-1},w_{2n_k})}{3}\right)).\end{aligned}$$

Similarly

$$\begin{aligned}p(v_{2m_k},v_{2n_k+1}) &=\tfrac{1}{3}F(\psi\,(p(u_{2m_k-1},u_{2n_k})+p(v_{2m_k-1},v_{2n_k})+p(w_{2m_k-1},w_{2n_k}))\,,\\ &\quad\varphi\left(\tfrac{p(u_{2m_k-1},u_{2n_k})+p(v_{2m_k-1},v_{2n_k})+p(w_{2m_k-1},w_{2n_k})}{3}\right)).\end{aligned}$$

and

$$\begin{aligned}p(w_{2m_k},w_{2n_k+1}) &=\tfrac{1}{3}F(\psi\,(p(u_{2m_k-1},u_{2n_k})+p(v_{2m_k-1},v_{2n_k})+p(w_{2m_k-1},w_{2n_k}))\,,\\ &\quad\varphi\left(\tfrac{p(u_{2m_k-1},u_{2n_k})+p(v_{2m_k-1},v_{2n_k})+p(w_{2m_k-1},w_{2n_k})}{3}\right)).\end{aligned}$$

so,

$$\psi(p(u_{2m_k},u_{2n_k+1})+p(v_{2m_k},v_{2n_k+1})+p(w_{2m_k},w_{2n_k+1}))$$
$$\leq F(\psi(p(u_{2m_k-1},u_{2n_k})+p(v_{2m_k-1},v_{2n_k})+p(w_{2m_k-1},w_{2n_k})),$$
$$\varphi(\frac{p(u_{2m_k-1},u_{2n_k})+p(v_{2m_k-1},v_{2n_k})+p(w_{2m_k-1},w_{2n_k})}{3})$$

Letting $k \to \infty$ and using (2.13) and (2.14), we have
$\psi(\frac{\epsilon}{2}) \leq F(\psi(\frac{\epsilon}{2}), \varphi(\frac{\epsilon}{6}))$. So, $\psi(\frac{\epsilon}{2})$ or, $\varphi(\frac{\epsilon}{6})$. Hence $\varepsilon = 0$. It is a contradiction.

Hence $\{u_{2n}\}, \{v_{2n}\}$ and $\{w_{2n}\}$ are Cauchy sequences in the metric space (X, p^s). Letting $n, m \to \infty$ in

$$|p^s(u_{2n+1}, u_{2m+1}) - p^s(u_{2n}, u_{2m})| \leq p^s(u_{2n+1}, u_{2n}) + p^s(u_{2m+1}, u_{2m})$$

we get $\lim_{n,m\to\infty} p^s(u_{2n+1}, u_{2m+1}) = 0$.
Letting $n,m \to \infty$ in

$$|p^s(v_{2n+1}, v_{2m+1}) - p^s(v_{2n}, v_{2m})| \leq p^s(v_{2n+1}, v_{2n}) + p^s(v_{2m+1}, v_{2m})$$

we get $\lim_{n,m\to\infty} ps(v_{2n+1}, v_{2m+1}) = 0$.
Letting $n, m \to \infty$ in

$$|p^s(w_{2n+1}, w_{2m+1}) - p^s(w_{2n}, w_{2m})| \leq p^s(w2n+1,w2n) + ps(w2m+1,w2m)$$

we get $\lim_{n,m\to\infty} ps(w2n+1,w2m+1) = 0$.
Thus $\{u_{2n+1}\}, \{v_{2n+1}\}$ and $\{w_{2n+1}\}$ are Cauchy sequences in the metric space (X, p^s).
Hence $\{u_n\}, \{v_n\}$ and $\{w_n\}$ are Cauchy sequences in the metric space (X, p^s).
Hence we have that

$$\lim_{n,m\to\infty} p^s(u_m,u_n) = \lim_{n,m\to\infty} p^s(v_m,v_n) = \lim_{n,m\to\infty} p^s(w_m,w_n) = 0. \tag{2.15}$$

Now from definition of ps and from (2.4), we have

$$\lim_{n,m\to\infty} p(u_m,u_n) = \lim_{n,m\to\infty} p(v_m,v_n) = \lim_{n,m\to\infty} p(w_m,w_n) = 0. \tag{2.16}$$

Suppose $f(X)$ is complete.
Since $\{u_{2n+1}\} \subseteq f(X)$, $\{v_{2n+1}\} \subseteq f(X)$ and $\{w_{2n+1}\} \subseteq f(X)$ are Cauchy sequences in the complete metric space $(f(X), p^s)$, it follows that the sequences $\{u_{2n+1}\}$, $\{v_{2n+1}\}$ and $\{w_{2n+1}\}$ are convergent in $(f(X), p^s)$.
Thus $\lim_{n\to\infty} p^s(u_{2n+1}, \alpha) = 0$, $\lim_{n\to\infty} p^s(v_{2n+1}, \beta) = 0$ and $\lim_{n\to\infty} p^s(w_{2n+1}, \gamma) = 0$ for some α, β and γ in $f(X)$.
There exist $x, y, z \in X$ such that $\alpha = fx$, $\beta = fy$ and $\gamma = fz$.
Since $\{u_n\}, \{v_n\}$ and $\{w_n\}$ are Cauchy sequences in X and $\{u_{2n+1}\} \to \alpha, \{v_{2n+1}\} \to \beta$ and $\{w_{2n+1}\} \to \gamma$, it follows that $\{u_{2n}\} \to \alpha, \{v_{2n}\} \to \beta$ and $\{w_{2n}\} \to \gamma$. From Lemma 1.6(2), we have

$$p(\alpha, \alpha) = \lim p(u_{2n}, \alpha) = \lim p(u_{2n+1}, \alpha) = \lim p(u_n, u_m). \tag{2.17}$$
$$p(\beta, \beta) = \lim p(v_{2n}, \beta) = \lim p(v_{2n+1}, \beta) = \lim p(v_n, v_m) \tag{2.18}$$

and

$$p(\gamma, \gamma) = \lim p(w_{2n}, \gamma) = \lim p(w_{2n+1}, \gamma) = \lim p(w_n, w_m). \tag{2.19}$$

From (2.16), (2.17), (2.18) and (2.19) we have

$$p(\alpha, \alpha) = \lim p(u_{2n}, \alpha) = \lim p(u_{2n+1}, \alpha) = 0. \tag{2.20}$$
$$p(\beta, \beta) = \lim p(v_{2n}, \beta) = \lim p(v_{2n+1}, \beta) = 0 \tag{2.21}$$

and

$$p(\gamma, \gamma) = \lim p(w_{2n}, \gamma) = \lim p(w_{2n+1}, \gamma) = 0. \tag{2.22}$$

Now

$$\begin{aligned}\psi p(S(x,y,z),\alpha)) &\le \psi(p(S(x,y,z),u_{2n+1})+p(u_{2n+1},\alpha)-p(u_{2n+1},u_{2n+1}))\\ &\le \psi(p(S(x,y,z),T(x_{2n+1},y_{2n+1},z_{2n+1})))+\psi(p(u_{2n+1},\alpha))\\ &\le \tfrac{1}{3}F(\psi\left(p(fx,gx_{2n+1})+p(fy,gy_{2n+1})+p(fz,gz_{2n+1})\right),\\ &\varphi\left(\tfrac{p(fx,gx_{2n+1})+p(fy,gy_{2n+1})+p(fz,gz_{2n+1})}{3}\right)+\psi(p(u_{2n+1},\alpha))\\ &\le \tfrac{1}{3}F(\psi\left(p(\alpha,u_{2n})+p(\beta,v_{2n})+p(\gamma,w_{2n})\right),\\ &\varphi\left(\tfrac{p(\alpha,u_{2n})+p(\beta,v_{2n})+p(\gamma,w_{2n})}{3}\right))+\psi(p(u_{2n+1},\alpha)).\end{aligned}$$

Letting $n\to\infty$, it follows that $S(x, y, z) = \alpha$. Similarly $S(y, z, x) = \beta$ and $S(z, x, y) = \gamma$.

Thus $\alpha = fx = S(x, y, z), \beta = fy = S(y, z, x)$ and $\gamma = fz = S(z, x, y)$.

Since (f, S) is w-compatible we have

$$S(\alpha,\beta,\gamma) = f\alpha,\ S(\beta,\gamma,\alpha) = f\beta \text{ and } S(\gamma,\alpha,\beta) = f\gamma.$$

From Remark 5(2) and (2.17),(2.18) and (2.19),we have

$$\lim p(f\alpha, u_{2n}) = p(f\alpha,\alpha),\ \lim p(f\beta, v_{2n}) = p(f\beta,\beta) \text{ and } \lim p(f\gamma, w_{2n}) = p(f\gamma,\gamma).$$

Now, we shall prove that $f\alpha = \alpha, f\beta = \beta$ and $f\gamma = \gamma$.

From $(p_4)$and (i), we have

$$\begin{aligned}\psi p(f\alpha,\alpha)) &\le \psi(p(f\alpha,u_{2n+1})+p(u_{2n+1},\alpha)-p(u_{2n+1},u_{2n+1}))\\ &\le \psi(p(S(\alpha,\beta,\gamma),T(x_{2n+1},y_{2n+1},z_{2n+1})))+\psi(p(u_{2n+1},\alpha))\\ &\le \tfrac{1}{3}F\begin{pmatrix}\psi\left(p(f\alpha,u_{2n})+p(f\beta,v_{2n})+p(f\gamma,w_{2n})\right),\\ \varphi\left(\tfrac{p(f\alpha,u_{2n})+p(f\beta,v_{2n})+p(f\gamma,w_{2n})}{3}\right)\end{pmatrix}+\psi(p(u_{2n+1},\alpha))\end{aligned}$$

Similarly

$$\begin{aligned}\psi(p(f\beta,\beta)) \le \tfrac{1}{3}F(&\psi(p(f\alpha,u_{2n})+p(f\beta,v_{2n})+p(f\gamma,w_{2n})),\\ &\varphi(\tfrac{p(f\alpha,u_{2n})+p(f\beta,v_{2n})+p(f\gamma,w_{2n})}{3}))+\psi(p(v_{2n+1},\beta))\end{aligned}$$

and

$$\begin{aligned}\psi(p(f\gamma,\gamma)) &\le \tfrac{1}{3}F(\psi\left(p(f\alpha,u_{2n})+p(f\beta,v_{2n})+p(f\gamma,w_{2n})\right),\\ &\varphi\left(\tfrac{p(f\alpha,u_{2n})+p(f\beta,v_{2n})+p(f\gamma,w_{2n})}{3}\right))+\psi(p(w_{2n+1},\gamma))\end{aligned}$$

Thus

$$\begin{aligned}\psi\ &(p(f\alpha,\alpha)+p(f\beta,\beta)+p(f\gamma,\gamma))\\ &\le F\left(\psi\left(p(f\alpha,u_{2n})+p(f\beta,v_{2n})+p(f\gamma,w_{2n})\right),\varphi\left(\tfrac{p(f\alpha,u_{2n})+p(f\beta,v_{2n})+p(f\gamma,w_{2n})}{3}\right)\right)\\ &+p(u_{2n+1},\alpha)+p(v_{2n+1},\beta)+p(w_{2n+1},\gamma)\end{aligned}$$

Letting $n\to\infty$, we have

$$\begin{aligned}&\psi(p(f\alpha,\alpha)+p(f\beta,\beta)+p(f\gamma,\gamma))\\ &\le F\left(\psi\left(p(f\alpha,\alpha)+p(f\beta,\beta)+p(f\gamma,\gamma)\right),\varphi\left(\tfrac{p(f\alpha,\alpha)+p(f\beta,\beta)+p(f\gamma,\gamma)}{3}\right)\right)\end{aligned}$$

By the definition of C - class function, we have that

$$\alpha = f\alpha = S(\alpha,\beta,\gamma),\ \beta = f\beta = S(\beta,\gamma,\alpha) \text{ and } \gamma = f\gamma = S(\gamma,\alpha,\beta). \tag{2.23}$$

Since $S(X\times X\times X)\subseteq g(X)$, there exist $r,s,t\in X$ such that

$$\alpha = S(\alpha,\beta,\gamma) = gr,\ \beta = S(\beta,\gamma,\alpha) = gs \text{ and } \gamma = S(\gamma,\alpha,\beta) = gt.$$

Consider

$$\begin{aligned}\psi(p(\alpha, T(r,s,t))) &= \psi(p(S(\alpha,\beta,\gamma), T(r,s,t))))\\ &\leq \tfrac{1}{3}F(\psi\left(p(f\alpha, gr) + p(f\beta, gs) + p(f\gamma, gt)\right), \varphi\left(\tfrac{p(f\alpha,gr)+p(f\beta,gs)+p(f\gamma,gt)}{3}\right))\\ &= \tfrac{1}{3}F(\psi\left(p(\alpha,\alpha) + p(\beta,\beta) + p(\gamma,\gamma)\right), \varphi\left(\tfrac{p(\alpha,\alpha)+p(\beta,\beta)+p(\gamma,\gamma)}{3}\right))\\ &= 0 \text{ from (2.20), (2.21) and (2.22).}\end{aligned}$$

It follows that $T(r, s, t) = \alpha = gr$.

Similarly $T(s, t, r) = \beta = gs$ and $T(t, r, s) = \gamma = gt$.

Since (g, T) is w - compatible, we have

$$T(\alpha,\beta,\gamma) = g\alpha,\ T(\beta,\gamma,\alpha) = g\beta \text{ and } T(\gamma,\alpha,\beta) = g\gamma.$$

Now we prove that $g\alpha = \alpha$, $g\beta = \beta$ and $g\gamma = \gamma$.

$$\begin{aligned}\psi(p(\alpha, g\alpha)) &= \psi(p(S(\alpha,\beta,\gamma), T(\alpha,\beta,\gamma)))\\ &\leq \tfrac{1}{3}F(\psi\left(p(f\alpha, g\alpha) + p(f\beta, g\beta) + p(f\gamma, g\gamma)\right), \varphi\left(\tfrac{p(f\alpha,g\alpha)+p(f\beta,g\beta)+p(f\gamma,g\gamma)}{3}\right))\\ &= \tfrac{1}{3}F(\psi\left(p(\alpha, g\alpha) + p(\beta, g\beta) + p(\gamma, g\gamma)\right), \varphi\left(\tfrac{p(\alpha,g\alpha)+p(\beta,g\beta)+p(\gamma,g\gamma)}{3}\right)).\end{aligned}$$

Similarly

$$\psi(p(\beta, g\beta)) \leq \tfrac{1}{3}F(\psi\left(p(\alpha, g\alpha) + p(\beta, g\beta) + p(\gamma, g\gamma)\right), \varphi\left(\tfrac{p(\alpha,g\alpha)+p(\beta,g\beta)+p(\gamma,g\gamma)}{3}\right))$$

and

$$\psi(p(\gamma, g\gamma)) \leq \tfrac{1}{3}F(\psi\left(p(\alpha, g\alpha) + p(\beta, g\beta) + p(\gamma, g\gamma)\right), \varphi\left(\tfrac{p(\alpha,g\alpha)+p(\beta,g\beta)+p(\gamma,g\gamma)}{3}\right)).$$

Thus

$$\psi(p(\alpha, g\alpha) + p(\beta, g\beta) + p(\gamma, g\gamma)) \leq F(\psi\left(p(\alpha, g\alpha) + p(\beta, g\beta) + p(\gamma, g\gamma)\right), \varphi\left(\tfrac{p(\alpha,g\alpha)+p(\beta,g\beta)+p(\gamma,g\gamma)}{3}\right)).$$

By the definitin of C - condition, we have that

$$\alpha = g\alpha = T(\alpha,\beta,\gamma), \beta = g\beta = T(\beta,\gamma,\alpha) \text{ and } \gamma = g\gamma = T(\gamma,\alpha,\beta). \tag{2.24}$$

Hence from (2.23) and (2.24), (α, β, γ) is common 3-tupled fixed point of S, T, f and g.

To prove uniqueness, let $(\alpha^*, \beta^*, \gamma^*)$ is another common 3-tupled fixed point of S, T, f and g such that $\alpha \neq \alpha^*$ or $\beta \neq \beta^*$ or $\gamma \neq \gamma^*$.

Consider

$$\begin{aligned}\psi(p(\alpha, \alpha^*)) &= \psi(p(S(\alpha,\beta,\gamma), T(\alpha^*,\beta^*,\gamma^*)))\\ &\leq \tfrac{1}{3}F(\psi\left(p(f\alpha, g\alpha^*) + p(f\beta, g\beta^*) + p(f\gamma, g\gamma^*)\right), \varphi\left(\tfrac{p(f\alpha,g\alpha^*)+p(f\beta,g\beta^*)+p(f\gamma,g\gamma^*)}{3}\right)\\ &= \tfrac{1}{3}F(\psi\left(p(\alpha, \alpha^*) + p(\beta, \beta^*) + p(\gamma, \gamma^*)\right), \varphi\left(\tfrac{p(\alpha,\alpha^*)+p(\beta,\beta^*)+p(\gamma,\gamma^*)}{3}\right).\end{aligned}$$

Similarly

$$\psi(p(\beta, \beta^*)) \leq \tfrac{1}{3}F(\psi\left(p(\alpha, \alpha^*) + p(\beta, \beta^*) + p(\gamma, \gamma^*)\right), \varphi\left(\tfrac{p(\alpha,\alpha^*)+p(\beta,\beta^*)+p(\gamma,\gamma^*)}{3}\right).$$

and

$$\psi(p(\gamma, \gamma^*)) \leq \tfrac{1}{3}F(\psi\left(p(\alpha, \alpha^*) + p(\beta, \beta^*) + p(\gamma, \gamma^*)\right), \varphi\left(\tfrac{p(\alpha,\alpha^*)+p(\beta,\beta^*)+p(\gamma,\gamma^*)}{3}\right).$$

Thus

$$\psi(p(\alpha, \alpha^*) + p(\beta, \beta^*) + p(\gamma, \gamma^*)) \leq F(\psi\left(p(\alpha, \alpha^*) + p(\beta, \beta^*) + p(\gamma, \gamma^*)\right), \varphi\left(\tfrac{p(\alpha,\alpha^*)+p(\beta,\beta^*)+p(\gamma,\gamma^*)}{3}\right)).$$

So, $\psi(p(\alpha, \alpha^*) + p(\beta, \beta^*) + p(\gamma, \gamma^*)) = 0$, or, $\varphi(p(\alpha, \alpha^*) + p(\beta, \beta^*) + p(\gamma, \gamma^*)) = 0$.

Hence $p(\alpha, \alpha^*) + p(\beta, \beta^*) + p(\gamma, \gamma^*) = 0$ It is a contradiction.

Hence

$$\alpha = \alpha^*, \beta = \beta^* \text{ and } \gamma = \gamma^*.$$

Therefore (α, β, γ) is unique common 3-tupled fixed point of S, T, f and g.

Now we claim that $\alpha = \beta = \gamma$. Suppose $\alpha \neq \beta$ or $\beta \neq \gamma$ or $\gamma \neq \alpha$.

$$\begin{aligned}\psi(p(\alpha,\beta)) &= \psi(p(S(\alpha,\beta,\gamma),T(\beta,\gamma,\alpha)))\\ &\leq \tfrac{1}{3}F(\psi\,(p(f\alpha,g\beta)+p(f\beta,g\gamma)+p(f\gamma,g\alpha)),\varphi\left(\tfrac{p(f\alpha,g\beta)+p(f\beta,g\gamma)+p(f\gamma,g\alpha)}{3}\right))\\ &= \tfrac{1}{3}F(\psi\,(p(\alpha,\beta)+p(\beta,\gamma)+p(\gamma,\alpha)),\varphi\left(\tfrac{p(\alpha,\beta)+p(\beta,\gamma)+p(\gamma,\alpha)}{3}\right)).\end{aligned}$$

$$\begin{aligned}\psi(p(\beta,\gamma)) &= \psi(p(S(\beta,\gamma,\alpha),T(\gamma,\alpha,\beta)))\\ &\leq \tfrac{1}{3}F(\psi\,(p(f\beta,g\gamma)+p(f\gamma,g\alpha)+p(f\alpha,g\beta)),\varphi\left(\tfrac{p(f\beta,g\gamma)+p(f\gamma,g\alpha)+p(f\alpha,g\beta)}{3}\right))\\ &= \tfrac{1}{3}F(\psi\,(p(\alpha,\beta)+p(\beta,\gamma)+p(\gamma,\alpha)),\varphi\left(\tfrac{p(\alpha,\beta)+p(\beta,\gamma)+p(\gamma,\alpha)}{3}\right)).\end{aligned}$$

and

$$\begin{aligned}\psi(p(\gamma,\alpha)) &= \psi(p(S(\gamma,\alpha,\beta),T(\alpha,\beta,\gamma)))\\ &\leq \tfrac{1}{3}F(\psi\,(p(f\gamma,g\alpha)+p(f\alpha,g\beta)+p(f\beta,g\gamma)),\varphi\left(\tfrac{p(f\gamma,g\alpha)+p(f\alpha,g\beta)+p(f\beta,g\gamma)}{3}\right))\\ &= \tfrac{1}{3}F(\psi\,(p(\alpha,\beta)+p(\beta,\gamma)+p(\gamma,\alpha)),\varphi\left(\tfrac{p(\alpha,\beta)+p(\beta,\gamma)+p(\gamma,\alpha)}{3}\right)).\end{aligned}$$

Thus

$$\begin{aligned}\psi(p(\alpha,\beta)+p(\beta,\gamma)+p(\gamma,\alpha)) \;&\leq F(\psi\,(p(\alpha,\beta)+p(\beta,\gamma)+p(\gamma,\alpha)),\\ &\quad ,\varphi\left(\tfrac{p(\alpha,\beta)+p(\beta,\gamma)+p(\gamma,\alpha)}{3}\right)).\end{aligned}$$

So, $\psi(p(\alpha, \beta) + p(\beta, \gamma) + p(\gamma, \alpha)) = 0$, or, $\varphi(p(\alpha, \beta) + p(\beta, \gamma) + p(\gamma, \alpha)) = 0$. Hence $p(\alpha, \beta) + p(\beta, \gamma) + p(\gamma, \alpha) = 0$ It is a contradiction. Hence $\alpha = \beta = \gamma$.

Thus S, T, f and g have unique common 3-tupled fixed point of the form (α, α, α) in $X \times X \times X$.

Example 2.2. *Let $X = [0,1]$, the mappings $S,T : X \times X \times X \to X$ and $f,g : X \to X$ be defined by $S(x, y, z)$ $\frac{x^2+y^2+z^2}{5}$, and $S(x, y, z)$ $\frac{x+y+z}{15}$, respectively and $p : X \times X \to [0, \infty)$ by $p(x, y) = \max\{x, y\}$. Let $\psi, \varphi : [0, \infty) \to [0, \infty)$ be defined by $\psi(x) = x$ and $\varphi(x) = \frac{1}{4}$ also define the C - class function $F : [0, \infty)^2 \to \mathbb{R}$* by *$F(s, t)$ $\frac{s}{1+t}$ Clearly the conditions (ii), (iii) and (iv) are satisfied.*

Now

$$\begin{aligned}\psi(p(S(x,y,z),T(u,v,w))) &= \max\left\{\tfrac{x^2+y^2+z^2}{5},\tfrac{u+v+w}{15}\right\}\\ &= \max\left\{\tfrac{x^2}{5}+\tfrac{y^2}{5}+\tfrac{z^2}{5},\tfrac{u}{15}+\tfrac{v}{15}+\tfrac{w}{15}\right\}\\ &\leq \max\left\{\tfrac{x^2}{5},\tfrac{u}{15}\right\}+\max\left\{\tfrac{y^2}{5},\tfrac{v}{15}\right\}+\max\left\{\tfrac{z^2}{5},\tfrac{w}{15}\right\}\\ &= \tfrac{1}{5}\left(\max\left\{x^2,\tfrac{u}{3}\right\}+\max\left\{y^2,\tfrac{v}{3}\right\}+\max\left\{z^2,\tfrac{w}{3}\right\}\right)\\ &= \tfrac{1}{3}\tfrac{3}{5}(\,p(fx,gu)+p(fy,gv)+p(fz,gw))\\ &\leq \tfrac{1}{3}\tfrac{4}{5}(\,p(fx,gu)+p(fy,gv)+p(fz,gw))\\ &= \tfrac{1}{3}F\left(\psi\,(p(fx,gu)+p(fy,gv)+p(fz,gw)),\varphi\left(\tfrac{p(fx,gu)+p(fy,gv)+p(fz,gw)}{3}\right)\right).\end{aligned}$$

Hence all conditions of Theorem 2.1 are satisfied and (0, 0, 0) is unique common 3-tupled fixed point of S, T, f and g.

Example 2.3. *Let $X = [0,1]$, the mappings $S, T : X \times X \times X \to X$ and $f, g : X \to X$ be defined by $S(x, y, z) = \frac{x^2+y^2+z^2}{6}$, $T(x, y, z) = \frac{x+y+z}{12}$, $f(x) = x^2$ and $g(x)$ $\frac{x}{2}$ respectively and $p : X \times X \to [0, \infty)$ by $p(x, y) = \max\{x, y\}$. Let $\psi, \varphi : [0, \infty) \to [0, \infty)$ be defined by $\psi(x) = x$ and $\varphi(x) = \frac{x}{2}$, also define the C - class function $F : [0, \infty)^2 \to \mathbb{R}$ by $F(s, t) = s - t$ Clearly all conditions of Theorem 2.1 are satisfied and (0,0,0) is unique common 3-tupled fixed point of S,T,f and g.*

REFERENCES

1. T. Abdeljawad, E. Karapınar, K. Tas, *Existence and uniqueness of a common fixed point on partial metric spaces, Appl. Math. Lett.* **24** (11), 1894–1899(2011).
2. I. Altun, F. Sola and H. Simsek, *Generalized contractions on partial metric spaces, Topology and its Applications.* **157** (18) (2010), 2778–2785.
3. I. Altun and A. Erduran, *Fixed point theorems for monotone mappings on partial metric spaces, Fixed Point Theory and Applications*, Volume 2011, Article ID 508730, 10 pages, doi: 10.1155/2011/508730.
4. H. Aydi, *Some coupled fixed point results on partial metric spaces, International Journal of Mathematics and Mathematical Sciences*, Volume 2011, Article ID 647091, 11 pages, doi:10.1155/2011/647091.
5. H. Aydi, *Fixed point results for weakly contractive mappings in ordered partial metric spaces, Journal of Advanced Mathematical and Studies*, Volume **4**, Number 2, (2011).
6. V. Berinde and M. Borcut, *Tripled fixed point theorems for contractive type mappings in partially ordered metric spaces, Nonlinear Analysis*, **74**(15), 4889–4897 (2011).
7. T. G. Bhaskar and V. Lakshmikantham, *Fixed point theorems in partially ordered metric spaces and applications, Nonlinear Analysis.* **65** (2006), 1379–1393.
8. R. Heckmann, *Approximation of metric spaces by partial metric spaces, Appl. Categ. Structures*, **no.1-2, 7**, 1999, 71-83.
9. D. Ili´c, V. Pavlovi´c, V. Rako˘cevi´c, *Some new extensions of Banach's contraction principle to partial metric spaces, Appl. Math. Letters*, doi:10.1016/j.aml.2011.02.025.
10. E. Karapınar, I. M. Erhan, *Fixed point theorems for operators on partial metric spaces, Applied Mathematics Letters*, **24** (11),1900-1904 (2011), 10.1016 /j.aml. 2011.05.013.
11. E. Karapınar, *Weak ϕ-contraction on partial contraction, J. Comput. Anal. Appl.* (in press).
12. E. Karapınar, *Weak ϕ-contraction on partial contraction and existence of fixed points in partially ordered sets, Mathematica Aeterna*, **1**(4)237-244(2011).
13. E. Karapınar, *Generalizations of Caristi Kirk's Theorem on Partial metric Spaces, Fixed Point Theory and. Appl.* 2011:4, doi:10.1186/1687-1812-2011-4.
14. R. Kopperman, S.G. Matthews, and H. Pajoohesh, *What do partial metrics represent?, Spatial representation: discrete vs. continuous computational models, Dagstuhl Seminar Proceedings*, **No. 04351**, Internationales Begegnungs- und Forschungszentrum fu¨r Informatik (IBFI), Schloss Dagstuhl, Germany, (2005). MR 2005j:54007.
15. H. P. A. Ku¨nzi, H. Pajoohesh, and M.P. Schellekens, *Partial quasi-metrics, Theoret. Comput. Sci.* **365 no.3** (2006) 237-246. MR 2007f:54048.
16. V. Lakshmikantham and Lj. Ćirić,*Coupled fixed point theorems for nonlinear contractions in partially ordered metric spaces, Nonlinear Analysis.* **70** (2009) 4341–4349.
17. S. G. Matthews. *Partial metric topology. Research Report 212. Dept. of Computer Science*. University of Warwick, 1992.
18. S. G. Matthews, *Partial metric topology, in Proceedings of the 8th Summer Conference on General Topology and Applications*, **vol. 728**, pp. 183–197, Annals of the New York Academy of Sciences, 1994.
19. S. Oltra and O. Valero, *Banach's fixed point theorem for partial metric spaces, Rendiconti dell'Istituto di Matematica dell'Universita` di Trieste.* **36** (1–2) (2004) 17–26.
20. S. J. ONeill, *Two topologies are better than one, Tech. report, University of Warwick, Coventry*, UK, http://www.dcs.warwick.ac.uk/reports/283.html, (1995).
21. K. P. R. Rao and G. N. V. Kishore, *A unique common 3 - tupled fixed point, theorem for four maps in partial metric spaces, South East Asian Bulletin of Mathematics*, 37, (2013) 565 578.
22. S. Romaguera and M. Schellekens, *Weightable quasi-metric semigroup and semilattices, Electronic Notes of Theoretical computer science*, Proceedings of MFCSIT, **40**, Elsevier, (2003).
23. S. Romaguera, *A Kirk type characterization of completeness for partial metric spaces, Fixed Point Theory and Applications*, Volume 2010, Article ID 493298, 6 pages, 2010.
24. M. Schellekens, *The Smtth comletion: a common foundation for denotational semantics and complexity analysis, Electronic Notes in Theoretical Computer Science*, **vol. 1**, 1995, 535 556.
25. M.P. Schellekens, *A characterization of partial metrizability: domains are quantifiable, Topology in computer science (Schlo Dagstuhl, 2000), Theoretical Computer Science* **305** no. 1-3 (2003) 409–432. MR 2004i:54037.
26. O. Valero, *On Banach fixed point theorems for partial metric spaces, Applied General Topology*. **6** (2) (2005) 229–240.
27. P. Waszkiewicz, *Quantitative continuous domains, Applied Categorical Structures*, **vol. 11**, no. 1, 2003, 41–67.
28. P. Waszkiewicz, *Partial metrizebility of continuous posets, Mathematical Structures in Computer Sciences*, **vol. 16**, no. 2, 2006, 359–372.
29. C. Chen, C. Zhu, Fixed point theorems for weakly C-contractive mappings in partial metric spaces, Fixed Point Theory Appl. (2013) doi: 10.1186/1687-1812-2013-107
30. A. H.Ansari,Note on" φ–ψ-contractive type mappings and related fixed point",The 2 nd Regional Conference onMathematics And Applications,PNU,September 2014 ,pages 377–380
31. M. S. Khan, M. Swaleh, and S. Sessa, Fixed point theorems by altering distances between the points, Bulletin of the Australian Mathematical Society, 30 (1) (1984) 1–9.

Impending Inquisitions in Humanities and Sciences (ICIIHS-2022) – Dr. Mohan Varkolu et al. (eds)

A Two-warehouse Inventory Model with an Advanced Payment Mechanism for Stock-Dependent Demand, Deteriorating Items with a Fixed Shelf Life and Partial Backlogging

60

Ashfar Ahmed[1]
Department of Mathematics, School of Science, GITAM-Hyderabad Campus, Hyderabad, India
Department of Mathematics, Malla Reddy Engineering College (AUTONOMOUS), Main Campus, Hyderabad, Telangana, India

Puppala Bala Sai Manikanta Sandeep[2]
Department of Computer Science Engineering, School of Technology, GITAM-Hyderabad Campus, Hyderabad, India

Motahar Reza[3], **Krishna Kummari**[4]
Department of Mathematics, School of Science, GITAM-Hyderabad Campus, Hyderabad, India

ABSTRACT The most important factor in maximising profit margins in the expanding global market of today is inventory management. One of the most critical constraints in inventory management is the fixed shelf life of each item. Due to space constraints, a second warehouse may be required to store inventory for large-scale productions. The warehouse inventory is utilized to make installment payments for advance payments. Stock shortages are also possible with a partial backlog. This research looked at a two warehouse inventory model of deteriorating commodities with an advance payment mechanism and partial backlogging. We also consider that the demand rate is stock dependent, that one warehouse is owned, and that another is rented. The backlog rate is assumed to be time-dependent. Based on the assumptions, the cost function of this problem is a highly nonlinear constraint optimization problem. Mathematica is used to tackle this nonlinear optimization problem. A numerical example of this model has been simulated for various parameters, and the results have been physically interpreted.

KEYWORDS Inventory, Two-warehouse, Partial backlogging, Fixed-shelf life, Advance payment

1. Introduction

The main issue that people face on a daily basis is the inventory problem and the term "*inventory*" refers to a collection of usable commodities. The inventory contains a variety of commodities, including ingredients, works in process and products that have been completed. The management of many types of inventories is a huge concern in the present day. For this reason, a lot of commercial companies are emphasising the need for effective inventory management in order to operate their business effectively. This article covers a variety of inventory management issues, including one with a two warehouse inventory problem with an advance payment technique. Among the most commonly employed techniques is advance payment method. In advance payment facilities the suppliers and traders essentially attempt to draw customers. They seek compensation in full or in part

[1]ahmedashfaq02@gmail.com, [2]sandeeppuppala248@gmail.com, [3]mreza@gitam.edu, [4]krishna.maths@gmail.com

DOI: 10.1201/9781003489436-60

before sending the merchandise. They provide us with a price break on the item or another type of refund on the purchase price in exchange for this payment. These amenities enable shops to purchase an increasing number of goods. Advance payments and two warehouse have just a relationship with one another. Retailers will require additional storage capacity if they purchase more goods. These two criteria were taken into account simultaneously in this study.

For business houses, unforeseen unique circumstances may arise frequently, especially when reacting to seasonal requests, providing discounts to boost sales, importing essential facilitators to address various technological and enterprise issues, etc. Companies frequently wish to purchase a huge number of goods in account of all these aspects. In a single warehouse or owned warehouse, a large amount of goods cannot be kept because of limitations of capacity. Therefore, additional space for storage is needed for holding the extra or surplus goods. As a result, properly managing inventory storage is crucial to a company's success. Therefore, having adequate warehouse facilities is crucial for executing business operations successfully and ensuring the continuous flow of output supplies. From an economic aspect, owned warehouses generally offer client demand. However, a rented warehouse seems to have a higher holding cost compared to one that is owned. As a result, firms constantly seek to move their inventory from rented to owned warehouses in order to meet customer demand. Various scholars have studied the two-warehouse inventory system while taking the degradation impact into account. Degradation is defined as deterioration or damage, and the idea of deterioration must be introduced in the inventory models. The lot of the goods degrade while they are stored in the warehouse. The management of deteriorating goods in warehousing systems has drawn the interest of a large number of academics and researchers for this reason.

Gupta and Vrat [1] created an inventory model using a consumption rate that depends on stock. Later, by considering demand as a value of the level of immediate inventory, this model was modified by Baker and Urban [2]. In their research, Mandal and Phaujdar [3] looked at an inventory model for deteriorating goods that took stock dependence for consumption rate. A new model developed by Vrat and Padmanabhan [4] that incorporates the stock-dependent demand rate and the influence of inflation.

Various inventory issues with deteriorating goods and shortages are discussed in a model by Datta et al. [5] and that uses stock dependence as the demand rate. Furthermore, Padmanabhan and Vrat [6] performed their research using three models, i.e., a variety of backorders for deteriorating products. A two-component demand rate and shortages in inventory models have received a lot of recent study attention. (See, instance Karabi et al. [7], Zhou and Yang [8], Maiti et al. [9], Goyal et al. [10], Yang et al. [11], Bhunia et al. [12], Kumar et al. [13], Pal [14], and Tiwari et al. [15]). A number of other models are also produced by Shaikh et al. [16], Mashud et al. [17], Shah et al. [18], Shaikh et al. [19] , Panda [20] and C´ardenas-Barr´on et.al. [21].

Inventory model includes a lot of perishable commodities like explosive materials,berries, cereal, veggies, medications etc. An EOQ model with diminishing demand rate that is based on technical notes with a defined Shelf-life has bee proposed by Avinadav and Arponen [22]. A shortage-based inventory model with ramp-type demand and a fixed shelf life was presented by Chuang [23] , Ukil et al. [24], Muriana [25], and many more others.

In [26], Hartley illustrates an inventory model with two warehouses based on the assumption that the "*RW*" holding cost is greater than the "*OW*" holding cost. Later,by adding infinite replenishment rate to the two warehouse concept an inventory model is discussed by Sarma [27]. Further, Goswami and Chaudhuri [28] extended the use of non degrading items while taking time-dependent factors like shortages and demand rate into account and also they consider the price of transportation. In [29], Pakkala and Achary introduced a two ware housing systems with decaying items. Yang [30] considered two ware housing systems with the effects of inflation and shortages. Later, Chung and Haung [31] offered a different approach for non-deteriorating commodities with no shortages during a permitted payment delay. A two ware house inventory system with demand rate as a stock dependent combination was developed by Zhou and Yang [32]. For more information about a two warehouse inventory model one can refer Lee and Husu [33], Kumar et.al. [34] and Rastogi et.al. [35].

According to our knowledge, only a few research projects using a single ware house system have been funded in advance. A two-warehouse inventory system has yet to include the advance payment feature. Inventory analysis is significantly influenced by two warehousing systems. A large showroom in a highly competitive marketplace could be quite challenging to identify. Retailers need to rent an additional store area because there isn't enough room in a busy marketplace. Retailers cannot disregard this technique as a result of the globalisation of marketing strategy. This is an actual issue that exists in the corporate world. In fact, we have implemented an advance payment option for the first time in a two-warehouse inventory system. In this study, We therefore present a fixed Shelf in a two-warehouse inventory model with an option for advance payment. Finally we discuss a numerical example with the help of Mathematica Software. Below is a summary of the major contributions:

- Fixed Shelf-life in a Two-Warehouse System with Demand dependent on Stock.

- Prior to receiving the product, pay an equal installment.
- Inventory system for two warehouses that depends on stock levels to meet demand and deteriorate when there is some backlog.
- Partially backlogged shortages at a steady rate.
- The product's demand is stock-dependent.
- Rate of deterioration is constant.

The remaining sections are arranged as follows: Different symbols and Notations used in this study were covered in Section 2 of the paper. In Section 3, the model's mathematical formulation is described. In Section 4, we examine the Optimal solution of the model by applying different parameters. Finally in Section 5, the Conclusion of the paper is provided.

2. Notations

The following notations have been taken into consideration while we developed the inventory model:

Notations	Units	Description
A	$/Order	Ordering Cost
η	Units	Backlogging unit ($0 < \eta < 1$)
S	Units	Total Inventory level
a	Constant	Demand rate's coefficient part ($a > 0$)
M	yr	Enterprise's lead time for paying prepayments
b	Constant	Demand rate Constant for price ($b > 0$)
α	Constant	Rate of Deterioration at Owned warehouse
β	Constant	Rate of Deterioration at Rented warehouse
n	Constant	Prepayments evenly spaced throughout the Lead period
W_1	Units	Level of Inventory at Owned warehouse
C_p	Units	Unit Cost of Purchase
k	Constant	An amount that needs to be paid in multiple installments ($0 < k < 1$)
R	Units	Backlogged units
f_1	Constant	Fixed Shelf-life
t_a	yr	The time where the inventory level in the rented warehouse (*RW*) falls to zero
c	Unit	Cost of shortage, Rupee/unit
t_b	yr	The time where the inventory level in the owned warehouse (*OW*) falls to zero
q	Unit	Rupees per unit time spent in owned warehouse (*OW*)
γ	Constant	In an own warehouse (*OW*), the deterioration rate's value exists between (0, 1)
θ	Constant	In a rented warehouse (*RW*), the deterioration rate's value exists between (0, 1)
D_1	Unit	Rupees per unit time in a rented warehouse (*RW*) for the cost of deterioration
D_2	Unit	Rupees per unit time in a owned warehouse (*OW*) for the cost of deterioration
p	Unit	Rupees per unit time spent in rented warehouse (*RW*)
h	Constant	Holding Cost Constant that does not depend on Time
T	yr	Total length of an inventory cycle,hence $T = t_a + t_b$

3. Problem Definition

Suppose a scenario where a company place a request for $(S + R)$ units of a given product and pays for a number "k" of the purchase price by making "n" uniform payments at uniform periods over the lead time "M" before paying the balance at time $t = 0$ to receive the lot. The on-hand inventory level changes to "S" shortly after "R" are used to partially satisfy the backlogged demand. The remaining portion $(S - W_1)$ is saved in "*RW*" while "W_1" units are kept in "*OW*". The holding cost in the "*RW*" is apparently higher than that in the "*OW*" due to the "*RW's*" better facilities, and as a result, the "*RW*"*'s* products will reportedly be taken first. In the time interval $[0, f_1]$, the inventory decreases due to customer demand. But in the time interval $[f_1, t_a]$ the inventory level decreases due to both the constant depreciation rate "β" and the customer demand $D(p)$. It becomes zero in *RW* at the time $t = t_a$. The inventory level in *OW*, however, drops as a result of a constant rate of deterioration "α" within the range $[f_1, t_a]$. The levels of inventory are zero for rented warehouse (*RW*) and greater than zero for own warehouse (*OW*) in the time interval $[t_a, t_b]$. Individual time intervals may be considered to be $[0, f_1]$, $[f_1, t_a]$, $[t_a, t_b]$ and $[t_b, T]$.

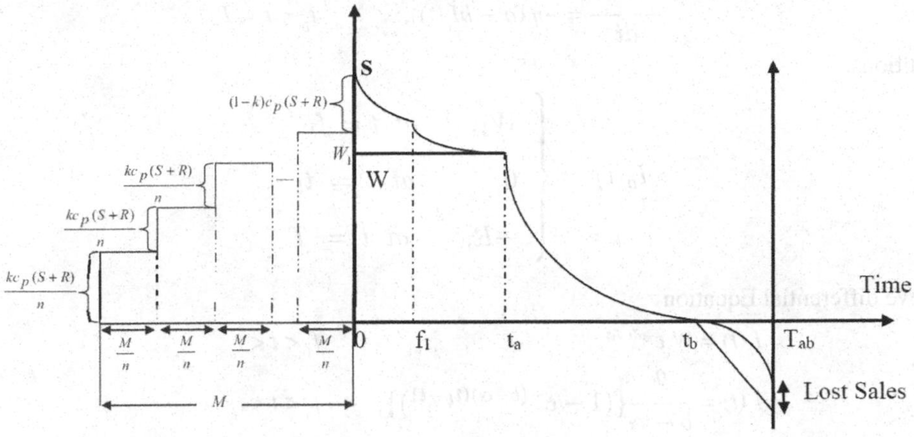

Fig. 59.1 A Visual illustration of the two warehouse system's inventory level

Here there will be two cases, the first case is the deterioration rate starts when the level of inventory of rented warehouse (*RW*) is in use in the interval $(0 \le f_1 \le t_a)$, and in the second case the deterioration rate starts when the level of inventory of own warehouse (*OW*) is in use in the interval $(t_a \le f_1 \le t_b)$.

3.1 For CASE I: $(0 \le f_1 \le t_a)$

Following are the differential equations that explain how the levels of inventory for Owned warehouses (*OW*) and Rental warehouses (*RW*) differ:

For various time periods, differential equations for a rented warehouse are:

$$\frac{dI_r(t)}{dt} = -(a - bI_r(t)), \qquad 0 < t < f_1 \tag{1}$$

$$\frac{dI_r(t)}{dt} = -(a - bI_r(t)) - \beta I_r(t). \qquad f_1 < t < t_a \tag{2}$$

subject to the conditions:

$$I_r(t) = \begin{cases} S - W_1, & at\ t = 0 \\ 0, & at\ t = t_a \end{cases} \tag{3}$$

On solving the above differential Equations:

$$I_r(t) = \frac{a}{b} - \left\{ \frac{a}{b} - (S - W_1) \right\} e^{bt}, \qquad 0 < t < f_1 \tag{4}$$

$$I_r(t) = \frac{a}{b - \beta}\{(1 - e^{-(b-\beta)(t_a - t)})\},, \qquad f_1 < t < t_a \tag{5}$$

Furthermore, the following differential Equations can be used to represent the inventory level $I_o(t)$ in *OW* at any instant "*t*"

$$\frac{dI_o(t)}{dt} = -\alpha I_o(t), \qquad f_1 < t < t_a \tag{6}$$

$$\frac{dI_o(t)}{dt} = -(a - bI_o(t)) - \alpha I_o(t), \quad t_a < t < t_b \tag{7}$$

$$\frac{dI_o(t)}{dt} = -\eta\,(a - bI_o(t)), \qquad t_b < t < T \tag{8}$$

subject to the conditions:

$$I_o(t) = \begin{cases} W_1, & at\ t = f_1 \\ 0, & at\ t = t_b \\ -R, & at\ t = T \end{cases} \tag{9}$$

On solving the above differential Equation

$$I_o(t) = W_1 e^{-\alpha(t-f_1)}, \qquad f_1 < t < t_a \tag{10}$$

$$I_o(t) = \frac{a}{b-\alpha}\{(1 - e^{-(b-\alpha)(t_b-t)})\}, \qquad t_a < t < t_b \tag{11}$$

$$I_o(t) = \frac{a}{b} - \left\{\frac{a}{b} + R\right\} e^{-\eta b(T-t)}, \qquad b < t < T \tag{12}$$

By considering the continuity at $t = f_1$, $t = t_a$ and $t = t_b$, we can write:

$$S = W_1 - \frac{a}{2-e^{bf_1}}\left[\frac{1}{b-\beta}\left\{1 - e^{-(b-\beta)(t_a-f_1)}\right\} + \frac{1}{b}e^{bf_1}\right] \tag{13}$$

$$t_b = t_a - \frac{1}{b-\alpha}\left[\ln\left(1 - \frac{W_1(b-\alpha)}{a}e^{-\alpha(t_a-f_1)}\right)\right] \tag{14}$$

$$R = \frac{a}{b}\left[e^{\eta b(T-t_b)} - 1\right] \tag{15}$$

Here, we discuss how the model's inventory-related costs originated based on the assumptions:

(a) Ordering Cost: A

(b) Purchase Cost: $C_p(S + R) = C_p\left(\frac{a\left((b-\beta)\exp\left(b\eta\left(\frac{\log\left(\frac{W_1(\alpha-b)e^{\alpha(f_1-t_a)}}{a}+1\right)}{b-\alpha} + T - t_a\right)\right)\right)}{b(b-\beta)} + \frac{a\left(\beta e^{-bf_1} - be^{-bt_a-\beta f_1+\beta t_a}\right)}{b(b-\beta)} + W_1\right)$

(c) Holding Cost:

$$\int_0^{f_1} h(t)I_r(t)\,dt + \int_{f_1}^{t_a} h(t)\,I_r(t)\,dt + \int_0^{f_1} h(t)I_o(t)\,dt + \int_{f_1}^{t_a} h(t)I_o(t)\,dt + \int_{t_a}^{t_b} h(t)I_o(t)\,dt$$

$$= \frac{-\frac{e^{bf_1}(a+b(W_1-S))(b(f_1p+h)-p)}{b^2} + \frac{(bh-p)(a+b(W_1-S))}{b^2} + \frac{1}{2}af_1{}^2p + af_1h}{b}$$
$$- \frac{a\left(\frac{e^{(b-\beta)(f_1-t_a)}(p-(b-\beta)(f_1p+h))}{(b-\beta)^2} + \frac{(b-\beta)(h+pt_a)-p}{(b-\beta)^2} + \frac{f_1{}^2p}{2} + f_1h - ht_a - \frac{pt_a{}^2}{2}\right)}{b-\beta}$$
$$- \frac{a\left(\frac{e^{(b-\alpha)(t_a-t_b)}(q-(b-\alpha)(h+qt_a))}{(b-\alpha)^2} + \frac{(b-\alpha)(h+qt_b)-q}{(b-\alpha)^2} + ht_a - ht_b + \frac{qt_a{}^2}{2} - \frac{qt_b{}^2}{2}\right)}{b-\alpha}$$
$$+ \frac{W_1\left(-e^{\alpha(f_1-t_a)}(\alpha h + \alpha q t_a + q) + \alpha f_1 q + \alpha h + q\right)}{\alpha^2} + \frac{1}{2}f_1W_1(f_1q + 2h)$$

(d) Deterioration Cost:

$$D_1\int_{f_1}^{t_a} \theta I_r(t)\,dt + D_2\int_{f_1}^{t_a} \gamma I_o(t)\,dt + D_2\int_{t_a}^{t_b} \gamma I_o(t)\,dt$$

$$= \frac{aD_1\theta\left(e^{(b-\beta)(f_1-t_a)} + b(t_a - f_1) + \beta f_1 - \beta t_a - 1\right)}{(b-\beta)^2} + \frac{\gamma D_2W_1\left(e^{\alpha(f_1-t_a)} - e^{\alpha(f_1-t_b)}\right)}{\alpha}$$
$$- \frac{\gamma D_2W_1\left(e^{\alpha(f_1-t_a)} - 1\right)}{\alpha}$$

(e) Shortage Cost:

$$-c\int_{t_b}^{T} I_o(t)\,dt = -\frac{c\left((a+bR)e^{b\eta(t_b-T)} + a(b\eta(T-t_b)-1) - bR\right)}{b^2\eta}$$

(f) Capital Cost:

$$I_c\left[\frac{kC_p(S+R)}{n}\,\frac{M}{n}\,(1+2+3+\dots\dots\dots+n)\right]$$

$$= \frac{C_pI_ckM(n+1)\left(\frac{a\left((b-\beta)\exp\left(b\eta\left(\frac{\log\left(\frac{\mathrm{w1}(\alpha-b)e^{\alpha(f_1-t_a)}}{a}+1\right)}{b-\alpha}+T-t_a\right)\right)+\beta e^{-bf_1}-be^{-bt_a-\beta f_1+\beta t_a}\right)}{b(b-\beta)} + W_1\right)}{2n}$$

Consequently, the total cyclic cost per unit of time is

$$TC1 = \frac{1}{T}\left[\begin{array}{c}\langle Ordering\,Cost\rangle + \langle Purchase\,Cost\rangle + \langle Holding\,Cost\rangle \\ + \langle Deterioration\,Cost\rangle + \langle Shortage\,Cost\rangle + \langle Capital\,Cost\rangle\end{array}\right]$$

$$= \frac{1}{T}\left[A + \frac{1}{2}f_1W_1(f_1q + 2h) + \frac{-\frac{e^{bf_1}(a+b(W_1-S))(b(f_1p+h)-p)}{b^2} + \frac{(bh-p)(a+b(W_1-S))}{b^2} + \frac{1}{2}af_1{}^2p + af_1h}{b}\right]$$

$$-\frac{1}{T}\left[\frac{a\left(\frac{e^{(b-\alpha)(t_a-t_b)}(q-(b-\alpha)(h+qt_a))}{(b-\alpha)^2} + \frac{(b-\alpha)(h+qt_b)-q}{(b-\alpha)^2} + ht_a - ht_b + \frac{qt_a{}^2}{2} - \frac{qt_b{}^2}{2}\right)}{b-\alpha}\right]$$

$$+\frac{1}{T}\left[\frac{W_1\left(-e^{\alpha(f_1-t_a)}(\alpha h + \alpha q t_a + q) + \alpha f_1 q + \alpha h + q\right)}{\alpha^2}\right]$$

$$-\frac{1}{T}\left[\frac{a\left(\frac{e^{(b-\beta)(f_1-t_a)}(p-(b-\beta)(f_1p+h))}{(b-\beta)^2} + \frac{(b-\beta)(h+pt_a)-p}{(b-\beta)^2} + \frac{f1^2p}{2} + f_1h - ht_a - \frac{pt_a{}^2}{2}\right)}{b-\beta}\right]$$

$$+\frac{1}{T}\left[C_p\left(\frac{a\left((b-\beta)\exp\left(b\eta\left(\frac{\log\left(\frac{W_1(\alpha-b)e^{\alpha(f_1-t_a)}}{a}+1\right)}{b-\alpha} + T - t_a\right)\right)\right) + \beta e^{-bf_1} - be^{-bt_a-\beta f_1+\beta t_a}}{b(b-\beta)} + W_1\right)\right.$$

$$+\frac{1}{T}\left[\frac{C_pI_ckM(n+1)\left(\frac{a\left((b-\beta)\exp\left(b\eta\left(\frac{\log\left(\frac{W_1(\alpha-b)e^{\alpha(f_1-t_a)}}{a}+1\right)}{b-\alpha}+T-t_a\right)\right)\right)+\beta e^{-bf_1}-be^{-bt_a-\beta f_1+\beta t_a}}{b(b-\beta)}+W_1\right)}{2n}\right]$$

$$-\frac{1}{T}\left[\frac{\gamma D_2 W_1\left(e^{\alpha(f_1-t_a)} - e^{\alpha(f_1-t_b)}\right)}{\alpha} + \frac{\gamma D_2 W_1\left(e^{\alpha(f_1-t_a)} - 1\right)}{\alpha}\right]$$

$$+\left[\frac{ac\left(\exp\left(b\eta\left(\frac{\log\left(\frac{W_1(\alpha-b)e^{\alpha(f_1-t_a)}}{a}+1\right)}{b-\alpha} + T - t_a\right)\right)\right)}{Tb^2\eta}\right.$$

$$-\frac{ac\left(\exp\left(b\eta\left(\frac{\log\left(\frac{W_1(\alpha-b)e^{\alpha(f_1-t_a)}}{a}+1\right)}{b-\alpha}-t_a+t_b\right)\right)\right)}{Tb^2\eta}\Bigg]$$

$$+\frac{1}{T}\left[\frac{acb\eta(t_b-T)}{b^2\eta}\right]+\frac{1}{T}\left[\frac{aD_1\theta\left(e^{(b-\beta)(f_1-t_a)}+b(t_a-f_1)+\beta f_1-\beta t_a-1\right)}{(b-\beta)^2}\right]$$

3.2 For CASE II: ($t_a \le f_1 \le t_b$)

For various time periods, differential equations for a rented warehouse are:

$$\frac{dI_r(t)}{dt} = -(a - bI_r(t)), \qquad 0 < t < t_a$$

subject to the conditions:

$$I_r(t) = \begin{cases} S - W_1, & at\ t = 0 \\ 0, & at\ t = t_a \end{cases}$$

On solving the above differential Equations:

$$I_r(t) = \tfrac{a}{b}\left[1 - e^{-b(t_a - t)}\right] \qquad 0 < t < t_a$$

Furthermore, the following differential Equations can be used to represent the inventory level $I_o(t)$ in *OW* at any instant "*t*"

$$\frac{dI_o(t)}{dt} = -(a - bI_o(t)), \qquad t_a < t < f_1$$

$$\frac{dI_o(t)}{dt} = -(a - bI_o(t)) - \alpha I_o(t), \qquad f_1 < t < t_b$$

$$\frac{dI_o(t)}{dt} = -\eta\,(a - bI_o(t)) \qquad t_b < t < T$$

subject to the Conditions:

$$I_o(t) = \begin{cases} W_1, & at\ t = t_a \\ 0, & at\ t = t_b \end{cases}$$

On solving the above differential Equations:

$$I_o(t) = \tfrac{a}{b} - \left[\tfrac{a}{b} - W_1\right] e^{b(t - t_a)} \qquad t_a < t < f_1$$

$$I_o(t) = \frac{a}{b-\alpha}\{(1 - e^{-(b-\alpha)(t_b - t)})\}, \qquad f_1 < t < t_b$$

$$I_o(t) = \frac{a}{b} - \left\{\frac{a}{b} + R\right\} e^{-\eta b(T - t)} \qquad t_b < t < T$$

By considering the continuity at $t = t_a$, $t = f_1$ and $t = t_b$, we can write:

$$S = W_1 + \tfrac{a}{b}\left[1 - e^{-bt_a}\right]$$

$$R = \tfrac{a}{b}\left[e^{\eta b(T-t_b)} - 1\right]$$

$$t_b = f_1 - \tfrac{1}{b-\alpha}\left[\ln\left(1 - \tfrac{b-\alpha}{b} + \left(\tfrac{b-\alpha}{b} + \tfrac{(b-\alpha)W_1}{a}\right)e^{b(f_1-t_a)}\right)\right]$$

Here, we discuss how the model's inventory-related costs originated based on the assumptions:

(a) Ordering Cost: A

(b) Purchase Cost: $C_p(S + R)$

$$= C_p\left[\frac{a\left(\exp\left(-b\eta\left(-\frac{\log\left(\frac{e^{-bt_a}\left(a\alpha\left(e^{bt_a}-e^{bf_1}\right)+abe^{bf_1}-bW_1(b-\alpha)e^{bf_1}\right)}{ab}\right)}{b-\alpha}+f_1-T\right)\right)-1\right)}{b}\right]$$

$$+ C_p\left[\frac{a - ae^{-bt_a}}{b} + W_1\right]$$

t

(c) Holding Cost:

$$\int_0^{t_a} h(t)\, I_r(t)dt + \int_{t_a}^{f_1} h(t)I_o(t)\,dt + \int_{f_1}^{t_b} h(t)I_o(t)\,dt$$

$$= \frac{\frac{(bW_1-a)e^{b(f_1-t_a)}(b(f_1q+h)-q)}{b^2} - \frac{(bW_1-a)(b(h+qt_a)-q)}{b^2} + \frac{1}{2}af_1^2q + af_1h - aht_a - \frac{1}{2}aqt_a^2}{b}$$

$$- \frac{a\left(\frac{e^{-bt_a}(p-bh)}{b^2} + \frac{bh+bpt_a-p}{b^2} - \frac{1}{2}t_a(2h+pt_a)\right)}{b}$$

$$- \frac{a\left(\frac{e^{(b-\alpha)(f_1-t_b)}(q-(b-\alpha)(f_1q+h))}{(b-\alpha)^2} + \frac{(b-\alpha)(h+qt_b)-q}{(b-\alpha)^2} + \frac{f_1^2q}{2} + f_1h - ht_b - \frac{qt_b^2}{2}\right)}{b-\alpha}$$

(d) Deterioration Cost:

$$D_2\int_{f_1}^{t_b} \gamma I_o(t)dt = \frac{a\gamma D_2\left(e^{(b-\alpha)(f_1-t_b)} + b(t_b - f_1) + \alpha f_1 - \alpha t_b - 1\right)}{(b-\alpha)^2}$$

(e) Shortage Cost:

$$-c\int_{t_b}^{T} I_o(t)dt = -\frac{c\left((a+bR)e^{b\eta(t_b-T)} + a(b\eta(T-t_b)-1) - bR\right)}{b^2\eta}$$

(f) Capital Cost:

$$I_c\left[\frac{kC_p(S+R)}{n}\frac{M}{n}\left(1+2+3+\dots\dots\dots+n\right)\right]$$

$$= \frac{C_pI_ckM(n+1)\left(\dfrac{a\left(\exp\left(-b\eta\left(-\dfrac{\log\left(\dfrac{e^{-bta}\left(a\alpha\left(e^{bta}-e^{bf_1}\right)+abe^{bf_1}-bW_1(b-\alpha)e^{bf_1}\right)}{ab}\right)}{b-\alpha}+f_1-T\right)\right)-1\right)}{b}+S\right)}{2n}$$

Consequently, the total cyclic cost per unit of time is

$$TC2=\frac{1}{T}\left[\begin{array}{l}\langle Ordering\,Cost\rangle+\langle Purchase\,Cost\rangle+\langle Holding\,Cost\rangle\\ +\langle Deterioration\,Cost\rangle+\langle Shortage\,Cost\rangle+\langle Capital\,Cost\rangle\end{array}\right]$$

$$=\frac{1}{T}\left[A+\frac{C_pI_ckM(n+1)\left(\dfrac{a\left(\exp\left(-b\eta\left(-\dfrac{\log\left(\dfrac{e^{-bta}\left(a\alpha\left(e^{bta}-e^{bf_1}\right)+abe^{bf_1}-bW_1(b-\alpha)e^{bf_1}\right)}{ab}\right)}{b-\alpha}+f_1-T\right)\right)-1\right)}{b}+S\right)}{2n}\right]$$

$$-\frac{1}{T}\left[\frac{a\left(\frac{e^{-bt_a}(p-bh)}{b^2}+\frac{bh+bpt_a-p}{b^2}-\frac{1}{2}t_a(2h+pt_a)\right)}{b}\right]$$

$$+\frac{1}{T}\left[\frac{\frac{(bW_1-a)e^{b(f_1-t_a)}(b(f_1q+h)-q)}{b^2}-\frac{(bW_1-a)(b(h+qt_a)-q)}{b^2}+\frac{1}{2}af_1{}^2q+af_1h-aht_a-\frac{1}{2}aqt_a{}^2}{b}\right]$$

$$-\frac{1}{T}\left[\frac{a\left(\frac{e^{(b-\alpha)(f_1-t_b)}(q-(b-\alpha)(f_1q+h))}{(b-\alpha)^2}+\frac{(b-\alpha)(h+qt_b)-q}{(b-\alpha)^2}+\frac{f_1{}^2q}{2}+f_1h-ht_b-\frac{qt_b{}^2}{2}\right)}{b-\alpha}\right]$$

$$+\frac{1}{T}\left[\frac{a\gamma D_2\left(e^{(b-\alpha)(f_1-t_b)}+b(t_b-f_1)+\alpha f_1-\alpha t_b-1\right)}{(b-\alpha)^2}\right]$$

$$-\frac{1}{T}\left[\frac{ac\exp\left(-b\eta\left(-\frac{\log\left(\frac{e^{-bt_a}\left(a\alpha\left(e^{bt_a}-e^{bf_1}\right)+abe^{bf_1}-bW_1(b-\alpha)e^{bf_1}\right)}{ab}\right)}{b-\alpha}+f_1-T\right)\right)}{b^2\eta}\right]$$

4. Numerical Examples

4.1 For CASE I: When ($0 \le f_1 \le t_a$)

Considering the values as $A = 500$, $S = 115$, $f_1 = 0.1912$, $h = 12$, $q = 12$, $W_1 = 100$, $p = 12$, $b = 0.5$, $t_a = 0.2228$, $t_b = 0.8818$, $\alpha = 0.1$, $C_p = 10$, $\beta = 0.08$, $I_c = 0.25$, $k = 0.4$, $M = 0.25$, $a = 200$, $\theta = 0.08$, $D_1 = 200$, $D_2 = 200$, $T = 1.1046$, $\gamma = 0.06$, $\eta = 0.8$, $c = 100$, $n = 15$.

After calculation the optimum value of $TC1 = 4605.9745$

Table 59.1 Analyzing holding cost for own warehouse with sensitivity

Variation in parameter (%)	TC1	Cost change in (%)	TC2	Cost change in (%)
-20	2563.70	-2.0423	1119.3334	-0.9795
-10	3433.31	-1.1727	1620.5915	-0.4782
10	6171.9858	1.566	3419.6445	1.3208
20	8247.0011	3.641	4892.6669	2.7938

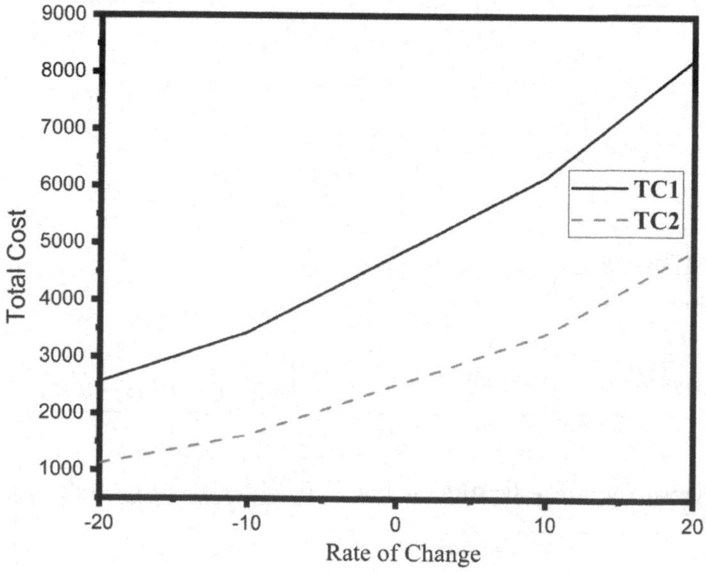

Observations from Table (59.1):

- Table 59.1 shows that the variation in the holding costs in its own warehouse causes the cost per unit of time to change (increase/decrease).
- The graph makes it abundantly obvious that the cost per unit time in these two situations (both) rapidly rises when the holding cost of the own warehouse increases in value.

- However, whether the holding cost rises or falls, *TC1's* value is always higher than *TC2's* value.
- It is absolutely shown that maintaining a longer fixed Self-life is beneficial.

4.2 For CASE II: ($t_a \leq f_1 \leq t_b$)

Considering the values as $A = 500$, $S = 115$, $f_1 = 0.3567$, $h = 12$, $q = 12$, $W_1 = 100$, $p = 12$, $b = 0.5$, $t_a = 0.2228$, $t_b = 0.8818$, $\alpha = 0.1$, $C_p = 10$, $\beta = 0.08$, $I_c = 0.25$, $k = 0.4$, $M = 0.25$, $a = 200$, $\theta = 0.08$, $D_1 = 200$, $D_2 = 200$, $T = 1.1046$, $\gamma = 0.06$, $\eta = 0.8$, $c = 100$, $n = 15$.

After calculation the optimum value of $TC2 = 2098.8231$

Table 59.2 Analyzing holding cost for rented warehouse with sensitivity

Variation in parameter (%)	TC1	Cost change in (%)	TC2	Cost change in (%)
-20	2290.7990	-2.3152	1015.4402	-1.0834
-10	3248.3251	-1.3576	1433.6596	-0.6652
10	5791.0131	1.1850	3017.0274	0.9182
20	7716.9747	3.111	4304.9340	2.2061

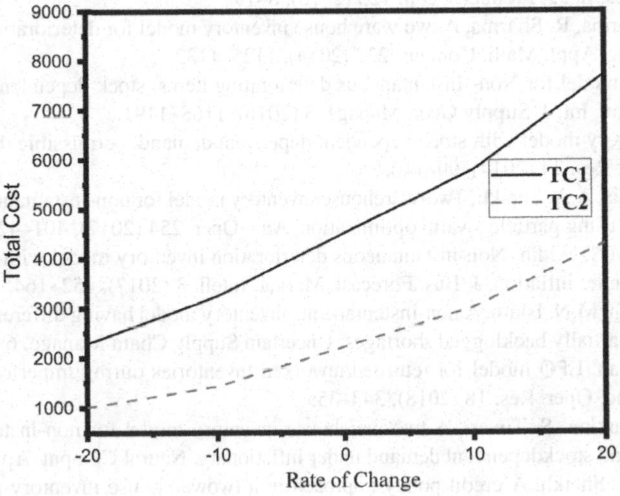

Observations from Table (59.2):

- Table 59.2 shows that the variation in the holding costs in its rented warehouse causes the cost per unit of time to change (increase/decrease).
- The graph makes it abundantly obvious that the cost per unit time in these two situations (both) rapidly rises when the holding cost of the rented warehouse increases in value.
- However, whether the holding cost rises or falls, *TC1's* value is always higher than *TC2's* value.
- It is absolutely shown that maintaining a longer fixed Self-life is beneficial.

5. Conclusion

This study looked at a two-warehouse inventory model of goods that get worse over time with advance payments and some backlog. We also considered that the rate of demand depends on the amount of stock, that one warehouse is owned and the other is rented. The backlog rate changes over time. Based on the assumption, the cost function of this problem is a highly nonlinear constraint optimization problem. Mathematica is used to solve this problem of nonlinear optimization. A numerical example of this model has been simulated with different parameters, and the results have been examined physically.

REFERENCES

1. R. Gupta, P. Vrat, Inventory model for stock dependent consumption rate, Op search, 23 (1986), 19–24.
2. R.C. Baker, T.L. Urban, A deterministic inventory system with an inventory level dependent rate, J. Oper. Res. Soc. 39 (1988), 823–831.
3. B.N. Mandal, S. Phaujdar, An inventory model for deteriorating items and stock dependent consumption rate, J. Oper. Res. Soc. 40 (1989), 483–488.
4. P. Vrat, G. Padmanabhan, An inventory model under inflation for stock-dependent consumption rate items, Eng. Costs Product. Econ. 33 (1990), 71–82.
5. T.K. Datta, A.K. Pal, A note on an inventory model with inventory level dependent demand rate, J. Oper. Res. Soc. 41 (1990), 971–975.
6. G. Padmanabhan, P. Vrat, EOQ models for perishable items under stock dependent selling rate, Eur. J. Oper. Res. 86 (1995), 281–292.
7. Karabi Paul, T.K. Datta, K.S. Chaudhuri, A. K. Pal, Inventory model with twocomponent demand rate and shortages, J. Oper. Res. Soc. 47 (1996), 1029–1036.
8. Y.W. Zhou, S.L. Yang, A two-warehouse inventory model for items with stock-leveldependent demand rate, Int. J. Product. Econ. 95 (2005), 215–228.
9. A. K. Maiti, M. K. Maiti, M. Maiti, Two storage inventory model with random planning horizon, Appl. Math. Comput. 183 (2006), 1084–1097.
10. S.K. Goyal, C.T. Chang, Optimal ordering and transfer policy for an inventory with stock-dependent demand, Eur. J. Oper. Res. 196 (2009), 177–185.
11. H.L. Yang, J.T. Teng, M.S. Chern, An inventory model under inflation for deteriorating items with stock- dependent consumption rate and partial backlogging shortages, Int. J. Product. Econ. 123 (2010), 8–19.
12. A.K. Bhunia, C.K. Jaggi, A. Sharma, R. Sharma, A two ware house inventory model for deteriorating items under permissible delay in payment with partial backlogging, Appl. Math. Comput. 232 (2014), 1125–1137.
13. N. Kumar, S. Kumar, Inventory model for Non- Instantaneous deteriorating items, stock dependent demand, partial backlogging, and inflation over a finite time horizon, Int. J. Supply Oper. Manage. 3 (2016), 1168–1191.
14. M. Pal, A periodic review inventory model with stock dependent dependent demand, permissible delay in payment and price discount on backorders, Yugoslav J. Oper. Res. 24 (2014), 99–110.
15. S. Tiwari, C.K. Jaggi, A.K. Bhunia, A.A. Shaikh, Two-warehouse inventory model for non-instantaneous deteriorating items with stock-dependent demand and inflation using particle swarm optimization, Ann. Oper. 254 (2017), 401–423.
16. A.A. Shaikh, A.H.M. Mashud, M.S. Uddin, Non-instantaneous deterioration inventory model with price and stock dependent demand for fully backlogged shortages under inflation, J. Bus. Forecast. Market. Intell. 3 (2017), 152–164.
17. A. Mashud, M. Khan, M.S. Uddin, M.N. Islam, A non-instantaneous inventory model having different deterioration rates with stock and price dependent demand under partially backlogged shortages, Uncertain Supply Chain Manage. 6 (2018), 49–64.
18. N.H. Shah, D.G. Patel, D.B. Shah, EPQ model for returned/reworked inventories during imperfect production process under price-sensitive stock-dependent demand, Oper. Res. 18 (2018), 343–359.
19. A.A. Shaikh, L.E. C´ardenas-Barr´on, S. Tiwari, A two-warehouse inventory model for non-instantaneous deteriorating items with interval-valued inventory costs and stockdependent demand under inflationary, Neural Comput. Appl. 31 (2019), 1931–1948.
20. G.C. Panda, M.A.A. Khan, A.A. Shaikh, A credit policy approach in a twowarehouse inventory model for deteriorating items with price-and stock-dependent demand under partial backlogging, J. Ind. Eng. 15 (2019), 147–170.
21. L.E. C´ardenas-Barr´on, A.A. Shaikh, S. Tiwari, G. Trevin˜o-Garza, An EOQ inventory model with nonlinear stock dependent holding cost, nonlinear stock dependent demand and trade credit, Computers Ind. Eng. 139 (2020), 105557.
22. Tal Avindav and Teijo Arponen, An EOQ Model for items with a Fixed Shelf-Life and a Declining Demand Rate Based on Time to Expiry Technical note, Asia-Pacific Journal of Operational Research , (2009) 26(6) 759–767
23. K.W. Chuang, C.N. Lin, Replenishment policies for deteriorating items with ramp type demand and a fixed shelf-life under shortage, J. Networks. 10 (2015), 470–476
24. S.I. Ukil, M.E. Islam, M.S. Uddin, A production inventory model of power demand and constant production rate where the products have fixed shelf-life, J. Serv. Sci. Manage. 8 (2015), 874–885.
25. C. Muriana, An EOQ model for perishable products with fixed shelf life under stochastic demand conditions, Eur. J. Oper. Res. 255 (2016), 388–396.
26. Hartley, R.V., Operation Research—A Managerial Emphasis, Good Year Pub.Comp, California pp. 315–317 (1976)
27. K.V.S. Sarma, A deterministic order level inventory model for deteriorating items with two storage facilities, Eur. J. Oper. Res. 29 (1987), 70–73.
28. Adrijit GoswamiK.S.Chaudhuri, Variations of order-level inventory models for deteriorating items,International Journal of Production Economics 27(2), 1992, 111–117.
29. T. P. M. Pakkala, K.K.Achary, A deterministic inventory model for deteriorating items with two warehouses and finite replenishment rate, European Journal of Operational Research 57(1), 1992, 71–76.
30. Yang, H. L. Two-warehouse inventory models for deteriorating items with shortages under inflation. European Journal of Operational Research, 157(2), 2004, 344–356.

31. K. J. Chung, Y.F. Huang, Optimal replenishment policies for EOQ inventory model with limited storage capacity under permissible delay in payments, Opserach, 41 (2004), 16–34.
32. Y. W. Zhou, S.L. Yang, A two-warehouse inventory model for items with stock-level dependent demand rate, Int. J. Product. Econ. 95 (2005), 215–228.
33. C.C. Lee, S.-L. Hsu, A two-warehouse production model for deteriorating inventory items with time- dependent demands, Eur. J. Oper. Res. 194 (2009), 700–710.
34. S. B. Kumar, B. Sarkar, A. Goswami, A two-warehouse inventory model with increasing demand and time varying deterioration, Sci. Iran. 19 (2012), 1969–1977.
35. M. Rastogi, S.R. Singh, P. Kushwah, S. Tayal, Two warehouse inventory policy with price dependent demand and deterioration under partial backlogging, Decision Sci. Lett. 6 (2017), 11–22.

Note: All the figures and tables in this chapter were made by the authors.

Impending Inquisitions in Humanities and Sciences (ICIIHS-2022) – Dr. Mohan Varkolu et al. (eds)

Flame Retardant Efficiency of Oxyanion in the Interlayer of MgAl-LDHs/PP Composite

61

Rajathsing Kalusulingam[1,2,4], Paulmanickam Koilraj[1,2], Shanthana Lakshmi Duraikkannu[1,3], Kannan Srinivasan[1,2,*]

[1]Inorganic Materials and Catalysis Division, CSIR-Central Salt and Marine Chemicals Research Institute, Council of Scientific and Industrial Research (CSIR), Bhavnagar, Gujarat, India

[2]Academy of Scientific and Innovative Research (AcSIR), Ghaziabad, Uttar Pradesh, India

[3]Membrane Science and Separation Technology Division, CSIR-Central Salt and Marine Chemicals Research Institute, Council of Scientific and Industrial Research (CSIR), Bhavnagar, India

[4]Laboratory of Functional Nanomaterials Technology, Institute of Nanotechnologies, Electronics and Electronic Equipment Engineering, Southern Federal University, Taganrog, Russia

ABSTRACT Selective oxyanions intercalated into the interlayer gallery of Mg and Al-containing LDHs; borate, carbonate, molybdate, phosphate, and silicate were prepared through anion exchange method. The oxyanion-LDHs are known for their potential usage in the flame-retardant filler. MgAl-oxyanion-LDHs/PP composite was prepared with polypropylene by micro compounding method, and the influence of the oxyanions on the thermal stability, crystallization behaviors, and flame retardancy were evaluated by TGA, DSC, and limiting oxygen index and cone calorimetry. Incorporation of the oxyanions enhanced the thermal stability, crystallization rate, limiting oxygen index, and flame retardant properties of PP. Limiting oxygen index of MgAl-oxyanion-LDHs/PP increased to 18.9% compared to that of PP (17.8%). Cone calorimetry studies revealed a 23% reduction in total smoke production. Owing to the presence of interlayer oxyanions, the heat release rate, and total heat release reduced by 12 and 10% respectively for the MgAl-oxyanion-LDHs/PP composite.

KEYWORDS layered double hydroxide, oxyanions, polypropylene, polymer composite, crystallization, and flame retardant

GRAPHICAL ABSTRACT

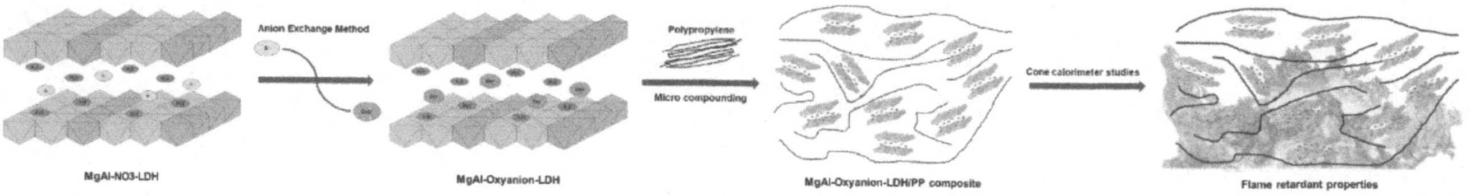

Flame retardant efficiency of oxyanion in the interlayer of MgAl-LDHs/PP composite

*Corresponding author: skannan@csmcri.res.in

DOI: 10.1201/9781003489436-61

1. Introduction

The combination of polymers and inorganic materials basic attracted more attention in recent years in industrial and academic featuring. Such former composites contain highly dispersed inorganic materials on a polymer material that often significant enhancement in the fire resistance. (Jin et al. 2020) The global annual consumption of flame retardants is estimated to nearly \$5.4 billion with a growth rate of 5.4% per annum. One of the viable methods to reduce the flammability of polymer-inorganic composite materials is quite popular from the 20th century. The performance of polymer nanocomposites strongly depends on the nature of the filler, (Gilman et al. 2000; Palacios et al. 2016; Wicklein et al. 2016) whereas size, the degree of dispersion and ratio towards the polymer and in turn influence on the various applications barrier properties, (Choudalakis and Gotsis 2009) flammability, (Porter et al. 2000; Zhang and Horrocks 2003) optical properties, (Beecroft and Ober 1997) rheology, (Saikia et al. 2016) and thermo-mechanical properties. (Wang J et al. 2012).

Polypropylene (PP) is one of the most widely used commodity thermoplastics, and has exceptional physical properties, including high stiffness and tensile strength. It is, however, a combustible substance with little thermal stability. Several inorganic nanofillers, such as magnesium hydroxide, (Wang M et al. 2015) aluminum hydroxide (Feng Y et al. 2017), metal oxyanions materials such as borate, (Shi et al. 2005; Wang Qiang et al. 2013) carbonate, (Gao Yanshan et al. 2013; Gao Y. et al. 2014) phosphate, antimony and layered double hydroxides (Matusinovic and Wilkie 2012; Gao Yanshan et al. 2014; Wang X, Spörer Y, et al. 2015; Yue et al. 2019) have been incorporated to enhance its flame-retardant properties. According to earlier research, researchers are interested in two-dimensional materials as nanofillers for polymers. Earlier days metal hydroxides such as magnesium hydroxide, (Wang M et al. 2015) and aluminum hydroxide, (Manzi-Nshuti et al. 2008; Li et al. 2015) were widely used as a commercial flame retardant. In the last couple of decades layered double hydroxide and their based polymer composites exhibit better flame retardant performance than aluminum and magnesium hydroxides due to their layered nanostructure that plays a key role.(Matusinovic and Wilkie 2012; Wang Q. and O'Hare 2012; Gao Y. et al. 2016) Layered double hydroxides (LDH) an anionic synthetic clay-based layered materials, have a positive hydroxyl main brucite layers separated by interlayer anionic species along with water molecules. It has high charge density and tuneable anion exchange properties in the interlayer space; strong interaction between the main brucite layer because of the unique structural properties and endothermic decomposition in the higher temperature during the time of the release of water and carbon dioxide. Hence these phenomena absorb a huge amount of heat and prevent the oxygen supply to fire makes them more effective for it's as a flame retardant. A recent trend in the development of nanotechnology-based on LDH composite, flame retardancy was evaluated by binary/ternary composition of the LDHs, (Wang X, Spörer Y, et al. 2015) effects of metal ions divalent, (Manzi-Nshuti, Chen, et al. 2009) trivalent, (Manzi-Nshuti, Wang, et al. 2009), anions, (Shi et al. 2005; Manzi-Nshuti, Songtipya, et al. 2009; Wang Qiang et al. 2013; Gao Y. et al. 2014; Hajibeygi et al. 2015; Kalali et al. 2015; Li et al. 2015; Wang X, Kalali EN, et al. 2015; Zhou et al. 2017) organic dye, (Kang et al. 2013) and surfactants, (Araújo et al. 2013; Qiu et al. 2018) into LDHs are reported to be able to suppress the fire, and which means that anion intercalated LDH-based materials are a promising material for flame retardancy.

The flame-retardant research was mainly focused on various functionalization with selective elements of B, C, P, Si, and Mo. These types of functionalized materials have tunable specific retardancy properties such as char promoting agent, smoke suppressant, radical scavenger, protective thermal insulation barrier, and especially resistant to heat, corrosion, and radiation. Based on the above-mentioned properties, it is advisable to use the selective oxy anions borate, carbonate, phosphate, silicate, and molybdate (BO_3, CO_3, PO_4, SiO_4, and MoO_4) as their oxyanion intercalation into layered double hydroxides (Shi et al. 2005; Feng YJ et al. 2006; Wang Qiang et al. 2013; Gao Y. et al. 2014; Xu et al. 2016). Oxyanion flame retardants act simultaneously to form a solid layer on the surface of the substrate to resist the effect of the oxygen and heat by cooling polymer via endothermic decomposition and reduction of the pyrolysis on the solid surface. The flame-retardant efficiency of LDHs for specific oxyanion can be improved by intercalating suitable interlayer oxyanion. MgAl-BO_3-LDH, MgAl-CO_3-LDH, MgAl-MoO_4-LDH, MgAl-PO_4-LDH, and MgAl-SiO_4-LDH.

In the present investigation, selective oxyanion-intercalated LDHs were prepared through anion exchange method. Oxyanion-LDHs were melted with PP to prepare MgAl-oxyanion-LDHs/PP composite. Subsequently, we demonstrated the influence of oxyanions on the thermal stability, crystallization rate, and LOI and flame retardancy of the composite.

2. Experimental Section

2.1 Materials

Magnesium nitrate hexahydrate ($Mg(NO_3)_2.6H_2O$), aluminum nitrate nonahydrate ($Al(NO_3)_3.9H_2O$), sodium hydroxide (NaOH), sodium nitrate ($NaNO_3$), sodium carbonate ($NaCO_3$), sodium molybdate (Na_2MoO_4), potassium hypophosphate (K_2HPO_4), sodium metasilicate ($Na_2SiO_3.9H_2O$) and disodium tetraborate ($Na_2B_4O_7 \cdot 10H_2O$), were purchased from S.D. fine Chem Ltd. Poly Propylene (PP) was purchased from Otto chemicals Ltd and used without any further purification.

Synthesis of MgAl-oxyanion-LDHs

The synthesis of MgAl-LDH having (M^{2+}/M^{3+}) atomic ratio 1:2 with nitrate as an interlayer anion was prepared by the co-precipitation method. Typically for the preparation of MgAl-LDH by taking the mixture of 1 M of $Mg(NO_3)_2.6H_2O$ and 1 M of $Al(NO_3)_3.9H_2O$ solution and the mixture of 2 M NaOH and 0.2 M $NaNO_3$ solution separately. Both solutions were added simultaneously to approximately 80 ml deionized water under vigorously stirring. During the synthesis, the pH of the medium was maintained constant at 9.5 ± 0.2 for MgAl-NO_3-LDH respectively. To avoid the interference of atmospheric carbon dioxide, the synthesis was carried out under a nitrogen environment, de-carbonate water was used for the complete synthesis to minimize contamination. After the complete addition of the metal solution under the N_2 atmosphere, the resulting slurry was aged at 65°C for 20 h in an oil bath. Subsequently, the slurry was filtered and washed repeatedly with decarbonated water until the pH of the filtrate became neutral. A part of the wet cake was dried under vacuum and referred to as MgAl-NO_3-LDH.

Oxyanions were intercalated on MgAl-NO_3-LDH through anion exchange method. In general, 5.0 g of MgAl-NO_3-LDH was dispersed in separately into 250 ml of each 0.25 M aqueous oxyanions solution (pH was adjusted to > 9.5), After that N_2 purged into the solution for creating an inert atmosphere, the resulting mixture was aged at 60 °C, 250 rpm for 6 h in an oil bath. Then the oxyanion exchange LDH was obtained by filtration and washing with decarbonated water until the pH of the filtrate became neutral and the powder was dried under vacuum. The oxyanion exchanged samples here are referred to as MgAl-BO_3-LDH, MgAl-CO_3-LDH, MgAl-MoO_4-LDH, MgAl-PO_4-LDH, and MgAl-SiO_4-LDH.

Preparation of the MgAl-oxyanion-LDHs/PP composite

Polymer composite was prepared by micro compounding method, 200 g of polypropylene (PP), and 10 g of LDHs were homogeneously mixed at 190°C, 150 rpm for 2 min in micro compounder. The homogeneous mixture obtained after the mixing was removed from micro compounder and compressing molded at the same condition.

2.2 Characterization Techniques

PXRD was carried out in a Rigaku - Miniflex II system using Cu Kα radiation (λ = 1.5406 Å) with a step size of 0.04° and a step time of 2 seconds. FT-IR absorption spectra of the samples were recorded on a Perkin-Elmer FT-IR spectrometer FT-1730 by mixing with KBr and spectra were recorded with a nominal resolution of 4 cm^{-1}. Thermogravimetric analysis (TGA) was performed in Mettler TGA/SDTA 851e and the crystallization and melting behaviors of neat PP and nanocomposite were analyzed using differential scanning calorimeter (DSC) were carried out in Mettler Toledo DSC 822 The melt crystallization temperature (T_{mc}) was measured. TGA and DSC thermal experiments were carried out at a nitrogen flow rate of 50 ml min^{-1} at a heating rate of 10°C/min. and the data were processed using Stare software. Morphology of the material was characterized by Field Emission-Scanning Electron Microscope (FESEM - JEOL JSM 7100M). Limiting oxygen index (LOI) measurements were conducted on an oxygen index model instrument (IMDEA Materials Institute, Spain) with a dimension of specimen: 100 mm × 6.5 mm × 3.2 mm according to ASTM D2863-97. Cone calorimeter tests were carried out on an FTT model (IMDEA Materials Institute, Spain) cone calorimeter, according to the procedures in ISO 5660-1. Each specimen with dimensions of 100 mm × 100 mm × 4 mm was irradiated horizontally at a heat flux of 60 kW m^{-2}. The samples were mounted in aluminum foil and placed on a holder with the thermocouple under the sample. An infrared thermometer (Optris CT laser MT-CF3) was used to monitor the surface temperature of the samples, and the system accuracy was ± 1%. All measurements were repeated at least three times, and the results were averaged. From the Cone calorimeter tests, time to ignition (TTI), heat release rate (HRR), total heat release (THR), total smoke production (TSP), and char residues were measured.

3. Results and Discussion

3.1 Characterization of as-synthesized and Different Anion-exchanged LDHs

The synthesized MgAl-oxyanion-LDHs were initially subjected to characterization using powder X-ray diffraction (PXRD) analysis. The resulting diffraction patterns are depicted in Fig. 61.1. The present study involved the observation of powerful

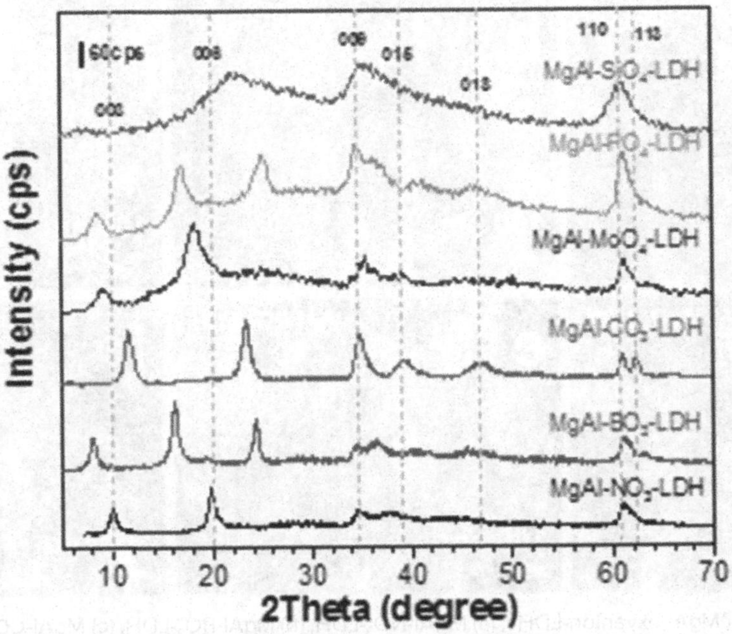

Fig. 61.1 (a) PXRD pattern of MgAl-oxyanion-LDHs (MgAl-NO_3-LDH, MgAl-BO_3-LDH, MgAl-CO_3-LDH, MgAl-MoO_4-LDH, MgAl-PO_4-LDH, and MgAl-SiO_4-LDH)

distinctive reflections (003, 006, 009, 015, 018, 110, and 113) for the corresponding layered double hydroxides (LDHs). The interlayer distance was determined, and it was shown that the spicing value of the 003 plane of LDHs strongly suggests that intercalated LDHs possess a high level of crystallinity and have a layered structure. A variation in the interlayer spacing was detected among different anions, attributable to disparities in their dimensions and charge carriers. Upon the introduction of the nitrate anion (NO_3-) as a guest, the interlayer distance was measured to be approximately 8.94 Å. In a similar manner, the order reflections also undergo a shift towards lower angles, which suggests that anions have been successfully intercalated into the interlayer galleries of MgAl-LDH, resulting in an observed increase in the interlayer spacing. The basal reflection of oxyanion-layered double hydroxides (LDHs) exhibits a crystalline structure, with the exception of MgAl-SiO_4-LDH, which displays a comparatively lower degree of crystallinity.

The morphologies of MgAl-oxyanion-LDHs were subsequently determined through investigation utilising field emission scanning electron microscopy (FE-SEM), as shown in Fig. 61.2. The MgAl-NO_3-LDH exhibited platelet morphologies with uniform shapes, a characteristic often found in inorganic anionic intercalated LDH structures. The nanoparticles in the layer possess a diameter in the range of a few hundred nanometers and a thickness spanning several tens of nanometers. Nevertheless, it is worth noting that regular nanoplates are not seen in MgAl-NO_3-LDH, MgAl-BO_3-LDH, MgAl-MoO_4-LDH, MgAl-PO_4-LDH, and MgAl-SiO_4-LDH. This observation suggests that the crystalline structure of these LDHs is comparatively inferior to that of MgAl-CO_3-LDH. The obtained result demonstrates a strong correlation with the analysis conducted using Powder X-ray Diffraction (PXRD).

3.2 Characterization of MgAl-oxyanion-LDHs/PP Composite

PXRD is commonly employed for the identification of layered structures, namely intercalated or exfoliated composites of MgAl-oxyanion-LDHs/PP. The diffraction peak associated with the d_{003} spacing was not observed in the polypropylene composite with a 5.0 wt% loading of oxyanion-layered double hydroxides (LDHs). There are two primary assertions that can be proposed to explain the lack of a diffraction peak in PXRD analysis. Firstly, it is possible that the absence of a diffraction peak is due to the full separation of the layers in the MgAl-oxyanion-LDHs/PP composite, resulting in delamination. Secondly, it is also plausible that the LDH layers within the PP matrix are disordered, while maintaining the same d-spacing. The formation of an intercalated polypropylene (PP) composite with a 5.0 wt% loading of oxyanion-layered double hydroxides (LDHs), synthesised by micro compounding composite, is evident from the X-ray diffraction patterns. This observation confirms the successful incorporation of oxyanion-LDHs into the PP matrix, as depicted in Fig. 61.3.

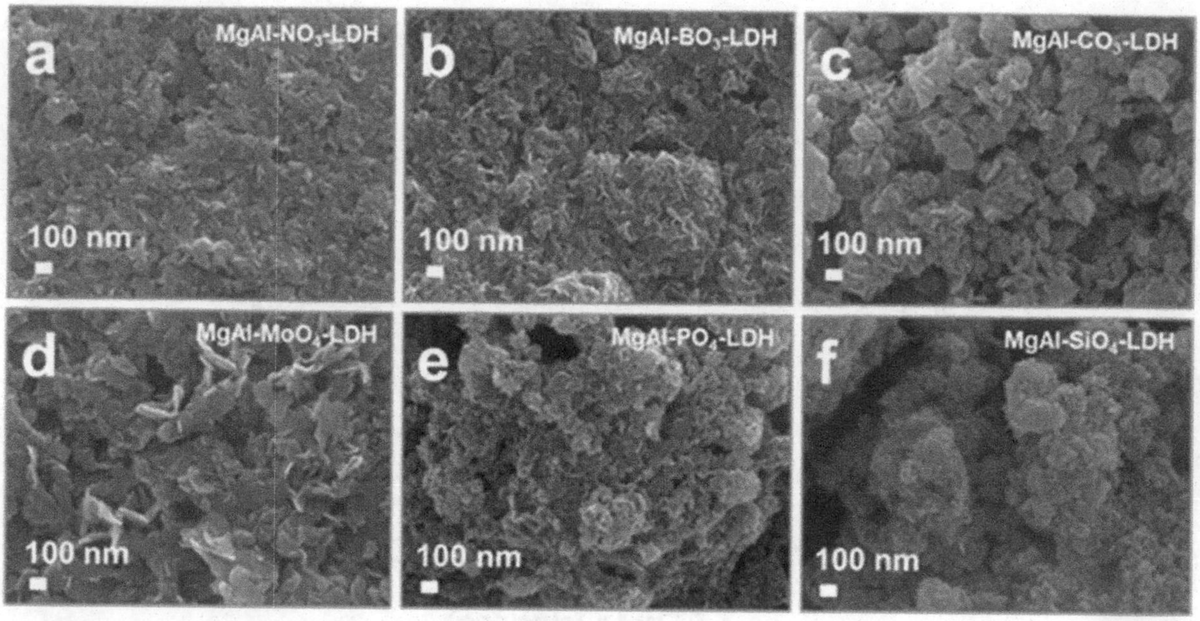

Fig. 61.2 FE-SEM images of MgAl-oxyanion-LDHs, (a) MgAl-NO_3-LDH, (b) MgAl-BO_3-LDH, (c) MgAl-CO_3-LDH, (d) MgAl-MoO_4-LDH, (e) MgAl-PO_4-LDH, and (f) MgAl-SiO_4-LDH. Horizontal bars indicate 100 nm

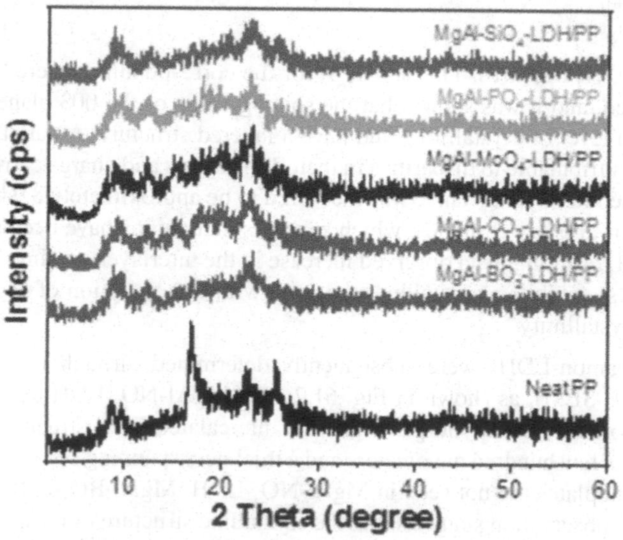

Fig. 61.3 (a) PXRD pattern of Neat PP and MgAl-oxyanion-LDHs/PP composites.

The potential impact of LDH fillers on the thermal stability of polypropylene (PP) has been investigated. Specifically, the influence of oxyanion LDHs on the thermal stability of LDH/PP composites was examined using thermogravimetric analysis (TGA). The TGA data, presented in Table 61.1 and visually represented in Fig. 61.4, provide insights into the observed effects. The onset of weight loss in all composites is reported to occur at approximately 350°C. The highest level of weight loss, referred to as the maximum weight loss (T_{max}), is found at a temperature of 432 °C for polypropylene (PP). The thermal stability of the MgAl-oxyanion-LDHs/PP composite is significantly influenced by the presence of oxyanions. This is evident from the observation that the composite exhibited the highest level of breakdown at temperatures ranging from 447 to 452°C. The residual mass of polypropylene (PP) after exposure to a temperature of 800 °C was found to be 2.9%. However, when oxyanions were intercalated into layered double hydroxides (LDHs) and combined with PP to form composites, the residual mass after 800°C increased. For instance, the MgAl-BO_3-LDH/PP composite exhibited a residual mass of 6.1%, the MgAl-CO_3-LDH/

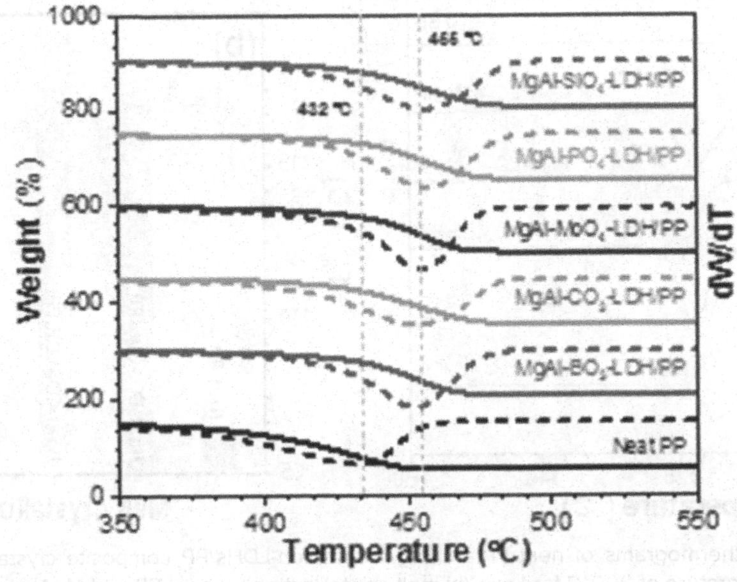

Fig. 61.4 (a) TG curves of Neat PP and MgAl-oxyanion-LDHs/PP composite.

Table 61.1 TG results of Neat PP and MgAl-oxyanion-LDHs/PP composite

Sample	$T_{0.1\%}$ (°C)	$T_{0.5\%}$ (°C)	T_{max} (°C)	Residue mass (wt%)
Neat PP	374	419	432	2.9
MgAl-BO_3-LDH/PP	411	447	452	6.1
MgAl-CO_3-LDH/PP	413	448	452	7.6
MgAl-MoO_4-LDH/PP	410	449	455	4.2
MgAl-PO_4-LDH/PP	419	452	456	5.7
MgAl-SiO_4-LDH/PP	419	452	456	7.6

PP composite had a residual mass of 7.6%, the MgAl-MoO_4-LDH/PP composite had a residual mass of 4.2%, the MgAl-PO_4-LDH/PP composite had a residual mass of 5.7%, and the MgAl-SiO_4-LDH/PP composite had a residual mass of 7.6%. The char residue percentages of the composites including MgAl-oxyanion-LDH/PP were found to be greater compared to those of pure PP. This observation suggests that the inclusion of oxyanions in the composites leads to a considerable enhancement in their thermal-oxidative resistance. In terms of enhancing thermal stability, the efficiency of oxyanions-LDHs can be ranked as follows: MgAl-PO_4-LDH and MgAl-SiO_4-LDH exhibit higher efficiency compared to MgAl-MoO_4-LDH, MgAl-CO_3-LDH, and MgAl-BO_3-LDH.

In order to investigate the impact of oxyanion LDH on the rate of crystallisation of PP, the temperature at which melt crystallisation occurs (referred to as T_{mc}) was calculated by the use of DSC. Fig. 61.5a displays the cooling thermograms obtained from DSC analysis of two samples: the pure PP and a composite material consisting of 5.0 wt% MgAl-oxyanion-LDHs loaded into the PP matrix. The LDHs used in the composite include borate (BO_3), carbonate (CO_3), molybdate (MoO_4), phosphate (PO_4), and silicate (SiO_4). The T_{mc} values have frequently been employed as a means of quantifying the rate of crystallisation in polymers. An increase in the T_{mc} value corresponds to an increase in the crystallisation rate of the polymer. Fig. 61.5a displays the cooling curves obtained from DSC measurements conducted on both pure PP and its corresponding nanocomposites, with a cooling rate of 10 °C per minute. The neat PP exhibits a melting temperature (T_{mc}) of around 106.4°C. However, the T_{mc} values for the MgAl-oxyanion-LDHs/PP composite, regardless of the specific oxyanion LDH used, shift to 117 ± 2 °C, as shown in Fig. 61.5b. The increased T_{mc} observed in the MgAl-oxyanion-LDHs/PP composite, in comparison to plain PP, suggests the occurrence of crystallisation of MgAl-oxyanion-LDHs/PP.

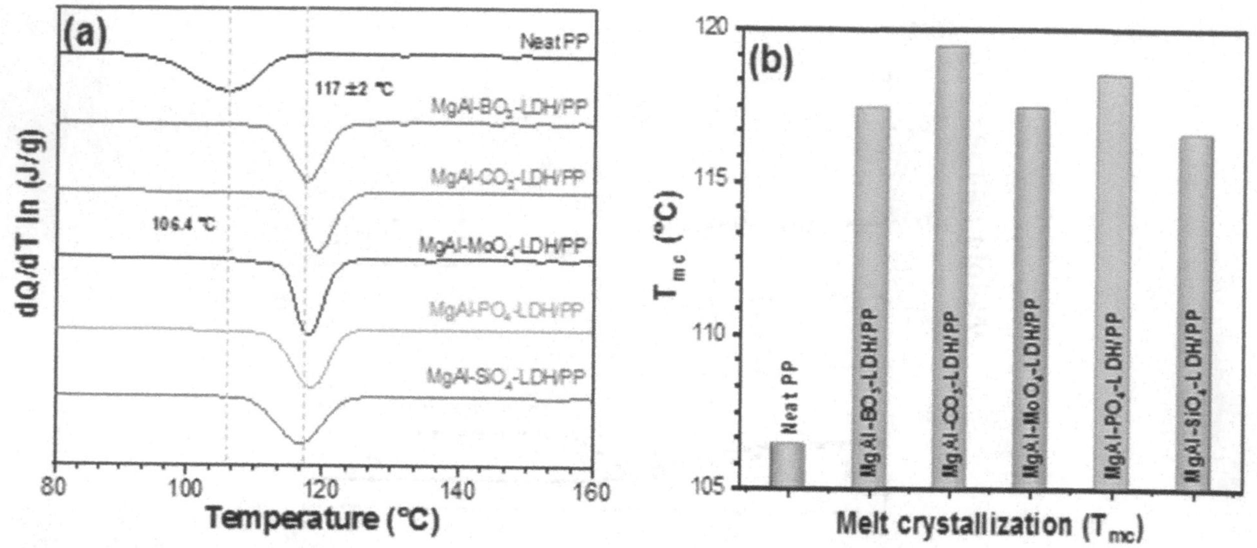

Fig. 61.5 (a) DSC cooling thermograms of neat PP and MgAl-oxyanion-LDHs/PP composite crystallized non-isothermal after melting at a temperature of 200°C for 1 min. (b) melt crystallization for neat PP and MgAl-oxyanion-LDHs/PP composite

3.3 Cone Calorimeter - Combustion Behavior of the Composite Materials

The cone calorimeter (CC) is a comparative method employed to assess the combustion characteristics of composite materials. This technology operates on the fundamental principle of oxygen combustion. The aforementioned methodology was effectively employed as a first assessment to evaluate the flame-retardant characteristics of the composite material consisting of polypropylene and magnesium aluminium oxyanion layered double hydroxides.

The rate of heat release is key-value due to express the fire intensity. The PHRR profiles of the different LDH/PP composites are shown in Fig. 61.6a, and their results are summarized in Table 61.2. There is one peak in the HRR curve of the neat PP and the composites associated with LDH degradation as shown in Fig. 61.6a. The heat release rate profile of neat PP showed a maximum PHRR 1069 kW m^{-2}, which is due to the rapid burning of PP after ignition. In contrast, the oxyanion-LDH/PP composite burned a relatively lower rate and showed PHRR values of 954 kW m^{-2} (11%), 905 kW m^{-2} (15%), 927 kW m^{-2} (13%), 982 kW m^{-2} (8%), and 946 kW m^{-2} (12%) for MgAl-BO_3-LDH/PP, MgAl-CO_3-LDH/PP, MgAl-MoO_4-LDH/PP, MgAl-PO_4-LDH/PP, and MgAl-SiO_4-LDH/PP composites respectively, which indicates a good deceleration effect of the related oxyanions composites. This is particularly important because of the physical barrier effect of composite and LDH during combustion. The lower PHHR is due to the isolation of oxygen from the evolved combustible gases during combustion. It should be noted that the PHRR of the all the composites is lower than PP, which is advantageous for flame retardants.

The gradient of the THR curve may be regarded as a metric for assessing the rate of flame propagation. Fig. 61.6b illustrates the thermal degradation behaviour of pure PP and composites of MgAl-oxyanion-LDHs with PP. The corresponding data are shown in Table 61.2. The intercalation of oxyanion in the LDHs resulted in a reduction of the THR values of all the composites. In comparison to the PP material with an energy value of 117 MJ m^{-2}, the THR (thermal heat release) values of MgAl-BO_3-LDH/PP, MgAl-CO_3-LDH/PP, MgAl-MoO_4-LDH/PP, MgAl-PO_4-LDH/PP, and MgAl-SiO_4-LDH/PP exhibited reductions of 28%, 26%, 34%, 23%, and 21% correspondingly. The cone colorimetric test assessed several composites, and it was shown that the silicate-LDH/PP composite exhibited a significantly lower THR. This outcome can perhaps be attributed to the interaction between the silicate LDH and PP components. Additionally, the presence of a condensed phase can facilitate the development of a char layer that exhibits condensed phase behaviour, resulting in the release of oxygen and inhibiting the combustion of the mixture.

The flame-retardant capabilities of PP/LDH composite are demonstrated in Fig. 61.6c and Table 61.2, where the TSP is recognized as a significant parameter. The TSP values of the composites consisting of MgAl-oxyanion-LDHs and PP were much lower compared to the TSP values of pure PP. The well-organized particulate matter had the greatest TSP value of 16 m^2 kg^{-1}. Following the incorporation of composite materials, a reduction in smoke emission was seen regardless of the specific oxyanion. The TSP values were measured as 13, 12, 12, 12, and 13 m^2 kg^{-1} for the composites MgAl-BO_3-LDH/

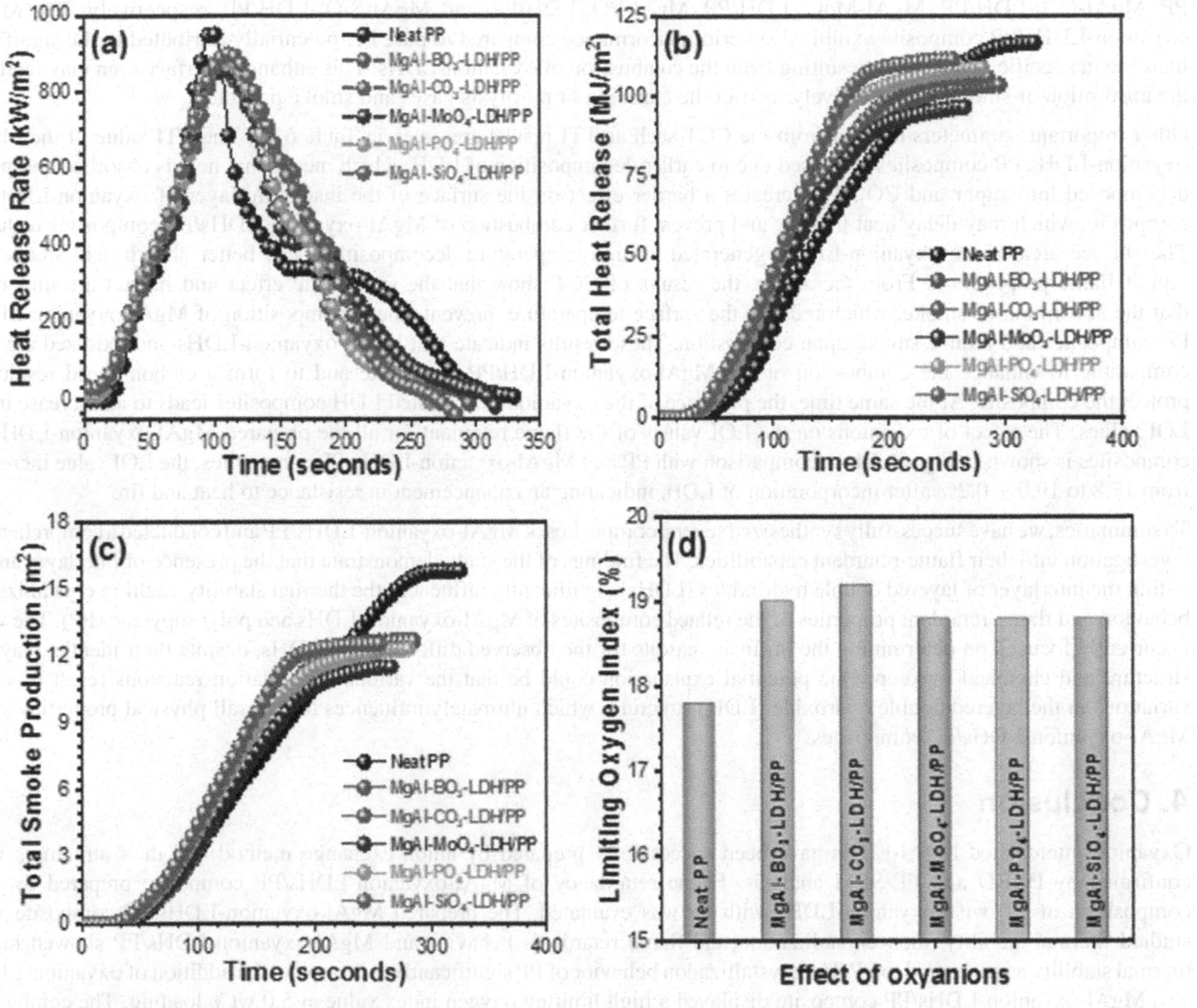

Fig. 61.6 (a) Heat release rate, (b) total heat release, and (c) total smoke production rate versus time curves of PP and MgAl-oxyanion-LDHs/PP composite from cone calorimeter tests and (d) Effect of oxyanion on limiting oxygen index of MgAl-oxyanion-LDHs /PP composite

Table 61.2 Cone calorimeter data of MgAl-oxyanion-LDHs/PP composite under the heat flux

Code	PHRR (kW m^{-2})	THR (MJ m^{-2})	TSP (m^2 kg^{-1})	TTI (s)	Limiting Oxygen Index (LOI)
Neat PP	1070	117	16	27	17.8
MgAl-BO_3-LDH/PP	954	103	13	17	19.0
MgAl-CO_3-LDH/PP	905	106	12	19	19.2
MgAl-MoO_4-LDH/PP	927	97	12	19	18.8
MgAl-PO_4-LDH/PP	982	109	12	20	18.8
MgAl-SiO_4-LDH/PP	946	112	13	20	18.8

Summary of the cone calorimetric (at 60 kW m^{-2}), LOI results for MgAl-oxyanion-LDHs/PP composite.

PP, MgAl-CO_3-LDH/PP, MgAl-MoO_3-LDH/PP, MgAl-PO_4-LDH/PP, and MgAl-SiO_4-LDH/PP, respectively. The MgAl-oxyanion-LDHs/PP composite exhibited superior performance compared to pure PP, potentially attributed to the significant increase in specific surface area resulting from the combustion of oxyanion-LDHs. This enhanced surface area may facilitate the absorption of smoke and effectively restrict the emission of pyrolysis gases and smoke particles.

Other important parameters derived from the CCT such as TTI results are given in Table 61.2. The TTI value of the MgAl-oxyanion-LDHs/PP composites increased due to earlier decomposition of LDH, which means that heat is absorbed and partly decomposed into vapor and CO_2. This creates a barrier effect on the surface of the insulating layer of oxyanion-LDHs/PP composite, which may delay heat transfer and prevent further combustion of MgAl-oxyanion-LDHs/PP composites material. The surface area of the oxyanion-LDHs, generated in high-temperature decomposition, can better absorb acid smoke and inhibit flame propagation. From the above, the results of CCT show that the synergistic effect and impact are important, that the absorption of smoke, which reduces the surface temperature, prevents the decomposition of MgAl-oxyanion-LDHs/PP composites and reduce smoke upon combustion. These results indicate that MgAl-oxyanion-LDHs and oxidized are well compatible to enhance the combustion of the MgAl-oxyanion-LDH/PP composite and to form a carbon-based residue to protect the composite. At the same time, the presence of the oxyanion intercalated LDH composites leads to an increase in the LOI values. The effect of oxyanions on the LOI values of the flame retardant for all the prepared MgAl-oxyanion-LDHs/PP composites is shown in Fig. 61.6d. In comparison with PP and MgAl-oxyanion-LDHs/PP composites, the LOI value increased from 17.8 to 19.0 ± 0.2% after incorporation of LDH, indicating an enhancement in resistance to heat and fire.

To summaries, we have successfully synthesized several composites of MgAl-oxyanion-LDHs/PP and conducted a comprehensive investigation into their flame-retardant capabilities. The findings of the study demonstrate that the presence of interlayer anions within the interlayer of layered double hydroxides (LDHs) significantly influences the thermal stability, melting crystallization behavior, and flame-retardant properties of the related composites of MgAl-oxyanion-LDHs and polypropylene (PP). The work is currently focused on determining the intrinsic reasons for the observed differences in LDHs, despite their identical layered structure and chemical makeup. One potential explanation could be that the various intercalation reactions result in subtle variations in the layered double hydroxide (LDH) structure, which ultimately influences the overall physical properties of the MgAl-oxyanion-LDHs/PP composites.

4. Conclusion

Oxyanion intercalated MgAl-LDHs have been successfully prepared by anion exchange method and their structures were confirmed by PXRD and FE-SEM analysis. Flame retardancy of MgAl-oxyanion-LDHs/PP composite prepared by melt composition of 5.0 wt% oxyanion-LDHs with PP was evaluated. The prepared MgAl-oxyanion-LDHs/PP composite were studied thermal stability, melt crystallization and flame retardants behavior and MgAl-oxyanion-LDHs/PP showed higher thermal stability as compared to PP. The crystallization behavior of PP significantly increased in the addition of oxyanion-LDHs. The MgAl-oxyanion-LDHs/PP composite displayed a high limiting oxygen index value at 5.0 wt% loading. The combustion behavior was investigated with cone calorimeter; the results revealed that intercalated oxyanion-LDHs are very efficient in the reduction of the flammability of PP. Thermal gravimetric analysis and differential scanning calorimeter demonstrated that the intercalation of various anions into LDHs can considerably enhance the thermal stability of PP. The addition of MgAl-oxyanion-LDHs accomplished the highest temperature enhances in thermal decomposition (24°C) and glass transition temperature (12°C). According to the cone calorimeter investigation, the effectiveness of the addition of MgAl-oxyanion-LDHs to PP can enhance its flame-retardant performance and the type of intercalate anions had a significant impact on this efficiency. PHRR reduction for the MgAl-BO_3-LDH/PP polymer nanocomposites, MgAl-CO_3-LDH/PP, MgAl-MoO_4-LDH/PP, MgAl-PO_4-LDH/PP, and MgAl-SiO_4-LDH/PP polymer nanocomposites were 11, 15, 13, 8 and 12%, respectively. Similarly, THR, and TSP were also significantly condensed after addition of MgAl-oxyanion-LDHs. These results indicate that the insertion of oxyanion into LDHs enhances the retardancy of PP. In summary, herein we proved that MgAl-oxyanion-LDHs/PP composite had reliably high thermal stability, better crystallization behavior, and improved flame retardancy performance; these properties highly depend on the type of interlayer anion. Further, we can tune the flame-retardant properties of MgAl-oxyanion-LDHs/PP composite by changing interlayer oxyanion and LDH compositions. These composites can be used as fireproof flooring material wire cables etc.

5. Acknowledgment

CSIR-CSMCRI Communication No. 77/2020. Authors are thankful to CSIR, New Delhi, India, for funding through Network Project on "CSC-0135". Rajathsing Kalusulingam thanks to AcSIR for enrollment in Ph.D. We are especially thankful for Prof. Dr. De-Yi Wang, High-Performance Polymers Nanocomposite and Fire Retardants Lab, IMDEA Materials Institute, Spain, and also thankful for Analytical Science Discipline and Centralized Instrumentation Facility of the institute and the kind support.

REFERENCES

1. Araújo RS, Grafova I, Marques MdFV, Kemell M, Leskelä M, Grafov A. 2013. Properties and nanoscale structure of polypropylene-layered double hydroxide composites prepared by compatibilizer-free way. Journal of Applied Polymer Science. 130(4):2429–2438.
2. Beecroft LL, Ober CK. 1997. Nanocomposite Materials for Optical Applications. Chemistry of Materials. 9(6):1302–1317.
3. Choudalakis G, Gotsis AD. 2009. Permeability of polymer/clay nanocomposites: A review. European Polymer Journal. 45(4):967–984.
4. Feng Y, Hu J, Xue Y, He C, Zhou X, Xie X, Ye Y, Mai Y-W. 2017. Simultaneous improvement in the flame resistance and thermal conductivity of epoxy/Al2O3 composites by incorporating polymeric flame retardant-functionalized graphene. Journal of Materials Chemistry A. 5(26):13544–13556.
5. Feng YJ, Williams GR, Leroux F, Taviot-Gueho C, O'Hare D. 2006. Selective Anion-Exchange Properties of Second-Stage Layered Double Hydroxide Heterostructures. Chemistry of Materials. 18(18):4312–4318.
6. Gao Y, Wang Q, Wang J, Huang L, Yan X, Zhang X, He Q, Xing Z, Guo Z. 2014. Synthesis of highly efficient flame retardant high-density polyethylene nanocomposites with inorgano-layered double hydroxides as nanofiller using solvent mixing method. ACS applied materials & interfaces. 6(7):5094–5104.
7. Gao Y, Wu J, Wang Q, Wilkie CA, O'Hare D. 2014. Flame retardant polymer/layered double hydroxide nanocomposites [10.1039/C4TA01030B]. Journal of Materials Chemistry A. 2(29):10996.
8. Gao Y, Wu J, Zhang Z, Jin R, Zhang X, Yan X, Umar A, Guo Z, Wang Q. 2013. Synthesis of polypropylene/Mg3Al–X (X = CO32–, NO3–, Cl–, SO42–) LDH nanocomposites using a solvent mixing method: thermal and melt rheological properties. Journal of Materials Chemistry A. 1(34):9928.
9. Gao Y, Zhang Y, Williams GR, O'Hare D, Wang Q. 2016. Layered double hydroxide-oxidized carbon nanotube hybrids as highly efficient flame retardant nanofillers for polypropylene. Sci Rep. 6:35502.
10. Gilman JW, Jackson CL, Morgan AB, Harris R, Manias E, Giannelis EP, Wuthenow M, Hilton D, Phillips SH. 2000. Flammability Properties of Polymer–Layered-Silicate Nanocomposites. Polypropylene and Polystyrene Nanocomposites. Chemistry of Materials. 12(7):1866–1873.
11. Hajibeygi M, Shabanian M, Moghanian H, Khonakdar HA, Häußler L. 2015. Development of one-step synthesized LDH reinforced multifunctional poly(amide–imide) matrix containing xanthene rings: study on thermal stability and flame retardancy [10.1039/C5RA05565B]. RSC Advances. 5(66):53726–53735.
12. Jin L, Zeng H-Y, Du J-Z, Xu S. 2020. Intercalation of organic and inorganic anions into layered double hydroxides for polymer flame retardancy. Applied Clay Science. 187:105481.
13. Kalali EN, Wang X, Wang D-Y. 2015. Functionalized layered double hydroxide-based epoxy nanocomposites with improved flame retardancy and mechanical properties [10.1039/C5TA00010F]. Journal of Materials Chemistry A. 3(13):6819–6826.
14. Kang NJ, Wang DY, Kutlu B, Zhao PC, Leuteritz A, Wagenknecht U, Heinrich G. 2013. A new approach to reducing the flammability of layered double hydroxide (LDH)-based polymer composites: preparation and characterization of dye structure-intercalated LDH and its effect on the flammability of polypropylene-grafted maleic anhydride/d-LDH composites. ACS applied materials & interfaces. 5(18):8991–8997.
15. Li C, Wan J, Kalali EN, Fan H, Wang D-Y. 2015. Synthesis and characterization of functional eugenol derivative based layered double hydroxide and its use as a nanoflame-retardant in epoxy resin [10.1039/C4TA05740F]. Journal of Materials Chemistry A. 3(7):3471–3479.
16. Manzi-Nshuti C, Chen D, Su S, Wilkie CA. 2009. The effects of intralayer metal composition of layered double hydroxides on glass transition, dispersion, thermal and fire properties of their PMMA nanocomposites. Thermochimica Acta. 495(1-2):63–71.
17. Manzi-Nshuti C, Songtipya P, Manias E, Jimenez-Gasco MM, Hossenlopp JM, Wilkie CA. 2009. Polymer nanocomposites using zinc aluminum and magnesium aluminum oleate layered double hydroxides: Effects of LDH divalent metals on dispersion, thermal, mechanical and fire performance in various polymers. Polymer. 50(15):3564–3574.
18. Manzi-Nshuti C, Wang D, Hossenlopp JM, Wilkie CA. 2008. Aluminum-containing layered double hydroxides: the thermal, mechanical, and fire properties of (nano)composites of poly(methyl methacrylate) [10.1039/B802553C]. Journal of Materials Chemistry. 18(26):3091.
19. Manzi-Nshuti C, Wang D, Hossenlopp JM, Wilkie CA. 2009. The role of the trivalent metal in an LDH: Synthesis, characterization and fire properties of thermally stable PMMA/LDH systems. Polymer Degradation and Stability. 94(4):705–711.

20. Matusinovic Z, Wilkie CA. 2012. Fire retardancy and morphology of layered double hydroxide nanocomposites: a review [10.1039/C2JM33179A]. Journal of Materials Chemistry. 22(36):18701.
21. Palacios E, Leret P, De La Mata MJ, Fernandez JF, De Aza AH, Rodriguez MA, Rubio-Marcos F. 2016. Self-Forming 3D Core-Shell Ceramic Nanostructures for Halogen-Free Flame Retardant Materials. ACS applied materials & interfaces. 8(14):9462–9471.
22. Porter D, Metcalfe E, Thomas MJK. 2000. Nanocomposite fire retardants: a review. Fire and Materials. 24(1):45–52.
. Qiu L, Gao Y, Zhang C, Yan Q, O'Hare D, Wang Q. 2018. Synthesis of highly efficient flame retardant polypropylene nanocomposites with surfactant intercalated layered double hydroxides [10.1039/C7DT03477F]. Dalton Trans. 47(9):2965–2975.
24. Saikia P, Gautam A, Goswamee RL. 2016. Synthesis of nanohybrid alcogels of SiO2 and Ni–Cr/Mg–Cr–LDH: study of their rheological and dip coating properties [10.1039/C6RA23475E]. RSC Advances. 6(113):112092–112102.
25. Shi L, Li D, Wang J, Li S, Evans DG, Duan X. 2005. Synthesis, flame-retardant and smoke-suppressant properties of a borate-intercalated layered double hydroxide. Clay Clay Miner. 53(3):294–300.
26. Wang J, Cheng Q, Tang Z. 2012. Layered nanocomposites inspired by the structure and mechanical properties of nacre [10.1039/C1CS15106A]. Chemical Society reviews. 41(3):1111–1129.
27. Wang M, Han X-W, Liu L, Zeng X-F, Zou H-K, Wang J-X, Chen J-F. 2015. Transparent Aqueous Mg(OH)2 Nanodispersion for Transparent and Flexible Polymer Film with Enhanced Flame-Retardant Property. Industrial & Engineering Chemistry Research. 54(51):12805–12812.
28. Wang Q, O'Hare D. 2012. Recent advances in the synthesis and application of layered double hydroxide (LDH) nanosheets. Chemical reviews. 112(7):4124–4155.
29. Wang Q, Undrell JP, Gao Y, Cai G, Buffet J-C, Wilkie CA, O'Hare D. 2013. Synthesis of Flame-Retardant Polypropylene/LDH-Borate Nanocomposites. Macromolecules. 46(15):6145–6150.
30. Wang X, Kalali EN, Wang D-Y. 2015. Renewable Cardanol-Based Surfactant Modified Layered Double Hydroxide as a Flame Retardant for Epoxy Resin. ACS Sustainable Chemistry & Engineering. 3(12):3281–3290.
31. Wang X, Spörer Y, Leuteritz A, Kuehnert I, Wagenknecht U, Heinrich G, Wang D-Y. 2015. Comparative study of the synergistic effect of binary and ternary LDH with intumescent flame retardant on the properties of polypropylene composites [10.1039/C5RA15565G]. RSC Advances. 5(96):78979–78985.
32. Wicklein B, Kocjan D, Carosio F, Camino G, Bergström L. 2016. Tuning the Nanocellulose–Borate Interaction To Achieve Highly Flame Retardant Hybrid Materials. Chemistry of Materials. 28(7):1985–1989.
33. Xu W, Xu B, Li A, Wang X, Wang G. 2016. Flame Retardancy and Smoke Suppression of MgAl Layered Double Hydroxides Containing P and Si in Polyurethane Elastomer. Industrial & Engineering Chemistry Research. 55(42):11175–11185.
34. Yue X, Li C, Ni Y, Xu Y, Wang J. 2019. Flame retardant nanocomposites based on 2D layered nanomaterials: a review. J Mater Sci. 54(20):13070–13105.
35. Zhang S, Horrocks AR. 2003. A review of flame retardant polypropylene fibres. Progress in Polymer Science. 28(11):1517–1538.
36. Zhou X, Chen H, Chen Q, Ling Q. 2017. Synthesis and characterization of two-component acidic ion intercalated layered double hydroxide and its use as a nanoflame-retardant in ethylene vi-nyl acetate copolymer (EVA) [10.1039/C7RA08859K]. RSC Advances. 7(84):53064–53075.

Impending Inquisitions in Humanities and Sciences (ICIIHS-2022) – Dr. Mohan Varkolu et al. (eds)

Adaptable Synthetic Routes—from Aldehydes to Nitriles

62

Madhavi Latha Duda, Amarnath Velidandi*
Department of Chemistry, Chaitanya Deemed to be University, Warangal, Telangana State, India

ABSTRACT Nitrile intermediates are having crucial role at Organic synthesis, for the preparation of acid, amides, and amines. In the synthesis of nitrile derivatives from aldehydes, many methods are possible. In the current illustration, some important recent advances, improvements, simple, cheaper prepration methods are discussed.

KEYWORDS Catalysis, Nitriles, Aldehydes

1. Introduction

Nitriles are intermediates for many organic compounds, one of the way to prepare these compounds is converting aldehydes to nitriles, in literature various methods arc available, many of them were having their limitations, advantages depends on 1) aromatic/aliphatic 2) electron-donating/electron-withdrawing groups 3) effect of solvent 4) effect of reagent 5) effect of catalyst.

$$\underset{\text{Aromatic (or) Unsaturated}}{\text{R-CHO}} \xrightarrow[\text{THF / R.T}]{\text{aq } NH_3/\ I_2} \text{R-CN}$$

Scheme-I

Various substitutions on cyano quinolines are prepared in a short period with a higher yield. Using aqueous ammonia, Tetrahydrofuran as a solvent, I_2, and aldehydes with (electron-acceptor (or) electron-donor groups) stirred at room temperature. The development of reaction checked with Thin Layer Chromatography and ethyl acetate used as solvent for extraction of compound mentioned by Shraddaupadyay et al. (2009). To convert the aryl aldehydes into required nitriles, the reaction mixture which consists of aryl aldehydes, hydroxylamine hydrochloride, a process with easier, effective, higher-yielding, eco-friendly, and solvent-free reaction with an acid catalyst (p-toluenesulfonic acid), microwave radiation, after that cool the reaction mixture at 25°C and extracting with Dichloromethane. The pure nitriles extracted with the column chromatography.

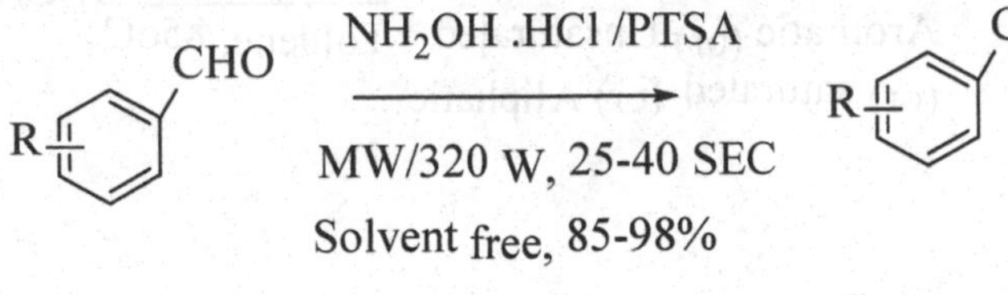

*Corresponding author: velidandi@yahoo.co.in

DOI: 10.1201/9781003489436-62

Scheme-II

The advantage of this method is to prevent the use of toxic organic solvents, faster, and efficient process implemented by Madhusudana Reddy et al. (2010).

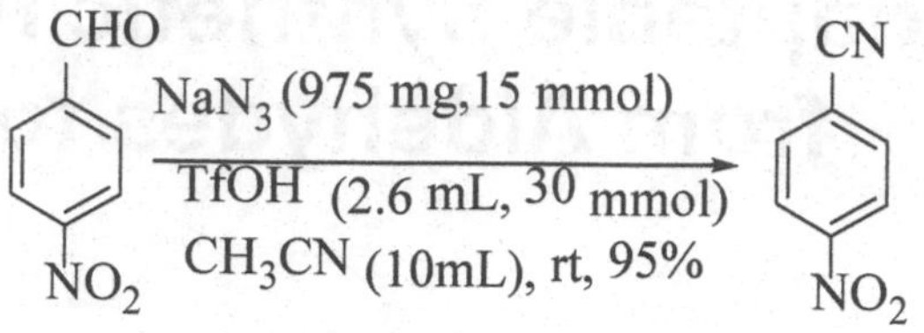

Scheme-III

Intramolecular chemoselective transformation of aldehyde to nitrile:

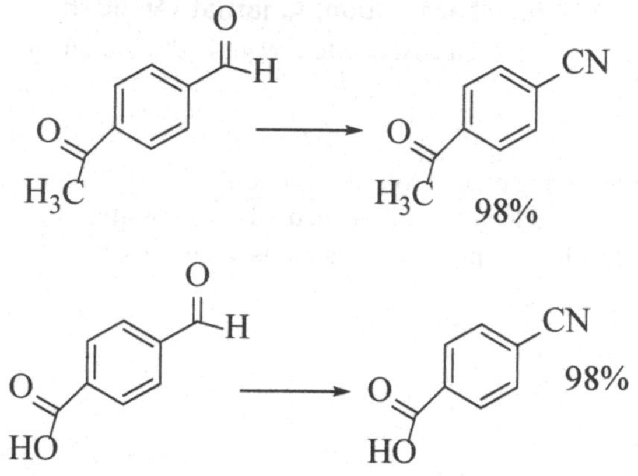

A quick, versatile Nitrile formation from aldehyde compound, without formation of formamide as the bi-product is observed in this procedure.

Plausible Mechanism:

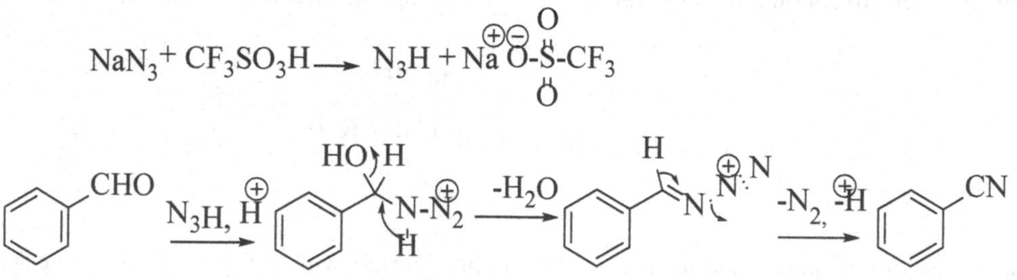

The Schmidt reaction applied to produce nitriles from corresponding aldehydes. In present scheme NaN_3, TfOH and aldehydes used to form Azido alcohol.

The elimination of H_2O molecules from Azido alcohol and nitrogen leads to the formation of Nitrile. It is a huge advantage of a Chemo-selective reaction with high yield indicated by Balaji et al. (2012).

R-CHO —DPPH / Toluene, 85oC→ R-CN

Aromatic (or) Unsaturated (or) saturated (or) Aliphatic

Mechanism:

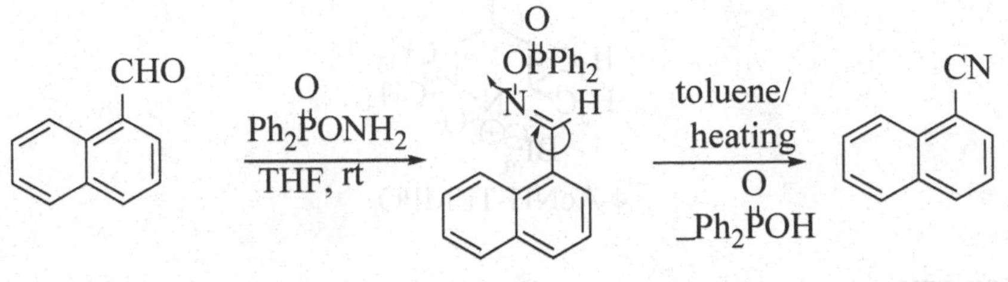

Scheme–IV

The aldehyde, DPPH (2,2-diphenyl-1-picrylhydrazyl) dissolved in tetrahydro furan and this mixture stirred at room temperature for three hours, than boiled to 85°C in toulene without using any oxidizing agent. After completion of reaction, the pure nitrile extracted from crude residue using column chromatography presented by Sébastien et al. (2012).

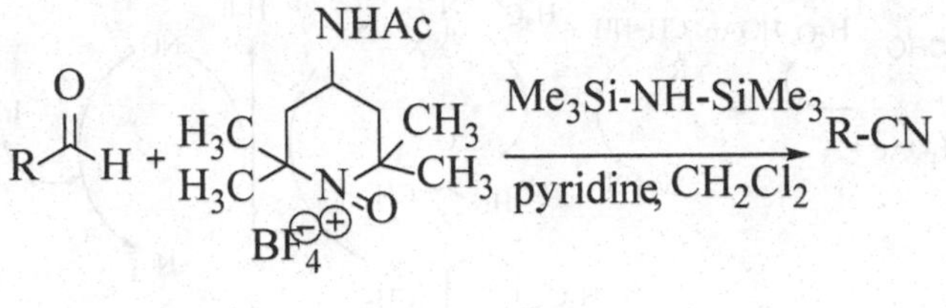

Scheme–V, Mechanism:

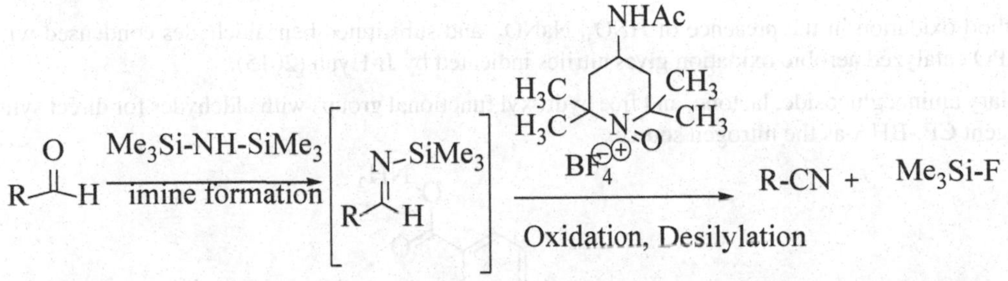

The higher yields of nitriles achived from aldehydes, HMDS, and (reusable oxidant) oxo ammonium salt. In this method HMDS role as a "N" source, polycyclic and (a variety of sulphur, oxygen, and nitrogen) heterocyclic nitriles formation also possible. The formation and oxidation of the intermediate N-trimethylsilyl imine can influence the rate of the reaction. The steps involved in the formation of nitrile, step (1) silyl-imine formation, step (2) oxo ammonium salt mediated oxidation of silyl-imine, step (3) desilylation work done by Christopher (2015).

Synthesis of Nitriles from aldehydes, amides, and oximes (aliphatic (or) aromatic (or) saturated, and unsaturated) by using dehydrating and oxidizing agents with good yields discussed in various papers. Nitriles prominent medical motifs for the preparation of a large number of pharmacologically active compounds with a wide range of activities but also are versatile intermediates in the preparation of various valuable chemicals.

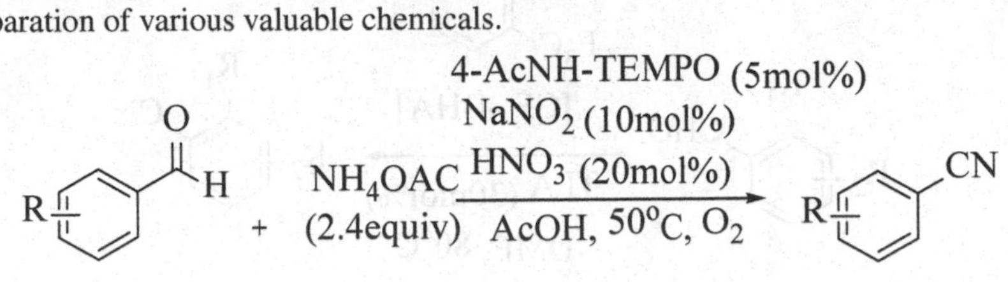

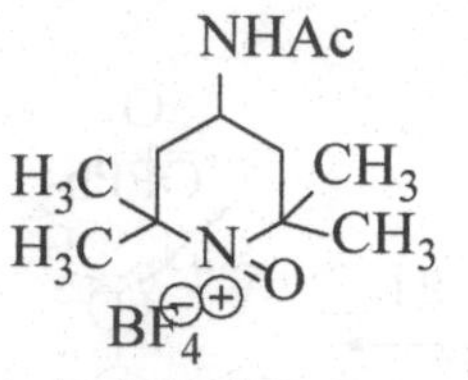

4-AcNH-TEMPO

Scheme–VI, Mechanism:

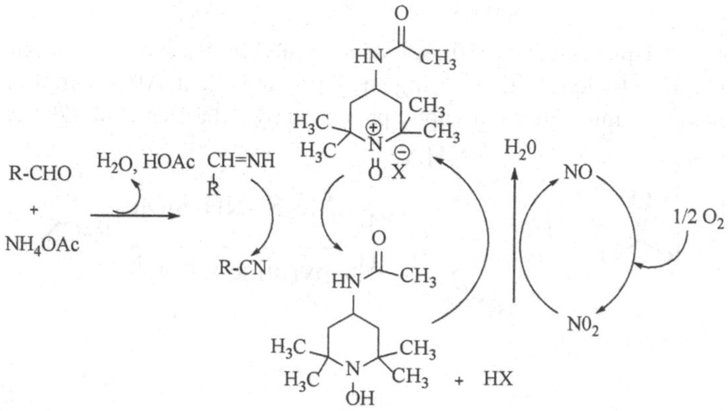

In one-pot method oxidation in the presence of HNO_3, $NaNO_2$, and substituted benzaldehydes condensed with the NH_4OAc, 4-AcNH-TEMPO catalyzed aerobic oxidation gives nitriles indicated by Ji-Hyun (2015).

The olefin, tertiary amine, glucoside, lactone, and free hydroxyl functional groups with aldehydes for direct synthesis of nitriles with a new reagent CF_3-BHA as the nitrogen source.

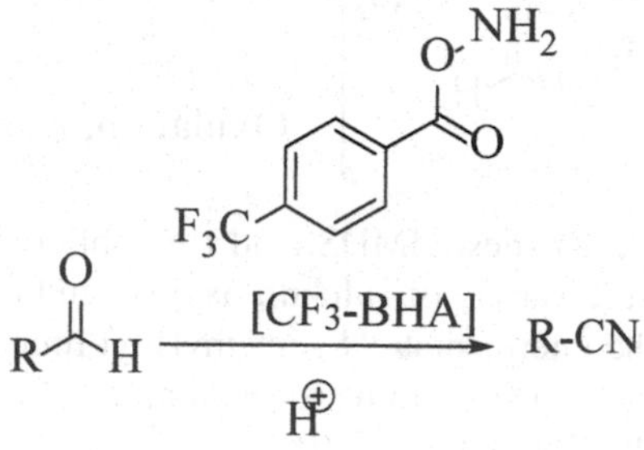

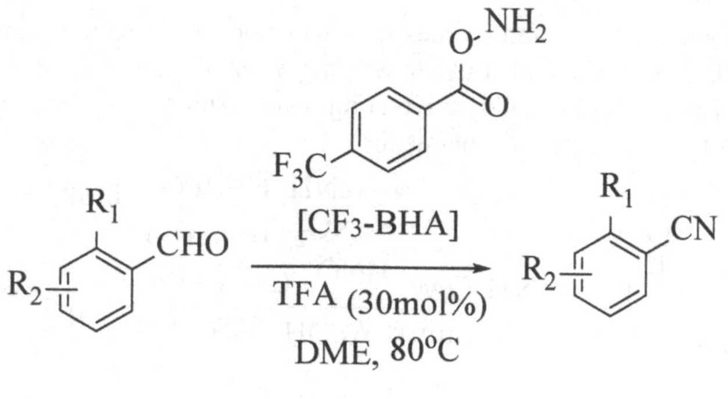

Scheme-VII

Plausible Mechanism:

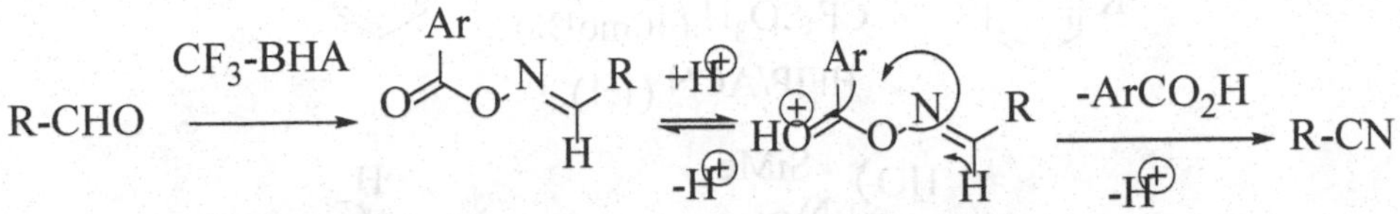

Mild conditions, higher yields, and good functional group tolerance, O-(4-CF_3-benzoyl)-hydroxylamine CF_3-BHA (source of nitrogen) reacts with aldehydes in the presence of Brønsted acid generating in situ (O-acyl oximes), it produces corresponding nitriles discribed by Xiao-DAn (2015). A fast, gentle and effective way to convert different types of aldehydes into Nitriles formed in combination with aqueous ammonia in acetonitrile at room temperature using ionic liquid reagents (Hexamethylene bis(N-methylimidazolium) bis(dichloroiodate) HMBMIBDCI.

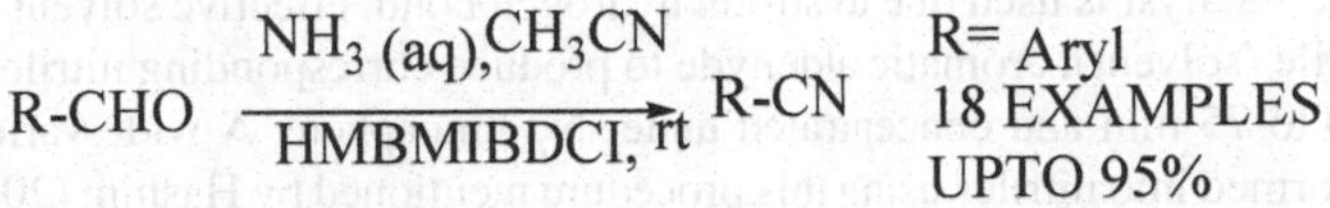

Scheme-VIII

Quick, easy work-up and more yield with simple purification are the main benefits of this process. This has been shown to reuse the ionic liquid oxidant in three times without substantial oxidation failure, without major losses mentioned by Rahman (2015). The possible mechanism is without formation of oxime, with the sustainable, effective, and modest chemical conversion of aldehydes into high yielded nitriles.

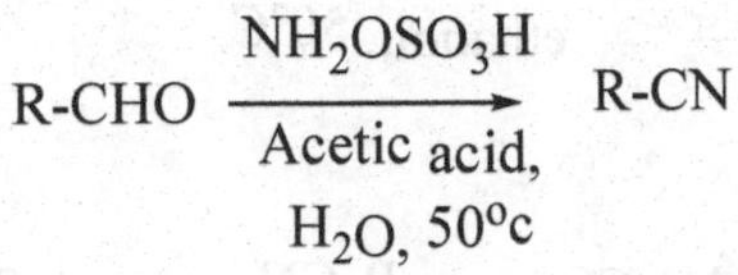

Scheme-IX

A nitrogen source is [(NH_2OSO_3H) hydroxylamine-O-sulfonic acid], the reaction takes place in acetic acid along with very simple conditions proposed by Dylan (2016).

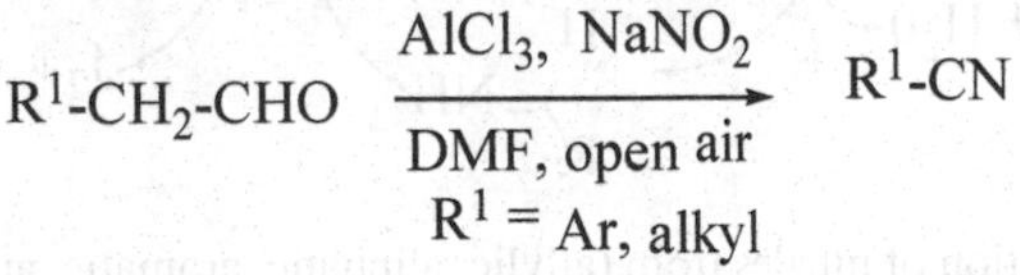

Scheme–X, Mechanism:

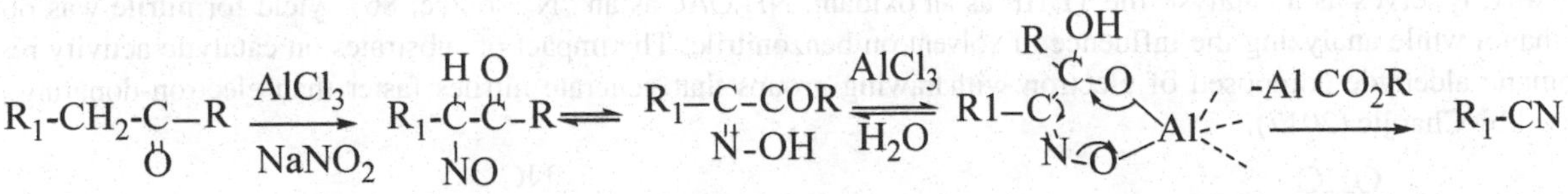

The first time without using transition metal, conversion of aldehydes to nitriles by oxidation occurred with $NaNO_2$ and it acts as an N_2 source. Aluminum chloride is cheaper and easily available lewis acid used for the "C-C" bond cleavage. Various nitriles produced with high yields for aryl, heteroaryl, alkyl, and alkenyl groups could be easily obtained by this method from carbonyl compounds work done by Jing (2016). Nitriles were obtained in higher yields from the carbonyl group by Schmidt reaction applying a sub-stoichiometric amount of triflic acid.

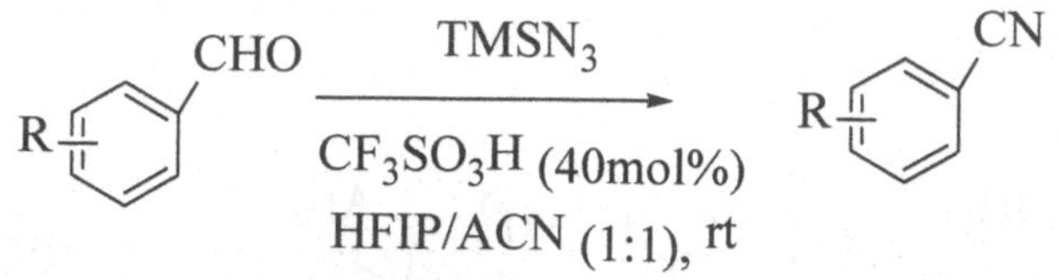

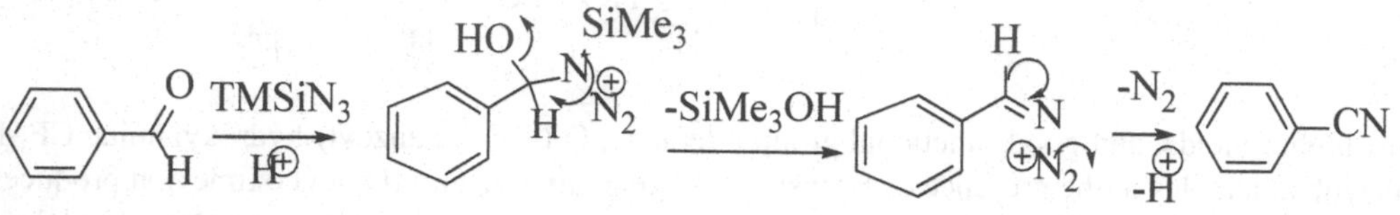

Scheme-XI

A small amount of triflic acid as catalyst is used due to strong hydrogen bond, effective solvent hexafluoro-2-propanol (HFIP), a solution of $TMSN_3$, acetonitrile (solvent), aromatic aldehyde to produce corresponding nitriles without water washings. The stirred reaction mixture for 20 to 75 min and concentrated under N_2 atmosphere. A wide variety of aldehydes with different functional groups can be transformed into nitriles using this procedure mentioned by Hashim (2016). Fast and efficient synthesis of nitriles from carbaldehyde and ammonium acetate, when saturated aliphatic aldehyde used 94% yield obtained for nitrile and from Halogen substituted aldehyde 93% yield produced for nitriles.

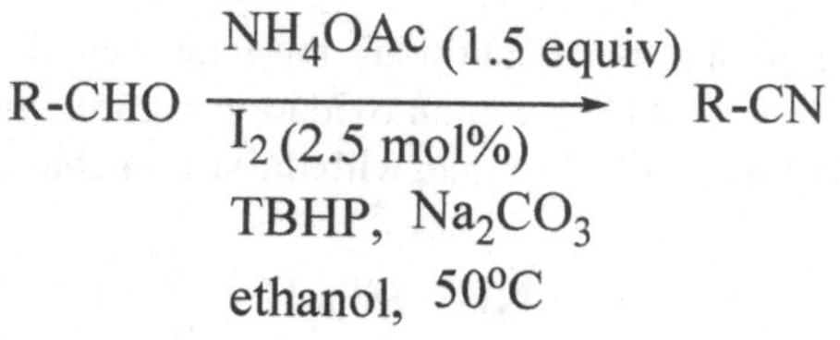

Scheme-XII, Mechanism:

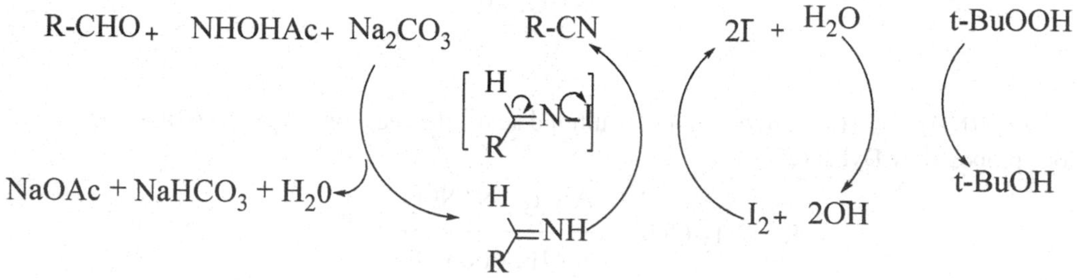

The effective and eco-friendly preparation of nitriles from (allylic, aliphatic, aromatic, and hetero-aromatic) aldehydes utilizing NH_4OAc and I_2 / TBHP.

In this way, I_2 serves as a catalyst, the TBHP as an oxidant, NH_4OAc as an "N" source. 86% yield for nitrile was observed with ethanol while analyzing the influence of solvent on benzonitrile. The impact of substrates on catalytic activity results in an aromatic aldehyde composed of electron withdrawing groups that generate nitriles faster than electron-donating groups discussed by Chaojie (2017).

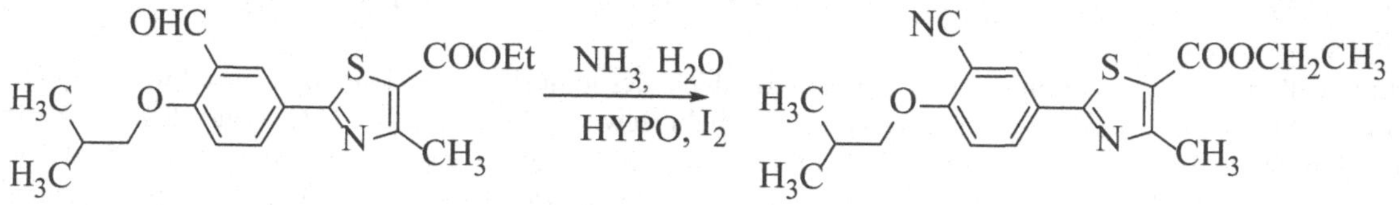

Scheme–XIII

The transformation of the aldehyde group into nitrile by ammonia (as nitrogen source), water as a solvent, I_2 / hypo (sodium thiosulphate) as oxidant. Alkyl (or) aryl cyano group produced in presence F, Cl, Br, NO_2, ester, and -OR groups mentioned by Chen (2019).

Table 62.1 Comparison of discussed methods of nitriles synthesis

Scheme Number	Method	Advantage	Limitation
I.	Ammonia, Tetrahydrofuran as a solvent, I_2	Low Cost	
II.	Hydroxylamine hydrochloride / (p-toluenesulfonic acid) / Dichloromethane	High-yield	
III.	NaN_3, TfOH	Chemo-selective reaction	
IV.	DPPH (2,2-diphenyl-1-picrylhydrazyl), tetrahydro furan	simple mixing at room temperature	column chromatography Purification
V.	HMDS(Hexamethyldisilazane)	Heterocyclic nitriles formation	
VI.	NH_4OAc, 4-AcNH-TEMPO	one-pot method	High Cost
VII.	O-(4-CF_3-benzoyl)-hydroxylamine CF_3-BHA	Good functional group tolerance	
VIII.	(Hexamethylene bis(N-methylimidazolium) bis(dichloroiodate) HMBMIBDCI	Simple purification	High Cost
IX.	(NH_2OSO_3H) Hydroxylamine-O-sulfonic acid	Very simple conditions	
X.	$NaNO_2$, Aluminum chloride	Cheaper and easily available	
XI.	$TMSN_3$	Good functional group tolerance	
XII.	NH_4OAc and I_2 / TBHP	High yield	
XIII.	Ammonia, water, I_2, hypo	Good functional group tolerance	

2. Conclusions

Nitrile functional group is commonly used in organic synthesis and is often used in pharmaceutical and agrochemical processing. Therefore, the development of new, eco-friendly safe and highly efficient synthetic protocols for the preparation of nitriles is a very attractive topic of study in modern organic chemistry. We conclude by hoping that this review will help in giving a comprehensive understanding of nitrile synthesis and will stimulate further research in this area of nitrile synthesis.

3. Funding

Authors did not receive any funding from any organisation.

4. Acknowledgments

The authors, sincerely thanking to Prof. P. V. Srilakshmi, Department of Chemistry, and National Institute of Technology Warangal for encouragement and suggestions in the research work.

REFERENCES

1. Balaji V. Rokade and Kandikere Ramaiah Prabhu (2012). Chemo selective Schmidt Reaction Mediated by Triflic Acid: Selective Synthesis of Nitriles from Aldehydes. J. Org. Chem 77(12) 5364–5370. https://doi.org/10.1021/jo3008258.
2. Chaojie Fang, Meichao Li, Xinquan Hu, Weimin Mo, Baoxiang Hu, Nan Sun, LiqunJin and Zhenlu Shen (2017). A Practical Iodine-Catalyzed Oxidative Conversion of Aldehydes to Nitriles. RSCAdv7 1484–1489. https://doi.org/10.1039/C6RA26435B.

3. Chen H, Sun S, Xi H, Hu K, Zhang N, Qu J, Zhou Y (2019). Catalytic oxidative conversion of aldehydes into nitriles using $NH_3 \cdot H_2O$/ $FeCl_2$/NaI/$Na_2S_2O_8$: A practical approach to febuxostat. Tetrahedron Lett 60 1434–1436. doi: 10.1016/j.tetlet.2019.04.043.
4. Christopher B. Kelly, Kyle M. Lambert, Michael A. Mercadante, John M. Ovian, William F. Bailey and Nicholas E. Leadbeater (2015). Access to Nitriles from Aldehydes Mediated by an Oxoammonium Salt. AngewandteChemie 544241–4245. DOI: 10.1002/anie.201412256.
5. Dylan J. Quinn, Graham J. Haun and Gustavo Moura-Letts(2016). Direct Synthesis of Nitriles from Aldehydes with Hydroxylamine O-Sulfonic Acid in Acidic Water. Tetrahedron Letters 57 3844–47. http://dx.doi.org/10.1016/j.tetlet.2016.07.047.
6. Hashim F. Motiwala,Qin Yin (2016). Azidotrimethylsilane in 1,1,1,3,3,3-Hexafluoro-2-propanol. Molecules, 21 45; doi: 10.3390/molecules21010045.
7. Ji-Hyun Noh and Jinho Kim (2015). Aerobic Oxidative Conversion of Aromatic Aldehydes to Nitriles Using a Nitroxyl/NOx Catalyst System. J.Org. Chem 80(22) 11624–11628. https://doi.org/10.1021/acs.joc.5b02333.
8. Jing-Jie Ge, Chuan-Zhi Yao, Mei-Mei Wang,Hong-Xing Zheng,Yan-Biao Kang, and Yadong Li (2016). Transition-Metal-Free Deacylative Cleavage of Unstrained C(sp3)–C(sp2) Bonds: Cyanide-Free Access to and Jeffrey Aubé, Improved Schmidt Conversion of Aldehydes to Nitriles Using Aryl and Aliphatic Nitriles from Ketones and Aldehydes. *Org. Lett* 18 228–231. DOI: 10.1021/acs.orglett.5b03367
9. M.B. Madhusudana Reddy and M. A. Pasha (2010). Efficient and High-Yielding Protocol For The Synthesis of Nitriles From Aldehydes. Synthetic Communications 403384–89. https://doi.org/10.1080/00397910903419894
10. Rahman Hosseinzadeh, Hamid Golchoubian & Mahboobe Nouzarian (2015). A mild and efficient method for the conversion of aldehydes into nitriles and thiols into disulfides using an ionic liquid oxidant. *Research on Chemical Intermediates*. 41 4713–4725. Doi:10.1007/s11164-014-1562-4.
11. Sébastien Laulhé, Sadakatali S. Gori and Michael H. Nantz (2012). Chemoselective one-Pot transformation of Aldehydes to Nitriles. J. Org. Chem 77(20) 9334–9337. doi.org/10.1021/jo301133y.
12. Shraddaupadyay, Atish Chandra and Radhey M Singh (2009). A One Pot Method of Conversion of Aldehydes into Nitriles Using Iodine in Ammonia Water: Synthesis of 2-Chloro-3-Cyanoquinolines. Indian Journal of Chemistry 48 152–154.
13. Xiao-DAn and Shouyun Yu (2015). Direct Synthesis of Nitriles from Aldehydes Using an O-BenzoylHydroxylamine (BHA) as the Nitrogen Source. Org. Lett17 5064–5067. https://doi.org/10.1021/acs.orglett.5b02547.

Impending Inquisitions in Humanities and Sciences (ICIIHS-2022) – Dr. Mohan Varkolu et al. (eds)

Analytical Method Development and Validation of Metformin HCl of Different Market Brands by UV-Spectroscopy

63

Kishore More*

HOD, Department of Pharmaceutical Analysis, Vikas College of Pharmaceutical Sciences, Suryapet, Telangana, India

Banothu Bhadru

Department of Pharmaceutical Analysis, CMR College of Pharmacy, Kandlakoya, Hyderabad, Telangana, India

Natesh Gunturu, G. Mahesh, Nimmala Shanthi, Peddinti Naveen Kumar

Vikas College of Pharmaceutical Sciences, Suryapet, Telangana, India

ABSTRACT Method development and validation of Metformin Hcl conducted in API and tablet formulations by UV-Visible spectroscopic method. The λ_{max} was found that 236 nm, The linearity was carried out for the concentrations of 1,2,3,4, 5, 6, 7, 8, 9µg/ml with regression coefficient (r^2) 0.9986. The percentage of drug content is found to be 99.7016 ± 1.4996 for Metformin HCL in API formulation. The method was found to be accurate and precise, as indicate by recovery studies as recoveries were close to 100% and %RSD is less than 2%. Inter day Precision results for different brands like okamet shows %RSD of 0.3382, glycomet shows %RSD of 2.1933, glyciphage shows %RSD of 2.8841and Intraday variations was found to be okamet shows %RSD of 1.520, glycomet shows %RSD of 1.009, glyciphage shows %RSD of 2.396. Accuracy results for different brands like okamet at 80% shows %RSD of 0.696268, at 100% shows %RSD of 0.081663, at 120% shows %RSD of 0.411235, Glycomet at 80% shows %RSD of 0.183157 at 100% shows %RSD of 0.128092 at 120% shows %RSD of 0.190224 and Glyciphage- at 80% shows %RSD of 0.18149, at100% shows %RSD of 0.166053217, at 120% shows %RSD of 0.196933465. Robustness for Okamet 235nm shows %RSD of 2.98217216, at 236nm shows %RSD of 2.264530905, at 237nm shows %RSD of 6.430021 Glycomet at 235nm shows %RSD of 3.595899474, at 236nm shows %RSD of 3.177528527, at 237nm shows %RSD of 13.34564475 Glyciphage at 235nm shows %RSD of 10.5995176, at 236nm shows %RSD of 5.634288012, at 237nm shows %RSD of 8.09250084).

KEYWORDS Method development, Validation, Metformin Hcl, Uv-Vis spectroscopy

1. Introduction

Analysis can be devided into two parts

(i) Qualitative Analysis

(ii) Quantitative Analysis

In Qualitative analysis we can estimate the unknown sample, while quantitative analysis helpful for the exact concentration of the sample. By using quantitative analysis we will perform different validation parameters.

*morekishore.pharma@gmail.com

DOI: 10.1201/9781003489436-63

2. Importance of Method Validation

Method validation is the process of establishing the performance characteristics of an analytical method to ensure that it is fit for its intended purpose. The importance of method validation in UV-Visible spectroscopy can be explained in the following ways:

Ensuring accuracy and precision: Method validation provides a means of ensuring that the UV-Visible spectroscopy method is accurate and precise. Validation helps to identify and correct any potential sources of error in the analytical method, such as instrumental or analytical errors.

Ensuring reliability: Method validation also ensures that the analytical method is reliable, meaning that it produces consistent results over time and across different analysts or instruments. This is particularly important for UV-Visible spectroscopy, where small variations in the instrument or analytical conditions can affect the accuracy and precision of the results.

Meeting regulatory requirements: Method validation is often required by regulatory bodies, such as the US Food and Drug Administration (FDA) and the European Medicines Agency (EMA), to ensure that analytical methods used for quality control or product release meet certain standards of accuracy, precision, and reliability.

Ensuring method suitability: Method validation can also help to determine the suitability of the UV-Visible spectroscopy method for its intended use, such as quantitative analysis, impurity profiling, or stability testing. This can help to ensure that the analytical method is fit for purpose and provides meaningful results.[1-4]

Reatul Karim et.al A precise UV spectrophotometric was created and approved for evaluation of Metformin hydrochloride in bulk and in tablet formulations. The linearity chart built at 233nm was straight and linear of 1-25µg/ml with a correlation coefficient of 0.9998. The strategy was approved according to ICH rules. Linearity, precision, accuracy (Interday and intraday), Robustness and ruggedness were viewed as 0.2226µg/ml and 0.6745µg/ml separately. Hence, the proposed technique is reasonable and can be shown assurance of Metformin hydrochloride from drug measurements structure in routine quality control examination. "Advancement and approval of uv spectroscopic estimation of Metformin hydrochloride in tablet measurement structure IJPSR (2012), Vol. 3, Issue 09"

Yuvraj Dilip Dange et al., Create UV/VIS strategy for Metformin Hydrochloride in bulk and tablets dosage forms. Technique: API is dissolved in methanol, ethanol, acetonitrile and phosphate buffer and different concentrations were measured, Among various solvents water has showed improved results. Metformin Hydrochloride showed most extreme absorbance at 234 nm. The technique was basic, advantageous and reasonable for the estimation of Metformin Hydrochloride in bulk and tablet dosage forms. "Method development and validation of UV-Spectrophotometric methods for the estimation of Metformin in bulk and Tablet dosage forms. Indian journal of Drug guidance and discovery | Vol 51 | Issue 4S | Oct-Dec (Suppl), 2017"

2.1 Drug Profile

Name: Metformin hydrochloride

Structure

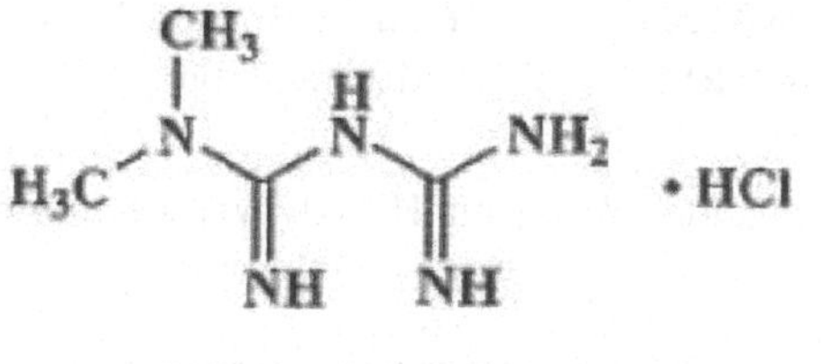

Metformin Hydrochloride

Weight: 165

Chemical Formula: $C_4H_{12}ClN_5$

3. Methodology

Estimation of marketed formulations:

Weigh accurately 20 tablets of Metformin Hcl tablets, take mean of them, crush it by using mortar and pestle, weigh equivalent weight of 100mg of drug. Now transfer into 100ml volumetric flask, dilute it with water and make up to 100ml, sonicated it for 20 minutes, After that filter through whatman filter paper, Now make up the volume to 100mL with water, Now the concentration becomes 1000μg/mL. Prepare the series of concentrations 1 μg/mL, 2 μg/mL, 3 μg/mL, 4 μg/mL, 5 μg/mL, 6 μg/mL, 7 μg/mL, 8 μg/mL, and 9 μg/mL by taking 0.1mL, 0.2mL, 0.3mL, 0.4mL, 0.5mL, 0.6mL, 0.7mL, 0.8mL, and 0.9mL from stock solution and make the volume to 100mL with ditilled water. The λ_{max} is found that 236 nm.[5-8]

3.1 Validation

Linearity:

Prepare the standard stock solutions of API from 1 to 9μg/mL. By taking concentration on X-axis and absorbance on Y-axis, plot the graph and find out the regression value.

Precision:

Precision was performed for Interday and Intraday evaluation, find out the %RSD value for the above.

Accuracy:

Accuracy studies were carried out for 80%, 100%, and 120% by addition of standard API as per the label claim of 500mg of Metformin Hcl

Limit of Detection:

The LOD is typically calculated using the following equation:

LOD = 3.3 x (standard deviation of the response of the blank) / slope of the calibration curve

where:

- 3.3 is a constant that represents the signal-to-noise ratio of 3:1, which is commonly used to define the LOD.
- The standard deviation of the response of the blank is the standard deviation of a series of measurements of the sample matrix that does not contain the analyte of interest.
- The slope of the calibration curve is the slope of the straight line that relates the response of the analytical method to the concentration or amount of the analyte of interest in a series of calibration standards.

Limit of Quantitation:

The LOQ is typically calculated using the following equation:

LOQ = 10 x (standard deviation of the response of the blank) / slope of the calibration curve

where:

- 10 is a constant that represents the signal-to-noise ratio of 10:1, which is commonly used to define the LOQ.
- The standard deviation of the response of the blank is the standard deviation of a series of measurements of the sample matrix that does not contain the analyte of interest.
- The slope of the calibration curve is the slope of the straight line that relates the response of the analytical method to the concentration or amount of the analyte of interest in a series of calibration standards.

Robustness:

By evaluating the robustness of an analytical method, the analyst can identify the key parameters that affect the method performance and determine the range of variation that can be tolerated without affecting the accuracy and precision of the method. This information can be used to establish the method conditions that are most robust and to optimize the method parameters to improve the method's performance and reliability.[9-15]

4. Results and Discussion

4.1 Selection of Wavelength

Solution is scanned between 200nm to 400nm wavelength, the absorption maxima is found that 236 nm.

4.2 Method Validation

Linearity:

Linearity was obtained by plotting 1-9μg/mL of metformin. Hcl linear regression equation was $y = 0.076x - 0.0533$ with correlation coefficient 0.9986

Table 63.1 Linearity, LOD & LOQ of Metformin

CONC (μg/mL)	ABS	FOUND CONC(μg/ml)	%RECOVERY
1	0.0226	0.994736842	99.47368421
2	0.0959	1.959210526	97.96052632
3	0.1726	2.968421053	98.94736842
4	0.2487	3.969736842	99.24342105
5	0.3259	4.985526316	99.71052632
6	0.4098	6.089473684	101.4912281
7	0.4867	7.101315789	101.4473684
8	0.5641	8.119736842	101.4967105
9	0.6142	8.778947368	97.54385965
MEAN			99.76163255
STANDARD DEVIATION			1.499620673
STANDARD ERROR OF INTERCEPT			0.006090054
SD OF INTERCEPT	SE OF INTERCEPT*n		0.054810486
LOD	3*3(SD OF INTERCEPT/SLOPE)		6.490715485
LOQ	10*(SD OF INTERCEPT/SLOPE)		0.721190609

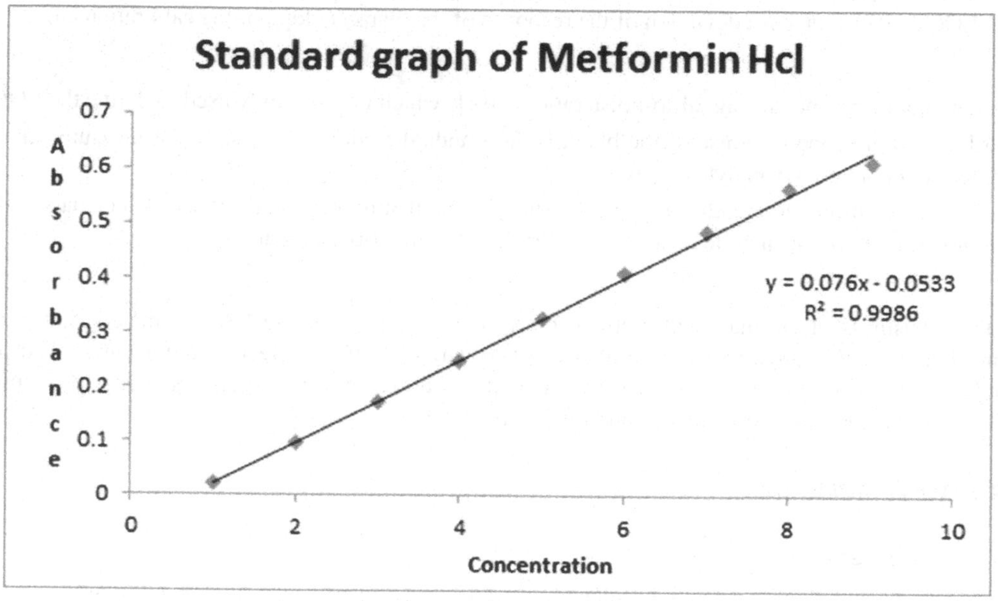

Fig. 63.1 Precision of Metformin

INTRADAY

OKAMET(CIPLA)							
TIME	**CONC µg/ml**	**A1**	**A2**	**A3**	**AVG**	**FOUND CONC**	**%RECOVERY**
10:30AM	5	0.3158	0.3185	0.3095	0.3146	4.83684211	96.73684211
12:30PM	5	0.3168	0.3178	0.283	0.3058	4.72105263	94.42105263
3:30PM	5	0.3145	0.3085	0.2905	0.3045	4.70394737	94.67894737
MEAN							95.07894737
SD							1.445932302
%RSD							1.520770204

GLYCOMET(USV)							
TIME	**CONC µg/ml**	**A1**	**A2**	**A3**	**AVG**	**FOUND CONC**	**%RECOVERY**
10:30AM	5	0.3251	0.3198	0.3154	0.311267	4.792982456	95.85964912
12:30PM	5	0.3169	0.2876	0.2896	0.291967	4.539035088	90.78070175
3:30PM	5	0.3098	0.3245	0.3054	0.295167	4.581140351	91.62280702
MEAN							92.75438596
SD							2.22250306
%RSD							2.396116407

GLYCIPHAGE(TRI STAR)							
TIME	**CONC µg/ml**	**A1**	**A2**	**A3**	**AVG**	**FOUND CONC**	**%RECOVERY**
10:30AM	5	0.3059	0.3125	0.3154	0.3112667	4.792982456	95.85964912
12:30PM	5	0.2987	0.2876	0.2896	0.2919667	4.539035088	90.78070175
3:30PM	5	0.2847	0.2954	0.3054	0.2951667	4.581140351	91.62280702
MEAN							92.75438596
SD							2.22250306
%RSD							2.396116407

INTERDAY

OKAMET(CIPLA)							
DATE	**CONC µg/ml**	**A1**	**A2**	**A3**	**AVG**	**FOUND CONC**	**%RECOVERY**
5-04-2022	5	0.3099	0.311	0.307	0.3093	4.76710526	95.34210526
6-04-2022	5	0.3076	0.3086	0.3085	0.3085	4.75657895	95.13157895
7-04-2022	5	0.3502	0.3104	0.3069	0.3069	4.73552632	94.71052632
MEAN							95.06140351
SD							0.3215584259
%RSD							0.338291091

GLYCOMET(USV)							
DATE	CONC µg/ml	A1	A2	A3	AVG	FOUND CONC	%RECOVERY
5-04-2022	5	0.3249	0.3188	0.322	0.3219	4.932894737	98.65789474
6-04-2022	5	0.3169	0.3155	0.3155	0.3156	4.85	97
7-04-2022	5	0.3045	0.3031	0.3101	0.3059	4.722368421	94.44736842
MEAN							96.70175439
SD							2.121048278
%RSD							2.1933917225

GLYCIPHAGE (TRI STAR)							
DATE	CONC µg/ml	A1	A2	A3	AVG	FOUND CONC	%RECOVERY
5-04-2022	5	0.3198	0.3156	0.312	0.3158	4.852631573	97.05263158
6-04-2022	5	0.321	0.2806	0.2974	0.2974	4.610526316	92.21052632
7-04-2022	5	0.3406	0.3242	0.2986	0.2986	4.626315789	92.52631579
MEAN							93.92982456
SD							2.709035529
%RSD							2.884105821

5.2 Accuracy

Accuracy of the strategy is as curtained by standard expansion technique at 3 levels. Standard amount comparable to 80%, 100 percent, and 120% is to be included as example.

ACCURACY OF METFORMIN

OKAMET(CIPLA):

okamet (Cipla)					
S.NO	CONC	ABS	MEAN	SD	%RSD
1	80%	0.562	0.566233333	0.00394	0.69627
2		0.5698			
3		0.5669			
4	100%	0.6819	0.6818	0.00056	0.08166
5		0.6823			
6		0.6812			
7	120%	0.7038	0.703233333	0.00289	0.41124
8		0.7001			
9		0.7058			

GLYCOMET(USV):					
S.NO	CONC	ABS	MEAN	SD	%RSD
1	80%	0.568	0.5674	0.0010392	0.18316
2		0.5662			
3		0.568			
4	100%	0.6646	0.663966667	0.0008505	0.12809
5		0.663			
6		0.6643			
7	120%	0.713	0.713733333	0.0013577	0.19022
8		0.7129			
9		0.7153			

GLYCIPHAGE(TRI STAR):

S.NO	CONC	ABS	MEAN	SD	%RSD
1	80%	0.5938	0.5942667	0.001078579	0.18149753
2		0.5935			
3		0.5955			
4	100%	0.6718	0.6706	0.001113553	0.16605322
5		0.6696			
6		0.6704			
7	120%	0.7109	0.7109	0.0014	0.19693346
8		0.7095			
9		0.7123			

ROBUSTNESS OF METFORMIN

OKAMET(CIPLA):				
S.NO	CONC (μg/ml)	231nm	236nm	241nm
1	5	0.2918	0.3154	0.2945
2		0.2978	0.3254	0.2746
3		0.2899	0.3146	0.2845
4		0.2789	0.3164	0.2689
5		0.2759	0.3325	0.2897
6		0.2812	0.3258	0.2456
	MEAN	0.285916667	0.321683333	0.2763
	SD	0.008526527	0.007284618	0.017766
	%RSD	2.98217216	2.264530905	4.430021

GLYCOMET(USV):				
S.NO	CONC (μg/ml)	231nm	236nm	241nm
1	5	0.3015	0.3215	0.2987
2		0.3158	0.3263	0.2165
3		0.3256	0.3197	0.2756
4		0.3174	0.3249	0.2455
5		0.3345	0.3485	0.2685
6		0.3275	0.3282	0.2145
	MEAN	0.320383333	0.328183333	0.253216667
	SD	0.011520663	0.010428119	0.03379337
	%RSD	3.595899474	3.177528527	4.34564475

GLYCIPHAGE(TRI STAR):				
S.NO	CONC (µg/ml)	231nm	236nm	241nm
1	5	0.2458	0.3158	0.2368
2		0.2145	0.3257	0.2345
3		0.2547	0.3658	0.2685
4		0.2687	0.3457	0.2185
5		0.2156	0.3189	0.2574
6		0.2758	0.3358	0.2658
MEAN		0.24585	0.334616667	0.24691667
SD		0.026058914	0.018853267	0.01998173
%RSD		3.5995176	0.634288012	4.09250084

6. Conclusion

Metformin HCL tablet formulations (Okamet, Glycomet, Glyciphage) were validated according to ICH guidelines performed linearity, range, LOD, LOQ, Accuracy, precision (Intraday and Interday) and Robustness.

Linearity found that 1-9 µg/ml which obeys Beer lamberts law and the regression found that $R^2 = 0.9986$

Metformin shown in intraday precision okamet shows %RSD less than 2%(1.5207) but the glycomet and glyciphage shown slightly more than 2%RSD.

Interday precision okamet shown %RSD <2 (0.3382) but the glycomet and glyciphage shown slightly more than 2% RSD.

In accuracy data all the 3 formulations (okamet, glycomet, glyciphage) shown <2% RSD which was acceptable.

In robustness all the formulations shown %RSD within the limit.

By the above data we can conclude that all three formulations obey all parameters.

The validation study shows that the developed UV method is linear, accurate, rapid, precise, robust and inexpensive with acceptable correlation co-efficient, % RSD and standard deviations which make it versatile and valuable for determination of Metformin in bulk or pharmaceutical dosage forms.

REFERENCES

1. "Development and validation of a UV-spectrophotometric method for the determination of metformin hydrochloride in pharmaceutical formulations." Authors: Shah DA, Patel NB, Patel NM. Journal: Journal of Analytical Chemistry. Year: 2010. DOI: 10.1134/S1061934810040052.
2. "Analytical methods for the determination of metformin in pharmaceutical formulations." Authors: Önal A, Ağçam E. Journal: Acta Chromatographica. Year: 2009. DOI: 10.1556/AChrom.21.2009.4.4.
3. "Development and validation of a spectrophotometric method for the determination of metformin hydrochloride in bulk and pharmaceutical dosage forms." Authors: Sahoo HB, Sahoo SK, Sahoo SK. Journal: International Journal of Pharmaceutical Sciences and Drug Research. Year: 2012. DOI: Not available.
4. "Development and validation of spectrophotometric and HPLC methods for determination of metformin HCl in the presence of sitagliptin and saxagliptin." Authors: Kassem MG, Naguib IA, Abu-Hashem AA, Serag A. Journal: Journal of Chromatography B. Year: 2013. DOI: 10.1016/j.jchromb.2012.12.019.
5. "Analytical method development and validation of metformin hydrochloride by UV-spectrophotometry in bulk and pharmaceutical dosage forms." Authors: Visweswaramurthy D, Muppa SK, Bongu SK. Journal: Journal of Chemical and Pharmaceutical Research. Year: 2011. DOI: Not available.
6. "Development and validation of a UV-visible spectrophotometric method for determination of metformin hydrochloride in bulk and tablet dosage forms." Authors: Jadhav SD, Agarwal SP, Vaidya VV, Pore YV. Journal: Der Pharma Chemica. Year: 2010. DOI: Not available.

7. "Development and validation of a UV-spectrophotometric method for estimation of metformin hydrochloride in bulk and tablet dosage form." Authors: Gupta MK, Sharma S, Agarwal S. Journal: Research Journal of Pharmacy and Technology. Year: 2017. DOI: 10.5958/0974-360X.2017.00168.X.
8. "Analytical method development and validation of metformin hydrochloride in pharmaceutical dosage form by UV spectrophotometry." Authors: Patel DS, Patel JA, Patel DB. Journal: International Journal of Pharmacy and Pharmaceutical Sciences. Year: 2011. DOI: Not available.
9. "Development and validation of a UV-visible spectrophotometric method for determination of metformin hydrochloride in bulk and pharmaceutical dosage forms." Authors: Mahadik MV, Suhagia BN, Shah SA. Journal: Scientia Pharmaceutica. Year: 2007. DOI: 10.3797/scipharm.2007.74.77.
10. "Analytical method development and validation of metformin hydrochloride in bulk and pharmaceutical dosage form by UV-spectrophotometry." Authors: Shah H, Savale S, Joshi H. Journal: Asian Journal of Pharmaceutical Analysis. Year: 2013. DOI: Not available.
11. "Development and validation of UV-visible spectrophotometric method for determination of metformin hydrochloride in bulk and tablet dosage form." Authors: Rathore AS, Vardhan H, Rathore MS, Singhai AK. Journal: International Journal of ChemTech Research. Year: 2010. DOI: Not available.
12. "Development and validation of UV-spectrophotometric method for determination of metformin hydrochloride in pharmaceutical dosage forms." Authors: Gandhi SV, Panchal RB, Shah DA, Shah BK. Journal: Indian Journal of Pharmaceutical Sciences. Year: 2008. DOI: 10.4103/0250-474X.41456.
13. "Development and validation of UV-spectrophotometric method for estimation of metformin hydrochloride in bulk and tablet dosage form." Authors: Bhanu P, Raju M, Anusha P, Chakradhar D. Journal: Journal of Chemical and Pharmaceutical Research. Year: 2011
14. "Analytical method development and validation for determination of metformin hydrochloride by using UV-visible spectrophotometry in bulk and tablet dosage form." Authors: Abbagani S, Saibaba D, Neerada G, Ramesh K. Journal: International Journal of Pharmacy and Pharmaceutical Sciences. Year: 2014.
15. "Development and validation of a UV-spectrophotometric method for determination of metformin hydrochloride in bulk and tablet dosage form." Authors: Chawla N, Kalia AN. Journal: International Journal of Pharma and Bio Sciences. Year: 2011.

Impending Inquisitions in Humanities and Sciences (ICIIHS-2022) – Dr. Mohan Varkolu et al. (eds)

Chemical Reduction Synthesis Method of Nano-ZnO Particles, Characterisation and its Application of Anti-Microbial Activity with Gram-Positive Bacteria Gram-Negative Bacteria

64

Madan Kumar Gundala*, M. Sujatha
KL (Deemed to be University), Guntur, Andhra Pradesh, India
G. Kalpana Ginne, M. Ravinder
Chaitanya (Deemed to be University), Hanamakonda, Warangal, Telangana, India
Gundekari Sreedhar
KL (Deemed to be University), Hyderabad, Telangana, India

ABSTRACT In the current work, we used the chemical reduction approach to create nano-Zinc Oxide particles (ZnO NPs) in a range of sizes. Zinc nitrate is a metal precursor, and sodium borohydride a reducing agent, are the raw ingredients used to create nano-Zinc Oxide particles. The particular chemical composition of Nano-Zinc Oxide powders was also determined utilizing Fourier Transform Infrared Spectroscopy (FTIR), X-ray diffraction (XRD), Scanning Electron Microscopy (SEM), and Transmission Electron Microscopy (TEM). The Nano-Zinc Oxide particle was characterized for composition, shape, size, and crystallinity. The Nano-Zn Oxide particles were 100 nm in size and had a wurtzite hexagonal structure. The Anti-bacterial test approach for Gram-positive and Gram-negative bacteria allowed researchers to observe the anti-microbial activity of the nano-Zn Oxide particles that were generated.

KEYWORDS Nano-Zinc Oxide particles, Chemical reduction, FTIR, XRD, SEM, TEM, and Anti-Microbial activity

1. Introduction

Carbon nanomaterials are the initial and finest clear embodiment of nanotechnology, while metal nanosized particles, of which Nano-Zn Oxide particles are one of the greatest examples. The size range of the nano-zinc oxide particles is between 1 and 100 nm. Metal nanoparticles fluctuate from those bulk materials in terms of properties depending on the designated pioneering applications (Matthew E. et al., 2012). Additionally, Nano-Zn Oxide materials belong to the II-VI group and have a huge energy band gap of $E0 = 3.4$ eV and high excitation energy of 60 eV, which allows them to withstand strong electric fields, high temperatures, and high-power operations (Kusuma, URS., et al., 2018).

Nano-Zn Oxide Particles are employed in a variety of applications, including gas sensors, luminous oxides, rubber, ceramics, paints, and others because of their physical and chemical properties (Nagarajan. P and Rajagopalan. V, 2008). Nano-Zn Oxides are suggested for the food, cosmetic, and pharmaceutical industries due to their high purity and anti-microbial action (Hernández. R et al., 2020) Many physical and chemical processes, such as the Sol-Gel technique, Co-precipitation, Ball milling, laser ablation, Solvothermal, Hydrothermal synthesis methods, etc., have been used to create nano-Zn oxide particles in addition to chemical reduction (Mazhdi, M and Tafreshi, M. J. 2018, Rajendran, S. et al., 2017, Abdolhoseinzadeh, A and Sheibani, S. 2019 and Rajendran, S. et al., 2017).

*Corresponding author: madankumar.gundala@gamil.com

DOI: 10.1201/9781003489436-64

FTIR, XRD, SEM, and TEM are just a few of the spectroscopic and microscopic methods used to examine and characterize nano-Zn Oxide particles (Patil, B.N. et al., 2016) Gram-positive and gram-negative bacteria were more effectively inhibited by the nano-Zn oxide particles studied in the other study (Wang, L and Muhammed, M, 1999).

2. Materials

We need Zn $(NO_3)_2$ (min 98.0%) as the beginning reactant, $NaBH_4$ (min 98.0%) as a reducing agent, Sodium alginate (min 98.0%) as a stabilizing agent, THF (99.0%) as a solvent, and Deionized water purchased from a nearby Marchant. All analytical quality reagents are from S.D. fine chemicals.

2.1 Experimental

The chemical reduction approach was used to generate the nano-Zinc Oxide particle [10]. We developed 100 ml of 0.1 M Zn (NO3)2 solution by dissolving it in deionized water, followed by 100 ml of 0.1 M NaBH4 solution by dissolving it in THF, and subsequently prepared 100 ml of 10% sodium alginate solution by dissolving it in deionized water for the manufacture of nano-Zinc Oxide particles.

50 ml of 0.1 M Zn (NO_3)2 solution should be placed in an RB flask, and it should be heated to the temperature range of 60 to 70 °C while being continuously stirred with a magnetic stirrer for around 60 minutes. This standard operating procedure must be carried out under the varied dropping durations and the reaction temperature, then 20 ml of 10% sodium alginate solution drops in fall with steady stirring is added after 50 ml of 0.1 M NaBH4 solution (dropping for 20–30 minutes). Now, for 180 minutes, heat the entire reaction mixture at 60–70 °C on a magnetic stirrer. Extract the nano-Zinc oxide precipitate with a light-yellow tint at this time.

3. Characterization

3.1 Scanning Electron Microscopy (SEM)

The 3-D pictures of the Nano-zinc Oxide particles in the SEM micrographs display the chemical's exterior shape and composition. A concentrated stream of high-energy electrons from the JSM-6510 SEM equipment is used to identify these discoveries at various magnification levels (20X to 30000X).

The concentrated electron beam that interacts with the Nano-zinc oxide sample when it is put in an SEM has affected its electrons. The SEM generates a variety of signals, including the contrast in composition and crystalline structure of Nano-zinc oxide particles can be illustrated by backscattered and diffracted backscattered electrons and the signals containing secondary electrons are useful for demonstrating surface morphology to convert the signals into detectable 3-D images that contain information about the morphology and chemical composition of Nano-zinc oxide particles.

According to the provided SEM data, the nano-zinc oxide particle size is 100 nm. Dynamic Image Analysis (DIA), a contemporary particle characterization technique, was used to calculate size distributions and shape parameters. As shown in Figs. 64.1(a), (b), (c), and 64.1(d), their surface shape is spherical and amorphous. We confirmed the surface structure of nano-zinc oxide particles and the size of zinc oxide nanoparticles using the chemical reduction method, according to the data. Particle size has a considerable impact on the anti-bacterial and anti-microbial activities of zinc nanoparticles.

3.2 Transmission Electron Microscopy (TEM)

The JOEL JEM-2200 FS 300 kV Transmission electron microscope is used to conduct in-depth morphological and size studies. A TEM microscope operates on the same principles as ordinary microscopy. Instead of using a light source to operate, TEM uses an electron source. Diffraction and imaging modes are utilized in the TEM to collect data about nanoparticles.

Dip a 300-mesh carbon grid covered with zinc into a sonicated toluene solution to prepare a TEM sample of nano-zinc oxide particles. Before inspecting the sample of Nano-zinc particles, the film on the TEM grid is allowed to dry 24/7 in a fume hood and preserved in a vacuum container. The extra solution is then blotted away using blotting paper.

The sample is put through a TEM examination, which provides information about the Nano-zinc oxide particles in the form of pictures, once the full creation of the Nano-zinc oxide film has been established. Images are processed with ImageJ to determine the average particle size and provide a histogram of particle diameter. Electron diffraction from a TEM makes it possible to determine the orientation of nano-zinc oxide and the components present.

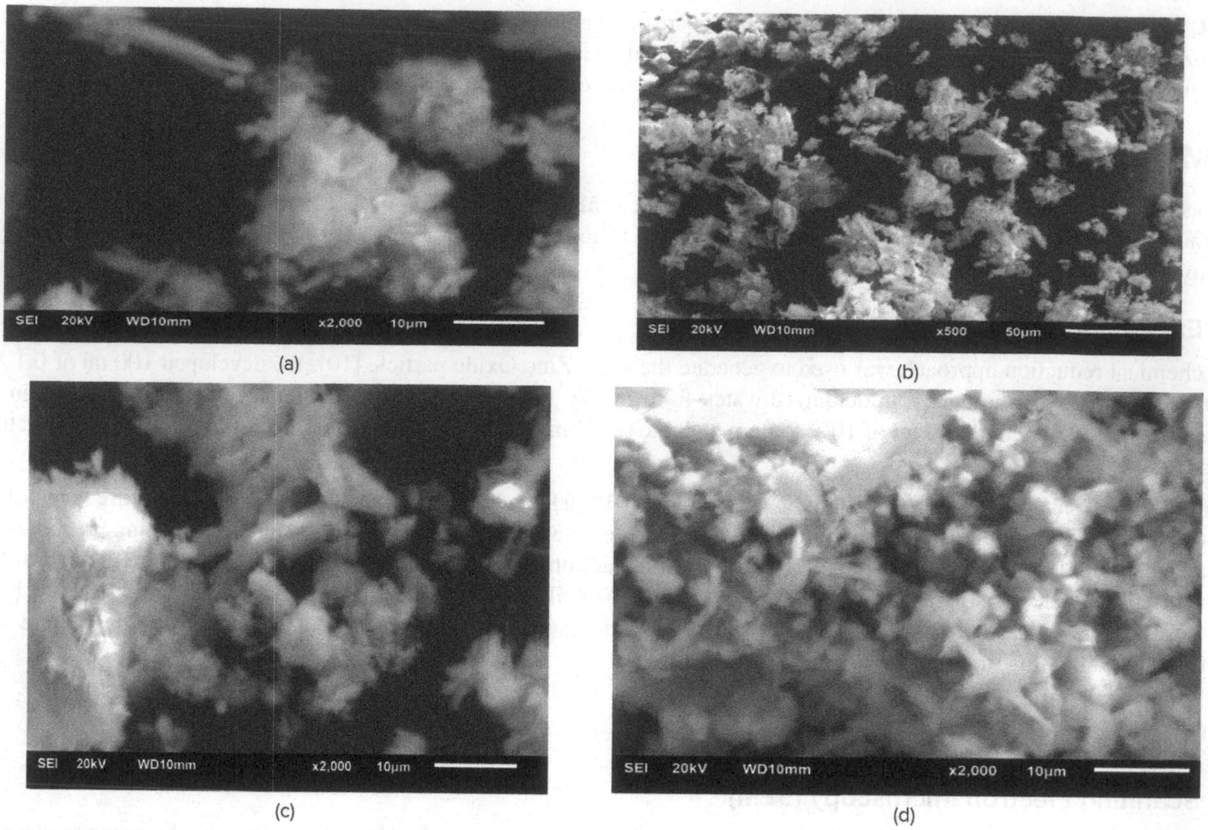

Fig. 64.1 SEM-Images of Nano-Zinc Oxide particles

According to the study above, the information on nano-zinc oxide particles demonstrates that the morphological findings are spherical and amorphous, and the size of the particles was 100 nm as shown in Figs. 64.2. Finally, we conclude that the chemical reduction procedure causes the particle size to decrease.

3.3 X-Ray Diffraction (XRD):

An adaptable, non-corrosive quantitative method for determining the phase of crystalline material is the X-ray diffractometer. It provides information about the crystalline size and unit cell dimensions. Using XRD (model XPERT-PRO X-ray diffraction) functioning at 45 kV and a current of 40 mA with Cu Kα radiation in a θ - 2θ configuration, we can analyze Nano-Zinc Oxide material. The powdered Nano-Zinc Oxide materials that have been subjected to analysis are homogenized and finely ground, and their average bulk composition is ascertained using X-rays emitted by a cathode ray tube and thereafter created monochromatically by filtering.

Figure 64.3 displays the results of an X-ray diffractometer that verified the existence of a nano-sized zinc oxide powder and crystalline structure. The results of Fig. 64.3 show that the peaks are at 31.77°, 34.44°, and 47.60°. The produced Nano-Zinc Oxide particle architecture was proven to be the hexagonal phase of Zinc Oxide based on the evidence and peaks. The peaks in the XRD patterns were widened, which had an impact on the solid nano-Zinc Oxide material. The full width at half the maximum (FWHM) of the Diffraction peaks was used to determine the mean crystalline size of Nano-Zinc Oxide powdered material by the Scherrer equation, , The XRD pattern revealed that the ZnO nanoparticles were either a cubic or a hexagonal wurtzite crystal with an average crystallite size of 30-36 nm, depending on the reaction time.

3.4 FTIR (Fourier Transform Infrared) Spectroscopy

According to the components and types of bonds, FTIR spectroscopy can measure the vibration of chemical bonds in Nano-Zinc Oxide materials at various Hz. It can be demonstrated that the bond's vibration increases and leads to a transition between the ground state and a variety of excited states, which may be observed by looking for electromagnetic radiation. FTIR may

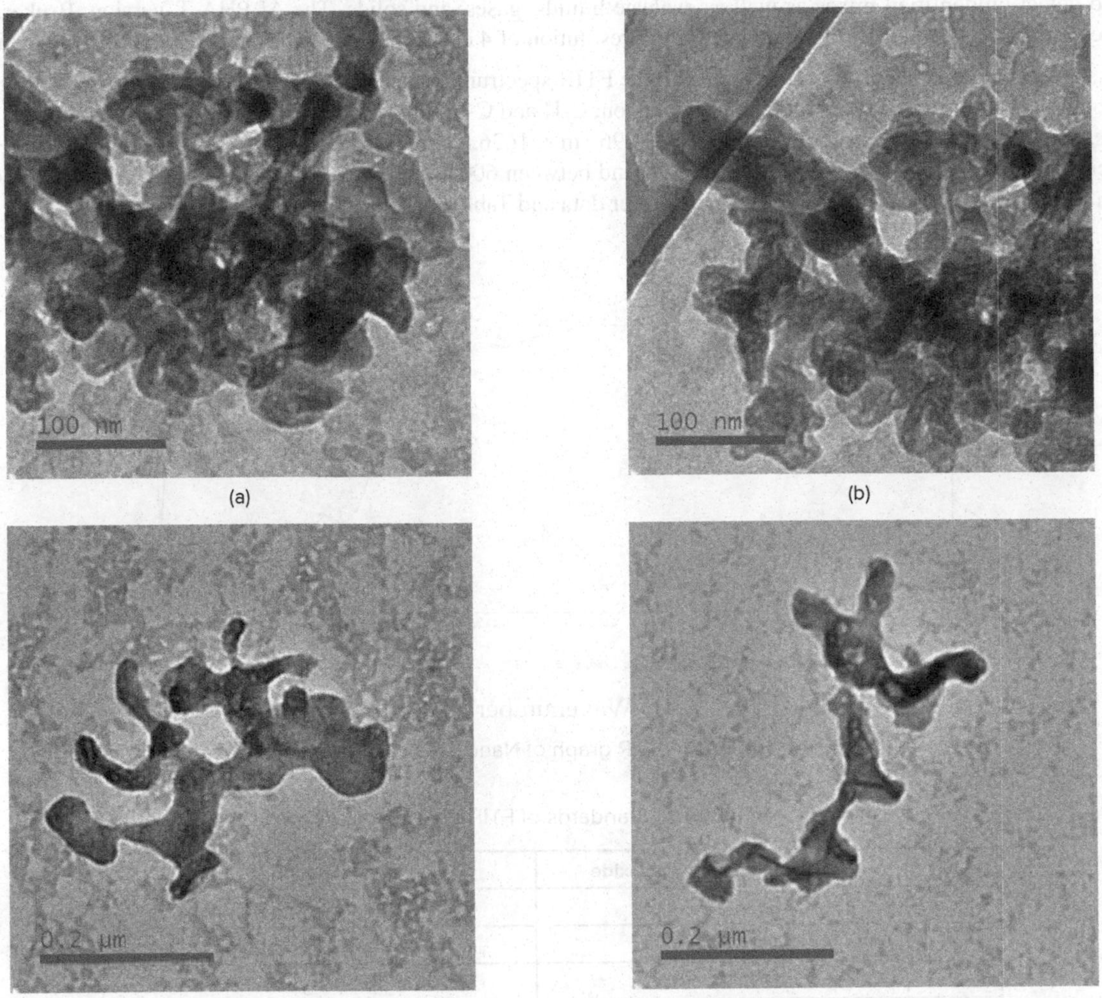

Fig. 64.2 TEM-Images of Nano-Zinc Oxide particles

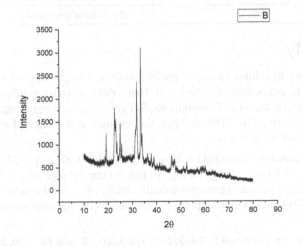

Fig. 64.3 X-ray diffraction graph of Nano-Zinc Oxide particles

be used to detect unidentified mixes as well as evaluate liquids, gases, and solids. The ALPHA-T Version, Bruker, Germany spectrometer employs a range of 400-4000 cm^{-1} and a resolution of 4 cm^{-1} to get spectra for wavelength and transmittance.

Fig. 64.4 shows the broad absorption band peaks in the FTIR spectrum of nano-zinc correspond to the stretching frequencies of the strong hydroxyl group, C=O group, ammonium ion, C-H and C-O single bond stretching vibrations, and B-H stretching vibrations, respectively. These peak positions are 3434.96 cm^{-1}, 1626.60 cm^{-1}, 1384.20 cm^{-1}, 2800-3000 cm^{-1}, 1111.06 cm^{-1}, and 994 cm^{-1}. Additionally, it supports delineating a band between 600 and 720 cm^{-1} as C-H bending vibrations and the peak at 469 cm^{-1} as a nano-zinc oxide particle confirms as per data and Table 64.1.

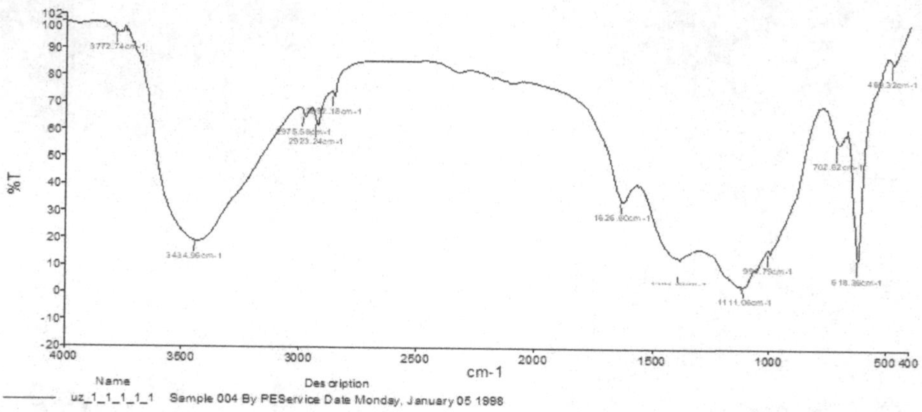

Wavenumber cm^{-1}

Fig. 64.4 FTIR graph of Nano-ZnO particles

Table 64.1 Standards of FTIR spectroscopy

S.No	IR values of Nano-zinc oxide	Obligation
1.	3434.96 cm^{-1}	Strong OH stretching vibrations
2.	1626.60 cm^{-1}	C=O stretching vibrations
3.	1384.20 cm^{-1}	N-H stretching vibrations
4.	2800-3000 cm^{-1}	C-H Stretching vibrations
5.	1111.06 cm^{-1}	C-O single bond stretching vibrations
6.	600-720 cm^{-1}	C-H bending vibrations
7.	469 cm^{-1}	The peak of zinc oxide nanoparticles

4. Anti-Microbial Activity

This influences an antibiotic's ability to inhibit pathogen growth and the dosage required to treat a patient. Using Muller-Hinton agar bacterial inoculum, the antimicrobial activity of Nano-ZnO particles produced by chemical reduction was quantitatively tested against Gram +ve bacteria of P. aeruginosa, S. Pyogenes, and Gram-negative Klebsiella. The production of nano ZnO particle powder was confirmed by FTIR analysis, the resultant adsorbent had an average pore size of 2.527 nm and a 113.751 $m^2 g^{-1}$ N2 physisorption surface area.

For measuring bacterial activity, we generate Nano-ZnO solutions with moles 0.0012, 0.0024, 0.0037, 0.0049, 0.0061, 0.0074, 0.0086, 0.0096, 0.0111, and 0.0123, or 10 mg, 20 mg, 30 mg, 40 mg, 50 mg, 60 mg, 70 mg, 80 mg, 90 mg, and 100 mg. The three bacterial agar inoculums solidified in Petri plates individually. Wells of 7 mm diameter were formed on each agar Petri plate, and 10 various doses of Nano-ZnO solution were put into indicated wells. Inhibition zones were measured after 24 hours at 37°C.

P. aeruginosa determined inhibitory zones are mm 0.1, 0.8, 3, 5, 6, 7, 8.5, 10, 12, and 14 at 10, 20, 30, 40, 50, 60, 70, 80, 90, and 100 mg. S. Pyogenes' zone values are 0.0, 0.5, 1.5, 2.7, 3.5, 5, 6, 7, 9, and 11 mm starting from 30 mg, and Klebsiella's are 0.0,

0.2, 0.7, 2, 3, 5.5, 7, 8.3, 9, and 10 mm starting from 40 mg. shows in Table 64.2. According to the data, Nano-ZnO particles have strong bacterial activity, and P. aeruginosa has the most shown in Fig. 64.5, Fig. 64.6, and Fig. 64.7.

Table 64.2 Zone of inhibition standards of anti-microbial activity

S. No	Conc. of Nano-ZnO		Zone of inhibition in mm		
	In Moles	In mg	P. aeruginosa	S. Pyogenes	Klebsiella
1.	0.0012	10	0.1	0.0	0.0
2.	0.0024	20	0.8	0.5	0.2
3.	0.0037	30	3	1.5	0.7
4.	0.0049	40	5	2.7	2
5.	0.0061	50	6	3.5	3
6.	0.0074	60	7	5	5.5
7.	0.0086	70	8.5	6	7
8.	0.0096	80	10	7	8.3
9.	0.0111	90	12	9	9
10.	0.0123	100	14	11	10

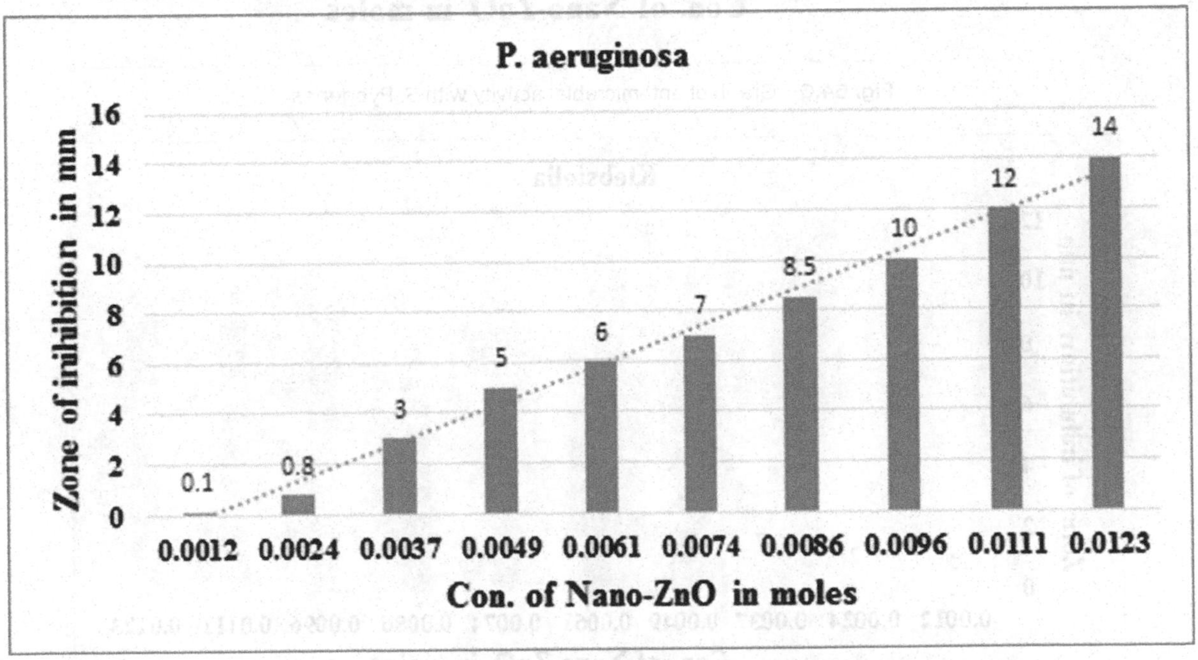

Fig. 64.5 Graph of anti-microbial activity with P. Aeruginosa

5. Conclusion

In this investigation of the chemical reduction technique utilized to create Nano-ZnO particles, it was shown that this technique generates NPS of smaller size and on a bigger scale while minimizing waste. The difficult issue, in this case, is utilizing the SEM and TEM to determine the 100 nm size, surface morphology, and topology of Zn NPs. We checked and confirmed that from the FTIR spectrum data peak of 469 cm^{-1} stretching frequency of Zn NPs, which is utilized for defining the crystalline phase by XRD. The nano ZnO particles are very lesser size and have good bacterial activity compared with commercial zinc materials. And the ZnO NPs have strong antimicrobial action against P. aeruginosa, S. pyogenes, and Klebsiella, with P. aeruginosa having the maximum activity, with a 14 mm zone of inhibition for 100 mg of Zn NPs in comparison to the other two bacteria.

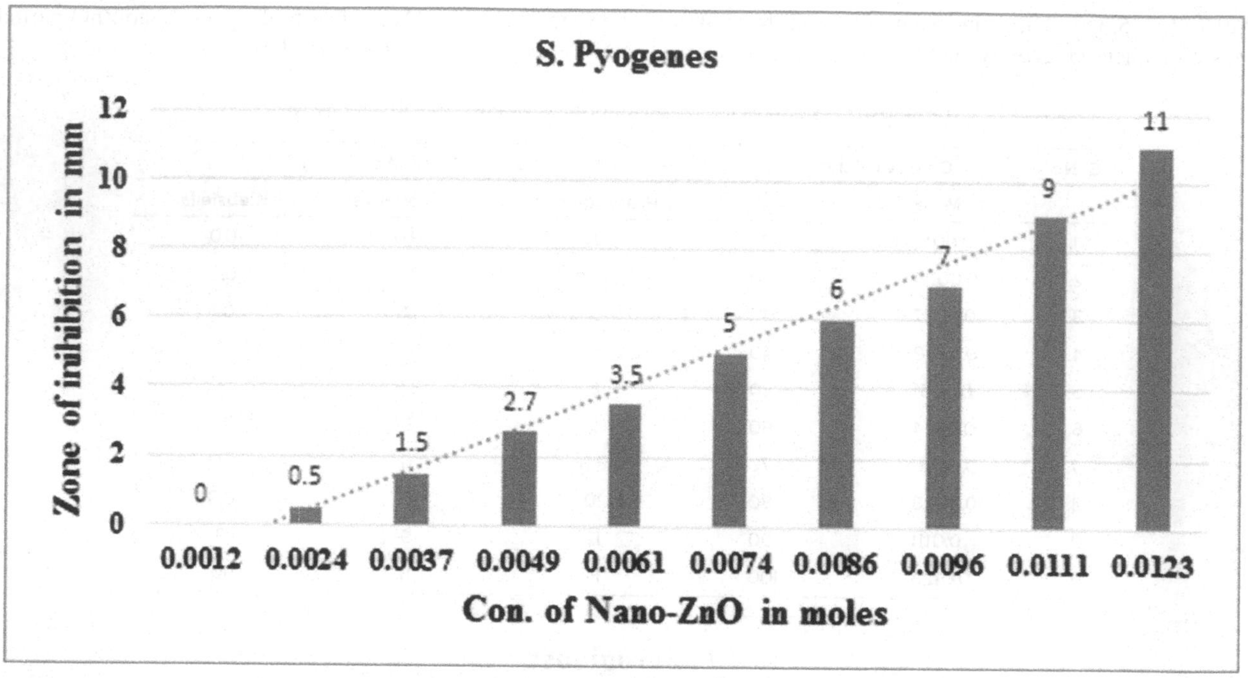

Fig. 64.6 Graph of anti-microbial activity with S. Pyogenes

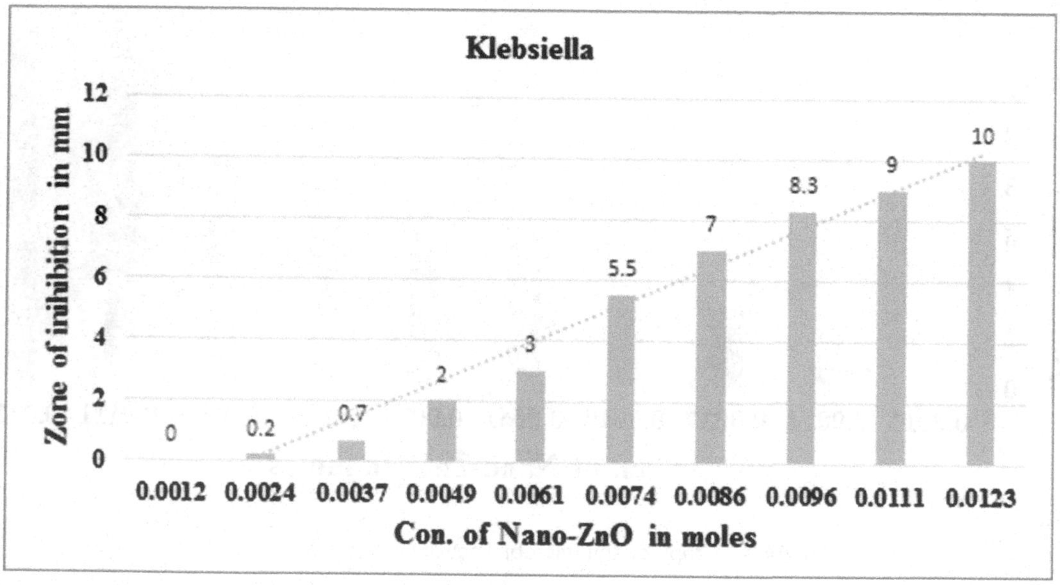

Fig. 64.7 Graph of anti-microbial activity with Klebsiella

REFERENCES

1. Matthew E. Edwards., Ashok K. Batra., Ashwith K. Chilvery., Padmaja, G., Michael, Curley., Mohan, D., Aggarwal., (2012). Pyroelectric Properties of PVDF: MWCNT Nanocomposite Film for Uncooled Infrared Detectors. Mater. sci. appl. 3(12), 851–855.
2. Kusuma, URS., Bhat, S.V., and Vinayak, K., (2018). On exceeding the solubility limit of Cr+3 dopants in SnO2 nanoparticles-based dilute magnetic semiconductors. J. Appl. Phys. 123(16):161518–161524.
3. Nagarajan, P., Rajagopalan, V., (2008). Enhanced bioactivity of ZnO nanoparticles and antimicrobial study. Sci. Technol. Adv. Mater. 9(3):035004–035012.

4. Hernández, R., Hernández-Reséndiz, J.R., Martínez-Chávez, A., Velázquez-Castillo, R., Escobar-Alarcón, L., Esquivel, K., (2020). Au-TiO2 Synthesized by a Microwave- and Sonochemistry-Assisted Sol-Gel Method: Characterization and Application as Photocatalyst. J. Sol-Gel Sci. Technol. 10(9):1052–1069.
5. Mazhdi, M., Tafreshi, M. J., (2018). The effects of gadolinium doping on the structural, morphological, optical, and photoluminescence properties of zinc oxide nanoparticles prepared by co-precipitation method. Appl. Phys. A., 124(12), 863–871.
6. Kołodziejczak-Radzimska, A., Jesionowski, T., (2014). Zinc Oxide—From Synthesis to Application: A Review. Materials, 7(4), 2833–2881.
7. Abdolhoseinzadeh, A., Sheibani, S., (2019). The enhanced photocatalytic performance of Cu2O nano-photocatalyst powder was modified by ball milling and ZnO. Adv. Powder Technol. S0921883119303498–.
8. Rajendran, S., Kandasamy, S., and Rajendran., (2017). Synthesis and Characterization of Zinc Oxide and Iron Oxide Nanoparticles Using Sesbania Grandiflora Leaf Extract as Reducing Agent. J. Nanosci. 2017: 1–7.
9. Patil, B.N., Taranath, T.C., Limonia acidissima, L., (2016). Limonia acidissima L. leaf mediated synthesis of zinc oxide nanoparticles: A potent tool against Mycobacterium tuberculosis. Int. J. Mycobacterial. S2212553116300115–.
10. Wang, L., Muhammed, M., (1999). Synthesis of zinc oxide nanoparticles with controlled morphology. J. Mater. Chem. 9(11):2871–2878.

Impending Inquisitions in Humanities and Sciences (ICIIHS-2022) – Dr. Mohan Varkolu et al. (eds)
© 2024 Taylor & Francis Group, London, ISBN 978-1-032-78829-6

Polymer-Based Poly Aniline Copper Oxide Nanocomposites by Sol-Gel Approach Method Characterization and their Photocatalytic Activity

65

Kalpana Ginne[1], S. Kavitha[2]
Chaitanya (Deemed to be University), Hanamakonda, Telangana, India
Madan Kumar Gundala
K L (Deemed to be University), Guntur, Andhra Pradesh, India

ABSTRACT $CuNO_3$ and aniline are utilized as precursors, together with $NH_4S_2O_8$, Glacial acetic acid, and DMF, to create Polymer matrix Polyaniline Copper Oxide (PACO) nanocomposites utilizing the bottom-up Sol-Gel approach technique. The synthesized Poly aniline CuO nanocomposites were analyzed by XRD, SEM, and TEM to demonstrate the results of a crystalline shape, size, and chemical composition. Here, the Scherrer equation is employed to determine the size, and the FTIR value provides the confirmation peak of CuO. By absorbing UV light, it was possible to determine the photocatalytic activity for CuO nanocomposites at varied concentrations of 0.1, 0.2, 0.3,0.4, and 0.5 moles at RT at different time durations of 20, 40, 60, 80, and 100 minutes using the $KMnO_4$ dye.

KEYWORDS Polyaniline-Copper Oxide nanocomposites, Sol-Gel method, Microscopic analysis, Spectroscopic analysis, and Photocatalytic activity

1. Introduction

Polyaniline copper oxide (PACO) nanocomposites exhibit both novel and photocatalytic activity. These materials must yet satisfy harsh application criteria. Conducting Polyaniline Nickel Oxide Nanocomposites (PANO) is an advantageous option for photocatalytic activity applications due to their recently observed photocatalytic activity, low cost of synthesis, thermal stability up to 3000°C, and the release of previously unrevealed physical characteristics (Andrew T. Smith et al., 2016, Alan G., MacDiarmid., 2001, Zheng, H.Y. et al., 2015, Shirakawa, H. et al., 1997).

In this survey, PACO nanocomposites were produced using the sol-gel technique, and their size, shape, and crystallinity were analyzed. Degrading organic dye, the PACO nanoparticles' photocatalytic activity was also measured (Heeger. A.J. 1997).

PACO nanocomposites include mostly CuO and Cu_2O oxides. Cupric oxide is a monoclinic p-type semiconductor with an energy band gap of 1.2 to 1.7 eV at room temperature. Its lattice parameters are c=1=1288 Å, b=3=4226 Å, a=4=6837 Å, and a=99=54 Å. Cu_2O is a cubic p-type semiconductor with direct band energy (Armelao, L. et al., 2003, Marabelli, F. et al., 1995). In addition to their potential application as innovative micro and nano-electronic device building blocks because of their band gap, PACO nanocomposites have recently been shown to have the potential to serve as functional cores of such devices (Di Mauro, A. et al., 2017, Thomas, D.J., 2018, Gervasio, M. and Lu, K., 2017, Mallakpour, S. and Behranvand, V., 2016, Loste, J., 2019).

Corresponding author: [1]kalpana.gundala@gmail.com, [2]kavithavbr@gmail.com

DOI: 10.1201/9781003489436-65

Using photocatalytic materials, solar energy has the potential to be an effective solution to environmental degradation and serve as a sustainable energy source. This is one of the primary motivations for cutting-edge research and development (Wang, X., 2018).

Sol-Gel synthesis was used to create PACO nanocomposites in this study. Sol-gel synthesis entails creating a colloidal suspension (sol) and then using the gelatin of the sol to create a network in a continuous liquid phase (gel). Currently, nanocomposite materials containing Carbon nanofibers (CNF) are created via the sol-gel technique. X-ray diffraction (XRD), Fourier transform infrared spectroscopy (FTIR), Scanning electron microscopy (SEM), and Transmission electron microscopy (TEM) were all used to learn about PACO Nanoparticles' properties. Under ultraviolet (UV) and visible light (VIS) irradiation, the polymer nanocomposite photocatalytic activity was measured by the degradation of $KMnO_4$ dye. Dye concentrations of 6 and 30 mg/L were determined to have the highest predicted degradation efficiencies RB (Rhodamine B) and AR 57 (Acid red 57) dyes respectively. It has also been observed that the chemical interaction of RB and AR57 dyes during photocatalysis follows a generalized process.

2. Experimental

2.1 Materials

After undergoing a twofold distillation procedure, Aniline (S. D. Fine-Chem Ltd., 99.5%) was utilized. Other chemicals utilized were starch, N, N-dimethylformamide, Ethyl ethylene glycol ($C_4H_{10}O_2$) (99% purity), Ammonium persulfate ($NH_4S_2O_8$) (99%), Copper nitrate trihydrate (Cu $(NO_3)_2.3H_2O$) (99%), Con. HNO_3 others that were purchased from a local shop. The ammonia of the AR grade was used. This investigation was carried out using de-ionized water.

2.2 Synthesis Method of PACO Nanocomposites

Step-1 Preparation of CuO Nanoparticles

100 ml of starch solution and 0.1 M (Cu $(NO_3)_2.3H_2O$ were combined, and the concoction was continuously stirred while ammonia was added in drops. The samples color changed from blue to white after the ammonia had been copiously dissolved in the mixture and the solution had been allowed to settle overnight. Whatman filter paper is used to filter acquired samples. Simply wash a poised sample of CuO nanoparticles in a solution of Ethyl ethylene glycol and deionized water to abolish any contaminants. It was then dried in hot air furnaces at a temperature of 650°C.

Step-2 Preparation of Poly aniline Copper Oxide Nanocomposites

In-situ chemical oxidative polymerization creates PACO matrix nanocomposites from synthesized CuO nanoparticles. Dissolve 10 ml 0.5 M of HNO_3 in 10 ml aniline. Stir ingredients on a magnetic stirrer for 30 minutes. Add this obtained mixture to the calcinated CuO solution. Stir the reaction mixture for 2 hours. Then add potassium persulfate with continuously stirring for three hours. A shift in color from white to yellow indicates the polymerization and creation of PACO nanocomposites. After keeping the solution overnight in a dark place, a sample was collected. Clean the sample using ethylene glycol and deionized water. Polyaniline and copper oxide nanocomposites varying in moles were produced (0.1, 0.2, 0.3, 0.4, and 0.5).

Scheme-1 Schematic presentation of PACO nanocomposite preparation

See Fig. 65.1.

3. Result and Discussions

3.1 Transmission Electron Microscopy (TEM)

In situ, oxidation may make PACO nanocomposite filler with a morphology based on the polymer nanocomposite structure, and the size and shape of PACO-generated sol-gel processes can be estimated by TEM. Electrons are commonly created using an electron cannon on top of a TEM and accelerated to 100 to 300 kV, as specified by the user, using a 300 kV TEM with a 2 Å point resolution. The electron emitter's work function is low. PACO Nanocomposites were analyzed using TEM. Fig. 65.2(a) and Fig. 65.2(b) establish different PACO Nanocomposites morphologies. Due to its capacity to image materials at the nanoscale scale, aniline's polymerization rate impacts the generation of sphere-shaped nanoparticles from polyaniline copper oxide nanocomposites and provides the easiest way to examine PACO nanocomposites' exfoliation phases shown in Fig. 65.2(c). Despite a few minor agglomerates, polymer nanocomposites were distributed evenly throughout the matrix. Based on TEM images in Fig. 65.2(c) and Fig. 65.2(d), we estimate polyaniline copper oxide nanocomposites are 70 nm.

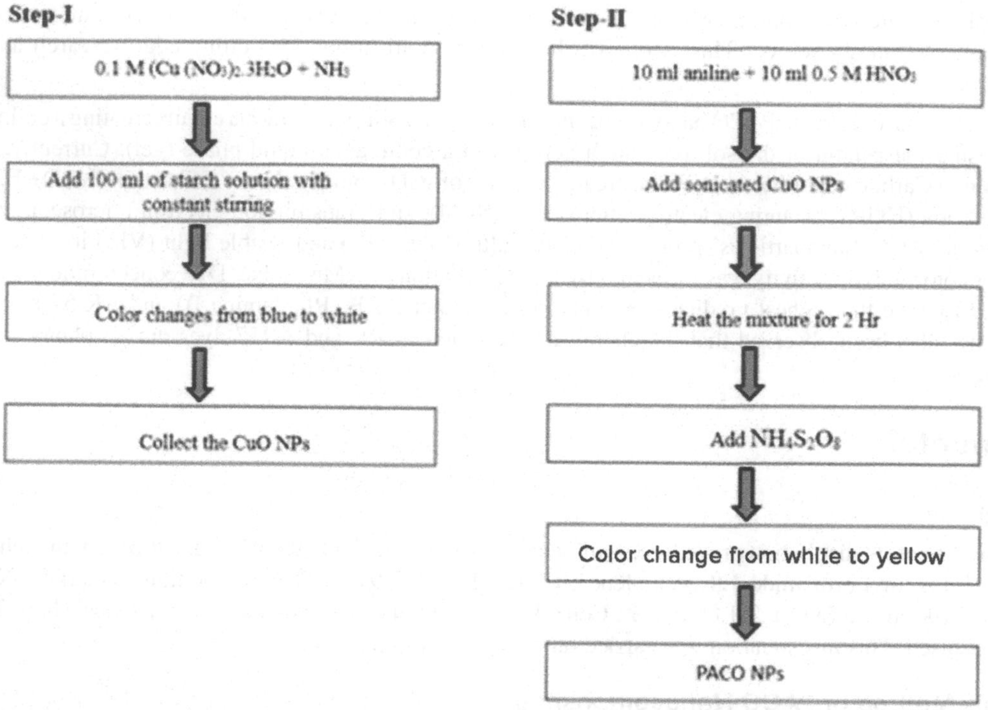

Fig. 65.1 Schematic presentation of PACO nanocomposite preparation

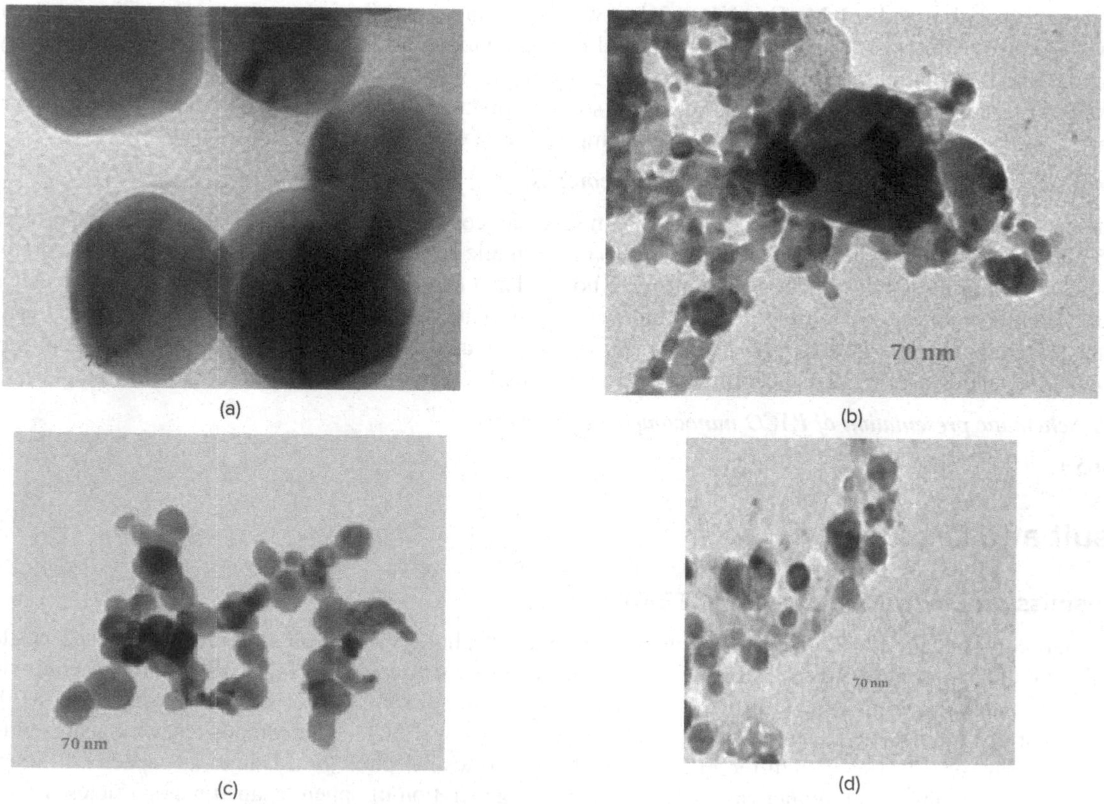

Fig. 65.2 TEM-images of polymer-based Polyaniline Copper Oxide (PACO) nanocomposites

3.2 Scanning Electron Microscopy (SEM)

Figure 65.3(a) shows an SEM study of Polyaniline-CuO nanocomposites. Polyaniline-CuO nanocomposites display CuO's monoclinic space group C2/c crystal structure with octahedral Cu^{2+} and O_2 sites. Figure 65.3(b) shows the PACO nanocomposites produced on the surface. 100 nm polymer-based PACO nanocomposites were examined by SEM. Figure 65.3(c) shows a polyaniline copper oxide nanocomposite. SEM shows that nanocomposite PACO particles are clean and white. SEM images of polymer-based Copper Oxide nanocomposites show uniform particle size. Figure 65.3(d) shows a more consistent size distribution when the concentration was medium. Lower particle agglomeration in the samples causes this.

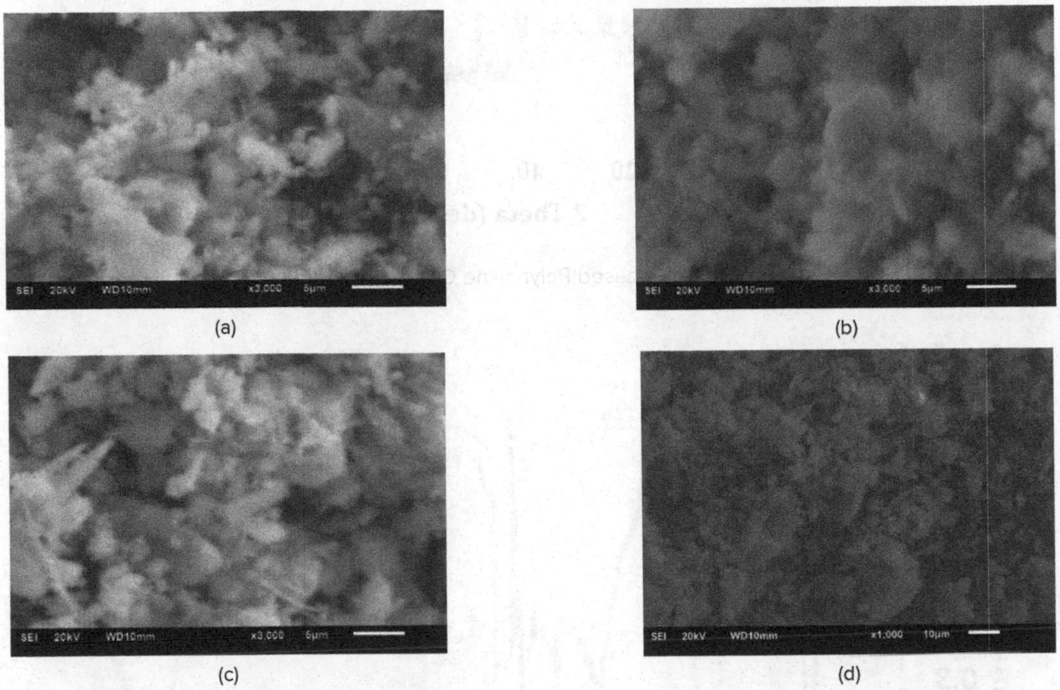

Fig. 65.3 SEM images of polymer-based Polyaniline Copper Oxide (PACO) nanocomposites.

3.3 X-ray Diffraction Studies (XRD)

Both metal nanoparticles and polymers may have their morphologies revealed by X-ray diffraction (XRD) imaging. Using distinctive peaks in XRD patterns, the polymer and metal nanoparticles may be isolated from one another. We may check the crystallinity of the nanocomposite using this method. The XRD patterns of PACO nanocomposites are shown in Fig. 65.4. Copper oxide contains two crystalline peaks at $2\theta = 36°$ and $38°$ which have been linked to CuO's crystal structure in the monoclinic space group C2/c. PACO nanocomposites feature two separate broad peaks at $2\theta = 50.50°$ and $73.23°$. The partly crystalline structure of the PACO nanocomposites was confirmed by analysis; the crystallite size was found to be 70 nm. X-ray diffraction (XRD) patterns for PACO nanocomposites may provide evidence of interfacial interactions between crystalline peaks of PACO and crystalline peaks of CuO.

3.4 Fourier Transform Infrared Spectroscopy (FTIR)

FTIR spectroscopy was used to record the locations of the absorption and emission peak positions for PACO polymer nanocomposites. The generated nanocomposites' IR spectra are shown in Fig. 65.5 and Table 65.1 and the findings demonstrate that they do not significantly alter the polymer matrix. Finally, using FTIR measurements, we verified that the intensity of the band at 750 cm^{-1} rises as the quantity of copper in the polymer nanocomposites increases. Aniline's N-H bond stretching vibration frequency is 3360 cm^{-1}, the primary amine's N-H bond stretching frequency is 1619 cm^{-1}, the primary amine's C-N stretching vibration is 1320 cm^{-1}, and the primary amine's C-H stretching vibration is 3027 cm^{-1}.

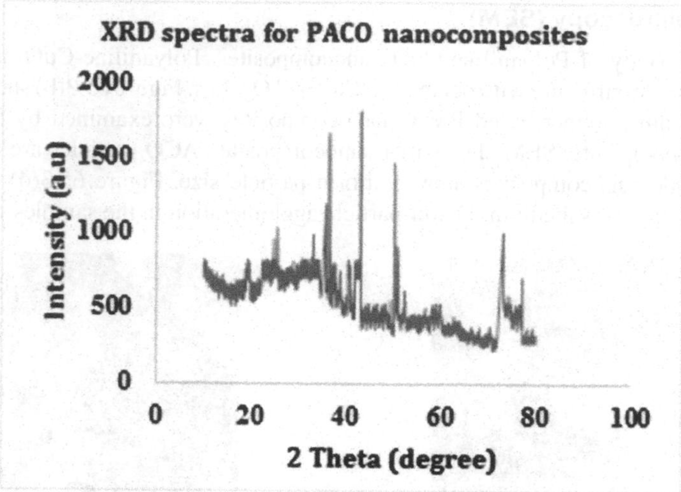

Fig. 65.4 XRD-Spectra of polymer-based Polyaniline Copper Oxide (PACO) nanocomposites

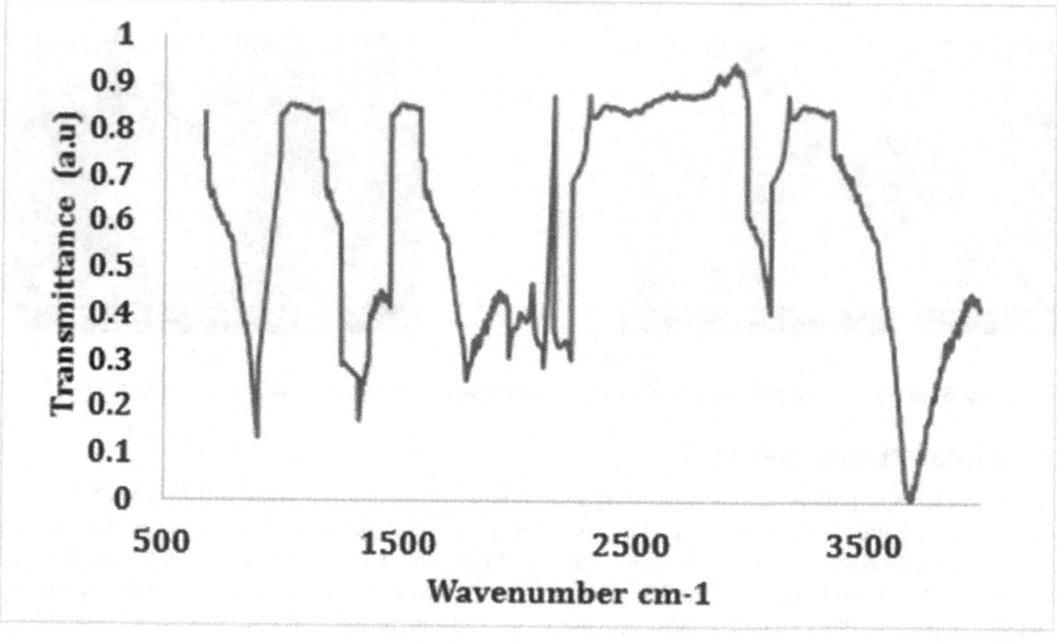

Fig. 65.5 FTIR spectra of PACO

Table 65.1 FTIR absorption characteristics of PACO nanocomposites

S.No	PACO Nanocomposite in cm^{-1}	Assignments
1.	750	Band of Copper nanoparticle
2.	3360	N-H Stretching vibration
3.	1619	N-H Stretching frequency
4.	1320	C-N Stretching vibration
5.	3027	C-H Stretching vibration

3.5 PACO Nanocomposites' Ability to act as Photocatalysts

It was found that the synthesis of PACO nanocomposites that displayed strong photocatalytic activity was brought about as a consequence of the photo-degradation of a concentration of 0.2 M dye utilizing potassium permanganate as the light source. Potassium permanganate is a common chemical that can be purchased at a reasonable price and is not difficult to get in research facilities. The potassium permanganate was first transferred to a Petri plate before being dissolved in the distilled water that had been used. After these two steps, the potassium permanganate was discarded. You should receive the following findings after making numerous samples of PACO nanocomposite in varied concentrations, adding them to a solution of ($KMnO_4$), and then exposing them to UV light sources on each of the samples: Fig. 65.6 is a representation of the connection that exists between the absorbance of $KMnO_4$ and UV-irradiation. This relationship is displayed as a function of the duration in minutes. (10, 20, 30,40,50,60,70,80,90,100, and 120 minutes). The sample's absorbance was evaluated after it had been annealed in PACO nanocomposites at 650°C. This finding indicated in Fig. 65.6 that the regularity and integrity of the PACO nanocomposite structure both significantly impact the pace at which photodegradation occurs. A lower number of photo-generated electron-hole pairs were created on the surface of the polymer nanocomposite catalyst, which has poor crystallinity and an imperfect crystal structure.

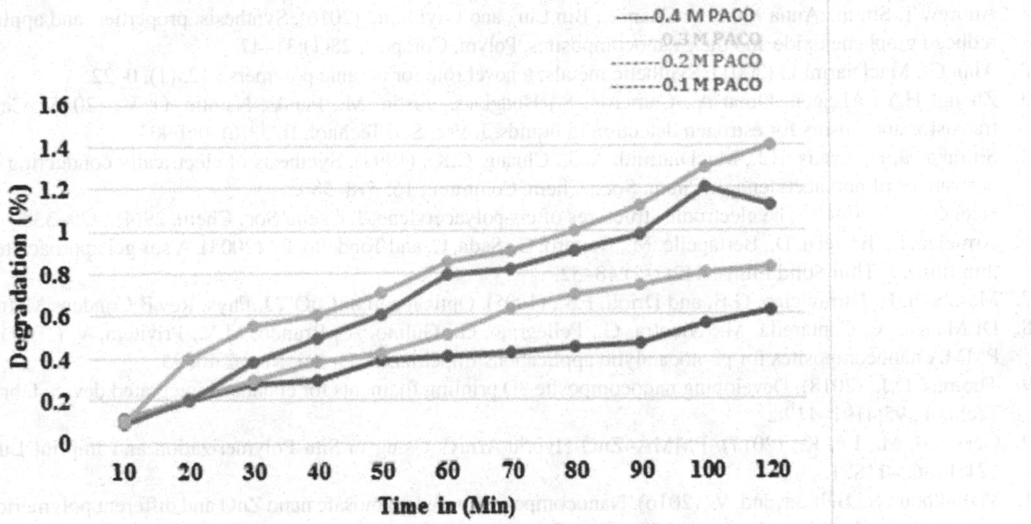

Fig. 65.6 Image of photocatalysis with respective time (Min)

Table 65.2 Values obtained from experiments about the use of PACO nanocomposites in photocatalysis

S.No	Time in (Min)	Con. 0.1 M	Con. 0.2 M	Con. 0.3 M	Con. 0.4 M
1.	10	0.12	0.11	0.09	0.1
2.	20	0.2	0.21	0.2	0.4
3.	30	0.28	0.3	0.38	0.55
4.	40	0.3	0.38	0.49	0.6
5.	50	0.39	0.42	0.6	0.71
6.	60	0.41	0.51	0.79	0.85
7.	70	0.42	0.63	0.82	0.9
8.	80	0.45	0.7	0.9	1
9.	90	0.47	0.74	0.98	1.12
10.	100	0.56	0.8	1.2	1.3
11.	120	0.62	0.83	1.12	1.4

4. Conclusion

The type of particles to be used and their qualities determine the many pathways that may be taken to build nano-polymer composites with a high degree of homogeneity. The present study showed that it is possible to make polymer nanocomposite materials based on CuO nanocomposites using a Sol-Gel process that is both straightforward and economical. This result was accomplished by using a heterogeneous assortment of methods in concert with one another. Through the use of TEM and SEM, this research was able to determine that the morphology and topology of PACO nanocomposites with a size of 70 nm are the most important criteria to consider. The technique of X-ray diffraction was used so that researchers could prove that nanocomposites had a crystal structure in the monoclinic space group C2/c. The use of FTIR enabled the confirmation that the chemical grafting of the PACO nanocomposite to the polymer was successful. This was made possible thanks to technology. According to the results obtained from the photocatalytic activity, there is sufficient absorption for 0.4 M PACO nanocomposites for ten minutes. In addition to this, it was shown to possess photocatalytic activity, which qualified it as a suitable component material for a semiconducting device.

REFERENCES

1. Andrew T. Smith., Anna Marie LaChance., Bin Liu., and Luyi Sun., (2016), Synthesis, properties, and applications of graphene oxide/ reduced graphene oxide and their nanocomposites, Polym. Compos., 28(1):31–47.
2. Alan G., MacDiarmid., (2001). Synthetic metals: a novel role for organic polymers., 125(1), 0–22.
3. Zheng, H.Y., Alsager., Omar A., Cameron. S., Hodgkiss., Justin, M., Plank., Natalie, O. V., (2015). Carbon nanotube field effect transistor apt sensors for estrogen detection in liquids. J. Vac. Sci. Technol. B. 33(6), 06F904.
4. Shirakawa, H., Louis, E.J., MacDiarmid, A.G., Chiang, C.K., (1997). Synthesis of electrically conducting organic polymers: halogen derivatives of polyacetylene, J. Chem. Soc., Chem. Commun., 16: 578–580.
5. Heeger. A.J., (1997). The electronic structures of cis-polyacetylene, J. Chem. Soc. Chem. 29(4): 329–333.
6. Armelao, L., Barreca, D., Bertapelle, M., Bottaro, G., Sada, C. and Tondello, E., (2003). A sol-gel approach to monophasic copper oxide thin films. J. Thin Solid Films. 442(1-2):48–52.
7. Marabelli, F., Parraviciny, G.B. and Drioli, F.S., (1995). Optical gap of CuO", J. Phys. Rev.B Condens Matter., 52(3):1433–1436.
8. Di Mauro, A., Cantarella, M., Nicotra, G., Pellegrino, G., Gulino, A., Brundo, M.V., Privitera, V. (2017). Novel synthesis of ZnO/ PMMA nanocomposites for photocatalytic applications Impellizzeri. G. Sci. Rep. 7:40895.
9. Thomas, D.J., (2018). Developing nanocomposite 3D printing filaments for enhanced integrated device fabrication. Int. J. Adv. Manuf. Technol., 95:4191–4198.
10. Gervasio, M., Lu, K., (2017). PMMA–ZnO Hybrid Arrays Using in Situ Polymerization and Imprint Lithography. J. Phys. Chem. 121:11862–11871.
11. Mallakpour, S., Behranvand, V., (2016). Nanocomposites based on biosafe nano ZnO and different polymeric matrixes for antibacterial, optical, thermal, and mechanical applications. Eur. Polym. J. 84: 377–403.
12. Loste, J.; Lopez-Cuesta, J.-M.; Billon, L.; Garay, H.; Save, M. Prog. (2019). Transparent polymer nanocomposites: An overview of their synthesis and advanced properties. Prog. Polym. Sci. 89: 133–158.
13. Wang, X., Cheng, X., Yu, X., Quan, X., (2018). Study on Surface-Enhanced Raman Scattering Substrate Based on Titanium Oxide Nanorods Coated with Gold Nanoparticles. J. Nanotech. 2018: 1–9.

Impending Inquisitions in Humanities and Sciences (ICIIHS-2022) – Dr. Mohan Varkolu et al. (eds)

A Review on Strategic Development of Gastroretentive Drug Delivery Systems and its Challenges in Providing Better Treatment Options

66

Sushma Desai*
GITAM School of Pharmacy, GITAM deemed to be University, Rudraram, Hyderabad, India

Kumar Shiva Gubbiyappa, Naveen Amgoth, Baru Chandrasekhar Rao, Aenugu Jyothi Reddy, T. Rajithasree
Chilkur Balaji College of Pharmacy, Moinabad, Hyderabad, India

ABSTRACT Every year millions of people report for suffering with gastro related health issues and undergo diagnosis for it. Creating demand for novel health systems with oral as preferred route of administration. Drug absorption as Prime requisite dosage forms designed for regions in gastric and intestinal. Utmost drugs get absorbed in gastric region and upper part of small intestine. Oral conventional dosage forms (tablets, capsules, pellets, oral liquids) found to have total transit time of 5.8 hours. Rapid gastrointestinal transit time leads to incomplete drug absorption & efficacy. Filling the gap for the above conditions gastroretentive drug delivery systems serve as a best suitable treatment options for patients. Gastro retentive dosage forms market share expected to reach USD 19.7 billion by 2028. Designing of gastric retention delivery systems (GRDDS) involves thorough knowledge of 1) Anatomical & physiological properties of gastrointestinal tract (GIT). 2) physico-chemical properties. 3) Biochemical processes. Various technologies developed for gastroretentive dosage forms such as A) swelling & expanding systems. B) Bioadhesive systems. C) Raft formers. D) High density delivery systems. Compelling part of GRDDS are polymers of various origin such as natural, synthetic and semi-synthetic are as Stallers of release rate, buoyancy enhancers, inert fatty materials and hydrocolloids. Continuous efforts are put to develop these dosage forms for giving better treatment to patients.

KEYWORDS Gastroretentive drug delivery systems, Raft formers, Bioadhesive systems

1. Introduction

Gastroretentive type of drug delivery systems stands as the commercially successful novel oral pharmaceuticals developed as drug release forestallers (Elsherbeeny W et al. 2022). The challenge to develop into an efficient drug delivery system began in 20th century detailed by Warren & Marshall with Helicobacter Pylori (bacterium responsible for peptic ulcer).

Primary site of absorption after Ingestion (food, nutraceuticals and drugs) through oral route is stomach. The main principle involved in gastroretentive systems is drug retaining within the delivery system prolonging its residence time giving scope for maximum absorption there by achieving target site for localized and overall systemic effect (Rajendra Awasti et al. 2014).

Here we discuss the things in interest of drug to formulate into gastro retentive drug delivery systems as follows:

(a) Should have primary absorption site as stomach.
(b) Drugs having insolubility in alkaline pH conditions.

*Corresponding author: d.sushmapharma@gmail.com

DOI: 10.1201/9781003489436-66

(c) NAW (narrow absorption window) Drugs.
(d) Colon degrading drugs.
(e) drugs which interfere with inhabitant colonic microbes.
(f) Drugs possessing rapid git absorption.
(g) Drugs with targeted organ as git (gastro-intestinal tract) itself.

1.1 Anatomical and Physiological Conditions of Stomach

Average length: 20cm

Diameter: 15cm

Absorption mechanism: passive diffusion and connective transport.

Transit time of food: 0.25-3hr.

The performance of the drug delivery system depends on during which particular phase that is going on referred to as MMC (migrating motor complex) that is responsible for the cyclic events of motility in stomach Hence the fate of drug dependent on the time of administration of phase is detailed below.

Table 66.1 Phase of event occurring in Migrating Motor Complex (MMC)

S.NO	PHASE	TIME SPAN (min)	EVENTS
1.	I	30-60	Referred as quiescent period due to lack of secretory, electrical & contractions.
2.	II	20-40	Bile secretion and mucus discharge starts with contractions.
3.	III	10-20	Increased mucus discharge with increased intensity of contractile force responsible for sweeping undigested food.
4.	IV	0-5	Referred as transition period.

Source: https://pubmed.ncbi.nlm.nih.gov

1.2 Factors Affecting Gastric Retention

1. Density <1.004 g/ml (showing good floating in gastric fluids).
2. Dosage size >7.5mm (size specifications to delay its passing through the antrum part of stomach).
3. Shape preferably tetrahedron (complex shapes can prolong the drug release).
4. Simple or multiple unit formulations (suitable for multiple dosage regimens).
5. Patient fed/unfed state to assess the gastric residence time (in fed state it is short and unfed state is long).
6. Nature of meal consumed by the patient (fatty acids or indigestible polymer interfering with stomach motility pattern delaying gastric emptying.

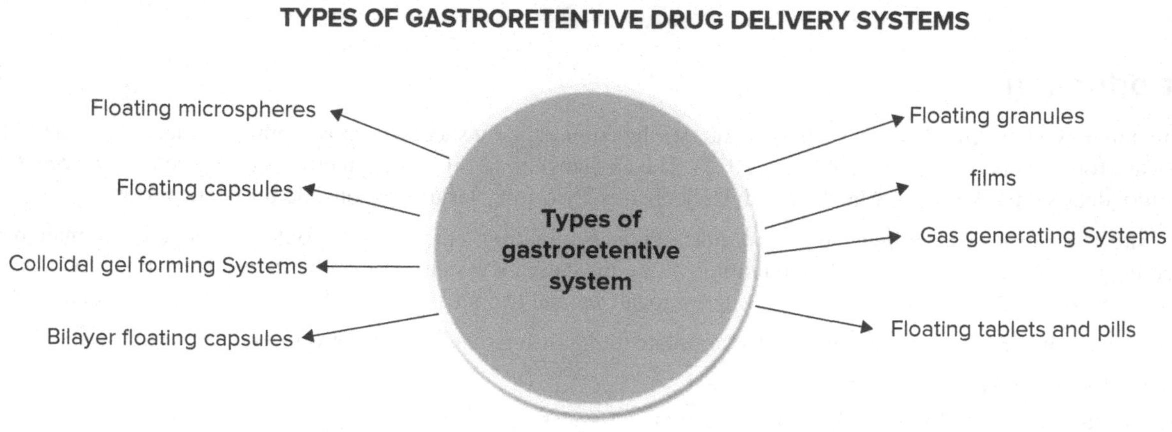

Fig. 66.1 Gastroretentive dosage forms

Source: https://www.itmedicalteam.pl

Polymers: polymers applicability are providing controlled release of drug can be of diffusion controlled, solvent controlled or chemically controlled systems to increase the residence time in gastric region explained as below (Jagtap. et al., 2012):

- Capsule shell made of controlled release polymers.
- Heterogeneous dispersion of drug particles made with biodegradable or non-biodegradable polymers.
- Polymer coating as laminate of drug molecules.
- Liquid – liquid encapsulation of the drug.
- Low density polymer coating micropellets.
- Chemical bonding between polymer and drug
- Formation of polymer-drug macromolecular structures.

Polymers being the core of these dosages can be categorized depending upon their nature and also their category.

Natural polymers: Natural polymers selected for incorporation in grdds for their beneficial in availability, extraction, eco-friendly, compatible to gastric conditions and safe and combination of hydrophobic and hydrophilic polymers are used & a lot of research required in establishing them over the synthetic polymers.

Examples of natural polymers having swelling, controlled effect are locust bean gum,mimosa pudica gum, guar gum, psyllium husk, tamarind gum, colacasia esculenta gum, limonia acidissima gum, karaya gum, xanthum gum, tara gum, okra gum, carrageenan, chitosan, pectin, peanut husk powder etc. and with biodegradable nature type polysaccharides are alginates, arabinogalactan, arabinoxylan, galactomnam, chondroitin sulphate etc.

With the supportive role of polymers apart from controlling drug release it provides protection and stability to the dosage form.

Semi synthetic polymers: these are chemical variants of natural polymers like carboxy methy ethyl cellulose, sodium alginate.

Synthetic polymers: examples of lactic acid derivatives are macrogol, polylactide etc. and acrylic acid derivatives are eudragit and carbopol.

Effervescent agents: sodium carbonate, citric acid, tartaric acid,

Grdds specifically designed to retain in stomach for prolonged time thereby improving drug bioavailability, reducing drug going waste and in view of patient benefits (pocket friendly, reduced frequency of the drug regimen prevalent in conventional dosage forms) with pharmacokinetic & pharmacodynamics advantages. Drugs passing the complex gastrointestinal tract reaching target site with appropriate concentrations is the achievement of therapeutic action. Many antibiotics therapy developed successfully for treatment of disorders like gastroenteritis, Cholecystitis, typhoid fever, perirectal abscess, peritonitis etc.

Brief scenario of stomach: the primary site of absorption of drug is low (in stomach) compared to the entire gastrointestinal tract. Stomach having capacity of 0.5-1 litre capacity with contents as pepsinogen, mucus and hydrochloric acid. pH environment is 1-3and varies (after heavy meal).irrespective of the fed state gastric emptying occurs. Few stomach specific biomarkers are cagA, vac A, babA2, IL-1, PG-1, Ig-G etc.

Listing the physical and the physiological factors affecting drug absorption across the gastrointestinal tract are as follows:

Physical factors:

- Molecular size,
- Partition coefficient,
- type of delivery system
- Chemical degradation,
- Dissolution rate,
- Solubility.

Physiological factors:

- Gastric and intestinal transit,
- Regional pH,
- Complexation,
- Hepatic and luminal metabolism,
- Intestinal permeability.

Common gastrointestinal related diseases are grouped as follows:

Hepatic diseases:

- Amoebic liver abscess
- Biliary strictures
- Ascites,
- Liver cirrhosis
- Cholecystitis

Gastric diseases:

- Carcinoma
- Gastric erosion
- Oesophagitis
- Gastric ulcer
- Peptic ulcer

Orogenital infections:

- Pelvic inflammatory disease

Lympho-reticular system disease:

- Hodgkin's disease.

Large intestine disease:

- Appendicitis

Diagnosis: In treatment of gastrointestinal related disorders the diagnosis plays a major role in viewing internal organs and tissues,bile ducts and assess the blood flow to it, food residence time and its emptying time, manometric tests to measure the electrical & stomach muscular activity related issues are listed as follows:

- Chromoendoscopy
- Liver scan
- Colonoscopy,
- Peroral cholangioscopy
- Sigmoidoscopy
- Anoscopy
- Pancreas scan
- Wireless esophageal pH TEST (Bravo test)
- Ultrasound endoscopy
- MRI scan
- CT scan
- Esophageal /gastric manometry
- Lab tests (fecal occult test, stool culture)

Evaluation: To know the formulation performance and the patient physiological factors that affect the gastric retention time, evaluation tests are given in table as follows:

2. Conclusion

In this review article we put efforts as start way in better understanding the challenges of the gastrointestinal system its common ailments and diagnosis for effective treatment thereby developing the dosage forms that can be successfully given by this oral route targeting desired organ of choice overcoming maximum drawbacks of conventional dosage forms.

Table 66.2 Various strategic developments of gastroretentive drug delivery systems mechanisms briefed as follows:

S.NO	Type of system	Mechanism
1.	Floating/lower density dosage forms	Floating/lower density type dosage forms also known as hydro-dynamically balanced system are developed as low density formulations remain floating on gastric fluids controlling their drug kinetics and thereby provide controlled drug release (Rumpa. B et al., 2021).
	• **Effervescence generating drug delivery systems:**	These systems provide buoyancy in gastric contents by liberation of co2 gas using swellable polymers.
	1. Gas evolving type of dosages:	Principle involved is the reaction between the carbonate salts and citric acid to release co2 and achieve buoyancy in less than 10 min with floating time over 5-6 hrs. (Ichikawa et al.).
	2. Volatile type liquid dosages:	These systems are designed to have a two compartment with drug in first and liquid in second part designed to evaporate at body temperature as inflatable chamber forming a reversible deformable unit i.e., converts from collapsible to expand state and vice versa required for ejection from stomach (Michaels et. al.).
	• **No effervescence generating drug delivery systems:**	Systems developed using polymers of wide range characteristic in providing gastric retention by mechanisms of swelling and bioadhesion to gastric mucosa.
	1. Colloidal Gel Barrier type formers:	These systems consist of more than one gel forming polymers of varied type. These polymers hydrate in presence of gastric fluids forming a colloidal gel barrier (Sheth & Tossounian).
	2. Bilayer tablets with Floating property:	These are developed giving dual release pattern as first layer giving faster effect and second layer as controlled release by forming colloidal gel from gastric fluid absorption (Sheth & Tossounian).
	3. Micro-porous dosages:	These systems consist of encapsulated drug reservoir within its microporous chamber having apertures by which gastric fluids enters, drug dissolves & eventually drug absorption occurs across intestinal tract by transport carriers (Yuasa et al.).
	4. Floating Beads/Alginate solid Beads:	These are the spherical beads formed by the cross linkage between alginate polymer and calcium chloride measuring in rage of 2-5mm diameter. In this calcium alginate is precipitated out from sodium alginate and calcium chloride solution by freezing in liquid nitrogen at −40°C for 24 hours. The prepared beads are superior to the solid beads with 12 hr retention time (Whitehead et al.).
	5. Microballoons/Hollow Microspheres:	Preparation of these dosage forms require unique technique emulsion solvent diffusion method with polymer solution and the gas phase is the result of removal of organic solvent by evaporation forming dispersed particle i.e., forming internal cavity in microsphere (Roy et al.).
2.	**Bioadhesive systems**	bioadhesive/mucoadhesive drug delivery systems developed to provide poorly soluble drug with improved bioavailability and stability bypassing first pass effect. In this bioadhesion occurs by three mechanisms **1. Hydrated mediated adhesion:** hydrophilic polymers absorb water and stick to the stomach wall giving prolonged release. **2. Bonding mediated adhesion:** physical & chemical bonds formed by the deposition of adhesion material in the folds of mucus and hydrogen bonds respectively. **3. Receptor triggering adhesion:** A strong adhesive polymer with this special ability directly binds to targets cell receptor on surface (Park & Robinson).
3.	**Raft forming systems**	These are specially designed to counter the gastric reflux caused due to acidity and infections where a thick gel like formed on surface of gastric contents also called raft and floats for a quiet time providing desired prolonged effect.
4.	**Swelling and expandable systems**	These also called as 'plug type system' upon ingestion reaching stomach gets swelled in stomach contents unfolding and expand approximately 12-18 mm cannot easily pass through the pylorus hence a controlled action obtained giving maximum time for absorption (Kaewroek. K. et. al., 2022).
5.	**Magnetic systems**	Magnetic type of grdds formulated in as magnetic materials in small gastroretentive capsules which are retained in stomach by external application of magnet exactly on the bod surface of stomach region and is continuing research in improving its convenience.
6.	**High density delivery systems:**	High density system possessing density 3g/cm3 and remain settled to the stomach bottom providing controlled release of the drug resisting peristaltic movements in stomach (Deshpande et al.).

Source: https://www.tandfonline.com

Table 66.3 Evaluation factors affecting the gastroretentive formulations

S.No.	Evaluation Factors	Parameters
1.	Formulation factors	i. Diameter, density, shape and flexibility of matrices ii. Floating time iii. Content uniformity iv. Hardness v. Friability vi. Dissolution.
2.	Physiological factors	I. Age II. Sex III. Posture IV. Food V. Bioadhesion.
3.	In vivo evaluation tests	1. Radiology 2. Scintigraphy 3. Gastroscopy 4. Magnetic marker monitoring 5. Ultrasonography 6. ^{13}c octanoic breathe test.

Source: https://www.ncbi.nlm.nih.gov

REFERENCES

1. Elsherbeeny W, El-Gogary R, Nasr M, Sammour. (2014). Current progress of oral site specific dosage forms: Emphasis on gastroretentive drug delivery systems. *Archives of Pharmaceutical Sciences Ain Shams University*, Dec 1; 6(2); 221–38.
2. Awasthi R, Kulkari GT. (2016). Decades of research in drug targeting to the upper gastrointestinal tract using gastroretention technologies: where do we stand drug delivery? Feb 12; 23(2): 378–94.
3. Rajamane A, Trivedi R, Nandgude T. (2022). A Novel approach to Enhance Gastric Retention for better Therapeutic Activity: Gastro Retentive Drug Delivery System. *Research Journal of Pharmacy and Technology*. Jul 29; 15(7): 3324–30.
4. Badoni A, Ojha A, Gnanarajan G, Kothiyal P. (2006). Review on gastro retentive drug delivery system. *The pharma innovation*. Oct 1; 1(8, Part A): 32.
5. Streubel A, Siepmann J, Bodmeier R. Gatroretentive drug delivery systems. (2006). *Expert opinion on drug delivery*. Mar 1; 3(2): 217–33.
6. Makwana A, Sameja K, Parekh H, Pandya Y. (2012). Advancements in controlled release gastroretentive drug delivery system: A review. *Journal of Drug Delivery and Therapeutics*. May 14; 2(3).
7. Lopes CM, Bettencourt C, Rossi A, Buttini F, Barata P. (2016). Overview on gastroretentive drug delivery systems for improving drug bioavailability. *International journal of pharmaceutics*. Aug 20; 510(1):144–58.
8. Madal UK, Chatterjee B, Senjoti FG. (2016). Gastro-retentive drug delivery systems and their in vivo success: A recent update. *Asian journal of pharmaceutical sciences*. Oct 1; 11(5): 575–84.
9. Khan R. Gastroretentive drug delivery system-a review. (2013). *Int J Pharm Bio Sci*. 4(2)630–46.
10. Kalla U, Gohil P, Jain H, Meshram DB. (2022). Micro balloons: As a gastro retentive drug delivery system. *Gradiva Review-Journal*. 8 (4):341–7.
11. Saini S, Asija, Goyal A.(2022). Floating microsphere as gastroretentive drug delivery system: an updated review. *Tropical Journal of Pharmaceutical and Life Sciences*. Apr 27; 9(2): 21–9.
12. Nurhalifa N, Sundawan PD, Veronita SC, Destria SI, Nuryamah S, Yuniarsih N. (2022). Literature Review Article: drug delivery system held in stomach (gastroretentive). *Journal of social research*. Dec 10; 2(1)912–9.
13. Andrew A. (2022). A review on raft forming drug delivery system-Mechanism and its significance. Australasian Medical Journal. 15(2): 336–7.
14. Zanke AA, Gangurde HH, Ghonge AB, Chavan PS. (2022). Recent advance in gastroretentive drug delivery system (GRDDS). *Asian Journal of Pharmaceutical Research*. 12(2):143–9.
15. Jagtap, Y.M., Bhujbal, R.K. Ranade, A.N. and Ranpise, N.S. (2012). Effect of various polymers concentrations on physiochemical properties of floating microspheres. *Indian Journal of Pharmaceutical Sciences*, 74(6).p.512.
16. Kumar A and Srivastava. R. (2021). In vitro in vivo studies on floating microspheres for gastroretentive drug delivery system; a review. *Asian Journal of Pharmaceutical and Clinical Research*. pp. 13–26.

16. Birajdar. AA, Deshmukh, M.T and Shete. R.V. (2021). A review on gastro-retentive floating microspheres. *Journal of Drug Delivery and Therapeutics*. 11(1-s). pp. 131–26.
17. Rumpa, Tanmay. M, Sujit. D and Suhasis B. (2021). Recent advances in the development of floating microspheres for the treatment of hypertension.
18. Agarwal. S, Thakur, A and Sharma. A. (2022). Development and evaluation of ketoprofen loaded floating microspheres for sustained delivery. *Materials Today Proceedings*.

Impending Inquisitions in Humanities and Sciences (ICIIHS-2022) – Dr. Mohan Varkolu et al. (eds)
 ISBN 978-1-032-78829-6

Novel Enzyme by Prodrug Therapy for Treating Cancer—Deadly Disease

67

Sushma Desai*
Gitam school of pharmacy, Rudraram, Hyderabad, India
Ayesha Sara, Pally Mahesh, Baru Chandrasekhar Rao, Anumula Rama Rao, Srujana Gandrathi
Chilkur Balaji College of Pharmacy, Moinabad, Hyderabad, India

ABSTRACT Since its development in cancer chemotherapy in 1980, investigational research in developing pro-drug therapies for cancer treatment has been progressing. Where 1-9 million people worldwide with more than 200 types of cancer are diagnosed every year and survival chances are found to be very slim (5.4%). The variant cancers like leukaemia, carcinoma, lymphoma, myeloma, brain, spinal cord, and sarcoma are found to be known as extensive cancers. Cancers require novel targeted therapy for their treatment to overcome disadvantages that are evident from conventional chemotherapy. This novel enzyme pro-drug is an investigational therapy receiving promising results found in Ovarian, brain, spinal cord, keratinocyte and urological cancers in facing calamitous conditions are observed in conventional chemotherapies. In this article, we discuss the comparative profile of typical conventional and non-conventional chemotherapy and in-detailed information about various enzyme pro-drug therapies available (its ongoing research and the mechanism of action) in giving effective treatment.

KEYWORDS Enzyme, Pro-drug, Cancer

1. Introduction

Yearly 9-10 million of deaths are cognate to cancer. The existing conventional therapies have prolonged & extensive side effects. One of which is the destruction of healthy cells along with cancer cells which instigates serious side effects. To deal with these calamitous problems of non-conventional therapies, advanced enzyme prodrug therapies came into existence. This only targets tumor cells and no harm to healthy cells avoiding ominous effects. Hence a demand arisen for advanced targeted therapies for giving better treatment options and gaining patience compliance.

CANCER: The abnormal cell growth which escalates to nearby tissues resulting anomalous condition called cancer which is life-threatening if untreated in early stages. The main route through which the cancer spreads is Lymphatic system or Blood stream.

Lymphatic System: To lymph nodes in groin, neck, under arm regions

Blood Stream: To brain, lungs, bones, liver. (Cancer.Net Editorial Board, Aug, 2019).

Cancers are available in centuplicate types. Few cancers which are in investigational trials are briefly discussed in the given table below. (Cancer.Net Editorial Board, Aug, 2019).

*Corresponding author: d.sushmapharma@gmail.com

DOI: 10.1201/9781003489436-67

Table 67.1 Types of cancers

S. No.	Disease	Organs Effected	Symptoms	Age Group
1.	**Leukaemia:** Leukemia emerges when there are possible genetic mutations in DNA leading to an anomaly rise in white blood cells. It effects bone marrow and blood	Bone Marrow, Blood spreads to liver, Spleen, CNS, lymph nodes	Recurring infections, fatigue, fever and chill.	**>65 Years**
2.	**Lymphoma:** effects the lymphatic system when the abnormal growth of white blood cells is non-fatal and turns to cancer hence, effecting the immune system.	Lymph nodes, bone marrow, spleen, thymus gland	Fatigue, cachexia, swelling of lymph nodes, sudden bleeding.	**(15-39 yrs) Teens and Adults >75 years**
3.	**Myeloma:** The abnormal growth of carcinogenic cells force out the normal cells in the bone marrow i.e. plasma cells, which makes anti bodies and effects immune system	Immune system, bones, WBC, RBC count, kidneys	Anorchia, delirium, osteomyelitis, fever, nausea, cachexia, frequent infection.	**>60 years**
4.	**Sarcoma:** Cancer which begins in the bone and soft tissue and spreads to other various body structures.	Muscles, blood vessels, tendons, nerves, bones, joints, fat	Pain associate with bones, Weak bones causing minor, Fractures, abdominal pain.	**>50 years**
5.	**Carcinoma Cancer:** Cancer which effects the epithelial tissue lining of internal pathway in body and skin due to mutations in DNA , which instructs cells to perform specific functions	Skin, colour, stomach, lungs, rectums, pancreas etc.	Fatigue, changes in skin, abnormal, weight changes, development of lumps, or thick ness in area, bursting and blooding.	**177.5/100,000 men and 128-7/100,000 women per year**
6.	**Brain & spinal cord cancer:** tumor originating from brain and spreading to other parts known as primary tumor (Gilomas) and tumor originating from another part of body and spreads to brain is known as secondary tumor (metastatic) is brain &spinal cord cancer	Brain and spinal cord	Seizures, anorexia, bipolar disorder, anxlety elusory, vertigo, nausea vomiting's frequently, hearing, speech, vision problem, and headache: - usually go away after the vomit, generalized weakness.	**Child, adult >65 years**

Source: https://pubmed.ncbi.nlm.nih.gov , https://www.cancer.gov/

CAUSES OF CANCER: Cancer is characterized as multi stage process of transformation of normal cells to malicious tumours cells takes place and the risk factors triggering them are given as follows:

1. Physical: ionization radiations and ultraviolet radiations.
2. Biological: infections from virus, parasites, bacteria
   Ex: hepatitis B and C viruses, HPV virus causes liver and cervical cancers.
   HIV: increases risk of cervical and Kaposi sarcoma.
3. Chemical: smoking tobacco, alcohol drinking, asbestos, arsenic and aflatoxins.
4. Age relates factors: characterized by impaired cellular repair mechanisms as body ages. (WHO, 2022, 3 Feb)

Effect on Life Span of Humans

According to W.H.O. the average life expectancy in a healthy individual is 73.4 yrs. The patients suffering from cancer has shorter "TELOMERES" than normal person of same age. This maybe result of intense regime of chemotherapy & radiation therapy. (Ananya Mandal (2017)).

Table 67.2 No of Cancer cases reported and deaths reported. (WHO, 2022, 3 Feb)

S.no	Cancer's reported	Reported cancer death
1.	Stomach cancer (1.09 million)	Stomach cancer (769,000)
2.	Breast cancer (2.26 million)	Breast cancer (685,000)
3.	Skin cancer (1.20 million)	Skin cancer (767,000)
4.	Lung cancers (2.21 million)	Lung cancers (1.80 million)
5.	Prostate cancer (1.41 Million)	Prostate cancer (375,000)
6.	Colon & rectum (1-93 Million))	Colon & rectum (916,00)

Source: https://www.who.int/

Diagnosis of cancer: The symptoms of cancer might be similar to symptoms of other infections so, Cancer diagnosis require a thorough screening of patient's history by genetic screening, lab test, tumor biopsy, endoscopy and imaging listed in table below

Table 67.3 Diagnosis of cancer (Stanford Medicine)

BIOPSY	IMAGING	ENDOSCOPY	LAB TEST'S	GENETIC SCREENING
Skin Biopsy	X-ray	Cystoscopy	CBC (Complete Blood picture)	Tumour DNA sequencing
Bone marrow Biopsy	Bone scan	Colonoscopy	Urinalysis	
Endoscopic Biopsy	CT Scan	ERCP	Tumour marker	
Excisional or incisional biopsy	Mammogram	Upper Endoscopy (EGD)		
Fine needle aspiration Biopsy	Lymphangiogram	Sigmoidoscopy		
Shave Biopsy	Ultrasound MRI			
Punch Biopsy				

Source: https://med.stanford.edu

Conventional Therapies and their Mechanism

1. **Chemotherapy:** A drug regime is followed. Anti-cancer drugs are given either alone or in combination. The inhibition of synthesis of proteins and nucleic acids (both RNA & DNA) by affecting function of neoplastic cells and macromolecules by chemotherapy agents leads to improper functioning of molecules (whole body is exposed to chemotherapy effecting healthy cells also) (Muhammed T, et at. Feb.27.2003)
2. **Surgery:** The mass cells (tumor) and few surrounding tissues are removed. Surgery is also done to decrease the effects of side effects due to tumors.
3. **Radiation:** in this X-rays are used to quell tumor cells.
   A) X-rays: Given from outside of the body (external beam)
   B) Radioactive seeds: Placed near tumor. Route of administration is IV or pill and delivers radiation inside the body (internal beam)

   Radiations help in breaking of DNA into small segments in the cell preventing cancer cells from growing and resulting in death of the cells.

   This is a planned therapy so only the affected part is focused but it does also affect the healthy cells (Abshire, et al., March.01.2000).

Non-Conventional Chemo Therapy

1. **Immunotherapy:** this therapy helps in energizing the immune system to get rid of cancer cells by slowing down its division rate & growth. This effect is totally dependent on immune system so this therapy doesn't work for everyone. The immunotherapy drugs also called as immune checkpoint works by inhibiting the binding of checkpoints in the proteins with concerned proteins. This then prevents (off signals) sending and resulting in death of cancer cells by T cells. (National Cancer Institute, April, 2022)

   Ex: CTLA-4 protein, PD-1 protein and its partner proteins PD-L1 inhibited by immune checkpoint inhibitors.
2. **Hormonal therapy:** in this therapy body's natural hormone production is decreased or stopped using drugs to stop the division of cancer cells .ex: breast, prostate, ovarian cancer. The drugs involves in disruption of mechanism of abnormal hormonal production in the body which inhibit the growth of cancer cells. (National Cancer Institute, April, 2022)
3. **TARGETED THERAPY:** This therapy involves targetting of specific cancer cells without backing out normal cells. A drug is designed to target the protein cells, preventing further growth of cancer.

ENZYME PRO DRUG THERAPY: In this the delivery of drug activates the enzyme gene or the particular functional protein to targeted tissue then a prodrug follows resulting in destruction of cancer cells (Bagshawe, et at., 1994)

PRODRUG: prodrug (which is a pharmacologically inactive substance) transforms through enzymatic reactions into a pharmacologically active drug in the body.

Table 67.4 Conventional and non-conventional therapies. (National Cancer Institute, 2019)

THERAPY	ADVANTAGES	DISADVANTAGES	SIDE EFFECTS	SUCCESS RATE	COST
Surgery	Solid tumors can be removed completely	Along with tumor surrounding lymph nodes and healthy tissue are also moved or damaged	infection and pain at the site of injury, blood clots & risk of bleeding, reactions due to anesthesia& other drugs at surgery time.	30%, morality rate is higher in men's than women's	2,80,000 INR (4,000 USD)- 10,50,000 INR (15,000 USD)
CHEMO THERAPY	Shrink cancer and slow down its growth, chemotherapy after surgery and other therapies is effective in eradication cancer completely,	Damages healthy cells, unpleasant and long lasting side effects.	Alopecia, lethargy, xerosis, mucositis (40%), constipation and diarrhea (44%), anorexia, bleeding and brusing, cachexia, anaemia, Breathing problems, amnesia, insomnia, sex & fertility issues (45% males & 50% females), nausea & vomiting (80%), nervepain-CIPN (30-40%), depression.	1. Breast cancer-26% at stage 4 with 5 years survival rate 2. Bladder cancer-5yrs survival rate is 77% 3. Colorectal cancer - 68% in stage 3 & 12% in stage 4. 4. lung cancer - 19% 5. Prostate & testicular cancer - upto 99% with surgery.	1,000 USD - 48,000 USD
RADIATION THERAPY	Prevent cancer from recurrence, palliative treatment, slow or stop the growth of cancer.	Healthy cells are also effected	1. alopecia 2. edema 3. changes in skin 4. infertility problems	90% can be achieved in early stages	60,000 INR (772 USD) - 2,25,000 INR (2895 USD)
IMMUNOTHERAPY	Cancer is less likely to return, tumor size can be decreased.	Negative reactions may occur, harm other organs, working is limited to certain people, body can become resistance to these drugs	rev up with immune system and causes: 1. flu, fever, fatigue & chills 2. Edema 3. Weight gain 4. Diarrhea 5. Heart palpitations.	15%–50% after therapy	1-1.5 lacs INR\ session
HORMONAL THERAPY	Decline the cancer division and reduce the chances of recurrence, mitigate cancer symptoms	Long term use may show certain side effects	1. Visceral fat growth 2. dementia 3. Fatigue 4. Hot flashes 5. Low sex drives 6. Weaker bones	43%–88%	52,000 INR (673 USD)
TARGETED THERAPY	Necrosis of cancer cells, push immune system to fight infection, there are no serious side effects reported	Cells can become resistant in very rare cases	There are fewer side effects: fatigue, rashes, proteinuria etc.	Upto 95%	INR 2-3 lacs per session

Source: https://www.cancer.gov/

Enzyme pro drug therapy involves three consecutive stages and repeated as sessions explained as flowchart (see Fig. 67.1):

Enzyme Prodrug Therapies and Their Mechanisms

ADEPT Technology: ADEPT principle involves antibody directed at the vector associates antigen to the targeted vector enzyme to tumour site by restricting the cytotoxic effects. It has 3 steps 1. Antibody of enzyme conjugates administration 2. Clearing agents are added to clear any left unbound conjugates from bloodstream 3. Pro drug administration (the drug reaches all the parts but is only released at the targeted site (Kenet D Bagshawe, et al., Feb. 23, 2005).

GDEPT Technology: also referred as suicidal genes therapy or Gene directing enzyme prodrug treatment. The principle constitute of 3 processes 1. First a gene (mainly coding gene) is cloned with a vector and targeted to tumour cell. 2. The gene is then translated into an enzyme after it is transcribed into mRNA. 3. Administered prodrug transforms to a cytotoxic drug. The favourable conversion of prodrug to cytotoxic drug takes place only in tumour cell (targeted cells) resulting in minute exposure to healthy cells. This makes the "GDEPT" a desirable therapy for cancer treatment. Achieve better clinical response bystander

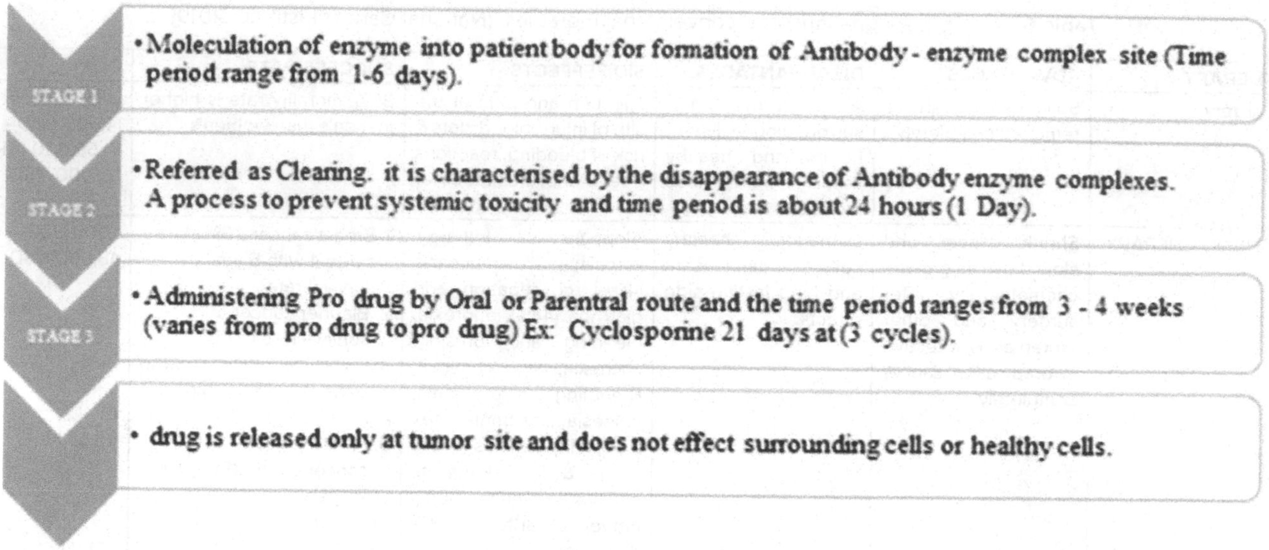

STAGE 1
•Moleculation of enzyme into patient body for formation of Antibody - enzyme complex site (Time period range from 1-6 days).
STAGE 2
•Referred as Clearing. it is characterised by the disappearance of Antibody enzyme complexes. A process to prevent systemic toxicity and time period is about 24 hours (1 Day).
STAGE 3
•Administering Pro drug by Oral or Parentral route and the time period ranges from 3 - 4 weeks (varies from pro drug to pro drug) Ex: Cyclosporine 21 days at (3 cycles).
• drug is released only at tumor site and does not effect surrounding cells or healthy cells.

Fig. 67.1 Mechanism of enzyme prodrug therapy (M.P. Napier, et al., March. 01, 2001)

Source: https://pubmed.ncbi.nlm.nih.gov

effect is achieved through various mechanisms. (Jin Zhang, et al., oct 23, 2015)

POLYMER DIRECTED ENZYME PRO DRUG THERAPY: in this therapy first a polymeric prodrug is administered which then combines with an activating conjugate of polymer - enzyme then it is generated into cytotoxic drug at the targeted site (tumour cell).

BDEPT Technology: The administered prodrugs are activated by the bacteria in the tumour which encourage the destruction of cancerous cells. The anaerobic bacteria are converted into cytotoxic prodrugs, the bacteria proliferation takes place only in hypoxic and necrotic regions and the enzyme is limited to only tumour. Spores aerobic bacteria can only multiply, spread and reach to oxygen starved tissues when oxygenated side of tumour are contacted hence they are totally safe and innocuous to rest of the body.

Table 67.5 Details of different types of Enzyme prodrug therapies (Schellmann, et at., Jun 7, 2001)

THERAPY	ENZYME	PRODRUG	ADVANTAGES	DISADVANTAGES
ADEPT	1. carboxypeptidase G2 (CPG2) 2. B-Lactamase 3. alkaline phosphate	1. Glucoronidate 2. cephalosporin vinca 3. phenol mustard, etoposide, mitomycin, doxorubin phosphate	1. Rapid clearance from blood is achieved 2. even small cytotoxic agent developed diffuses more than antibodies	1. Failure in expressing target antigen 2. Immunogenicity of antibody enzyme conjugate
VDEPT	1. thymidine kinase 2. cystosine deaminase 3. nitrireductase 4. cytochrome P450	1. Ganciclovir 2. 5-flurocystosine (5-FC) 3. CB1954 4. cyclophosphamide ifosfamide	1. Therapeutic index is much higher than cancer 2. minimal exposure to surrounding healthy cells	1. high infusibility may limit concentration in targeted cells 2. in order to spread virus mostly rely to convection to spread in tumour 3. Cannot be eliminated if any complications arise.
PDEPT	HPMA copolymercathepsin B	N-(2-hydroxypropyl) methacrylamide HPMA	Decreased toxicity activity	Retention is increased if lymphatic drainage is absent in tumour site.
BDEPT	1. Nitroreductase 2. cystosine deaminase 3. HSV-(TK)thymidine kinase	1. CB1954 2. 5-flurocystosine 3. ganciclovir	1. Bacteria are motile and can swim in a diffusion gradient and pressure created in tumour. 2. if complications are arised, they can be killed with antibodies during treatment	Doesn't spread to all parts of malignant tissues, so combination therapy is needed

Source : www.researchgate.net , https://europepmc.org , https://pubmed.ncbi.nlm.nih.gov

2. Case Studies

ANTIBODY DIRECTING ENZYME PRODRUG THERAPY: Based on ADEPT working principle first trial was conducted in which the treatment was carried out using calicheamicin and 1gG4 protein (humanized protein) in targeting acute myeloid leukemia which resulted in less severe side effects compared to conventional chemotherapy. The yielded product was MYLOTARG which was approved in 2001 and FDA approved in 2017.In one of the other study 11 patients were administered a fixed dose of 2-3 MFECP1 cycles & increasing dose (prodrug). 2 cycles as maximum dose was set at 1200mg/m². Efficacy was assessed using FDG-PET as relation to prodrug dose. 9 patients were evaluated and 4 of them showing partial response with prodrug dose ≥900mg/m². Additionally stability in a condition was achieved by 69% of patients reported by RECIST criteria. Other favourable agents are ZEVALIN AND BEXCAR (anti - CD20 mAbs which are conjugated to Y and I) are in clinical trials yet waiting. For FDA approval for treatment of B cell lymphomas. (PF Ross, et al., May, 2018).

TARGET THERAPY IN MELANOMA:

In 2019, combinations of Dabrafenib + trametinib, targeted therapy drugs were given to 35 patients before surgery who were on stage 3 melanoma. These 2 drugs are BRAF protein inhibitor and MEK inhibitor. This therapy involved blocking the division and growth of targeted cancer cells. The result showed 46% patients showed not a single sign of cancer and 80% of patients tumour shrunk. This made the removal of tumour in surgery easier. Hence treatment of melanoma with BRAF V600 mutation with combination of these two drugs was approved by FDA. (Brielle Gregory, Feb 4, 2000)

3. Conclusion

In this review article an attempt done to put forth in detail the latest developments in cancer treatment technologies which are achieving success in their trials and can be implemented as soon as possible for making difference in the cancer affected patients in terms of cost, time and sufferings.

REFERECES

1. Ananya Mandal (2017), Cancer Survivors have shorter life span finds new study, *News Medical life sciences*. Dec.18
2. Muhammed T. Amjad, Anusha Chidharlal, Anup Kasi (2023). Cancer chemotherapy, *National library of Medicine*. Feb.27
3. Kennet D Bagshawe, Surinder K Sharma, Richard HJ Begent (2005), Antibody-directed enzyme prodrug therapy (ADEPT) for Cancer, *Taylor & Francis*. Feb 23
4. M. P. Napier, S. K. Sharma, C. J. Springer, K. D. Bagshawe, A. J. Green, J. Martin, S. M. Stribbling, N. Cushen, D. O'Malley, R. H. J. Begent (2000). Antibody-directed enzyme prodrug therapy: Efficacy and Mechanism of Action in Colorectal Carcinoma, *AACR journals*. March.01
5. Abshire, D., & Lang, M. K. (2018), The evolution of radiation therapy in treating cancer. In Seminars in oncology nursing, may, (Vol. 34, No. 2, pp. 151–157). WB Saunders
6. P F Bross, J Beitz, G Chen, X H Chen, E Duffy, L Kieffer, S Roy, R Sridhara, A Rahman, G Williams, R Pazdur (2001). Approval summary: gemtuzumab ozogamicin in relapsed acute myeloid leukemia, Pubmed, Nationaql library of Medicine. Jun, 7(6):1490–6.
7. Brielle Gregory, ASCO staff, (2020). Improvements in Surgery for Cancer: The 2020 Advance of the Year, Cancer.Net. Feb 4
8. Schellmann, N., Deckert, P. M., Bachran, D., Fuchs, H., & Bachran, C. (2010). Targeted enzyme prodrug therapies. Mini reviews in medicinal chemistry, 10(10), 887–904.
9. Bagshawe, K. D., Sharma, S. K., Springer, C. J., & Rogers, G. T. (1994). Antibody directed enzyme prodrug therapy (ADEPT): A review of some theoretical, experimental and clinical aspects. Annals of Oncology, 5(10), 879–891.
10. National cancer institute, (2019). Cancer Therapies, Dec. 27
11. Natioal cancer institute (2022). Immune Checkpoit Inhibitor, April, 2022
12. Jin Zhang, Vijay Kale, Mingnan Chen. (2015). Gene Directed Enzyme Prodrug Therapy, National Library of Medicine, Pubmed, Oct.23
13. Cancer.Net Editorial Board, 2019, What is cancer, Aug.
14. World Healtrh Organisation, 2022, Feb.23
15. Stanford Medicine Health Care

Impending Inquisitions in Humanities and Sciences (ICIIHS-2022) – Dr. Mohan Varkolu et al. (eds)

A Review on Tele-Pharmacy and E-Pharmacy as Digital Revolutionary in Indian Healthcare System

68

Sushma Desai*
GITAM School of Pharmacy, Rudraram, Patencheru, Hyderabad, India

Sasasvi Dharmapuri, Pathlavath Thulsiram, Bestha Lakshmi Kalyani, M. Shiroja, Nelluri Swarna
Chilkur balaji college of pharmacy, Moinabad, Hyderabad, India

ABSTRACT Tele pharmacy and e-pharmacy are the digital revolutionaries found advancing healthcare connecting patient's medical practitioner and pharmacist in giving informed decision providing many advantages outnumbering disadvantages. Forthwith online services in demand digital revolution have greatly improved access to quality services for public. Overcoming the miss interpretation caused due to physical prescription Pharmacy profession in India since independence customizing to the dire needs of the healthcare system keeping abreast with the technology, knowledge to reach out common man in all aspects. Need has risen for each hospital being equipped with highly qualified pharmacy professionals giving quick responses and handling them during emergencies have been evident during covid-19 pandemic. This review aims to give an insight to the necessity of Tele pharmacy and E-pharmacy emergence its advantages and its implementation bought changes in healthcare system, acceptance by the people getting precise, comfortable and cost-effective treatment saving time in finding pharmacy, receiving and travelling in emergencies.

KEYWORDS Tele pharmacy, E-pharmacy, Emergency, Digital revolution

1. Introduction

Tele pharmacy services started in year 2015 and in India Apollo was the first company to start Tele pharmacy Tele pharmacy is communication of the patient from a location where they cannot directly contact with the pharmacist but can communicate via telecommunication (it is provision of use of tele commutations and information technology (internet of things) in pharmaceutical care for remote area patients. Tele pharmacy services include patient counselling, prior authorization, drug therapy monitoring, monitoring of formulary of teleconferencing or tele communications. Whereas Epharmacy is an online pharmacy that is operated over internet and which sends information to patient through mails & shipping companies (It's an online pharmacy service that provides ability to fill prescription medication online. The medication can be delivered without any issues or difficulty)[1][2][4][12]19]

*Corresponding author: d.sushmapharma@gmail.com

DOI: 10.1201/9781003489436-68

2. Tele Pharmacy and E Pharmacy Companies

Names	Location
Icebreaker health	San Francisco (United States)
Nurx	San Francisco (US)
Pocket pills	Surrey (Canada) Gurgaon (India)
Aspenrxhealth	Tampa(US)
Lyopds pharmacy	Coventry (UK)
Pharmesay	India
Midlife	India
1mg	India
Myra	India
Netmeds	India
Apollo pharmacy	India
Sun pharmaceutical industries limited	India
Karnataka antibiotics and pharmaceuticals private limited	India

Source: https://tracxn.com

3. Procedure of Telepharmacy[19][22]

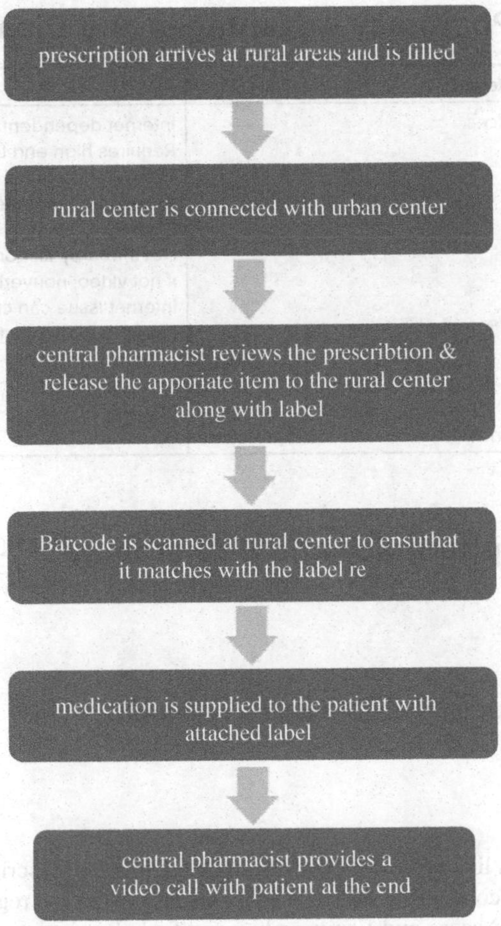
prescription arrives at rural areas and is filled
rural center is connected with urban center
central pharmacist reviews the prescribtion & release the apporiate item to the rural center along with label
Barcode is scanned at rural center to ensuthat it matches with the label re
medication is supplied to the patient with attached label
central pharmacist provides a video call with patient at the end

Source: https://www.ncbi.nlm.nih.gov

4. Procedure of E-Pharmacy[19][17][14]

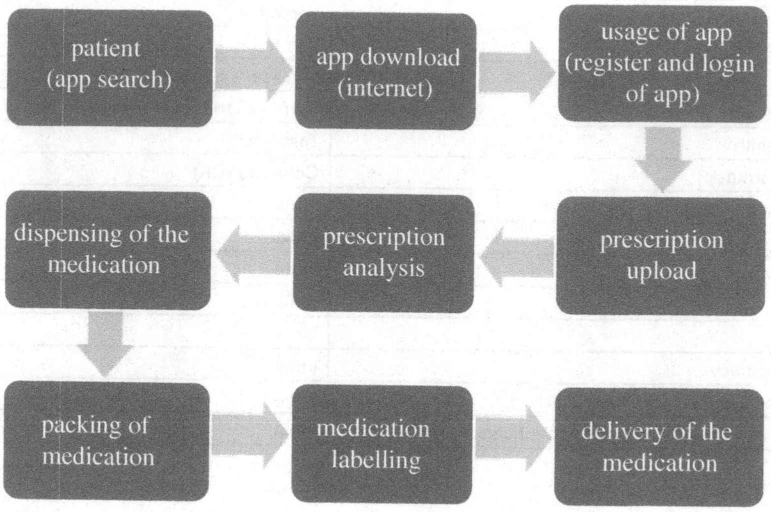

Source: https: www.lexology.com

5. Telepharmacy and E-Pharmacy Advantages and Disadvantages[21][16][13]

Advantages	Disadvantages
Helpful in providing rural areas services. Cost effective. Clears myths about illness. Follow-ups made easy. Time is saved. Convention. Reduces cancellations. Reach more patients Remove geographic barriers. Quick response in emergencies. paper work not there. Better triage. A better way to receive medical information without misinterpretation.	Internet dependent. Requires high end technical devices. Often Generational differences observed. Assessment is limited. Security. Not afford by all doctors lost clinical information. If not video, nonverbal clues are missed. Internet issue can create treatment difficulty. Lesser opportunity for clinical difficulty.

Source: https://www.ncbi.nlm.nih.gov

6. Requirements for Telepharmacy and E-Pharmacy[15][14][9][6]

- Android phone.
- Good internet connection.
- Proper camera condition.
- Having proper apps in phone for tele pharmacy like pharma easy, Apollo.
- Certified pharmacist.

2. Conclusion

Despite of the challenges faced in areas like technical(checking authenticity of e-prescription, drugs shortage, supply, financial recovery is less), privacy of patient is a concerning matter in cyber security, legal and regulatory issues are to be dealt as raising concern need to be solved. Regarding Access and Equity only a certified pharmacist can be accountable. For quality of care Pharmacist is entrusted member, serving healthcare community. Their role in the Tele pharmacy and E-pharmacy services had widely been utilized and an overwhelming response found since its implementation and continuing with updating knowledge and technological advances with collaboration of technology, medicine, pharmacist providing and quality healthcare services to people.

REFERENCE

1. Ibrahim OM, Ibrahim RM,Z A1 Meslamani A, A1 Mazrouei N.(2019) tele pharmacy in pharamacist counselling to coronavirus diseased patient and medication dispensing errors. *Journal of telemedicine and telecare*. 2020 oct15:1357633X20964347
2. Le T, Toscani M, Colaizzi J (2020) Telepharmacy: a new paradigm for our journal of pharmacy practice. Apr:33(2):176–82.
3. Baldoni S, Amenta F, Ricci G. (2019). Telepharmacy services: present status and future prepectives: a review Medicina Jul 1;55(7):327.
4. Schneider PJ (2013). Evaluating the impact of telepharmacy. *Amercican journal of health-system pharmacy*. Dec;70(23):2130–5
5. Poudel A, Nissen LM. (2016). Telepharmacy: a pharmacist's prepective on the clinical benefits and challenges. Integrated pharmacy Research & Practice. 5:75.
6. Pathak S, Haynes M, Qato DM, Urick BY. (2013–2019). Peer Reviewed: Telepharmacy and Quality of Medication Use in Rural Areas, Preventing Chronic Disease. 2020–17.
7. Misbah S, Ahmad A, Butt MH, Khan YH, Alotaibi NH, Mallhi TH. (2020). A systematic analysis of studies on corona virus disease 19 (COVID-19) from viral emergence to treatment J Coll Physicians Surg Pak. Jun 1;30(6):9–18.
8. Mohamed Ibrahim O, Ibrahim RM, Abdel-Qader DH, A1 Meslamani AZ, A1 Mazrouei N. (2021). Evaluation of telepharmacy services in light of COVID-19 Telemedicine and e-health Jun 1;27(6):649–56.
9. Kimber MB, Peterson GM. (2006). Telepharmacy-enabling technology to provide quality pharmacy services in rural and remote communities. *Journal of pharmacy Practice and Research*. Jun;36(2):128–33.
10. Cimperman M, Brencic MM, Trkman P, Stanonik MD. (2013). Older adults' preceptions of home telehealth services. Telemedicine and e-Health. Oct 1;19(10):786–90.
11. Crico C, Renzi C, Graf N, Buyx A, Kondylakis H, Koumakis L, Praventtoni G. (2018). Ehealth and telemedicine apps:in search of a common regulation.Ecancer medical science 2018;12.
12. Asseri AA, Manna MM, Yasin IM, Moustafa MM, Roubie FM, E1-Anssasy SM, Baqqwie SK, A1saeed MA. (2020). Implementation and evaluation of telepharmacy during COVID-19 pandemic in an academic medical city in the Kingdom of Saudi Arabia: Paving the way for telepharmacy. *World Journal of Advanced Research ans Reviews*. 2020;7(2):218–26.
13. Alexander E, Butler CD, Darr A, Jenkins MT, Long RD, Shipman CJ, Stratton TP. ASHP statement on telepharmacy. *American journal of Health-system Pharmacy*. 2017 May 1;74(9):e236–41.
14. Guadamuz JS, McCormick CD, Choi S, Urick B, Alexander GC, Qato DM. (2021). Telepharmacy and medication adherence in urban areas. *Journal of the American Pharmacists Association*. 2021 Mar 1;61(2):e100–13.
15. Angaran DM Telemedicine and telepharmacy: (1999) current status and future implications. *American Journal of Health-system pharmacy*. 1999 Jul 15;56(14):1405–26.
16. Killeen RM, Grindrod K, Ong SW. (2020). Innovations in practice: Telepharmcy's time has arrived. *Canadian Pharmacists Journal/ Revue des Pharmaciens du Canada*. 2020 Sep; 153(5):252–5.
17. Kosmisky DE, Everhart SS, Griffiths CL. (2019). Implemantation, evolution and impact of ICU telepharmacy services across a health care system. Hospital pharmacy. 2019 Aug; 54(4); 232–40.
18. Unni EJ, Patel K, Beazer IR, Hung M. Telapharmacy during COVID-19: a scoping review. pharmacy. 2021 Nov 11;9(4):183.
19. Lam AY, Rose D. (2009). Telepharmacy services in n urban community health clinic system. *Journal of the American Pharmacists Association*. 2009 Sep 1;49(5):652–9.
20. Frenzel J, Porter A. (2021). The Need to Educate Pharmacy Students in Telepharmacy and Telehealth. *American Journal of Pharmaceutical Education*. 2021 Sep 1;85(8).
21. Langbecker D, Caffery LJ, Gillespie N, Smith AC. (2017). Using survey methods in telehealth research. A practical guide. *Journal of Telemedicine and telecare*. 2017 Oct; 23(9):770–9.
22. Bashshur R, Doarn CR, Frenk JM, Kvedar JC, Woolliscroft JO. (2020). Telemedicine and the COVID-19 pandemic, lessons for the furture.Telemedicine and e-Health. 2020 May 1; 26(5):571–3.

Impending Inquisitions in Humanities and Sciences (ICIIHS-2022) – Dr. Mohan Varkolu et al. (eds)

A Glance on Medication Triggered Rare Disease—Stevens Johnson Syndrome as Global Challenge in Healthcare System

69

Sushma Desai*
GITAM School of Pharmacy, Rudraram, Patencheru, Hyderabad, India

Neha Yasmeen, Naveen Amgoth, Anumula Rama Rao, Beebireddy Vidhya, Jestadi Ragaswetha
Chilkur Balaji College of Pharmacy, Moinabad, Hyderabad, India

ABSTRACT In this review article we try to understand Stevens Johnson syndrome (sjs) as a rare disease affecting millions of people every year. It is characterized by a type-IV serious skin rash causing hypersensitivity disease with 1-5 % of mortality chances. With no defined universal diagnostic criteria presently diagnosis of this disease is based on studying affected persons medical history and observing the symptoms suffering with. The case observations reveal the causes as multiple medical over toxicity, infections, sudden genetic changes, family hereditary history and immune compromised patients. Standard therapies above all include symptomatic treatment by topical care, wound healing, fluid replacement with concerted specialists along with other healthcare professionals and psychosocial support required from family. Hence a need for close monitoring required for avoiding causative drugs in the treatment and thereby checking for any chances for counterpoise medication & educating patients for its possible risks and mitigation measures to be taken.

KEYWORDS Stevens johnson syndrome, Multiple medications, Mitigation, Therapies

1. Introduction

Stevens Johnson syndrome (sjs) is a severe allergic with potentially fatal reaction of skin, a characteristic rash that affects skin and mucous membrane such as buccal mucosa, genital areas and conjunctiva. The upper layer affected skin layer showing shedding then dies and the only begins to heal after several days. The sjs development seen in 4 to 30 days (usually takes 8 weeks). The causative agent first exposure with adverent rechallenge is seen in cases very rarely and non contagious. sjs was put forth by Albert Mason Stevens & Frank Chambliss Johnson, an American paediatricians who reported and published 2 cases in children having symptoms of unknown conditions in New York city, 1922.

SJS found incidence rate as 1-2 / lakhs as medical emergency.

2. Observed Symptoms

- Fever
- Sore throat and mouth
- Cough
- Joint pain
- Burning eyes

*Corresponding author: d.sushmapharma@gmail.com

DOI: 10.1201/9781003489436-69

As the condition progresses observations include blisters on skin & mucous membrane of nose, eyes and genitals with unexplained skin pain and ulcers & other lesions start appearing in mucous membrane.

Possible Causes for Trigerring SJS are

1. Infections: mycoplasma pneumonia, herpes virus, HIV, hepatitis A
2. Genetic variation: specific gene called human leukocyte antigen-B (HLA-B)
3. Graft versus host disease: Bone marrow donation, Peripheral Blood Stem Cells
4. Treatment of radiation
5. Vaccination: 2 reported cases of SJS due to covid-19
6. Allergic reaction to SJS
7. Medications
   - Antibacterial sulfonamides like sulfasalazine
   - Analgesics: acetaminophen
   - Medicines of arthritis like allopurinol
   - Antiepileptics like carbamazepine, phenytoin, lamotrigine, phenobarbital
   - NSAIDS: piroxicam, diclofenac, nevirapine
   - Medications that treat mental illness like anticonvulsants, antipsychotics.

3. Scorten Scale

Scorten scale is an assessment to evaluate severity of the conditions starting from mild to severe range. In 2000, scorten as an evaluating scale was introduced. The scale can also be used to assess sufferers of SJS, burn victims, cutaneous drug reactions or exfoliative wounds. It has 7 independent risk factors. scaling mortality rate and the presence of each factor is given by 1 point where during first 24 hours of admission was evaluated.

Prognostic Factors are:

- Age of Patient > 40 years
- Neoplasia / cancer
- Serum Bicarbonate level < 20mmol/L
- Serum Urea level > 10mmol/L
- Body surface area > 10%
- Heart rate > 120bpm
- Blood glucose level > 14mmol/L

The higher the Scorten score, higher is the mortality.

3.1 Scorten Score Interpretation

SCORTEN SCORE	CHANCES OF MORTALITY (%)
0-1	3
2	12
3	35
4	58
>5	90

Source: https://www.ncbi.nlm.nih.gov

Long Term Complications Observed in SJS

- Xerosis (dryness of skin and mucous membrane)
- Alopecia (hair loss)
- Hyperhidrosis (excess sweating)
- Koilonychia (abnormal growth of nails)

- Dyspigmentation and bumps (permanent skin damage)
- Visual impairment
- Dehydration
- Scarring of mucosal surface
- Sepsis (blood infection)
- Pigmentation (changes in skin colour)

Examples of few reported cases for the medication triggered sjs are:

1. LAMOTRIGENE + VALPROIC ACID

Age: 35 yrs

Gender: Female

Complaints: painful ulcers in mouth, bleeding lips, body rashes,high fever (39°C) since 4 days

Past history: suffering from seizure for past 4 years and was on medication valproate 600mg/day.

(S Kavitha et al., 2015, august 7.)

2. OXCARBAZEPINE

Age: 18 years

Gender: female

Complaints: maculopapular rashes, orogenital lesions and lymphadenopathy

Past history: The patient was on tab **OXCARBAZEPINE** 200mg OD and tab **CLONAZEPAM** 5mg OD for treating seizures. (Bhavi S. Trivedi; et al., Dec, 2016 (unique ID number 23455))

3. PHENYTOIN

Age: 65 years

Gender: male

Complaints: **Skin Lesions** for the past 2 days and **Oral Lesions** for 30 days

Past history: patient was suffering from sub arachnoid hemorrage and was on tab phenytoin 100mg BD

(Bhavi S. Trivedi; et, al., Dec, 2016 (unique ID number 17000)).

4. CARBAMAZEPINE

Age: 21 years

Gender: male

Complaints: fever, watering from eyes, vomiting, erosions on lips and mucosal candidiasis, bleachable erythematous rashes on neck & face, macular rashes on chest, upper & lower limbs and erythematous.

Past history: The patient had a history of epileptic attacks and was on tab carbamazepine 200mg TDS

(Bhavi S. Trivedi; et al., Dec, 2016 (unique ID number 00497)).

5. PARACETAMOL

Age: 38 years

Gender: female

Complaints: painful lesion in oral cavity, redness in eye and purulent discharge

Past history: patient was suffering from fever & coughs since 5 days and was on tab paracetamol 650mg.

(Neethu George, et al. 2016).

6. CEFAZOLIN

Age: 10 years

Gender: male

Complaints: bilateral conjunctivitis, oedema of lips and eyelids, haemorrhagic crusts of lips, superficial erosions of hard palate, widespread bullous, erythematous pruritis eruption and high fever.

Past history: the patient was suffering from acute pharyngitis for which he administered cefazolin 1gm IM

(M Yazicioglu, Jan 9, 2010)

7. URSODEOXYCHOLIC ACID

Age: 24 years

Gender: female

Complaints: whole body observed maculopapular rashes with oromucocutaneous erosions and skin desquamation.

Past history: 10 days prior to presenting features patient administered UDCA 300mg BD along with FDC'S **Omeprazole & Domperidone** QD owing to deranged liver function test.

(Shatavia Mukherjee, et al., Dec, 2020.)

8. DICLOFENAC

Age: 12 years

Gender: male

Complaints: ulceration of mouth, lips, sore throat, high fever since 4 dys, reddish purple maculopapular lesions in stomach and forearms.

Past history: patient had pain in the thigh region for which diclofenac sodium 500mg BD was given.

(Satish Kumar Reddy. K, et al., 2018)

9. TETRACYCLINE

Age: 14 years

Gender: male

Complaints: 4 days of increasing dysphagia, dysuria, photophobia, nacular rash extending from trunk towards extremities.

Past history: patient was using tetracycline for last 2 weeks as a treatment for facial acne. (Matthew Smelik, 2002)

10. NIMESULIDE AND ALLOPURINOL

Age 53 years

Gender: male

Complaints: pain and burning mouth on examination there was haemorrhagic crusts evident in vermillion of lips and buccal mucosa.

Past history: patient was on medication nimesulide, allopurinol, omeprazole, simvastatin for treatment of conjunctivitis. (JAA Arruda et al., 2014)

Treatment

- Depending upon the severity, hospitalization is required for treating SJS like dermatology unit, burn unit and intensive care unit.
- Rule out the cause of the disease.
- Stop all the non-essential medications.
- Supportive care: It includes wound care, eye care, fluid replacement and nutrition, genital care, respiratory support, monitoring the infected skin (non-adhesive dressing), maintenance of room temperature between 86.6F–89.6F
- Immune modulated treatment: Immunomodulatory agents like glucocorticoids, IV immunoglobulin, immunosuppressants or combination of these.
- Medications: Topical steroids to deal with inflammation, Antibiotics for infection control, Pain relievers to ease discomfort.

Mitigation Measures to be taken for sjs:

- If found to have SJS history, avoid the medication causing it to prevent recurrence.
- The genetic testing for HLA-B gene to be done prior treatment.

2. Conclusion

With the above given information about the Stevens Johnson syndrome its symptoms, long term effects, treatment and the mitigation measures taken. A need to understand that the highest reported cases of sjs are due to medications rather than other causes and its delayed treatment leading to severeity hence utmost care to be taken in medication prescribing to prevent this global challenge.

REFERENCES

1. Fouchard N, Bertocchi M, Roujeau JC, Revuz J, Wolkenstein P, Bastuji-Garin S. (2000). SCORTEN: a severity of illness score for toxic epidermal necrolysis. *Journal of Investigative Dermatology*. Aug 1;115(2):149–53.
2. Worswick S, cotliar J. (2011). Stevens-Johnson syndrome and toxic epidermal necrolysis, a review of treatment options. *Dermatologic therapy*. Mar 24(2):207–18.
3. Dash S, Sirka CS, Mishra S Viswan P. (2021). Vaccine induced Stevens -Johnson syndrome. *Clinical and Experimental Dermatology*. Dec;46(8):1615–7.
4. Shanbhag SS, Chodosh J, Fathy C, Goverman J, Mitchell C, Saeed HN. (2020). Multidisciplinary care in Stevens-Johnson syndrome. *Therapeutic Advances in Chronic Disease*. Apr;11:2040622319894469.
5. Halevy S, Ghislain PD, Mockenhaupt M, Fagot JP, Bavinck JN, Sidoroff A, Naldi L, Dunant A, Viboud C, Roujeau JC, EuroSCAR study Group. (2008). Allopurinol is the most common cause of Stevens-Johnson syndrome and toxic epidermal necrolysis in Europe and Israel. *Journal of the American Academy of Dermatology*. Jan 1;58(1):25–32.
6. Olson D, Watkins LK, Demirjian A, Lin X, Robinson CC, Pretty K, Benitez AJ, Winchell JM, Diaz MH, Miller LA,Foo TA. (2015). Outbreak of Mycoplasma pneumoniae-associated Stevens-Johnson syndrome. pediatrics. Aug; 136(2):e386–94.
7. Miliszewski MA, Kirchoff MG, Sikora S, Papp A, Dutz JP. (2016). Stevens-Johnson syndrome and toxic epidermal necrolysis: an analysis of triggers and implications for improving prevention. *The American journal of medicine*. Nov 1;129(11):1221–5.
8. Saeed H, Mantagos IS, Chodosh J. (2016). Complications of Stevens-Johnson syndrome beyond the eye and skin. *Burns*. Feb 1;(1):20–7.
9. Bianchine JR, Macaraeg Jr PV, Lasagna L, Azarnoff DL, fred Brunk S, Hvdberg EF, Owen Jr JA. (1968). Drugs as etiologic factors in the Stevens-Johnson syndrome. *The American Journal of Medicine*. Mar 1;44(3):390–405.
10. Kim HI, Kim SW, Park GY, Kwon EG, Kim HH, Jeong JY, Chang HH, Lee JM, Kim NS. (2012). Causes and treatment outcomes of Stevens-Johnson syndrome and toxic epidermal necrolysis in 83 adult patients. *The Korean journal of internal medicine*. Jun;27(2):203.
11. Lin LC, Lai PC, Yang SF, Yang RC. (2009). Oxcarbazepine-induced Stevens-Johnson syndrome: a case report. *The Kaohsiung journal of medical sciences*. Feb 1;25(2):82–6.
12. Mockenhaupt M. (2014). Stevens-Johnson syndrome and toxic epidermal necrolysis: clinical patterns, diagnostic considerations, and therapeutic management. *Semin Cutan Med Surg*. Mar 1;33(1):10–6.
13. Abtahi-Naeini B, Dehghan MS, Paknazar F, Shahmoradi Z, Faghihi G, Sabzghabaee AM, Akbari M, Haidan M, Momen T. (2022). Clinical and epidemiological features of patients with drug-induced Stevens-Johnson and toxic epidermal necrolysis in Iran: different points of children from adults. *International Journal of pediatrics*. Feb 8.
14. de Bustros P, Baldea A, Sanford A, Joyce C, Adams W, Bouchard C. (2022). Review of culprit drugs associated with patients admitted to the burn unit with the diagnosis of Stevens-Johnson syndromeand toxic epidermal necrolysis syndrome. *Burns*. Nov 1;48(7):1561–73.
15. Lian BS, Lee HY. (2022). Managing the ADR of Stevens-Johnson syndrome/toxic epidermal necrolysis. *Expert opinion on Drug Safety*. Aug 3;21(8):1039–46.
16. Pollard MC, Le LM, Dhaliwal DK. (2022). Stevens-Johnson syndrome in corneal Emergencies. *Springer*, Singapore. (pp. 325–337).
17. Dobry A, Himed S, Waters M, Kaffenberger BH. (2022). Scoring Assessments in Stevens-Johnson syndrome and Toxic Epidermal Necrolysis. *Frontiers in Medicine*. Jun 16:1766.
18. Hoffman M, Chansky PB, Bashyam AR, Boettler MA, Challa N, Dominguez A, Estupinan B, Gupta R, Hennessy K, Huckell SN, Hylwa- Deufel S. (2021). Long-term physical and psychological outcomes of stevens-johnson syndrome/toxic epidermal necrolysis. *JAMA dermatology*. Jun 1;157(6): 712–5.
19. Mansouri P, Farshi S. (2022). A case of Stevens-Johnson syndrome after COVID-19 vaccination. *Journal of cosmetic dermatology*. Apr;21 (4):1358–60.
20. Lee HY, Walsh SA, Creamer D. (2017). Long-term complications of Stevens-Johnson syndrome/toxic epidermal necrolysis (SJS/TEN): the spectrum of chronic problems in patients who survive an episode of SJS/TEN necessitates multidisciplinary follow-up. *British Journal of Dermatology*. Oct;177(4):924–35.
21. Kavitha S, Anbuchelvan T, Mahalakshmi V, Sathya R, Sabarinath TR, Gururaj N, Kalaivani S. (2015). Stevens-johnson syndrome induced by a combination of lamotrigine and valproic acid. *Journal of pharmacy and Bioallied sciences*. Aug;7 (suppl 2):S756.
22. Trivedi BS, Darji NH, Malhotra SD, Patel PR. (2016). Antiepileptic drugs-induced Stevens -Johnson syndrome: A case series. *Journal of Basic and Clinical pharmacy*. Dec;8 (1); 42.

23. Mukherjee S, Saha D, Dasgupta S, Tripathi SK. (2020). Case report: Suspected case of Stevens-Johnson syndrome and toxic epidermal necrolysis overlap due to ursodeoxycholic acid. *Dermatology*. Dec
24. SD SK, NC VR. (2018). A case report on diclofenac induced Stevens Johnson syndrome. *Journal of Basic and Clinical Pharmacy*. 9(2).
25. JAA Arruda JA, Sampaio GC. (2014). Stevens-Johnson syndrome with allopurinol and nimesulide: case report. Revista de Cirurgia e Traumatologia Buco-maxilo-facial. Sep;14(3): 59-64.
26. Yazicioglu M. (2010). Stevens-Johnson syndrome in a child: case report. *Reactions*. Jan 9; 1283:9.

Impending Inquisitions in Humanities and Sciences (ICIIHS-2022) – Dr. Mohan Varkolu et al. (eds)

A Comprehensive Review on Impact of Nanomaterials on Human Health

70

T. Indira Priyadarshini*
Assistant Professor, Department of Pharmaceutics,
Chilkur Balaji College of Pharmacy, Hyderabad

Chandrasekhar Rao Baru
Professor, Department of Pharmaceutics,
Chilkur Balaji College of Pharmacy, Hyderabad

Srujana Gandrathi
Assistant Professor, Department of Pharmaceutics,
Chilkur Balaji College of Pharmacy, Hyderabad

ABSTRACT Nanomaterials have been widely used for their biomedical applications like diagnosis, treatment of diseases, artificial cells and tissue engineering through which they have direct access into human body. They are used as materials for coatings of containers and food packages, from which they may leach into the contents and reach the human body. Also, a wide range of consumer products utilize nanotechnology for better results such as beauty solutions, weight loss drinks. Generally, nanoparticles are long circulating in the systemic circulation and are excreted into the urine and take their path into the environment and waterbodies. Therefore, humans can be exposed to nanomaterials through a known source such as drugs or foods, or inadvertent exposure through the environment. Nanomaterials due to their ability to interact with the human cells at molecular level especially possess some inherent toxicity, which is carefully evaluated by means of nanotoxicity assays. There is a thorough need to device methods for estimating toxicity of various types of nanomaterials. This review focuses on various applications for which nanomaterials are used, and resulting toxicity and methods of estimating toxicity and suggested methods to overcome. Also, it outlines the environmental toxicity resulting from the usage of nanomaterials.

KEYWORDS Nanomaterials, Drug delivery, Toxicity, Diagnosis

1. Introduction

Increasing use of nanomaterials in medicine, cosmetics and other nano enabled consumer products like biomolecules in nutrition products, tattoo inks, textiles, household cleaning products, building materials like water repellent paints, agrochemicals, are all sources of NPs lead to a situation that nanotechnology has penetrated considerably into the human world, to an extent as much as it became inseparable. As the matter is scaled down to nanoscale, the properties of the matter change with respect to the bulk counter parts, due to which reactivity of nanomaterials is quite high. The same novel properties like extremely small particles size, large surface area to volume ratio, that makes nanoparticles attractive tend to make them potentially toxic to the human body as well as other living organisms. Aquatic organisms when tested, were found to be loaded with considerable amounts of nanoparticles, which resulted in various levels of deleterious effects on the environment.

*Corresponding author: indira7taragaturi@gmail.com

DOI: 10.1201/9781003489436-70

The large surface area, and thus resulting reactivity of nanoparticles, may facilitate various kinds of reactions in the environment and also may result transfer of toxic materials in the environment. The size and chemical composition of few nanostructures may cause biological harm because of the way they are taken up by the cells and their ability to catalyse in-vivo reactions. Nanostructures are capable of self-assembly in the laboratory. Thus, chances of self-assembly and aggregation when they are disposed into the environment also cannot be ruled out. Therefore, the fate of nanostructures reaching the environment and their impact on human health need to be studied. In addition, nanotechnology led to societal changes that influence transportation, urban development, information management, and other activities of our society that directly had an impact on human health and wellbeing. Few areas where nanotechnology is applied like in the treatment of dreadful diseases like cancer and new frontiers in material sciences have revolutionized human life, whose benefit outweighs the risk to many folds. But slow elimination of nanoparticles from human body and persistent nanostructures in the environment may have few effects on human health which are discussed in this review.

Various Types of Nanoparticles and their Applications

Inorganic nanoparticles (metal nanoparticles)

- **Silver nanoparticles:** AgNPs have been used extensively in the health/personal care industry, house-hold utensils, in food storage, environmental, and biomedical applications owing to their proven antibacterial, antifungal, antiviral, anti-inflammatory, anti-cancer, and anti-angiogenic properties. Extensive research has been carried out and vast amount of literature is available on usage of silver nanoparticles. (Zhang XF, 2016)
- **Gold nanoparticles:** AuNps are well known for their optical properties, therefore used in colorimetric sensors in many diagnostic applications, of which an example is pregnancy test kit. They are used for targeted drug delivery in many types of cancer treatments, also as theranostics. They also find applications in catalysis and electronics. (Ines Hammami, 2021)
- **Titanium dioxide nanoparticles:** Titanium dioxide nanoparticles are most commonly used nanomaterials in biomedicine, cosmetics and food supplements in candies, chewing gums and in toothpastes.
- **Silica nanoparticles:** Products containing silica nanoparticles are widely used. It is used as an additive for the manufacture of rubber and plastics; as a strengthening filler for concrete and other construction materials. It is also a stable and non-toxic platform for biomedical applications such as drug delivery and theranostics.
- **Ferrous ion nanoparticles:** Magnetic properties of ferrous nanoparticles enable them to be used in laboratory applications like magnetic data storage and resonance imaging (MRI), targeting drugs to a specific site inside the body. They are also used for treating industrial sites contaminated with chlorinated organic compounds and also to treat many types of ground contamination such as grounds contaminated by polychlorinated biphenyls (PCBs), organochlorine pesticides, and chlorinated organic solvents. They also find use as catalyst and in certain alloy applications.
- **Copper nanoparticles:** Copper nanoparticles are proven to have "contact killing", phenomenon and therefore used to develop antibacterial and antiviral combination. Essa A.M.M, 2016 demonstrates the mechanism of copper nanoparticles antibacterial activities in bacterial cells.

Semiconductor nanoparticles

- **Ceramic nanoparticles:** Ceramic nanoparticles have been successfully used as drug delivery systems against a number of diseases, such as bacterial infections, glaucoma, etc., and most widely, against cancer.
- **Quantum dots:** Quantum dots (QDs) are semiconductors-based nanomaterials with numerous biomedical applications such as drug delivery, live imaging, and medical diagnosis, in addition to other applications beyond medicine such as in solar cells.

Organic nanoparticles

- **Carbon nanotubes:** Carbon nanotubes are used as additives in plastic to prepare composites with enhanced mechanical strength and electrical conductivity. They are used as vehicles for targeted drug-delivery and nerve cell regeneration. Other applications of CNTs are gene therapies, tissue regeneration, biosensor diagnosis, enantiomer separation of chiral drugs, extraction and analysis of drugs and pollutants.
- **Carbon dots:** Specific advantages of carbon dots are biocompatible, non-toxic, photostable, and easily functionalized with good photoluminescence and water solubility. Owing to these unique properties, they are used broadly in live cell imaging, catalysis, electronics, biosensing, power, targeted drug delivery, and other biomedical applications in sensors, bioimaging, solar cells and photocatalysis, etc

Polymeric nanoparticles: These are promising type of nanoparticles for the drug delivery, due to non-poisonous and biocompatible nature.

Solid lipid nanoparticles: SLNs and nanostructured lipid carriers (NLCs) are suitable for parenteral, dermal, pulmonary and topical delivery of drugs. These products are developed to reduce toxic side effects of the highly potent drugs administered intravenously, and also to improve bioavailability of drugs.

Toxicity of various types of nanoparticles:

Metal nanoparticles are often proven to have toxic effects to aquatic microbiota. Also these nanoparticles may reach the cells in human body through any route, when ingested intentionally like consumption of foods, medicines and other nano enabled products or accidentally, through water contaminated with nanoparticles from water bodies. Also occupational exposure is another possibility, where nano dust of the product being prepared could be inhaled by the person.

Toxicity of nanoparticles is accounted for a variety of reasons. The decreased size of the particle results in increased surface volume ratio, which means the number of atoms present on the surface to the inside of the particle is relatively greater compared to a particle of bigger size. This results in increased reactivity, increased electrostatic interaction, resulting in easy uptake of the nanoparticle by a living cell. Once inside the biological system, the nanoparticle due to increased reactivity may participate in a variety of biochemical reactions and interrupt them. They generate reactive oxygen species, resulting in increased oxidative stress, which is substantiated by the results of many studies. Inhaled nanoparticles, orally ingested nanoparticles and topically-applied nanoproducts, the majority are stuck in the exposure organs, and elicit hazardous effects before being gradually cleared out the body, unlike intravenously administered nanoparticles (Cui, X., 2020).

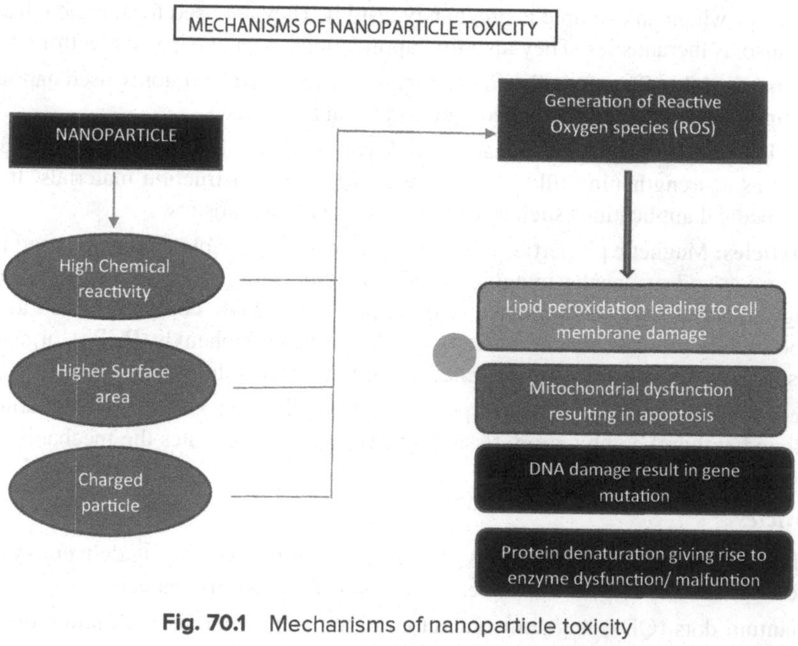

Fig. 70.1 Mechanisms of nanoparticle toxicity

Once the nanoparticles reach the environment, they loose their pristine nano state through various physical, chemical and biological transformation processes occurring in variety of environments (Lowry, 2012). The processes that the nanoparticles undergo in the environment depends on the properties such as size, shape, surface functionality of the Nps. For example, silver nanoparticles in aquatic environment undergo sulfidation and oxidation ultimately releasing the toxic metal ions (Brunetti, 2015). Gold nanoparticles produce more amounts of oxidative stress on the aquatic organisms through production of ROS (Domingos, 2013).

It was found that carbon-based nanomaterials like carbon nanotubes (CNTs) and graphene can to combine with the different gases and water vapors and change which change their reactivity (Yu J.G, 2015). This atmosphere release of nanomaterials was found to produce global climate changes through possible interactions with gaseous pollutants (John. 2017)

Nanoparticles are found to have potential impact on aquatic microbiota. Photosynthetic activity of the aquatic organisms is found to be inhibited by nano-TiO2 and nano CeO2 nanoparticles. Exposure to Gram-positive (Bacillus subtilis) and Gram-negative (Escherichia coli) bacteria to Ag NPs at environmentally relevant concentration (5 mg/L) for 8 h inhibited their growth and changed the growth kinetics (Solivers, 2016). A retardation in growth rate is found in other bacterial such as Azotobacter, nitrifying, phosphorus-and potassium-solubilizing bacteria upon exposure to metal oxide NPs due to the inhibition of enzymatic activities (Wu et al., 2018). Greater oxidative stress and gill damage induced by nano TiO2 to zebrafish larvae suggests toxicity of nano TiO2 towards humans (Honghui Guo, 2022)

Apart from oxidative stress nanoparticles are capable of inducing genotoxicity. Genotoxicity with silver nanoparticles was observed in chromosomal aberration studies both in vivo and invitro, but more effect was observed in invitro studies. It was found that genotoxicity of silver nanoparticles was dependant on size and shape of nanoparticles (Rodriguez 2020).

Neurotoxicity of nanoparticles is evident in many cases because of their ability to cross the blood brain barrier and ability to generate reactive oxygen species (Yongmei Zhao, 2022).

Toxicity to lungs is a direct result of inhalation of nanoparticles. Nanoparticles of natural occurrence in smoke and volcano dust are well known cause of pulmonary diseases like asthma. Silica (SiO2) nanoparticles are investigated for their lung toxicity in mice and found to increase the IL-4, an inflammatory mediator, level in lungs (Han B, 2011). According to a study, cobalt nanoparticles are found to induce inflammatory response in mice (Rong Won, 2017). Polymeric nanoparticles may result in toxic monomer aggregation, and toxic degradation process (Singh N et al., 2017). Cytotxicity of polymeric nanoparticles is observed to be enhanced with various polymers used for preparation of polymeric nanoparticles.

Few findings on toxicity risks associated with various types of nanoparticles discussed in this manuscript are summarized in the Table 70.1.

Table 70.1 Toxicity risks associated with various types of nanoparticles

Nanoparticle type	Possible Toxicity	Mechanism
Silver nanoparticles	cytotoxicity, genotoxicity	oxidative Stress
TiO2 nanoparticles	cytotoxicity, genotoxicity	oxidative Stress
Gold nanoparticles	cytotoxicity	oxidative Stress
carbon nanotubes	cytotoxicity, genotoxicity	oxidative Stress
Silica nanoparticles	cytotoxicity, lung toxicity	Increased IL-4 production
Cobalt nanoparticles	cytotoxicity, lung toxicity	Increased IL-4 production
Polymeric nanoparticles	cytotoxicity, genotoxicity	dependent on quantum size effects, oxidative stress

Methods for assessment of toxicity:

Nanotoxicology is a science that deals with the study and a better understanding of the toxicity of nanoparticles, which is a vast research area currently. Toxicity of nanoparticles is assessed by means of different invitro and in vivo models

In Vitro Assays: the advantage of these studies is that they do not require the use of animals, which avoids any form of ethical issues. Cell lines are usually to be conducted during a toxicity evaluation on monocultures of long-lived cell lines of expected in-vivo target organ and is usually carried out for single incubation period after exposure of nanoparticles to the cells (Hillegass, 2010), (Jones and Grainger, 2009). Then the toxicity is evaluated by various biochemical assays such as enzymatic immunoassays (Nel et al., 2013) or by staining the cell lines with fluorescent labelled molecular probes (Soenen et al., 2011) These studies aim to mimic cellular events inside the human body after being exposed to any toxic substance. The drawback of invitro assay is that it cannot mimic the toxicity outcome observed in a physiological system.

In Vivo Assay: Various animal models such as rats, zebra fishes, earthworms and plants are used as invio models based on the respective environmental impact to be studied. Effects of nanoparticle reaching the water bodies are studies using a fresh water fish, zebrafish, whose genetic makeup is 70% similar to that of human beings (Zhang, C, 2003). Nanoparticles accumulated in the soil are taken up by the plants, due to which plants as well as earthworms are used to study the toxic effects of nanoparticles. Drosophila melanogaster is another best in-vivo model organism for assessing the genotoxicity and cytotoxicity of nanoparticles, due to its similarity to human genome (Vecchio, 2015). The cellular uptake, distribution, metabolism, and the removal pathway can be studied in this technique. Histopathology is after nanoparticle exposure is observed to understand the level of toxicity of nanoparticles in various tissues of organs such as brain, heart, , kidney, etc (Kumar, 2017).

In Silico Assay: In silico assay predicts the toxicity of any molecular compound using the available experimental data and further interpolating using mathematical models (Irini Furxhi, 2020). In silico prediction is advantageous because it is a rapid and cost-efficient tool and avoids any ethical conflicts, as there is no use of animals. However, defining the toxicological end points after the exposure requires experimental verification

Measures to be taken to overcome toxicity issues:

Measures need to be taken for proper disposal of waste nanomaterials to prevent them from entering the environment contaminate the natural resources. Presently tons of nanomaterials are reaching the environment on a continuous basis, which need to be circumvented with appropriate check points, otherwise the environment would be flushed with nano-waste, exposing every single person to nanotoxicity (Kabir E, 2018).

Nanowaste that are produced from a factory (dry process) should be collected, stored separately and can be utilized as an external coat by adding required ingredients (Saravanan, 2017). ROS production in cells can be overcome by introducing ascorbic acid, which is an antioxidant capable of scavenging free radicals. Surface modification of nanoparticles is also suggested as a way to limit the nanotoxicity (Khanna, 2015). Incineration to destroy the nanowaste can be carried out to completely prevent the nanomaterial from reaching the ecosystem (Holder, A.L, 2013). Phytoremediation is another approach wherein certain types of plants accumulate the metallic nanoparticles, preventing them from contaminating the environment (Andreotti., F, 2015).

Redisgn of nanoparticles based on the major mode of toxicity can also be considered to address the issue of toxicity. For example interactions with the cell surface can be reduced by imparting a negative surface charge, by use of ligands such as polyethylene glycol that reduce protein binding, or create a morphology that discourages binding with a cell surface. Dissolution of nanoparticles to toxic ions can be reduced by replacing the toxic species with less toxic elements that have similar properties, the nanoparticle can be capped with a shell material. Also the morphology of the nanoparticle can be chosen to minimize surface area and thus minimize dissolution, or a chelating agent can be co-introduced or functionalized onto the nanomaterial's surface. Production of reactive oxygen species can be overcome by tuning the band gap of the material either by using different elements or by doping, or antioxidant molecules can be tethered to the nanoparticle surface (Buchman JT et al., 2019).

The redesigning strategy should be adequately tested that it actually reduces toxicity to organisms from relevant environmental compartments. It is also necessary to confirm that the nanomaterial still demonstrates the critical physicochemical properties that are required for its performance.

2. Conclusion

Nanotechnology has been a boon to the world, which tremendously improved the human wellbeing in almost all the aspects of life, especially healthcare, where dreadful diseases like cancer can be treated more effectively. Also, gene therapy and artificial tissues have given solutions for problems that were once counted as untreatable for a long period. But nanotechnology is a double-edged sword, which carries risks associated with the benefits. Even though immediate benefits with the usage of nanomaterials outweigh the long-term risks associated, there is a need to carry out extensive research to estimate the potential long-term risks and also to come up with proper measures that minimize the risks associated. There are policies in place to address the problem, but growing usage of nanomaterials suggest continual updating of existing ones and their adequate enforcement to effectively circumvent the problem and protect the overall health of a normal human being.

REFERENCES

1. Andreotti, F., Mucha, A.P., Caetano, C., Rodrigues, P., Gomes, C.R., Almeida, C.M.R., 2015. Interactions between salt marsh plants and Cu nanoparticles–effects on metal uptake and phytoremediation processes. Ecotoxicol. Environ. Saf. 120, 303–309.
2. Brunetti, G., Donner, E., Laera, G., Sekine, R., Scheckel, K. G., Khaksar, M., … Lombi, E. (2015). Fate of zinc and silver engineered nanoparticles in sewerage networks. Water Research, 77, 72–84
3. Cui, X., Bao, L., Wang, X., & Chen, C. (2020). The nano–intestine interaction: Understanding the location-oriented effects of engineered nanomaterials in the intestine. Small, 16
4. Domingos, R. F., Rafiei, Z., Monteiro, C. E., Khan, M. A. K., & Wilkinson, K. J. (2013). Agglomeration and dissolution of zinc oxide nanoparticles: Role of pH, ionic strength and fulvic acid. Environmental Chemistry, 10(4), 306–312.
5. Essa A.M.M., Khallaf M.K. Antimicrobial potential of consolidation polymers loaded with biological copper nanoparticles. *BMC Microbiol.* 2016;16:144.
6. Han B., Guo J., Abrahaley T., Qin L., Wang L., Zheng Y., Li B., Liu D., Yao H., Yang J., et al. Adverse effect of nano-silicon dioxide on lung function of rats with or without ovalbumin immunization. *PLoS One.* 2011;6:e17236.
7. He H, Pham-Huy LA, Dramou P, Xiao D, Zuo P, Pham-Huy C. Carbon nanotubes: applications in pharmacy and medicine. Biomed Res Int. 2013.

8. Hillegass, J.M., Shukla, A., Lathrop, S.A., MacPherson, M.B., Fukagawa, N.K., Mossman, B.T., 2010. Assessing nanotoxicity in cells in-vitro. Wiley Interdiscip. Rev. Nanomed. 2, 219–231
9. Honghui Guo, Yu Kuang, Kang Ouyang, Ce Zhang, Hui Yang, Siqi Chen, Rong Tang, Xi Zhang, Dapeng Li, Li Li, Ammonia in the presence of nano titanium dioxide (nano-TiO2) induces greater oxidative damage in the gill and liver of female zebrafish, Ecotoxicology and Environmental Safety, Volume 236, 2022, 113458.
10. Holder, A.L., Vejerano, E.P., Zhou, X., Marr, L.C., 2013. Nanomaterial disposal by incineration. Environ. Sci. Process Impacts 15, 1652–1664.
11. Inès Hammami, Nadiyah M. Alabdallah, Amjad Al jomaa, Madiha kamoun, Gold nanoparticles: synthesis properties and applications, Journal of King Saud University - Science, Volume 33, Issue 7, 2021.
12. Irini Furxhi, Finbarr Murphy, Martin Mullins, Athanasios Arvanitis & Craig A. Poland (2020). Nanotoxicology data for *in silico* tools: a literature review, Nanotoxicology, 14:5, 612–637.
13. John, A. C., Kupper, M., Manders-Groot, A. M. M., Debray, B., Lacome, J. M., & € Kuhlbusch, T. A. J. (2017). Emissions and possible environmental Implication of engineered nanomaterials (ENMs) in the atmosphere. Atmosphere (Basel), 8, 1–29
14. Jones, C.F., Grainger, D.W., 2009. In-vitro assessments of nanomaterial toxicity. Adv. Drug Deliv. Rev. 61, 438–456.
15. Kabir E, Kumar V, Kim KH, Yip ACK, Sohn JR. Environmental impacts of nanomaterials. J Environ Manage. 2018 Nov 1;225:261–271. doi: 10.1016/j.jenvman.2018.07.087. Epub 2018 Aug 7. PMID: 30096714.
16. Khanna, P., Ong, C., Bay, B.H., Baeg, G.H., 2015. Nanotoxicity: an interplay of oxidative stress, inflammation and cell death. Nanomaterials 5, 1163–1180.
17. Kumar, V., Sharma, N., Maitra, S.S., 2017. In-vitro and in-vivo toxicity assessment of nanoparticles. Int. Nano Lett. 7, 243–256.
18. Lowry, G. V., Gregory, K. B., Apte, S. C., & Lead, J. R. (2012). Transformations of nanomaterials in the environment. Environmental Science & Technology, 46(13), 6893–6899.
19. Nel, A., Xia, T., Meng, H., Wang, X., Lin, S., Ji, Z., Zhang, H., 2013. Nanomaterial toxicity testing in the 21st century: use of a predictive toxicological approach and highthroughput screening. Acc. Chem. Res. 46, 607–621.
20. Rodriguez-Garraus A, Azqueta A, Vettorazzi A, López de Cerain A. Genotoxicity of Silver Nanoparticles. Nanomaterials (Basel). 2020 Jan 31;10(2):251.
21. Saravanan, J., Karthickraja, R., Vignesh, J., 2017. Nanowaste. International Journal of Civil Engineering and Technology Volume 8, Issue 12, December 2017, pp. 483–491.
22. Soenen, S.J., Rivera-Gil, P., Montenegro, J.M., Parak, W.J., De Smedt, S.C., Braeckmans, K., 2011. Cellular toxicity of inorganic nanoparticles: common aspects and guidelines for improved nanotoxicity evaluation. Nano Today 6, 446–465.
23. Soliveres, S., Van Der Plas, F., Manning, P., Prati, D., Gossner, M. M., Renner, S. C., … Allan, E. (2016). Biodiversity at multiple trophic levels is needed for ecosystem multifunctionality. Nature, 536(7617), 456–459.
24. Vecchio, G., 2015. A fruit fly in the nanoworld: once again Drosophila contributes to environment and human health. Nanotoxicology 9, 135–137.
25. Wan R, Mo Y, Zhang Z, Jiang M, Tang S, Zhang Q. Cobalt nanoparticles induce lung injury, DNA damage and mutations in mice. Part Fibre Toxicol. 2017 Sep 18;14(1):38.
26. Yongmei Zhao, Qiongxia Yang, Dong Liu, Tianqing Liu, Lingyan Xing, Neurotoxicity of nanoparticles: Insight from studies in zebrafish, Ecotoxicology and Environmental Safety, Volume 242, 2022.
27. Yu, J. G., Yu, L. Y., Yang, H., Liu, Q., Chen, X. H., Jiang, X. Y., … Jiao, F. P. (2015). Graphene nanosheets as novel adsorbents in adsorption, preconcentration and removal of gases, organic compounds and metal ions. Science of the Total Environment, 502, 70–79.
28. Zhang, C., Willett, C. and Fremgen, T. (2003), Zebrafish: An Animal Model for Toxicological Studies. Current Protocols in Toxicology, 17: 1.7.1-1.7.18.
29. Zhang XF, Liu ZG, Shen W, Gurunathan S. Silver Nanoparticles: Synthesis, Characterization, Properties, Applications, and Therapeutic Approaches. Int J Mol Sci. 2016 Sep 13;17(9):1534. doi: 10.3390/ijms17091534. PMID: 27649147; PMCID: PMC5037809.

Impending Inquisitions in Humanities and Sciences (ICIIHS-2022) – Dr. Mohan Varkolu et al. (eds)

A Special Focus on Paraquat Poisoning as a Leading Worldwide Concern for its Ban

71

Sushma Desai*
GITAM School of Pharmacy, Rudraram, Patencheru, Hyderabad, India
Mohammed Dawood Shareef, Fareesa Khan, Beebireddy Vidhya, Permula Praveen Kumar, Vadladi Nikhila
Chilkur Balaji of Pharmacy, Moinabad, Hyderabad, India

ABSTRACT A world widely used Paraquat herbicide by farmers since 55yrs is sold under various trade names by approx 377 certified companies receiving a world-wide concern for its ban despite of its numerous benefits to farming in economy, environment and time but found fatal to human life due to accidental and suicidal ingestion leaving very less survival chances (as single sip causing aggravate generation of reactive oxygen leading to multiorgan failure) and no effective antidote available for its treatment hence became social concern for its ban by the families of the farmers for accidental exposure and of overall suicidal reported cases 20-55 % occupied by herbicidal suicides and in it paraquat occupies as first choice due its availability rather than intentional. This review aims to understand the reasons for its raising concern as fatal herbicide, and its ongoing research for treatments and to provide safety education to the handlers.

KEYWORDS benefits, paraquat, herbicide, treatment

1. Introduction

Paraquat is one of the most popular herbicide used in agriculture due to its good herbicidal properties it is highly toxic and shows severe toxic effects. It has been banned in around 32 countries currently the largest manufacture is China, produces around 1 lakh tons of paraquat. Paraquat is extremely toxic and can cause poisoning in every possible way through inhalation, dermal contact etc. many suicide cases are reported every year. Paraquat has been reported as utmost active biologically & toxic to living (both plants & animals) usage by us. Environment protection agency. And also reported as ecotoxic to aquatic environment by New Zealand environment risk management authority. It was reported about 13 countries has become resistant to paraquat with over 22 various weed species. Its absorption, metabolism and toxicity found as follows:

Absorption by skin (intact): 0.5%

Absorption by broken skin: lead to death.

Metabolism is usually limited.

Excretion: 69-96% in faeces which is unchanged. Microbial degradation is showed in gut and is highly concentrated in lungs.

Toxicity: rapidly absorbed through inhalation and by intestines.

The acute and chronic toxicities effect observations and the level of toxicities with dose through route of administration are tabulated in table 1 & 2 respectively.

*Corresponding author: d.sushmapharma@gmail.com

DOI: 10.1201/9781003489436-71

Table 71.1 Acute and chronic toxicity observations

ACUTE TOXICITY EFFECTS	CHRONIC TOXICITY EFFECT
1. Extensive toxic if inhaled 2. Acute respiratory distress 3. Burns in mouth 4. Dermal damage 5. Eye injury 6. Giddiness 7. Vertigo 8. Headaches 9. Shortness of breath 10. Abdominal pain 11. Nausea & vomiting 12. Diarrhea 13. Nose bleeds 14. Loosing nails 15. Myalgia 16. lethargy	1. Accelerate development of parkinsons disease. 2. Can penetrate nervous system and cause brain dysfunction. 3. Prolonged exposure in pregnancy alter brain development in foetus 4. Possible carcinogenic 5. Foetal death before birth and neonatal death after delivery. 6. Cause endocrine disruption 7. Causes birth defects 8. Effects immune system.

Source: https://www.ncbi.nlm.nih.gov

Environmental Effects: paraquat has been reported as extremely toxic being biologically active to both animals & plants.

Effects showing in amphibians:

1. Shown teratogenicity in fishes.
2. Genotoxicity in tadpole due to hormonal disturbance in frogs

Effects in floral environment:

1. Planktonic algae are highly sensitive to paraquat.
2. Paraquat alters with species composition and cause biodiversity loss.
3. Results in weed algae or dangerous algae blooming, variant diseases thus results in decreased fisheries population.

Effects in birds: it causes endocrine disturbances in birds which effects reproduction system and also egg hatchability when exposed to paraquat. Ruffled feathers, lack of co-ordination, imbalance, wing shiver and death in 3-20 hrs after administration is seen.

Environmental fate

In soil:

1. It increases soil pathogen reproduction. It shows toxicity in soil's bacteria & fungi.
   Ex: Gaeumannomyces graminisvar tritici.
2. Shows high binding capacity in soil hence stays there for longer time but in inactive form.
   This later can be debsorb and become active again
3. Half life is nearly 20 yrs in soil. It was estimated as 0.05 mg/kg findings of 11.4-25.4 in soil segment.

In water: half life found to be >800 yrs under mid European conditions with factors like depth and sunlight reaching inside water.

It is immobile in soil, so cannot reach underground water but can be found in exposed water and drinking water.

ECONOMIC IMPACT ON BANNING PARAQUAT: with evident resources (data available in FAOSTAT) which is available investigated literature based on 6-8 years of study have proven that there is no evidence of economic impact on agriculture based countries by banning it for crop yielding.

CHALLENGES AND OPPURTUNITIES OF BANNING PARAQUAT: Challenges: Despite banning paraquat, IT was believed by farmers as negative impact on their livelihoods & crop yield which is opposite to the fact (study conducted by sethi et.al.2022) In Kerala, India, south Korea, Taiwan revealed that banning paraquat and other 14 deadly pesticides found no effect on crop yield (6-8 years of study). As Effective alternatives of paraquat are available (carfentrazone-ethyl).

Table 71.2 Toxicity with dose levels & route of administration

TOXICITY	ORAL		DERMAL		INHALATION		OTHERS
	DOSE	EFFECT	DOSE	EFFECT	DOSE	EFFECT	EFFECTS
ACUTE TOXICITY	LD50 human = 40-60 mg/kg	Dehydration, piloerection, splay reflex is decreased, stains near mouth and nose, breathing problems, hypithermia, skin discoloration around mouth and nose.	LD50 rat = >660 mg/kg	Skin becomes think, severeitching, blistering of skin and irritation.	Inhalation dose in rat LC50 = 0.6-1.4 mg.	Lungs are congested, edema in kidney is seen, haemorrhages is reported	EYES: Discharge in eyes, redness, swelling , shedding and thickening of external layers, corneal opacity
SUB CHRONIC	0.45 mg/kg bw/day seen in dogs,	Collapsed Alveolar, irregular heartbeat, loss of appetite, weight loss, large lesions increases weight in lungs	1.15 mg/kg bw/day seen in rabbits	Carcinogenic cell proliferation, serious inflammation, exucation and thickening, ulceration.	10 μg/m³ in rats,	Ulcers and inflammation in larynx, nasal discharge, alveolar wall thickening, white blood cells get aggregated.	Myocardium, reccurenthaemolytic anaemia
CHRONIC		Oral carcinogenic risks		squamous cell carcinoma with combined bipyridines and sunlight exposure.		Pheochromocytoma & adenomas	parkinson disease, birth defects and genotoxicity, endocrine disruption

Source: https://emergency.cdc.gov/agent/paraquat/basics/facts.asp

2. Handling and Precautions of Paraquate Poisoning

1. Pesticide should all ways keep in well secured place all the time. The us people have died just after having small quantity of thinking it was soft drink which was packed in container
2. Well trained and certified person have to handle & use it. For this training is available in online of no cost. This course can be completed in an hour
3. Water must be spared on the field before applying the paraquat. 5 gallons are applied by air better result are seen if 10 gallons of water is applied on ground of 1 acre.
4. To illegal appalled right quantity and proper directions must be followed as on the bottle pre-harvest methods must be respected before under cutting and combining of the crop
5. Medium size droplets size must be preferred large amount must not be used as paraquat as paraquat is sensitive to most plants.
6. Medium size droplets size must be preferred large amount must not be used as paraquat as paraquat is sensitive to most plants.

3. Labelling Information

- Always follow the directions that are on the label don't not store the product in food or drinking products.
- The labelling must contain 'ONE SIP CAN KILL' the label have to contain skull and crossbones on the labelling.

BENEFITS OF PARAQUAT CHLORIDE: Paraquat has very unique mode of action as

1. *Agro-environmental benefits:* Paraquat decreases the effort and need of preparing the land for crop growth with which there will be saving in machinery tools, labour and fuel. Preparing of the land for crop growth that is losing on the soil makes a well healthy condition for the crop to grow properly. Losing of soil also reduces the soil erosion as particles settle properly it also improves physical condition of the crop and wildlife. It also increases the air quality and decrease the air free dust also decreases their emission of the greenhouse gases by binding carbon in organic matter.
2. *Socio economic benefits:* Paraquat increases the yield constituents and controls weeds when weeds decrease the crop yields rise. If also decrease the framer work which the farmer removes the weeds by hand weeding. It also saves time of farmer where they spend hours away from the family to remove weeds by hand.
3. *Time saving:* Paraquat also controls the weeds even if rains 15 mins to 30 mins for spraying

4. Treatment Options for Paraquat

- Elimination of toxin: is done with haemodialysis, hemofiltration & immunosuppressants.
- Anti-oxidants in paraquat poisoning: n-acetyl cysteine, vitamin c, Vitamin E, Deferoxamine, Salicyclic acid.

5. Case Report

An 18-year female patient admitted with alleged attempted history of suicidal attempt with unknown quantity consumed I liquid form at her residence. Initially patient at local hospital managed with IV fluids, H2 blockers and Antiemetics and was brought to Super speciality hospital after 24 hrs. She has difficulty in mouth opening & urine output decreasing observed. There was no history of vomiting, abdominal pain, seizures, loose stools & fever reported and patient was conscious & oriented was summarized.

Conclusion: with the above discussed facts about its potential herbicidal profile and failing to sustain due to the fatal case reports received worldwide and couldn't save the lives of the people affected with over toxicity either occupational or intentional hence need of the hour in dealing with the usage of paraquat is by giving strict training in handling and safety precautions as no immediate effective antidote available.

REFERENCE

1. Xiao Q, Wang W, Qi H, Gao X, Zhu B, Li J, Wang P. (2020). Continuous hemo-perfusion relieves pulmonary fibrosis in patients with acute mild and moderate paraquat poisoning. *The Journal of Toxicological Sciences*. 45(10):611–7.
2. Wang JW, Yang X, Ning BY, Yang ZY, Luo LH, Xiao H, (2019) Ning Z. The successful treatment of systemic toxic induced paraquat poisoning by skin absorption: case reports and a literature review. *International journal of clinical and experimental pathology*. 12(9):3662.
3. Pateiro-Moure M, Martínez-Carballo E, Arias-Estévez M, Simal-Gándara J. (2008) Determination of quaternary ammonium herbicides in soils: comparison of digestion, shaking and microwave-assisted extractions. *Journal of Chromatography*; Jul 4 1196:110–6.
4. Kervégant M, Merigot L, Glaizal M, Schmitt C, Tichadou L, de Haro (2013) L. Paraquat poisonings in France during the European ban: *experience of the Poison Control Center in Marseille. Journal of medical toxicology*. June 9(2):144–7.
5. Taylor PJ, Salm P, Pillans PI 2001. A detection scheme for paraquat poisoning: validation and a five-year experience in Australia. *Journal of analytical toxicology*. Sep 1 25(6):456–60.
6. Lock EA, Wilks MF (2010). Paraquat. *In Hayes' Handbook of Pesticide Toxicology Academic Press*. Jan 1 (pp. 1771–1827).
7. Bullivant CM. Accidental poisoning by paraquat: Report of two cases in man. *British Medical Journal*. 1(5498):1272.
8. Delirrad M, Majidi M, Boushehri B (2015). Clinical features and prognosis of paraquat poisoning: *a review of 41 cases. International journal of clinical and experimental medicine*. 8(5):8122.
9. Kolilekas L, Ghizopoulou E, Retsou S, Kourelea S, Hadjistavrou C (2006). Severe paraquat poisoning. *A long-term survivor. Respiratory Medicine Extra*. Jan 1; 2(2):67–70.
10. Ong ML, Glew S (1989): Paraquat poisoning: per vagina. *Postgrad Med J*. 65(769): 835–836.
11. Soloukides A, Moutzouris DA, Kassimatis T, Metaxatos G, Hadjiconstantinou V (2007). *A fatal case of paraquat poisoning following minimal dermal exposure. Renal failure*. Jan 1 29(3):375–7.
12. Chen HW, Tseng TK, Ding LW (2009). Intravenous paraquat poisoning. *Journal of the Chinese Medical Association*. Oct 1 72(10):547–50.
13. Roh HK, Oh BJ, Suh JH, Kim JS. Fatal inhalation poisoning after diluted paraquat spray (2006). *Inclinical Toxicology 325 Chestnut St, Suite 800, Philadelphia, PA 19106 USA: Taylor & Francis Inc*. Jan 1 (Vol. 44, No. 5, pp. 670–670).
14. Yoon SC. (2009) Clinical outcome of paraquat poisoning. *The Korean Journal of Internal Medicine*. Jun 24(2):93.
15. Gawarammana IB, Buckley NA. Medical management of paraquat ingestion. *British journal of clinical pharmacology*. 72(5):745–57.
16. Wesseling C, Corriols M, Bravo V (2005). Acute pesticide poisoning and pesticide registration in Central America. *Toxicology and applied pharmacology*. Sep 1 207(2):697–705.
17. Sukumar CA, Shanbhag V, Shastry AB (2019). Paraquat: The poison potion. Indian Journal of Critical Care Medicine: Peer-reviewed, *Official Publication of Indian Society of Critical Care Medicine*. Dec 23(Suppl 4):S263.
18. Raghu K, Mahesh V, Sasidhar P, Reddy PR, Venkataramaniah V, Agrawal A (2013). Paraquat poisoning: A case report and review of literature. *Journal of family & community medicine*. Sep 20(3):198.
19. Tominack R, Pond S. Herbicides. *Goldfrank'stoxicologic emergencies*.
20. Wright N, Yeoman WB, Hale KA (1978). Assessment of severity of paraquat poisoning. *British Medical Journal*. Aug 8 2(6134):396.
21. Sriperumbuduru VP (2014). Paraquat Poisoning in Clinical and Medico-Legal Perspective a Case Report and Over View. *Indian Journal of Forensic Medicine & Toxicology*. Jan 18(1).
22. Sandhu JS, Dhiman A, Mahajan R, Sandhu P (2003). *Outcome of paraquat poisoning-a five year study. Indian J Nephrol*. Apr 1 13:64–8.
23. Eddleston M, Wilks MF, Buckley NA (2003). Prospects for treatment of paraquat-induced lung fibrosis with immunosuppressive drugs and the need for better prediction of outcome: *A systematic review*. Qjm. Nov 1; 96(11):809–24.
24. Lee EY, Hwang KY, Yang JO, Hong SY (2002). Predictors of survival after acute paraquat poisoning. *Toxicology and industrial health*. 18(4):201–6.

Impending Inquisitions in Humanities and Sciences (ICIIHS-2022) – Dr. Mohan Varkolu et al. (eds)

Synthesis and Characterization of Novel 1,3-indanedione Pyrazole Derivatives as Potential Antimicrobial, Antioxidant Agents

72

S. Sandhya Rani*
Department of Pharmaceutical Chemistry,
Chilkur Balaji college of Pharmacy, Moinabad, R. R. Dist., India

N. Swarna, B. Lakshmi Kalyani
Department of Pharmaceutical Analysis,
Chilkur Balaji college of Pharmacy, Moinabad, R. R. Dist., India

ABSTRACT In quest of elucidating the diverse pharmacological properties of 1,3-indanedione derivatives, we have synthesized a new series of 1,3-indanedione derivatives via Knoevenagel condensation reaction mechanism, by condensation of 1,3-indanedione and different aldehydes to form a styrylated indanediones as first step products, synthesized derivatives are cyclized at their chalcone moieties with the hydrazine hydrate and glacial acetic acid quickly to form indanedione pyrazole derivatives. Structures of newly synthesized compounds were characterized by 1H and 13C NMR, FT-IR and Mass spectral analysis. The synthesized compounds are evaluated for their pharmacological activities being like antioxidant, antimicrobial and antifungal activities. The newly synthesized compounds are evaluated for their antimicrobial activity (by using Cup- Plate method) agonistic to selected bacterial strains amongst S. aureus, E. coli among total compounds, compound II4, II6, II7, II10 are found to have higher activity against selected strains and the results were found to be as moderate activity, for anti-fungal activity- compound II2 is more potent, and compounds are evaluated for their antioxidant NO scavenging compound II8 is having more activity, for DPPH scavenging compound II9 is having more activity, and inhibition of lipid peroxidation compounds II4 and II10 are having more effective activity, for anti-oxidative activity moderate outcome has demonstrated. The potential importance of the pharmacophore, styryl, pyrazole notices a role in the development of new potential anti- bacterial, anti-fungal, and anti-oxidative agents.

KEYWORDS Indanediones, Indanedione pyrazole derivatives, Antibacterial agents, Ant oxidative agents, Antifungal agents

1 Introduction

Indane-1,3-dione is the compound having benzene fused with cyclopentane-1,3-dione. The last two decades have witnessed profound changes in indane-1,3-dione chemistry in both quality and quantity. Synthesis of compounds in a number up to now unexplored fields has been developed.1,3-Indanedione, a potent pharmacophore. An aromatic nucleus has gained prominence in medicinal chemistry during the years. The β-di carbonyl moiety of the Indanedione is established as an important starting material in various organic transformations because of its cost effective, eco-friendliness and operational simplicity, easy to handle, harmless properties and affording higher yields of corresponding products (Asadi *et al.*, 2017). 1,3-Indanediones were found to have variety of pharmacological activities including ani-coagulant, anti-convulsant, anti-microbial (Durden *et al.*, 1975), anti-inflammatory (Rosini *et al.*, 1976), anti-oxidative and also cytotoxic activity etc. Their diverse biological and

*Corresponding author: ranisandhya543@gmail.com

DOI: 10.1201/9781003489436-72

chemical application was created interest in researchers who have prepared derivatives on pharmacophore based synthesis.1,3-Indanedione itself is an anti-coagulant but it can possess pharmacophore based activities with its pyrazole derivatives (Naim *et al.*, 2016). At 2nd position of the 1,3-indanedione structure serves as a key group for the synthesis of structurally complex compounds via condensation. The styryl moiety which is introduced by the condensation of Indanedione (the active methylene group) and different aldehydes were proved with antimicrobial (Jeyachandran *et al.*, 2011), anti-oxidative activities (Van Den Berg *et al.*, 1975). The pyrazole moiety containing a novel indanedione derivatives were prepared by cyclizing the styrylated Indanedione derivatives containing chalcone moiety and phenyl hydrazine along with the glacial acetic acid as a solvent mix and at suitable atmospheric conditions, prepared compounds are ready for the screening of different biological activities. In this present work, we have designed and synthesized a novel series of 1,3-indanedionepyrazole derivatives and evaluated for their pharmacological activities such as antioxidant, antimicrobial and antifungal activities.

2. Materials and Methods

2.1 Reaction Scheme for the Synthesis of Novel 1,3-Indanedione Pyrazole Derivatives

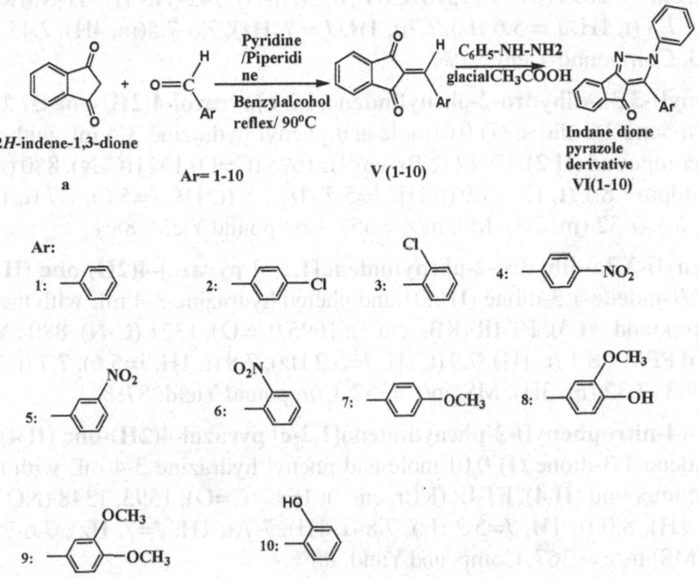

All the chemicals used in the synthesis of the intermediates and final derivatives are of analytical grade and obtained from S.D fine chem. Limited (Mumbai). Purity of the synthesized compounds was analyzed by using TLC plate, silica gel-G as adsorbent and solvent system (or) mobile phase was used with various ratios of hexane, chloroform, ethanol, ethyl acetate appropriately. Rf values produced for each compound were correlating with the literature and assumed to be pure. Characterizations of synthesized compounds were interpreted by FT-IR, 1H-NMR, 13C-NMR and Mass Spectroscopy.

2.2 Procedure for the Synthesis of novel 1,3-indanedione Pyrazole Derivatives

Step-1: Synthesis of 2-arylidene-2H-indene-1,3-dione compounds I (1-10) styrylated indanedione derivatives: Condensation reaction of substituted benzaldehydes with indanedione is a classic general method for the preparation of styrylated indanedione derivatives. In this reaction a respective aldehyde 0.1mole, indanedione 0.73g (0.01 mole) and benzyl alcohol 15 mL required in presence of a base piperidine 2-3 ml refluxed at 90°C for 10-12 hrs. with a knoevenagel condensation reaction (Dubey *et al.*, 2011), the active methylene group is reacted initially with the aldehyde to form styrylated indanedione derivatives I (1-10), in where Ar = (1) Phenyl; (2) 4-chlorophenyl; (3) 2-chlorophenyl; (4) 4-nitrophenyl; (5) 3-nitrophenyl; (6) 2-nitrophenyl; (7) 4-methoxy phenyl; (8) 4-hydroxy, 3-methoxy phenyl; (9) 3, 4-dimethoxyphenyl and (10) 2-hydroxy phenyl. Synthesized compounds were purified by using column chromatography by running the mobile phase containing hexane and methanol, purified products were dried, used for further synthesis.

Synthesis of 2-benzylidene-2H-indene-1,3-dionecompound (I1): 1,3-Indanedionestyryl derivative **I1** was synthesized by condensation of 1,3-indanedione (0.73g) and benzaldehyde 0.1 mole b by using piperidine (1 mL) as base, benzyl alcohol as a solvent (15mL) with a knoevenagel condensation reaction (Sandhya Rani et al., 2018; Sandhya Rani *et al.*, 2019) by refluxing 4 -5 hr at 90oC temperature. The reaction mixture was cooled and the compound was filtered and air dried. Further the obtained compound purified by column chromatography, in that the compound containing chloroform fraction was collected and evaporate the solvent to get the pure compound. FT-IR (KBr, cm^{-1}): 3066 (=C-H), 1726 (C=O). ^{1}H-NMR (300 MHz, (CDC13, PPM): 8.67 (s, 1H), 8.44 (s, 1H), 8.35 (t, 2H), 8.05-7.9 (m, *J*=5.5, 3H), 7.86 (d, 1H), 7.27 (d, 1H). MS: m/z =233. Compound Yield: 90%.

Step-2: Synthesis of indanedione pyrazole derivatives II (1-10):

In the formation of indanedione pyrazole derivatives, reaction was carried out by 1,3-indanedione styryl derivative compound I (1-10) 0.01 mole and phenyl hydrazine 3-4 mL with the solvent glacial acetic acid 10 mL by cyclization to form heterocyclic compounds II (1-10). Formed compounds were purified by column chromatography by using mobile phase hexane and methanol with 1:2 ratios (Vitinzelinskaite *et al.*, 2014).

Synthesis of 3,3a-dihydro-2,3-diphenylindeno[1,2-c] pyrazol-4(2H)-one (II 1): Reaction was carried out by 2-benzylidene-2H-indene-1,3-dione (**I**) 0.01 mole and phenyl hydrazine 3-4 mL with the solvent glacial acetic acid 10mL to form a heterocyclic compound (**II 1**). FT-IR (KBr, cm^{-1}): 1695 (C=O), 1346 (C-N), 833(Ar-H), 742 (Ar- H). ^{1}H-NMR (300 MHz, $CDCl_3$, dppm):8.0 (t, 1H), 7.9(t, 1H, J = 5.2 Hz), 7.8 (t, 1H, *J* = 5.6 Hz), 7.7(t, 1H, *J* = 7. Hz), 7.6-7.5(m, 4H), 7.45 -7.4 (m, 3H, *J* = 7.6 Hz), 7.3.-7.22 (m, 3H). MS: m/z = 323. Compound Yield: 91%.

Synthesis of 3-(4-chlorophenyl)-3,3a-dihydro-2-phenylindeno[1,2-c]pyrazol-4(2H)-one (II 2): Reaction was carried out by 2-(4-chlorobenzylidene)-2*H*-indene-1,3-dione (**I**) 0.01mole and phenyl hydrazine 3-4 mL with the solvent glacial acetic acid 10mL to form a heterocyclic compound (**II 2**). FT-IR (KBr, cm^{-1}): 1695 (C=O) 1321(C-N), 880 (Ar-H), 742 (Ar-H).762 (C-Cl). ^{1}H NMR (300 MHz, $CDCl_3$, dppm): 8.1 (t, 1H), 7.9 (t, 1H, *J*=5.2 Hz),7.8 (t, 1H, *J*=5.6), 7.7 (t, 1H, *J*=7. Hz), 7.6-7.5 (m, 3H), 7.45 -7.4 (m, 3H, *J*=7.6 Hz), 7.3.-7.32 (m, 3H). MS: m/z = 357. Compound Yield: 88%.

Synthesis of 3-(2-chlorophenyl)-3,3a-dihydro-2-phenylindeno[1,2-c] pyrazol-4(2H)-one (II 3): Reaction was carried out by 2-(2-chlorobenzylidene)-2*H*-indene-1,3-dione (**I**) 0.01 and phenyl hydrazine 3-4 mL with the solvent glacial acetic acid 10 mL to form a heterocyclic compound (**II 3**). FT-IR (KBr, cm^{-1}): 1695 (C=O), 1321 (C-N), 880 (Ar-H), 742 (Ar- H),762 (C-Cl). ^{1}H-NMR (300 MHZ, $CDCl_3$, d PPM): 8.1 (t, 1H), 7.9 (t, 1H, *J*=5.2 Hz), 7.8 (t, 1H, J=5.6), 7.7 (t, 1H, J=7. Hz), 7.6-7.5 (m, 3H), 7.45-7.4 (m, 3H, *J*=7.6 Hz), 7.3.-7.32 (m, 3H). MS: m/z =357. Compound Yield: 87%.

Synthesis of 3,3a-dihydro-3-(4-nitrophenyl)-2-phenylindeno[1,2-c] pyrazol-4(2H)-one (II 4): Reaction was carried out by 2-(4-nitrobenzylidene)-2*H*-indene-1,3-dione (**I**) 0.01 mole and phenyl hydrazine 3-4 mL with the solvent glacial acetic acid 10mL to form a heterocyclic compound (**II 4**). FT-IR (KBr, cm^{-1}): 1695 (C=O); 1595, 1348 (NO2), 1350 (C-N). ^{1}H-NMR (300 MHz, $CDCl_3$, dppm: 8.12 (t, 1H), 8.0 (t, 1H, *J*=5.2 Hz), 7.8 (t, 1H), 7.7(t, 1H, *J*=7. Hz), 7.6-7.5 (m, 3H), 7.45 -7.4 (m, 3H, *J*=7.6 Hz), 7.3.-7.3 (m, 3H). MS: m/z = 367. Compound Yield: 88%.

Synthesis of 3,3a-dihydro-3-(3-nitrophenyl)-2-phenylindeno[1,2-c] pyrazol-4(2H)-one (II 5): Reaction was carried out by 2-(3-nitrobenzylidene)-2*H*-indene-1,3-dione (**I**) 0.01 mole and phenyl hydrazine 3-4 mL with the solvent glacial acetic acid 10 mL to form a heterocyclic compound (**II 5**). FT-IR (KBr, cm^{-1}):1695 (C=O), 1595, 1375 (NO2); 1350 (C-N); ^{1}H-NMR (300 MHz, $CDCl_3$, dppm): 8.12(t, 1H), 8.0 (t, 1H, *J*=5.2 Hz), 7.8 (t, 1H), 7.7 (t, 1H, *J*=7. Hz), 7.6-7.5 (m, 3H), 7.45 -7.4 (m, 3H, *J*=7.6 Hz), 7.3.-7.32(m, 3H). MS: m/z = 367. Compound Yield: 86%.

Synthesis of3,3a-dihydro-3-(2-nitrophenyl)-2-phenylindeno[1,2-c] pyrazol-4(2H)-one (II 6): Reaction was carried out by 2-(2-nitrobenzylidene)-2*H*-indene-1,3-dione (**I**) 0.01 mole and phenyl hydrazine 3-4 mL with the solvent glacial acetic acid 10 mL to form a heterocyclic compound (**II 6**). FT-IR (KBr, cm^{-1}): 1695 (C=O); 1595, 1375 (NO2); 1350 (C- N). ^{1}HNMR (300 MHz, $CDCl_3$, dppm): 8.12(t, 1H), 8.0 (t, 1H, *J*=5.2 Hz), 7.8(t, 1H), 7.7 (t, 1H, *J*=7. Hz), 7.6-7.5 (m, 3H), 7.45 -7.4 (m, 3H, *J*=7.6 Hz), 7.3.-7.32(m, 3H). MS: m/z = 367. Compound Yield: 85%.

Synthesis of 3,3a-dihydro-3-(4-methoxyphenyl)-2-phenylindeno[1,2-c] pyrazol-4(2H)- one (II 7): Reaction was carried out by 2-(4-metoxy benzylidene)-2*H*-indene-1,3-dione (**I**) 0.01 mole and phenyl hydrazine 3-4 mL with the solvent glacial acetic acid 10 mL to form a heterocyclic compound (**II 7**). FT-IR (KBr, cm^{-1}): 2808 (OCH3), 1695 (C=O), 1350 (C-N). ^{1}H-NMR (300 MHz, $CDCl_3$, dppm): 8.12(t, 1H, *J*=2.5), 8.0 (t, 1H, *J*=5.2 Hz), 7.8(t, 1H, *J*=5.6), 7.7(t, 1H, *J*=7. Hz), 7.6-7.5 (m, 3H), 7.45 -7.4 (m, 3H), 7.3.-7.32 (m, 3H). MS: m/z = 353. Compound Yield: 85%.

Synthesis of 3,3a-dihydro-3-(4-hydroxy-3-methoxyphenyl)-2-phenylindeno[1,2-c] pyrazol-4(2H)-one (II 8): Reaction was carried out by 2-(4-hydroxy, 3-methoxy benzylidene)-2*H*-indene-1,3-dione (**I**) 0.01 mole and phenyl hydrazine 3-4 mL with the solvent glacial acetic acid 10mL to form a heterocyclic compound (**II 8**). FT-IR (KBr, cm^{-1}): 3437 (OH), 2974 (OCH3), 1631(C=C), 1348 (C-N). ^{1}H-NMR (300 MHz, $CDCl_3$, dppm): 8.14(t, 1H, *J*=2.5), 7.8(t, 1H, *J*=5.2), 7.7(t, 1H,

J=5.6), 7.58(m, 2H), 7.52-7.57(m, 3H),7.4(s, 1H), 7.34-7.47 (m, 3H), 5.05 (s, 1H, OH), 3.49 (s, 3H, CH3). MS: m/z = 371. Compound Yield: 84%.

Synthesis of 3,3a-dihydro-3-(3,4-dimethoxyphenyl)-2-phenylindeno[1,2-c] pyrazol- 4(2H)-one (II 9): Reaction was carried out by 2-(3,4-dimethoxy benzylidene)-2*H*-indene-1,3-dione (**I**) 0.01 mole and phenyl hydrazine 3-4 mL with the solvent glacial acetic acid 10 mL to form a heterocyclic compound (**II 9**). FT-IR (KBr, cm^{-1}): 2856, 2914(OCH3),1730(C=O),1591 (C=C), 1350 (C-N). ^{1}H-NMR (300 MHz, $CDCl_3$, d ppm): 8.04(t, 1H, *J*=2.5), 7.9(t, 1H), 7.6(t, 1H, *J*=5.2), 7.58(t, 1H, *J*=5.6), 7.5(m, 4H), 7.57-7.52(m, 3H),7.4(s, 1H) 3.65 (s, 3H); 3.73 (s, 3H). MS: m/z = 381. Compound Yield: 84%.

Table 72.1 Physical characteristic data of 1,3-indanedione pyrazole derivatives.

S.No	Compound	Molecular formula	Structure	Yield	Melting Point
1	II1	$C_{22}H_{14}N_2O$	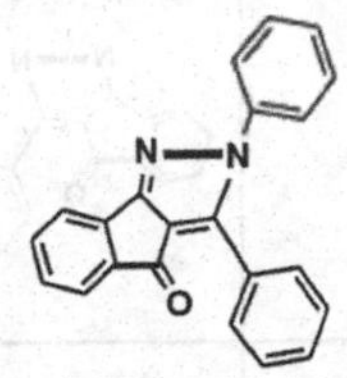	91%	176-178°C
2	II2	$C_{22}H_{13}ClN_2O$	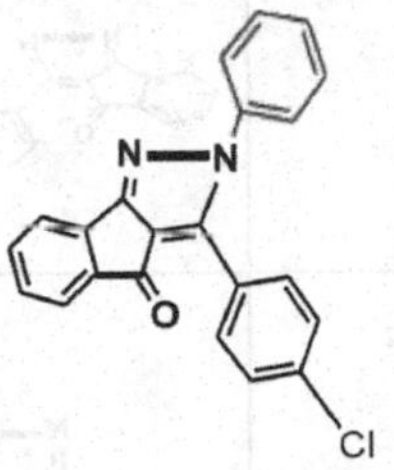	88%	180-184°C
3	II3	$C_{22}H_{13}ClN_2O$	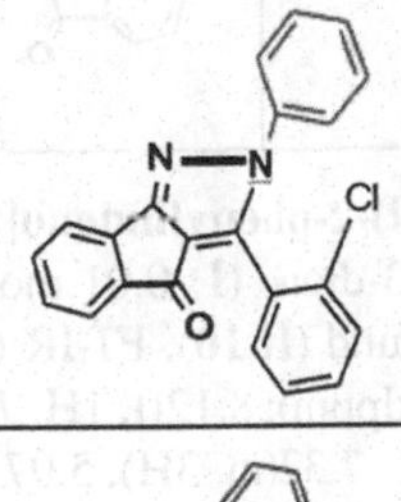	87%	180-183°C
4	II4	$C_{22}H_{13}N_3O_3$	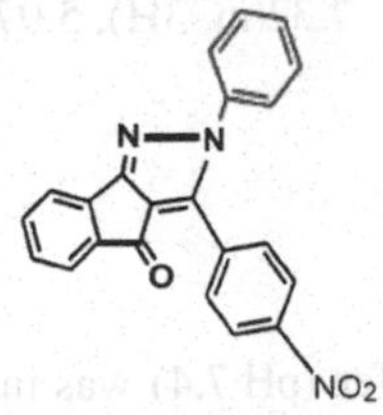	88%	183-186°C
5	II5	$C_{22}H_{13}N_3O_3$	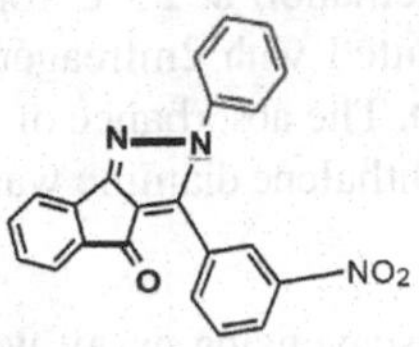	86%	185-187°C
6	II6	$C_{22}H_{13}N_3O_3$	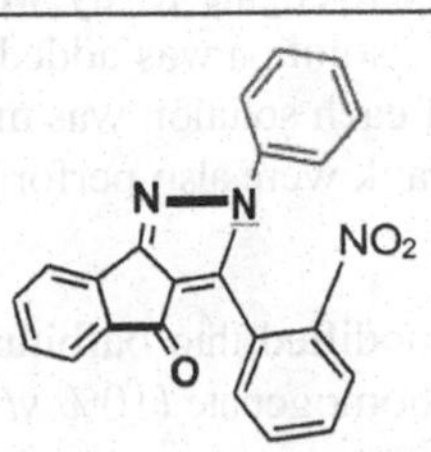	85%	178-182°C

S.No	Compound	Molecular formula	Structure	Yield	Melting Point
7	II7	$C_{23}H_{16}N_2O_2$	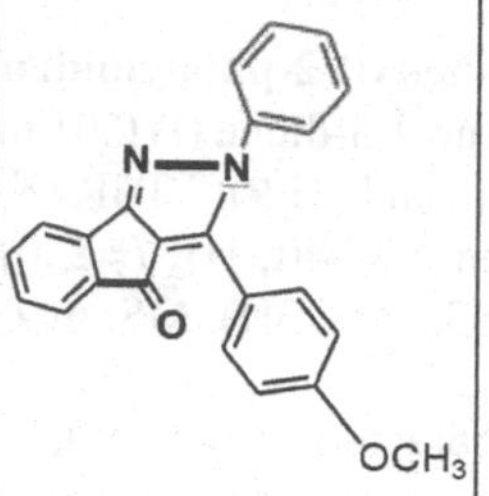	85%	178-182°C
8	II8	$C_{23}H_{16}N_2O_3$	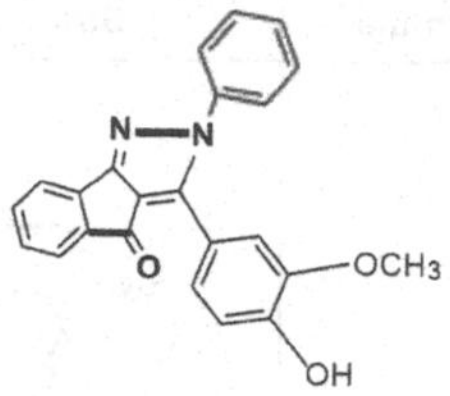	84%	180-183°C
9	II9	$C_{24}H_{18}N_2O_3$	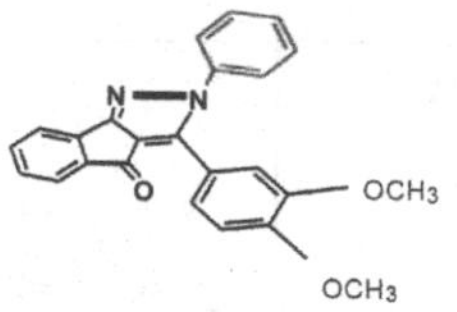	84%	180-185°C
10	II10	$C_{22}H_{14}N_2O_2$	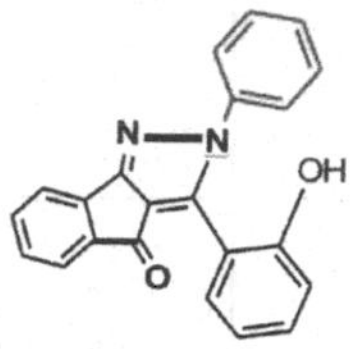	84%	178-185°C

Synthesis of 3,3a-dihydro-3-(2-hydroxyphenyl)-2-phenylindeno[1,2-c] pyrazol-4(2H)-one (II 10): Reaction was carried out by 2-(2-hydroxy benzylidene)-2*H*-indene-1,3-dione (**I**) 0.01 mole and phenyl hydrazine 3-4 mL with the solvent glacial acetic acid 10 mL to form a heterocyclic compound (**II 10**). FT-IR (KBr, cm^{-1}): 3500(OH), 1668 (C=C), 1695 (C=O), 1350 (C-N), 992 (Ar-H). ^{1}H-NMR (300 MHz, $CDCl_3$, dppm): 8.12(t, 1H, *J*=2.5), 8.0 (t, 1H, *J*=5.2 Hz), 7.8(t, 1H, *J*=5.6), 7.7 (t, 1H, *J*=7. Hz), 7.6-7.5 (m, 3H), 7.45-7.4 (m, 3H), 7.3.- 7.32(m, 3H), 5.07 (s, 1H). MS: m/z = 340. Compound Yield: 84%.

2.3 Biological Activity

In vitro Antioxidant activity

Assay of Nitric Oxide scavenging activity

Sodium nitroprusside (10 mM) in phosphate buffer (pH 7.4) was incubated with thenewly synthesized compounds (100 μM) by dissolving in a suitable solvent (dioxane/methanol) at 25°C for 2 hr. Control experiment was conducted under similar conditions. 2 ml of incubated solution was diluted with 2mlreagent (Grriess) (1% sulfanilamide, 2% H_3PO_4 and 0.1% N-(1-naphthyl) ethylene diamine dihydrochloride)). The absorbance of the chromophore, which is formed during diazotization of nitrite with sulphanilamide and subsequent Naphthalene diamine was recorded at 546 nm (Boora *et al.*, 2014).

Interaction with stable free radical DPPH

Di phenyl picryl hydrazyl(DPPH) free radical scavenging assay was carried out by adding 100µl of test compound to the buffer, methanol solution. To this100µl of DPPH solution was added, the mixture was incubated at 37°C for 30 minutes after that without exposing to light, the absorbance of each solution was measured at 540 nm. Control was performed with 100µl of DMSO and DPPH. Sample blank and control blank were also performed (Kedare *et al.*, 2011).

Lipid peroxidation

Lipid peroxide formation was measured by a modified thio barbituric acid-reactive species (TBARS) assay using egg yolk homogenate (lipid rich medium). 0.5ml of egg homogenate (10% v/v) and 0.1ml of the test compounds were added to a test tube and made up to 1ml with distilled water. Lipid peroxidation was induced by adding $FeSO_4$ (0.07M) to the mixture and incubated for 30 min. To this, 1.5ml of 20% acetic acid (pH adjusted to 3.5 with NaOH), 1.5 ml of 0.8% (w/v) TBA in 1.1% sodium dodecyl sulphate and 0.5ml 20% TCA were added and vortexed and heated for 60 min at 95°C. Butanol (5 ml) was added to each tube after cooling it centrifuged at 3000 rpm for 10 min. The absorbance of the upper layer (organic layer) was measured at 532 nm (Upadhyay *et al.*, 2014).

Anti-microbial activity

The antibiotic potency of the newly synthesized derivatives of all the schemes of compounds was determined by using Cup-Plate method against *Pseudomonas aureus* (Gram positive), and *Escherichia coli* (Gram negative) as test organisms. Initially, the prepared nutrient agar medium was sterilized by autoclaving at 15 lbs pressure and 121°C for 25 min. Agar media was cooled to 25°C temperature after the organism was inoculated to the media. From that 15 ml of media was transferred to the petri plates aseptically. Synthesized compounds were properly dissolved in solvent and diluted to get 10mg/ml of concentration, whereas, streptomycin was used as standard drug at a concentration of 10μg/ml. The cultured plates were incubated at 37°C for 24 hrs. The zones of inhibition produced by test compounds were recorded in mm (Kalyani *et al.*, 2017).

Anti-fungal activity

Anti-fungal activity of the newly synthesized derivatives of all the schemes was evaluated by Disc diffusion method. Sabour and dextrose agar plates (5-6 mm) were prepared aseptically and were dried at 37°C before inoculation. *Candida albicans* were inoculated into sabour and dextrose agar plates by using sterile inoculation loop and were incubated at 37°C for about 24hrs. Ketoconazole (10μg/disc) was used as standard. What man No. 2 filter paper disc (5 mm diameter) was sterile and it was soaked into synthesized compounds (20μg/disc) separately and evaporated to dryness and placed on the media. One more disc carefully immersed into dimethyl sulphoxide and placed on the media as control. The petri dishes were incubated at 37°C for 24 hrs, cooled them for an hour in a refrigerator to facilitate uniform diffusion.

3. Results and Discussion

A series of 10 novel 1,3-indanedione pyrazole derivatives were synthesized by 1,3-indanedione condensed with different aldehydes to form the styrylated indanedione derivatives, which can further have cyclized in the presence of phenyl hydrazine and glacial acetic acid. Formed products were separated using column chromatography and purified by recrystallization method. All the Synthesized compounds were characterized by IR, NMR and Mass spectrometric data.

Biological activity: All the Synthesized compounds II (1-10) were evaluated for antioxidant, antimicrobial activities.

3.1 Antioxidant Activity of 1,3-indanedione Pyrazole Derivatives

The antioxidant activity of newly synthesized1,3-indanedione pyrazole derivatives have been determined in terms of % scavenging of NO and DPPH and % inhibition of lipid peroxidation and presented in Table 72.2 and Fig. 72.1. Among all

Table 72.2 Antioxidant activity of newly synthesized 1,3-indanedione derivatives II (1-10).

Compound	Percentage Scavenging Nitric Oxide activity	Percentage DPPH Scavenging	Percentage Inhibition of Lipid Peroxidation
II 1	14.5	16.75	40.99
2	12.54	18.56	41.24
3	17.35	16.65	42.75
4	18.48	8.9	52.85
5	22.21	19.29	49.34
6	25.38	19.12	50.65
7	20.82	18.25	43.66
8	29.9	23.41	41.63
9	22.7	25.16	48.23
10	16.5	20.23	50.98

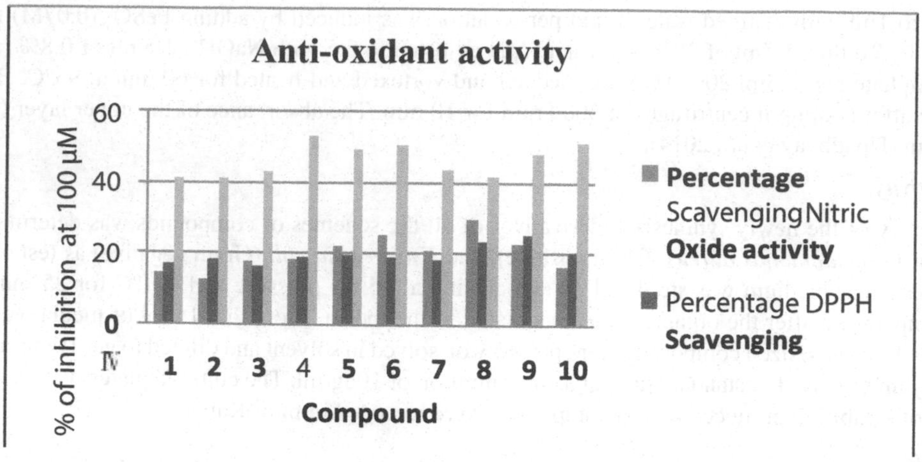

Fig. 72.1 Antioxidant activity of synthesized 1,3 indanedione derivatives II (1-10)

the tested compounds few are exhibited moderate and remaining are to be exhibited low scavenging activity towards NO and DPPH, and exhibited substantial inhibition of lipid peroxidation. NO scavenging is ranged between 12.54% (compound 2) and 29.9% (compound 8), whereas DPPH scavenging is ranged between 16.65% (compound 3) and 25.16% (compound 9). The antioxidant activity in terms inhibition of lipid peroxidation is ranged from 40.99%(compound 1) to 52.85% (Compound 4). **Standard (Vitamin-C)** action was found to be 44.78% for scavenging of NO while 49.56% for scavenging of DPPH and the %inhibition of lipid peroxidation 66.78%.

3.2 Antibacterial and Antifungal activities

The antibacterial and antifungal activities of the newly synthesized 1,3-indanedione derivatives (scheme-3) against bacterial/fungal strains have been determined and are shown in the Table 72.3 and Fig. 72.2. Zone of inhibition the bacterial strain against *P. aureus* is ranged from 10 mm (compounds 9, 10) to 13 mm (compounds 6), whereas, against *E. coli* is ranged from 16 mm (compound 3, 5, 8, 9) to 18 mm (compounds 4, 7, 10). Compound 2 has showed as highest antifungal activity 13 mm against *Candida. albicans*, followed by (compounds 8, 6) 12 mm from Table 72.3 and Fig. 72.2. Zone of inhibition values of standard streptomycin is 15 mm and 21 mm against *P. aureus* and *E. coli* respectively for antibacterial and ketoconazole is 14 mm for antifungal activities.

Table 72.3 Antibacterial and antifungal activities of novel 1,3-indanedionederivatives II (1-10)

Compound	*P. aureus*	Zone	of inhibition (mm) *E. coli*	*C. albicans*
II 1	11		16	10
2	12		17	13
3	11		16	11
4	12		18	11
5	12		16	10
6	13		17	12
7	12		18	11
8	11		16	12
9	10		16	11
10	10		18	10
Standard	15		21	14

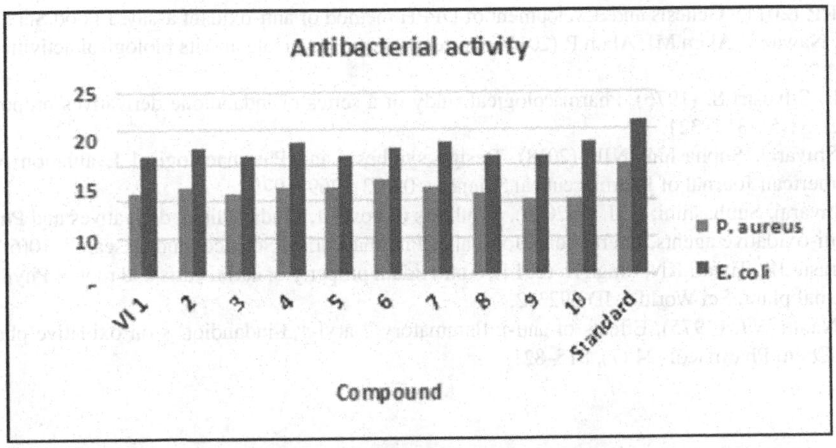

Fig. 72.2 Antibacterial activity of novel 1,3-indanediones II (1-10)

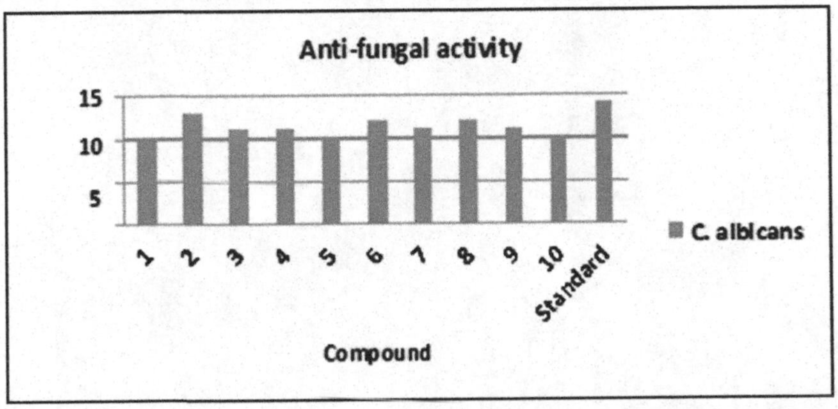

Fig. 72.3 Antifungal activity of novel 1,3-indanedione derivatives II (1-10)

4. Conclusion

1,3-Indanedione derivatives exhibited significant activities at moderate doses. Novel series of 1,3-indanedione derivatives **1,3-indanedione pyrazole derivatives** were evaluated for their pharmacological potentials in terms of antioxidant, antibacterial and antifungal activities. Compounds with phenolic group as the terminal substitution have shown to higher antioxidant activity among the tested compounds, and remaining compounds have shown minimum to moderate activities at moderate doses. Compounds with para substitution with chloro, ortho position with –nitro, para-hydroxy group were showed with higher antibacterial and antifungal activities compared to other substitutions remaining compounds, which have shown minimum to moderate activities.

REFERENCES

1. Asadi S, Ziarani GM. (2016), The molecular diversity scope of 1,3-indandione in organic synthesis, Mol Divers. 20(1), 111–152.
2. Boora F, Chirisa E, Mukanganyama S. (2014), Evaluation of nitrite radical scavenging properties of selected Zimbabwean plant extracts and their phyto constituents, J Food Proc. ID 918018.
3. Dubey PK, Venkatanarayana M. (2011), Synthesis of 2-Arylideneindane-1,3-diones by Knoevenagel Condensation, J. Chem. Sci. 123, 609.
4. Durden JA, Biocidal activity of 1,3-indanedione and related compounds. (1975), Med Chem. 1975, 10, 143.
5. Jeyachandran M, Ramesh P. (2011), Synthesis, Antimicrobial, and Anticoagulant activities of 2-(Aryl sulfonyl) indane-1,3-diones. Org Chem Int. ID 360810.
6. Kalyani G, Srinivas B, Sastry KV, Vijaya K. (2017), Synthesis, characterization and pharmacological evaluation of new mannich bases with coumarin derivatives, Der Pharma Chemica. 9 (13): 61–64.

7. Kedare SB, Singh RP. (2011), Genesis and development of DPPH method of anti-oxidant assay, J Food Sci Technol. 48(4), 412–422.
8. Naim MJ, Alam O, Nawaz F, Alam MJ, Alam P. (2016), Current status of pyrazole and its biological activities, J Pharm Bio allied Sci. 8(1), 2–17.
9. Rosini S, Trallori L, Silvestri S. (1976), Pharmacological study of a series of indandione derivatives proposed as anti-inflammatory agents, Farmaco Sci. 31(5), 315–321.
10. Sandhya Rani S, Shivaraj, Subhashini NJP. (2018), Design synthesis and Pharmacological Evaluation of novel 1,3-Indonedione derivatives, Indo American Journal of Pharmaceutical Sciences. 05(03)1969–1979.
11. Sandhya Rani S, Shivaraj, Subhashini. N. J. P. (2019), Synthesis of novel 1,3-indanedione derivatives and Pharmacological evaluation as antimicrobial, anti-oxidative agents, International Journal of Pharmaceutical Sciences and Research. 10(6)2838–2844.
12. Upadhyay R, Chaurasia JK, Tiwari KN, Singh K. (2014), Antioxidant property of aerial parts and root of Phyllanthus fraternus Webster, an important medicinal plant, Sci World J. ID 692392.
13. Van Den Berg G, Nauta WT. (1975), Effects of anti-inflammatory 2-aryl-1,3-indandiones on oxidative phosphorylation in rat liver mitochondria, Bio. Chem Pharmacol. 24 (7), 815–821.

Impending Inquisitions in Humanities and Sciences (ICIIHS-2022) – Dr. Mohan Varkolu et al. (eds)
© 2024 Taylor & Francis Group, London, ISBN 978-1-032-78829-6

An Empirical Study of Management's Perceptions towards Quality Management in Technical Education

73

K. V. S. Prasad*
Associate Professor, GMR Institute of Technology, RAJAM, Andhra Pradesh, India

ABSTRACT This paper aims to present the opinions of management members of various technical education institutions about management of quality in technical education. This study is an attempt to understand how the opinions differ among several factors such as gender, age, qualification, and experience in teaching of the management members of selected technical education institutes. This study collected data from 50 respondents of selected 7 technical education institutions in the State of Andhra Pradesh and SPSS was used to analyse their observations. This research will add to the body of knowledge in this field and will serve as a catalyst for future research.

KEYWORDS Quality, Technical Education, Quality Management, Management's Perception

1. Introduction

The progress of a country does not depend on the prevailing means but on effective deployment of these resources. We essential a technically equipped human force for the effective operation of increasing economic prosperity by the application of Science and Technology to information technology, communication, other industries and also agriculture sector etc. The practise of development cannot be strengthened until competent technical education is offered to the nation's youth; consequently, engineering education plays an important part in the growth of national production, prosperity, and self-sufficiency. We want improved and competent technical education structure in India and it should have finest set-up to create the best engineering professionals. The Indian Government and private institutions has been taken initiative to form good institutes in our country.

2. Objectives of the Study

- To study the perception of management members of engineering institutions towards quality management.
- To study the difference between gender, age groups, educational qualification and teaching experience among various management members of engineering institutions.

3. Research Methodology

The study was conducted for fifty management members of seven selected engineering institutions in Srikakulam and Vizianagarm Districts of Andhra Pradesh. The management members include Directors, Principals, Vice-Principals, Deans, HODs etc. Primary data collected from the sample respondents' through questionnaire. This comprehensive questionnaire comprises some key quality aspects i.e. Quality Management, effective leadership, providing quality learning resources, teaching

*prasad.kvs@gmrit.edu.in, prasadkvs3@gmail.com

DOI: 10.1201/9781003489436-73

leaning, evaluation process etc. These quality aspects are sub-divided and scattered randomly to the sample respondents and requested them to give ranks for their answers on Likert five points scale. Secondary data collected through Government reports, journals, magazines, books, research papers, various handbooks of AICTE, various official websites and the data was analysed with SPSS.

4. Review of Literature

Oliveira, Oliveira and Costa [2012] conducted a study on Students' and teachers' views about quality of engineering education in Portugal and their key conclusion is that both students and teachers refer to the similar characteristics that must exist in the quality of education. Irfan A. Gulbarga, Soma. V. Chetty, J. P. Ganjigatti and Suniel Prakash [2012] described that quality is important aspect in all institutions mainly technical education. Nair, Patil and Mertova [2011] conducted a study on enhancing the quality of engineering education by exploiting student feedback. R. Chakka and G. T. Kulkarni [2010] expressed the stress on enrichment of teaching quality and learning processes through total quality management and described the evaluation of teaching quality by peer reviewing, the methods to achieve teaching quality, student feedback and evaluation of learning process. Rao & Pandi [2007] made significant contributions through their study on the topic of 'Quality Enhancement in engineering Institutions through Knowledge Management and Total Quality Management'. Mitra[2007] conducted a study on Graph Theory Approach for TQM Evaluation of Technical Institutions and said that Infrastructure, Faculty, curriculum, stakeholder and System and policies have an effect on quality education. Sakthivel et al., [2005] have concluded from the views of students' that the ISO 9001:2000 certified engineering institutions are moving towards the path of TQM offering better quality of service than the non-ISO certified institutions.

5. Empirical Results

Table 73.1 Perceptive score difference among sample management respondents based on their qualification

Statement	Highest Qualification	[N]	[Mean]	[Std. Deviation]	[Standard Error]	[f-value]	[p-value]
Management & Leadership (ML)	Degree	8	53.25	8.031	2.839	0.397	0.675
	PG	20	50.25	8.632	1.930		
	Above P.G	22	50.82	7.682	1.638		
	Total	50	50.98	8.029	1.136		
Infrastructure and Learning Resources (ILR)	Degree	8	45.25	10.306	3.644	1.386	0.260
	PG	20	37.65	15.608	3.490		
	Above P.G	22	35.41	14.295	3.048		
	Total	50	37.88	14.451	2.044		
Teaching Learning Process and Research (TLPR)	Degree	8	46.88	14.466	5.115	0.360	0.700
	PG	20	50.40	9.185	2.054		
	Above P.G	22	50.23	10.076	2.148		
	Total	50	49.76	10.380	1.468		

The table 73.1 represents the perceived score difference among different sample management members based on their qualification. It is observed that for ML, the average score of degree qualified respondents found highest than the other i.e. Above P.G and P.G. i.e. 50.82 and 50.25 respectively. In this distribution of mean values the tested f-value is 0.397 which is not significant as the p-value is 0.675 which is higher than 0.05. Hence, it indicates that degree qualified respondents found more satisfied than the others for ML. The average perceived score of management persons towards ILR shows that the average value of degree qualified respondents is 45.25 and found it is higher than P.G. (37.65) and Above P.G. (35.41) qualified respondents. The f-value of 1.386, on the other hand, is not significant because the p-value is 0.260, which is more than 0.05. Hence, it infers that the ILR gives higher satisfaction to the degree qualified respondents. It is found that the average score of P.G. qualified respondents is 50.40 regarding the TLPR in their organization, which is higher than above P.G. (50.23) and degree (46.88) qualified respondents, because the p-value 0.700 is more than 0.05, the f-value 0.360 is not significant. The results indicate that P.G. qualified respondents are greater level of attitude about the TLP and the research in their institutes.

Table 73.2 Perceptive score differences among sample management respondents based on their gender

Statement	Gender	[N]	[Mean]	[Std. Deviation]	[Standard Error]	[f-value]	[p-value]
Management & Leadership (ML)	Male	36	49.97	7.792	1.299	1.396	0.177
	Female	14	53.57	8.336	2.228		
Infrastructure and Learning Resources (ILR)	Male	36	35.00	14.652	2.442	2.652*	0.013
	Female	14	45.29	11.276	3.014		
Teaching Leaning Process and Research TLPR)	Male	36	51.06	7.375	1.229	1.065	0.303
	Female	14	46.43	15.590	4.166		

*significant @5% level

A comparison of the average perceptive scores of sample management members based on their gender details is presented in the table 73.2. It is observed that mean value for female respondents is 53.57, which is higher than the male respondents i.e. 49.97 for ML. Because the p-value is 0.177, the estimated t-value for these differences in mean values is 1.396, which is not significant. It implies that female management members in the sample are extra positive about quality of' ML. The average score of female respondents in the 'infrastructure and learning resources' category is 45.29, which is higher than male respondents (35.00). The tested t-value in this distribution of mean values is 2.652, and it is significant at the 5% level because the p-value is 0.013, which is less than 0.05. Therefore, it indicates that sample female management members are more favourable about the ILR. It is found that the sample male respondents are perceived an average score of 51.06 for 'teaching leaning process and research' which is higher than the sample female respondents i.e. of 46.43. The estimated t-value is 1.065, and the p-value is 0.303, which is more than 0.05, thus it is not significant. This shows that the sample male management members are positive towards the TLPR.

Table 72.3 Perceptive score difference among the sample management respondents based on their age group

Statement	Age (in years)	[N]	[Mean]	[Std. Deviation]	[Standard Error]	[f-value]	[p-value]
Management and Leadership (ML)	Below 30	9	53.33	4.42	1.47	1.042	0.383
	30-40	14	48.50	10.95	2.93		
	40-50	22	52.18	7.69	1.64		
	Above 50	5	48.40	0.55	0.25		
	Total	50	50.98	8.03	1.14		
Infrastructure and Learning Resources (ILR)	Below 30	9	43.22	13.76	4.59	7.295**	0.000
	30-40	14	30.00	14.77	3.95		
	40-50	22	44.32	11.09	2.36		
	Above 50	5	22.00	0.71	0.32		
	Total	50	37.88	14.45	2.04		
Teaching Learning Process and Research (TLPR)	Below 30	9	52.44	5.13	1.71	0.822	0.489
	30-40	14	51.86	8.93	2.39		
	40-50	22	47.23	13.34	2.84		
	Above 50	5	50.20	2.68	1.20		
	Total	50	49.76	10.38	1.47		

**Significant @1%level

The table 73.3 represents the perceptive score difference among sample management members based on their age group. It is noticed that regarding the 'management and leadership' the average perceived score of below 30 years age group (53.33) of sample management members is found higher than the other age group respondents i.e. 40-50 years (52.18), 30-40 years

(48.50) and above 50 years (48.40). The 'f' value of 1.042 is not significant in this distribution of mean values because the calculated 'p' value is 0.383. Hence, it indicates that below 30 years age group respondents are highly satisfied about the ML. The mean score for the 40-50 years age group (44.32) is significantly higher than the other age groups, i.e. below 30 years (43.22), 30-40 years (30.00), and above 50 years (22.00). Because of the p-value 0.000, the f-value 7.295 is notable at the 1% level due to these differences in mean values. This shows that the ILR gives more satisfaction to the respondents between the age group of 40-50 years. The perceptive mean score value of below 30 years of age group respondents is 52.44 which is higher than the 30-40 years (51.86), 40-50 years (47.23) and above 50 years (50.20) for 'Teaching Learning Process and Research'. Because the p-value 0.489 is more than 0.05, the calculated f-value of 0.822 is not significant. It shows that respondents under the age of 30 provided a good answer to TLPR.

Table 73.4 Perceptive score difference among sample management respondents by their experience in teaching

Statement	Experience in Teaching	[N]	[Mean]	[Std. Deviation]	[Standard Error]	[f-value]	[p-value]
Management & Leadership (ML)	< 5 years	11	55.45	7.738	2.333	2.134	0.109
	5 - 10 years	20	51.30	8.791	1.966		
	Above 10 years	14	47.93	7.364	1.968		
	No Experience	5	48.40	0.548	0.245		
	Total	50	50.98	8.029	1.136		
Infrastructure and Learning Resources (ILR)	< 5 years	11	48.18	11.026	3.324	7.495**	0.000
	5 - 10 years	20	41.60	13.578	3.036		
	Above 10 years	14	29.07	12.413	3.317		
	No Experience	5	25.00	6.164	2.757		
	Total	50	37.88	14.451	2.044		
Teaching Learning Process and Research (TLPR)	< 5 years	11	53.45	6.330	1.909	0.756	0.524
	5 - 10 years	20	49.85	10.698	2.392		
	Above 10 years	14	47.29	11.512	3.077		
	No Experience	5	48.20	13.535	6.053		
	Total	50	49.76	10.380	1.468		

** Significant @ 1% level

The table 73.4 shows the perceptive score difference among sample management members by their experience in teaching. It can be found that the average perceptive score of below 5 years of experience is 55.45 and it is found higher than the other respondents i.e. 5-10 years and above 10 years with a mean score value of 51.30 and 47.93 respectively. The mean score of the respondents with no experience is 48.40. Because the p-value 0.109 is more than 0.05, the estimated f-value 2.134 is not significant. Hence, we can conclude that most of the respondents who are below 5 years of experience have more positive towards the quality in ML. Further it is found that the ILR measures that the average perceived score of below 5 years of experience members is 48.18, which is found higher while compare with the other respondents i.e. 5-10 years (41.60), above 10 years (29.07) and no experience (25.00). At the 1% level, the f-value of 7.495 is statistically significant because the p-value of 0.000 is less than 0.01. It shows that the 'infrastructure and learning resources' gave more satisfaction for the respondents who have below 5 years of experience in teaching. Regarding the TLPR, it shows the mean score of below 5 years of experience is 53.45 and it is highest than the other respondents i.e. 5-10 years (49.85), above 10 years (47.29) and no experience (48.20). Because the p-value 0.524 is more than 0.05, the f-value 0.756 is not considered significant. With a mean score of 53.45, it reveals that respondents with less than 5 years of experience are extremely satisfied with the 'teaching learning process and research'.

6. Findings and Conclusion

It can be derived from the above study, It found that the perception score difference among sample management members based on their age group the statements 'Management and Leadership' and 'Teaching Learning Process and Research' not found any significant difference, but 'Infrastructure and Learning Resources' found significant. Regarding gender it is found that the significant at 5 percent level for 'Infrastructure and Learning Resources' and no significant for 'Management and Leadership' and 'Teaching Leaning Process and Research'. Regarding the educational qualification of the sample management members, majority of the respondents satisfied with the 'Management & Leadership' and 'Infrastructure & Learning Resources' are having degree qualification, and P.G. qualified respondents have more satisfaction with the 'Teaching Learning Process and Research'. It also found that, the sample management members are highly satisfied with the 'infrastructure and learning resources' as it is found significant at one percent level (7.495), in which below 5 years of experience respondents gave positive response.

REFERENCES

1. Garvin, D.A. (1998), Managing Quality: The Strategic and Competitive Edge, New York, NY: Free Press.
2. Feigenbaum, A.V. (1956): Total Quality Control, Harvard Business Review, 34(6), 93–101.
3. Juran, J. M. (1974), Quality Control Handbook, New York, McGraw-Hill.
4. Crosby, P. B. (1979), Quality is free, New York, McGraw-Hill.
5. Deming. W. E. (1986), Out of the Crisis, Cambridge: Massachusetts Institute of Technology, Center for Advanced Engineering Study.
6. Taguchi, G. (1994), Introduction to Quality Engineering: Designing Quality into Products and Processes, Tokyo: Asian Productivity Organization Ho, S.K.M., Is the ISO9000 Series for Total Quality Management?, International Journal of Quality and Reliability Management, 11(9), 74–89.
7. Flynn, B.B., Schroder, R.G., and Sakakibara, S. (1994): A framework for Quality Management Research and an Associated Measurement Instrument, Journal of Operations Management, 11(4), 339–366.
8. Rao S, Pandi. (2007): Quality Enhancement in Engineering Institutions through Knowledge Management and Total Quality Management', The Journal of Engineering Education, India, XX(3), 10–15.
9. Mitra, R. M (2007): Graph Theory Approach for TQM Evaluation of Technical Institutions, The Journal of Engineering Education, 96–110.
10. Sakthivel, P.B., Rajendran, G. and Raju, R. (2005): TQM implementation and students' satisfaction of academic performance, The TQM Magazine, 17(6), 573–589.
11. Khan F. ((2010), Developing a Total Quality Management Framework for Public Sector Universities in Pakistan, Ph.D. Thesis.
12. Patel, S.G. (2013), TQM in Higher Education Institutions (HEIs).
13. Ali M. and Shastri, R.K. (2010): Implementation of Total Quality Management in Higher Education, Asian Journal of Business Management, 2 (1), 9–16.
14. Oliveira, C. G., Oliveira, P. C., and Costa, N. (2012): Students' and teachers' perspectives about quality of engineering education in Portugal. European Journal of Engineering Education, 37(1), 49–57.
15. Irfan A. Gulbarga, Soma. V. Chetty, J.P. Ganjigatti and Suniel Prakash (2012): Assessing Technical Institutions through the Principles of Total Quality Management: The Empirical Study, International Journal of Scientific and Research Publications, 2(8), 1–9.
16. Nair, C., Patil, A., and Mertova, P. (2011): Enhancing the quality of engineering education by utilizing student feedback. European Journal of Engineering Education, 36(1), 3–12.
17. R. Chakka and G.T. Kulkarni (2010): Total Quality Management In Pedagogy: An Update, Indian J. Pharm. Educ. Res., 44(4).
18. Prasad, K.V.S. (2020), A Case Study of Parent's Perception Towards Quality Management In Engineering Education Institutions. International Journal of Scientific & Technology Research, Vol. 9(4), 47–50.
19. Prasad, K.V.S. (2018), Quality Management in Technical Education: A Case Study on Perceptions of Students. International Journal of Pure and Applied Mathematics, 119(16), 1315–3121.
20. Prasad, K.V.S. (2018), An Empirical Study of Total Quality Management in Engineering Education Institutions: Perspective of Management. International Journal of Mechanical Engineering and Technology, 9(2), 547–555.
21. Prasad, K.V.S. (2018), Teachers' Perception towards Total Quality Management Practices in Technical Education. Jour of Adv Research in Dynamical & Control Systems, 10 (8), 318–322.

Impending Inquisitions in Humanities and Sciences (ICIIHS-2022) – Dr. Mohan Varkolu et al. (eds)

Identification of Plant Diseases Using AI

74

Ravi Boda*, Aysuh Dasgupta, Raj Parikh, VedhaSree G., Sai Nikitha G., Akshitha Mancharla
Department of ECE, Koneru Lakshmiah Educational Foundation, Aziznagar, Hyderabad

ABSTRACT This article proposed building a convolutional neural network (CNN) model that can identify plant diseases. We use image processing with a convolutional neural network (CNN) to detect plant diseases. Farmers typically do not detect new diseases in crops and plants. Plants are therefore not specially treated for any disease or virus. By using the CNN algorithm, plant disease can be accurately identified. Using image processing to detect plant diseases is the best way to predict and obtain accurate results. The accuracy results show that this model outperforms any conventional framing. Learn how to solve this real-world problem using existing tools such as TensorFlow and Keras. Create a CNN model using a dataset from Kaggle or any open source. Then the model is trained and tested.

KEYWORDS CNN, Deep learning, TensorFlow, Pre-processing, Kaggle data-set

1. Introduction

Agriculture is one of the best relations between earth and farmer who can knows nature as an art and science of cultivating the crops and raising livestock. But every month we're observed that, there may be big amount of crop receives damage because of horrific weather, viruses and exceptional vegetation sicknesses. In a few cases, environmental conditions is likewise have an effect on the plant growing. Environmental stress can cripple a plant and makes it extra exposed to sickness or insect assault. Generally, farmer fails to find new diseases on vegetation and flora. So, plant does no longer get specific treatment for specific disorder or viruses. commonly many farmers can't come up with the money for the specialists advises due to lack of cash and other instances like travelling lengthy distance to get the help, and the time-consuming techniques.to overcome all this problem, we've got amazing gear in technology. To make sure great yield we sincerely want to use exceptional techniques and to be had technology. Using photograph technique with machine learning algorithms offer higher result. Right here we have chosen to put into effect neural inter network the use of TensorFlow to come across different sicknesses on plant leaves.

2. Literature Review

In order to find accurate way to predict plant diseases and to find best fit AI algorithm, we went to through various papers and different authors who published papers in this regard. It gave us idea on how various implementations can be done to get the maximum accuracy possible. Collected information and tabulate the data Collected information and tabulated the data.

*Corresponding author: raviou2015@klh.edu.in

DOI: 10.1201/9781003489436-74

S.NO	Journal Name and Publications Year	Title	Technology	Working
1	International Conference On Internet of Things and Intelligence System Analysis [IEEE][2018]	Krishimitr(Farmer Friend): Using Machine Learning to Identify Diseases in plants.	Tensor Flow Framework by CNN model	Calculate CNN model
2	International Conference for Convergence in Technology [IEEE] [2018]	Mango leaf Deficiency Detection using digital image processing and Machine learning.	Image Processing	Calculate only leaf area
3	International Research Journal of Engineering and Technology [IEEE][2018]	Plant Disease Detection Using Machine Learning	Canny Edge Detection Algorithm	Diseases in crop Mostly on leaves
4	ARXIV Organization Of Computer Science [11-04-2016]	Prasanna Mohanty Using Deep Learning	Image Based CNN model	To detect disease in Plants by training a CNN
5	International Journal of Technical Research and Applications [May, 2015]	Malvika Ranjan Using ANN	ANN model	To Capture image of diseased leaf
6	Information Processing In Agriculture ELSEVIER [march, 2017]	Kulkarni Using ANN	Image Processing ANN model	Classified by Gabor filter for feature Extraction

Fig. 74.1 Review on plant disease identification

3. Methodology

Algorithm Introduction

Convolutional neural network CNN is a type of artificial neural network, which is used in Machine learning Fig. 1. Review on Plant Disease Identification algorithms, perceptron, for Supervised learning. CNN is applied to natural language processing, image processing and some kinds of cognitive tasks. There are 5 layers in Convolutional Neural Network. They are: Input layer, Convolution RELU layer, Max pooling layer, Fully Connected layer and Output layer. From the above layers Convolution RELU layer, Max pooling layer are Feature Learning, fully Connected layer is classification. Three important layers are convolutional RELU, Max pooling and fully connected layers.

Convolutional RELU: Convolution layer is core building block for CNN. This layer performs dot product between two matrices. It is main portion for network's computational load. It acts as the activation function applied elementwise to the output of the convolutional operation, allowing the network to learn and extract meaningful features from the input images

Max pooling: Pooling layer replaces output of networks at certain locations by deriving summary statistic of the nearby outputs. It helps in reducing spatial size of the representation, which decreases the required amount of computation and weights. It has been partially replaced in favor of other techniques like stride convolution. Also, this aims to retain more spatial information in the network while reducing the risk of losing relevant features.

Fully Connected layers: the fully connected layers, also known as dense layers, typically appear towards the end of the network architecture. These layers are responsible for high-level reasoning and understanding based on the features learned by earlier convolutional and pooling layers.

The foundational component of CNN is the convolution layer. The dot product between two matrices is performed by this layer. It is essential for computational demand on the network. Pooling to a maximum layer substitutes network output at certain areas with obtaining a statistical summary of the neighboring outputs. It aids in decreasing the representation's spatial size, which declines the necessary number of calculations and weights.

Finally, the neurons in this layer are completely linked with every neuron in the layer before and after it. It is a bias effect that follows matrix multiplication it aids to translate the input and output representations. CNN's primary building block is layer. This layer functions between two matrices and the dot product. It provides a structured approach to train and optimize CNNs for various tasks, ensuring that the model learns meaningful representations from the data and performs well on unseen samples. Adjusting and refining these steps based on specific requirements and challenges of the task is common in CNN development. CNNs leverage these principles to learn and extract meaningful features from images, enabling them to perform tasks like image classification, object detection, segmentation, and more, mirroring human visual perception to a certain extent and achieving state-of-the-art performance in various vision tasks using CNNs, the identification of diseases becomes faster, more accurate, and cost-effective, ultimately aiding in better crop management and reducing agricultural losses.

$$H1 = x1 \star w1 + x2 \star w2 + b \tag{1}$$

H1= Hidden layer

x1, x2=inputs

w1, w2=Weights

b=bias

$$[(n + 2p - fi/s + 1]X [(n + 2p - f)/s + 1 \tag{2}$$

n: Inputs

p: padding

s: strides

f: filter size.

INCEPTION V3: Inception v3 has less computational power and is modified from previous Inception architectures. Inception and adaptation of different network cases turns out to be problem due to uncertainty of new network efficiency. It optimizing several techniques like regularization, Convolutions, dimension reductions and parallelized computations. This have been suggested to loosen the constraints for easier model adaptations. Batch normalization helped in stabilizing and accelerating the training process, while factorized convolutions decomposed larger convolutions into smaller ones to reduce computational cost and increase efficiency. Inception layers in plant disease detection CNNs facilitates the extraction of diverse and informative features from plant images, enabling the network to accurately classify and identify various diseases. The multi-scale feature extraction and efficient utilization of computational resources are key advantages offered by these modules in this context, using CNNs, the identification of diseases becomes faster, more accurate, and cost-effective, ultimately aiding in better crop management and reducing agricultural losses. Inception architectures stack multiple inception modules, allowing for deeper and more expressive models while maintaining computational efficiency. The depth of the network allows for the extraction of hierarchical representations from low-level features like edges and textures to higher-level concepts and patterns. The Inception layer primary goal was to strike a balance between model complexity and computational efficiency, enabling the development of deeper networks while controlling computational costs. This architectural innovation has influenced subsequent designs in convolutional neural networks, paving the way for improved performance in various computer vision tasks. These modules perform convolutions at multiple scales concurrently by using filters of various sizes (1x1, 3x3, 5x5) and pooling operations the Inception architecture's innovation in utilizing diverse filter sizes, pooling operations, and managing computational complexity has significantly advanced the effectiveness and efficiency of CNNs in capturing complex features from images, leading to improved performance in computer vision tasks.

There are five steps in inception Architecture.

1. Factorized Convolutions: They enhance computational efficiency and help to cut down on the amount of parameters used in networks. It watches over the network.
2. Smaller convolutions: This enables for faster
3. Asymmetric Convolutions: It was proposed that there are perhaps more parameters than the Convolution that is asymmetric.
4. Auxiliary Classifier: It is administered to integrate a little CNN between layers while training. The loss is in addition to the total network loss. Inception an extra classifier in version three serves as a regularize.

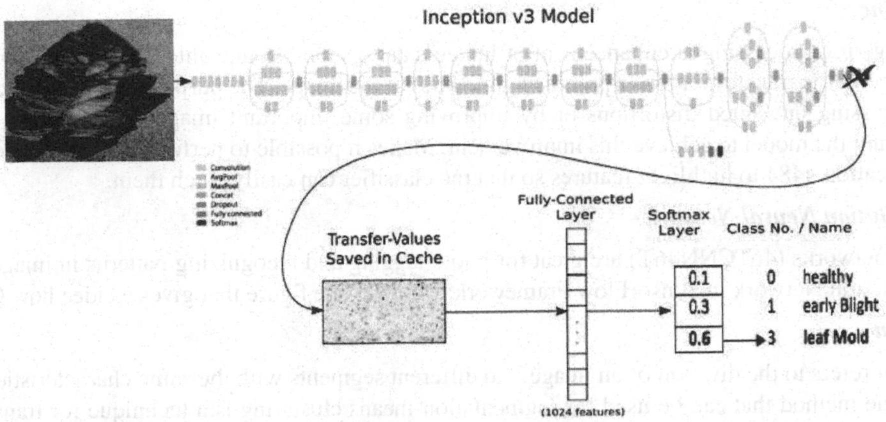

Fig. 74.2 Inception V3

5. Grid size reduction: This more appropriate methodology is recommended to address computational cost limitations. It is typically carried out by pooling firms. An extra classifier in related process serves as a regularize. Here, a more effective method has been suggested. It is typically carried out by pooling businesses.

Block Diagram

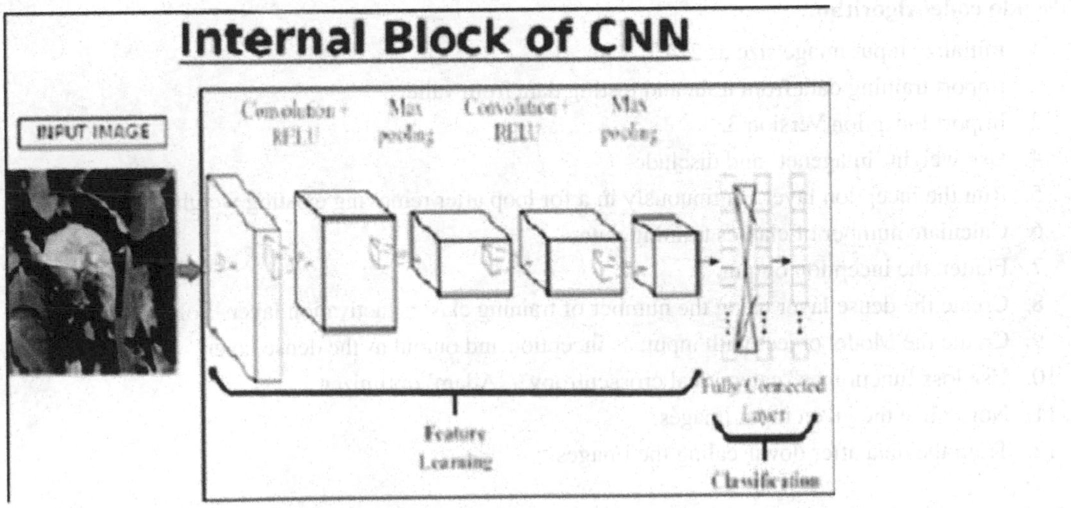

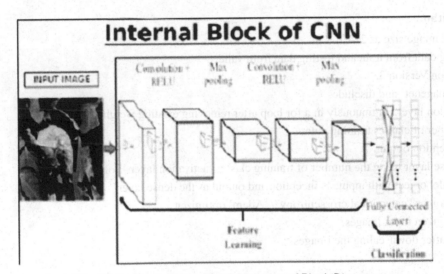

Fig. 74.3 Block diagram 3.3 Description of Block Diagram

Description of Block Diagram

Image Acquisition

To train the model, we gathered a variety of leaf photos. Therefore, we gather high-quality photographs from many sources and group them according to various categories. Right now, we will make use of the Kaggle Plant Village dataset. There are up to fifteen separate directories, with up to nine thousand photos. Hence Using this dataset, we can easily train our model and use it for forecasts. Later, we may also add additional photos and Retrain the model to improve it. In order to train the model, we gathered a variety of leaf photos. Consequently, we obtain high-quality photos from several sources. And classify them into several groups. Right now, we will make use of the Kaggle dataset. There are up to fifteen separate directories, with up to fifteen thousand photos.

Image Pre-processing

The purpose of image pre-processing is enhancement of image data to suppress unwanted distortion or to improve important image features by geometric transformation (z processing method) of the image. The purpose of image processing is to improve the image by suppressing unwanted distortions or by improving some important image features, thus improving the data (features) and allowing the model to achieve this improvement. Makes it possible to perform calculations on the data using the data. We use modification 4484 to highlight features so that the classifier can easily catch them.

Configuring Convolution Neural Network

Convolution neural networks (46TCNN46T) are great for photo tagging and recognizing patterns in image data. Here we will build Convolution Neural Network in TensorFlow Framework. Below is the figure that gives us idea how CNN work.

Image Segmentation

Picture segmentation refers to the division of an image into different segments with the same characteristics or some degree of similarity. Otsu is one method that can be used for segmentation-means clustering is a technique for translating RGB images into such as this model. When an image is segmented, it is divided into multiple components with the same qualities or some degree of similarity. The Several techniques, including otsu, can be used to segment data.

Feature Extraction

Feature extraction is essential for object identification. Numerous image processing tools make use of feature extraction morphology, edge, color, texture, etc. are characteristics that can be used to identify plant diseases. 3.3.6 Display Images: Finally diseased leaf is displayed among all other leaves from the dataset. On checking various features, diseased leaves will be categorized.

Pseudo code/Algorithm:

1. Initialize input image size as 224 X 224.
2. Import training data from train and testing data from valid.
3. Import Inception Version 3.
4. Use weight 'imagenet' and disclude.
5. Run the inception layer continuously in a for loop after removing existing weights.
6. Calculate number of classes training data.
7. Flatten the inception output.
8. Create the dense layer using the number of training classes, activation layer 'SoftMax'.
9. Create the Model object with input. as inception and output as the dense layer.
10. Use loss function as" categorical crossentropy", 'Adam' optimizer.
11. Normalize the given input images.
12. Train the data after downscaling the images.

4. Conclusion

Our report is all about finding the accuracy of our proposed model in identifying disease of various plants. This comprehensive report delves into the meticulous evaluation of our proposed model designed for the precise identification of plant diseases, leveraging the formidable capabilities of the Convolutional Neural Network (CNN) with Inception layers. The pivotal stages of our study encompassed the meticulous preprocessing of images from our dataset, followed by rigorous training. In pursuit of heightened accuracy, we harnessed the power of VGG-16 as a feature extractor, enhancing the model's capabilities for robust classification and identification of diverse plant diseases. The primary objective of this paper is to provide an insightful exploration of various plant diseases and elucidate the intricacies of our model's identification process.

The integration of CNN Inception layers and VGG-16 feature extraction contributes significantly to the model's effectiveness in discerning and categorizing diseases accurately. Furthermore, our work serves as a foundation that can be extended for maximum accuracy, paving the way for broader applications in the realm of plant disease identification. The paper thus encapsulates a valuable contribution to the understanding of plant diseases, offering a robust methodology and a roadmap for future advancements in this critical domain. We attained an accuracy of **95 %** using Inception layer.

Category	Original Image	Contrast	Zoom and Crop	Central Zoom
Healthy				
Early Blight	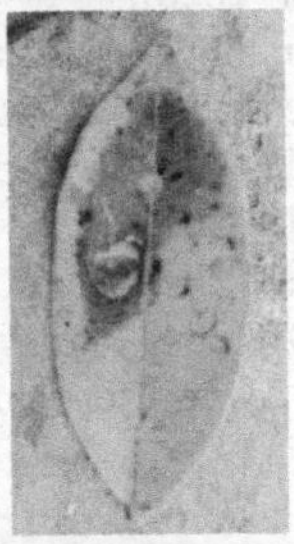	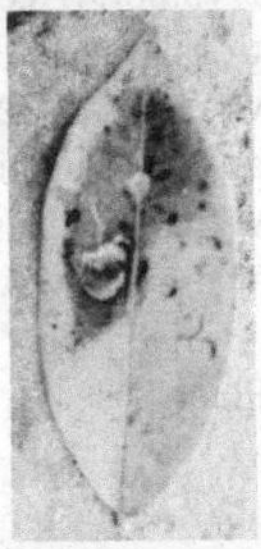	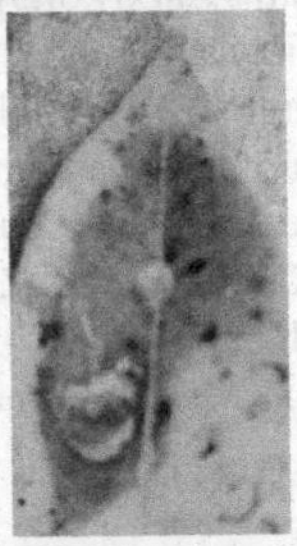	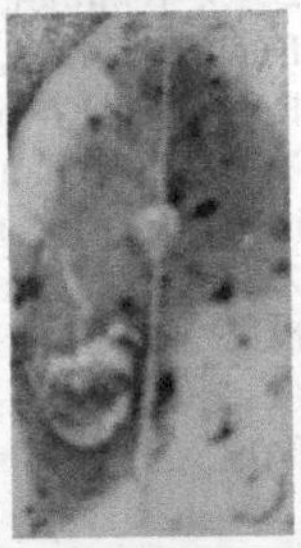
Late Blight	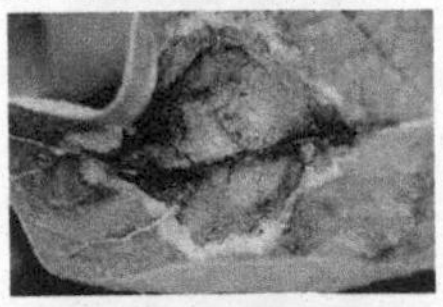	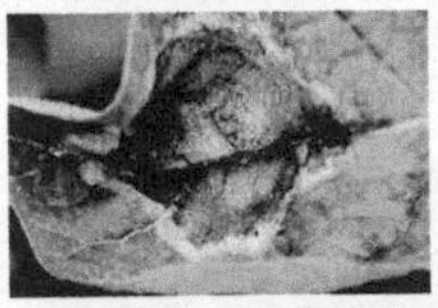	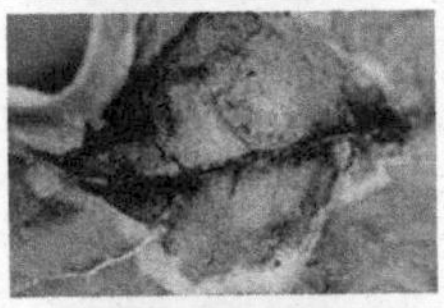	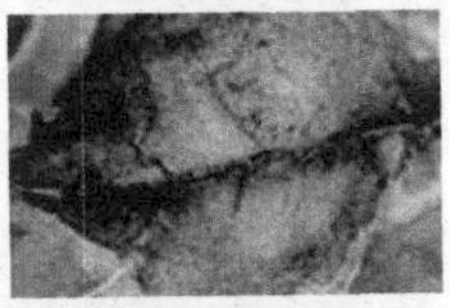
Leaf Mould		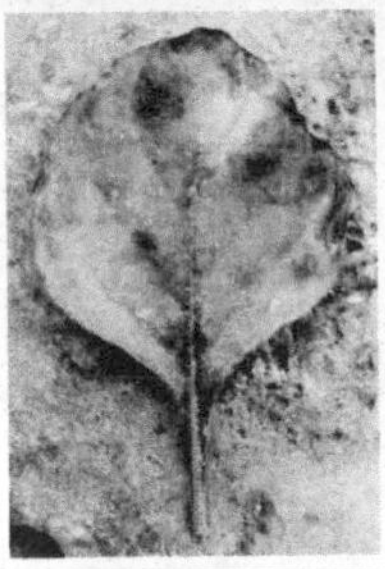	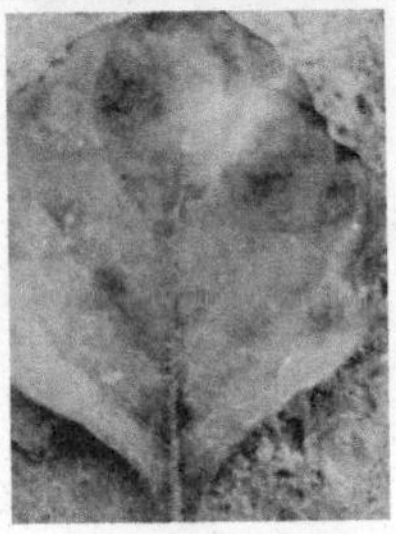	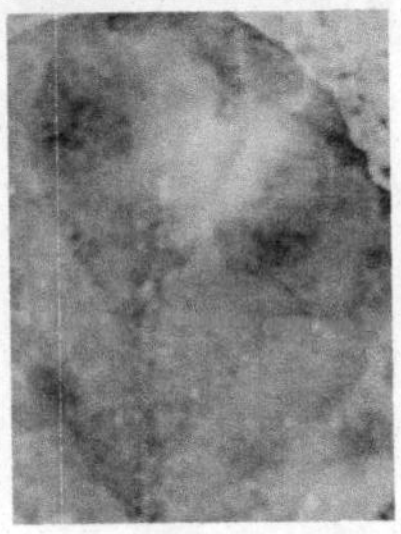

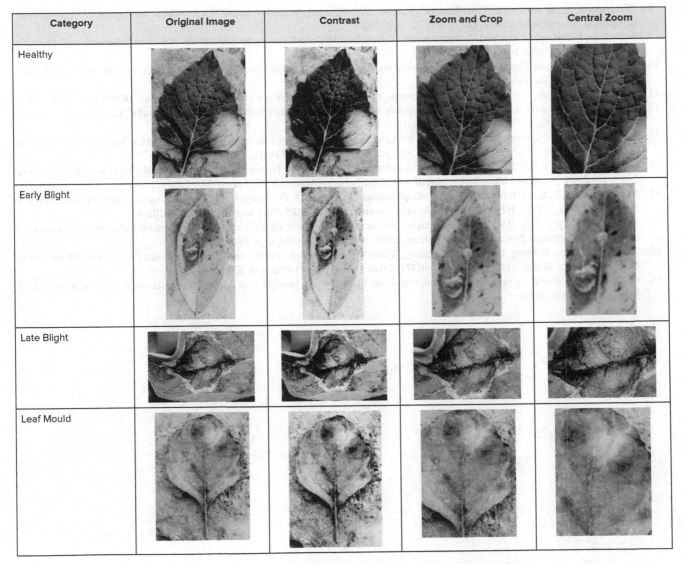

Fig. 74.4 Attention embedded residual CNN for disease detection in leaves.

Adding more images needs further improvement in training. Further, this paperwork can be extended to obtain maximum accuracy and for wide applications.

REFERENCES

1. M. Hunter, R. Smith, M. Schipanski, L. Atwood and D. Mortensen,"Agriculture in 2050: Recalibrating Targets for Sustainable Intensification", BioScience, vol. 67, no. 4, pp. 386-391, 2017.Available: 10.1093/biosci/bix010
2. "Plant Village", Plantvillage.psu.edu, 2020. [Online]. Available: https://plantvillage.psu.edu/. [Accessed: 31- Jan- 2020].
3. D. Klauser, "Challenges in monitoring and managing plant diseases in developing countries", Journal of Plant Diseases and Protection, vol.125, no. 3, pp. 235-237, 2018. Available: 10.1007/s41348-018-0145-9.
4. A. Muimba-Kankolongo, Food crop production by smallholder farmers in Southern Africa. Elsevier, 2018, pp. 23-27.
5. J. Boulent, S. Foucher, J. Théau and P. St-Charles, "Convolutional Neural Networks for the Automatic Identification of Plant Diseases", Frontiers in Plant Science, vol. 10, 2019. Available: 10.3389/fpls.2019.00941.
6. "iNaturalist", iNaturalist, 2020. [Online]. Available: https://www.inaturalist.org/. [Accessed: 30- Jan- 2020].
7. I. Ltd., "Success-stories - PlantSnap: Training the world's largest plant recognition classifier. | Image Technologies Ltd.", Imagga.com, 2020. [Online]. Available: https://imagga.com/success-stories/plantsnap-case-study. [Accessed: 31- Jan- 2020].

8. "Smartphone users worldwide 2020 | Statista", Statista, 2020. [Online]. Available: https://www.statista.com/statistics/330695/number-of-smartphone-users-worldwide/. [Accessed: 22- Apr- 2020].
9. K. Liakos, P. Busato, D. Moshou, S. Pearson and D. Bochtis, "Machine Learning in Agriculture: A Review", Sensors, vol. 18, no. 8, p. 2674,2018. Available: 10.3390/s18082674.
10. Y. Toda and F. Okura, "How Convolutional Neural Networks Diagnose Plant Disease", Plant Phenomics, vol. 2019, pp. 1-14, 2019. Available:10.34133/2019/9237136.
11. Atabay, H. A. 2016b. A convolutional neural network with a new architecture applied on leaf classification. IIOAB J 7(5):226–331
12. A. Krizhevsky, I. Sutskever and G. Hinton, "ImageNet classification with deep convolutional neural networks", Communications of the ACM, vol. 60, no. 6, pp. 84-90, 2017. Available: 10.1145/3065386.
13. S. Jadhav, "Convolutional Neural Networks for Leaf Image-Based Plant Disease Classification", IAES International Journal of Artificial Intelligence (IJ-AI), vol. 8, no. 4, p. 328, 2019. Available:10.11591/ijai. v8. i4. pp328-341.
14. C. Szegedy and S. Ioffe, "Batch Normalization: Accelerating Deep Network Training by Reducing Internal Covariate Shift", Arxiv.org, 2015. [Online]. Available: https://arxiv.org/pdf/1502.03167v3.pdf. [Accessed: 10- Apr- 2020].
15. D. Wu, Y. Wang, S. Xia, J. Bailey and X. Ma, "Skip Connections Matter: On the Transferability of Adversarial Examples Generated with Resets", arXiv.org, 2020. [Online]. Available: https://arxiv.org/abs/2002.05990. [Accessed: 10- Apr- 2020].
16. M. Ji, L. Zhang and Q. Wu, "Automatic grape leaf disease identification via United Model based on multiple convolutional neural networks", Information Processing in Agriculture, 2019. Available: 10.1016/j.inpa.2019.10.003.
17. X. Zhang, Y. Qiao, F. Meng, C. Fan and M. Zhang, "Identification of Maize Leaf Diseases Using Improved Deep Convolutional Neural Networks", IEEE Access, vol. 6, pp. 30370-30377, 2018. Available:10.1109/access.2018.2844405.
18. C. Buche, S. Whaleling and M. Polceanu, "Soybean Plant Disease Identification Using Convolutional Neural Network", 2018. [Accessed22 April 2020].

Impending Inquisitions in Humanities and Sciences (ICIIHS-2022) – Dr. Mohan Varkolu et al. (eds)
© 2024 Taylor & Francis Group, London, ISBN 978-1-032-78829-6

Parametric and Mond-Weir Duality for Fractional Interval-valued Optimization Problem with Vanishing Constraints

75

Krishna Kummari*
Department of Mathematics, School of Science, GITAM-Hyderabad Campus, Hyderabad, India
B. Japamala Rani[1]
Department of Mathematics, St. Ann's College for Women-Mehdipatanam, Hyderabad, India

ABSTRACT This paper serves and analyses the duality theorems associated with intervalvalued fractional optimization issues concerning vanishing constraints (FVC). Through the construction of Mond-Weir-type dual and the parametric dual models, we prove the weak, strong, and strict converse duality findings between (FVC) and its considerate dual paradigms, presuming convexity and generalized convexity.

KEYWORDS Interval-valued fractional optimization problem, LU-efficient solution, Parametric duality, Mond-Weir duality, Vanishing constraints

1. Introduction

In recent years, numerous academics have delved deeply into mathematical optimization problems concerning vanishing constraints. This understands from the presence of optimization problems with vanishing constraints originating from problems of the topology of mechanical structures [1] and possess considerable applicability in a variety of domains, including robot motion planning problem based on logic communication [15], the nonlinear mixed-integer optimal control problem [17] and the economic dispatch problem with disjoint feasible regions [12]. Various speculative properties and distinctive numerical techniques can be found in [1, 10, 11, 16].

In recent years, the problems involving interval values have shown a growing interest in optimization; see, e.g., [5, 18, 19, 20]. Interval-valued optimization problems handle inexact linear programming problems in a simplified approach. Extensive research has been studied to investigate both single and multiobjective scenarios based on various configurations of differentiability and convexity assumptions that have been examined in diverse contexts, one can see the works among Ben-Israel and Robers [3], Wu [21, 22, 23], Zhou and Wang [24], Bhurjee and Panda [4], Chalco-Cano et al. [5], Ahmad et al. [2], as well as the references therein.

Researchers have recently shown a growing interest in Fractional Programming problems. (see, e.g., [8, 25]). These problems arise in cutting and stock, decision-making applications, game theory, and science management see, e.g., [26]. Debnath and Gupta [6] provide some optimality criteria for nondifferentiable fractional interval-valued programming with convex numerators

*Corresponding author: krishna.maths@gmail.com
[1]bjapamalarani@gmail.com

DOI: 10.1201/9781003489436-75

and concave denominators for the objective function. Subsequent work has recently expanded these findings to fractional interval-valued multiobjective problems (refer, [28, 27, 7]).

Numerous authors have aided the evolution of duality theory for fractional programming problems under the presumptions of convexity, invexity, generalized convexity, and generalized invexity. We refer to ([26, 7, 13]) for more details about duality theory for fractional programming problems.

This article proposes the parametric duality and Mond-Weir dual model for the fractional interval-valued optimization problem concerning vanishing constraints under modified Abadie constraint qualification. The following is the paper's outline: In Section 2, we provide basic definitions and fundamental concepts. In Section 3, based on the concept of convexity, optimality conditions for (FVC) are analyzed. Section 4 deals with conceptualizing parametric duality, followed by the duality results involving convexity and generalized convexity. Section 5 provides the Mond-Weir type dual model for (FVC) and usual duality findings. We conclude the discussion in Section 6.

2. Preliminaries

This preliminary section serves several fundamental concepts and suitable lemmas utilized throughout the paper. We use the notation R^n to denote the n-dimensional Euclidean space, and R^n_+ to denote its non-negative orthant. To advance our discussion, it is essential to introduce the subsequent fundamental concepts of interval mathematics:

Assume $\frac{\Lambda}{\Gamma} = \left[\frac{\alpha_1^L}{\gamma_1^L}, \frac{\alpha_1^U}{\gamma_1^U}\right]$ and $\frac{\chi}{\Phi} = \left[\frac{\alpha_2^L}{\gamma_2^L}, \frac{\alpha_2^U}{\gamma_2^U}\right]$ be fractional closed intervals with $\frac{\alpha_1^L}{\gamma_1^L} \leq \frac{\alpha_1^U}{\gamma_1^U}$ and $\frac{\alpha_2^L}{\gamma_2^L} \leq \frac{\alpha_2^U}{\gamma_2^U}$, $\gamma_1^L, \gamma_1^U, \gamma_2^L, \gamma_2^U \neq 0$.

(i) $\frac{\Lambda}{\Gamma} + \frac{\chi}{\Phi} = \left[\frac{\alpha_1^L}{\gamma_1^L} + \frac{\alpha_2^L}{\gamma_2^L}, \frac{\alpha_1^U}{\gamma_1^U} + \frac{\alpha_2^U}{\gamma_2^U}\right]$,

(ii) $\frac{-\Lambda}{\Gamma} = \left[\frac{-\alpha_1^U}{\gamma_1^U}, \frac{-\alpha_1^L}{\gamma_1^L}\right]$,

(iii) $\frac{\Lambda}{\Gamma} - \frac{\chi}{\Phi} = \frac{\Lambda}{\Gamma} + \left(\frac{-\chi}{\Phi}\right) = \left[\frac{\alpha_1^L}{\gamma_1^L} - \frac{\alpha_2^U}{\gamma_2^U}, \frac{\alpha_1^U}{\gamma_1^U} - \frac{\alpha_2^L}{\gamma_2^L}\right]$,

(iv) $\beta\left(\frac{\Lambda}{\Gamma}\right) = \begin{cases} \left[\frac{\alpha_1^L}{\gamma_1^L}, \frac{\alpha_1^U}{\gamma_1^U}\right], & \text{if } \beta \geq 0, \\ \left[\frac{\alpha_1^U}{\gamma_1^U}, \frac{\alpha_1^L}{\gamma_1^L}\right], & \text{if } \beta < 0. \end{cases}$

The order relation $\leq \mathcal{LU}$ is defined between $\frac{\Lambda}{\Gamma}$ and $\frac{\chi}{\Phi}$ as

(i) $\frac{\Lambda}{\Gamma} \leq_{LU} \frac{\chi}{\Phi}$ iff $\frac{\alpha_1^L}{\gamma_1^L} \leq \frac{\alpha_2^L}{\gamma_2^L}$ and $\frac{\alpha_1^U}{\gamma_1^U} \leq \frac{\alpha_2^U}{\gamma_2^U}$.

(ii) $\frac{\Lambda}{\Gamma} < \frac{\chi}{\Phi}$ iff $\frac{\Lambda}{\Gamma} \leq \frac{\chi}{\Phi}$ and $\frac{\Lambda}{\Gamma} \neq \frac{\chi}{\Phi}$, equivalently

$$\begin{cases} \frac{\alpha_1^L}{\gamma_1^L} < \frac{\alpha_2^L}{\gamma_2^L} \\ \frac{\alpha_1^U}{\gamma_1^U} \leq \frac{\alpha_2^U}{\gamma_2^U} \end{cases}, \text{ or } \begin{cases} \frac{\alpha_1^L}{\gamma_1^L} \leq \frac{\alpha_2^L}{\gamma_2^L} \\ \frac{\alpha_1^U}{\gamma_1^U} < \frac{\alpha_2^U}{\gamma_2^U} \end{cases}, \text{ or } \begin{cases} \frac{\alpha_1^L}{\gamma_1^L} < \frac{\alpha_2^L}{\gamma_2^L} \\ \frac{\alpha_1^U}{\gamma_1^U} < \frac{\alpha_2^U}{\gamma_2^U}. \end{cases}$$

Definition 2.1 *Let $\bar{\varsigma} \in \mathcal{F}$ is a $\mathcal{LU}$ optimal solution for (FVC) iff there is no other feasible point $\varsigma \in \mathcal{F}$ satisfying*

$$\left[\frac{\alpha^L}{\gamma^L}(\varsigma), \frac{\alpha^U}{\gamma^U}(\varsigma)\right] \prec_{\mathcal{LU}} \left[\frac{\alpha^L}{\gamma^L}(\bar{\varsigma}), \frac{\alpha^U}{\gamma^U}(\bar{\varsigma})\right].$$

Now, we define some well-known fundamentals of convexity and generalized convexity for differentiable real-valued functions.

Definition 2.2 *A function $\Theta : X \subseteq R^n \to R$ be continuously differentiable. Then Θ is considered as a convex function at $\varsigma^* \in X$ if for any $\varsigma \in X$, we hold*

$$(\varsigma - \varsigma^*)^T \nabla\Theta(\varsigma^*) \leq \Theta(\varsigma) - \Theta(\varsigma^*).$$

Furthermore, the inequality is strict if Θ is strictly convex with $\varsigma \neq \varsigma^*$.

Definition 2.3 *A function $\Theta : X \subseteq R^n \to R$ be continuously differentiable.*
Then Θ becomes a (strictly) pseudoconvex at $\varsigma^* \in X$ if for any $\varsigma \in X$, we hold

$$(\varsigma - \varsigma^*)^T \nabla\Theta(\varsigma^*) \geq 0 \Rightarrow \Theta(\varsigma)(>) \geq \Theta(\varsigma^*),$$

equivalently

$$\Theta(\varsigma)(\leq) < \Theta(\varsigma^*) \Rightarrow (\varsigma - \varsigma^*)^T \nabla\Theta(\varsigma^*) < 0.$$

Definition 2.4 *Let $\Theta : X \subseteq R^n \to R$ be a continuously differentiable function. Then Θ becomes a quasiconvex at $\varsigma^* \in X$ if any point $\varsigma \in X$, holds*

$$\Theta(\varsigma) \leq \Theta(\varsigma^*) \Rightarrow (\varsigma - \varsigma^*)^T \nabla\Theta(\varsigma^*) \leq 0,$$

equivalently

$$(\varsigma - \varsigma^*)^T \nabla\Theta(\varsigma^*) > 0 \Rightarrow \Theta(\varsigma) > \Theta(\varsigma^*).$$

We formulate the subsequent optimization problem of fractional intervalvalued with vanishing constraints:

$$\text{(FVC)} \qquad \min \left[\frac{\zeta^L(\varsigma), \zeta^U(\varsigma)}{\delta^L(\varsigma), \delta^U(\varsigma)}\right]$$

$$\begin{aligned}
&\text{subject to}\\
&\phi_\rho(\varsigma) \leq 0, \forall\ \rho = 1, 2, \cdots, \omega,\\
&\psi_s(\varsigma) = 0, \forall\ s = 1, 2, \cdots, p,\\
&G_\rho(\varsigma) \geq 0, \forall\ \rho = 1, 2, \cdots, \ell,\\
&H_\rho(\varsigma) G_\rho(\varsigma) \leq 0, \forall\ \rho = 1, 2, \cdots, \ell,
\end{aligned}$$

where functions $\varphi : R^n \to R^\omega$, $\psi : R^n \to R^p$, $H, G : R^n \to R^\ell$ are continuously differentiable.

Consider the simplified form of (FVC) as follows

$$\min \left[\frac{\zeta^L(\varsigma)}{\delta^U(\varsigma)}, \frac{\zeta^U(\varsigma)}{\delta^L(\varsigma)}\right]$$

$$\begin{aligned}
&\text{subject to}\\
&\phi_\rho(\varsigma) \leq 0, \forall\ \rho = 1, 2, \cdots, \omega,\\
&\psi_s(\varsigma) = 0, \forall\ s = 1, 2, \cdots, p,\\
&G_\rho(\varsigma) \geq 0, \forall\ \rho = 1, 2, \cdots, \ell,\\
&H_\rho(\varsigma) G_\rho(\varsigma) \leq 0, \forall\ \rho = 1, 2, \cdots, \ell,
\end{aligned}$$

where $\zeta^L(\varsigma)$, $\zeta^U(\varsigma) \geq 0$, $\delta^L(\varsigma)$, $\delta^U(\varsigma) > 0$, and $\varphi : R^n \to R^\Omega$, $\psi : R^n \to R^p$, H, $G : R^n \to R^\ell$ are continuous functions on X. Set $\zeta^L = \alpha^L$, $\delta^U = \gamma^L$, $\zeta^U = \alpha^U$, $\delta^L = \gamma^U$, then, the above problem reduces to

$$\text{(FVC)} \qquad \min \left[\frac{\alpha^L}{\gamma^L}(\varsigma), \frac{\alpha^U}{\gamma^U}(\varsigma)\right]$$

$$\begin{aligned}
&\text{subject to}\\
&\phi_\rho(\varsigma) \leq 0, \forall\ \rho = 1, 2, \cdots, \omega,\\
&\psi_s(\varsigma) = 0, \forall s = 1, 2, \cdots, p,\\
&G_\rho(\varsigma) \geq 0, \forall\ \rho = 1, 2, \cdots, \ell,\\
&H_\rho(\varsigma) G_\rho(\varsigma) \leq 0, \forall\ \rho = 1, 2, \cdots, \ell.
\end{aligned}$$

Let $\mathcal{F}$ indicate the feasible region for the problem (FVC) and is defined as

$$\mathcal{F} := \Big\{ \varsigma \in R^n \mid \phi_\rho(\varsigma) \leq 0, \ \forall\ \rho = 1, 2, \cdots, \omega,$$
$$\psi_s(\varsigma) = 0, \ \forall s = 1, 2, \cdots, p,$$
$$G_\rho(\varsigma) \geq 0, \ \forall\ \rho = 1, 2, \cdots, \ell,$$
$$H_\rho(\varsigma) G_\rho(\varsigma) \leq 0, \ \forall\ \rho = 1, 2, \cdots, \ell \Big\}.$$

3. Optimality Conditions

Let $\varsigma^* \in F$ be any feasible solution of the (FVC). In the subsequent, we discuss the index sets.

$$\Omega_\phi = \{\rho \in \{1, 2, ..., \omega\} \mid \phi_\rho(\varsigma^*) = 0\},$$
$$\Omega_\psi = \{1, 2, ..., p\},$$
$$\Omega_+ = \{\rho \in \{1, 2, ..., \ell\} \mid G_\rho(\varsigma^*) > 0\},$$
$$\Omega_0 = \{\rho \in \{1, 2, ..., \ell\} \mid G_\rho(\varsigma^*) = 0\}.$$

Additionally, the index set Ω_+ can be partitioned into the subsequent subsets

$$\Omega_{+0} = \{\rho \in \{1, 2, ..., \ell\} \mid G_\rho(\varsigma^*) > 0, H_\rho(\varsigma^*) = 0\},$$
$$\Omega_{+-} = \{\rho \in \{1, 2, ..., \ell\} \mid G_\rho(\varsigma^*) > 0, H_\rho(\varsigma^*) < 0\}.$$

Likewise, the index set Λ_0 can be partitioned in the subsequent manner

$$\Omega_{0+} = \{\rho \in \{1, 2, ..., \ell\} \mid G_\rho(\varsigma^*) = 0, H_\rho(\varsigma^*) > 0\},$$
$$\Omega_{00} = \{\rho \in \{1, 2, ..., \ell\} \mid G_\rho(\varsigma^*) = 0, H_\rho(\varsigma^*) = 0\},$$
$$\Omega_{0-} = \{\rho \in \{1, 2, ..., \ell\} \mid G_\rho(\varsigma^*) = 0, H_\rho(\varsigma^*) < 0\}.$$

For $\varsigma \in S$, we also define the succeeding index sets:

$$\begin{cases}
\Omega_\phi^+(\varsigma) = \{\rho = 1, 2, ..., \Omega \mid \lambda_\rho > 0\} \\
\Omega_\psi^+(\varsigma) = \{s \in \Omega_\psi(\varsigma) \mid \beta_s > 0\} \\
\Omega_\psi^-(\varsigma) = \{s \in \Omega_\psi(\varsigma) \mid \beta_s < 0\} \\
\Omega_{0+}^+(\varsigma) = \{\rho \in \Omega_{0+}(\varsigma) \mid \alpha_\rho^G > 0\} \\
\Omega_{0+}^-(\varsigma) = \{\rho \in \Omega_{0+}(\varsigma) \mid \alpha_\rho^G < 0\} \\
\Omega_{00}^+(\varsigma) = \{\rho \in \Omega_{00}(\varsigma) \mid \alpha_\rho^G > 0\} \\
\Omega_{+0}^+(\varsigma) = \{\rho \in \Omega_{+0}(\varsigma) \mid \alpha_\rho^G > 0\} \\
\Omega_{+-}^+(\varsigma) = \{\rho \in \Omega_{+-}(\varsigma) \mid \alpha_\rho^G > 0\} \\
\Omega_{0-}^+(\varsigma) = \{\rho \in \Omega_{0-}(\varsigma) \mid \alpha_\rho^G > 0\} \\
\Omega_{+0}^{++}(\varsigma) = \{\rho \in \Omega_{+0}(\varsigma) \mid \alpha_\rho^H > 0\} \\
\Omega_{+-}^{++}(\varsigma) = \{\rho \in \Omega_{+-}(\varsigma) \mid \alpha_\rho^H > 0\}
\end{cases}$$

Definition 3.1 *The modified Abadie constraint qualification (FVC-MACQ) is considered to sustain at $\bar{\varsigma} \in S$ iff $L^{VC}(\bar{\varsigma}) \subseteq \%(\bar{\varsigma})$, wherein*

$$\varrho(\bar{\varsigma}) = \left\{ d_c \in R^n \mid \exists\{\varsigma^k\} \subseteq X, \exists\{t_k\} \downarrow 0, \ \varsigma^k \to \bar{\varsigma} \ and \ \frac{\varsigma^k - \bar{\varsigma}}{t_k} \to d_c \right\}$$

is the standard tangent cone of the (FVC) at ς*,and*

$$\mathcal{L}^{VC}(\bar{\varsigma}) = \Big\{ d_c \in R^n \;\Big|\; \nabla\phi_\rho(\bar{\varsigma})^T d_c \le 0,\ \rho \in \Omega_\phi(\bar{\varsigma}), \\ \nabla\psi_s(\bar{\varsigma})^T d_c = 0,\ s = 1,2,\ldots,p, \\ \nabla G_\rho(\bar{\varsigma})^T d_c = 0,\ \rho \in \Omega_{0+}(\bar{\varsigma}), \\ \nabla G_\rho(\bar{\varsigma})^T d_c \ge 0,\ \rho \in \Omega_{00}(\bar{\varsigma}) \cup \Omega_{0-}(\bar{\varsigma}), \\ \nabla H_\rho(\bar{\varsigma})^T d_c \le 0,\ \rho \in \Omega_{+0}(\bar{\varsigma}) \cup \Omega_{00}(\bar{\varsigma}) \Big\}$$

is the corresponding vanishing constraints linearized cone of the (FVC) at ς*.*

We assert the succeeding theorem (Karush-Kuhn-Tucker(KKT) type necessary optimality conditions) from Kummari and Rani [14].

Theorem 3.1 (KKT necessary optimality conditions) *Let the (FVC-MACQ), defined in Definition* (3.1)*, be satisfied at* $\bar{\varsigma} \in \mathcal{F}$*. If* $\bar{\varsigma} \in \mathcal{F}$ *be a* $\mathcal{LU}$ *optimal solution of the (FVC), then there exist* $v^L, v^U \in R_+$, $\lambda_\rho \in R$, $(\rho = 1, 2, \ldots, \omega)$, $\beta_s \in R$, $(s \in J_\psi)$, α^G_ρ, $\alpha^H_\rho \in R$, $(\rho = 1, 2, \ldots, \ell)$*, such that*

$$\nabla\alpha^L(\bar{\varsigma}) - v^L\nabla\gamma^L(\bar{\varsigma}) + \nabla\alpha^U(\bar{\varsigma}) - v^U\nabla\gamma^U(\bar{\varsigma}) + \sum_{\rho=1}^{\omega}\lambda_\rho\nabla\phi_\rho(\bar{\varsigma}) \\ + \sum_{s=1}^{p}\beta_s\nabla\psi_s(\bar{\varsigma}) - \sum_{\rho=1}^{\ell}\alpha_\rho^G\nabla G_\rho(\bar{\varsigma}) + \sum_{\rho=1}^{\ell}\alpha_\rho^H\nabla H_\rho(\bar{\varsigma}) = 0, \tag{1}$$

and

$$\alpha^L(\bar{\varsigma}) - v^L\gamma^L(\bar{\varsigma}) = 0, \tag{2}$$

$$\alpha^U(\bar{\varsigma}) - v^U\gamma^U(\bar{\varsigma}) = 0, \tag{3}$$

$$\psi_s(\bar{\varsigma}) = 0,\ s \in \Omega_\psi(\bar{\varsigma}), \tag{4}$$

$$\lambda_\rho \ge 0,\ \varphi_\rho(\bar{\varsigma}) \le 0,\ \lambda_\rho\varphi_\rho(\bar{\varsigma}) = 0,\ \rho = 1,2,\ldots\omega,$$

$$\alpha_\rho^G = 0, \rho \in \Omega_+(\bar{\varsigma}),\ \alpha_\rho^G \ge 0,\ \rho \in \Omega_{00}(\bar{\varsigma}) \cup \Omega_{0-}(\bar{\varsigma}),\ \alpha_\rho^G \text{ is free } \rho \in \Omega_{0+}(\bar{\varsigma}), \tag{5}$$

$$\alpha_\rho^H = 0,\ \rho \in \Omega_{0+}(\bar{\varsigma}) \cup \Omega_{0-}(\bar{\varsigma}) \cup \Omega_{+-}(\bar{\varsigma}),\ \alpha_\rho^H \ge 0, \tag{6}$$

$$\rho \in \Omega_{00}(\bar{\varsigma}) \cup \Omega_{+0}(\bar{\varsigma}), \tag{7}$$

$$\alpha_\rho^G G_\rho(\bar{\varsigma}) = 0,\ \alpha_\rho^H H\rho(\bar{\varsigma}) = 0,\ \forall\rho = 1, 2, \ldots, \ell. \tag{8}$$

4. Parametric Duality

This section examines the parametric dual formulation corresponding to the primal problem (FVC).

$$\text{(PD)}\ \max\ v = \left[v^L, v^U\right]$$

subject to

$$\nabla\alpha^L(\varrho) - v^L\nabla\gamma^L(\varrho) + \nabla\alpha^U(\varrho) - v^U\nabla\gamma^U(\varrho) + \sum_{\rho=1}^{\Omega}\lambda_\rho\nabla\phi_\rho(\varrho) \\ + \sum_{s=1}^{p}\beta_s\nabla\psi_s(\varrho) - \sum_{\rho=1}^{\ell}\alpha_\rho^G\nabla G_\rho(\varrho) + \sum_{\rho=1}^{\ell}\alpha_\rho^H\nabla H_\rho(\varrho) = 0, \tag{9}$$

$$\alpha^L(\varrho) - v^L\gamma^L(\varrho) \ge 0, \tag{10}$$

$$\alpha^U(\varrho) - v^U\gamma^U(\varrho) \ge 0, \tag{11}$$

$$\lambda_\rho \ge 0, \lambda_\rho\phi_\rho(\varrho) \ge 0,\ \forall\rho = 1,2,\ldots,\omega, \tag{12}$$

$$\beta_s \in R, \beta_s\psi_s(\varrho) = 0, \ \forall s = 1, 2, ..., p, \tag{13}$$
$$\alpha_\rho^H H_\rho(\varrho) \geq 0, \ \forall \rho = 1, 2, ..., \ell, \tag{14}$$
$$\alpha_\rho^H = \eta_\rho G_\rho(\varrho), \ \eta_\rho \geq 0, \ \forall \rho = 1, 2, ..., \ell, \tag{15}$$
$$-\alpha_\rho^G G_\rho(\varrho) \geq 0, \ \forall \rho = 1, 2, ..., \ell, \tag{16}$$
$$\alpha_\rho^G = \varpi_\rho - \eta_\rho H_\rho(\varrho), \ \varpi_\rho \geq 0, \ \forall \rho = 1, 2, ..., \ell, \tag{17}$$

where $\varrho \in R^n$, $v \in R_+$, $\lambda \in R_+^{\,\rho}$, $\beta \in R^p$, $\alpha^G \in R^\ell$, $\alpha^H \in R^\ell$, $\eta \in R_+^{\,\ell}$ and $\rho \in R_+^{\,\ell}$. We indicate by $\mathcal{W}(\varsigma)$ the set of all possible feasible solutions of (PD)(ϱ) and let $pr\mathcal{W}(\varsigma) = \{\varrho \in R^n | (\varrho, v, \lambda, \beta, \alpha^G, \alpha^H, \eta, \varpi) \in \mathcal{W}(\varsigma)$ be the projection of the set $\mathcal{W}(\varsigma)$ on R^n.

To ensure independence from the primal problem (FVC), we introduce an alternative dual problem presented below:

$$\text{(PD)} \qquad \max \quad v = \left[v^L, v^U\right]$$
$$\text{subject to}$$
$$(\varrho, v, \lambda, \beta, \alpha^G, \alpha^H, \eta, \varpi) \in \bigcap_{\varsigma \in \mathcal{F}} \mathcal{W}(\varsigma).$$

The feasible solutions set of the (PD) are indicated by $\mathcal{W} = \bigcap_{\varsigma \in \mathcal{F}} \mathcal{W}(\varsigma)$, the projection of the set $\mathcal{W}$ on R^n is indicated by $pr\mathcal{W}$.

Next, we establish duality between (FVC) and (PD) under specific presumptions of generalized convexity imposed on the relevant functions.

Theorem 4.1 (Weak Duality) *Let $\varsigma \in \mathcal{F}$ and $(\varrho, v, \lambda, \beta, \alpha^G, \alpha^H, \eta, \varpi) \in \mathcal{W}$. Further, assume that*

(i) $\alpha^L(.) - v^L\gamma^L(.) + \alpha^U(.) - v^U\gamma^U(.)$ *is convex function at % on $\mathcal{F} \cup pr\mathcal{W}$,*

(ii) $\sum_{\rho=1}^{\omega} \lambda_\rho \phi_\rho(.) + \sum_{s=1}^{p} \beta_s \psi_s(.) - \sum_{\rho=1}^{\ell} \alpha_\rho^G G_\rho(.) + \sum_{\rho=1}^{\ell} \alpha_\rho^H H_\rho(.)$ *is a convex function at ϱ on $\mathcal{F} \cup pr\mathcal{W}$,*

then $\Psi(\varsigma) \nprec_{LU} v$

Proof: Supposing on the contradiction, let us suppose that $\Psi(\varsigma) \prec_{LU} v$, subsequently with the application of (10) and (11) , we obtain

$$\alpha^L(\varsigma) - v^L\gamma^L(\varsigma) < \alpha^L(\varrho) - v^L\gamma^L(\varrho),$$
$$\alpha^U(\varsigma) - v^U\gamma^U(\varsigma) < \alpha^U(\varrho) - v^U\gamma^U(\varrho),$$

or

$$\alpha^L(\varsigma) - v^L\gamma^L(\varsigma) < \alpha^L(\varrho) - v^L\gamma^L(\varrho),$$
$$\alpha^U(\varsigma) - v^U\gamma^U(\varsigma) \leq \alpha^U(\varrho) - v^U\gamma^U(\varrho),$$

or

$$\alpha^L(\varsigma) - v^L\gamma^L(\varsigma) \leq \alpha^L(\varrho) - v^L\gamma^L(\varrho),$$
$$\alpha^U(\varsigma) - v^U\gamma^U(\varsigma) < \alpha^U(\varrho) - v^U\gamma^U(\varrho),$$

the above inequalities yield,

$$\alpha^L(\varsigma) - v^L\gamma^L(\varsigma) + \alpha^U(\varsigma) - v^U\gamma^U(\varsigma) < \alpha^L(\varrho) - v^L\gamma^L(\varrho) + \alpha^U(\varrho) - v^U\gamma^U(\varrho),$$

which by convexity of $\alpha^L(.) - v^{-L}\gamma^L(.) + \alpha^U(.) - v^{-U}\gamma^U(.)$ at ϱ on $\mathcal{F} \cup pr\mathcal{W}$, one can get

$$(\varsigma - \varrho)^T \left[[\nabla\alpha^L(\varrho) - v^L\nabla\gamma^L(\varrho)] + [\nabla\alpha^U(\varrho) - v^U\nabla\gamma^U(\varrho)] \right] < 0. \tag{18}$$

Since $\varsigma \in \mathcal{F}$ and $(\varrho, v, \lambda, \beta, \alpha^G, \alpha^H, \eta, \rho) \in \mathcal{W}$, it follows that

$$\begin{aligned}
&\phi_\rho(\varsigma) \leq 0,\ \varsigma_\rho \geq 0,\ \forall \rho = 1, 2, ..., \omega,\\
&\psi_s(\varsigma) = 0,\ \gamma_s \in R,\ \forall s = 1, 2, ..., p,\\
&-G_\rho(\varsigma) < 0,\ \alpha^G_\rho \geq 0,\ \forall \rho \in \Omega_+,\\
&-G_\rho(\varsigma) = 0,\ \alpha^G_\rho \in R,\ \forall \rho \in \Omega_0,\\
&H_\rho(\varsigma) > 0,\ \alpha^H_\rho = 0,\ \forall \rho \in \Omega_{0+},\\
&H_\rho(\varsigma) = 0,\ \alpha^H_\rho \geq 0,\ \forall \rho \in \Omega_{00} \cup \Omega_{+0},\\
&H_\rho(\varsigma) < 0,\ \alpha^H_\rho \geq 0\ ,\forall \rho \in \Omega_{0-} \cup \Omega_{+-},
\end{aligned}$$

that is,

$$\sum_{\rho=1}^{\omega} \lambda_\rho \phi_\rho(\varsigma) + \sum_{s=1}^{p} \beta_s \psi_s(\varsigma) - \sum_{\rho=1}^{\ell} \alpha^G_\rho G_\rho(\varsigma) + \sum_{\rho=1}^{\ell} \alpha^H_\rho H_\rho(\varsigma) \leq 0.$$

By (12), (13), (14) and (16), above inequality implies that

$$\sum_{\rho=1}^{\omega} \lambda_\rho \phi_\rho(\varsigma) + \sum_{s=1}^{p} \beta_s \psi_s(\varsigma) - \sum_{\rho=1}^{\ell} \alpha^G_\rho G_\rho(\varsigma) + \sum_{\rho=1}^{\ell} \alpha^H_\rho H_\rho(\varsigma)$$

$$\leq \sum_{\rho=1}^{\omega} \lambda_\rho \phi_\rho(\varrho) + \sum_{s=1}^{p} \beta_s \psi_s(\varrho) - \sum_{\rho=1}^{\ell} \alpha^G_\rho G_\rho(\varrho) + \sum_{\rho=1}^{\ell} \alpha^H_\rho H_\rho(\varrho),$$

which by convexity of $\sum_{\rho=1}^{\omega} \lambda_\rho \phi_\rho(.) + \sum_{s=1}^{p} \beta_s \psi_s(.) - \sum_{\rho=1}^{\ell} \alpha^G_\rho G_\rho(.) + \sum_{\rho=1}^{\ell} \alpha^H_\rho H_\rho(.)$ at ϱ on $\mathcal{F} \cup pr\mathcal{W}$, we obtain

$$(\varsigma - \varrho)^T \left[\sum_{\rho=1}^{\omega} \lambda_\rho \nabla \phi_\rho(\varrho) + \sum_{s=1}^{p} \beta_s \nabla \psi_s(\varrho) - \sum_{\rho=1}^{\ell} \alpha^G_\rho \nabla G_\rho(\varrho) \right.$$

$$\left. + \sum_{\rho=1}^{\ell} \alpha^H_\rho \nabla H_\rho(\varrho) \right] \leq 0. \tag{19}$$

On adding (18) and (19), we get

$$(\varsigma - \varrho)^T \left[[\nabla \alpha^L(\varrho) - v^L \nabla \gamma^L(\varrho)] + [\nabla \alpha^U(\varrho) - v^U \nabla \gamma^U(\varrho)] + \sum_{\rho=1}^{\omega} \lambda_\rho \right.$$

$$\left. \nabla \phi_\rho(\varrho) + \sum_{s=1}^{p} \beta_s \nabla \psi_s(\varrho) - \sum_{\rho=1}^{\ell} \alpha^G_\rho \nabla G_\rho(\varrho) + \sum_{\rho=1}^{\ell} \alpha^H_\rho \nabla H_\rho(\varrho) \right] < 0,$$

which is contradicting (9).

Theorem 4.2 (Weak Duality) *Let $\varsigma \in \mathcal{F}$ and $(\varrho, v, \lambda, \beta, \alpha^G, \alpha^H, \eta, \varpi) \in \mathcal{W}$. Further, assume that*

(i) *$\alpha^L(.) - v^L\gamma^L(.) + \alpha^U(.) - v^U\gamma^U(.)$ is pseudoconvex function at % on $\mathcal{F} \cup pr\mathcal{W}$,*

(ii) *$\sum_{\rho=1}^{\omega} \lambda_\rho \phi_\rho(.) + \sum_{s=1}^{p} \beta_s \psi_s(.) - \sum_{\rho=1}^{\ell} \alpha^G_\rho G_\rho(.) + \sum_{\rho=1}^{\ell} \alpha^H_\rho H_\rho(.)$ is a quasiconvex function at ϱ on $\mathcal{F} \cup pr\mathcal{W}$,*

then $\Psi(\varsigma) \nless \mathcal{LU}^v$.

Proof: Proof is comparable to the proof of Theorem (4.1).

Theorem 4.3 (Strong Duality) *Let $\bar{\varsigma}$ be a $\mathcal{LU}$ optimal solution of (FVC) and* (FVC-MACQ) *is satisfied at $\bar{\varsigma}$. Then there exist, $\bar{v} \in R_+$, $\bar{\lambda} \in R_+^{\omega}$, $\bar{\beta} \in R^p$, $\bar{\alpha}^G \in R^{\ell}$, $\bar{\alpha}^H \in R^{\ell}$, $\bar{\eta} \in R^{\ell}_+$ and such that ($\bar{\varrho}$, $\bar{v}$, $\bar{\lambda}$, $\bar{\beta}$, $\bar{\alpha}^G$, $\bar{\alpha}^H$, $\bar{\eta}$, $\bar{\varpi}$) is a feasible solution of (PD) and the both objective values are equal. Further, if all the presumptions of the weak duality (Theorem* (4.1) *or* (4.2)*) are fulfilled, then the point ($\bar{\varrho}$, $\bar{v}$, $\bar{\lambda}$, $\bar{\beta}$, $\bar{\alpha}^G$, $\bar{\alpha}^H$, $\bar{\eta}$, $\bar{\varpi}$) is a $\mathcal{LU}$ optimal solution of (PD).*

Proof: By presumption $\bar{\varsigma}$ is a $\mathcal{LU}$ optimal solution of (FVC) and (FVC-MACQ) condition is satisfied at this point, then by Theorem (3.1), there exists $\bar{v} \in R_+$, $\bar{\lambda} \in R_+^{\omega}$, $\bar{\beta} \in R^p$, $\bar{\alpha}^G \in R^{\ell}$, $\bar{\alpha}^H \in R^{\ell}$, $\bar{\eta} \in R^{\ell}_+$ such that the conditions (1) – (8) are satisfied. Thus, ($\bar{\varrho}$, $\bar{v}$, $\bar{\lambda}$, $\bar{\beta}$, $\bar{\alpha}^G$, $\bar{\alpha}^H$, $\bar{\eta}$, $\bar{\varpi}$) is a feasible solution of (PD). Furthermore, the corresponding objective values of (FVC) and (PD) are equal. Moreover, if ($\bar{\varrho}$, $\bar{v}$, $\bar{\lambda}$, $\bar{\beta}$, $\bar{\alpha}^G$, $\bar{\alpha}^H$, $\bar{\eta}$, $\bar{\varpi}$) is not a $\mathcal{LU}$ optimal solution of (PD), then there exists a feasible solution ($\bar{\varrho}$, $\bar{v}$, $\bar{\lambda}$, $\bar{\beta}$, $\bar{\alpha}^G$, $\bar{\alpha}^H$, $\bar{\eta}$, $\bar{\varpi}$) for (PD), such that the following inequality is satisfied

$$\Psi(\varsigma) \prec_{LU} v.$$

Which contradicts (Theorem (4.1) or (4.2)). Hence ($\bar{\varrho}$, $\bar{v}$, $\bar{\lambda}$, $\bar{\beta}$, $\bar{\alpha}^G$, $\bar{\alpha}^H$, $\bar{\eta}$, $\bar{\varpi}$) becomes a $\mathcal{LU}$ optimal solution of (PD).

Theorem 4.4 (Strict Converse Duality) *Let $\bar{\varsigma} \in \mathcal{F}$ and ($\tilde{\varrho}$, $\tilde{v}$, $\tilde{\lambda}$, $\tilde{\beta}$, $\tilde{\alpha}^G$, $\tilde{\alpha}^H$, $\tilde{\eta}$, $\tilde{\varpi}$) $\in \mathcal{W}_1$ be a $\mathcal{LU}$ optimal solution of (FVC) and (PD). Assume that the presumptions of Theorem (4.2) are fullled. Further, assume that*

(i) $\alpha^L(.) - v^L\gamma^L(.) + \alpha^U(.) - v^U\gamma^U(.)$ *is a strictly convex function at $\tilde{\varrho}$ on $\mathcal{F} \cup pr\mathcal{W}_1$,*

(ii) $\sum_{\rho=1}^{\omega} \lambda_\rho \phi_\rho(.) + \sum_{s=1}^{p} \beta_s \psi_s(.) - \sum_{\rho=1}^{\ell} \alpha_\rho^G G_\rho(.) + \sum_{\rho=1}^{\ell} \alpha_\rho^H H_\rho(.)$ *is a convex function at $\tilde{\varrho}$ on $\mathcal{F} \cup pr\mathcal{W}_1$,*

then $\bar{\varsigma} = \tilde{\varrho}$.

Proof: Presume, contrary to the result, that $\bar{\varsigma} = \tilde{\varrho}$. By Theorem (4.2), we possess

$$\left[\frac{\alpha^L}{\gamma^L}(\bar{\varsigma}), \frac{\alpha^U}{\gamma^U}(\bar{\varsigma})\right] = [v^L, v^U],$$

which implies

$$\alpha^L(\bar{\varsigma}) - v^L\gamma^L(\bar{\varsigma}) = 0,$$
$$\alpha^U(\bar{\varsigma}) - v^U\gamma^U(\bar{\varsigma}) = 0,$$

by using the above equations, we obtain

$$[\alpha^L(\bar{\varsigma}) - v^L\gamma^L(\bar{\varsigma}) + \alpha^U(\bar{\varsigma}) - v^U\gamma^U(\bar{\varsigma})] = 0. \tag{20}$$

Since $\bar{\varsigma} \in \mathcal{F}$ and ($\tilde{\varrho}$, $\tilde{v}$, $\tilde{\lambda}$, $\tilde{\beta}$, $\tilde{\alpha}^G$, $\tilde{\alpha}^H$, $\tilde{\eta}$, $\tilde{\varpi}$) $\in \mathcal{W}_1$, it follows that

$$\begin{aligned}
&\phi_\rho(\bar{\varsigma}) \le 0,\ \varsigma_\rho \ge 0,\ \forall \rho = 1, 2, ..., \omega,\\
&\psi_s(\bar{\varsigma}) = 0,\ \gamma_s \in R,\ \forall s = 1, 2, ..., p,\\
&-G_\rho(\bar{\varsigma}) < 0,\ \alpha_\rho^G \ge 0,\ \forall \rho \in \Omega_+,\\
&-G_\rho(\bar{\varsigma}) = 0,\ \alpha_\rho^G \in R,\ \forall \rho \in \Omega_0,\\
&H_\rho(\bar{\varsigma}) > 0,\ \alpha_\rho^H = 0,\ \forall \rho \in \Omega_{0+},\\
&H_\rho(\bar{\varsigma}) = 0,\ \alpha_\rho^H \ge 0,\ \forall \rho \in \Omega_{00} \cup \Omega_{+0},\\
&H_\rho(\bar{\varsigma}) < 0,\ \alpha_\rho^H \ge 0\ , \forall \rho \in \Omega_{0-} \cup \Omega_{+-},
\end{aligned}$$

that is,

$$\sum_{\rho=1}^{\omega} \lambda_\rho \phi_\rho(\bar{\varsigma}) + \sum_{s=1}^{p} \beta_s \psi_s(\bar{\varsigma}) - \sum_{\rho=1}^{\ell} \alpha_\rho^G G_\rho(\bar{\varsigma}) + \sum_{\rho=1}^{\ell} \alpha_\rho^H H_\rho(\bar{\varsigma}) \le 0.$$

By (12), (13), (14) and (16), above inequality implies that

$$\sum_{\rho=1}^{\omega} \lambda_\rho \phi_\rho(\bar{\varsigma}) + \sum_{s=1}^{p} \beta_s \psi_s(\bar{\varsigma}) - \sum_{\rho=1}^{\ell} \alpha_\rho^G G_\rho(\bar{\varsigma}) + \sum_{\rho=1}^{\ell} \alpha_\rho^H H_\rho(\bar{\varsigma})$$

$$\le \sum_{\rho=1}^{\omega} \lambda_\rho \phi_\rho(\tilde{\varrho}) + \sum_{s=1}^{p} \beta_s \psi_s(\tilde{\varrho}) - \sum_{\rho=1}^{\ell} \alpha_\rho^G G_\rho(\tilde{\varrho}) + \sum_{\rho=1}^{\ell} \alpha_\rho^H H_\rho(\tilde{\varrho}),$$

which by convexity of $\sum_{\rho=1}^{\omega}\lambda_\rho\phi_\rho(.)+\sum_{s=1}^{p}\beta_s\psi_s(.)-\sum_{\rho=1}^{\ell}\alpha_\rho^G G_\rho(.)\ +\sum_{\rho=1}^{\ell}\alpha_\rho^H H_\rho(.)$ at $\tilde{\varrho}$ on $\mathcal{F}\cup prW_1$, we obtain

$$(\tilde{\varsigma}-\tilde{\varrho})^T\left[\sum_{\rho=1}^{\omega}\lambda_\rho\nabla\phi_\rho(\varrho)+\sum_{s=1}^{p}\beta_s\nabla\psi_s(\tilde{\varrho})-\sum_{\rho=1}^{\ell}\alpha_\rho^G\nabla G_\rho(\tilde{\varrho})+\sum_{\rho=1}^{\ell}\alpha_\rho^H\nabla H_\rho(\tilde{\varrho})\right]\leq 0. \tag{21}$$

By hypothesis (i), $\alpha^L(.)-v^L\gamma^L(.)+\alpha^U(.)-v^U\gamma^U(.)$ is a strictly convex function at %˜ on $\mathcal{F}\cup prW_1$. Then, by Definition (3.1), for any $\varsigma^\sim\in\mathcal{F}$, we have

$$[\alpha^L(\tilde{\varsigma})-v^L\gamma^L(\tilde{\varsigma})+\alpha^U(\tilde{\varsigma})-v^U\gamma^U(\tilde{\varsigma})]-[\alpha^L(\tilde{\varrho})-v^L\gamma^L(\tilde{\varrho})+\alpha^U(\tilde{\varrho})-v^U\gamma^U(\tilde{\varrho})]>(\tilde{\varsigma}-\tilde{\varrho})^T[\nabla\alpha^L(\tilde{\varrho})-v^L\nabla\gamma^L(\tilde{\varrho})]+[\nabla\alpha^U(\tilde{\varrho})-v^U\nabla\gamma^U(\tilde{\varrho}) \tag{22}$$

On adding (21) and (22), we get

$$\begin{aligned}&[\alpha^L(\varsigma)-v^L\gamma^L(\tilde{\varsigma})+\alpha^U(\tilde{\varsigma})-v^U\gamma^U(\tilde{\varsigma})]-[\alpha^L(\tilde{\varrho})-v^L\gamma^L(\tilde{\varrho})\\ &+\alpha^U(\varrho)-v^U\gamma^U(\tilde{\varrho})]>(\tilde{\varsigma}-\tilde{\varrho})^T\Big[[\nabla\alpha^L(\tilde{\varrho})-v^L\nabla\gamma^L(\tilde{\varrho})]+[\nabla\alpha^U(\tilde{\varrho})\\ &-v^U\nabla\gamma^U(\tilde{\varrho})]+\sum_{\rho=1}^{\omega}\lambda_\rho\nabla\phi_\rho(\tilde{\varrho})+\sum_{s=1}^{p}\beta_s\nabla\psi_s(\tilde{\varrho})-\sum_{\rho=1}^{\ell}\alpha_\rho^G\nabla G_\rho(\tilde{\varrho})\\ &+\sum_{\rho=1}^{\ell}\alpha_\rho^H\nabla H_\rho(\tilde{\varrho})\Big],\end{aligned}$$

which by using (9), (10) and (11), yields

$$[\alpha^L(\tilde{\varsigma})-v^L\gamma^L(\tilde{\varsigma})+\alpha^U(\tilde{\varsigma})-v^U\gamma^U(\tilde{\varsigma})]>0,$$

which is contradicting (20).

5. Mond-Weir Duality

Considering the Mond-Weir dual problem for (FVC).

$$(MD)(\varsigma)\qquad \max\ \Psi(\varrho)=\left[\frac{\alpha^L}{\gamma^L}(\varrho),\frac{\alpha^U}{\gamma^U}(\varrho)\right]$$

subject to

$$\left.\begin{array}{l}\nabla\alpha^L(\varrho)-\Psi^L(\varrho)\nabla\gamma^L(\varrho)+\nabla\alpha^U(\varrho)-\Psi^U(\varrho)\nabla\gamma^U(\varrho)\\ +\sum_{\rho=1}^{\omega}\lambda_\rho\nabla\phi_\rho(\varrho)+\sum_{s=1}^{p}\beta_s\nabla\psi_s(\varrho)-\sum_{\rho=1}^{\ell}\alpha_\rho^G\nabla G_\rho(\varrho)\\ +\sum_{\rho=1}^{\ell}\alpha_\rho^H\nabla H_\rho(\varrho)=0,\\ \alpha^L(\varrho)-\Psi^L(\varrho)\gamma^L(\varrho)\geq 0,\\ \alpha^U(\varrho)-\Psi^U(\varrho)\gamma^U(\varrho)\geq 0,\\ \lambda_\rho\geq 0,\lambda_\rho\phi_\rho(\varrho)\geq 0,\ \forall\rho=1,2,...,\omega,\\ \beta_s\in R,\beta_s\psi_s(\varrho)=0,\ \forall\rho=1,2,...,p,\\ \alpha_\rho^H H_\rho(\varrho)\geq 0,\ \forall\rho=1,2,...,\ell,\\ \alpha_\rho^H=\eta_\rho G_\rho(\varrho),\ \eta_\rho\geq 0,\ \forall\rho=1,2,...,\ell,\\ -\alpha_\rho^G G_\rho(\varrho)\geq 0,\ \forall\rho=1,2,...,\ell,\\ \alpha_\rho^G=\varpi_\rho-\eta_\rho H_\rho(\varrho),\ \varpi_\rho\geq 0,\ \forall\rho=1,2,...,\ell.\end{array}\right\} \tag{23}$$

We denote by

$$pr\mathcal{F}_{MD}(\varsigma) = \{\varrho \in R^n \mid (\varrho, \lambda, \beta, \alpha^G, \alpha^H, \eta, \varpi) \in \mathcal{F}_{MD}(\varsigma)$$

the projection of the set $\mathcal{F}_{MD}(\varsigma)$ on R^n.

Comparable to the parametric dual, we also examine an alternative dual problem designated as MD, which is expressed as follows:

$$\text{(MD)} \quad \max \Psi(\varrho)$$
$$\text{subject to}$$
$$(\varrho, \lambda, \beta, \alpha^G, \alpha^H, \eta, \varpi) \in \bigcap_{\varsigma \in \mathcal{F}} \mathcal{F}_{MD}(\varsigma)$$

The feasible points set of (MD) is indicated by $\mathcal{F}_{MD} = \bigcap_{x \in \mathcal{F}} \mathcal{F}_{MD}(\varsigma)$ and the projection set $\mathcal{F}_{MD}$ on R^n is denoted by $pr\mathcal{F}_{MD}$.

Definition 5.1 *A feasible point* $(\bar{\varrho}, \bar{\lambda}, \bar{\beta}, \bar{\alpha}^G, \bar{\alpha}^H, \bar{\eta}, \bar{\varpi}) \in \mathcal{F}_{MD}$ *is considered to be a* $\mathcal{LU}$ *optimal solution for* (MD) *if and only if there is no other feasible point* $(\varrho, \lambda, \beta, \alpha^G, \alpha^H, \eta, \varpi) \in \mathcal{F}_{MD}$ *satisfying* $\Psi(\varrho) \prec_{LU} \Psi(\bar{\varrho})$. *That is,*

$$\left[\frac{\alpha^L}{\gamma^L}(\varrho), \frac{\alpha^U}{\gamma^U}(\varrho)\right] \prec_{\mathcal{LU}} \left[\frac{\alpha^L}{\gamma^L}(\bar{\varrho}), \frac{\alpha^U}{\gamma^U}(\bar{\varrho})\right].$$

The subsequent section outlines the duality findings between (FVC) and (MD).

Theorem 5.1 (Weak Duality) *Let* $\varsigma \in \mathcal{F}$ *and* $(\varrho, \lambda, \beta, \alpha^G, \alpha^H, \eta, \varpi) \in \mathcal{F}_{MD}$ *be feasible points for the (FVC) and the (MD), respectively. Additionally, if any of the following conditions hold:*

(i) $\alpha^L(.) - \Psi^L(\varrho)\gamma^L(.) + \alpha^U(.) - \Psi^U(\varrho)\gamma^U(.)$ *is pseudoconvex function and*

$$\sum_{\rho=1}^{\omega} \lambda_\rho \phi_\rho(.) + \sum_{s=1}^{p} \beta_s \psi_s(.) - \sum_{\rho=1}^{\ell} \alpha_\rho^G G_\rho(.) + \sum_{\rho=1}^{\ell} \alpha_\rho^H H_\rho(.)$$

is a quasiconvex function at % on $\mathcal{F} \cup pr\mathcal{F}_{MD}$, *respectively;*

(ii) $\alpha^L(.) - \Psi^L(\%)\gamma^L(.) + \alpha^U(.) - \Psi^U(\%)\gamma^U(.)$ *is pseudoconvex function and* $\phi_\rho (\rho \in \Omega^+_\phi(\varsigma)), \psi_s (s \in \Omega^+_\psi(\varsigma)), -\psi_s (s \in \Omega^-_\psi(\varsigma))$, $-G_\rho (\rho \in \Omega^+_{+0}(\varsigma) \cup \Omega^+_{+-}(\varsigma) \cup \Omega^+_{00}(\varsigma) \cup \Omega^+_{0-}(\varsigma) \cup \Omega^+_{0+}(\varsigma)), -G_\rho (\rho \in \Omega^-_{0+}(\varsigma))$, $H\rho$ $(\rho \in \Omega^{++}_{+0}(\varsigma) \cup \Omega^{++}_{+-}(\varsigma))$ *are quasiconvex functions at* ϱ *on* $\mathcal{F} \cup pr\mathcal{F}_{MD}$, *respectively;*

then the following inequality holds:

$$\left[\frac{\alpha^L}{\gamma^L}(\varsigma), \frac{\alpha^U}{\gamma^U}(\varsigma)\right] \nprec_{\mathcal{LU}} \left[\frac{\alpha^L}{\gamma^L}(\varrho), \frac{\alpha^U}{\gamma^U}(\varrho)\right].$$

Proof: Supposing the contradiction, let us presume that

$$\left[\frac{\alpha^L}{\gamma^L}(\varsigma), \frac{\alpha^U}{\gamma^U}(\varsigma)\right] \prec_{\mathcal{LU}} \left[\frac{\alpha^L}{\gamma^L}(\varrho), \frac{\alpha^U}{\gamma^U}(\varrho)\right].$$

that is

$$\begin{cases} \frac{\alpha^L}{\gamma^L}(\varsigma) < \frac{\alpha^L}{\gamma^L}(\varrho) \\ \frac{\alpha^U}{\gamma^U}(\varsigma) \le \frac{\alpha^U}{\gamma^U}(\varrho) \end{cases}, \text{ or } \begin{cases} \frac{\alpha^L}{\gamma^L}(\varsigma) \le \frac{\alpha^L}{\gamma^L}(\varrho) \\ \frac{\alpha^U}{\gamma^U}(\varsigma) < \frac{\alpha^U}{\gamma^U}(\varrho) \end{cases}, \text{ or } \begin{cases} \frac{\alpha^L}{\gamma^L}(\varsigma) < \frac{\alpha^L}{\gamma^L}(\varrho) \\ \frac{\alpha^U}{\gamma^U}(\varsigma) < \frac{\alpha^U}{\gamma^U}(\varrho) \end{cases},$$

which implies

$$\alpha^L(\varsigma) - v^L\gamma^L(\varsigma) < \alpha^L(\varrho) - v^L\gamma^L(\varrho),$$
$$\alpha^U(\varsigma) - v^U\gamma^U(\varsigma) \le \alpha^U(\varrho) - v^U\gamma^U(\varrho),$$

or

$$\alpha^L(\varsigma) - \upsilon^L\gamma^L(\varsigma) \leq \alpha^L(\varrho) - \upsilon^L\gamma^L(\varrho),$$
$$\alpha^U(\varsigma) - \upsilon^U\gamma^U(\varsigma) < \alpha^U(\varrho) - \upsilon^U\gamma^U(\varrho),$$

or

$$\alpha^L(\varsigma) - \upsilon^L\gamma^L(\varsigma) < \alpha^L(\varrho) - \upsilon^L\gamma^L(\varrho),$$
$$\alpha^U(\varsigma) - \upsilon^U\gamma^U(\varsigma) < \alpha^U(\varrho) - \upsilon^U\gamma^U(\varrho),$$

the above inequalities yield,

$$\alpha^L(\varsigma) - \upsilon^L\gamma^L(\varsigma) + \alpha^U(\varsigma) - \upsilon^U\gamma^U(\varsigma)$$
$$< \alpha^L(\varrho) - \upsilon^L\gamma^L(\varrho) + \alpha^U(\varrho) - \upsilon^U\gamma^U(\varrho). \tag{24}$$

(i) Since $\varsigma \in \mathcal{F}$ and $(\varrho, \lambda, \beta, \alpha^G, \alpha^H, \eta, \varpi) \in \mathcal{F}_{MD}$, it follows that

$$\phi_\rho(\varsigma) \leq 0,\ \varsigma_\rho \geq 0,\ \forall \rho = 1, 2, ..., \omega,$$
$$\psi_s(\varsigma) = 0,\ \gamma_s \in R,\ \forall s = 1, 2, ..., p,$$
$$-G_\rho(\varsigma) < 0,\ \alpha_\rho^G \geq 0,\ \forall \rho \in \Omega_+(\varsigma),$$
$$-G_\rho(\varsigma) = 0,\ \alpha_\rho^G \in R,\ \forall \rho \in \Omega_0(\varsigma),$$
$$H_\rho(\varsigma) > 0,\ \alpha_\rho^H = 0,\ \forall \rho \in \Omega_{0+}(\varsigma),$$
$$H_\rho(\varsigma) = 0,\ \alpha_\rho^H \geq 0,\ \forall \rho \in \Omega_{00}(\varsigma) \cup \Omega_{+0}(\varsigma),$$
$$H_\rho(\varsigma) < 0,\ \alpha_\rho^H \geq 0\ , \forall \rho \in \Omega_{0-}(\varsigma) \cup \Omega_{+-}(\varsigma).$$

By (23), it implies that

$$\sum_{\rho=1}^{\omega} \lambda_\rho \phi_\rho(\varsigma) + \sum_{s=1}^{p} \beta_s \psi_s(\varsigma) - \sum_{\rho=1}^{\ell} \alpha_\rho^G G_\rho(\varsigma) + \sum_{\rho=1}^{\ell} \alpha_\rho^H H_\rho(\varsigma)$$
$$\leq \sum_{\rho=1}^{\omega} \lambda_\rho \phi_\rho(\varrho) + \sum_{s=1}^{p} \beta_s \psi_s(\varrho) - \sum_{\rho=1}^{\ell} \alpha_\rho^G G_\rho(\varrho) + \sum_{\rho=1}^{\ell} \alpha_\rho^H H_\rho(\varrho).$$

Combining the quasiconvexity of $\sum_{\rho=1}^{\omega} \lambda_\rho \phi_\rho(.) + \sum_{s=1}^{p} \beta_s \psi_s(.) - \sum_{\rho=1}^{\ell} \alpha_\rho^G G_\rho(.) + \sum_{\rho=1}^{\ell} \alpha_\rho^H H_\rho(.)$, one has

$$(\varsigma - \varrho)^T \Bigg[\sum_{\rho=1}^{\omega} \lambda_\rho \nabla\phi_\rho(\varrho) + \sum_{s=1}^{p} \beta_s \nabla\psi_s(\varrho) - \sum_{\rho=1}^{\ell} \alpha_\rho^G \nabla G_\rho(\varrho)$$
$$+ \sum_{\rho=1}^{\ell} \alpha_\rho^H \nabla H_\rho(\varrho) \Bigg] \leq 0.$$

Using the above inequality and from (23), one has

$$(\varsigma - \varrho)^T[\nabla\alpha^L(\varrho) - \Psi^L(\varrho)\nabla\gamma^L(\varrho) + \nabla\alpha^U(\varrho) - \Psi^U(\varrho)\nabla\gamma^U(\varrho)] \geq 0.$$

By the pseudoconvexity of $\alpha^L(.) - \Psi^L(\varrho)\gamma^L(.) + \alpha^U(.) - \Psi^U(\varrho)\gamma^U(.)$, it implies that

$$\alpha^L(\varsigma) - \upsilon^L\gamma^L(\varsigma) + \alpha^U(\varsigma) - \upsilon^U\gamma^U(\varsigma) \geq \alpha^L(\varrho) - \upsilon^L\gamma^L(\varrho) + \alpha^U(\varrho) - \upsilon^U\gamma^U(\varrho).$$

which is contradicting (24).

(ii) By $\varsigma \in \mathcal{F}$ and $(\varrho, \lambda, \beta, \alpha^G, \alpha^H, \eta, \varpi) \in \mathcal{F}_{MD}$, it follows that

$$\varphi_\rho(\varsigma) \le \phi_\rho(\varrho),\ \rho \in \Omega^+_\phi(\varsigma),\ \psi_s(\varsigma) = \psi_s(\varrho), s \in \Omega^+_\psi(\varsigma) \cup \Omega^-_\psi(\varsigma),$$
$$-G_\rho(\varsigma) \le -G_\rho(\varrho), \rho \in \Omega^+_{+0}(\varsigma) \cup \Omega^+_{+-}(\varsigma) \cup \Omega^+_{00}(\varsigma) \cup \Omega^+_{0-}(\varsigma) \cup \Omega^+_{0+}(\varsigma),$$
$$-G_\rho(\varsigma) \ge -G_\rho(\varrho), \rho \in \Omega^-_{0+}(\varsigma),\ H_\rho(\varsigma) \le H_\rho(\varrho), \rho \in \Omega^{++}_{+0}(\varsigma) \cup \Omega^{++}_{+}-(\varsigma),$$

By the quasiconvexity of $\phi_\rho(\rho \in \Omega^+_\phi(\varsigma))$, $\psi_s\ (s \in \Omega^+_\psi(\varsigma))$, $-\psi_s\ (s \in \Omega^-_\psi(\varsigma))$, $-G_\rho\ (\rho \in \Omega^+_{+0}(\varsigma) \cup \Omega^+_{+-}(\varsigma) \cup \Omega^+_{00}(\varsigma) \cup \Omega^+_{0-}(\varsigma) \cup \Omega^+_{0+}(\varsigma))$, $-G_\rho$
$(\rho \in \Omega^-_{0+}(\varsigma))$, $H_\rho\ (\rho \in \Omega^{++}_{+0}(\varsigma) \cup \Omega^{++}_{+-}(\varsigma))$, it implies that

$$(\varsigma - \varrho)^T \nabla \phi_\rho(\varrho) \le 0,\ \rho \in \Omega^+_\phi(\varsigma),\ (\varsigma - \varrho)^T \nabla \psi_s(\varrho) \le 0, s \in \Omega^+_\psi,$$
$$(\varsigma - \varrho)^T \nabla \psi_s(\varrho) \ge 0, s \in \Omega^-_\psi,\ -(\varsigma - \varrho)^T \nabla G_\rho(\varrho) \le 0, \rho \in \Omega^+_{+0}(\varsigma) \cup \Omega_+{}^+{}_-(\varsigma) \cup \Omega^+_{00}(\varsigma) \cup \Omega^+_{0-}(\varsigma) \cup \Omega^+_{0+}(\varsigma),$$
$$-(\varsigma - \varrho)^T \nabla G_\rho(\varrho) \ge 0, \rho \in \Omega^-_{0+},\ (\varsigma - \%)^T \nabla H_\rho(\varrho) \le 0, \rho \in \Omega^{++}_{+0} \cup \Omega^{++}_{+-}.$$

From the above inequalities and from the above-mentioned index sets, one obtain

$$(\varsigma - \varrho)^T \left[\sum_{\rho=1}^{\omega} \lambda_\rho \nabla \phi_\rho(\varrho) + \sum_{s=1}^{p} \beta_s \nabla \psi_s(\varrho) - \sum_{\rho=1}^{\ell} \alpha^G_\rho \nabla G_\rho(\varrho) + \sum_{\rho=1}^{\ell} \alpha^H_\rho \nabla H_\rho(\varrho) \right] \le 0.$$

Combining the aforementioned inequality and (23), one has

$$(\varsigma - \varrho)^T [\nabla \alpha^L(\varrho) - \Psi^L(\varrho) \nabla \gamma^L(\varrho) + \nabla \alpha^U(\varrho) - \Psi^U(\varrho) \nabla \gamma^U(\varrho)] \ge 0.$$

By the pseudoconvexity of $\alpha^L(.) - \Psi^L(.)\gamma^L(.) + \alpha^U(.) - \Psi^U(.)\gamma^U(.)$, it implies that

$$\alpha^L(\varsigma) - v^L\gamma^L(\varsigma) + \alpha^U(\varsigma) - v^U\gamma^U(\varsigma) \ge \alpha^L(\varrho) - v^L\gamma^L(\varrho) + \alpha^U(\varrho) - v^U\gamma^U(\varrho).$$

which is contradicting (24).

Theorem 5.2 (Strong Duality) *Let $\bar{\varsigma}$ be a* LU *optimal solution of (FVC) and* (FVC-MACQ) *is fulfilled at $\bar{\varsigma}$. Then there exist $\tilde{\lambda} \in R^\omega_+$, $\tilde{\beta} \in R^p$, $\tilde{\alpha}^G$, $\tilde{\alpha}^H$, $\tilde{\eta}$, $\tilde{\varpi} \in R^\ell$ such that $(\tilde{\varrho}, \tilde{\lambda}, \tilde{\beta}, \tilde{\alpha}^G, \tilde{\alpha}^H, \tilde{\eta}, \tilde{\varpi})$ is a feasible solution of (MD) and the value of objective functions are equal. Additionally, if all the suppositions of the weak duality Theorem* (5.1) *are fulfilled, then the point $(\tilde{\varrho}, \tilde{\lambda}, \tilde{\beta}, \tilde{\alpha}^G, \tilde{\alpha}^H, \tilde{\eta}, \tilde{\varpi})$ is a* LU *optimal solution of (MD).*

Proof: By presumption $\bar{\varsigma}$ is a $\mathcal{LU}$ optimal solution of (FVC) and (FVCMACQ) condition is ascertained at this point, then by Theorem (3.1), there exists $\tilde{\lambda} \in R^\omega_+$, $\tilde{\beta} \in R^p$, $\tilde{\alpha}^G$, $\tilde{\alpha}^H$, $\tilde{\eta}$, $\tilde{\varpi} \in R^\ell$ such that the conditions (1)–(8) are satisfied. Thus, the feasible solution of (PD) is $(\tilde{\varrho}, \tilde{\lambda}, \tilde{\beta}, \tilde{\alpha}^G, \tilde{\alpha}^H, \tilde{\eta}, \tilde{\varpi})$, Further, the objective values of both the (FVC) problem and the (MD) problem are equal. Furthermore, if $(\tilde{\varrho}, \tilde{\lambda}, \tilde{\beta}, \tilde{\alpha}^G, \tilde{\alpha}^H, \tilde{\eta}, \tilde{\varpi})$ is not a $\mathcal{LU}$ optimal solution of (MD), then there is a feasible solution $(\tilde{\varrho}, \tilde{\lambda}, \tilde{\beta}, \tilde{\alpha}^G, \tilde{\alpha}^H, \tilde{\eta}, \tilde{\varpi})$ for (MD), such that the following inequality is satisfied

$$\left[\frac{\alpha^L}{\gamma^L}(\tilde{\varsigma}), \frac{\alpha^U}{\gamma^U}(\tilde{\varsigma})\right] \prec_{\mathcal{LU}} \left[\frac{\alpha^L}{\gamma^L}(\tilde{\varrho}), \frac{\alpha^U}{\gamma^U}(\tilde{\varrho})\right].$$

This contradicts the weak duality theorem (5.1). Hence $(\tilde{\varrho}, \tilde{\lambda}, \tilde{\beta}, \tilde{\alpha}^G, \tilde{\alpha}^H, \tilde{\eta}, \tilde{\varpi})$ is a $\mathcal{LU}$ optimal solution of (MD).

Theorem 5.3 (Strict Converse Duality) *Let $\bar{\varsigma} \in \mathcal{F}$ be a $\mathcal{LU}$ optimal solution of (FVC) such that the (FVC-ACQ) at $\bar{\varsigma}$. Assume the conditions of Theorem* (5.2) *hold and $(\bar{\varrho}, \bar{\lambda}, \bar{\beta}, \bar{\alpha}^G, \bar{\alpha}^H, \bar{\eta}, \bar{\varpi}) \in \mathcal{F}_{MD}$ be a $\mathcal{LU}$ optimal solution of (MD). If any of the following conditions hold;*

(i) $\alpha^L(.) - \Psi^L(\bar{\varrho})\gamma^L(.) + \alpha^U(.) - \Psi^U(\bar{\varrho})\gamma^U(.)$ *is strictly pseudoconvex and*

$\sum_{\rho=1}^{\omega} \lambda_\rho \phi_\rho(.) + \sum_{s=1}^{p} \beta_s \psi_s(.) - \sum_{\rho=1}^{\ell} \alpha^G_\rho G_\rho(.) + \sum_{\rho=1}^{\ell} \alpha^H_\rho H_\rho(.)$ *is quasiconvex at $\bar{\varrho}$ on $\mathcal{F} \cup pr\mathcal{F}_{MD}$,*

(ii) $\alpha^L(.) - \Psi^L(\bar{\varrho})\gamma^L(.) + \alpha^U(.) - \Psi^U(\bar{\varrho})\gamma^U(.)$ *is strictly pseudoconvex and $\phi_\rho\ (\rho \in \Omega^+_\varphi(\varsigma))$, $\psi_s\ (s \in \Omega^+_\psi(\varsigma))$, $-\psi_s\ (s \in \Omega^-_\psi(\varsigma))$, $-G_\rho\ (\rho \in \Omega^+_{+0}(\varsigma) \cup \Omega^+_{+-}(\varsigma) \cup \Omega^+_{00}(\varsigma) \cup \Omega^+_{0-}(\varsigma) \cup \Omega^+_{0+}(\varsigma))$, $-G_\rho\ (\rho \in \Omega^-_{0+}(\varsigma))$, $H_\rho\ (\rho \in \Omega^{++}_{+0}(\varsigma) \cup \Omega^{++}_{+-}(\varsigma))$ are quasiconvex functions at $\bar{\varrho}$ on $\mathcal{F} \cup pr\mathcal{F}_{MD}$,*

then $\bar{\varsigma} = \bar{\varrho}$.

Proof: (i) Supposing on the contradiction, let us presume that $\bar{\varsigma} \ne \bar{\varrho}$. By Theorem (5.2), there exist $\bar{\lambda} \in R^\omega_+$, $\bar{\beta} \in R^p$, $\bar{\alpha}G \in R^\ell$, $\bar{\alpha}^H \in R^\ell$, $\bar{\eta} \in R^\ell$, $\bar{\rho} \in R^\ell$ such that $(\bar{\varrho}, \bar{\lambda}, \bar{\beta}, \bar{\alpha}^G, \bar{\alpha}^H, \bar{\eta}, \bar{\varpi})$ be a $\mathcal{LU}$ optimal solution of (MD). Hence

$$\alpha^L(\varsigma) - \Psi^L(\varrho)\gamma^L(\varsigma) + \alpha^U(\varsigma) - \Psi^U(\varrho)\gamma^U(\varsigma) = \alpha^L(\varrho) - \Psi^L(\varrho)\gamma^L(\varrho) + \alpha^U(\varrho) - \Psi^U(\varrho)\gamma^U(\varrho) \tag{25}$$

Since $\varsigma \in \mathcal{F}$ and $(\bar{\varrho}, \bar{\lambda}, \bar{\beta}, \bar{\alpha}^G, \bar{\alpha}^H, \bar{\eta}, \bar{\varpi}) \in \mathcal{F}_{MD}$, it follows that

$$\begin{aligned}
&\phi_\rho(\bar{\varsigma}) \le 0,\ \bar{\varsigma}_\rho \ge 0,\ \forall \rho = 1, 2, ..., \omega,\\
&\psi_s(\bar{\varsigma}) = 0,\ \bar{\gamma}_s \in R,\ \forall s = 1, 2, ..., p,\\
&-G_\rho(\bar{\varsigma}) < 0,\ \bar{\alpha}^G_\rho \ge 0,\ \forall \rho \in \Omega_+(\bar{\varsigma}),\\
&-G_\rho(\bar{\varsigma}) = 0,\ \bar{\alpha}^G_\rho \in R,\ \forall \rho \in \Omega_0(\bar{\varsigma}),\\
&H_\rho(\bar{\varsigma}) > 0,\ \bar{\alpha}^H_\rho = 0,\ \forall \rho \in \Omega_{0+}(\bar{\varsigma}),\\
&H_\rho(\bar{\varsigma}) = 0,\ \bar{\alpha}^H_\rho \ge 0,\ \forall \rho \in \Omega_{00}(\bar{\varsigma}) \cup \Omega_{+0}(\bar{\varsigma}),\\
&H_\rho(\bar{\varsigma}) < 0,\ \bar{\alpha}^H_\rho \ge 0\ , \forall \rho \in \Omega_{0-}(\bar{\varsigma}) \cup \Omega_{+-}(\bar{\varsigma}).
\end{aligned}$$

By (23), it implies that

$$\begin{aligned}
&\sum_{\rho=1}^{\omega} \lambda_\rho \phi_\rho(\bar{\varsigma}) + \sum_{s=1}^{p} \beta_s \psi_s(\bar{\varsigma}) - \sum_{\rho=1}^{\ell} \alpha^G_\rho G_\rho(\bar{\varsigma}) + \sum_{\rho=1}^{\ell} \alpha^H_\rho H_\rho(\bar{\varsigma})\\
&\le \sum_{\rho=1}^{\omega} \lambda_\rho \phi_\rho(\bar{\varrho}) + \sum_{s=1}^{p} \beta_s \psi_s(\bar{\varrho}) - \sum_{\rho=1}^{\ell} \alpha^G_\rho G_\rho(\bar{\varrho}) + \sum_{\rho=1}^{\ell} \alpha^H_\rho H_\rho(\bar{\varrho}).
\end{aligned}$$

Combining the quasiconvexity of $\sum_{\rho=1}^{\omega} \lambda_\rho \phi_\rho(.) + \sum_{s=1}^{p} \beta_s \psi_s(.) \ - \sum_{\rho=1}^{\ell} \alpha^G_\rho G_\rho(.) + \sum_{\rho=1}^{\ell} \alpha^H_\rho H_\rho(.)$, one has

$$(\bar{\varsigma} - \bar{\varrho})^T \left[\sum_{\rho=1}^{\omega} \lambda_\rho \nabla \phi_\rho(\bar{\varrho}) + \sum_{s=1}^{p} \beta_s \nabla \psi_s(\bar{\varrho}) - \sum_{\rho=1}^{\ell} \alpha^G_\rho \nabla G_\rho(\bar{\varrho}) + \sum_{\rho=1}^{\ell} \alpha^H_\rho \nabla H_\rho(\bar{\varrho}) \right] \le 0.$$

Using the above inequality and from (23), one has

$$(\bar{\varsigma} - \bar{\varrho})^T [\nabla \alpha^L(\bar{\varrho}) - \Psi^L(\bar{\varrho}) \nabla \gamma^L(\bar{\varrho}) + \nabla \alpha^U(\bar{\varrho}) - \Psi^U(\bar{\varrho}) \nabla \gamma^U(\bar{\varrho})] \ge 0.$$

In view of strict pseudoconvexity of $\alpha^L(.) - \Psi^L(\bar{\varrho})\gamma^L(.) + \alpha^U(.) - \Psi^U(\bar{\varrho})\gamma^U(.)$ at $\bar{\varrho}$ on $\mathcal{F} \cup pr\mathcal{F}_{MD}$, the above inequality gives

$$\begin{aligned}
&\alpha^L(\bar{\varsigma}) - \Psi^L(\bar{\varrho})\gamma^L(\bar{\varsigma}) + \alpha^U(\bar{\varsigma}) - \Psi^U(\bar{\varrho})\gamma^U(\bar{\varsigma})\\
&\succ_{\mathcal{LU}} \alpha^L(\bar{\varrho}) - \Psi^L(\bar{\varrho})\gamma^L(\bar{\varrho}) + \alpha^U(\bar{\varrho}) - \Psi^U(\bar{\varrho})\gamma^U(\bar{\varrho}),
\end{aligned}$$

which contradicts (25). This completes the proof.

(ii) By $\varsigma \in \mathcal{F}$ and $(\bar{\varrho}, \bar{\lambda}, \bar{\beta}, \bar{\alpha}^G, \bar{\alpha}^H, \bar{\eta}, \bar{\varpi}) \in \mathcal{F}_{MD}$, it follows that

$$\begin{aligned}
&\phi_\rho(\bar{\varsigma}) \le \phi_\rho(\bar{\varrho}), \rho \in \Omega^+_\phi(\bar{\varsigma}),\\
&\psi_s(\bar{\varsigma}) = \psi_s(\bar{\varrho}), s \in \Omega^+_\psi(\bar{\varsigma}) \cup \Omega^-_\psi(\bar{\varsigma}),\\
&-G_\rho(\bar{\varsigma}) \le -G_\rho(\bar{\varrho}), \rho \in \Omega^+_{+0}(\bar{\varsigma}) \cup \Omega^+_{+-}(\bar{\varsigma}) \cup \Omega^+_{00}(\bar{\varsigma}) \cup \Omega^+_{0-}(\bar{\varsigma})\\
&\qquad \cup \Omega^+_{0+}(\bar{\varsigma}),\\
&-G_\rho(\bar{\varsigma}) \ge -G_\rho(\bar{\varrho}), \rho \in \Omega^-_{0+}(\bar{\varsigma}),\\
&H_\rho(\bar{\varsigma}) \le H_\rho(\bar{\varrho}), \rho \in \Omega^{++}_{+0}(\bar{\varsigma}) \cup \Omega^{++}_{+-}(\bar{\varsigma}),
\end{aligned}$$

By the quasiconvexity of $\varphi_\rho (\rho \in \Omega^+_\varphi(\bar{\varsigma}))$, $\psi_s (s \in \Omega^+_\psi(\bar{\varsigma}))$, $-\psi_s (s \in \Omega^-_\psi(\bar{\varsigma}))$, $-G_\rho (\rho \in \Omega^+_{+0}(\bar{\varsigma}) \cup \Omega^+_{+-}(\bar{\varsigma}) \cup \Omega^+_{00}(\bar{\varsigma}) \cup \Omega^+_{0-}(\bar{\varsigma}) \cup \Omega^+_{0+}(\bar{\varsigma}))$, $-G_\rho(\rho \in \Omega^-_{0+}(\bar{\varsigma}))$, $H_\rho (\rho \in \Omega^{++}_{+0}(\bar{\varsigma}) \cup \Omega^{++}_{+-}(\bar{\varsigma}))$, it implies that

$$(\bar{\varsigma} - \bar{\varrho})^T \nabla \varphi_\rho(\bar{\varrho}) \leq 0,\ \rho \in \Omega^+_\varphi(\bar{\varsigma}),$$
$$(\bar{\varsigma} - \bar{\varrho})^T \nabla \psi_s(\bar{\varrho}) \leq 0, s \in \Omega^+_\psi,$$
$$(\bar{\varsigma} - \bar{\varrho})^T \nabla \psi_s(\bar{\varrho}) \geq 0, s \in \Omega^-_\psi,$$
$$-(\bar{\varsigma} - \bar{\varrho})^T \nabla G_\rho(\bar{\varrho}) \leq 0, \rho \in \Omega^+_{+0}(\bar{\varsigma}) \cup \Omega^+_{+-}(\bar{\varsigma}) \cup \Omega^+_{00}(\bar{\varsigma}) \cup \Omega^+_{0-}(\bar{\varsigma}) \cup \Omega^+_{0+}(\bar{\varsigma}),$$
$$-(\bar{\varsigma} - \bar{\varrho})^T \nabla G_\rho(\bar{\varrho}) \geq 0, \rho \in \Omega^-_{0+},$$
$$(\bar{\varsigma} - \bar{\varrho})^T \nabla H_\rho(\bar{\varrho}) \leq 0, \rho \in \Omega^{++}_{+0} \cup \Omega^{++}_{+-}.$$

From the above inequalities and by index sets, one has

$$(\bar{\varsigma} - \bar{\varrho})^T \left[\sum_{\rho=1}^{\omega} \lambda_\rho \nabla \phi_\rho(\bar{\varrho}) + \sum_{j=1}^{p} \beta_s \nabla \psi_s(\bar{\varrho}) - \sum_{\rho=1}^{\ell} \alpha^G_\rho \nabla G_\rho(\bar{\varrho}) + \sum_{\rho=1}^{\ell} \alpha^H_\rho \nabla H_\rho(\bar{\varrho}) \right] \leq 0.$$

We obtain the following expression by combining the aforementioned inequality with equation (23).

$$(\bar{\varsigma} - \bar{\varrho})^T [\nabla \alpha^L(\bar{\varrho}) - \Psi^L(\bar{\varrho}) \nabla \gamma^L(\bar{\varrho}) + \nabla \alpha^U(\bar{\varrho}) - \Psi^U(\bar{\varrho}) \nabla \gamma^U(\bar{\varrho})] \geq 0.$$

By the strict pseudoconvexity of $\alpha^L(.) - \Psi^L(.)\gamma^L(.) + \alpha^U(.) - \Psi^U(.)\gamma^U(.)$, it implies that

$$\alpha^L(\bar{\varsigma}) - \Psi^L(\bar{\varrho})\gamma^L(\bar{\varsigma}) + \alpha^U(\bar{\varsigma}) - \Psi^U(\bar{\varrho})\gamma^U(\bar{\varsigma})$$
$$\succ_{\mathcal{LU}} \alpha^L(\bar{\varrho}) - \Psi^L(\bar{\varrho})\gamma^L(\bar{\varrho}) + \alpha^U(\bar{\varrho}) - \Psi^U(\bar{\varrho})\gamma^U(\bar{\varrho}),$$

which is contradicting (25).

6. Conclusion

This article introduces a parametric dual (PD) for the optimization problem with fractional interval-valued objectives concerning vanishing constraints (FVC). It establishes the weak, strong, and strict converse duality findings between (FVC) and (PD). A Mond-Weir framework is formulated, and its duality results are explored under the assumption of generalized convex functions.

REFERENCES

1. Achtziger, W. and Kanzow, C. (2008): Mathematical programs with vanishing constraints: Optimality conditions and constraints qualifications. Math. Program. 114(1), 6999.
2. Ahmad, I., Singh, D. and Ahmad, B. (2016): Optimality conditions for invex interval-valued nonlinear programming problems involving generalized H-derivative, Filomat, 30, 21212138.
3. Ben-Israel, A. and Robers, P. D. (1969/1970): A decomposition method for interval linear programming, Manage. Sci., 16, 374387.
4. Bhurjee, A. and Panda, G. (2012): Efficient solution of interval optimization problem, Math. Methods Oper. Res., 76, 273288.
5. Chalco-Cano, Y., Lodwick, W. A. and Rufian-Lizana, A. (2013): Optimality conditions of type KKT for optimization problem with interval-valued objective function via a generalized derivative. Fuzzy Optim. Decis. Mak., 12, 305322.
6. Debnath, I.P. and Gupta S.K. (2019): Necessary and sufficient optimality conditions for fractional interval-valued optimization problems. In: K. Deep, M. Jain, S. Salhi, (eds) Decision Science in Action. Asset Analytics, pp. 155173, Springer.
7. Dar, B. A., Jayswal, A., Singh, D. (2021): Optimality, duality and saddle point analysis for interval-valued nondifferentiable multiobjective fractional programming problems, Optimization 70, 12751305.
8. Dubey, R. and Gupta, S. K. (2017): On duality for a second-order multiobjective fractional programming problem involving type-I functions, Georgian Math. J., 26(3) 393–404.
9. Ghosh, D. (2017): Newton method to obtain efficient solutions of the optimization problems with interval-valued objective functions. J. Appl. Math. Comput. 53, 709731.
10. Hoheisel, T., Kanzow, C., Outrata, J.V. (2010): Exact penalty results for mathematical programs with vanishing constraints. Nonlinear Anal. Theory Methods Appl. 72(5), 25142526.
11. Hu, Q.J., Chen, Y., Zhu, Z.B., Zhang, B.S. (2014): Notes on some convergence properties for a smoothing regularization approach to mathematical programs with vanishing constraints. Abstr. Appl. Anal. (1), 17.
12. Jabr, R.A. (2012): Solution to economic dispatching with disjoint feasible regions via semidefinite programming. IEEE Trans. Power Syst. 27(1), 572573.

13. Kim, D. S. (2004): Optimality and Duality for Multiobjective Fractional Programming with Generalized Invexity. MCDM, Whistler, B.C. Canada.
14. Kummari, K. and Rani, B. J.: Optimality conditions and saddle point criteria for fractional interval-valued optimization problem with vanishing constraints using generalized convexity. (Communicated)
15. Kirches, C., Potschka, A., Bock, H.G., Sager, S., (2013): A parametric active set method for quadratic programs with vanishing constraints. Pac. J. Optim. 9(2), 275299.
16. Laha, V., Kumar, R., Singh, H.N., Mishra, S.K. (2020): On minimax programming with vanishing constraints. In: Laha, V., Marechal, P., Mishra, S.K. (eds.) Optimization, Variational Analysis and Applications. IFSOVAA 2020. Springer Proceedings in Mathematics & Statistics, vol. 355 , pp. 247263. Springer, Singapore.
17. Michael, N.J., Kirches, C., Sager, S. (2013): On perspective functions and vanishing constraints in mixed integer nonlinear optimal control. In: J unger, M., Reinelt, G. (eds.) Facets of Combinatorial Optimization, pp. 387417. Springer, Berlin.
18. Singh, A. D. and Dar, B. A. (2015): Optimality conditions in multiobjective programming problems with interval-valued objective functions. Control Cybern., 44, 1945.
19. Tung, L. T. (2019): KarushKuhnTucker optimality conditions and duality for semi-infinite programming with multiple interval-valued objective functions. J. Nonlinear Funct. Anal.
20. Tung, L. T. (2020): KarushKuhnTucker optimality conditions and duality for convex semi-infinite programming with multiple interval-valued objective functions. J. Appl. Math. Comput., 62, 6791.
21. WU, H. C. (2007): The KarushKuhnTucker optimality conditions in an optimization problem with interval-valued objective functions. Eur. J. Oper. Res., 176, 4659.
22. WU, H. C. (2009): The KarushKuhnTucker optimality conditions in multiobjective programming problems with interval-valued objective functions. Eur. J. Oper. Res., 196, 4960.
23. WU, H. C. (2009): The KarushKuhnTucker optimality conditions for multi-objective programming problems with fuzzy-valued objective functions. Fuzzy Optim. Decis. Mak., 8, 128.
24. H. C. Zhou and Y. J. Wang, (2009): Optimality conditions and mixed duality for interval-valued optimization, Fuzzy Info. and Eng., 2, 13151323.
25. M. A. Hejazi and S. Nobakhtian, (2020): Optimality conditions for multiobjective fractional programming via convexificators, J. Ind. Manag. Optim., 16(2) 623-631.
26. I. M. Stancu-Minasian, (2019): A ninth bibliography of fractional programming, Optimization, 68(11) 2125–2169
27. I. P. Debnath and S. K. Gupta, (2020): The KarushKuhnTucker conditions for multiple objective fractional interval-valued optimization problems, RAIRO Oper. Res. 54, 11611188.
28. T. D. Chuong, D. S. Kim, (2016): A class of nonsmooth fractional multiobjective optimization problems, Ann. Oper. Res. 244, 367383.

Impending Inquisitions in Humanities and Sciences (ICIIHS-2022) – Dr. Mohan Varkolu et al. (eds)

Reducing the Number of Trainable Parameters does not Affect the Accuracy of Res-Net18 on Cervical Cancer Images

76

K. Sreenivasa Rao
Department of Computer Science and Engineering,
Koneru Lakshmaiah Education Foundation Hyderabad, India
Priyadarshini Chatterjee[1]
Research Scholar, Department of Computer Science and Engineering,
Koneru Lakshmaiah Education Foundation, Hyderabad, India
And
Assistant Professor, Department of Computer Science and Engineering,
B V Raju Institute of Technology, Narsapur, Medak. Telengana

ABSTRACT Res-Net18 is one amongst the most important tool for medical image processing. It is used both for image segmentation and classification. There are CNN that are used to train very deep network using the 'technique of batch normalization. However, these CNNs cannot tackle the problem of degradation. Researchers have found out an improved version of CNN called Res-Net with shortcut connection. The residual unit of this u-net helps in improving the flow of information. It makes the layers to learn according to the residual function and according to the input given in the layers. The short circuit connection makes the output and the input dimensions similar. There are various evolved versions of res u-net. This paper has considered res-net18 whose number of trainable parameters is eleven million with a standard accuracy of 94% with respect to the three datasets of cervical cancer. We have tried to reduce the number of trainable parameters of Res-Net18 by using an additional max-pooling layer in the architecture of the Res U-Net. We have successfully reduced the trainable parameters to eight million, keeping the accuracy to 94% on a cervical cancer data set.

KEYWORDS Deep learning, CNN, Residual network, Segmentation

1. Introduction

The high-performance computing facility has made the growth of deep learning a promiscuous field. The performance of deep learning involves analyzing the structured data to unstructured data. As the name implies it uses many layers to process the features of large volume of data. Each layer is capable of extracting features and pass it to the next layer. Layers that are present at the initial level is capable of extracting low level features and the successive layers combine these features to perceive the complete representation. The evolution of deep learning models are as follows:

The Artificial Neural of first generation comprising of perceptron in their layers were of limited computations. There was an improvement in the second generation of ANN's that calculated the error and back propagated the error. Then Boltzmann

*Corresponding author: jinipriya@klh.edu.in; priyadarshini.ch@bvrit.ac.in

DOI: 10.1201/9781003489436-76

machine was discovered but it was in its restricted version, though learning was made easier. Then there was the evolution of decision tress in 1966, Recurrent Neural Network in 1973, Support Vector Machine in 1979 followed by Convolutional Neural Network and so on and so forth. Each of these evolutions was an improvement of the previous versions. The deep learning approaches are supervised learning, unsupervised learning, reinforcement learning, hybrid learning.

Convolutional Neural Network is a type of deep learning model that is used for analysis of images. It is a multiplelayer deep learning blueprint that is based on the biological systems of living beings. The CNN uses different fields of computer vision and natural language processing. CNNs became popular in 2012 after the performance of Alex Net, though they are initiated much before. The CNN is inspired by the philosophy that cells of the visual cortex of the animals are able to recognize light in a small receptive field. In 1980 Kunihiko Fukusima proposed Neocognitron which is the first theoretical model of CNN. Then with the advancement of other versions of CNN, these networks were able to solve complex computation problems.

With the passage of time, it was observed that convolutional neural network was degrading in performance. The need was the requirement of a network that will be deeper and richer. This need pushed the development of a network called Residual Net. The Residual Net uses shortcut connection to improve the degradation. It is also used to improve the flow of information across the layers. It also makes the layers to reformulate as per the input layer. A deep residual network consists of a residual block. The layers in the residual block consists of stacked layers of batch normalization, a weight layer and ReLu activation. Short connections are those connections that skips one or more layers in the neural network. The short connection also maintains the dimension of the main convolutional block. The encoder and the decoder can be built by stacking the residual unit. There is total four stages in the encoder and decoder path and each stage consists of a residual block. This paper tries to improve the performance of ResNet18 by reducing the number of trainable parameters. The proposed algorithm is described in details in the third section of this paper.

Instance segmentation of glands and nuclei has become a very important part in computational pathology for diagnosing cancer. As the availability of large number of public data sets has increased, there is a surge in automated computation of predicting the abnormality in a suspected gland or nuclei. There is a race to prove oneself best in the domain specific challenges. It is also proved by researchers that accurate segmentation always leads to correct prediction of the disease. With the advent of technology, the segmentation process is been made robust. There are various models of convolutional network used for instance segmentation within the paradigm of computer vision. Residual Network is one such model that improves upon the problem of degradation. There are various uses of ResNet like image denoising, segmentation, feature extraction to name a few. This paper aims to improve the performance of Res-Net by modifying its architecture and thereby using the number of trainable parameters.

2. Literature Review

In the basic U-Net, there is a large use of data augmentation for improving the training and testing strategy. The architecture of the basic U-Net consists of contracting path and an expanding path. The contracting path is used for understanding the reference and the expanding path provides with the exact location. This network has outperformed previous methods and has shown promising results using very few images also [1,2].

Spine parsing is another very recent technology to diagnose spinal diseases. There is a SpineParseNet that does a system driven spine parsing. It uses region pooling to project the image and thus constructs the graph. The node of the graph denotes a specific spinal structure. The graph is analyzed by the graph convolution [3].

The attenuation correction in a mixed PET/MRI system has always been a challenging task. There is an efficient method to predict the CT images from T2 and T1 weighted data and that also uses limited data is used in recent times. This method uses improved neighborhood anchored regression that calculates the matrices to predict the false patches [4,5].

The noises in the images makes the images of low contrast and of poor quality. There is an improved fuzzy algorithm approach on the basis traditional Pas. S.K algorithm. This algorithm helps to preserve the important information after changing the planar space of the image to the fuzzy space.

Medical Imaging is one of the most important aspects in the field of medical image processing. A clear image always gives a much better result at the time of diagnosis. Medical Imaging tries to reveal the inner structure that is generally hidden by bones and skins. However, these images might not show accurate result because of image degradation. So, it is the responsibility of the researchers to improve the quality of the image. There are various algorithms and tools in today's time to process these images so that they give better results [6].

It is seen that medical images are grey in color, because of which the information that was required to be color loses important messages. These images also fail to analyze the features that are in the deeper side. There are algorithms that iteratively colors a grey scale image. These algorithms are based on deep neural network. There is a Y-Loss that preserves the content of the target and colorized medical image [7].

The M-PAT can resolve tissue color distribution that is based on spectral un-mixing. It identifies the absorption spectrum through a sequence of color images. These images are acquired using multiple illuminated source and wavelength. Due to this kind of acquisition, a dataset of images is created. This results in a huge data set. In order to reduce the volume of the data set sparce sampling method is used. There is a unique scanning-based image acquiring scheme that has a sparse detector array that rotates as the illumination wavelength changes [8].

It becomes necessary to pin the anatomical marks in a medical image. There are algorithms based on fully convolutional neural network that provides a global to local solution of highlighting a region of interest in a medical image. This algorithm is divided into parts, firstly a global FCNN brings at the forefront multiple patches. It performs classification and regression simultaneously. Subsequently a local FCNN analyzes the images highlighted by the global FCNN [9,10].

It has always been a problem for the researchers to denoise a tomographic image and save them from image quality degradation. The algorithm called as Discrete Wavelet Transform helps to denoise tomographic images in practice. This DWT is a filter that quantizes the 3D images. This algorithm also minimizes the hardware requirements for the process. The data is presented in fixed point format. This method does the work by reducing the bit width from 15% to 70% and also comprises of fewer energy costs [11].

Deformation in medical images is very common because of metallic installations or limited view of MRI images. This hinders the image analysis task. There is a new generative framework that does the medical image inpainting. This algorithm in paint the deformed regions without prior pinning the region of interest. They also help in image translation if it is required at the time of inpainting [12].

It is necessary now adays to analyze the images in the context of big data. The Gabor filter algorithm analyzes images in the context of Big Data. This filter is tested on 320 mammographic images that provided of precision of 75% and a recall of 33 percent. When this algorithm is performed on 1000 data, the precision is 70 percent and the recall remains as 33 percent. It is also proved that the Gabor filter does the analysis more concisely and accurately [13].

Not only deep learning is used for medical image processing, but also artificial intelligence plays a major role in processing medical images. There are many states of the art algorithms that helps in managing medical data and is also used for clinical researches.

Predicting the volume and size of future tumors using convolutional network has outperformed the traditional mathematical models. These algorithms use deep learning-based data driven approach and shows higher rate of accuracy. Current algorithms are not sufficient to analyze 4D longitudinal data. There is an algorithm that analyzes a tumor's static imaging appearances. Then it captures the dynamic time invariant changes. This algorithm shows a dice score of 83.2% [14].

Noises in the images are always ubiquitous, especially the Gaussian noise and stripe noise. There is a two-stage convolutional neural network that treats the image and the noise component equally. There is a noise subnetwork that separates the noise component and there is an image subnetwork that is guided by the noise subnetwork so that different noise distribution and noise levels can be analyzed for better [15].

3. Data and Variables

3.1 Data Preprocessing

The DSB data set of cervical cancer available in 2018 Data Science Bowl is divided into separate folders. The test and the train folder. The train folder contains seventy percent of 1340 images and the test folder contains thirty percent of 1340 images. The train folder is pre-processed using keras.preprocessing.image library before they are provided as input to the batch Normalize function. The algorithm is run on google-colab notebook with a support of 1 GPU and tensorflow 2.12.0 and keras 2.12.0 versions.

4. Methodology and Model Specifications

Residual Network is an improvement over the convolutional neural network models. In CNN, the researchers faced a difficulty in training the deep layers. It was believed that neural networks provide faster result if the number of layers is more, but experimental analysis did not prove the same. The reason behind this degradation in performance of the Res-Net was increasing the number of layers and uneven flow of information across all the deeper layers. So, Res-Net was made to use skip connections that helps to reduce the diminishing gradient problem. We have further modified the ResNet18 by reducing the trainable parameter to eight million from 11 million using the public data set of cervical cancer. It is also seen that the accuracy with the same data set remains to be 94% percent by using the unmodified ResNet18 as compared to using the modified ResNet18. In the following paragraph we provide with the algorithm that is used in the modification of ResNet18.

The algorithm can be briefly analyzed as the input X goes under batch normalization and the ReLu activation function is administered. This is then fed as input to the Residual Block. The output of the Residual Block, denoted as Y is applied a maxpool and the output of this is denoted as M. The output of the Conv2D () function is assigned as S. The output of this encoder block (M+S) denoted as E is given as input into the decoder block. The decoder block using keras decoder_block() function returns the output denoted as D. The build_resnet() function returns the output on D.

A. Algorithm

START

Encoder Block

1. Input X
2. Batch Normalize X
3. ReLu on Batch Normalized X
4. Step 1 to 3 is given as input to the Residual Block
5. Output of Step 4 is Y
6. mpool(Y) - MaxPooling2D of Keras and Tensorflow
7. Step 6 output is denoted as M
8. S is the output of Conv2D () of Keras and Tensorflow
9. The output of the encoder block is denoted as E=M + S
10. The decoder block takes the input E and uses the decoder_block function of Keras to return D
11. The build_resnet(input_shape) returns the output on input D.

END

After we implement the model summary of this modified ResNet18, we get the trainable parameters as eight million and with an accuracy of 94% on cervical cancer public data set. The unmodified version of the ResNet18 when implemented on the same data set shows a trainable parameter of eleven million. We have also tried to capture the evaluation results on accuracy, jaccard, recall, F1 and precision. The experimental results are based on TensorFlow version 2.9 with 1340 images in the data set. We have performed the experiment using ten epochs both for training and testing. Trainable parameters mean is the count of the learnable quantities by the filter. They are the weights that are learnt during training. They help to improve the predictive power of the model and they change during the back propagation. By introducing the Maxpool layer we have reduced the number of trainable parameters to eight million than the traditional ResNet18 which shows a trainable parameter count of eleven million on the same cervical cancer data set. The Fig. 76.1 shows the modified ResNet1.

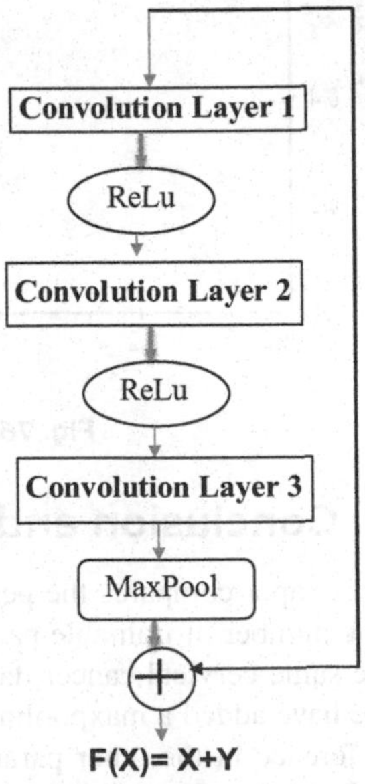

Fig. 76.1 Modified ResNet18

5. Empirical Results

5.1 Tabular Analysis

The Table 76.1 below shows the comparison analysis of ResNet 18 and the modified ResNet-18. The table clearly shows that by adding an extra maxpooling layer, in the new ResNet-18, the number of trainable parameters got reduced from 11 million in case of ResNet18 to 8 million in case of modified ResNet-18. The accuracy of ResNet18 is 94% and

Table 76.1 Comparison of ResNet18 and modified ResNet18

Method	Accuracy	Jaccard	F1	Precision
ResNet-18	0.94003	0.8845	0.8365	0.85543
Modified ResNet-18	0.94223	0.8934	0.8422	0.8625

the accuracy of modified ResNet-18 has improved to 94.223%. The table below also shows there is an improvement in metrics (Jaccard, F1 and Precision) of modified ResNet-18 to that of ResNet18.

5.2 Result Analysis

From the Table 76.1 above we see that the accuracy of ResNet18 without modification is almost similar to the modified ResNet18. The trainable parameter of the unmodified ResNet18 is 11,556,121 and the trainable parameter of the modified ResNet18 is 8,220,993. The Maxpooling function is used to diminish the dimension of the input image. ResNet18 is a dense net that has many layers, so the number of feature count is high. One way to minimize the number of trainable parameters is to normalize the data. In ResNet18, we are using Batch Normalization, that helps to reduce the trainable parameter. Maxpooling layer contributes a lot to reduce the trainable parameter as it down sample the data. In the modified ResNet18, we are using Maxpooling layer in the encoder side. Along with Batch Normalization, Maxpooling layer helps to reduce the trainable parameter from eleven million to eight million.

5.3 Graphical Analysis

Figure 76.2 and 76.3 below shoes the comparison of accuracy, Jaccard, F1 and precision value of ResNet18 and modified ResNet18. Below plots Fig. 76.2 and Fig. 76.3 is a graphical comparison of ResNet18 and modified ResNet-18 in terms of the metrics (Accuracy, Jaccard, Precision and Recall). The x-axis and the y-axis are in a scale of 0.2. As the improvement in these parameters of modified ResNet-18 in comparison to ResNet18 is in scale of 0.1, so we get the following graphs.

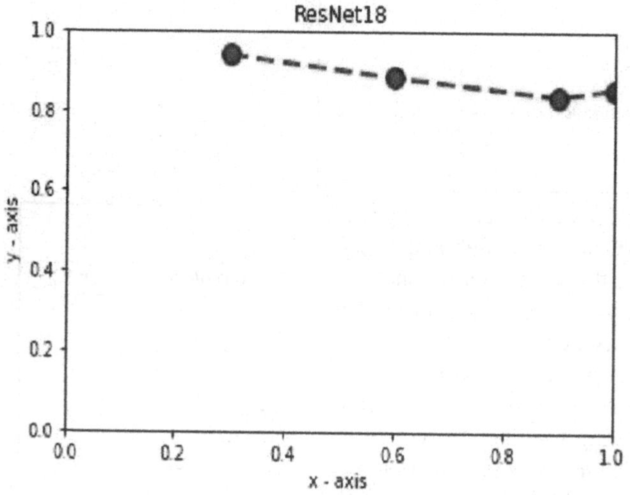

Fig. 76.2 ResNet18

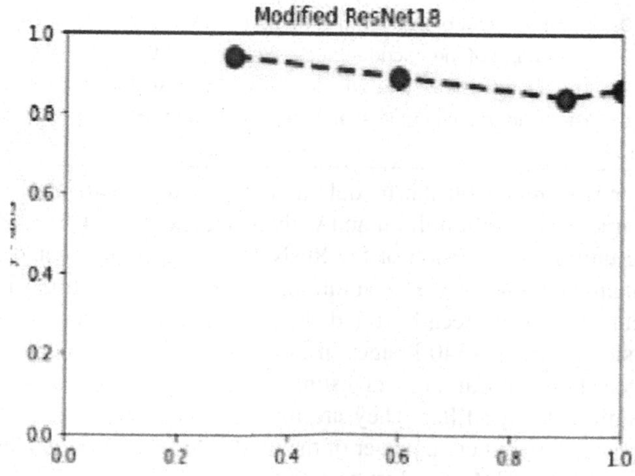

Fig. 76.3 Modified RsNet18

6. Conclusion and Future Work

This paper compares the performance of ResNet18 and modified ResNet18 in terms of accuracy, jaccard, F1, and precision. The number of trainable parameters of ResNet18 without modification is found out be eleven million when implemented on the same cervical cancer data set as compared to the modified ResNet18 that shows a trainable parameter of eight million. We have added a maxpooling layer at the encoder side. The accuracy of both the versions of ResNet18 shows 94% with little difference in the other parameters. In future we will work towards increasing the accuracy of ResNet18 after reducing the trainable parameter further.

REFERENCES

1. M. Karaman, M. A. Kutay, and G. Bozdagi, "An adaptive speckle suppression filter for medical ultrasonic imaging," IEEE Trans. Med. Imag., vol. 14, no. 2, pp. 283–292, Jun. 1995. (references) Cella, C. (2009). Institutional investors and corporate investment. United Sates: Indiana University, Kelley School of Business.
2. T. Loupas, W. N. McDicken, and P. L. Allan, "An adaptive weighted median filter for speckle suppression in medical ultrasonic images," IEEE Trans. Circuits Syst., vol. 36, no. 1, pp. 129–135, Jan. 1989.
3. F. Attivissimo, G. Cavone, A. M. L. Lanzolla, and M. Spadavecchia, "A technique to improve the image quality in computer tomography," IEEE Trans. Instrum. Meas., vol. 59, no. 5, pp. 1251–1257, May 2010.
4. M. Georgiev, R. Bregovi´c, and A. Gotchev, "Fixed-pattern noise modeling and removal in time-of-flight sensing," IEEE Trans. Instrum. Meas., vol. 65, no. 4, pp. 808–820, Apr. 2016.
5. F. Russo, "An image-enhancement system based on noise estimation," IEEE Trans. Instrum. Meas., vol. 56, no. 4, pp. 1435–1442, Aug. 2007.
6. A. Mencattini, M. Salmeri, R. Lojacono, M. Frigerio, and F. Caselli, "Mammographic images enhancement and denoising for breast cancer detection using dyadic wavelet processing," IEEE Trans. Instrum. Meas., vol. 57, no. 7, pp. 1422–1430, Jul. 2008.
7. L. Yan, T. Wu, S. Zhong, and Q. Zhang, "A variation-based ring artifact correction method with sparse constraint for flat-detector ct," Phys. Med. Biol., vol. 61, no. 3, p. 1278, 2016.
8. W. Zhao and H. Lu, "Medical image fusion and denoising with alternating sequential filter and adaptive fractional order total variation," IEEE Trans. Instrum. Meas., vol. 66, no. 9, pp. 2283–2294, Sep. 2017.
9. M. Elad and M. Aharon, "Image denoising via sparse and redundant representations over learned dictionaries," IEEE Trans. Image Process., vol. 15, no. 12, pp. 3736–3745, Dec. 2006.
10. S. Li and L. Fang, "Signal denoising with random refined orthogonal matching pursuit," IEEE Trans. Instrum. Meas., vol. 61, no. 1, pp. 26–34, Jan. 2012.
11. S. Li, L. Fang, and H. Yin, "An efficient dictionary learning algorithm and its application to 3-D medical image denoising," IEEE Trans. Biomed. Eng., vol. 59, no. 2, pp. 417–427, Feb. 2012.
12. Q. Xu, H. Yu, X. Mou, L. Zhang, J. Hsieh, and G. Wang, "Low-dose X-ray CT reconstruction via dictionary learning," IEEE Trans. Med. Imaging, vol. 31, no. 9, pp. 1682–1697, Sep. 2012.
13. S. Li, H. Yin, and L. Fang, "Group-sparse representation with dictionary learning for medical image denoising and fusion," IEEE Trans. Biomed. Eng., vol. 59, no. 12, pp. 3450–3459, Dec. 2012.
14. K. Dabov, A. Foi, V. Katkovnik, and K. Egiazarian, "Image denoising by sparse 3-D transform-domain collaborative filtering," IEEE Trans. Image Process., vol. 16, no. 8, pp. 2080–2095, Aug. 2007.

Impending Inquisitions in Humanities and Sciences (ICIIHS-2022) – Dr. Mohan Varkolu et al. (eds)

Optimality Conditions and Saddle Point Criteria for Fractional Interval-Valued Optimization Problem with Vanishing Constraints Using Generalized Convexity

77

Krishna Kummari[1]

Department of Mathematics, School of Science,
GITAM-Hyderabad Campus, Hyderabad, India

B. Japamala Rani[2]

Department of Mathematics, School of Science,
GITAM-Hyderabad Campus, Hyderabad, India

Department of Mathematics, St. Ann's College for Women-Mehdipatanam,
Hyderabad, India

ABSTRACT Using the concept of convexifactors, we discuss the optimality conditions of a fractional interval-valued optimization issue with vanishing constraints (FIVVC). The existence of LU efficient solution to the problem is investigated. Sufficient optimality conditions are established using the concept of generalized convex functions. In addition, concerning the problem, the saddle point analysis of a Lagrangian function in fractional interval-valued is developed for (FIVVC).

KEYWORDS Fractional interval-valued optimization problem, LU efficient solution, Lagrange functions, Saddle point, Vanishing constraints

1. Introduction

In optimization theory, the mathematical program with vanishing constraints (MPVC) is significant because it frequently appears in applications of applied science, economics, and engineering. Several illustrations of (MPVC) appear in structural and topology optimization, and references to recent articles are provided to the reader [12, 18, 15]. Numerous studies have been conducted and are accessible in the literature addressing interval optimization issues, developing optimality criteria, and duality results, see [5, 16, 14, 3]. Very recently, Hoheisel and Kanzow [8, 9, 10] provided an inclusive study on the set of optimality conditions in the idea of MPVCcontext and presented a necessary and sufficient optimality criterion of the second-order for MPVCs, as well as a new sufficient criterion of the first order. Specifically, these papers examine in greater depth the Abadie, Guignard, and weak constraint qualifications for a particular kind of optimization issue. The numerical results of Achtziger et al. [1] and Izmailov and Solodov [11] are analyzed using relaxation, smoothing, and regularization concepts.

As a result, the present work is devoted to examining the sufficient optimality criteria for the *LU* optimal solution of a fractional interval-valued optimization problem by constraining convex functions. Furthermore, we define a Lagrangian interval-valued function and establish a few saddle point results for the aforementioned Lagrangian function.

The following is the overview of the study. Section 2 includes fundamental definitions and basic notions of the operations on intervals, which will be necessary later in the sequel. Section 3 determines the Karush-Kuhn-Tucker necessary and sufficient

[1]krishna.maths@gmail.com, [2]bjapamalarani@gmail.com

DOI: 10.1201/9781003489436-77

optimality conditions for a fractional interval-valued optimization issue with vanishing constraints under convex functions. The relationship between a saddle point of the Lagrange function and a *LU* optimum solution of the given fractional interval-valued optimization issue with vanishing constraints is further established in Section 4. Section 5, ends with concluding remarks.

2. Preliminaries

This section provides essential definitions and lemmas which will be employed throughout the paper.

Let R^n_+ be the non-negative orthogonal component of an n-dimensional Euclidean space R^n. This article focuses primarily on Banach spaces. Let T^* be the topological dual of a given Banach space T with the canonical dual pairing $\langle .,. \rangle$.

The prefatory concepts of the operations on intervals are as follows:

Let $\frac{\mathcal{H}}{\mathcal{G}} = \left[\frac{\vartheta^L}{\rho^L}, \frac{\vartheta^U}{\rho^U}\right]$ and $\frac{\mathcal{N}}{\mathcal{D}} = \left[\frac{\gamma^L}{\upsilon^L}, \frac{\gamma^U}{\upsilon^U}\right]$ be two fractional closed intervals with $\frac{\vartheta^L}{\rho^L} \leq \frac{\vartheta^U}{\rho^U}$ and $\frac{\gamma^L}{\upsilon^L} \leq \frac{\gamma^U}{\upsilon^U}, \rho^L, \rho^U, \upsilon^L, \upsilon^U \neq 0.$

(i) $\frac{\mathcal{H}}{\mathcal{G}} + \frac{\mathcal{N}}{\mathcal{D}} = \left[\frac{\vartheta^L}{\rho^L} + \frac{\gamma^L}{\upsilon^L}, \frac{\vartheta^U}{\rho^U} + \frac{\gamma^U}{\upsilon^U}\right],$

(ii) $\frac{-\mathcal{H}}{\mathcal{G}} = \left[\frac{-\vartheta^U}{\rho^U}, \frac{-\vartheta^L}{\rho^L}\right],$

(iii) $\frac{\mathcal{H}}{\mathcal{G}} - \frac{\mathcal{N}}{\mathcal{D}} = \frac{\mathcal{H}}{\mathcal{G}} + \left(\frac{-\mathcal{C}}{\mathcal{D}}\right) = \left[\frac{\vartheta^L}{\rho^L} - \frac{\gamma^U}{\upsilon^U}, \frac{\vartheta^U}{\rho^U} - \frac{\gamma^L}{\upsilon^L}\right],$

(iv) $$\hat{\Psi}\left(\frac{\mathcal{H}}{\mathcal{G}}\right) = \begin{cases} \left[\frac{\vartheta^L}{\rho^L}, \frac{\vartheta^U}{\rho^U}\right], & \text{if } \hat{\Psi} \geq 0, \\ \left[\frac{\vartheta^U}{\rho^U}, \frac{\vartheta^L}{\rho^L}\right], & \text{if } \hat{\Psi} < 0. \end{cases}$$

The partial ordering relation $\preceq_{LU}$ among two intervals $\frac{\mathcal{H}}{\mathcal{G}}$ and $\frac{\mathcal{N}}{\mathcal{D}}$ are elucidated as

(i) $\frac{\mathcal{H}}{\mathcal{G}} \preceq_{\mathcal{L}\mathcal{U}} \frac{\mathcal{N}}{\mathcal{D}}$ iff $\frac{\vartheta^L}{\rho^L} \leq \frac{\gamma^L}{\upsilon^L}$ and $\frac{\vartheta^U}{\rho^U} \leq \frac{\gamma^U}{\upsilon^U}.$

(ii) $\frac{\mathcal{H}}{\mathcal{G}} \prec_{\mathcal{L}\mathcal{U}} \frac{\mathcal{N}}{\mathcal{D}}$ iff $\frac{\mathcal{H}}{\mathcal{G}} \leq \frac{\mathcal{N}}{\mathcal{D}}$ and $\frac{\mathcal{H}}{\mathcal{G}} \neq \frac{\mathcal{N}}{\mathcal{D}}$

$$\begin{cases} \frac{\vartheta^L}{\rho^L} < \frac{\gamma^L}{\upsilon^L} \\ \frac{\vartheta^U}{\rho^U} \leq \frac{\gamma^U}{\upsilon^U}, \end{cases} \text{or} \quad \begin{cases} \frac{\vartheta^L}{\rho^L} \leq \frac{\gamma^L}{\upsilon^L} \\ \frac{\vartheta^U}{\rho^U} < \frac{\gamma^U}{\upsilon^U}, \end{cases} \text{or} \quad \begin{cases} \frac{\vartheta^L}{\rho^L} < \frac{\gamma^L}{\upsilon^L} \\ \frac{\vartheta^U}{\rho^U} < \frac{\gamma^U}{\upsilon^U}. \end{cases}$$

Consider the fractional interval-valued optimization problem with vanishing constraints:

(FIVVC) $\quad \min \left[\frac{\xi^L(\varrho), \xi^U(\varrho)}{\zeta^L(\varrho), \zeta^U(\varrho)}\right]$

subject to

$\phi_\mu(\varrho) \leq 0, \forall\, \mu = 1,2,\cdots,\sigma,$

$\psi_\iota(\varrho) = 0, \forall \iota = 1,2,\cdots,t,$

$\varsigma_\mu(\varrho) \geq 0, \forall\, \mu = 1,2,\cdots,r,$

$\varphi_\mu(\varrho)\varsigma_\mu(\varrho) \leq 0, \forall\, \mu = 1,2,\cdots,r,$

where $\xi^L(\varrho), \xi^U(\varrho) \geq 0, \zeta^L(\varrho), \zeta^U(\varrho) > 0$, and $\phi : R^n \to R^\sigma, \psi : R^n \to R^t, \varphi, \varsigma : R^n \to R^r$ are continuously differentiable functions.

Consider the simplified form of (FIVVC) as follows

$$\min \left[\frac{\xi^L(\varrho)}{\zeta^U(\varrho)}, \frac{\xi^U(\varrho)}{\zeta^L(\varrho)}\right]$$

subject to

$$\phi_\mu(\varrho) \leq 0, \forall \mu = 1,2,\cdots,\sigma,$$
$$\psi_\iota(\varrho) = 0, \forall \iota = 1,2,\cdots,t,$$
$$\varsigma_\mu(\varrho) \geq 0, \forall \mu = 1,2,\cdots,r,$$
$$\varphi_\mu(\varrho)\varsigma_\mu(\varrho) \leq 0, \forall \mu = 1,2,\cdots,r,$$

Set $\xi^L = \kappa^L$, $\zeta^U = \tau^L$, $\xi^U = \kappa^U$, $\zeta^L = \tau^U$, then, the above problem reduces to

(FIVVC) $$\min \left[\frac{\kappa^L}{\tau^L}(\varrho), \frac{\kappa^U}{\tau^U}(\varrho)\right]$$

subject to

$$\phi_\mu(\varrho) \leq 0, \forall \mu = 1,2,\cdots,\sigma,$$
$$\psi_\iota(\varrho) = 0, \forall \iota = 1,2,\cdots,t,$$
$$\varsigma_\mu(\varrho) \geq 0, \forall \mu = 1,2,\cdots,r,$$
$$\varphi_\mu(\varrho)\varsigma_\mu(\varrho) \leq 0, \forall \mu = 1,2,\cdots,r.$$

Let $\mathcal{F}$ indicate the feasible region for the issue (FIVVC) and is defined as

$$\begin{aligned}\mathcal{F} := \Big\{ \varrho \in \mathcal{R}^n \mid & \phi_\mu(\varrho) \leq 0, \ \forall \mu = 1,2,\cdots,\sigma, \\ & \psi_\iota(\varrho) = 0, \ \forall \iota = 1,2,\cdots,t, \\ & \varsigma_\mu(\varrho) \geq 0, \ \forall \mu = 1,2,\cdots,r, \\ & \varphi_\mu(\varrho)\varsigma_\mu(\varrho) \leq 0, \ \forall \mu = 1,2,\cdots,r \Big\}.\end{aligned}$$

Definition 2.1 (Debnath and Gupta [6]) *A feasible point $\bar{\varrho} \in F$ is considered as a LU optimal solution for (FIVVC) if and only if there is no such feasible point $\varrho \in F$ satisfying*

$$\left[\frac{\kappa^L}{\tau^L}(\varrho), \frac{\kappa^U}{\tau^U}(\varrho)\right] \prec_{LU} \left[\frac{\kappa^L}{\tau^L}(\bar{\varrho}), \frac{\kappa^U}{\tau^U}(\bar{\varrho})\right]$$

Definition 2.2 *The modified Abadie constraint qualification (FIVVC-MACQ) is considered to hold at $\bar{\varrho} \in F$ if and only if $L^{VC}(\bar{\varrho}) \subseteq T(\bar{\varrho})$, where $L^{VC}(\bar{\varrho})$ and $T(\bar{\varrho})$ are linearized cone and tangent cone respectively.*

For the definition of linearized cone and tangent cone, one can refer (Achtziger and Kanzow[2]; Ahmad et al.[4]; Tung[17]).

For $\bar{\varrho} \in \mathcal{F}$, we define the subsequent index sets, which are useful in the sequel of the paper:

$$J_\phi(\bar{\varrho}) = \{\mu = 1,2,\ldots\sigma \mid \phi_\mu(\bar{\varrho}) = 0\}$$
$$J_\psi(\bar{\varrho}) = \{1,2,\ldots,t\}$$
$$J_+(\bar{\varrho}) = \{\mu \mid \varsigma_\mu(\bar{\varrho}) > 0\}$$
$$J_0(\bar{\varrho}) = \{\mu \mid \varsigma_\mu(\bar{\varrho}) = 0\}$$
$$J_{+0}(\bar{\varrho}) = \{\mu \mid \varsigma_\mu(\bar{\varrho}) > 0, \ \varphi_\mu(\bar{\varrho}) = 0\}$$
$$J_{+-}(\bar{\varrho}) = \{\mu \mid \varsigma_\mu(\bar{\varrho}) > 0, \ \varphi_\mu(\bar{\varrho}) < 0$$
$$J_{0+}(\bar{\varrho}) = \{\mu \mid \varsigma_\mu(\bar{\varrho}) = 0, \ \varphi_\mu(\bar{\varrho}) > 0\}$$
$$J_{00}(\bar{\varrho}) = \{\mu \mid \varsigma_\mu(\bar{\varrho}) = 0, \ \varphi_\mu(\bar{\varrho}) = 0\}$$
$$J_{0-}(\bar{\varrho}) = \{\mu \mid \varsigma_\mu(\bar{\varrho}) = 0, \ \varphi_\mu(\bar{\varrho}) < 0\}$$

For $\varrho \in F$, we also define the following index sets:

$$\begin{cases} \mathcal{J}_{\phi}^{+}(\varrho) = \{\mu = 1, 2, ..., \sigma \mid \hat{\lambda}_{\mu} > 0\} \\ \mathcal{J}_{\psi}^{+}(\varrho) = \{\iota \in \mathcal{J}_{\psi}(\varrho) \mid \hat{\Psi}_{\iota} > 0\} \\ \mathcal{J}_{\psi}^{-}(\varrho) = \{\iota \in \mathcal{J}_{\psi}(\varrho) \mid \hat{\Psi}_{\iota} < 0\} \\ \mathcal{J}_{0+}^{+}(\varrho) = \{\mu \in \mathcal{J}_{0+}(\varrho) \mid \hat{\alpha}_{\mu}^{\varsigma} > 0\} \\ \mathcal{J}_{0+}^{-}(\varrho) = \{\mu \in \mathcal{J}_{0+}(\varrho) \mid \hat{\alpha}_{\mu}^{\varsigma} < 0\} \\ \mathcal{J}_{00}^{+}(\varrho) = \{\mu \in \mathcal{J}_{00}(\varrho) \mid \hat{\alpha}_{\mu}^{\varsigma} > 0\} \\ \mathcal{J}_{+0}^{+}(\varrho) = \{\mu \in \mathcal{J}_{+0}(\varrho) \mid \alpha_{\mu}^{\varsigma} > 0\} \\ \mathcal{J}_{+-}^{+}(\varrho) = \{\mu \in \mathcal{J}_{+-}(\varrho) \mid \alpha_{\mu}^{\varsigma} > 0\} \\ \mathcal{J}_{0-}^{+}(\varrho) = \{\mu \in \mathcal{J}_{0-}(\varrho) \mid \alpha_{\mu}^{\varsigma} > 0\} \\ \mathcal{J}_{+0}^{++}(\varrho) = \{\mu \in \mathcal{J}_{+0}(\varrho) \mid \alpha_{\mu}^{\varphi} > 0\} \\ \mathcal{J}_{+-}^{++}(\varrho) = \{\mu \in \mathcal{J}_{+-}(\varrho) \mid \alpha_{\mu}^{\varphi} > 0\} \end{cases} \tag{1}$$

As we proceed, the concepts of pseudoconvexity and quasiconvexity have been used in the following discussion. There are also references to these generalized functions. To avoid lengthening the paper, we do not repeat the details of such definitions.

Definition 2.3 *A function* $\Lambda : T \subseteq R^n \to R$ *is a (strictly) convex at* $\varrho^* \in T$ *if for any* $\varrho \in T$, *we have*

$$\Lambda(\varrho) - \Lambda(\varrho^*)(>) \geq (\varrho - \varrho^*)^T \nabla\Lambda(\varrho^*).$$

3. Optimality Conditions

For the proposed feasible solution $\bar{\varrho}$, consider two independent fractional problems:

(FP1) $\quad \min v^L = \dfrac{\kappa^L}{\tau^L}(\varrho)$

subject to

$\phi_{\mu}(\varrho) \leq 0, \forall \mu = 1, 2, ..., \sigma,$

$\psi_{\iota}(\varrho) = 0, \forall \iota = 1, 2, ..., t,$

$\varsigma_{\mu}(\varrho) \geq 0, \forall \mu = 1, 2, \cdots, r,$

$\varphi_{\mu}(\varrho)\varsigma\mu(\varrho) \leq 0, \forall \mu = 1, 2, \cdots, r,$

$\dfrac{\kappa^U}{\tau^U}(\varrho) \leq \dfrac{\kappa^U}{\tau^U}(\bar{\varrho}),$

$\varrho \in T$

(FP1) $\quad \min v^U = \dfrac{\kappa^U}{\tau^U}(\varrho)$

subject to

$\phi_{\mu}(\varrho) \leq 0, \forall \mu = 1, 2, ..., \sigma,$

$\psi_{\iota}(\varrho = 0, \forall \iota = 1, 2, ..., t,$

$\varsigma_{\mu}(\varrho) \geq 0, \forall \mu = 1, 2, ..., r,$

$\varphi_{\mu}(\varrho)\varsigma_{\mu}(\varrho) \leq 0, \forall \mu = 1, 2, ..., r,$

$\dfrac{\kappa^L}{\tau^L}(\varrho) \leq \dfrac{\kappa^L}{\tau^L}(\bar{\varrho}),$

$\varrho \in T.$

On a similar note of Debnath and Gupta[6], the following Lemmas (3.1) and (3.2) are stated below.

Lemma 3.1 *If* $\bar{\varrho}$ *is a LU efficient solution of (FIVVC) if and only if* $\bar{\varrho}$ *is an efficient solution for both (FP1) and (FP2).*

Lemma3.2 *$\bar{\varrho}$ is a LU optimum of (FIVVC) if and only if $\bar{\varrho}$ minimizes $\frac{\kappa^L}{\tau^L}(\varrho)$ on the constraint set*

$$N = \Big\{ \varrho \in \mathcal{T} \,\Big|\, \frac{\kappa^U}{\tau^U}(\varrho) \leq \frac{\kappa^U}{\tau^U}(\bar{\varrho}),\ \phi_\mu(\varrho) \leq 0, \forall \mu = 1, 2, \cdots, \sigma,$$
$$\psi_\iota(\varrho) = 0, \forall \iota = 1, 2, \cdots, t,$$
$$\varsigma_\mu(\varrho) \geq 0, \forall \mu = 1, 2, \cdots, r,$$
$$\varphi_\mu(\varrho)\varsigma_\mu(\varrho) \leq 0, \forall \mu = 1, 2, \cdots, r \Big\}.$$

Considered the following single-objective fractional problem:

(DF) $\quad \min v = \frac{\kappa_1}{\tau_1}(\varrho)$

subject to

$\phi_\mu(\varrho) \leq 0, \forall \mu = 1, 2, \ldots, \sigma,$

$\psi_\iota(\varrho) = 0, \forall \iota = 1, 2, \ldots, t,$

$\varsigma_\mu(\varrho) \geq 0, \forall \mu = 1, 2, \ldots, r,$

$\varphi_\mu(\varrho)\varsigma_\mu(\varrho) \leq 0, \forall \mu = 1, 2, \ldots, r,$

where κ_1, τ_1 are continuous functions on T such that $\kappa_1(\varrho) \geq 0$ and $\tau_1(\varrho) > 0$, for all $\varrho \in T$.

On the lines of Theorem 3.1 of Kummari [13], we state the following theorem for (DF):

Theorem 3.1 *Let the (FIVVC-MACQ), defined in Definition (2.2), be satisfied at $\bar{\varrho} \in F$. If $\bar{\varrho} \in F$ is an optimal solution of the (DF), then there exists $v \in R_+$, $\lambda_\mu \in R_+$, ($\mu = 1,2, \ldots, \sigma$), $\beta_\iota \in R$, ($\iota \in J_\psi$), α_ς^μ, $\alpha^\varphi_\mu \in R$, ($\mu = 1, 2, \ldots, r$), such that*

$$\nabla\kappa_1(\bar{\varrho}) - v\nabla\tau_1(\bar{\varrho}) + \sum_{\mu=1}^{\sigma}\lambda_\mu\nabla\phi_\mu(\bar{\varrho}) + \sum_{\iota=1}^{t}\beta_\iota\nabla\psi_\iota(\bar{\varrho}) - \sum_{\mu=1}^{r}\alpha_\mu^\varsigma\nabla\varsigma_\mu(\bar{\varrho}) + \sum_{\mu=1}^{r}\alpha_\mu^\varphi\nabla\varphi_\mu(\bar{\varrho}) = 0, \tag{2}$$

and

$$\kappa_1(\bar{\varrho}) - v\tau_1(\bar{\varrho}) = 0, \tag{3}$$
$$\psi_\iota(\bar{\varrho}) = 0, \iota \in J_\psi(\bar{\varrho}), \tag{4}$$
$$\lambda_\mu \geq 0,\ \phi_\mu(\bar{\varrho}) \leq 0,\ \lambda_\mu\phi_\mu(\bar{\varrho}) = 0,\ \mu = 1,2,\ldots,\sigma, \tag{5}$$
$$\alpha^\varsigma_\mu = 0, \mu \in J_+(\bar{\varrho}),\ \alpha^\varsigma_\mu \geq 0, \mu \in J_{00}(\bar{\varrho}) \cup J_{0-}(\bar{\varrho}),\ \alpha^\varsigma_\mu \text{ is free}$$
$$\mu \in J_{0+}(\bar{\varrho}), \tag{6}$$
$$\alpha_\mu^\varphi = 0, \mu \in J_{0+}(\bar{\varrho}) \cup J_{0-}(\bar{\varrho}) \cup J_{+-}(\bar{\varrho}),\ \alpha_\mu^\varphi \geq 0,$$
$$\mu \in J_{00}(\bar{\varrho}) \cup J_{+0}(\bar{\varrho}), \tag{7}$$
$$\alpha^\varsigma_\mu\varsigma_\mu(\bar{\varrho}) = 0,\ \alpha^\varphi_\mu\varphi_\mu(\bar{\varrho}) = 0, \forall \mu = 1, 2, \ldots, r. \tag{8}$$

Theorem 3.2 (Necessary optimality conditions) *Let the (FIVVC-MACQ), defined in Definition(2.2), be satisfied at $\bar{\varrho} \in$ F. If $\bar{\varrho} \in$ F is a LU optimal solution of (FIVVC), then there exist v^L, $v^U \in R_+$, $\lambda_\mu \in R_+$, ($\mu = 1, 2, \ldots, \sigma$), $\beta_\iota \in R$, ($\iota \in J_\psi$), α^ς_μ, $\alpha^\varphi_\mu \in R$, ($\mu = 1, 2, \ldots, r$), such that*

$$\nabla\kappa^L(\bar{\varrho}) - v^L\nabla\tau^L(\bar{\varrho}) + \nabla\kappa^U(\bar{\varrho}) - v^U\nabla\tau^U(\bar{\varrho}) + \sum_{\mu=1}^{\sigma}\lambda_\mu\nabla\phi_\mu(\bar{\varrho}) + \sum_{\iota=1}^{t}\beta_\iota\nabla\psi_\iota(\bar{\varrho}) - \sum_{\mu=1}^{r}\alpha_\mu^\varsigma\nabla\varsigma_\mu(\bar{\varrho}) + \sum_{\mu=1}^{r}\alpha_\mu^\varphi\nabla\varphi_\mu(\bar{\varrho}) = 0, \tag{9}$$

and

$$\kappa^L(\bar{\varrho}) - v^L\tau^L(\bar{\varrho}) = 0, \tag{10}$$
$$\kappa^U(\bar{\varrho}) - v^U\tau^U(\bar{\varrho}) = 0, \tag{11}$$
$$\psi_\iota(\bar{\varrho}) = 0, \iota \in J_\psi(\bar{\varrho}), \tag{12}$$
$$\lambda_\mu \geq 0,\ \phi_\mu(\bar{\varrho}) \leq 0,\ \lambda_\mu\phi_\mu(\bar{\varrho}) = 0,\ \mu = 1,2,\ldots,\sigma, \tag{13}$$
$$\alpha^\varsigma_\mu = 0, \mu \in J_+(\bar{\varrho}),\ \alpha^\varsigma_\mu \geq 0, \mu \in J_{00}(\bar{\varrho}) \cup J_{0-}(\bar{\varrho}),\ \alpha^\varsigma_\mu \text{ is free}$$

$$\mu \in J_{0+}(\bar{\varrho}), \tag{14}$$

$$\alpha^{\varphi}_{\mu} = 0,\ \mu \in J_{0+}(\bar{\varrho}) \cup J_{0-}(\bar{\varrho}) \cup J_{+-}(\bar{\varrho}),\ \alpha^{\varphi}_{\mu} \geq 0,$$

$$\mu \in J_{00}(\bar{\varrho}) \cup J_{+0}(\bar{\varrho}), \tag{15}$$

$$\alpha^{\varsigma}_{\mu}\varsigma_{\mu}(\bar{\varrho}) = 0,\ \alpha^{\varphi}_{\mu}\varphi_{\mu}(\bar{\varrho}) = 0,\ \forall \mu = 1, 2, \ldots, r. \tag{16}$$

Proof: By presumption, $\bar{\varrho}$ is a LU optimal solution of (FIVVC), and a suitable abadie constraint qualification is fulfilled at $\bar{\varrho}$. Since $\bar{\varrho}$ is a LU optimal solution, by Lemma(3.1), $\bar{\varrho}$ is also a optimal solution for the problems (FP1) and (FP2). Hence, by Lemma(3.2), at $\bar{\varrho}$ the minimum value of $\frac{\kappa^L}{\tau^L}(\varrho)$ is obtained on the constraint set

$$\mathcal{N}^L = \Big\{ \varrho \in \mathcal{T} \;\Big|\; \frac{\kappa^U}{\tau^U}(\varrho) \leq \frac{\kappa^U}{\tau^U}(\bar{\varrho}),\ \phi_\mu(\varrho) \leq 0, \forall \mu = 1, 2, \cdots, \sigma,$$
$$\psi_\iota(\varrho) = 0, \forall \iota = 1, 2, \cdots, t,$$
$$\varsigma_\mu(\varrho) \geq 0, \forall \mu = 1, 2, \cdots, r,$$
$$\varphi_\mu(\varrho)\varsigma_\mu(\varrho) \leq 0, \forall \mu = 1, 2, \cdots, r \Big\},$$

and the minimum value of $\frac{\kappa^U}{\tau^U}(\varrho)$ is obtained at $\bar{\varrho}$ on the constraint set

$$\mathcal{N}^U = \Big\{ \varrho \in \mathcal{T} \;\Big|\; \frac{\kappa^L}{\tau^L}(\varrho) \leq \frac{\kappa^L}{\tau^L}(\bar{\varrho}),\ \phi_\mu(\varrho) \leq 0, \forall \mu = 1, 2, \cdots, \sigma,$$
$$\psi_\iota(\varrho) = 0, \forall \iota = 1, 2, \cdots, t,$$
$$\varsigma_\mu(\varrho) \geq 0, \forall \mu = 1, 2, \cdots, r,$$
$$\varphi_\mu(\varrho)\varsigma_i(\varrho) \leq 0, \forall \mu = 1, 2, \cdots, r \Big\}.$$

By Theorem(3.1), it ensures that there exist $v^L > 0 \in R_+$, $\lambda^L \in R_+$, β^L, $\alpha^{\varsigma L}$, $\alpha^{\varphi L} \in R$ and $v^U > 0 \in R_+$, $\lambda^U \in R_+$, β^U, $\alpha^{\varsigma U}$, $\alpha^{\varphi U} \in R$ such that

$$\nabla\kappa^L(\bar{\varrho}) - v^L\nabla\tau^L(\bar{\varrho}) + \sum_{\mu=1}^{\sigma}\lambda^L_\mu\nabla\phi_\mu(\bar{\varrho}) + \sum_{\iota=1}^{t}\beta^L_\iota\nabla\psi_\iota(\bar{\varrho}) - \sum_{\mu=1}^{r}\alpha^{\varsigma L}_\mu\nabla\varsigma_\mu(\bar{\varrho}) + \sum_{\mu=1}^{r}\alpha^{\varphi L}_\mu\nabla\varphi_\mu(\bar{\varrho}) = 0, \tag{17}$$

$$\kappa^L(\bar{\varrho}) - v^L\tau^L(\bar{\varrho}) = 0, \tag{18}$$

$$\psi_\iota(\bar{\varrho}) = 0,\ \iota \in J_\psi(\bar{\varrho}), \tag{19}$$

$$\lambda^L_\mu \geq 0,\ \phi_\mu(\bar{\varrho}) \leq 0,\ \lambda^L_\mu\phi_\mu(\bar{\varrho}) = 0,\ \mu = 1,2,\ldots,\sigma, \tag{20}$$

$$\alpha^{\varsigma L}_\mu = 0,\ \mu \in J_+(\bar{\varrho}),\ \alpha^{\varsigma L}_\mu \geq 0,\ \mu \in J_{00}(\bar{\varrho}) \cup J_{0-}(\bar{\varrho}),\ \alpha^{\varsigma L}_\mu \text{ is free}$$

$$\mu \in J_{0+}(\bar{\varrho}), \tag{21}$$

$$\alpha^{\varphi L}_\mu = 0,\ \mu \in J_{0+}(\bar{\varrho}) \cup J_{0-}(\bar{\varrho}) \cup J_{+-}(\bar{\varrho}),\ \alpha^{\varphi L}_\mu \geq 0,$$

$$\mu \in J_{00}(\bar{\varrho}) \cup J_{+0}(\bar{\varrho}), \tag{22}$$

$$\alpha^{\varsigma L}_\mu\varsigma_\mu(\bar{\varrho}) = 0, \alpha^{\varphi L}_\mu\varphi_\mu(\bar{\varrho}) = 0, \forall\mu = 1,2,\ldots,r. \tag{23}$$

and

$$\nabla\kappa^U(\bar{\varrho}) - v^U\nabla\tau^U(\bar{\varrho}) + \sum_{\mu=1}^{\sigma}\lambda^U_\mu\nabla\phi_\mu(\bar{\varrho}) + \sum_{\iota=1}^{t}\beta^U_\iota\nabla\psi_\iota(\bar{\varrho}) - \sum_{\mu=1}^{r}\alpha^{\varsigma U}_\mu\nabla\varsigma_\mu(\bar{\varrho}) + \sum_{\mu=1}^{r}\alpha^{\varphi U}_\mu\nabla\varphi_\mu(\bar{\varrho}) = 0, \tag{24}$$

$$\kappa^U(\bar{\varrho}) - v^U\tau^U(\bar{\varrho}) = 0, \tag{25}$$

$$\psi_j(\bar{\varrho}) = 0,\ \iota \in J_\psi(\bar{\varrho}), \tag{26}$$

$$\lambda^U_\mu \geq 0,\ \phi_\mu(\bar{\varrho}) \leq 0,\ \lambda^U_\mu\phi_\mu(\bar{\varrho}) = 0,\ \mu = 1, 2, \ldots, \sigma, \tag{27}$$

$$\alpha^{\varsigma U}_\mu = 0,\ \mu \in J_+(\bar{\varrho}),\ \alpha^{\mu}_{\varsigma U} \geq 0,\ \mu \in J_{00}(\bar{\varrho}) \cup J_{0-}(\bar{\varrho}),\ \alpha^{\varsigma U}_\mu \text{is free}$$

$$\mu \in J_0 + (\bar{\varrho}), \tag{28}$$

$$\alpha_\mu^{\varphi U} = 0, \mu \in J_{0+}(\bar{\varrho}) \cup J_{0-}(\bar{\varrho}) \cup J_{+-}(\bar{\varrho}), \alpha_\mu^{\varphi U} \geq 0,$$
$$\mu \in J_{00}(\bar{\varrho}) \cup J_{+0}(\bar{\varrho}), \tag{29}$$
$$\alpha_\mu^{\varsigma U} \varsigma_\mu(\bar{\varrho}) = 0, \alpha_\mu^{\varphi U} \varphi_\mu(\bar{\varrho}) = 0, \forall \mu = 1, 2, \ldots, r. \tag{30}$$

From (17) to (30), we obtain

$$\nabla\kappa^L(\bar{\varrho}) - v^L\nabla\tau^L(\bar{\varrho}) + \nabla\kappa^U(\bar{\varrho}) - v^U\nabla\tau^U(\bar{\varrho}) + \sum_{\mu=1}^{\sigma}(\lambda_\mu^L + \lambda_\mu^U)$$
$$\nabla\phi_\mu(\bar{\varrho}) + \sum_{\iota=1}^{t}(\beta_\iota^L + \beta_\iota^U)\nabla\psi_\iota(\bar{\varrho}) - \sum_{\mu=1}^{r}(\alpha_\mu^{\varsigma L} + \alpha_\mu^{\varsigma U})\nabla\varsigma_\mu(\bar{\varrho}) + \sum_{\mu=1}^{r}(\alpha_\mu^{\varphi L} + \alpha_\mu^{\varphi U})\nabla\varphi_\mu(\bar{\varrho}) = 0, \tag{31}$$

and

$$\kappa^L(\bar{\varrho}) - v^L\tau^L(\bar{\varrho}) = 0, \tag{32}$$
$$\kappa^U(\bar{\varrho}) - v^U\tau^U(\bar{\varrho}) = 0, \tag{33}$$
$$\psi_\iota(\bar{\varrho}) = 0, \iota \in J_\psi(\bar{\varrho}), \tag{34}$$
$$(\lambda_\mu^L + \lambda_\mu^U) \geq 0, \phi_\mu(\bar{\varrho}) \leq 0, (\lambda_\mu^L + \lambda_\mu^U)\phi_\mu(\bar{\varrho}) = 0, \mu = 1,2,\ldots,\sigma, \tag{35}$$
$$(\alpha_\mu^{\varsigma L} + \alpha_\mu^{\varsigma U}) = 0, \mu \in J_+(\bar{\varrho}), (\alpha_\mu^{\varsigma L} + \alpha_\mu^{\varsigma U}) \geq 0,$$
$$(\alpha_\mu^{\varsigma L} + \alpha_\mu^{\varsigma U}) = 0, \mu \in J_+(\bar{\varrho}), (\alpha_\mu^{\varsigma L} + \alpha_\mu^{\varsigma U}) \geq 0,$$
$$\mu \in J_{00}(\bar{\varrho}) \cup J_{0-}(\bar{\varrho}), \alpha_\mu^{\varsigma L} + \alpha_\mu^{\varsigma U} \text{ is free } \mu \in J_{0+}(\bar{\varrho}), \tag{36}$$
$$(\alpha_\mu^{\varphi L} + \alpha_\mu^{\varphi U}) = 0, \mu \in J_{0+}(\bar{\varrho}) \cup J_{0-}(\bar{\varrho}) \cup J_{+-}(\bar{\varrho}),$$
$$(\alpha_\mu^{\varphi L} + \alpha_\mu^{\varphi U}) \geq 0, \mu \in J_{00}(\bar{\varrho}) \cup J_{+0}(\bar{\varrho}), \tag{37}$$
$$(\alpha_\mu^{\varsigma L} + \alpha_\mu^{\varsigma U})\varsigma_\mu(\bar{\varrho}) = 0, (\alpha_\mu^{\varphi L} + \alpha_\mu^{\varphi U})\varphi_\mu(\bar{\varrho}) = 0, \forall \mu = 1,2,\ldots,l. \tag{38}$$

The equations from (31) - (38) along with $\lambda^L_\mu + \lambda^U_\mu = \lambda_\mu$, $\beta^L_j + \beta^U_j = \beta_j$, $\alpha_\mu^{\varsigma L} + \alpha_\mu^{\varsigma U} = \alpha_\mu^{\varsigma}$, $\alpha_\mu^{\varphi L} + \alpha_\mu^{\varphi U} = \alpha_\mu^{\varphi}$, yield

$$\nabla\kappa^L(\bar{\varrho}) - v^L\nabla\tau^L(\bar{\varrho}) + \nabla\kappa^U(\bar{\varrho}) - v^U\nabla\tau^U(\bar{\varrho}) + \sum_{\mu=1}^{\sigma}\lambda_\mu\nabla\phi_\mu(\bar{\varrho}) + \sum_{\iota=1}^{t}\beta_\iota\nabla\psi_\iota(\bar{\varrho}) - \sum_{\mu=1}^{r}\alpha_\mu^\varsigma\nabla\varsigma_\mu(\bar{\varrho}) + \sum_{\mu=1}^{r}\alpha_\mu^\varphi\nabla\varphi_\mu(\bar{\varrho}) = 0,$$

and

$$\kappa^L(\bar{\varrho}) - v^L\tau^L(\bar{\varrho}) = 0,$$
$$\kappa^U(\bar{\varrho}) - v^U\tau^U(\bar{\varrho}) = 0,$$
$$\psi_\iota(\bar{\varrho}) = 0, \iota \in J_\psi(\bar{\varrho}),$$
$$\lambda_\mu \geq 0, \phi_\mu(\bar{\varrho}) \leq 0, \lambda_\mu\phi_\mu(\bar{\varrho}) = 0, \mu = 1,2,\ldots,\sigma,$$
$$\alpha_\mu^\varsigma = 0, \mu \in J_+(\bar{\varrho}), \alpha_\mu^\varsigma \geq 0, \mu \in J_{00}(\bar{\varrho}) \cup J_{0-}(\bar{\varrho}), \alpha_\mu^\varsigma \text{ is free}$$
$$\mu \in J_{0+}(\bar{\varrho}), \alpha_\mu^\varphi = 0, \mu \in J_{0+}(\bar{\varrho}) \cup J_{0-}(\bar{\varrho}) \cup J_{+-}(\bar{\varrho}), \alpha_\mu^\varphi \geq 0,$$
$$\mu \in J_{00}(\bar{\varrho}) \cup J_{+0}(\bar{\varrho}), \alpha_\mu^\varsigma\varsigma_\mu(\bar{\varrho}) = 0, \alpha_\mu^\varphi\varphi_\mu(\bar{\varrho}) = 0, \forall \mu = 1,2,\ldots,r.$$

Hence the required proof.

Theorem 3.3 (Sufficient optimality conditions) *Let* $\tilde{\varrho} \in F$ *and there exist* $v^L, v^U \in R_+$, $\lambda_\mu \in R_+$, ($\mu = 1,2, \ldots, \sigma$), $\beta_\iota \in R$, ($\iota \in J_\psi$), α^ς_μ, $\alpha^\varphi_\mu \in R$, ($\mu = 1, 2, \ldots, l$), *such that* (9) - (16) *hold at* $\tilde{\varrho}$. *Further, assume that*

(i) $\kappa^L(.) - v^L\tau^L(.) + \kappa^U(.) - v^U\tau^U(.)$ *is convex function at* $\tilde{\varrho}$ *on* F,

(ii) $\sum_{\mu=1}^{\sigma}\phi_\mu(.)$, $\psi_\iota(.)$ ($j \in J^+_\psi$), $-\psi_\iota(.)$ ($\iota \in J^-_\psi$), $-\varsigma_\mu(.)$ ($\mu \in J^+_+ \cup J^+_0$), $\varsigma_\mu(.)$ ($\mu \in J^-_0$), $-\varphi_\mu(.)$ ($\mu \in J^-_{0+} \cup J^-_{00} \cup J^-_{+0}$), and $\varphi_\mu(.)$ ($\mu \in J^+_{00} \cup J^+_{0-} \cup J^+_{+0} \cup J^+_{+-}$), *are convex functions at* $\tilde{\varrho}$ *on* F, *then* $\tilde{\varrho}$ *is a* LU *optimal solution of* (*FIVVC*).

Proof: By assumption, (9) – (16) are satisfied at $\tilde{\varrho}$ with $v^L > 0$, $v^U > 0$, $\lambda_\mu \in R_+$, β_j, α_μ^ς, $\alpha_\mu^\varphi \in R$. It follows from (9), that

$$\nabla\kappa^L(\tilde{\varrho}) - v^L\nabla\tau^L(\tilde{\varrho}) + \nabla\kappa^U(\tilde{\varrho}) - v^U\nabla\tau^U(\tilde{\varrho}) + \sum_{\mu=1}^{\sigma}\lambda_\mu\nabla\phi_\mu(\tilde{\varrho}) + \sum_{\iota=1}^{t}\beta_\iota\nabla\psi_\iota(\tilde{\varrho}) - \sum_{\mu=1}^{r}\alpha_\mu^\varsigma\nabla\varsigma_\mu(\tilde{\varrho}) + \sum_{\mu=1}^{r}\alpha_\mu^\varphi\nabla\varphi_\mu(\tilde{\varrho}) = 0,$$

presume contrary to the result, that $\tilde{\varrho}$ is not a LU optimal solution for (FIVVC). Hence by Definition (2.1), there exist a feasible solution ϱ such that

$$\left[\frac{\kappa^L}{\tau^L}(\varrho), \frac{\kappa^U}{\tau^U}(\varrho)\right] \prec_{LU} \left[\frac{\kappa^L}{\tau^L}(\tilde{\varrho}), \frac{\kappa^U}{\tau^U}(\tilde{\varrho})\right]$$

that is

$$\begin{cases} \frac{\kappa^L}{\tau^L}(\varrho) < \frac{\kappa^L}{\tau^L}(\tilde{\varrho}) \\ \frac{\kappa^U}{\tau^U}(\varrho) \leq \frac{\kappa^U}{\tau^U}(\tilde{\varrho}) \end{cases}, \text{ or } \begin{cases} \frac{\kappa^L}{\tau^L}(\varrho) \leq \frac{\kappa^L}{\tau^L}(\tilde{\varrho}) \\ \frac{\kappa^U}{\tau^U}(\varrho) < \frac{\kappa^U}{\tau^U}(\tilde{\varrho}) \end{cases}, \text{ or } \begin{cases} \frac{\kappa^L}{\tau^L}(\varrho) < \frac{\kappa^L}{\tau^L}(\tilde{\varrho}) \\ \frac{\kappa^U}{\tau^U}(\varrho) < \frac{\kappa^U}{\tau^U}(\tilde{\varrho}) \end{cases},$$

which implies

$$\kappa^L(\varrho) - \nu^L(\tilde{\varrho})\tau^L(\varrho) < \kappa^L(\tilde{\varrho}) - \nu^L(\tilde{\varrho})\tau^L(\tilde{\varrho}),$$
$$\kappa^U(\varrho) - \nu^U(\tilde{\varrho})\tau^U(\varrho) \leq \kappa^U(\tilde{\varrho}) - \nu^U(\tilde{\varrho})\tau^U(\tilde{\varrho}),$$

or

$$\kappa^L(\varrho) - \nu^L(\tilde{\varrho})\tau^L(\varrho) \leq \kappa^L(\tilde{\varrho}) - \nu^L(\tilde{\varrho})\tau^L(\tilde{\varrho}),$$
$$\kappa^U(\varrho) - \nu^U(\tilde{\varrho})\tau^U(\varrho) < \kappa^U(\tilde{\varrho}) - \nu^U(\tilde{\varrho})\tau^U(\tilde{\varrho}),$$

or

$$\kappa^L(\varrho) - \nu^L(\tilde{\varrho})\tau^L(\varrho) < \kappa^L(\tilde{\varrho}) - \nu^L(\tilde{\varrho})\tau^L(\tilde{\varrho}),$$
$$\kappa^U(\varrho) - \nu^U(\tilde{\varrho})\tau^U(\varrho) < \kappa^U(\tilde{\varrho}) - \nu^U(\tilde{\varrho})\tau^U(\tilde{\varrho}).$$

From hypothesis (i), $\kappa^L(.) - \nu^L(\tilde{\varrho})\tau^L(.) + \kappa^U(.) - \nu^U(\tilde{\varrho})\tau^U(.)$ are convex functions at $\tilde{\varrho}$ on F, we obtain

$$(\varrho - \tilde{\varrho})^T[\nabla\kappa^L(\tilde{\varrho}) - \nu^L(\tilde{\varrho})\nabla\tau^L(\tilde{\varrho})] + [\nabla\kappa^U(\tilde{\varrho}) - \nu^U(\tilde{\varrho})\nabla\tau^U(\tilde{\varrho})] < 0, \tag{39}$$

For $\varrho \in F$, $\lambda_\mu \in R_+$, $\mu = 1, 2, ..., \sigma$, we have $\lambda_\mu\phi_\mu(\varrho) \leq 0$, $\mu = 1, 2, ..., \sigma$, which in view of (5) implies that

$$\sum_{\mu=1}^{\sigma} \lambda_\mu\phi_\mu(\varrho) \leq \sum_{\mu=1}^{\sigma} \lambda_\mu\phi_\mu(\tilde{\varrho}),$$

which by convexity of $\sum_{\mu=1}^{\sigma} \lambda_\mu\phi_\mu(.)$ at $\tilde{\varrho}$ on F, we obtain

$$(\varrho - \tilde{\varrho})^T \sum_{\mu=1}^{\sigma} \lambda_\mu\nabla\phi_\mu(\tilde{\varrho}) \leq 0. \tag{40}$$

By similar arguments, we get

$$(\varrho - \tilde{\varrho})^T\nabla\psi_\iota(\tilde{\varrho}) \leq 0,\ \forall\iota \in J_\psi^+,$$
$$-(\varrho - \tilde{\varrho})^T\nabla\psi_\iota(\tilde{\varrho}) \leq 0,\ \forall\iota \in J_\psi^-,$$
$$-(\varrho - \tilde{\varrho})^T\nabla\varsigma_\mu(\tilde{\varrho}) \leq 0,\ \forall\mu \in J_+^+ \cup J_0^+,$$
$$(\varrho - \tilde{\varrho})^T\nabla\varsigma_\mu(\tilde{\varrho}) \leq 0,\ \forall\mu \in J_0^-,$$
$$-(\varrho - \tilde{\varrho})^T\nabla\varphi_\mu(\tilde{\varrho}) \leq 0,\ \forall\mu \in J_{0+}^- \cup J_{00}^- \cup J_{+0}^-,$$
$$(\varrho - \tilde{\varrho})^T\nabla\varphi_\mu(\tilde{\varrho}) \leq 0,\ \forall\mu \in J_{00}^+ \cup J_{0+}^- \cup J_{+0}^+ \cup J_{+-}^+,$$

which by the concept of index sets, one possesses

$$(\varrho - \tilde{\varrho})^T\left[\sum_{\iota=1}^{t} \beta_\iota\nabla\psi_\iota(\tilde{\varrho}) - \sum_{\mu=1}^{r} \alpha_\mu^\varsigma\nabla\varsigma_\mu(\tilde{\varrho}) + \sum_{\mu=1}^{r} \alpha_\mu^\varphi\nabla\varphi_\mu(\tilde{\varrho})\right] \leq 0. \tag{41}$$

Combining (39), (40) and (41), we get

$$\nabla\kappa^L(\tilde{\varrho}) - \nu^L\nabla\tau^L(\tilde{\varrho}) + \nabla\kappa^U(\tilde{\varrho}) - \nu^U\nabla\tau^U(\tilde{\varrho}) + \sum_{\mu=1}^{\sigma} \lambda_\mu\nabla\phi_\mu(\tilde{\varrho}) + \sum_{\iota=1}^{t} \beta_\iota\nabla\psi_\iota(\tilde{\varrho}) - \sum_{\mu=1}^{r} \alpha_\mu^\varsigma\nabla\varsigma_\mu(\tilde{\varrho}) + \sum_{\mu=1}^{r} \alpha_\mu^\varphi\nabla\varphi_\mu(\tilde{\varrho}) < 0,$$

which is a contradiction to (9). Hence, $\tilde{\varrho}$ is a LU optimal solution of (FIVVC).

We consider the following paradigm to illustrate Theorem(3.3):

Example 1

$$\text{(FIVVC-1)} \qquad \min \left[\frac{\xi^L(\delta), \xi^U(\delta)}{\zeta^L(\delta), \zeta^U(\delta)} \right]$$

$$= \min \left[\frac{\delta_1^2 + 2\delta_2 + 1, 2 + \delta_1 + 2\delta_2^2 + 2\delta_2}{2 - \delta_1^2, \delta_2 + 4} \right]$$

subject to

$$\varsigma(\delta) = \delta_1 + \delta_2 \geq 0,$$

$$\varphi(\delta)\varsigma(\delta) = -\delta_2(\delta_1 + \delta_2) \leq 0,$$

Now, we rewrite (FIVVC-1) in the subsequent manner

$$\min \left[\frac{\delta_1^2 + 2\delta_2 + 1}{\delta_2 + 4}, \frac{2 + \delta_1 + 2\delta_2^2 + 2\delta_2}{2 - \delta_1^2} \right]$$

subject to

$$\varsigma(\delta) = \delta_1 + \delta_2 \geq 0,$$

$$\varphi(\delta)\varsigma(\delta) = -\delta_2(\delta_1 + \delta_2) \leq 0.$$

which is in the form of (FIVVC) with $n = 2, \sigma = t = 0, r = 1$ *and* $\frac{\kappa^L}{\tau^L}(\delta) = \frac{\delta_1^2 + 2\delta_2 + 1}{\delta_2 + 4}, \frac{\kappa^U}{\tau^U}(\delta) = \frac{2 + \delta_1 + 2\delta_2^2 + 2\delta_2}{2 - \delta_1^2}$. *The feasible region of (FIVVC-1) is* $F = \{\delta = (\delta_1, \delta_2) \in R^2 \mid \varsigma(\delta) = \delta_1 + \delta_2 \geq 0, \varphi(\delta)\varsigma(\delta) = -\delta_2(\delta_1 + \delta_2) \leq 0\}$. *Note that* $\tilde{\delta} = (0,0)$ *is a feasible solution of (FIVVC-1) and it is not difficult to see that there exist* $\alpha^\varsigma = 1$ *and* $\alpha^\varphi = 1 \in R$ *such that* (9) - (16) *are satisfied at* $\tilde{\delta} = (0,0)$ *for the problem (FIVVC-1). Further, it can be easily observe that the hypothesis (i) and (ii) of the Theorem* (3.3) *hold at* $\tilde{\delta} = (0,0)$. *Since all the suppositions of Theorem* (3.3) *are fulfilled, then* $\tilde{\delta} = (0,0)$ *is a* LU *optimal solution of (FIVVC-1).*

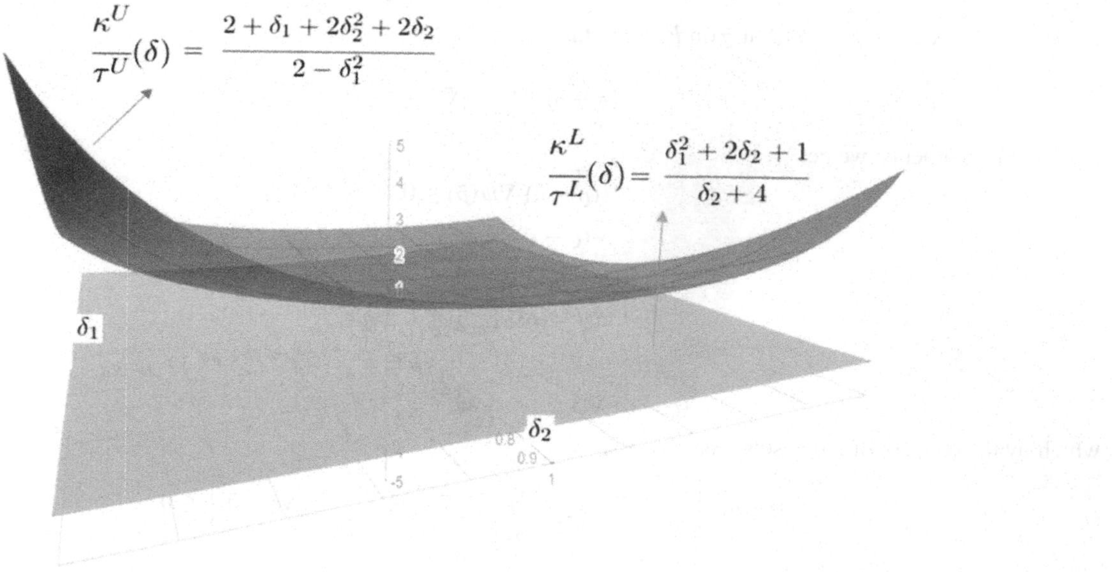

Fig. 77.1 3D plot of the objective function of (FIVVC-1)

Theorem 3.4 (Sufficient optimality conditions) *Let* $\tilde{\varrho} \in F$ *and there exist* $v^L, v^U \in R_+$, $\lambda_\mu \in R_+$, $(\mu = 1,2, \ldots, \sigma)$, $\beta_\iota \in R$, $(\iota \in J_\psi)$, $\alpha_\mu^\varsigma, \alpha_\mu^\varphi \in R$, $(\mu = 1,2, \ldots, r)$, *such that* (2) - (8) *hold at* $\tilde{\varrho}$. *Further, assume that*

(i) $\kappa^L(.) - v^L\tau^L(.) + \kappa^U(.) - v^U\tau^U(.)$ *is pseudoconvex function at* $\tilde{\varrho}$ *on* F

(ii) $\sum_{\mu=1}^{\sigma} \phi_\mu(.)$, $\psi_\iota(.)$ $(\iota \in J_\psi^+)$, $-\psi_\iota(.)$ $(\iota \in J_\psi^-)$, $-\varsigma_\mu(.)$ $(\mu \in J^+_+ \cup J_0^+)$, $\varsigma_\mu(.)$ $(\mu \in J_0^-)$, $-\varphi_\mu(.)$ $(\mu \in J_{0+}^- \cup J_{00}^- \cup J+0^-)$, and $\varphi_\mu(.)$ $(\mu \in J_{00}^+ \cup J_{0-}^+ \cup J_{+0}^+ \cup J_{+-}^+)$ are quasiconvex functions at $\bar{\varrho}$ on F.

Then $\bar{\varrho}$ *is a* LU *optimal solution of (FIVVC).*

Proof: The proof is comparable to the Theorem(3.3) proof.

4. Lagrangian Function and Saddle-point Analysis

In this section, we define the Lagrangian type function for (FIVVC) concerning the feasible point $\bar{\varrho} \in F$, as:

$$\mathcal{L}(\varrho,\lambda,\beta,\alpha^\varsigma,\alpha^H) = \left(\kappa^L(\varrho) - v^L(\bar{\varrho})\tau^L(\varrho)\right) + \left(\kappa^U(\varrho) - v^U(\bar{\varrho})\tau^U(\varrho)\right) + \sum_{\mu=1}^{\sigma}\lambda_\mu\phi_\mu(\varrho) + \sum_{\iota=1}^{t}\beta_\iota\psi_\iota(\varrho) - \sum_{\mu=1}^{r}\alpha_\mu^\varsigma\varsigma_\mu(\varrho) + \sum_{\mu=1}^{r}\alpha_\mu^\varphi\varphi_\mu(\varrho)$$

where $\varrho \in T$, $\lambda \in R_+$ and $\beta, \alpha^\varsigma, \alpha^\varphi \in R$.

Definition 4.1 *A point* $(\bar{\varrho}, \bar{\lambda}, \bar{\beta}, \bar{\alpha}^\varsigma, \bar{\alpha}^\varphi) \in T \times R_+ \times R \times R \times R$ *is said to be a saddle point for* $L(\varrho, \lambda, \beta, \alpha^\varsigma, \alpha^\varphi)$*, if*

$L(\bar{\varrho},\lambda,\beta,\alpha^\varsigma,\alpha^\varphi) \le L(\bar{\varrho}, \bar{\lambda}, \bar{\beta}, \bar{\alpha}^\varsigma, \bar{\alpha}^\varphi)$*, for all* $\lambda \in R_+, \beta, \alpha^\varsigma, \alpha^\varphi \in R$,

$L(\bar{\varrho},\bar{\lambda}, \bar{\beta}, \bar{\alpha}^\varsigma, \alpha^{-\varphi}) \le L(\varrho, \bar{\lambda}, \bar{\beta}, \bar{\alpha}^\varsigma, \bar{\alpha}^\varphi)$*, for all* $\varrho \in T$.

Theorem 4.1 *Let* $v^{-L} > 0$, $v^{-U} > 0$ *and* $(\bar{\varrho},\bar{\lambda}, \bar{\beta}, \bar{\alpha}^\varsigma,\bar{\alpha}^\varphi)$ *be a saddle point for* $L(\varrho,\lambda,\beta,\alpha^\varsigma,\alpha^\varphi)$*, then* $\bar{\varrho}$ *is a* LU *optimal solution of (FIVVC).*

Proof: Suppose, in contradiction to the result, that $\bar{\varrho}$ is not a LU optimal solution for (FIVVC). Hence, based on the Definition (2.1) , there exist a feasible solution ϱ such that

$$\left[\frac{\kappa^L}{\tau^L}(\varrho), \frac{\kappa^U}{\tau^U}(\varrho)\right] \prec_{\mathcal{LU}} \left[\frac{\kappa^L}{\tau^L}(\bar{\varrho}), \frac{\kappa^U}{\tau^U}(\bar{\varrho})\right]$$

that is

$$\begin{cases} \frac{\kappa^L}{\tau^L}(\varrho) < \frac{\kappa^L}{\tau^L}(\bar{\varrho}) \\ \frac{\kappa^U}{\tau^U}(\varrho) \le \frac{\kappa^U}{\tau^U}(\bar{\varrho}) \end{cases}, \text{ or } \begin{cases} \frac{\kappa^L}{\tau^L}(\varrho) \le \frac{\kappa^L}{\tau^L}(\bar{\varrho}) \\ \frac{\kappa^U}{\tau^U}(\varrho) < \frac{\kappa^U}{\tau^U}(\bar{\varrho}) \end{cases}, \text{ or } \begin{cases} \frac{\kappa^L}{\tau^L}(\varrho) < \frac{\kappa^L}{\tau^L}(\bar{\varrho}) \\ \frac{\kappa^U}{\tau^U}(\varrho) < \frac{\kappa^U}{\tau^U}(\bar{\varrho}) \end{cases},$$

which implies

$$\kappa^L(\varrho) - v^{-L}\tau^L(\varrho) < \kappa^L(\bar{\varrho}) - v^{-L}\tau^L(\bar{\varrho}),$$
$$\kappa^U(\varrho) - v^{-U}\tau^U(\varrho) \le \kappa^U(\bar{\varrho}) - v^{-U}\tau^U(\bar{\varrho}),$$

or

$$\kappa^L(\varrho) - v^{-L}\tau^L(\varrho) \le \kappa^L(\bar{\varrho}) - v^{-L}\tau^L(\bar{\varrho}),$$
$$\kappa^U(\varrho) - v^{-U}\tau^U(\varrho) < \kappa^U(\bar{\varrho}) - v^{-U}\tau^U(\bar{\varrho}),$$

or

$$\kappa^L(\varrho) - v^{-L}\tau^L(\varrho) < \kappa^L(\bar{\varrho}) - v^{-L}\tau^L(\bar{\varrho}),$$
$$\kappa^U(\varrho) - v^{-U}\tau^U(\varrho) < \kappa^U(\bar{\varrho}) - v^{-U}\tau^U(\bar{\varrho}).$$

By above inequalities and from $v^{-L} > 0$, $v^{-U} > 0$, we have

$$[\kappa^L(\varrho) - v^{-L}\tau^L(\varrho)] + [\kappa^U(\varrho) - v^{-U}\tau^U(\varrho)] < [\kappa^L(\bar{\varrho}) - v^{-L}\tau^L(\bar{\varrho})] + [\kappa^U(\bar{\varrho}) - v^{-U}\tau^U(\bar{\varrho})]. \tag{42}$$

Since $(\bar{\varrho}, \bar{\lambda}, \bar{\beta}, \bar{\alpha}^\varsigma, \bar{\alpha}^\varphi)$ is a saddle point for $L(\varrho, \lambda, \beta, \alpha^\varsigma, \alpha^\varphi)$, therefore by Definition(4.1) (i), we obtain

$$L(\bar{\varrho}, \lambda, \beta, \alpha^\varsigma, \alpha^\varphi) \le L(\bar{\varrho}, \bar{\lambda}, \bar{\beta}, \bar{\alpha}^\varsigma, \bar{\alpha}^\varphi),$$

that is,

$$\sum_{\mu=1}^{\sigma}\lambda_\mu\phi_\mu(\bar{\varrho}) + \sum_{\iota=1}^{t}\beta_\iota\psi_\iota(\bar{\varrho}) - \sum_{\mu=1}^{r}\alpha_\mu^\varsigma\varsigma_\mu(\bar{\varrho}) + \sum_{\mu=1}^{r}\alpha_\mu^\varphi\varphi_\mu(\bar{\varrho}) \le \sum_{\mu=1}^{\sigma}\bar{\lambda}_\mu\phi_\mu(\bar{\varrho}) + \sum_{\iota=1}^{t}\bar{\beta}_\iota\psi_\iota(\bar{\varrho}) - \sum_{\mu=1}^{r}\bar{\alpha}_\mu^\varsigma\varsigma_\mu(\bar{\varrho}) + \sum_{\mu=1}^{r}\bar{\alpha}_\mu^\varphi\varphi_\mu(\bar{\varrho}). \tag{43}$$

If we set $(\lambda, \beta, \alpha^{\varsigma}, \alpha^{\varphi}) = (0,0,0,0)$ in the above inequality, then we get

$$\sum_{\mu=1}^{\sigma} \bar{\lambda}_{\mu}\phi_{\mu}(\bar{\varrho}) + \sum_{\iota=1}^{p} \bar{\beta}_{\iota}\psi_{\iota}(\bar{\varrho}) - \sum_{\mu=1}^{r} \bar{\alpha}^{\varsigma}_{\mu}\varsigma_{\mu}(\bar{\varrho}) + \sum_{\mu=1}^{r} \bar{\alpha}^{\varphi}_{\mu}\varphi_{\mu}(\bar{\varrho}) \geq 0. \quad (44)$$

Using $\bar{\lambda} \in R_{+}, \bar{\beta}, \bar{\alpha}^{\varsigma}, \bar{\alpha}^{\varphi} \in R$

$$\bar{\lambda}_{\mu}\phi_{\mu}(\bar{\varrho}) \leq 0, \forall \mu = 1, 2, \cdots, \sigma, \quad (45)$$

$$\bar{\beta}_{\iota}\psi_{\iota}(\bar{\varrho}) = 0, \forall \iota = 1, 2, \cdots, p, \quad (46)$$

$$-\bar{\alpha}^{\varsigma}_{\mu}\varsigma_{\mu}(\bar{\varrho}) \geq 0, \forall \mu = 1, 2, \cdots, r, \quad (47)$$

$$\bar{\alpha}^{\varphi}_{\mu}\varphi_{\mu}(\bar{\varrho}) \leq 0, \forall \mu = 1, 2, \cdots, r, \quad (48)$$

which implies that

$$\sum_{\mu=1}^{\sigma} \bar{\lambda}_{\mu}\phi_{\mu}(\bar{\varrho}) + \sum_{\iota=1}^{t} \bar{\beta}_{\iota}\psi_{\iota}(\bar{\varrho}) - \sum_{\mu=1}^{r} \bar{\alpha}^{\varsigma}_{\mu}\varsigma_{\mu}(\bar{\varrho}) + \sum_{\mu=1}^{r} \bar{\alpha}^{\varphi}_{\mu}\varphi_{\mu}(\bar{\varrho}) \leq 0. \quad (49)$$

From the inequalities (44) and (49), we conclude that

$$\sum_{\mu=1}^{m} \bar{\lambda}_{\mu}\phi_{\mu}(\bar{\varrho}) + \sum_{\iota=1}^{t} \bar{\beta}_{\iota}\psi_{\iota}(\bar{\varrho}) - \sum_{\mu=1}^{r} \bar{\alpha}^{\varsigma}_{\mu}\varsigma_{\mu}(\bar{\varrho}) + \sum_{\mu=1}^{r} \bar{\alpha}^{\varphi}_{\mu}\varphi_{\mu}(\bar{\varrho}) = 0. \quad (50)$$

On other hand, Since $(\bar{\varrho}, \bar{\lambda}, \bar{\beta}, \bar{\alpha}^{\varsigma}, \bar{\alpha}^{\varphi})$ is a saddle point for $L(\varrho, \lambda, \beta, \alpha^{\varsigma}, \alpha^{\varphi})$, therefore by Definition(4.1) (ii), we obtain

$$L(\bar{\varrho}, \bar{\lambda}, \bar{\beta}, \bar{\alpha}^{\varsigma}, \bar{\alpha}^{\varphi}) \leq L(\varrho, \bar{\lambda}, \bar{\beta}, \bar{\alpha}^{\varsigma}, \bar{\alpha}^{\varphi}),$$

that is

$$[\kappa^{L}(\bar{\varrho}) - \bar{v}^{L}\tau^{L}(\bar{\varrho})] + [\kappa^{U}(\bar{\varrho}) - \bar{v}^{U}\tau^{U}(\bar{\varrho})] + \sum_{\mu=1}^{\sigma} \bar{\lambda}_{\mu}\phi_{\mu}(\bar{\varrho}) + \sum_{\iota=1}^{t} \bar{\beta}_{\iota}\psi_{\iota}(\bar{\varrho}) - \sum_{\mu=1}^{r} \bar{\alpha}^{\varsigma}_{\mu}\varsigma_{\mu}(\bar{\varrho}) + \sum_{\mu=1}^{r} \bar{\alpha}^{\varphi}_{\mu}\varphi_{\mu}(\bar{\varrho}) \leq [\kappa^{L}(\varrho) - \bar{v}^{L}\tau^{L}(\varrho)]$$

$$+ [\kappa^{U}(\varrho) - \bar{v}^{U}\tau^{U}(\varrho)] + \sum_{\mu=1}^{\sigma} \bar{\lambda}_{\mu}\phi_{\mu}(\varrho) + \sum_{\iota=1}^{t} \bar{\beta}_{\iota}\psi_{\iota}(\varrho) - \sum_{\mu=1}^{r} \bar{\alpha}^{\varsigma}_{\mu}\varsigma_{\mu}(\varrho) + \sum_{\mu=1}^{r} \bar{\alpha}^{\varphi}_{\mu}\varphi_{\mu}(\varrho)$$

Using the feasibility of ϱ of the problem (FIVVC) together with $\bar{\lambda} \in R_{+}, \bar{\beta}, \bar{\alpha}^{\varsigma}, \bar{\alpha}^{\varphi} \in R$ and (50), above inequality yields

$$[\kappa^{L}(\bar{\varrho}) - v^{-L}(\bar{\varrho})\tau^{L}(\bar{\varrho})] + [\kappa^{U}(\bar{\varrho}) - v^{-U}(\bar{\varrho})\tau^{U}(\bar{\varrho})] \leq [\kappa^{L}(\varrho) - v^{-L}(\bar{\varrho})\tau^{L}(\varrho)] + [\kappa^{U}(\varrho) - v^{-U}(\bar{\varrho})\tau^{U}(\varrho)].$$

This contradicts (42). Hence the required proof.

Theorem 4.2 *Let $\bar{\varrho}$ be a LU optimal solution to (FIVVC) and suppose the existence of f $\bar{v}^{L} > 0$, $\bar{v}^{U} > 0$, $\bar{\lambda} \in R_{+}, \bar{\beta}, \bar{\alpha}^{\varsigma}, \bar{\alpha}^{\varphi} \in R$, such that necessary conditions are fulfilled at $\bar{\varrho}$. Also assume that*

(i) *$\kappa^{L}(.) - v^{-L}\tau^{L}(.) + \kappa^{U}(.) - v^{-U}\tau^{U}(.)$ is convex function at $\bar{\varrho}$ on F*

(ii) *ϕ_{μ}, $\mu = 1, 2, \ldots, \sigma$, ψ_{ι} $(\iota \in J_{\psi}^{+})$, $-\psi_{\iota}$ $(\iota \in J_{\psi}^{-})$, $-\varsigma_{\mu}$ $(\mu \in J_{+}^{+} \cup J^{+}_{0})$, ς_{μ} $(\mu \in J^{-}_{0})$, $-\varphi_{\mu}$ $(\mu \in J^{-}_{0+} \cup J^{-}_{00} \cup J^{-}_{+0})$, and φ_{μ} $(\mu \in J_{00}^{+} \cup J^{+}_{0-} \cup J_{+0}^{+} \cup J^{+}_{+-})$ are convex functions at $\bar{\varrho}$ on F.*

Then $(\bar{\varrho}, \bar{\lambda}, \bar{\beta}, \bar{\alpha}^{G}, \bar{\alpha}^{\varphi})$ is a saddle point for $L(\varrho, \lambda, \beta, \alpha^{\varsigma}, \alpha^{\varphi})$.

Proof: Since, (9) – (16) hold at $\bar{\varrho}$ with $v^{-L} > 0$, $v^{-U} > 0$, $\bar{\lambda} \in R_{+}, \bar{\beta}, \bar{\alpha}^{\varsigma}, \bar{\alpha}^{\varphi} \in R$. As it follows from (9), that

$$\nabla\kappa^{L}(\bar{\varrho}) - v^{L}\nabla\tau^{L}(\bar{\varrho}) + \nabla\kappa^{U}(\bar{\varrho}) - v^{U}\nabla\tau^{U}(\bar{\varrho}) + \sum_{\mu=1}^{\sigma} \lambda_{\mu}\nabla\phi_{\mu}(\bar{\varrho}) + \sum_{\iota=1}^{t} \beta_{\iota}\nabla\psi_{\iota}(\bar{\varrho}) - \sum_{\mu=1}^{r} \alpha^{\varsigma}_{\mu}\nabla\varsigma_{\mu}(\bar{\varrho}) + \sum_{i=1}^{r} \alpha^{\varphi}_{\mu}\nabla\varphi_{\mu}(\bar{\varrho}) = 0,$$

From the hypothesis (*i*), $\kappa^{L}(.) - \bar{v}^{L}\tau^{L}(.) + \kappa^{U}(.) - \bar{v}^{U}\tau^{U}(.)$ is convex at $\bar{\varrho}$, we have

$$\kappa^{L}(\varrho) - \bar{v}^{L}\tau^{L}(\varrho) + \kappa^{U}(\varrho) - \bar{v}^{U}\tau^{U}(\varrho) - \kappa^{L}(\bar{\varrho}) - \bar{v}^{L}\tau^{L}(\bar{\varrho}) - \kappa^{U}(\bar{\varrho}) - \bar{v}^{U}\tau^{U}(\bar{\varrho}) \geq (\varrho - \varrho^{-})$$

$$\geq (\varrho - \bar{\varrho})^{T}[\nabla\kappa^{L}(\bar{\varrho}) - \bar{v}^{L}\nabla\tau^{L}(\bar{\varrho}) + \nabla\kappa^{U}(\bar{\varrho}) - \bar{v}^{U}\nabla\tau^{U}(\bar{\varrho})] \quad (51)$$

From hypothesis (*ii*), we get

$$\sum_{\mu=1}^{\sigma} \bar{\lambda}_{\mu}\phi_{\mu}(\varrho) - \sum_{\mu=1}^{\sigma} \bar{\lambda}_{\mu}\phi_{\mu}(\bar{\varrho}) \geq (\varrho - \bar{\varrho})^{T} \sum_{\mu=1}^{\sigma} \bar{\lambda}_{\mu}\nabla\phi_{\mu}(\bar{\varrho}). \quad (52)$$

By similar argument and by the definition of index sets, one has

$$\sum_{\iota=1}^{t}\bar\beta_\iota\psi_\iota(\varrho)-\sum_{\iota=1}^{t}\bar\beta_\iota\psi_\iota(\bar\varrho)-\left[\sum_{\mu=1}^{r}\bar\alpha^{\varsigma}_{\mu}\varsigma_\mu(\varrho)-\sum_{\mu=1}^{r}\bar\alpha^{\varsigma}_{\mu}\varsigma_\mu(\bar\varrho)\right]+\sum_{\mu=1}^{r}\bar\alpha^{\varphi}_{\mu}\varphi_\mu(\varrho)$$
$$-\sum_{\mu=1}^{r}\bar\alpha^{\varphi}_{\mu}\varphi_\mu(\bar\varrho)\geq(\varrho-\bar\varrho)^T\left[\sum_{\iota=1}^{t}\beta_\iota\nabla\psi_\iota(\bar\varrho)-\sum_{\mu=1}^{r}\alpha^{\varsigma}_{\mu}\nabla\varsigma_\mu(\bar\varrho)+\sum_{\mu=1}^{r}\alpha^{\varphi}_{\mu}\nabla\varphi_\mu(\bar\varrho)\right] \tag{53}$$

Combining (51)- (53), we have

$$\kappa^L(\varrho)-\bar\upsilon^L\tau^L(\varrho)+\kappa^U(\varrho)-\bar\upsilon^U\tau^U(\varrho)-\kappa^L(\bar\varrho)-\bar\upsilon^L\tau^L(\bar\varrho)-\kappa^U(\bar\varrho)-\bar\upsilon^U\tau^U(\bar\varrho)$$
$$+\sum_{\mu=1}^{\sigma}\bar\lambda_\mu\phi_\mu(\varrho)-\sum_{\mu=1}^{\sigma}\bar\lambda_\mu\phi_\mu(\bar\varrho)+\sum_{\iota=1}^{t}\bar\beta_\iota\psi_\iota(\varrho)-\sum_{\iota=1}^{t}\bar\beta_\iota\psi_\iota(\bar\varrho)-\left[\sum_{i=1}^{r}\bar\alpha^{\varsigma}_{i}\varsigma_\mu(\varrho)-\sum_{\mu=1}^{r}\bar\alpha^{\varsigma}_{\mu}\varsigma_\mu(\bar\varrho)\right]$$
$$+\sum_{\mu=1}^{r}\bar\alpha^{\varphi}_{\mu}\varphi_\mu(\varrho)-\sum_{\mu=1}^{l}\bar\alpha^{\varphi}_{\mu}\varphi_\mu(\bar\varrho)\geq(\varrho-\bar\varrho)^T\Big[[\nabla\kappa^L-\bar\upsilon^L\nabla\tau^L]+[\nabla\kappa^U-\bar\upsilon^U\nabla\tau^U]$$
$$+\sum_{\mu=1}^{\sigma}\lambda_\mu\nabla\phi_\mu(\bar\varrho)+\sum_{\iota=1}^{t}\beta_\iota\nabla\psi_\iota(\bar\varrho)-\sum_{\mu=1}^{r}\alpha^{\varsigma}_{\mu}\nabla\varsigma_\mu(\bar\varrho)+\sum_{\mu=1}^{r}\alpha^{\varphi}_{\mu}\nabla\varphi_\mu(\bar\varrho)\Big]$$

which by (9), yields

$$\kappa^L(\varrho)-\bar\upsilon^L\tau^L(\varrho)+\kappa^U(\varrho)-\bar\upsilon^U\tau^U(\varrho)+\sum_{\mu=1}^{\sigma}\bar\lambda_\mu\phi_\mu(\varrho)+\sum_{\iota=1}^{t}\bar\beta_\iota\psi_\iota(\varrho)$$
$$-\sum_{\mu=1}^{r}\bar\alpha^{\varsigma}_{\mu}\varsigma_\mu(\varrho)+\sum_{\mu=1}^{r}\bar\alpha^{\varphi}_{\mu}\varphi_\mu(\varrho)\geq\kappa^L(\bar\varrho)-\bar\upsilon^L\tau^L(\bar\varrho)+\kappa^U(\bar\varrho)-\bar\upsilon^U\tau^U(\bar\varrho)$$
$$+\sum_{\mu=1}^{\sigma}\bar\lambda_\mu\phi_\mu(\bar\varrho)+\sum_{\iota=1}^{t}\bar\beta_\iota\psi_\iota(\bar\varrho)-\sum_{\mu=1}^{r}\bar\alpha^{\varsigma}_{\mu}\varsigma_\mu(\bar\varrho)+\sum_{\mu=1}^{r}\bar\alpha^{\varphi}_{\mu}\varphi_\mu(\bar\varrho),$$

that is,

$$L(\bar\varrho,\bar\lambda,\bar\beta,\bar\alpha^{\varsigma},\bar\alpha^{\varphi})\leq L(\varrho,\bar\lambda,\bar\beta,\bar\alpha^{\varsigma},\bar\alpha^{\varphi}). \tag{54}$$

On the other hand, using the feasibility of $\bar\varrho$ of the problem (FIVVC) and the fact that $\lambda\in R_+$, β, α^{ς}, $\alpha^{\varphi}\in R$, we have

$$\lambda_\mu\phi_\mu(\bar\varrho)\leq 0,\ \forall\mu=1,2,\ldots,\sigma, \tag{55}$$
$$\beta_\iota\psi_\iota(\bar\varrho)=0,\ \forall\iota=1,2,\ldots,t, \tag{56}$$
$$-\alpha^{\varsigma}_{\mu}\varsigma_\mu(\bar\varrho)\geq 0,\ \forall\mu=1,2,\ldots,r, \tag{57}$$
$$\alpha^{\varphi}_{\mu}\varphi_\mu(\bar\varrho)\leq 0,\ \forall\mu=1,2,\ldots,r. \tag{58}$$

The above relations (55)–(58) together with optimality conditions (13) – (16), gives

$$L(\bar\varrho,\lambda,\beta,\alpha^{\varsigma},\alpha^{\varphi})\leq L(\bar\varrho,\bar\lambda,\bar\beta,\bar\alpha^{\varsigma},\bar\alpha^{\varphi}). \tag{59}$$

The inequalities (54) and (59) implies $(\bar\varrho,\bar\lambda,\bar\beta,\bar\alpha^{\varsigma},\bar\alpha^{\varphi})$ is a saddle point for $L(\varrho,\lambda,\beta,\alpha^{\varsigma},\alpha^{\varphi})$. Hence, the required proof.

5. Conclusion

In this article, we have obtained conditions under which a feasible solution is a LU optimal solution and established sufficient optimality conditions for (FIVVC) involving convex functions. Also, we provided an example to validate the results of sufficient optimality conditions. Ultimately, the equivalence between the saddle point criteria of a Lagrangian type function and a LU optimal solution of (FIVVC) in the presence of convexity is also investigated.

REFERENCES

1. Wolfgang Achtziger, Tim Hoheisel and Christian Kanzow, *A smoothing-regularization approach to mathematical programs with vanishing constraints*, Computational Optimization and Applications, Springer, vol. 55 no. 3: 733–767 2013.
2. Achtziger Wolfgang, and Christian Kanzow, *Mathematical programs with vanishing constraints: optimality conditions and constraint qualifications*, Mathematical Programming, 114 6999 (2008).
3. Ahmad Izhar, Anurag Jayswal and Jonaki Banerjee *On interval-valued optimization problems with generalized invex functions*, Journal of Inequalities and Applications, 2013, no. 1: 1–14 (2013).
4. Ahmad Izhar, Krishna Kummari and Al-Homidan S., *Sufficiency and duality for interval-valued optimization problems with vanishing constraints using weak constraint qualifications*, International Journal of Analysis and Applications, 18 no. 5: 784–798 (2020).
5. Bhurjee Ajay K., and Geetanjali Panda, *Sufficient optimality conditions and duality theory for interval optimization problem*, Annals of Operations Research, 243 335–348 (2016).
6. Debnath Indira P., and Shiv K. Gupta, *Necessary and sufficient optimality conditions for fractional interval-valued optimization problems*, Decision Science in Action: Theory and Applications of Modern Decision Analytic Optimisation, 155173 (2019).
7. Debnath Indira P., and Shiv K. Gupta, *The*
8. *Karush–Kuhn–Tucker conditions for multiple objective fractional interval valued optimization problems*, RAIROOperations Research, 54 no. 4: 1161–1188 (2020).
9. Hoheisel Tim, and Christian Kanzow, *On the Abadie and Guignard constraint qualifications for mathematical programmes with vanishing constraints*, Optimization, 58 no. 4: 431–448 (2009).
10. Hoheisel Tim, and Christian Kanzow, *Stationary conditions for mathematical programs with vanishing constraints using weak constraint qualifications*, Journal of Mathematical Analysis and Applications, 337 no. 1: 292–310 (2008).
11. Hoheisel Tim, and Christian Kanzow, *First-and second-order optimality conditions for mathematical programs with vanishing constraints*, Applications of Mathematics, 52 no. 6: 495–514 (2007).
12. Izmailov Alexey F., and Mikhail V. Solodov *Mathematical programs with vanishing constraints: optimality conditions, sensitivity, and a relaxation method*, Journal of Optimization Theory and Applications 142 no. 3: 501–532 (2009).
13. Jabr R. A., *Solution to economic dispatching with disjoint feasible regions via semidefinite programming*, IEEE Transactions on power systems, 27 no. 1: 572-573 (2011).
14. Kummari Krishna, *Optimality and duality in fractional programming with vanishing constraints*, (Communicated).
15. Li Lifeng, Sanyang Liu, and Jianke Zhang, *On interval-valued invex mappings and optimality conditions for interval-valued optimization problems*, Journal of Inequalities and Applications, 2015 no. 1: 1–19 (2015).
16. Rozvany G. I. N, *On design-dependent constraints and singular topologies*, Structural and Multidisciplinary Optimization, 21 164–172 (2001).
17. Sun Yuhua, Xiumei Xu, and Laisheng Wang, *Duality and saddle-point type optimality for interval-valued programming*, Optimization Letters, 8, no. 3: 1077–1091 (2014).
18. Tung L. Thanh, *Karush–Kuhn–Tucker optimality conditions and duality for multiobjective semi-infinite programming with vanishing constraints*, Annals of Operations Research, 311 no. 2: 1307–1334 (2022).
19. Verbart Alexander, Matthijs Langelaar and Fred V. Keulen, *Damage approach: A new method for topology optimization with local stress constraints*, Structural and Multidisciplinary Optimization, 53 1081–1098 (2016).

ed in the United States

Print